REVISED SECOND EDITION

HEATING AND COOLING OF BUILDINGS

Design for Efficiency

Mechanical Engineering Series
Frank Kreith & Roop Mahajan - Series Editors

REVISED SECOND EDITION

HEATING AND COOLING OF BUILDINGS

Design for Efficiency

JAN F. KREIDER

PETER S. CURTISS

ARI RABL

CRC Press
Taylor & Francis Group
Boca Raton London New York

CRC Press is an imprint of the
Taylor & Francis Group, an **informa** business

CRC Press
Taylor & Francis Group
6000 Broken Sound Parkway NW, Suite 300
Boca Raton, FL 33487-2742

© 2010 by Taylor and Francis Group, LLC
CRC Press is an imprint of Taylor & Francis Group, an Informa business

International Standard Book Number: 978-1-4398-1151-1 (Hardback)

Library of Congress Cataloging-in-Publication Data

Kreider, Jan F., 1942-
    Heating and cooling of buildings : design for efficiency / Jan F. Kreider, Peter S. Curtiss, Ari Rabl. -- Rev. 2nd ed.
        p. cm. -- (Mechanical engineering series)
    Includes bibliographical references and index.
    ISBN 978-1-4398-1151-1 (hardcover : alk. paper)
    1. Heating--Equipment and supplies--Design and construction. 2. Ventilation--Equipment and supplies--Design and construction. 3. Air conditioning--Equipment and supplies--Design and construction. I. Curtiss, Peter. II. Rabl, Ari. III. Title. IV. Series.

TH7345.K74 2010
697--dc22   2009041608

Visit the Taylor & Francis Web site at
http://www.taylorandfrancis.com

and the CRC Press Web site at
http://www.crcpress.com

Contents

Preface to the Revised Second Edition

The art and the science of building systems design evolve continuously as a worldwide cadre of designers, practitioners, and researchers all endeavor to improve the performance of buildings and the comfort and productivity of their occupants. This book, which presents the technical basis of building mechanical system and lighting system design, also has evolved due to the efforts of many in the building engineering profession. The authors decided to completely revise the first edition of this book to modernize topic coverage and to include much more reference information in electronic form.

Contents of the book have been expanded to include

- A chapter on economic analysis and optimization
- Heating and cooling load procedures and databases
- More than 200 new homework problems
- New and simplified procedures for ground coupled heat transfer calculations
- Many updates of fine points in all chapters
- Completely new Heating and Cooling of Buildings software
- All book graphics and appendices supplied on CD-ROM to assist with lecture preparation

One of the most noticeable differences in the second edition of this book is that many of the appendices from the first edition have been moved to the accompanying CD-ROM. This CD-ROM also contains an updated version of the HCB program and links to a Web site, www.HCBCentral.com. By packaging multimedia along with a textbook, we hope to provide resources to engineering students in a fashion that will eventually be standard practice. There are more than 1000 tables in the electronic appendixes that can be searched by major categories, a table list, or an index of topics. The CD-ROM also directs students to the HCBCentral Web site where several hundred links are maintained to help students find manufacturers' and government data, browse in newsgroups, and find any corrections and updates to the text and data tables. The tables are in HTML format, which means that data can be copied and pasted into applications directly from the electronic appendixes. In some cases, the data in the tables and the HCB software may be slightly different from other sources due to the process by which the data have been generated or converted. These differences are noted where appropriate.

The authors acknowledge the ongoing support of their families while this project was undertaken. Many students and professors shared their ideas with us so that the second edition would be improved in many ways. We particularly want to thank Professor J. Taylor Beard for his careful reporting of many suggested changes based on his use of the first edition for many years. The American Society of Heating, Refrigerating and Air Conditioning Engineers generously shared the data from their handbook series. Mike Johnston helped to update the solutions manual originally prepared by Wendy Hawthorne. Finally, Kreider and Rabl are very pleased to have their colleague, Dr. Peter S. Curtiss, PE, join the authorship of this edition.

Jan F. Kreider
Peter S. Curtiss
Ari Rabl
Boulder, Colorado

Acknowledgments

A book of this magnitude and scope represents the work of many in addition to the authors. First we thank our students, who have given us the luxury of testing the book in class. Their comments and ideas were essential in finalizing the organization and coverage of the book.

A number of reviewers took considerable time to examine the book closely, finding errors that the authors missed and suggesting improvements in the presentation. The review process began with helpful comments on the original outline by Louis Burmeister, University of Kansas; Thomas Hellman, New Mexico State University; Doug Hittle, Colorado State University; Ronald Howell, University of South Florida; Dennis O'Neal, Texas A&M University; Maurice Wildin, University of New Mexico; and Byron Winn, Colorado State University, and continued with reviews of the draft by John Lloyd, Michigan State University, and Trilochan Singh, Wayne State University. The exceptionally detailed and constructive comments by Professor Wildin on the entire manuscript are especially appreciated. Likewise, Wendy Hawthorne made a careful reading of the text and checked most of the examples, and Bill Shurcliff contributed a thorough review of several chapters. Various colleagues offered comments on parts of the book: Mike Brande-muehl (who also contributed novel end-of-chapter problems in the first half of the book), Manuel Collares-Pereira, Jeff Haberl, J. Y. Kao, John Littler, John Mitchell, Leslie K. Nordford, Mike Riley, Gideon Shavit, and Mike Scofield. Of course, any remaining errors are the sole responsibility of the authors.

Peter Curtiss wrote the software that accompanies this book. The thousands of lines of code that he wrote in producing the final product set a new standard for instructional software used in building systems education. Wendy Hawthorne prepared the solutions manual for the end-of-chapter problems.

Our book called on the work of many others. We are grateful to several people who generously provided material or data: R. Boehm, S. Burek, J. Harris, M.A. Piette, and R. Sullivan. We have tried to give credit to original sources wherever possible and have included complete reference lists in each chapter. Permissions were given liberally, and we thank the numerous original sources for their generosity.

Jan F. Kreider would like to express his personal thanks to Dottie for her patience during the apparently endless process of writing this book. Her good nature and support are very much appreciated. Kennon Stewart generously provided practical advice from his decades of HVAC design experience. The University of Colorado's College of Engineering provided the facilities used for preparing the book and for some of the experimental results contained within. Denis Clodic, director of the Centre d'Energétique of the Ecole des Mines de Paris, and Flavio Conti at the Joint Research Center of the European Community provided a gracious environment during his sabbatical for the final tasks needed to finish the book.

Ari Rabl would like to express his personal gratitude to several institutions and individuals. The Ecole des Mines offered a propitious setting for the writing, thanks to the encouragement of Jerome Adnot. The Joint Center for Energy Management of the University of Colorado provided support for summer visits. Robert H. Socolow invited him to join the Center for Energy and Environmental Studies of Princeton University and to focus on energy use of buildings. The work with colleagues at Princeton was a uniquely stimulating and productive experience to which this book is a tribute. And, of course, his most heartfelt thanks go to Brigitte for her moral support.

Jan F. Kreider
Ari Rabl

Authors

Jan F. Kreider is a professor of engineering and founding director of the Joint Center for Energy Management at the University of Colorado at Boulder. He received his BSME (magna cum laude) from Case Institute of Technology and his postgraduate degrees from the University of Colorado. Dr. Kreider is the author of 10 engineering textbooks and more than 200 technical articles and reports; he has directed more than $10 million in building energy–related research during the past decade. He is a fellow of the ASME, an active member of ASHRAE as well as a number of other technical societies, and a winner of ASHRAE's E.K. Campbell Award for excellence in building systems education. He is also the president of a consulting firm specializing in energy system design and analysis.

Ari Rabl is a research scientist at the Centre d'Energétique of the Ecole des Mines in Paris, as well as research professor at the University of Colorado at Boulder. He received a PhD in physics from the University of California at Berkeley and has worked at Argonne National Laboratory and at the Solar Energy Research Institute. From 1980 to 1989 he was research scientist and lecturer at the Center for Energy and Environmental Studies at Princeton University. His publications on energy-related topics include more than 50 journal articles, numerous review papers, and 10 patents. He has also written the book *Active Solar Collectors and Their Applications*. He is a member of the American Physical Society and ASHRAE.

Peter Curtiss is an adjunct professor at the University of Colorado and runs an engineering consulting business in Boulder. He received his BSCE from Princeton University and his postgraduate degrees from the University of Colorado. He has written more than 40 technical journal articles on subjects ranging from neural network modeling and control in buildings to solar radiation measurements. He has worked at research institutes in Israel, Portugal, and France as well as at a number of private engineering firms.

1

Introduction

1.1 A Bit of History

The quest for a safe and comfortable environment is older than the human species. Birds build nests, rabbits dig holes. Early human societies succeeded, in some cases, in creating remarkably pleasant accommodations. The cliff dwellings of the Pueblo Indians at Mesa Verde, Colorado, are an eminent example: Carved under an overhang to block the summer heat, yet accessible to the warming rays of the winter sun, with the massive heat capacity of the surrounding rocks, they are an early realization of the principles of passive solar architecture.

The ancient Greeks were quite conscious of the benefits to be obtained by good orientation of a building with respect to the sun, and they laid out entire settlements facing south (Butti and Perlin, 1980). Already in classical antiquity, fuel wood was scarce around the Mediterranean (our generation is not the first to be confronted with an energy shortage). The height of comfort in the classical world was achieved in some villas of the Roman Empire: The people who built them pioneered central heating, with a double floor through whose cavity the fumes of a fire were passed. Also in Roman times, the first translucent or transparent window coverings were introduced, made of materials such as mica or glass. Thus it became possible to admit light into a building without letting in wind, rain, or snow.

But it took a long time before buildings reached what we would consider comfortable conditions. When visiting castles built in Europe as late as the sixteenth century and looking at their heating arrangements, one shivers to think how cold winter must have been even for the wealthiest. In an antique store, one of the authors happened upon an old thermometer with a scale where 60.8°F (16°C) was marked as "room temperature." The comfort of heating to 68°F (20°C) or more is fairly recent.

Cooling is more difficult than heating. In the past the principal method was to ward off the sun, coupled with the use of heavy stonework for thermal inertia—actually quite effective in climates with cool nights. During the Middle Ages, architects of castles like the Alhambra supplemented that approach with skillful use of running water, providing some evaporative cooling. In certain parts of the world, some buildings take advantage of cool breezes that blow regularly from the same direction.

Nights used to be dim because candles and oil lamps were expensive, to say nothing of the quality of the light. In fact before the invention of electric lights, access to daylight was a primary criterion for the design of buildings. Daylight does not penetrate well into the interior; as a rule of thumb, adequate illumination on cloudy days cannot be provided to depths beyond 1.5 to 2 times the height of the upper window edge. This, together with the need for fresh air before the days of mechanical ventilation, explains the shape of buildings well into the first half of the twentieth century: No room could be far from the perimeter.

Total disregard for this constraint did not become practical until the availability of fluor-escent lighting (with incandescent lamps the cost of air conditioning would have been excessive).

Effective air conditioning had to await the development of mechanical refrigeration during the first decades of the twentieth century. Routine installation of central air con-ditioning systems dates from the 1960s. The oil crises of the 1970s stimulated intensive research on ways of reducing energy costs. The efficiency of existing technologies is being improved, and "new" technologies such as solar energy and daylighting are being tried. Some approaches are turning out to be more successful than others. This evolution is continuing at full speed, aided by advances in materials and computers. Elucidating these developments is one of the goals of this book.

1.2 Importance of Buildings in the U.S. Economy

People in modern times spend most of their time in buildings, and expend much of their wealth on these buildings. Even without trying for an accurate appraisal of the total real estate in the United States, we can get an idea by looking at the floor area, approximately 181×10^9 ft^2 ($\approx 18 \times 10^9$ m^2) in the residential and 60×10^9 ft^2 ($\approx 6 \times 10^9$ m^2) in the commer-cial sector (USDOE, 1999). Construction costs vary a great deal, of course, with the type and quality of building, but very roughly they are in the range of $100 per square foot ($1000 per square meter) ± 25 percent. Based on the replacement cost, the total value of the buildings in the United States is on the order of $\$20 \times 10^{12}$; the GNP, by comparison, was only $\$5 \times 10^{12}$ per year during the mid-1990s.

Obviously one cannot afford to replace the building stock very often. This ensures a certain continuity with the future: However little we may be able to foresee the future of our society, we can expect most of the buildings to last for decades, if not centuries. Rather than replace a building entirely, one may revamp its interior [currently each year some 2 percent of commercial buildings undergo a major retrofit (Brambley et al., 1988)]. This implies an awesome responsibility for city planners, architects, and engineers: *Do it well or the mistakes will haunt us for a long time.*

The demand for new construction was vigorous during the 1980s. Annual construction rates were around 1.5×10^9 ft^2 ($\approx 0.15 \times 10^9$ m^2) of residential and 1.2×10^9 ft^2 ($\approx 0.12 \times 10^9$ m^2) of commercial floor area; the residential stock was growing at about 1 percent, the commercial stock at about 2.5 percent per year. Single-family houses have accounted for about two-thirds of recent residential construction. Forecasting of trends is difficult, espe-cially in a sector that is as cyclical as the construction industry and as sensitive to the state of the economy. The residential market may have seen its peak of growth with the passing of the postwar baby boom, although there may be a continuing push for new and better housing as general living standards improve (a likely trend over the long term). The growing shift from industry to services implies an increasing need for commercial floor space.

It is interesting to consider the importance of cost components involved in a building, from design to operation. Brambley et al. (1988) state that one-time costs (design and construction) represent only about one-fifth of the total life cycle cost, with the remaining four-fifths being ongoing costs (operation and maintenance). Such a comparison is not without ambiguity, involving a choice of weights for present and future expenses (via the discount rate, as discussed in Chap. 15), to say nothing about differences between different

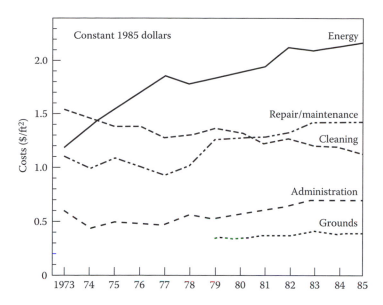

FIGURE 1.1
Average operating costs by component, per unit floor area, for office buildings, corrected for inflation. $1/ft^2 = $10.76/m^2$. (Courtesy of Brambley, M.R. et al., Advanced Energy Design and Operation Technologies Research: Recommendations for a U.S. Department of Energy Multiyear Program Plan, Battelle Pacific Northwest Laboratory, Richland, WA, Report PNL-6255 (December), 1988.)

types of buildings. We cite this figure merely to point out the dominant role of the cost of operation and maintenance. Of the latter, energy represents the lion's share, as can be seen from Fig. 1.1, where the ongoing costs, in dollars per unit floor area, are disaggregated according to the categories of administration, cleaning, repair and maintenance, security and grounds, and utilities (i.e., energy, since the cost of water does not amount to much). Since 1973 the importance of energy expenditures has increased in both relative and absolute terms. In the residential sector, energy expenditures tend to be smaller, somewhat less than the construction cost, but clearly they are an important item here, too. There is an important lesson: *Pay attention to energy costs at the design stage.*

Note that many of the data in Figs. 1.1 to 1.6 are estimated on the basis of limited statistical samples and can have uncertainties on the order of 10 percent; different reports may show different numbers.

To place the energy consumption of buildings in perspective, we show in Fig. 1.2 the evolution since 1973 of the primary energy consumption of the United States, broken down into the principal sectors: transportation, industry, commercial, and residential. Buildings, that is, the commercial and the residential sectors together, account for 36 percent of the total.

Several factors have contributed to the growth of energy use in buildings: The population has increased (from 180 million in 1960 to 275 million in 2000), comfort levels have improved, and there are more energy-using devices in the buildings. Figure 1.3 provides a more detailed view of this evolution. Part a shows the number of households and the consumption of primary energy per household; part b gives the analogous information for commercial buildings, in terms of floor area. Both follow the same pattern. While the number of households and the commercial floor area have grown steadily, the energy consumption per unit peaked in the early 1970s. Part of that may be due to population shifts to warmer climates, but most of it reflects gains in efficiency in response to the oil shocks.

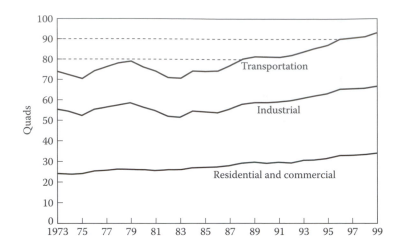

FIGURE 1.2

U.S. consumption of primary energy, by sector. 1 quad $= 1.055 \times 10^{18}$ J. (Courtesy of USDOE, *BTS Core Data Book*, Office of Building Technology, June 18, 1999; see also U.S. Department of Commerce, *Statistical Abstract of the United States*, USDOE, Washington, DC, 2001.)

A breakdown of energy consumption by end use is shown in Fig. 1.4. Space heating and water heating dominate in the residential sector, while lighting and space heating dominate in the commercial sector.

To highlight once more the importance of buildings and their energy consumption in the U.S. economy, we show in Fig. 1.5 the cost of constructing new buildings and the cost of the energy consumed by our building stock. Of the total GNP, the value of annual new construction accounts for some 10 percent and the annual energy bill for some 5 percent.

All these considerations indicate a major role for the heating, ventilating, and air conditioning (HVAC) design engineer, because his or her decisions determine, in large measure, the energy consumption. Yet, lest the reader conclude that energy consumption should be the dominant criterion for the design of commercial buildings, let us add the perspective of the value of the business transacted in the buildings. As an example, suppose an office worker requires 200 ft^2 ($\approx$20 m^2) of gross area and is paid \$30,000 per year (in exchange for services of roughly equivalent value). Thus the value of the services, per unit floor area, is \$150 per square foot ($\approx$\$1500 per square meter) per year, two orders of magnitude larger than the energy cost in Fig. 1.1. While this is merely an illustrative example, an even larger value of \$300 per square foot per year is cited for office buildings in the United States by Rosenfeld (1990). The value per unit floor area may be even larger, for instance, in retail stores, because of the profit to be made on the merchandise. Any reasonable assumptions about salary and density of workers lead to the same basic conclusion: In commercial buildings the energy costs are tiny in comparison with the value of the business that is transacted. Therefore, *do not risk a reduction in productivity while trying to save energy in commercial buildings*. A drop in productivity as small as 1 percent could more than wipe out any savings from energy.

As for energy forms used in commercial buildings, their relative contributions are shown in Fig. 1.6, from 1960 to 1998. One can see a steady and continuing growth of electricity relative to coal and petroleum. Two phenomena seem to be at play: a shift to electricity because it is more flexible and permits higher end-use efficiency and, as living standards

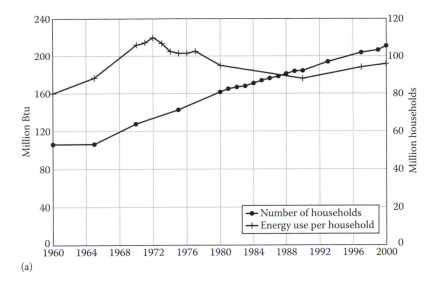

(a)

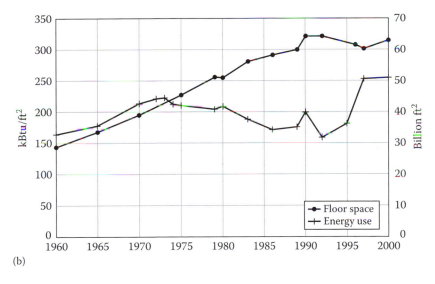

(b)

FIGURE 1.3

Building stock and consumption of primary energy. (a) Number of households and primary energy per house-hold. (b) Commercial floor area and primary energy per floor area. 1 MBtu = 1.055 GJ; 1 kBtu/ft^2 = 11.36 MJ/m^2. Decreases in commercial building area due to redefinition of commercial buildings. (Courtesy of USDOE, *BTS Core Data Book*, Office of Building Technology, June 18, 1999; see also U.S. Department of Commerce, *Statistical Abstract of the United States*, USDOE, Washington, DC, 2001.)

increase, a general trend from dirty to clean energy forms (clean at the site, not necessarily at the source). This evolution is likely to continue, including the growing use of heat pumps.

What may the future bring? Over the long term the cost of energy and materials tends to decline, thanks to technological progress. This trend is compensated to some extent by the exhaustion of resources and by the increasing costs of environmental protection. Occasion-ally the trend is punctuated by shocks, such as the oil crises of the 1970s. The "dash to gas"

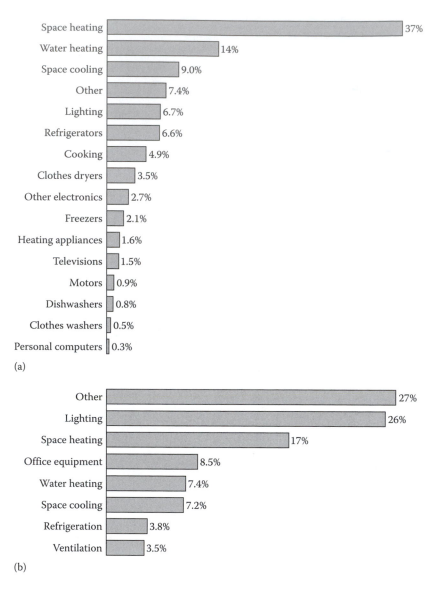

(a)

(b)

FIGURE 1.4

Breakdown of energy consumption by end use for 1997. (a) Residential sector. (b) Commercial sector. (Courtesy of USDOE, *BTS Core Data Book*, Office of Building Technology, June 18, 1999; see also U.S. Department of Commerce, *Statistical Abstract of the United States*, USDOE, Washington, DC, 2001.)

of recent years increases the risk of a shortage of gas, since most of the world's gas supply comes from politically unstable countries.

Concern about air pollution, both outdoor and indoor, has been increasing and is not likely to diminish soon. Since the late 1980s the risk of global impacts has captured worldwide attention, in particular the greenhouse effect and the depletion of ozone in the upper atmosphere (note that ozone in the upper atmosphere is good for us because it absorbs ultraviolet radiation from the sun, but at the earth's surface it is harmful to plants and to our health). Buildings are implicated in all these problems.

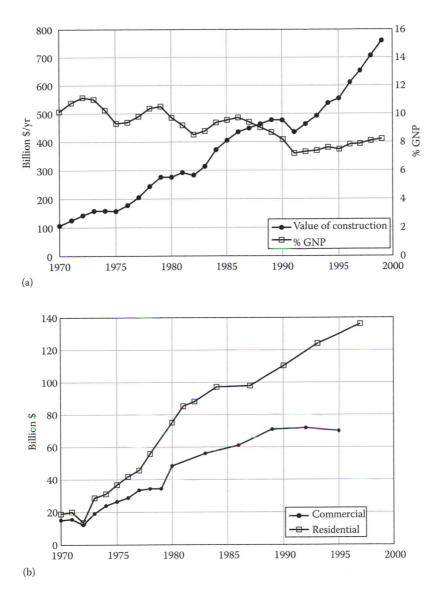

FIGURE 1.5
Cost of construction and cost of energy. (a) Value of annual new construction, in billion dollars and as percent of GNP. (b) Annual energy cost of building stock. (Courtesy of USDOE, *BTS Core Data Book*, Office of Building Technology, June 18, 1999; see also U.S. Department of Commerce, *Statistical Abstract of the United States*, USDOE, Washington, DC, 2001.)

All energy use, other than nuclear or renewable, contributes carbon dioxide (CO_2) to the greenhouse effect. Electricity production from coal or oil entails sizable emissions of NO_x, SO_2, and particulate matter. Natural gas is much cleaner, but the emission of NO_x remains a problem. Chlorofluorocarbons (CFCs), the major cause of ozone depletion, were widely used not only as working fluids in heat pumps and refrigeration equipment, but also in the manufacture of urethane foams for insulation and furniture. During the past decade the Montreal protocol of 1987 has taken effect, limiting the production of CFC11 and CFC12

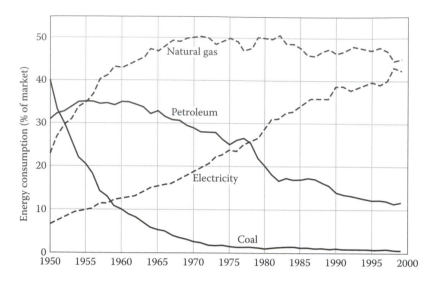

FIGURE 1.6
Relative share of energy sources for commercial buildings in the United States. (Courtesy of Brambley, M.R. et al., Advanced Energy Design and Operation Technologies Research: Recommendations for a U.S. Department of Energy Multiyear Program Plan, Battelle Pacific Northwest Laboratory, Richland, WA, Report PNL-6255 (December), 1988.)

and eventually banning CFCs altogether, with profound consequences for the design of air conditioning systems.

To emphasize the importance of environmental considerations for the HVAC engineer, we cite from the policy statement defining ASHRAE's concern for the environmental impact of its activities: "…ASHRAE's members will strive to minimize any possible deleterious effect on the indoor and outdoor environment of the systems and components in their responsibility while maximizing the beneficial effects these systems provide, consistent with accepted standards and the practical state of the art…."

1.3 Role of HVAC Design Engineer

The goal of rational building design is to provide an environment that is pleasant, comfortable, convenient, and safe—the best one can get for the money one is willing to spend. The design is a collaborative effort between architects and engineers. The domains of primary responsibilities are as follows:

- Architect and civil engineer for building shell and structure.
- Mechanical engineer for HVAC equipment, system integration, and controls.
- Electrical engineer for wiring, lights, and electric equipment.
- Other professionals for acoustics, interior design, security, and safety.

Since the oil crises of the 1970s, there has been a growing awareness of the influence of the building shell and the HVAC system on the energy bill. Significant savings can be realized by paying attention to energy consumption during the early design. Thus considerable feedback between architects and engineers is advisable, for example, to avoid large expanses of glass on south and west facades without control of solar heat gains. This is particularly important in the early steps of the design; otherwise it will be too late to make changes.[1]

The tasks of the mechanical engineer are to calculate the demands for heating, cooling, and ventilation; to choose the necessary equipment and controls; and to ensure that the components are correctly integrated into the building. Obviously the capacities of the equipment must be sufficient to maintain comfortable conditions even under extremes of summer heat and winter cold. Comfort depends not only on temperature but also on humidity. Excessive dryness parches the skin and irritates the throat and nose, even to the point of causing nosebleeds in sensitive individuals. The other extreme is well known to anyone who has visited England in winter or the southern United States in summer. High humidity interferes with the body's capacity for evaporative temperature regulation; it produces a sensation of clammy dampness in winter and of muggy, "sticky" heat in summer. During the 1980s, concern with indoor air quality added another dimension to the tasks of the design engineer.

The HVAC equipment should cost no more than necessary, and thus its capacity should not be excessive. It follows that the first job of the HVAC engineer is the calculation of peak loads. A building must not only function correctly, but also be economical. The annual energy cost is determined by the annual performance. To minimize the true total cost of a building, it is therefore necessary to calculate its annual heating and cooling loads as well, or, what is equivalent, its average loads. Thus the HVAC design process involves basically the following steps:

- Calculation of peak loads.
- Specification of equipment and system configuration.
- Calculation of annual performance.
- Calculation of costs.

The steps are iterative: Having determined the total cost of an initial design, one modifies some of its details and recalculates the total cost until one has come sufficiently close to an optimum.

During the past decades the architectural profession has developed a formal *building design procedure* consisting of several phases, as outlined in Table 1.1. During each phase, a specific level of detail is developed, and an estimate of project cost based on that design is prepared. The owner of the building reviews the output of each design phase, accepts the results, and authorizes the next phase with a formal, written approval of the current phase. The level of detail increases with each phase.

The first phase is *programming*. At this level, initial specifications of the building are made with respect to its volume, usable floor space, shape and orientation, operating schedule, interior temperatures, and ventilation rates.

---

[1] In the past, the load calculations were usually done only after most of the design was already set. Future generations of computers and software will make it easy to update the calculations of loads and costs automatically as the design evolves, thus allowing interactive optimization.

TABLE 1.1

The First Four Phases of the Building Design Procedure

Design Phase	Professional Services	Drawings or Specifications	Load Calculations	Cost Estimates
Programming	Architect Mechanical engineer	Sketches and models; no specifications	Rules of thumb	Rules of thumb
Schematic design	Architect Mechanical engineer Civil engineer	One-line diagrams; outline specifications	Peak and annual (rough)	Handbook values
Design development	Architect Mechanical engineer Civil engineer Electrical engineer Acoustics Lighting Landscaping	Intermediate level of detail; drawings and performance specifications	Peak and annual (final computer simulations)	Informal quotes
Construction documents	Same as design development	Final bid packages including detailed specifications and drawings	None	Formal bids

This is followed by the *schematic design* phase. Here the building loads are estimated, and the components of the HVAC system are specified. The first attempts are made toward optimizing the design.

The third phase is *design development*. Loads are calculated with precision, and all technical details are resolved, such as selection of equipment capacities and flow rates. Life cycle cost (i.e., equipment costs and operating costs, appropriately combined) should be calculated to find the optimum.

During the fourth phase, the *construction documents* are prepared. These involve full written specifications and drawings, including installation diagrams, control procedures, and equipment specifications. A standard format for this purpose has been developed by the Construction Specification Institute. Construction documents are the basis for the bidding phase (which we consider as part of the construction document phase).

The final phase is *contract administration*. The successful bidder is selected, and the building is constructed. As installation is completed, start-up of the equipment begins and a formal acceptance test is done, called *commissioning*. A user's manual is provided.

It is important to carry out the work in an orderly and systematic manner. Careful documentation is crucial, to minimize questions and problems that may arise later. It is best to develop good habits from the start; in particular, show job identification, the date, and your initials on each page.

This book may appear more concerned with commercial than residential buildings. There is a reason. Commercial buildings tend to be larger, and their HVAC systems are more complex than those of residential buildings. Furthermore, reliable performance is more important because of the high value of business transacted (the value per unit floor area is much higher than that in the residential sector). Commercial buildings, at least the larger ones, are usually custom-designed, while houses and their HVAC systems are more often based on mass production. Thus the HVAC designer will earn a living mostly with commercial buildings.

1.4 A Note on the Economics of Energy Efficiency

Making a design choice is easy when option A has a lower first cost (purchase and installation) *and* costs less to operate than an alternative option B. But usually an item that saves operating costs requires a higher initial investment. The optimal choice depends on the circumstances. Compare, for example, two refrigerators, both providing the same quantity and quality of service. Suppose model A costs $600 and draws electricity at an average rate of 150 W, while model B costs $700, but draws only 100 W; the price of electricity is 10¢/kWh. Noting that a year contains 8760 h, we find the difference in annual cost as

$$8760 \text{ h} \times [(0.15 - 0.10)] \text{ kW}] \times \$0.10/\text{kWh} = \$43.80 \text{ annual savings}$$

By paying the extra investment of $100, one saves $43.80 each year, for each of the roughly 20 years that a refrigerator can be expected to last. After $100/43.80 = 2.3$ years, the extra investment has been paid off, so to speak, and thereafter it yields pure profit year after year. This does not seem to be a bad deal when one compares it to typical interest payments from savings accounts. Is such a comparison appropriate? Yes, although with some complication, because at the end of its lifetime the refrigerator is worn out, whereas the money in a savings account remains.

In any case it is clear that choosing on the basis of first costs alone is very shortsighted. The cheap refrigerator, model A, would be very expensive in the long run. A rational decision takes all the costs into account that will be incurred over the lifetime of the item in question; in other words, it is based on an analysis of the *life cycle cost*. One would like to achieve the greatest benefit for the lowest life cycle cost. The building designer is constantly facing decisions of this kind: Is it worth paying the extra first cost for a design or a design element that has lower operating cost? Examples are extra insulation, high-performance glazing, efficient lighting, ducts with reduced pressure drops, and chillers with a high *coefficient of performance* (COP).

A rigorous analysis of such decisions is fairly involved, and we devote a complete chapter to it in Chapter 15. But for the benefit of readers who will not want to study it, we present at this point a very simple, albeit crude, decision tool. It is called *payback time*, defined as the ratio of initial investment to annual savings:

$$\text{Payback time} = \frac{\text{investment}}{\text{annual savings}} \tag{1.1}$$

In our example of the refrigerator, we have already calculated the payback time: It is 2.3 years. Payback time fails to take into account a number of effects, such as the difference in value between a dollar that is available today and one that will be available only in the future. But it has the advantage of simplicity. Roughly speaking, at the end of the payback time, the extra investment has paid for itself, and thereafter it brings a profit. The shorter the payback time, the better. *For investments without significant risk*, one can say as a rule of thumb that the profitability is *excellent* if the payback time is *less than one-third* of the lifetime of the investment; it is *good* if the payback time is *less than one-half* of the lifetime.

Example 1.1

Compare two alternatives for the lamp above your desk at home: an incandescent bulb and a compact fluorescent bulb, both giving light of the same quantity and quality. Make the following assumptions.

Given: First cost of incandescent bulb = $1

First cost of fluorescent bulb = $15

Power drawn by incandescent bulb = 60 W

Power drawn by fluorescent bulb = 15 W

Lifetime of incandescent bulb = 1000 h

Lifetime of fluorescent bulb = 10,000 h

Light is turned on 2000 h/yr

You have to pay for electricity at a price = 15¢/kWh.

Find: Payback time.

SOLUTION

Investment = difference in initial cost = $15−$1 = $14. There are savings in electricity and savings in lightbulb replacement. The difference in annual electricity cost = [(60−15) W] × 2000 h × $0.15/kWh = $13.50. The incandescent bulb has to be replaced twice a year, at a cost of $2 per year. Therefore the annual savings with the fluorescent bulb are $13.50 + $2 = $15.50. From Eq. (1.1) we have payback time = investment/annual savings = $14/$15.50 = 0.90 yr.
   The lifetime of the fluorescent bulb is 10,000 h/(2000 h/yr) = 5 yr.

COMMENTS

1. The payback time is less than one-third of the lifetime, and the extra investment of $14 for the fluorescent bulb is very profitable if there is no risk.
2. How about risk? In this case there is no significant technical risk because the technology is well established by now; compact fluorescent bulbs are reliable. But the utilization might be uncertain, depending on the circumstances. For instance, you might accept a job in an office where your employer will provide all the lighting, and you would no longer be using this lamp as much in future years. In that case the optimal choice depends on the circumstances.

This example illustrates some general aspects of the investment decisions that will confront the building designer. First, the decision involves a calculation of life cycle costs, or as a shortcut, an estimate of payback times. Second, it involves an examination of possible risks. Risks can depend on technology (e.g., how reliable is the equipment?) or on utilization (e.g., how will the building be used in the future?). In many cases, such as extra insulation or high-COP chillers, the risk is negligible because the technology is mature and there is no question about the utilization, so the rule of thumb about payback time and profitability can indeed be applied. Since the lifetimes are often on the order of 20 years, payback times of about 10 years may be quite interesting. This stands in sharp contrast to the very short payback times, often less than 2 years, that are demanded by many industrial decision makers. The explanation lies in the far greater risks of most industrial investments where it is difficult to predict the future more than a few years ahead. But in more stable situations, longer payback times can prevail; for example, many electric power plants have in effect payback times longer than 10 years.

Being concerned with energy in buildings, we mention a further point: the energy used in the process of constructing a building. Fabricating of the steel, glass, concrete, and all the other building materials requires a great deal of energy, and further energy is needed for transport and for construction. This topic, known as *embedded energy*, has attracted much attention. However, we do not believe it to be of immediate concern to the HVAC design engineer, as we explain with the following admittedly simplified argument.

In a free market economy, purchase decisions are based on money (and the communist societies have provided a warning example of what happens when one tries to ignore the principles of a free market). Therefore energy investments must be made on the basis of the monetary value of the energy, not on the basis of its physical quantity. And, of course, one must include all costs, not just the energy.

As for the embedded energy in a building, its cost is implicitly contained in the prices paid for the material and for the construction. For example, the charges for delivering a window to a site include the cost of the fuel for the transport vehicle. With costs one can account for energy, but with energy one cannot account for all costs. Embedded energy does not include the cost of labor; therefore it is not an appropriate criterion for investment decisions.

Of course, an efficient economy also presupposes that social costs are correctly taken into account. As an example of a social cost, suppose you burn a ton of coal without paying for the damage that will be caused by your contribution to acid rain and global warming: You impose a cost on society. The valuation of social costs is a difficult subject. Even more difficult is the implementation of laws, regulations, and taxes that will make people act in a way that is socially optimal. But enormous progress has been made, and continues to be made—that is part of our collective learning process. If policies are formulated correctly, they transmit the correct price signals to everybody, and the resulting decisions are indeed optimal.

It is not the purpose of this book to estimate social costs or to formulate policies. Nor would it be appropriate to try to do so during the design of one building. The designer is not prepared to evaluate the contribution that a building might make to the cost of global warming, but within the context of market prices she or he can and should optimize the design so as to provide the desired conditions of comfort for the lowest life-cycle cost. Developing the necessary tools for that job is the purpose of this book.

1.5 Units and Conversions

In almost all countries of the world, engineers use SI units (Système International d'Unités), a standardized and rational system based on metric units. In SI, all units are derived from a few fundamental units: mass, length, time, temperature, amount of substance, and luminous intensity. Prefixes for powers of 10 are added according to the table inside the cover of this book. This table also provides the conversion factors for the old English units, which are still in use in the United States (but no longer in England). These latter are also known as USCS (U.S. Customary System) or inch-pound units. In this book we use the labels "SI" and "US" in the equation numbers of dimensional equations to indicate the units.

In USCS, a multiplicity of units obscures the relation between equivalent quantities, and the need for conversion factors lurks at every step. For instance, to remove the heat produced by a 100-kW computer center, an engineer in a country with SI units specifies quite simply a chiller with a capacity of 100 kW; and if it is a gas-fired chiller with $COP = 1.1$, the engineer knows at once that the gas company will bill for $(100/1.1)$ kWh of fuel for each hour of full-load operation. The engineer's counterpart in the United States has to convert 100 kW first to 28.4 tons of refrigeration to specify the chiller size and then to Btu per hour because gas is metered in therms (1 therm $= 10^5$ Btu).

The *conversion of units* could be written in several slightly different ways. We have chosen to *multiply each quantity by a conversion factor* that is *equal to unity* and *whose units cancel the units we want to eliminate*. For instance, to convert 30 in to feet, multiply by $1 = 1$ ft/12 in and get

$$30 \text{ in} = 30 \text{ in} \times \frac{1 \text{ ft}}{12 \text{ in}} = 2.5 \text{ ft}$$

This method is systematic and safe.

Note that we *include the units with the numbers*, for good reason. If the units do not come out correctly, the result must certainly be wrong. Keeping the units, at least up to the point of verifying the correct units for the result, is a convenient trick for avoiding the most common errors to which we are prone in such calculations (forgetting a factor or dividing instead of multiplying).

Example 1.2

Water of density $\rho = 62.44$ lb$_m$/ft^3 (1000 kg/m^3) flows at rate $\dot{V} = 10.0$ gal/min (0.6308×10^{-3} m^3/s) through a pipe of interior diameter $D = 1.0$ in (0.0254 m). Find the velocity pressure which is given by the formula

$$p = \frac{\rho v^2}{2}$$

where $v =$ velocity.

    Given: $\rho = 62.44$ lb$_m$/ft^3 (1000 kg/m^3)

$$D = 1.0 \text{ in } (0.0254 \text{ m})$$

$$\dot{V} = 10.0 \text{ gal/min } (0.6308 \times 10^{-3} \text{ m}^3/\text{s})$$

    Assumptions: Assume for simplicity that all the water flows at the same velocity

$$V = \frac{\dot{V}}{D^2 \pi/4}$$

(This is not exact: The water at the center flows faster than the water near the wall.)

    Find: $p = \rho v^2/2$

SOLUTION

In SI units. Here everything is automatically consistent, and the quantities can be combined directly:

$$v = \frac{0.6308 \times 10^{-3} \text{ m}^3/\text{s}}{(0.0254 \text{ m})^2 \pi/4} = 1.245 \text{ m/s}$$

and

$$p = 0.5 \times 1000 \text{ kg/m}^3 \times (1.245 \text{ m/s})^2 = 775 \text{ kg}/(\text{m} \cdot \text{s}^2) = 775 \text{ Pa}$$

since the unit of pressure is $1\ Pa = 1\ N/m^2 = 1\ (kg \cdot m/s^2)/m^2$. This is about 0.8 percent of atmospheric pressure, the latter being 101 kPa under standard conditions.

In USCS units. First one must convert to consistent units:

$$\dot{V} = 10.0\ \text{gal/min} \times \frac{0.00223\ \text{ft}^3/\text{s}}{\text{gal/min}} = 0.0233\ \text{ft}^3/\text{s}$$

and

$$D = 1.0\ \text{in} = 1.0\ \text{in} \times \frac{1.0\ \text{ft}}{12.0\ \text{in}}$$

Hence

$$v = \frac{0.0223\ \text{ft}^3/\text{s}}{(\pi/4)(1/12.0)^2\ \text{ft}^2} = 4.084\ \text{ft/s}$$

Pressure $p = 0.5 \times 62.44\ \text{lb}_m/\text{ft}^3 \times (4.084\ \text{ft/s})^2 = 520.7\ \text{lb}_m/(\text{ft} \cdot \text{s}^2)$.

But this must be converted to USCS units for pressure, using the relation between pound-force and pound-mass. This is usually indicated by including the constant

$$g_c = 32.17\ \text{lb}_m \cdot \text{ft}/(\text{lb}_f \cdot \text{s}^2) \tag{1.2}$$

in the denominator of the equation for velocity pressure, as

$$p = \frac{\rho v^2}{2g_c}$$

In fact, g_c is a conversion factor that is equal to unity.

Dividing by g_c, we obtain

$$p = \frac{520.7\ \text{lb}_m/(\text{ft} \cdot \text{s}^2)}{32.17\ \text{lb}_m \cdot \text{ft}/(\text{lb}_f \cdot \text{s}^2)} = 16.2\ \text{lb}_f/\text{ft}^2$$

There is one more step because HVAC engineers have a predilection for psi ($= \text{lb}_f/\text{in}^2$) or for inWG (inches water gauge):

$$p = 16.2\ \text{lb}_f/\text{ft}^2 \times (1\ \text{ft}/12\ \text{in})^2 = 0.112\ \text{psi}$$

which in turn is equal to

$$p = 0.112\ \text{psi} \times 27.68\ \text{inWG/psi} = 3.11\ \text{inWG}$$

This is about 0.8 percent of standard atmospheric pressure, 14.70 psi.

1.6 Orders of Magnitude

When we start something new, it is a good idea to estimate orders of magnitude before embarking on a full-fledged analysis. The art of doing such back-of-the-envelope calculations is an essential skill of a scientist or an engineer. An understanding of orders of magnitude is a cornerstone of intuition. Often a decision can be made merely on the basis of a simple estimate. Sometimes the input is so uncertain as to render the accuracy of detailed calculations meaningless. Besides, a simple calculation, even if of limited accuracy or valid only under restricted circumstances, can often provide a valuable check for the reliability of the result of more complicated analysis (or point out a bug in a computer program).

A related point is the accuracy to which numbers should be stated. Computers can be quite deceptive when printing output to the nth significant figure. Presenting more figures than are meaningful is not just illusory but also confusing, making the output more difficult to read. Therefore, *round off a final result* to a point consistent with its accuracy, keeping just an extra figure in case of doubt. But note that the relevant criterion depends on the context. For instance, one might want to test the sensitivity of a result to a small variation in the input, even if the input itself is quite uncertain. Also, during an ongoing calculation one must carry a larger number of decimal points to minimize propagation of errors (fortunately, modern computers are so powerful that we rarely need to worry about rounding before the end).

Here is a simple example to illustrate these points and to develop your intuition about the order of magnitude of energy flows in everyday life.

Example 1.3

From what height do you have to jump into a bathtub to warm up the water by $\Delta T = 1°C$ (1.8°F), assuming all the kinetic energy goes into heating the water?

Given: As a reasonable estimate of the water in a bathtub, one can take

$M = 100$ kg (220.5 lb$_m$), and your mass might be $m = 70$ kg (154 lb$_m$).

Find: Height h such that $mgh = Mc_p\Delta T$.

Lookup values: $g = 9.80$ m/s^2 (32.17 ft/s^2),

$c_p = 4.186$ kJ/(kg · °C) [1.0 Btu/(lb$_m$ · °F)].

SOLUTION

In SI units. The required thermal energy is

$$Mc_p\Delta T = 100 \text{ kg} \times 4.186 \text{ kJ/(kg} \cdot °C) \times 1°C = 418.6 \text{ kJ}$$

The height is $h = Mc_p\Delta T/(mg) = 418.6$ kJ/(70 kg × 9.80 m/s^2).

Noting that 1 kJ $= 1$ kg × m^2/s^2, one obtains at once $h = 610$ m.

Since this is only an order-of-magnitude estimate, it is appropriate to state the result as 600 m.

However, if this number is to serve as input for further calculation, the full figures should be kept (e.g., if we want to convert it to USCS units for comparison with the result of the following calculation in USCS units).

In USCS units. The required thermal energy is

$$Mc_p\Delta T = 220.5 \text{ lb}_m \times 1.0 \text{ Btu}/(\text{lb}_m \cdot {}^\circ\text{F}) \times 1.8{}^\circ\text{F} = 396.9 \text{ Btu}$$

To proceed, one needs to know the equivalence between mechanical and thermal energy in USCS units; it is

$$1 \text{ Btu} = 778.16 \text{ ft} \cdot \text{lb}_f \qquad\qquad (1.3)$$

The height is found after adding the factor $g_c = 32.17 \text{ lb}_m \cdot \text{ft}/(\text{lb}_f \cdot \text{s}^2)$:

$$
\begin{aligned}
h &= \frac{Mc_p\Delta T g_c}{mg} \\
&= \frac{396.9 \text{ Btu} \times 778.16 \text{ ft} \cdot \text{lb}_f/\text{Btu} \times 32.17 \text{ lb}_m \cdot \text{ft}/(\text{lb}_f \cdot \text{s}^2)}{154 \text{ lb}_m \times 32.17 \text{ ft/s}^2} \\
&= 2005 \text{ ft} \approx 2000 \text{ ft}
\end{aligned}
$$

COMMENTS

In everyday life, warming up a bath by 1°C does not seem like a big deal. By contrast, climbing 600 m will take an average person around 2 h. In psychological terms, the magnitudes appear very different.

This relation between thermal and mechanical energy quantities of everyday life is quite typical. Thermal energy dominates the energy balance of buildings, while the contribution of gravitational energy is small or negligible. An intermediate role is played by the mechanical energy needed to overcome the friction of transporting fluids around a building. The HVAC designer does have to pay some attention to the energy consumption of pumps and fans. The latter can be a fairly important item, because of the large friction of air in typical air distribution systems, as illustrated by the following example.

Example 1.4

A building is to be cooled to $T_i = 24°C$ (75.2°F) by supplying air at flow rate $\dot{m}$ and temperature $T_s = 13°C$ (55.4°F). To move the air, the fan of the central air handler produces a pressure increase $\Delta p = 1000$ Pa (≈ 4 inWG). At what temperature $T'_s = T_s - \Delta T$ must the air enter the fan to compensate for the heat contributed by the fan?

Find: $\Delta T = T_s - T'_s$

Given: $\Delta p = 1000$ Pa

Assumptions: Here we anticipate the results of Section 5.4. According to Eq. 5.46, the work input by the fan is equal to

$$\dot{W}_{\text{fluid}} = \frac{\dot{m}\Delta p}{\rho}$$

and the resulting temperature rise ΔT is related to $\dot{W}_{\text{fluid}}$ by Eq. 5.51 (where for simplicity we neglect the fan efficiency by setting $\eta_f = 1$):

$$\dot{W}_{\text{fluid}} = \dot{m}c_p\Delta T$$

Combining these terms and canceling the flow rate $\dot{m}$, we have as the basic equation for this problem

$$\frac{\Delta p}{\rho} = c_p\Delta T$$

or

$$\Delta T = \frac{\Delta p}{\rho c_p}$$

Lookup values: The product of density and the specific heat of air is $\rho c_p = 1.2$ kJ/(m$^3 \cdot$ °C).

SOLUTION
Inserting $\Delta p = 1000$ N/m^2 and

$$\rho c_p = 1200 \ \text{N} \cdot \text{m}/(\text{m}^3 \cdot °\text{C})$$

we obtain

$$\Delta T = \frac{1000 \ \text{N/m}^2}{1200 \ \text{N} \cdot \text{m}/(\text{m}^3 \cdot °\text{C})} = 0.83°\text{C} \ (\approx 1.5°\text{F})$$

COMMENTS
The system supplies cooling at a rate $\dot{m}c_p(T_i - T_s)$, with $T_i - T_s = 11$ K. Thus the fan increases the cooling load by about 10 percent. These values are quite typical of large central air conditioning systems.

Problems

The problems in this book are arranged by topic. The approximate degree of difficulty is indicated by a parenthetic italic number from 1 to 10 at the end of the problem. Problems are stated most often in USCS units; when similar problems are presented in SI units, it is done with approximately equivalent values in parentheses. The USCS and SI versions of a problem are not exactly equivalent numerically. Solutions should be organized in the same order as the examples in the text: given, figure or sketch, assumptions, find, lookup values, solution. For some problems, the Heating and Cooling of Buildings (HCB) software included in a pocket inside the back cover of this book may be helpful. In some cases it is advisable to set up the solution as a spreadsheet, so the design variations are easy to evaluate.

The main purpose of the problems in Chapter 1 is to develop a sense of the order of magnitude of energy flows in and around buildings. For several of these problems you will need to make your own assumptions. State them clearly. Data on properties of materials can be found in the appendices of this book.

1.1 You are measuring electricity consumption by various end uses in a building. The sensors you use to make the measurements are accurate to about 1 percent of the reading. You are able to measure the following items separately and find that

- The telephone system uses 20 W.
- The lights use 10.3 kW.
- The computers and office equipment use 15.4 kW.
- The HVAC system uses 45 kW.

If these are the only electrical loads in the building, what is the total electrical consumption? How many figures are significant?   (1)

1.2 You are an American working with a European customer and must convert all USCS units to SI units. What are the equivalent SI units for the following?

- 13.4 MMBtu/hr
- 5 tons of refrigeration
- 25 psia
- 120 gallons per minute
- 0.5 Btu per hour per ft^2 per °F   (2)

1.3 You see the following numbers in an equipment catalog from a Japanese manufacturer.

3.5 dynes/cm^2

1.3 m^3 per second

23.1 N-m (torque)

1.2 kg/m^3

0.6 kJ/kg · °C

You want to sell them to a U.S. engineer. What are the equivalent USCS values?   (2)

1.4 A building with an interval volume of 12,450 ft^3 experiences natural infiltration of 0.333 air change per hour. What is the infiltration rate in cubic meters per second?   (2)

Helpful hint: 1 yr $= 8760$ h $= 3.1536 \times 10^7$ s (some find it convenient to remember as rule of thumb that 1 yr $\approx \pi \times 10^7$ s).

1.5 List the major pieces of energy-consuming equipment in your everyday life. For each, estimate the peak power, average power, and annual consumption.   (2)

1.6 Consider a residential furnace that is rated for a heat output of 50,000 Btu/h. Express the heat output in SI units. How many 100-W light bulbs would produce the same amount of heat?   (2)

1.7 Convert the horsepower rating of your car (or the car you would like to have) to kilowatts (the unit that is now officially used for cars in Europe).   (2)

1.8 Estimate the rate of energy transfer during the filling of the tank of a typical car at a gas station (heating value of gasoline $\approx 3.4 \times 10^7$ J/L). Compare with a typical all-electric residence (1 kW continuous average, without space heating).   (2)

1.9 Energy prices for fuels are usually quoted in terms of dollars per volume or per weight. Convert the following to cents per kilowatt hour and to dollars per MBtu.

(a) Gasoline at \$1.20 per gallon (heating value $\approx$ 21,000 Btu/lb)

(b) Crude oil at \$20 per barrel (heating value $\approx$ 18,000 Btu/lb)

(c) Coal at \$60 per ton (heating value $\approx$ 12,000 Btu/lb)

(d) Electricity at 10¢/kWh   (3)

1.10 A typical house in a temperate climate [on the order of 2500 K · days (4500°F · days), e.g., Albuquerque; Louisville, Kentucky; Washington, D.C.; or Paris] needs about 50 to 100 GJ of heat per year. Suppose this energy were collected as solar heat in summer at a temperature of 70°C (158°F) and stored. How large a storage tank would be needed if useful heat could be withdrawn as long as the tank is above 30°C (86°F)? Neglect losses from the tank. (Such seasonal storage has been proposed and tested in various places.)   (4)

1.11 Consider a house with a 1-ton air conditioner, running 500 h/yr at full power.

(a) What is the annual thermal energy delivered to the house?

(b) What is the corresponding electricity consumption, if the air conditioner delivers 2.0 J_t of cooling for 1.0 J_e of electricity [coefficient of performance (COP) = 2.0]?

(c) Suppose one stores winter ice in an ice tank for summer cooling. How large a volume is needed, if there are no losses from storage? Consider only the latent heat of melting, and take the density of ice as 0.9 t/m^3. (Such seasonal storage has been proposed and tested in various places.)

(d) How much is the cooling energy of 1 ton (2205 lb$_m$) of ice worth?

Note: The use of residential air conditioning depends on climate and on the construction of the building, but there is a strong behavioral component as well. In midlatitudes of the United States, the seasonal energy consumption for cooling in identical buildings can vary enormously, depending on how the occupants set the thermostat and how they control the heat gains (temperature preferences, utilization of lights and appliances, use of shades, and night ventilation).   (5)

1.12 A typical personal computer draws 100 W$_e$, while turned on, almost independent of how it is being used. Estimate the number of hours of utilization per year and the annual energy cost, if the price of electricity is 10¢/kWh$_e$.   (4)

1.13 The *Reynolds number* Re is a dimensionless group of variables that plays an important role in fluid mechanics. It represents the ratio of inertial and viscous forces in a fluid, and it is defined as

$$\text{Re} = \frac{vD}{v}$$

where v = velocity, D = characteristic linear dimension of system, and v = kinematic viscosity. Evaluate Re for air ($v = 15 \times 10^{-6}$ m^2/s) if $v = 10$ m/s and $D = 1$ m, in both SI and USCS units.   (3)

1.14 The average solar flux incident on a horizontal surface in the United States (averaged over 24 hours and 365 days) is about 180 W/m^2 [57.06 Btu/(h · ft^2)]. How much land would be needed to supply the total energy demand of the United States, approximately 80×10^{18} J (76 quad) per year, by means of photovoltaics (efficiency around 10%) and biomass (efficiency around 1%)? Compare with the land area of the United States, 9.16×10^6 km^2 (3.54×10^6 mi^2). (5)

1.15 A large coal or nuclear power plant has a peak output on the order of 1 GW$_e$. Over the year, its actual output is likely to average only 70% of that, due to the need for maintenance and due to an imperfect match between the supply and demand of electricity.

(a) What is the annual energy output, in kilowatthours, in joules, and in Btu?

(b) The year-round average residential electricity use amounts to about 1 kW$_e$ per housing unit in the United States. Assuming 3.2 residents per housing unit, estimate roughly how many power plants are needed to satisfy the residential demand of the New York metropolitan area (population around 20 million).

(c) The heating value of coal is approximately 12,000 Btu/lb, and the conversion efficiency of coal power plants is around 33%. Calculate the quantity of coal needed to satisfy the annual electricity consumption of a housing unit in (b). (5)

1.16 Suppose an air-to-air heat exchanger (for recovering heat from ventilation air) costs \$1200 and reduces the heating bill by \$200 per year. What is the payback time? (3)

1.17 Consider two choices for the furnace to heat a building: an ordinary gas furnace with efficiency 75% (seasonal average), costing \$1000, and a condensing furnace with efficiency 95% (seasonal average), costing \$2000. Suppose the annual heating load is 100 MBtu (105.5 GJ) and the fuel price \$7.00 per MBtu (\$6.64 per GJ). Calculate the payback time for the second furnace relative to the first. Is it a good investment, if both furnaces can be expected to last 15 yr? (4)

1.18 Suppose the annual heating bill of a building is \$1000. How much could you pay for a solar heating system that reduces this bill by one-half, if the payback time is to be no longer than 10 yr? (3)

References

ASHRAE (1976). *ASHRAE SI Metric Guide for Heating, Refrigerating, Ventilating and Air-Conditioning*. American Society of Heating, Refrigerating and Air-Conditioning Engineers, Atlanta, GA.

Brambley, M. R., D. B. Crawley, D. D. Hostetler, R. C. Stratton, M. S. Addison, J. J. Deringer, J. D. Hall, and S. E. Selhowitz (1988). Advanced Energy Design and Operation Technologies Research: Recommendations for a U.S. Department of Energy Multiyear Program Plan, Battelle Pacific Northwest Laboratory, Richland, WA, Report PNL-6255 (December).

Butti, K., and J. Perlin (1980). *A Golden Thread*. Cheshire Books, Palo Alto, Calif. A condensed version can be found in chap. 1 of J. F. Kreider and F. Kreith, eds., *Solar Energy Handbook,* McGraw-Hill, New York, 1980.

Rosenfeld, S. I. (1990). "Worker productivity: Hidden HVAC cost." *Heating, Piping and Air Conditioning*. September, pp. 117–119.

SERI (1981). *A New Prosperity—Building a Sustainable Energy Future*. Solar Energy Research Institute. Published by Brickhouse Publishing, Andover, Mass.

SERI (1985). *The Design of Energy-Responsive Commercial Buildings*. Solar Energy Research Institute. Published by Wiley, New York.

USDOC (1987). *U.S. Statistical Abstract*. Published annually by the U.S. Department of Commerce. Bureau of the Census. Government Printing Office, Washington, DC. Editions of 1979, 1987, and 1989.

USDOE (1999). *BTS Core Data Book*, Office of Building Technology, June 18; see also U.S. Department of Commerce, *Statistical Abstract of the United States*, USDOE, Washington, DC, 2001.

2

Elements of Heat Transfer for Buildings

2.1 Introduction to Heat Transfer

The transfer of heat is the principal mechanism by which environmental effects are manifested within buildings. Conduction of heat through a building's skin, transmission of solar radiation through windows, and cooling of occupants by ventilation are all examples of how heat transfer affects the thermal behavior of buildings and their occupants. This chapter reviews the key features of heat transfer that are needed to analyze and predict building thermal performance, heating and cooling loads, and equipment performance.

The goals of this chapter are to

- Review heat transfer essentials needed for HVAC load and system design and analysis.
- Provide basic reference material for the loads and systems chapters, Chaps. 6 to 11.
- Introduce *optional* material on building-to-earth heat transfer.

For those students who have had a first course in engineering heat transfer, we suggest that this chapter be used only for the second and third purposes. It need not be a part of the chapters formally covered in this course for students who have previously taken at least one heat transfer course, except for optional coverage of Sec. 2.2.3 on building-to-ground heat transfer.

Heat transfer is the science and art of predicting the rate at which heat flows through substances under various external conditions of imposed temperature and/or heat flow. The laws of heat transfer govern the rate at which heat energy must be supplied to or removed from a building to maintain the comfort of occupants or to meet other thermal requirements in buildings. This chapter is designed to be a review of the heat transfer fundamentals that one needs to calculate heating and cooling requirements of buildings and their HVAC equipment. Standard heat transfer texts listed in the References contain more details on each topic if needed by the reader (Holman, 1997; Incropera and Dewitt, 1996; Karlekar and Desmond, 1982).

There are three fundamentally different types of heat transfer that we will review generally in this chapter:

Conduction heat transfer is a result of molecular-level kinetic energy transfers in solids, liquids, and gases. There is a strong correlation between thermal conduction and electrical conduction in solids. Conduction heat flows occur in the direction of decreasing temperature. An example of the flow of heat by conduction is the heat loss through the opaque walls of buildings in winter.

Convection heat transfer is a result of larger-scale motions of a fluid, either liquid or gas. The higher the velocity of fluid flow, the higher the rate of convection heat transfer, in general. Convection heat loss occurs, e.g., when a cold wind blows over a person's skin and removes heat from it.

Radiation heat transfer is the transport of energy by electromagnetic waves. No material other than the surfaces exchanging energy need be present for radiation to occur. The sole requirement for radiation heat transfer to occur is the presence of two surfaces at different temperatures. Radiation must be absorbed by matter to produce internal energy. Energy is transported from the sun to earth by means of radiation, for example.

There are three heat quantities used in this book. First, the quantity of heat is denoted by Q, which has units of joules (or kilowatthours) or British thermal units. Second, the heat rate or power is denoted by $\dot{Q}$, with units of watts or British thermal units per hour. Third, the heat flux or heat rate per unit area is denoted by $\dot{q}$. Heat flux has units of watts per square meter or British thermal units per hour per square foot.

2.2 Conduction Heat Transfer

2.2.1 Fourier's Law and Steady Conduction

Joseph Fourier, in the early nineteenth century, set forth the law of the conduction heat transfer in his fundamental treatise, entitled *The Analytical Theory of Heat*. Fourier's law states that the rate of heat transfer by conduction $\dot{Q}$ is proportional to the temperature difference and the heat flow area, whereas $\dot{Q}$ is inversely proportional to the distance through which conduction occurs. Fourier's law is a property of matter, not a fundamental law of physics such as the conservation of mass. This law is similar to Ohm's law which governs the flow of electricity in electrical conductors. Current flow is proportional to the voltage difference and inversely proportional to the resistance of the material. We shall find the analogy between heat and current flows useful shortly.

As an equation, Fourier's law in one dimension with constant thermal properties is given by

$$\dot{Q} = -kA\frac{dT}{dx} \tag{2.1}$$

in which k is the thermal conductivity in units of Btu/[h $\cdot$ ft^2/($^\circ$F/ft)], which converts to Btu/(h $\cdot$ ft $\cdot$ $^\circ$F), or, in SI units, W/(m $\cdot$ K); A is the area through which heat flows; and the temperature gradient dT/dx is evaluated at point x where the rate of heat transfer is to be determined.

The second law of thermodynamics, which we review in the next chapter, requires that heat flow only from a high temperature to a low temperature. Therefore, a minus sign is needed in Eq. (2.1) since positive heat transfer is taken to occur in the positive x-coordinate direction. Figure 2.1 shows the temperature profile in a one-dimensional heat transfer problem with constant thermal conductivity. Heat flows to the right (positive heat flow) if the temperature decreases to the right.

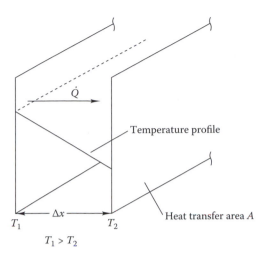

$T_1 > T_2$

FIGURE 2.1
Heat flow through a plane wall.

2.2.1.1 Steady Conduction in Plane Walls

Fourier's law can be integrated in Cartesian coordinates along a path of constant $\dot{Q}$. For the case shown in Fig. 2.1, if the temperature depends only on x, if there are no heat sources in the material, and if the thermal conductivity can be taken as a constant (not a bad assumption for most building materials, as noted below), then the result is

$$\dot{Q} = kA\frac{T_1 - T_2}{\Delta x} \tag{2.2}$$

where
$k =$ thermal conductivity, Btu/(h · ft · °F) [W/(m · K)]
$T_1 =$ higher temperature
$T_2 =$ lower temperature
$A =$ area through which conduction occurs
$\Delta x =$ thickness of material in which conduction occurs

Expressed slightly differently, this is

$$\dot{Q} = \frac{T_1 - T_2}{\Delta x/(kA)} \tag{2.3}$$

The denominator is often called the *resistance to heat transfer*

$$R = \frac{\Delta x}{kA} \ (\text{h} \cdot °\text{F})/\text{Btu} \ [\text{K/W}] \tag{2.4}$$

by analogy with electric resistance, which serves as the proportionality constant between voltage and current in Ohm's law. If one takes the temperature difference in the numerator of Eq. (2.3) to be the driving force analogous to voltage and the heat flow to be analogous to current flow, then R is the thermal analog of electric resistance.

The commonly used term *R value* is the *unit thermal* resistance:[1]

$$R_{th} = \frac{\Delta x}{k} = AR \ (h \cdot ft^2 \cdot {}^{\circ}F)/Btu[(m^2 \cdot K)/W] \tag{2.5}$$

One use of the thermal resistance concept is shown in Fig. 2.2a to c. Like electrical resistors, thermal "resistors" can be connected in series and in parallel or in combinations thereof. Figure 2.2b depicts a three-layer wall and shows the equivalent thermal model consisting of three resistors in series, whereas Fig. 2.2c shows a two-part wall, e.g., a residential wall consisting of studs (*B*) with insulation (*A*) between, through which heat flows by parallel paths, one through the studs and the other through the insulation.

Another convenient measure of thermal conductance is called the *unit conductance*, or *U value*. It is just the inverse of the *R* value:

$$U \equiv \frac{1}{R_{th}} \tag{2.6}$$

The use of R_{th} or U to solve a given problem depends on which is more convenient, as we see in the next two examples.

Example 2.1: *R* Value Calculation for a Building Wall

The outside wall of a home consists essentially of a 10-cm layer of common brick [$k=0.68$ W/(m · K)], a 15-cm layer of fiberglass insulation [$k=0.038$ W/(m · K)], and a 1-cm layer of gypsum board [($k=0.48$ W/(m · K)]. What is the overall *R* value? What is the heat flux through this wall if the interior surface temperature is 22°C and the exterior surface is 5°C?

Given: $k_1, k_2, k_3, \Delta X_1, \Delta X_2, \Delta X_3$

Figure: See Fig. 2.2b.

Assumptions: Steady-state, one-dimensional conduction is the only heat transfer mode; *k*'s are constant.

Find: Overall *R* value and $\dot{q}$

SOLUTION

The *R* values are all given by Eq. (2.5). The *R* value of the outer layer is

$$R_{th,1} = \frac{0.10 \text{ m}}{0.68 \text{ W}/(\text{m} \cdot \text{K})} = 0.147 \ (m^2 \cdot K)/W$$

The *R* value of the center layer is

$$R_{th,2} = \frac{0.15 \text{ m}}{0.038 \text{ W}/(\text{m} \cdot \text{K})} = 3.947 \ (m^2 \cdot K)/W$$

And the *R* value for the inner gypsum board layer is

$$R_{th,3} = \frac{0.01 \text{ m}}{0.48 \text{ W}/(\text{m} \cdot \text{K})} = 0.021 \ (m^2 \cdot K)/W$$

---

[1] Some authors use the opposite convention for R and R_{th}. However, to comply with the electrical analogy, we use the convention noted.

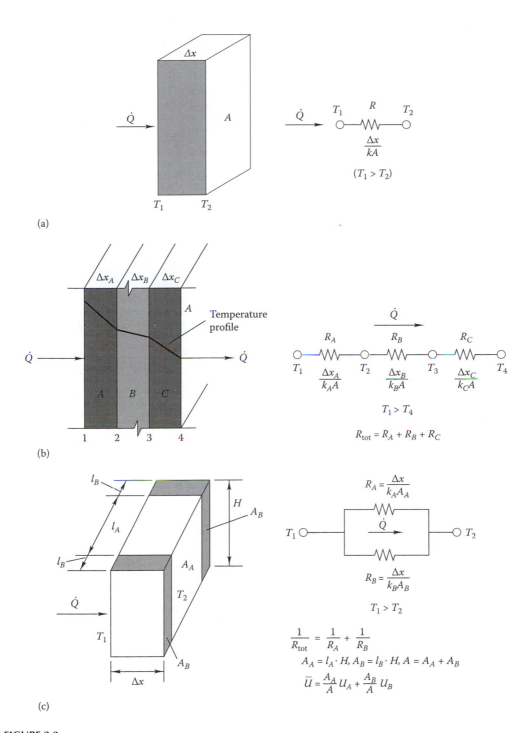

FIGURE 2.2
Electric resistance analog to (a) heat flow, (b) multiple-layer series-resistance heat flow, and (c) multiple-layer parallel-resistance heat flow.

The overall R value of the wall is just the sum of the three individual values in keeping with the electric series-resistance model: 4.115 $(m^2 \cdot K)/W$ or 23.3 $(ft^2 \cdot °F \cdot h)/Btu$.

One can find the heat flux by combining Eqs. (2.3) and (2.4) to get

$$\dot{q} = \frac{\Delta T}{R_{th}}$$

in which $\Delta T = T_{hot} - T_{cold}$. Since a unit wall area is used, by definition, to find the heat flux, R and R_{th} are the same numerically (but not dimensionally). Substituting values for R and the temperature difference, we find that the heat flux is

$$\dot{q} = \frac{(22 - 5)\ K}{4.115\ (m^2 \cdot K)/W} = 4.13\ W/m^2 \left[1.31\ Btu/(h \cdot ft^2)\right]$$

COMMENTS

The majority (96 percent) of the R value is due to the insulation. This is the thickest layer of the wall, but more importantly it is also the layer with the lowest thermal conductivity. In the future, do not routinely add R values to find the thermal resistance. This is correct only if the areas of all wall components are the same. Note that in this example we specified surface temperatures, not the interior and exterior air temperatures adjacent to the wall surfaces. This problem must await the discussion of thermal resistance due to convection in Sec. 2.3.

Finally, observe that the construction of the wall considered here is not practical since there is no structural connection between the outer (brick) and inner (gypsum board) parts. The physical connection must be accounted for by the studs (either metal or wood) whose heat transfer rate is higher than that of the insulation. Example 2.2 considers the *framing adjustment* that is needed.

Example 2.2: Effect of Studs on Wall Heat Loss

For structural reasons the wall described in Example 2.1 must have studs placed every 60 cm (24 in). The studs are fabricated from wood [conductivity 0.10 $W/(m \cdot K)$] and are 5 cm wide and 15 cm deep (note that this is slightly larger than the standard nominal 2-in by 6-in wood stud used in U.S. construction). Find the R value and heat flux $\dot{q}$, and compare with the results of Example 2.1 to quantify the effect of framing with studs ignoring the brick and gypsum board.

   Given: Data from Example 2.1 and stud dimensions

   $l_A = 60 - 5 = 55$ cm

   $l_B = 5$ cm

   $\Delta x = 15$ cm

   $k_{stud} = 0.10\ W/(m \cdot K)$

   Figure: See Fig. 2.2c.

   Find: $R, \dot{Q}$

SOLUTION

From Fig. 2.2c we see that the overall thermal resistance of the parallel network is given by (ignoring the gypsum and brick effects)

$$\frac{1}{R} = \frac{1}{R_{ins}} + \frac{1}{R_{stud}}$$

in which the thermal resistances are (for a unit wall height $H = 1$ m)

$$R_{\text{ins}} = \frac{\Delta x}{K_{\text{ins}}(H \times l_A)} = \frac{0.15 \text{ m}}{[0.038 \text{ W}/(\text{m} \cdot \text{K})](1 \text{ m} \times 0.55 \text{ m})} = 7.18 \text{ K/W}$$

and $$R_{\text{stud}} = \frac{\Delta x}{K_{\text{stud}}(H \times l_B)} = \frac{0.15 \text{ m}}{[0.10 \text{ W}/(\text{m} \cdot \text{K})](1 \text{ m} \times 0.05 \text{ m})} = 30.0 \text{ K/W}$$

The overall thermal resistance is

$$\frac{1}{R} = \frac{1}{7.18} + \frac{1}{30.0} = 0.173 \text{ W/K}$$

$$R = 5.79 \text{ K/W}$$

Finally, the heat flow through the 60-cm by 1-m wall is

$$\dot{Q} = \frac{(22 - 5) \text{ K}}{5.79 \text{ K/W}} = 2.93 \text{ W}$$

For comparison with Example 2.1, the heat flux $\dot{q}$ is needed:

$$\dot{q} = \frac{2.93 \text{ W}}{0.60 \text{ m} \times 1 \text{ m}} = 4.88 \text{ W/m}^2$$

The effective R value *averaged over the wall* is

$$R_{\text{th}} = \frac{\Delta T}{\dot{q}} = \frac{(22 - 5) \text{ K}}{4.88 \text{ W/m}^2} = 3.5 \text{ (K} \cdot \text{m}^2)/\text{W} \left[19.8 \text{ (h} \cdot \text{ft}^2 \cdot {}^\circ\text{F})/\text{Btu} \right]$$

COMMENTS

Because of the presence of wood studs, the heat flux through this wall is increased by 18 percent ($4.88/4.13 = 1.18$), even though the studs represent only 8 percent of the wall heat flow area, because the thermal conductivity of wood is about 2.5 times that of insulation. The stud is a simple example of a *thermal bridge,* a part of the building envelope with higher than average thermal losses. (See Sec. 2.4.6.) (To be very accurate recall that we ignored the 4 percent gypsum and brick effect.)

This example can be simplified by using U values. It is easy to show that the average U value of this composite wall is the area-weighted average of the U values of the components. For this example, using U values provides a shortcut:

$$U_{\text{ins}} = \frac{K_{\text{ins}}}{\Delta x_{\text{ins}}} = \frac{0.038 \text{ W}/(\text{m} \cdot \text{K})}{0.15 \text{ m}} = 0.253 \text{ W}/(\text{m}^2 \cdot \text{K})$$

$$U_{\text{stud}} = \frac{K_{\text{stud}}}{\Delta x_{\text{stud}}} = \frac{0.10 \text{ W}/(\text{m} \cdot \text{K})}{0.15 \text{ m}} = 0.667 \text{ W}/(\text{m}^2 \cdot \text{K})$$

$$\bar{U} = \left(\frac{5 \text{ cm}}{60 \text{ cm}} \right)(0.667) + \left(\frac{55 \text{ cm}}{60 \text{ cm}} \right)(0.253) = 0.288 \text{ W}/(\text{m}^2 \cdot \text{K})$$

The heat flow is then

$$\dot{Q} = \bar{U}A\Delta T = [0.288 \ \text{W}/(\text{m}^2 \cdot \text{K})](0.60 \ \text{m} \times 1 \ \text{m})[(22 - 5) \ \text{K}] = 2.93 \ \text{W}$$

confirming the previous calculation.

We have the following rule: *For parallel heat flow paths, the U values are area-averaged; for series heat flow paths, the R values are added* (when the heat flow areas are the same, as is usually the case in buildings).

The previous two examples are typical textbook examples in that perfect construction and installation have been intrinsically assumed. In practice, the HVAC engineer should exercise caution when calculating building heat losses to account for less than ideal construction. For example, insulation is often compressed or does not completely fill stud cavities in roofs and walls. As a result, the actual thermal resistances are lower than one would calculate by using the idealized equations given above.

The Appendix CD-ROM contains lengthy tables of wall, door, roof, and window $\bar{U}$ values for use in building load calculation (see Chap. 7).

2.2.1.2 Steady Conduction in Cylindrical Coordinates

Fourier's law applies to geometries other than plane walls. Of particular interest in buildings is the cylindrical geometry. Heat losses from piping in building HVAC systems are significant parasitic losses that merit attention. The heat transfer through a cylindrical solid such as that in Fig. 2.3a is also governed by Fourier's law. When one integrates Eq. (2.1) in cylindrical coordinates, using the same assumptions as before, the result is

$$\dot{Q} = \frac{T_1 - T_2}{\ln(r_o/r_i)/(2\pi kL)} \tag{2.7}$$

in which the geometric parameters are shown in Fig. 2.3a. The denominator in Eq. (2.7) is the thermal resistance for steady, cylindrical conduction.

The concept of resistance enables us to calculate the heat flow through layered, cylindrical walls much as we did for plane walls in Example 2.1. Figure 2.3b shows the method one would use. For example, layer A could represent a pipe carrying a fluid. Layer B could be an insulation layer, and layer C could be a protective jacket over the insulation.

2.2.1.3 Steady Conduction in Other Geometries

In building-related heat transfer analyses, there are often situations where the heat conduction is not strictly one-dimensional. An example is a pipe carrying a heated or cooled fluid from a central plant, underground to a building for heating or cooling. There are two methods for treating such cases. The first, which we discuss in this section, uses an expression called the *shape factor* to account for the two-dimensional effects. The second approach requires a full two-dimensional analysis and is not covered in detail in this book. However, results from such an analysis are presented in a later section for several

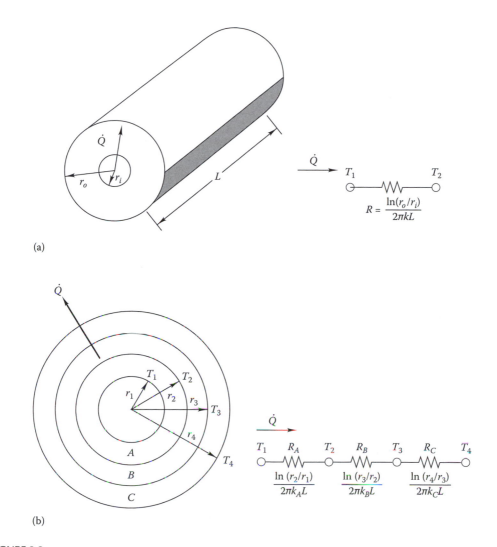

FIGURE 2.3
(a) Conduction in cylindrical coordinates with electric resistance analog circuit. (b) Multiple-layer conduction in cylindrical coordinates with electric resistance analog circuit.

cases having to do with the transfer of heat from the floors and basements of buildings to the earth.

The shape factor approach uses the following basic equation to find steady conduction heat transfer rates:

$$\dot{Q} = kS\,\Delta T \qquad (2.8)$$

in which S is the shape factor, given in Table 2.1, and k is the thermal conductivity. The following example shows how one uses the shape factor approach to find the loss of heat from a buried pipe carrying a hot fluid.

TABLE 2.1

Conduction Shape Factors

Physical System	Schematic	Shape Factor	Restrictions
Isothermal cylinder of radius r buried in semi-infinite medium having isothermal surface		$\dfrac{2\pi L}{\cosh^{-1}(D/r)}$	$L \gg r$
		$\dfrac{2\pi L}{\ln(2D/r)}$	$L \gg r$ $D > 3r$
		$\dfrac{2\pi L}{\ln\left(\dfrac{L}{r}\right)\left\{1 - \dfrac{\ln[L/2D)]}{\ln(L/r)}\right\}}$	$D \gg r$ $L \gg D$
Conduction between two isothermal cylinders buried in infinite medium		$\dfrac{2\pi L}{\cosh^{-1}\left(\dfrac{D^2 - r_1^2 - r_2^2}{2r_1 r_2}\right)}$	$L \gg r_1, r_2$ $L \gg D$
Conduction through two plane sections and the edge section of two walls of thermal conductivity k—inner and outer surface temperatures uniform		$\dfrac{al}{\Delta x} + \dfrac{bl}{\Delta x} + 0.54l$	

Source: Courtesy of Holman, J.P., *Heat Transfer*, 8th edn, McGraw-Hill, New York, 1997. With permission.

Example 2.3: Heat Loss from a Buried Pipe

A pipe with an outer surface temperature of 100°C and a radius of 15 cm is buried 30 cm deep in earth that has a thermal conductivity of 1.7 W/(m · K). If the surface temperature of the earth is 20°C, what is the heat loss for a pipe length of 10 m if the pipe is uninsulated?

Given: $k_{earth} = 1.7$ W/(m · K)

$T_{pipe} = 100°C$

$T_{earth} = 20°C$

$r_{pipe} = 15$cm

$D_{pipe} = 30$cm

Figure: See Table 2.1, upper panel.

Assumptions: There is steady, two-dimensional conduction; the thermal resistance between the pipe and earth is negligible. The pipe surface temperature remains constant over the 10-m length being considered.

Find: $\dot{Q}$

SOLUTION

The proper shape factor expression to be used from Table 2.1 depends on the ratio of the pipe length to the pipe radius. From the given data, the first equation in the table applies, i.e.,

$$S = \frac{2\pi L}{\cosh^{-1}(D/r)}$$

The inverse hyperbolic cosine value is

$$\cosh^{-1}\frac{D}{r} = \cosh^{-1}\frac{30}{15} = 1.32$$

from which the shape factor for the 10-m length of pipe is

$$S = \frac{2\pi(10)}{1.32} = 47.6 \text{ m}$$

and the heat loss is

$$\dot{Q} = kS\Delta T = [1.7 \text{ W/(m} \cdot \text{K)}](47.6 \text{ m})[(100 - 20)°\text{C}] = 6474 \text{ W } (22{,}096 \text{ Btu/h})$$

COMMENTS

Since the pipe is much longer than its diameter, the assumption of two-dimensional heat flow is appropriate. Note that the value of thermal conductivity to be used in such a calculation depends strongly on the soil moisture content, as discussed in a later section.

The two-dimensional heat flow situation shown in the bottom panel of Table 2.1 is often encountered in buildings. An example is the heat loss through insulated rectangular ducts carrying conditioned air. Note that the "corner" term $0.54l$ can have a nontrivial effect if the duct is relatively small with thick insulation. However, for large ducts, the effect is relatively smaller.

2.2.2 Thermal Conductivity

In previous sections, the thermal conductivity was seen to be the key physical property of materials governing the rate of conduction heat transfer. Table 2.2 shows the wide range of conductivities exhibited by common materials. Table 2.3 shows values of k in SI and USCS units for specific materials often used in buildings. The tables in the Appendix contain more complete tabulations of conductivities for many materials used in buildings.

The thermal conductivity of building materials is often assumed to be a constant, independent of temperature. Figure 2.4 shows why this is an excellent approximation.

TABLE 2.2

Representative Magnitudes of Thermal Conductivity

Material	Conductivity, Btu/(h · ft · °F)	Conductivity, W/(m · K)
Atmospheric-pressure gases	0.004–0.10	0.007–0.17
Insulating materials	0.02–0.12	0.034–0.21
Nonmetallic liquids	0.05–0.40	0.086–0.69
Nonmetallic solids (brick, stone, concrete)	0.02–1.50	0.034–2.6
Metal alloys	8–70	14–120
Pure metals	30–240	52–410

TABLE 2.3

Values of Thermal Conductivity for Building Materials

Material	k, Btu/(h · ft · °F)	T, °F	k, W/(m · K)	T, °C
Construction materials				
Asphalt	0.43–0.44	68–132	0.74–0.76	20–55
Cement, cinder	0.44	75	0.76	24
Glass, window	0.45	68	0.78	20
Concrete	1.0	68	1.73	20
Marble	1.2–1.7	—	2.08–2.94	—
Balsa	0.032	86	0.055	30
White pine	0.065	86	0.112	30
Oak	0.096	86	0.166	30
Insulating materials				
Glass fiber	0.021	75	0.036	24
Expanded polystyrene	0.017	75	0.029	24
Polyisocyanurate	0.012	75	0.020	24
Gases at atmospheric pressure				
Air	0.0157	100	0.027	38
Helium	0.0977	200	0.169	93
Refrigerant 12	0.0048	32	0.0083	0
	0.0080	212	0.0038	100
Oxygen	0.00790	−190	0.0137	−123
	0.02212	350	0.0383	175

Source: Courtesy of Karlekar, B. and Desmond, R.M., *Engineering Heat Transfer*, West Publishing, St. Paul, MN, 1982. With permission.

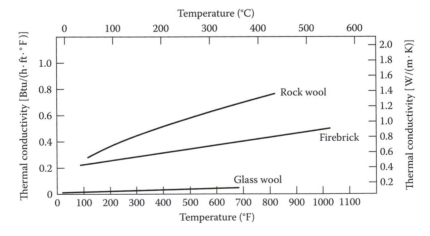

FIGURE 2.4

Thermal conductivity of several materials as a function of temperature.

Over the relatively small range of temperatures encountered in buildings during a year, it is usually adequate to assume that conductivities are constant. If data on the temperature dependence of conductivity are available for a particular material, use the conductivity at the average temperature since the value of k varies almost linearly with temperature.

In the next section on heat transfer by convection, we see that the conductivity temperature dependence for some liquids and gases is sometimes nonnegligible. In these cases we can account for this effect in a simple manner, as we shall see shortly.

2.2.3 Ground Coupling

The transfer of heat between the basement or floor slab of a building and the earth is called *ground coupling* and is essentially a problem of conduction heat transfer. A number of methods for calculating "ground-coupled" heat transfer have been suggested. Krarti (1999) provides a review of many of these methods. Unfortunately, the several common approaches sometimes disagree greatly in the values of heat transfer rates calculated (Claridge, 1986). This is particularly troublesome in residences and single-story buildings with large ratios of floor area to building volume. Further complicating the ground coupling calculations is the wide variation of soil properties. For example, the thermal conductivity of common soils varies between 0.3 and 1.4 Btu/(h·ft·°F) [0.5 and 2.5 W/(m·K)] depending on moisture and composition. Ground thermal coupling calculations have probably been the least accurate of any in building thermal analysis. Fortunately for designers of commercial buildings, basement and slab heat losses are a small portion of the total heat load.

2.2.3.1 ASHRAE Method

Heat transfer through below-grade floors and walls of residences has been the subject of study by a number of investigators (Mitalas, 1982; Krarti, 1987). According to ASHRAE (1999, 2001), the heat loss at peak conditions from *vertical basement walls* can be found by assuming that two-dimensional heat flow paths from a basement wall, when viewed in cross section, are circular arcs centered at the intersection of the wall and the earth's surface, as shown in Fig. 2.5. Based on this assumption, one uses the one-dimensional

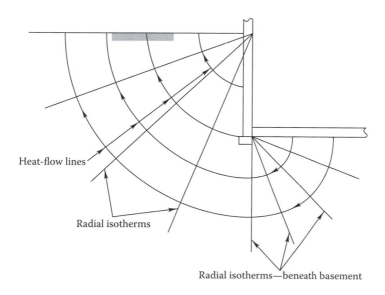

Heat-flow lines

Radial isotherms

Radial isotherms—beneath basement

FIGURE 2.5
Radial isotherms for basement heat loss assumed in the ASHRAE method.

TABLE 2.4a

Unit Heat Loss $\dot{Q}/(L\,\Delta T)$ through Basement Walls (ASHRAE Method), Btu/(h · ft · °F)

Depth of Wall Section, ft	Insulation Level			
	Uninsulated	R_{th}-4.17	R_{th}-8.34	R_{th}-12.5
0–1	0.410	0.152	0.093	0.067
1–2	0.222	0.116	0.079	0.059
2–3	0.155	0.094	0.068	0.053
3–4	0.119	0.079	0.060	0.048
4–5	0.096	0.069	0.053	0.044
5–6	0.079	0.060	0.048	0.040
6–7	0.069	0.054	0.044	0.037

Source: Courtesy of ASHRAE, *Handbook of Fundamentals*, American Society of Heating, Refrigerating and Air-Conditioning Engineers, Atlanta, GA, 1989. With permission.
[The values in the body of the table are heat loss per unit temperature difference (interior air temperature less soil design temperature) per unit length of building perimeter, assuming soil thermal conductivity of 0.8 Btu/(h · ft · °F). The R_{th} values have units of (ft^2 · h · °F)/Btu.]

TABLE 2.4b

Unit Heat Loss $\dot{Q}/(L\,\Delta T)$ through Basement Walls (ASHRAE Method), W/(m · K)

Depth of Wall Section, m	Insulation Level			
	Uninsulated	R_{th}-0.73	R_{th}-1.47	R_{th}-2.2
0–0.3	2.33	0.86	0.53	0.38
0.3–0.6	1.26	0.66	0.45	0.36
0.6–0.9	0.88	0.53	0.38	0.30
0.9–1.2	0.67	0.45	0.34	0.27
1.2–1.5	0.54	0.39	0.30	0.25
1.5–1.8	0.45	0.34	0.27	0.23
1.8–2.1	0.39	0.30	0.25	0.21

Source: Courtesy of ASHRAE, *Handbook of Fundamentals*, American Society of Heating, Refrigerating and Air-Conditioning Engineers, Atlanta, GA, 1989. With permission.
[The values in the body of the table are heat loss per unit temperature difference (interior air temperature less soil design temperature) per unit length of building perimeter, assuming soil thermal conductivity of 1.38 W/(m · K). The R_{th} values have units of (m^2 · K)/W.]

conduction equation to find the heat loss from basement walls. Table 2.4 contains unit heat loss data to be used with the ASHRAE method. These data are based on a soil conductivity of 0.8 Btu/(h · ft · °F) [1.38 W/(m · K)]. The appendix tables contain additional data on soil conductivity that are to be used to linearly adjust the values in Table 2.4 for other k values.

One uses the ASHRAE method to find peak heat loss as follows. First, the depth of any wall insulation must be known. Then from the table, sum the unit heat losses for the insulation level existing at each 1-ft depth of wall. Multiply the result by the building perimeter. The result is the total unit heat loss. Finally, multiply the total unit heat loss by the indoor air-to-ground temperature difference, to get the design heat loss. The design ground temperature $T_{g,des}$ is the average winter outdoor air temperature (see HCB software "weather data") minus the amplitude of the ground temperature swing ΔT_g read from Fig. 2.6. Equation (2.9) shows how to calculate ground heat losses.

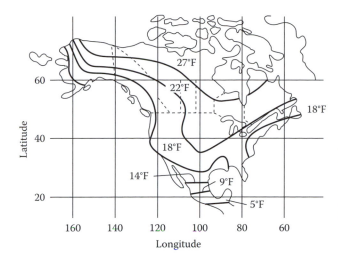

FIGURE 2.6
Map of amplitude of annual soil temperature swings used for ground coupling calculations (°F). For amplitude in SI units, divide by 1.8°F/K. (Courtesy of ASHRAE, *Handbook of Fundamentals*, American Society of Heating, Refrigerating and Air-Conditioning Engineers, Atlanta, GA, 1989. With permission.)

$$\dot{Q}_g = \left(\sum_{depth} \frac{\dot{Q}}{L\,\Delta T} \right) L(T_i - T_{g,des}) \tag{2.9}$$

where $\dot{Q}/(L\,\Delta T)$ is read from the table, L is the building perimeter, T_i is the building interior air temperature, and $T_{g,des}$ is the ground temperature.

Example 2.4: Partially Insulated Basement Wall Design Heat Loss

If a 6-ft-deep basement wall is insulated to a depth of only 2 ft with R-8.34 insulation and uninsulated below, find the design heat loss. The building perimeter is 250 ft, and $T_{g,des}$ is 50°F.

Given: Wall depth is 6 ft, and insulation R value is 8.34 for the top 2 ft.

Figure: Figure 2.5

Assumptions: Steady conduction; building interior temperature $T_i = 72$°F

Find: $\dot{Q}_g$

Lookup values: From Table 2.4a read the first two values in the R-8.34 column ($0.093 + 0.079$) and the next four values from the uninsulated column ($0.155 + 0.119 + 0.096 + 0.079$). The units on these quantities are Btu/(h · ft · °F).

SOLUTION

The total basement unit heat loss is found by summing these six values from Table 2.4a to get

$$\sum_{depth} \frac{\dot{Q}}{L\,\Delta T} = 0.621 \text{ Btu/(h · ft · °F)}$$

The total basement *wall* heat loss is the loss per unit length times the 250-ft perimeter times the temperature difference:

$$\dot{Q}_g = [0.621 \text{ Btu/(h · ft · °F)}](250 \text{ ft})[(72 - 50)°F] = 3416 \text{ Btu/h (1.0 kW)}$$

TABLE 2.5a

Heat Loss through Basement Floors (ASHRAE Method), U_{floor} Btu/(h · ft² · °F)

Depth of Wall below Grade, ft	Shortest Width of Building, ft			
	20	24	28	32
5	0.032	0.029	0.026	0.023
6	0.030	0.027	0.025	0.022
7	0.029	0.026	0.023	0.021

Source: Courtesy of ASHRAE, *Handbook of Fundamentals*, American Society of Heating, Refrigerating and Air-Conditioning Engineers, Atlanta, GA, 1989. With permission.
The values in the table are used to find floor heat loss $\dot{Q}_{floor} = U_{floor} A_{floor}(T_i - T_g)$.

TABLE 2.5b

Heat Loss through Basement Floors (ASHRAE Method), U_{floor} W/(m² · K)

Depth of Wall below Grade, m	Shortest Width of Building, m			
	6.0	7.3	8.5	9.7
1.5	0.18	0.16	0.15	0.13
1.8	0.17	0.15	0.14	0.12
2.1	0.16	0.15	0.13	0.12

Source: Courtesy of ASHRAE, *Handbook of Fundamentals*, American Society of Heating, Refrigerating and Air-Conditioning Engineers, Atlanta, GA, 1989. With permission.

Basement *floor* heat losses are calculated in a similar manner. Table 2.5 shows the unit heat loss (this time on a per-unit basement floor area, not perimeter, basis) as it depends on building size and wall depth. To find the basement heat loss, from the table read the unit heat loss and multiply by the floor area and same temperature difference as used for wall losses.

ASHRAE assumes that heat losses from *slabs on grade* are proportional to the perimeter of the slab, not the area of the slab, as for below-grade slabs. The governing equation based on work in the 1940s at the National Bureau of Standards is as follows:

$$\dot{Q}_{edge} = F_2 P(T_i - T_o) \tag{2.10}$$

For Eq. (2.10), the conductance F_2 is given in Table 2.6, P is the perimeter of the slab (the same physical quantity as L above, but P is used to remind the reader that slab edge losses are being determined), and the temperature difference is between the building interior air temperature and exterior air temperature. Note that ground temperatures are not involved since the slab is located above the ground. Insulation at the edge of a slab will decrease losses significantly. The data in Table 2.6 include this effect for the case of R-5.4 insulation applied from the top corner of the slab downward to the frost line. Essentially the same insulation effect is achieved if one insulates the edge of the slab and beneath the slab, instead of straight down along the foundation, for a distance equal to the frost line depth. Figure 2.7 shows one method of insulating a slab edge. Table 2.6 shows in its last line the significant effect of the poor practice of placing heating ducts in concrete walls. In the coldest climate shown, the heat loss is nearly 4 times that for a standard wall.

If slab edge losses are calculated by using Eq. (2.10), no other slab heat losses need be included. Slab and basement losses are usually ignored in cooling load calculations.

TABLE 2.6

Heat Loss Coefficients F_2 for Slab-on-Grade Floors (ASHRAE Method), Btu/(h · ft · °F) [W/(m · K)]

Construction	Insulation	Heating Degree-Days, °F · days (°C · days)		
		7433 (4130)	**5350 (2970)**	**2950 (1640)**
8-in (200-mm) block wall with brick	None	0.72 (1.24)	0.68 (1.18)	0.62 (1.07)
	R_{th}-5.4 (R_{th}-0.95)	0.56 (0.97)	0.50 (0.87)	0.48 (0.83)
4-in (100-mm) block wall with brick	None	0.93 (1.61)	0.84 (1.45)	0.80 (1.38)
	R_{th}-5.4 (R_{th}-0.95)	0.54 (0.94)	0.49 (0.85)	0.47 (0.81)
Metal-stud wall with stucco	None	1.34 (2.32)	1.20 (2.07)	1.15 (1.99)
	R_{th}-5.4 (R_{th}-0.95)	0.58 (1.00)	0.53 (0.92)	0.51 (0.88)
Poured-concrete wall with perimeter ducts	None	2.73 (4.72)	2.12 (3.67)	1.84 (3.18)
	R_{th}-5.4 (R_{th}-0.95)	0.90 (1.56)	0.72 (1.24)	0.64 (1.11)

Source: Courtesy of ASHRAE, *Handbook of Fundamentals*, American Society of Heating, Refrigerating and Air-Conditioning Engineers, Atlanta, GA, 1989. With permission.

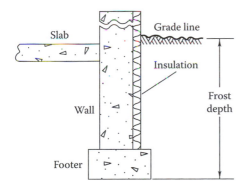

FIGURE 2.7
Schematic diagram of method of insulating slabs and foundations.

Recently, Bahnfleth and Pederson (1990) have questioned the simple model represented by Eq. (2.10). Their work, based on a detailed, transient, three-dimensional analysis, showed that Eq. (2.10) could be in error by up to 50 percent but that the time-varying part of the heat loss was well predicted by expressions similar to Eq. (2.10). However, the annual average slab heat loss about which the time variation occurs could not be found accurately by an equation of the form of Eq. (2.10). Instead they found that a power law in the slab perimeter-to-area ratio P/A gave excellent results for the annual mean part of the slab edge heat loss. For the annual part of slab edge loss, these authors suggest that F_2P be replaced by a term $c(P/A)^d$, where d is reported in the range [0.75, 0.90] depending on the ground temperature swing. The coefficient c has a value of 0.18 in USCS units and 1.0 in SI units. The improved version of Eq. (2.10) recommended for slab losses at both peak and nonpeak conditions in USCS units is

$$\dot{Q}_g = \underbrace{c\left(\frac{P}{A}\right)^d A(T_i - T_{\text{earth,av}})}_{\text{Annual average term}} + \underbrace{0.13P\Delta T_g \sin\left[(N_{\text{day}} + \phi)\frac{360}{365}\right]}_{\text{Time-varying term}} \text{ Btu/h} \qquad (2.11)$$

where ΔT_g is the ground temperature swing, read from Fig. 2.6, and ϕ is the lag between slab heat loss and day number (about 50 days for the example given by Bahnfleth and Pederson for Medford, Oregon). The method has yet to be generalized, but the results to date seriously call into question Eq. (2.10).

2.2.3.2 The Mitalas Method for Ground Losses

Another common method for estimating peak heat transfer from below-grade walls and floors is based on a two-dimensional analysis of Mitalas (1982). The wall and floor losses both are found from Eq. (2.12):

$$\dot{Q} = U_g A (T_i - T_g) \qquad (2.12)$$

in which U_g is the overall U value (Table 2.7 or 2.8) between interior air temperature T_i and ground temperature T_g and A is the wall or floor area. Table 2.7 summarizes the values of U_g to be used for various arrangements of insulation and wall depths.

TABLE 2.7

Overall Heat Transfer Coefficients U_g—Mitalas Method for Basement Walls, Btu/(h · ft^2 · °F) [W/(m^2 · K)]

Wall Depth, ft (m)	Uninsulated Wall	2-ft-Deep Insulation		Full-Depth Insulation	
		R_{th}-5 (R_{th}-0.88)	R_{th}-10 (R_{th}-1.76)	R_{th}-5 (R_{th}-0.88)	R_{th}-10 (R_{th}-1.76)
4	0.20	0.140	0.130	0.088	0.054
(1.22)	(0.035)	(0.025)	(0.023)	(0.015)	(0.010)
5	0.20	0.140	0.130	0.082	0.050
(1.52)	(0.032)	(0.025)	(0.023)	(0.014)	(0.009)
6	0.17	0.140	0.130	0.079	0.048
(1.83)	(0.030)	(0.025)	(0.023)	(0.014)	(0.008)
7	0.16	0.140	0.130	0.076	0.047
(2.13)	(0.028)	(0.025)	(0.023)	(0.013)	(0.008)

Source: Courtesy of Mitalas, G.P., Basement Heat Loss Studies at DBR/NRC, Division of Building Research Paper No. 1045, National Research Council, Ottawa, Canada, 1982. With permission.

TABLE 2.8

Overall Heat Transfer Coefficients U_g—Mitalas Method for Uninsulated Basement Floors, Btu/(h · ft^2 · °F) [W/(m^2 · K)]

Wall Uninsulated	Wall Insulated to 2-ft Depth	Fully Insulated Wall
0.045	0.036	0.025
(0.008)	(0.006)	(0.004)

Source: Courtesy of Mitalas, G.P., Basement Heat Loss Studies at DBR/NRC, Division of Building Research Paper No. 1045, National Research Council, Ottawa, Canada, 1982. With permission.

2.2.3.3 *True Two-Dimensional Methods for Long-Term Ground Coupling Losses*

The Mitalas and ASHRAE methods are meant primarily for peak heat load calculations of the sort needed in Chap. 7. However, for annual energy calculations (Chap. 8) one needs the long-term heat losses from slabs and below-grade basements in residential buildings. Figure 2.8 shows sample results for three annual average conditions in an uninsulated basement based on a full two-dimensional analysis (the problem is symmetric, so only one-half of the basement need be shown). In Fig. 2.8a, a 2-m-deep basement is located in earth which has a water table of 13°C at 3 m below the surface. Under these summer conditions (21°C earth surface temperature), most heat flow is from the earth's surface to the 13°C water table.

Figure 2.8b shows the same basement under different conditions. Here the basement is warm; the earth's surface is at an intermediate temperature of 17°C, and the water table remains at 13°C. The isotherms are closely spaced below the basement where most conduction occurs to the water table. In addition, losses occur from the basement wall to the water table. Finally, in Fig. 2.8c a cool-weather condition exists with the earth's surface at 13°C. In this case, the majority of conductive losses from the wall are to the earth's surface, and lines of constant flux (normal to the isotherms shown) are approximately circular, as assumed by the ASHRAE method. Hence, the exact two-dimensional solution supports the use of the ASHRAE method.

A complete dynamic analysis is beyond the scope of this book (see Sec. 2.4.4 for a description of annual load calculation), but a number of useful results appear in Krarti et al. (1988a, 1988b, and 1990). Figure 2.9 shows the effect of size, e.g., on annually varying heat loss for a slab ($2a$ m wide and $2c$ m long) on grade as calculated from a full, analytical three-dimensional analysis. Figure 2.9a shows the heat losses from various square slabs for a typical soil with a 50°F (10°C) water table located 16.4 ft (5 m) below the surface in a climate where the average soil temperature is 46°F (8°C). *The larger the slab, the lower the heat flux.* Figure 2.9b shows the effect of slab shape (width held fixed at $2a = 10$ m) for three slab lengths: $2c = 4$, 10, and 20 m. This figure reinforces the prior conclusion that larger slabs have smaller heat fluxes. Krarti et al. also showed, not surprisingly, that the effect of water table depth on heat loss decreases as the water table depth increases; and that larger soil conductivity–heat capacity ratios result in larger heat flux amplitudes through an annual cycle.

2.3 Convection Heat Transfer

2.3.1 Defining Equation for h_{con}

When a moving fluid contacts a surface at a different temperature, the resulting heat transfer is called *convection*. Convection heat transfer is always associated with large-scale (i.e., not molecular-scale) motion of a fluid—either liquid or gas—over a warmer or cooler surface. The higher the velocity of fluid flow, the higher the rate of convection heat transfer, in general.

Two kinds of convection exist: *natural* (or *free*) and *forced*. Free convection results from density differences in the fluid caused by contact with the surface to or from which the heat transfer occurs. The gentle circulation of air in a room caused by the presence of a solar-warmed window or wall is a manifestation of free convection. Buoyancy is the motive force in free convection. Heat is transferred from baseboard systems by natural convection, for example.

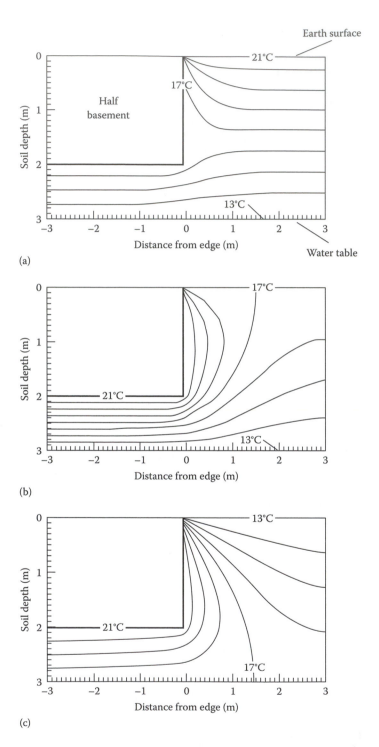

FIGURE 2.8
Isotherms near basement for three cases described in the text, with water table 9.8 ft (3 m) below the surface. (Courtesy of Krarti, M., New developments in ground coupling heat transfer, PhD dissertation, University of Colorado, Boulder, CO, 1987.)

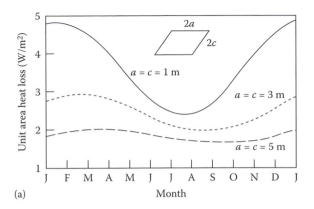

(a)

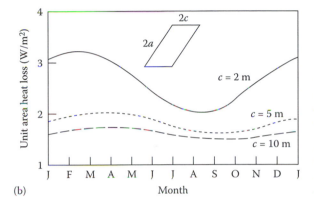

(b)

FIGURE 2.9
(a) Annual variation of heat loss from square slabs 2, 6, and 10 m on a side. (b) Annual variation of slab heat loss for rectangular slabs 20 m on one side and 4, 10, and 20 m on the other. (Courtesy of Krarti, M., Building foundation heat transfer, in Goswami, Y. and Boer, K. (eds), *Advances in Solar Energy*, vol. 13, Chapter 6, ASES, Boulder, CO, 242–308, 1999.)

Forced convection occurs when a force external to the problem (other than gravity or other body forces) moves a fluid past a warmer or cooler surface. Usually, the fluid velocities in forced convection are considerably higher than those in free convection, and the rate of heat transfer is generally greater. The improved heat transfer is at the expense of greater consumption of mechanical energy in the forced-flow case. Heat transfer from heating and cooling coils is primarily by forced convection, for example.

Although convection is a very common mode of heat transfer in buildings, a detailed analysis of it is complicated. However, Newton's law of cooling is a simplified approximation useful for both forced and free convection calculations. Simply stated, the rate at which heat is transferred by convection is proportional to the temperature difference and the heat transfer area. The equation expressing Newton's law of cooling is

$$\dot{Q} = h_{con}A(T_s - T_f) = h_{con}A\,\Delta T \tag{2.13}$$

where
h_{con} = convection coefficient, Btu/(h · ft^2 · °F) [W/(m^2 · K)]
A = surface area through which convection occurs, ft^2 (m^2)
T_s = surface temperature
T_f = fluid temperature well away from wall
$\Delta T = T_s - T_f$ = temperature difference

TABLE 2.9

Magnitude of Convection Coefficients

Arrangement	$W/(m^2 \cdot K)$	$Btu/(h \cdot ft^2 \cdot F)$
Air, free convection	6–30	1–5
Superheated steam or air, forced convection	30–300	5–50
Oil, forced convection	60–1800	10–300
Water, forced convection	300–6000	50–1000
Water, boiling	3000–60,000	500–10,000
Steam, condensing	6000–120,000	1000–20,000

The conversion between SI and USCS units is 5.678 $W/(m^2 \cdot K) = 1$ $Btu/(h \cdot ft^2 \cdot °F)$.

Standard heat transfer texts include dimensionless equations for finding h_{con}. These typically involve dimensionless numbers such as the Reynolds (forced convection), Grashof (natural convection), Prandtl, and Nusselt numbers [see, e.g., Holman (1997)]. In this design-oriented book, such equations are inconvenient to use since extensive physical property data tables are needed. We will use, instead, *dimensional* equations for all common situations needed by the HVAC designer, since such equations save considerable time.

Table 2.9 shows typical values of the convection coefficient. The largest values occur for boiling and condensing water, whereas the lowest values apply for free convection in gases.

2.3.2 Convection Thermal Resistance and *R* Value

In terms analogous to the definitions of thermal resistance and the *R* value for conduction, one can define both for convection. In these terms Eq. (2.13) expresses convection heat transfer by

$$\dot{Q} = \frac{\Delta T}{R} \tag{2.14}$$

from which the resistance to heat transfer for convection is

$$R = \frac{1}{h_{con}A} \tag{2.15}$$

The thermal resistance value R_{th} and its reciprocal, the *U* value, are given by

$$R_{th} = \frac{1}{h_{con}} \tag{2.16}$$

$$U \equiv \frac{1}{R_{th}} = h_{con} \tag{2.17}$$

The convection *R* and *U* values are used in a later section to simplify heat flow calculations when both conduction and convection modes coexist.

2.3.3 Convection Coefficients for Various Building System Configurations

2.3.3.1 External Flow Equations for Buildings

In this section we summarize convection equations for flows which occur in unconfined geometries. These are often called *external flows*. Examples include the airflow over the wall of a building or across a bank of tubes in a heat exchanger.

The equation to be used will depend on the type of flow—laminar or turbulent. In nearly all building-related convection situations where forced convection is involved, the flow will be turbulent (one exception is water flow in pipes for very cold liquid distribution systems). However, free convection can be either laminar or turbulent in buildings. For completeness, we will present expressions for laminar and turbulent free convection and forced convection.

The following equations (ASHRAE, 2001) are *dimensional*. In USCS, lengths are in feet, temperatures in degrees Fahrenheit, velocities in feet per second, and convection coefficients in Btu/(h · ft^2 · °F) unless otherwise noted. In the SI versions of the same equations, lengths are in meters, temperatures in degrees Celsius, velocities in meters per second, and convection coefficients in W/(m^2 · K). For those properties involving air, physical properties at 70°F (20°C) have been used. Two sets of equations are presented, one for each set of units, identified in the equation number.

Laminar free convection in air from a *tilted surface* is given by

$$h_{\text{con}} = 0.29 \left(\frac{\Delta T \sin \beta}{L} \right)^{1/4} \qquad (2.18\text{US})$$

$$h_{\text{con}} = 1.42 \left(\frac{\Delta T \sin \beta}{L} \right)^{1/4} \qquad (2.18\text{SI})$$

where ΔT is the temperature difference in Eq. (2.13), L is the length of the plate in the direction of the buoyancy-driven flow, and β is the surface tilt up from the horizontal. Equations (2.18) apply for surface tilts between 30° and 90°. Laminar flow occurs for $L^3 \Delta T < 63$ in USCS units or $L^3 \Delta T < 1.0$ in SI units.

Turbulent flow occurs if the preceding inequalities are reversed. The corresponding equations for turbulent free convection from *tilted surfaces* in air are

$$h_{\text{con}} = 0.19(\Delta T \sin \beta)^{1/3} \qquad (2.19\text{US})$$

$$h_{\text{con}} = 1.31(\Delta T \sin \beta)^{1/3} \qquad (2.19\text{SI})$$

Equations (2.18) and (2.19) apply to interior wall and window surfaces of buildings and to corresponding exterior surfaces in the absence of wind.

Laminar free convection from *horizontal pipes* and other horizontal cylinders in air is given by

$$h_{\text{con}} = 0.27 \left(\frac{\Delta T}{D} \right)^{1/4} \qquad (2.20\text{US})$$

$$h_{\text{con}} = 1.32 \left(\frac{\Delta T}{D} \right)^{1/4} \qquad (2.20\text{SI})$$

in which D is the cylinder's outer diameter (in feet for USCS units and meters for SI units). The criterion for laminar or turbulent flow is the same as that for flat plates given above (replace L with the diameter D in the inequality).

For turbulent free convection from horizontal cylinders in air, the convection coefficient is

$$h_{con} = 0.18(\Delta T)^{1/3} \tag{2.21US}$$

$$h_{con} = 1.24(\Delta T)^{1/3} \tag{2.21SI}$$

To find the laminar free convection coefficient for *warm horizontal surfaces facing up* (e.g., flat roofs of buildings warmed by the sun), one uses

$$h_{con} = 0.27\left(\frac{\Delta T}{L}\right)^{1/4} \tag{2.22US}$$

$$h_{con} = 1.32\left(\frac{\Delta T}{L}\right)^{1/4} \tag{2.22SI}$$

where L is the average length of the sides of the horizontal surface. This expression also applies to cold surfaces facing downward. An example is the inner surface of a plane skylight in the roof of a building in winter. The criterion for laminar free convection is the same as that in preceding cases in this section.

Turbulent free convection coefficients for warm surfaces facing up in turbulent flow are found from

$$h_{con} = 0.22(\Delta T)^{1/3} \tag{2.23US}$$

$$h_{con} = 1.52(\Delta T)^{1/3} \tag{2.23SI}$$

If a warmed surface faces downward, however, the laminar convection coefficient is reduced because of the stable stratification condition. Equations (2.22) can be used, but the dimensional multiplier is changed from 0.27 to 0.12 in USCS units. In SI units the multiplier becomes 0.59. This modified expression also applies for cooled flat surfaces (planes) facing upward (e.g., a horizontal skylight's outer surface in summer).

Example 2.5: Free Convection from a Hot Roof

Find the free convection coefficient at the 20-ft-square roof of a building. The roof's surface is heated by the sun to 110°F in air at 65°F.

Given: $T_s = 110°F$

$T_f = 65°F$

Find: h_{con}

SOLUTION

Either Eq. (2.22US) or (2.23US) applies to this case. We must determine which. Since the roof is square, $L = 20$ ft.

$$\Delta T = 110 - 65 = 45°F$$

Next check for laminar or turbulent flow:

$$L^3 \Delta T = 20^3 \times 45 > 63$$

Therefore, the turbulent flow equation, Eq. (2.23US), applies. Substituting into the equation, we have

$$h_{\text{con}} = 0.22(45)^{1/3} = 0.78 \text{ Btu}/(\text{h} \cdot \text{ft}^2 \cdot {}^{\circ}\text{F}) \left[4.4 \text{ W}/(\text{m}^2 \cdot \text{K})\right]$$

COMMENTS

Table 2.9 shows that free convection in air is a relatively ineffective heat transfer mechanism. This example further illustrates that one should expect small values of h_{con} in free convection.

Forced convection over planes does not depend on their orientation. For *laminar flow over planes*, the forced convection heat transfer coefficient is given by

$$h_{\text{con}} = 0.35 \left(\frac{v}{L}\right)^{1/2} \tag{2.24US}$$

$$h_{\text{con}} = 2.0 \left(\frac{v}{L}\right)^{1/2} \tag{2.24SI}$$

in which L is the length of the surface in the direction of flow [ft (m)] and v is the velocity [ft/s (m/s)]. Equation (2.24) applies if $vL < 15$ ft^2/s, indicating laminar flow ($vL < 1.4$ m^2/s in SI units).

For *turbulent flow over planes*, that is, $vL > 15$ ($vL > 1.4$ in SI), the convection coefficient is

$$h_{\text{con}} = 0.54 \left(\frac{v^4}{L}\right)^{1/5} \tag{2.25US}$$

$$h_{\text{con}} = 6.2 \left(\frac{v^4}{L}\right)^{1/5} \tag{2.25SI}$$

These equations apply, e.g., to wind blowing over the roof or wall of a building. The roughness of the exterior finish of a building can affect the heat transfer rate. Figure 2.10 shows that the forced convection coefficient will be twice as large for a rough, stucco wall as for a smooth surface such as glass. Equation (2.25) applies for smooth surfaces.

2.3.3.2 Internal Flow Equations for Building HVAC Systems

Flows of fluids confined by boundaries such as the sides of a duct are called *internal flows*. Since the mechanisms of convection are quite different for such flows vis-à-vis external flows, the expressions for h_{con} are also different. In this section we present equations for turbulent internal flow in water and air.

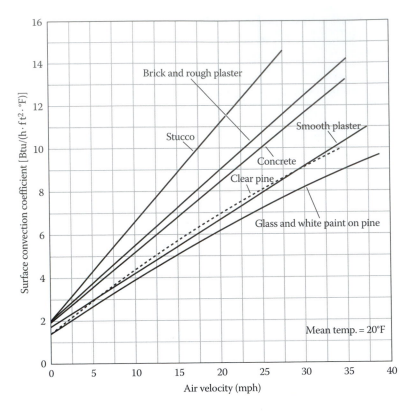

FIGURE 2.10
Surface forced convection coefficient h_{con} for various wall finishes. (For SI values of h_{con} multiply the values in the figure by 5.67.) (Courtesy of ASHRAE, *Handbook of Fundamentals*, American Society of Heating, Refrigerating and Air-Conditioning Engineers, Atlanta, GA, 1989. With permission.)

The forced convection coefficient for fully developed turbulent flow of *air through ducts and pipes* is given by

$$h_{con} = 0.5 \left(\frac{v^4}{D_h}\right)^{1/5} \qquad (2.26US)$$

$$h_{con} = 8.8 \left(\frac{v^4}{D_h}\right)^{1/5} \qquad (2.26SI)$$

in which the *hydraulic diameter* D_h in (cm) is defined as 4 times the ratio of the flow conduit's cross-sectional area divided by the perimeter of the conduit. For the familiar case of a round pipe with diameter D, the hydraulic diameter is

$$D_h = \frac{4(\pi D^2/4)}{\pi D} = D \qquad (2.27)$$

Convection coefficient values for fully developed turbulent *water flow through pipes* can be found from

$$h_{con} = 150(1 + 0.011T)\left(\frac{v^4}{D_h}\right)^{1/5} \qquad (2.28US)$$

In this equation, D_h is expressed in inches. The temperature term (in degrees Fahrenheit) accounts for the variation of water viscosity and conductivity with temperature.

In SI units,

$$h_{con} = 3580(1 + 0.015T)\left(\frac{v^4}{D_h}\right)^{1/5} \tag{2.28SI}$$

In this equation, D_h is in centimeters, and the temperature is in degrees Celsius. As in all SI dimensional equations in this section, v is in meters per second.

Dimensional equations for other fluids involved in building HVAC equipment are given in Mathur (1990).

2.3.3.3 Tables and Graphs of Convection Coefficients

For preliminary design, approximate values for convection coefficients are adequate. This is particularly true for heating and cooling load calculations. The thermal resistance due to convection is much smaller than that of the insulation in building walls and roofs. Therefore, use of a constant value of h_{con} instead of one depending on the parameters of the previous section is sufficient for the early phases of design. Appendix CD-ROM tables summarize conductance U values useful for load calculations.

Components of HVAC systems such as coils and piping that carry water can be exposed to subfreezing temperatures in winter. There are many methods of freeze protection, but one of the most common is to use an antifreeze solution in place of pure water. The standard antifreezes are either ethylene glycol or propylene glycol. Although these substances are well suited to freeze protection, they have lower convection heat transfer properties than pure water. Figure 2.11 shows the degradation of h_{con} for various concentrations of antifreeze (the proper concentration is discussed later). To find the convection coefficient with glycol $h_{con,g}$, the value for pure water is multiplied by F_g from the figure ($F_g = 1.0$ for pure water at 74°F); the temperature effect can also be found from the figure or from Eq. (2.28).

$$h_{con,g} = F_g h_{con,w} \tag{2.29}$$

Example 2.6: Effect of Glycol on Convection Coefficient

Find the convection heat transfer coefficient for a 50% solution of ethylene glycol at 40°F, flowing at 6 ft/s in a 0.6-in-diameter pipe.

Given: $v = 6$ ft/s

$D = 0.6$ in

Find: $h_{con,g}$

Lookup values: From Fig. 2.11 for 50% ethylene glycol, $F_g = 0.32$.

Solution

Equation (2.28US) can be used to find the convection coefficient for pure water:

$$h_{con,w} = 150(1 + 0.011 \times 40)\left(\frac{6.0^4}{0.6}\right)^{1/5} = 1003 \text{ Btu}/\left(\text{h} \cdot \text{ft}^2 \cdot {}^\circ\text{F}\right)$$

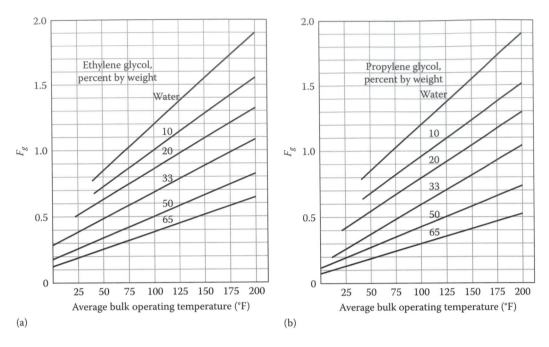

FIGURE 2.11
Glycol correction factor F_g for heat transfer in turbulent flow in pipes: (a) aqueous ethylene glycol solutions; (b) aqueous propylene glycol solutions. (Courtesy of Nussbaum, O.J., *Heat. Piping Air Cond.*, 62(1), 75, January 1990. With permission.)

The glycol coefficient is calculated from Eq. (2.29) as follows:

$$h_{con,g} = 0.32 \times 1003 = 321 \ \text{Btu}/\left(\text{h} \cdot \text{ft}^2 \cdot {}^\circ\text{F}\right)$$

COMMENTS

At low temperatures, where they are most needed for freeze protection, glycols have convection coefficients well below those for water. In this example the coefficient is 68 percent lower than that for water.

The designer of building retrofits must be aware of this problem if glycol is to used in a system formerly designed for water. Both heating and cooling capacity can be severely compromised. In addition to reduced heat transfer, glycol solutions require greater pumping power, as we will see in Chap. 5.

2.3.4 Heat Exchangers

Heat exchangers are devices in which two fluid streams, usually separated from each other by a solid wall, exchange thermal energy by convection. In the process of this exchange, one fluid is heated and the other is cooled. The fluids may be gases, liquids, or vapors (either condensing or boiling). For example, in buildings, heat exchangers are used to heat or cool air, to produce hot water from steam, or to reject heat from a chiller's condenser to the environment. The reader should already be familiar with heat exchanger analysis; therefore, this section reviews only equations useful for design.

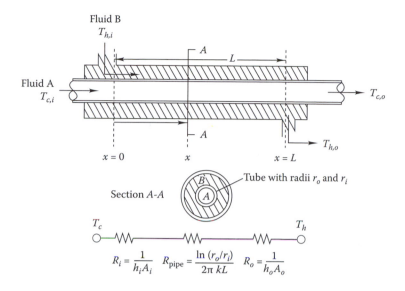

FIGURE 2.12

Schematic diagram of parallel-flow shell-and-tube heat exchanger showing fluid temperatures and equivalent thermal circuit.

Figure 2.12 shows a simple heat exchanger design consisting of two concentric pipes. If the fluids flow in the same direction, as shown, the heat exchanger design is called *parallel-flow*; if the flow direction of one stream is changed, the exchanger is of *counterflow* design. In this section we review the methods for computing heat exchanger performance for all designs commonly found in buildings.

Heat transfer from one fluid stream to the other in a heat exchanger involves two convection processes, one between each fluid and the surface between the streams. In addition, heat is conducted through the common tube surface. The overall heat transfer conductance $U_o A_o$ combines both modes of heat transfer by using the idea of thermal resistance that is shown in the equivalent circuit in Fig. 2.12 and discussed in earlier sections of this chapter. The overall conductance for this series heat transfer process is

$$U_o A_o = \frac{1}{R_i + R_{\text{pipe}} + R_o} \tag{2.30}$$

in which the resistances are given by the respective convection and conduction equations and are all based on the tube external area. To account for the buildup of surface films of fouling substances, a fouling resistance is sometimes added to the denominator of Eq. (2.30). Values can be found in standard heat transfer texts, listed in the References for this chapter.

2.3.4.1 Parallel Flow

The transfer of heat in a heat exchanger is governed by the first law of thermodynamics and the relevant conduction and convection heat transfer rate equations. A convenient method for predicting the heat transfer rate in heat exchangers is referred to as the ϵ–NTU, or *effectiveness number-of-transfer-units* approach. The reader is referred to Kays and London (1984) for details.

The *effectiveness* is defined as the ratio of actual to maximum possible heat transfer rates:

$$\epsilon \equiv \frac{\dot{Q}}{\dot{Q}_{max}} \tag{2.31}$$

The maximum possible rate of heat transfer is limited to the product of the maximum temperature difference existing across the heat exchanger and the minimum fluid capacitance rate $(\dot{m}c)_{min}$, where c is fluid specific heat. If one solves Eq. (2.31) for the actual heat rate, one obtains

$$\dot{Q} = \epsilon\,(\dot{m}c)_{min}(T_{h,i} - T_{c,i}) \tag{2.32}$$

in which the temperatures are shown in Fig. 2.12.

The ϵ–NTU approach uses an additional equation for effectiveness as a function of fluid capacitance rates and exchanger U value to complete the right-hand side of Eq. (2.32). Equation (2.32) applies for all heat exchangers, but the effectiveness equation is different for each.

For example, the effectiveness of single-pass *parallel-flow* shell-and-tube heat exchangers is given by

$$\epsilon = \frac{1 - \exp\left[-\text{NTU}\left(1 + \dot{C}_{min}/\dot{C}_{max}\right)\right]}{1 + \dot{C}_{min}/\dot{C}_{max}} \tag{2.33}$$

The capacitance rates denoted by $\dot{C}$ are the products of the mass flow rate $\dot{m}$ and the specific heat c for each stream. Also, $\dot{C}_{max}$ is the larger of the two capacitance rates, and $\dot{C}_{min}$ is the smaller.

The *number of transfer units* is defined as

$$\text{NTU} \equiv \frac{U_o A_o}{\dot{C}_{min}} \tag{2.34}$$

Figure 2.13 is a plot of parallel-flow effectiveness versus NTU. The figure shows a strong diminishing-returns effect with NTU values greater than 3.0 resulting in little improvement in effectiveness for larger NTUs. For a matched-capacitance situation, the effectiveness is limited to 0.5 in parallel-flow exchangers. The counterflow heat exchanger described next is not limited in this way.

2.3.4.2 Counterflow

Counterflow heat exchangers have larger effectiveness values than parallel-flow devices for given capacitance rates. This is a result of the fact that a larger average temperature difference exists between the two streams over a larger part of the heat exchanger than in the parallel-flow case. Figure 2.14 compares the two cases for the equal-capacitance-rate case $\dot{C}_{max} = \dot{C}_{min}$. The counterflow effectiveness (uppermost curve) is significantly greater than that for parallel flow (lowest curve).

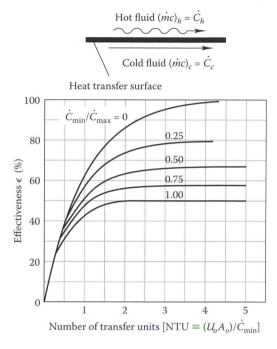

FIGURE 2.13

Parallel-flow heat exchanger effectiveness as a function of NTU.

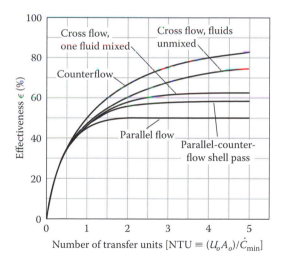

FIGURE 2.14

Comparison of effectiveness of several heat exchanger designs for equal hot- and cold-side capacitance rates, $\dot{C}_{min} = \dot{C}_{max}$.

The effectiveness for the counterflow heat exchanger is given by the second expression in Table 2.10. The same parameters appear in this equation as in Eq. (2.33), namely, NTU and the two capacitance rates. Figure 2.15 is a plot of counterflow heat exchanger effectiveness. Notice that effectivenesses are well above those shown in Fig. 2.13 for parallel-flow exchangers at identical capacitance and NTU values.

Example 2.7 illustrates the use of the effectiveness approach to calculate a number of thermal performance characteristics of a heat exchanger used to heat water in a building.

TABLE 2.10

Heat Exchanger Effectiveness Relations $N = \mathrm{NTU} = \dfrac{U_o A_o}{\dot{C}_{min}} \quad C = \dfrac{\dot{C}_{min}}{\dot{C}_{max}}$

Flow Geometry	Relation
Double pipe	
Parallel flow	$\varepsilon = \dfrac{1 - \exp\left[-N(1 + C)\right]}{1 + C}$
Counterflow	$\varepsilon = \dfrac{1 - \exp\left[-N(1 - C)\right]}{1 - C\exp\left[-N(1 - C)\right]}$
Crossflow	
Both fluids unmixed	$\varepsilon = 1 - \exp\left\{\dfrac{1}{Cn}\left[\exp(-NCn) - 1\right]\right\} \quad \text{where } n = N^{-0.22}$
Both fluids unmixed	$\varepsilon = N\left[\dfrac{N}{1 - \exp(-N)} + \dfrac{NC}{1 - \exp(-NC)} - 1\right]^{-1}$
$\dot{C}_{max}$ mixed, $\dot{C}_{min}$ unmixed	$\varepsilon = \dfrac{1}{C}\{1 - \exp\left[-C + C\ \exp(-N)\right]\}$
$\dot{C}_{max}$ unmixed, $\dot{C}_{min}$ mixed	$\varepsilon = 1 - \exp\left\{-\dfrac{1}{C}[1 - \exp(-NC)]\right\}$
Shell and tube	
One shell pass; two, four, six tube passes	$\varepsilon = 2\left[1 + C + \sqrt{1 + C^2}\dfrac{1 + \exp(-N\sqrt{1 + C^2})}{1 - \exp(-N\sqrt{1 + C^2})}\right]^{-1}$

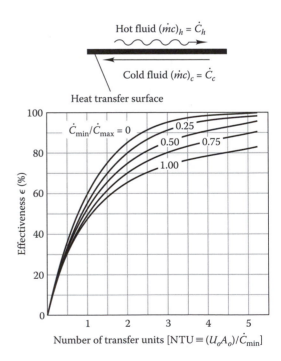

FIGURE 2.15
Counterflow heat exchanger effectiveness as a function of NTU.

Example 2.7: Counterflow Heat Exchanger

Potable service water is heated in a building from 20°C at a rate of 70 kg/min by using nonpotable pressurized water from a boiler at 110°C in a single-pass counterflow heat exchanger. Find the heat transfer rate if the hot-water flow is 90 kg/min. Also find the exit temperatures of both streams. The overall *U* value is 320 W/(m² · K), and the heat transfer area is 20 m².

Given: $T_{ci} = 20°C$

$T_{hi} = 110°C$

$U_o = 320 \text{ W}/(\text{m}^2 \cdot \text{K})$

$A_o = 20 \text{ m}^2$

$\dot{m}_c = 70 \text{ kg/min}$

$\dot{m}_h = 90 \text{ kg/min}$

Figure: See Fig. 2.12 (except the design is counterflow).

Assumptions: Jacket losses are negligible.

Find: ϵ, T_{co}, T_{ho}

Lookup values: $c = 4180 \text{ J}/(\text{kg} \cdot \text{K})$

SOLUTION

The cold-side capacitance rate is

$$\dot{C}_c = 70 \frac{\text{kg}}{\text{min}} \times 4180 \frac{\text{J}}{\text{kg} \cdot \text{K}} \times \frac{1 \text{ min}}{60 \text{ s}} = 4877 \text{ W/K}$$

and the hot-side capacitance rate is

$$\dot{C}_h = 90 \frac{\text{kg}}{\text{min}} \times 4180 \frac{\text{J}}{\text{kg} \cdot \text{K}} \times \frac{1 \text{ min}}{60 \text{ s}} = 6270 \text{ W/K}$$

The minimum capacitance is for the cold stream. The capacitance ratio is

$$\frac{\dot{C}_{min}}{\dot{C}_{max}} = \frac{4877}{6270} = 0.778$$

The NTU value is found from the definition

$$\text{NTU} = \frac{U_o A_o}{\dot{C}_{min}} = \frac{320 \times 20}{4877} = 1.31$$

Reading from Fig. 2.15 (or by using the second equation in Table 2.10), we find that the effectiveness is

$$\epsilon = 0.60$$

It is now a simple matter to find first the heat rate and then the outlet temperatures of each stream. From Eq. (2.32)

$$\dot{Q} = \epsilon \dot{C}_{min}(T_{hi} - T_{ci}) = (0.60 \times 4877 \text{ W/K})(110°C - 20°C) = 263.4 \text{ kW}$$

The heat rate for either stream is also given by the product of the mass flow, specific heat, and stream temperature rise:

$$\dot{Q} = \dot{m}c(T_i - T_o)$$

One now solves for the outlet temperatures. For the hot stream,

$$T_{ho} = T_{hi} - \frac{\dot{Q}}{\dot{C}_{max}}$$

$$= 110 - \frac{263{,}400}{6270} = 68.0°C$$

and for the cold stream,

$$T_{co} = 20 + \frac{263{,}400}{4877} = 74.0°C$$

COMMENTS

For improved accuracy, one could use the equation for effectiveness listed in Table 2.10. However, for design purposes, the curves are satisfactory. Alternatively, manufacturers' data can be consulted for effectiveness values, although many manufacturers use a different method for finding $\dot{Q}$, called the *log-mean temperature difference method,* for sizing heat exchangers. We discuss this approach in Chap. 9.

Since we made an effectiveness calculation at one temperature, the value applies for a relatively wide range of temperatures (if the flows remain fixed) since relevant water properties do not change much with temperature.

Table 2.10 contains a summary of effectiveness equations for all heat exchangers commonly used on buildings except for plate heat exchangers. For these devices, it is suggested that manufacturers' data be consulted since they are not true counter-flow, parallel-, or cross-flow heat exchangers and are constructed in many different variations.

Heat transfer in many types of heat exchangers can be enhanced by the use of fins. This topic is discussed in Chap. 9 where details of building heat exchangers are discussed.

2.4 Radiation Heat Transfer

Radiation heat transfer is important within conditioned spaces since it is one of the key determinants of both comfort and heat exchange in the zones of a building. Solar radiation is also important in analyzing the thermal behavior of both interiors and exteriors of buildings. It is so important, in fact, that we devote a separate chapter to it. Solar radiation and its effects on buildings are discussed in Chap. 6 and in Goswami, Kreith, and Kreider (2000). In this section we review nonsolar radiation heat transfer fundamentals as they apply to buildings.

There are three features of radiation which set it apart from conduction and convection. First, radiation heat transfer occurs by electromagnetic waves with a strong spectral dependence. Second, the microscale details (e.g., color) of surfaces involved in radiation have a first-order effect on the rate of heat transfer. Third, radiation is a highly nonlinear function of temperature.

2.4.1 Thermal Radiation Spectrum and the Stefan-Boltzmann Law

The fundamental physical law governing thermal radiation emission is the *Stefan-Boltzmann law*. This law states that the heat flux emitted by an ideal radiator called a *blackbody* is proportional to the absolute temperature (units of kelvins or degrees Rankine) to the fourth power. The proportionality constant denoted by σ is called the Stefan-Boltzmann constant. Equation (2.35) expresses the Stefan-Boltzmann law

$$\dot{q} = \sigma T^4 = E_b \tag{2.35}$$

The term E_b is called the *total blackbody emissive power*.

The value of the Stefan-Boltzmann constant in USCS units is 0.1714×10^{-8} Btu/(h · ft² · °R⁴), and in SI units it is 5.669×10^{-8} W/(m² · K⁴). Recall that absolute temperatures, in degrees Rankine or kelvins, must be used in radiation calculations.

Since radiation is transmitted by electromagnetic waves, it has a wavelength and frequency. In fact, radiation emitted by a surface has an entire spectrum of wavelengths. The wavelengths of radiation with which one is concerned in buildings range from a fraction of a micrometer to more than 100 μm. The upper curve in Fig. 2.16 shows how the *spectral emissive power* $E_{b\lambda}$ varies with wavelength λ for a surface near room temperature, 530°R (70°F). Note that there is a distinct peak to the distribution.

The Wien displacement law gives the wavelength at which the spectral blackbody emissive power is a maximum:

$$\lambda_{\max} T = 5216 \ \mu\text{m} \cdot °\text{R} \quad (2898 \ \mu\text{m} \cdot \text{K}) \tag{2.36}$$

In Fig. 2.16 the peak calculated from this expression is at 9.84 μm ($=5216/530$), in agreement with the upper curve on the figure.

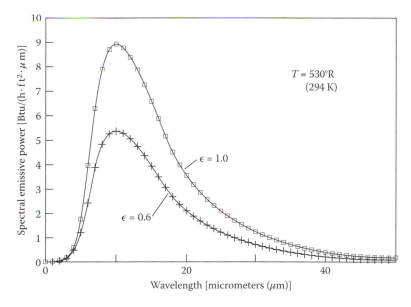

FIGURE 2.16

Example of thermal radiation spectra for black and gray ($\epsilon = 0.6$ in) surfaces at room temperature, 70°F (530°R; 294 K).

Other than this relation of maximum emission wavelength to temperature, we need not concern ourselves with the details of the spectral distribution. Standard heat transfer texts discuss the details.

2.4.2 Gray Surfaces

Real surfaces emit less radiation than ideal "black" ones. The ratio of actual emissive power E to the emissive power of a black surface at the same temperature E_b is called the *emissivity*. It is defined by

$$\epsilon = \frac{E}{E_b} \tag{2.37}$$

Although emissivity is a function of wavelength for many materials, a physically realistic simplification is often made by assuming that the emissivity is constant, at least over a limited range of wavelengths. For radiation calculations in buildings, this is an entirely satisfactory assumption. Surfaces for which the emissivity is constant are called *gray* surfaces. The lower curve in Fig. 2.16 shows the gray body emissive power for a gray surface at 530°R whose emissivity is 0.60.

Table 2.11 lists values for the emissivity of a number of building materials. We shall find such values useful in calculating heating and cooling loads on buildings in Chap. 7. As a rule of thumb, this table tells us that the emissivity of most building materials is approximately 0.9 (in the range of typical building temperatures).

2.4.3 Radiation Properties: Absorptivity, Transmissivity, and Reflectivity

In addition to emissivity, three additional properties of surfaces affect the rate of radiation heat transfer. Figure 2.17 shows how the three properties—absorptivity α, transmissivity τ, and reflectivity ρ—are related. Conservation of energy requires that the sum of these three properties be unity:

$$\alpha + \tau + \rho = 1 \tag{2.38}$$

Therefore, by knowing any two properties, the third can always be found. Although this equation is derived for a single wavelength, it is valid for piecewise gray surfaces if the wavelength range over which the three properties are calculated is the same.

Kirchhoff's identity, stating that absorptivity and emissivity for *gray surfaces* are equal, is another useful expression:

$$\alpha = \epsilon \tag{2.39}$$

This expression is also valid for nongray surfaces at a given wavelength.

2.4.4 Shape Factors

Since radiation is transferred by directional beams of radiation, the relative size and location of two (or more) surfaces interchanging radiation are important factors in quantifying the heat transfer rate. The *radiation shape factor* F_{12} includes all needed geometric information. The shape factor is the fraction of radiation leaving diffuse surface 1 that is

TABLE 2.11

Emissivities of Some Common Building Materials at Specified Temperatures

Surface	Temperature, °C	Temperature, °F	ϵ
Brick			
Red, rough	40	100	0.93
Concrete			
Rough	40	100	0.94
Glass			
Smooth	40	100	0.94
Ice			
Smooth	0	32	0.97
Marble			
White	40	100	0.95
Paints			
Black gloss	40	100	0.90
White	40	100	0.89–0.97
Various oil paints	40	100	0.92–0.96
Paper			
White	40	100	0.95
Sandstone	40–250	100–500	0.83–0.90
Snow	−12–−6	10–20	0.82
Water			
0.1 mm or more thick	40	100	0.96
Wood			
Oak, planed	40	100	0.90
Walnut, sanded	40	100	0.83
Spruce, sanded	40	100	0.82
Beech	40	100	0.94

Source: Courtesy of Sparrow, E.M. and Cess, R.D., *Radiation Heat Transfer,* augmented edn, Hemisphere, New York, 1978. With permission.

Incoming radiation (=1)

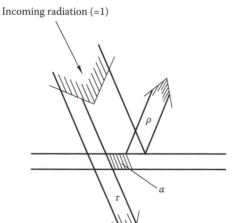

FIGURE 2.17
Schematic representation of transmissivity, absorptivity, and reflectivity. Incoming radiation ($=1$)

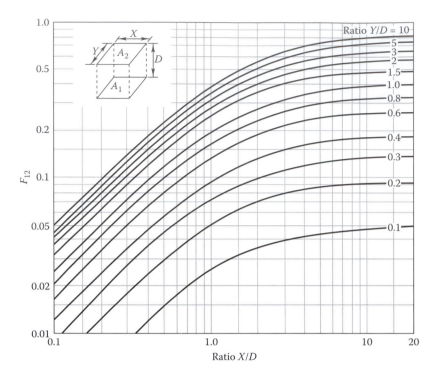

FIGURE 2.18
Shape factor F_{12} for two parallel planes.

intercepted (but not necessarily absorbed) by surface 2. The shape factor is strictly geometric and does not depend on surface properties such as emissivity or temperature. Figures 2.18 and 2.19 are plots of the shape factors for two parallel surfaces, e.g., between the floor and ceiling of a room, and two perpendicular surfaces, e.g., between a floor and an adjacent wall. These charts are plotted from equations given in Siegel and Howell (1981).

Two key relationships exist among shape factors. The first is called the *reciprocity relationship*, which is given by

$$A_1 F_{12} = A_2 F_{21} \qquad (2.40)$$

This can be shown to follow from the second law of thermodynamics.

The second relationship is easily derived from the first law of thermodynamics. It states that the sum of shape factors for a given surface must equal unity. This is easy to see since the radiation leaving a surface and intercepted elsewhere must just be equal to the total radiation leaving the surface. In equation form, the conservation of energy for j surfaces involved in radiant transport is given by

$$F_{11} + F_{12} + F_{13} + \cdots + F_{1j} = 1 \qquad (2.41)$$

The shape factor F_{11} is nonzero only for concave surfaces, i.e., surfaces which can "see" themselves.

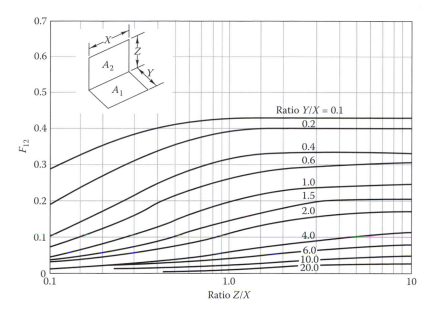

FIGURE 2.19
Shape factor F_{12} for two adjacent orthogonal planes.

It is important to recall that shape factor algebra described in this section applies only for diffusely emitting and reflecting surfaces. It cannot be used for surfaces such as mirrors which reflect specularly. However, nearly all surfaces in buildings are diffuse.

Equations (2.40) and (2.41), along with a few equations or charts of shape factors, can be used to calculate the shape factors needed for radiation problems within buildings. Example 2.8 illustrates how one calculates shape factors within a room. Example 2.9 shows how these results can be used to find the rate of radiation heat transfer between surfaces in such an enclosure.

Example 2.8: Radiation Shape Factors

One room of a building is 8 ft from floor to ceiling and 16 ft square. Find the shape factor from one of the walls to (1) the ceiling, (2) the adjacent wall, and (3) the opposite wall.

Given: $h = 8$ ft

$S = 16$ ft

Figure: See Fig. 2.20.

Assumptions: All surfaces are diffuse.

Find: F_{12}, F_{15}, F_{16}

SOLUTION

To use Figs. 2.18 and 2.19, one simply finds the ratios of lengths needed to enter the graphs. First, consider the shape factor between opposite walls. The ratio Y/D in the nomenclature of Fig. 2.18 is

$$\frac{Y}{D} = \frac{16 \text{ ft}}{16 \text{ ft}} = 1$$

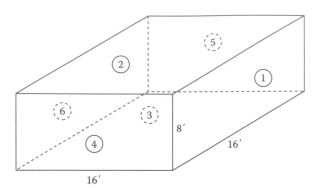

FIGURE 2.20
Isometric sketch of example room.

and the ratio X/D is

$$\frac{X}{D} = \frac{8 \text{ ft}}{16 \text{ ft}} = 0.05$$

From Fig. 2.18 it is an easy matter to read the shape factor between two opposing walls as

$$F_{16} = 0.12$$

For two adjacent walls, the shape factor is found from Fig. 2.19. In the nomenclature of this figure,

$$\frac{Y}{X} = \frac{16 \text{ ft}}{8 \text{ ft}} = 2$$

and

$$\frac{Z}{X} = \frac{16 \text{ ft}}{8 \text{ ft}} = 2$$

With these two values, one reads from Fig. 2.19

$$F_{15} = 0.14$$

Finally, for the wall-to-ceiling shape factor, we use Fig. 2.19 again with

$$\frac{Y}{X} = \frac{8 \text{ ft}}{16 \text{ ft}} = 0.5$$

and

$$\frac{Z}{X} = \frac{16 \text{ ft}}{16 \text{ ft}} = 1$$

The value of this shape factor is

$$F_{12} = 0.30$$

2.4.5 Radiative Exchange

With the tools developed in the three preceding sections, we are now able to find the rate of radiation heat transfer between surfaces. In this section, we will treat exchange between two surfaces. A similar method can be used if three or more surfaces are involved. If more than three surfaces are involved, the algebra becomes rather involved and matrix inversion methods are suggested (Siegel and Howell, 1981).

The rate of heat transport between two surfaces that see nothing else or that are configured so that one encloses the other is given by

$$\dot{Q}_{12} = \frac{A_1\left(\sigma T_1^4 - \sigma T_2^4\right)}{\rho_1/\epsilon_1 + 1/F_{12} + \rho_2 A_1/(\epsilon_2 A_2)} \tag{2.42}$$

We will use this expression in Example 2.9.

This equation applies to a number of cases of special interest such as radiant transport between two parallel planes spaced closely together. This condition can occur in buildings between the inner and outer surfaces of an airspace in a wall or in an attic radiant barrier. In these cases, the denominator of Eq. (2.42) simplifies since $F_{12} = 1$ and $A_1 = A_2 = A$, and we have

$$\dot{Q}_{12} = \frac{A_1\left(\sigma T_1^4 - \sigma T_2^4\right)}{1/\epsilon_1 + 1/\epsilon_2 - 1} \tag{2.43}$$

To get this result, we have used Eqs. (2.38) and (2.39) with $\tau = 0$ to eliminate the reflectivity. Since the surfaces are opaque, the transmissivity is zero.

A convenient shortcut for calculating radiation heat transfer in buildings is to define a radiation heat transfer coefficient h_{rad} as follows:

$$h_{\text{rad}} = \frac{\dot{q}_{12}}{T_1 - T_2} \tag{2.44}$$

This coefficient can be used just as one uses h_{con} in convection problems. If the two temperatures are close to each other (no more than 50°F difference), a good approximation to Eq. (2.44) for h_{rad} in two-surface situations in Eq. (2.43) is

$$h_{\text{rad}} = \frac{4\sigma \bar{T}^3}{1/\epsilon_1 + 1/\epsilon_2 - 1} \quad \text{with } \bar{T} \equiv \frac{T_1 + T_2}{2} \tag{2.45}$$

The denominator in this equation can be replaced with the denominator in Eq. (2.42) if the two surfaces are not parallel and of equal area.

Example 2.9 and Examples 2.10 and 2.11 in the next section illustrate how radiation calculations are performed to determine the heat transfer in several building-related situations.

Example 2.9: Room Radiation Heat Transfer

Assume that surface 1 in Example 2.8 is the exterior wall of a poorly insulated building and that its inner surface temperature is 60°F on a winter day. All other room surfaces are at 72°F. What is the heat loss from these surfaces to the cool wall? The emissivities of all walls are 0.85.

    Given: $\epsilon_1 = \epsilon_2 = 0.85$

Wall dimensions as in Example 2.8

$$T_1 = 60°\text{F} = 520°\text{R}$$

$$T_{n \neq 1}(=T_2) = 72°\text{F} = 532°\text{R}$$

    Find: $\dot{Q}$

    Figure: See Fig. 2.20.

SOLUTION

Although this problem involves six surfaces, five of the six are at the same temperature T_2 and can, therefore, be treated as one surface as far as radiation transport is concerned. In Eq. (2.42) we require values of the shape factor F_{12}, which, by inspection, is unity, since all radiation emitted by surface 1 is intercepted by the second surface.

The emissive powers in the numerator are

$$E_{b1} = \sigma T_1^4 = \left[0.1714 \times 10^{-8}\ \text{Btu}/(\text{h} \cdot \text{ft}^2 \cdot °\text{R}^4)\right](520°\text{R})^4 = 125.3\ \text{Btu}/(\text{h} \cdot \text{ft}^2)$$

and

$$E_{b2} = \sigma T_2^4 = (0.1714 \times 10^{-8})(532^4) = 137.3\ \text{Btu}/(\text{h} \cdot \text{ft}^2)$$

If one recalls that

$$\frac{\rho_1}{\epsilon_1} + \frac{1}{F_{12}} + \frac{1 - \epsilon_1}{\epsilon_1} + 1 = \frac{1}{\epsilon_1}$$

then Eq. (2.42) takes the following simplified form for this example:

$$\dot{Q}_{12} = \frac{A_1\left(\sigma T_1^4 - \sigma T_2^4\right)}{1/\epsilon_1 + \rho_2 A_1/(\epsilon_2 A_2)}$$

The total area of all walls except for 1 (i.e., surface 2) is 896 ft². Substituting numerical values, we have

$$\dot{Q}_{12} = \frac{(16\ \text{ft} \times 8\ \text{ft})(125.3 - 137.3)}{1/0.85 + 0.15(16 \times 8)/(0.85 \times 896)} = -1278\ \text{Btu/h}$$

The sign indicates that heat flows from surface 2 to surface 1.

2.4.6 Other Topics—Combined-Mode Heat Transfer and Thermal Bridges

2.4.6.1 Combined-Mode Heat Transfer

Nearly all heat transfer situations in buildings include more than one mode of heat transfer. For example, heat loss through a wall includes both conduction as treated in Examples 2.1 and 2.2 and convection from the inner and outer surfaces. Figure 2.21 shows the equivalent circuit for this case, and Example 2.10, based on Example 2.2, illustrates the calculation.

Likewise, convection and radiation often coexist in buildings. For example, reflective plastic sheets called *radiant barriers* are often used in residential walls or above residential building ceilings to reduce heat loss by controlling radiation.

Examples 2.10 and 2.11 illustrate how heat loss is computed when more than one heat transfer mode is present.

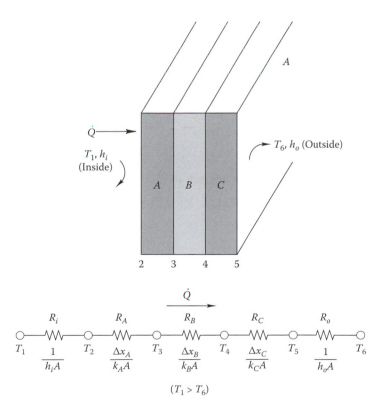

FIGURE 2.21
Solid insulated wall with inner- and outer-surface convection coefficients h_i and h_o.

Example 2.10: Effect of Convection on Wall R Value

Repeat Example 2.2 for a stud wall to include the effect of inner and outer surface convection coefficients. The inner surface coefficient has been calculated at 0.4 Btu/(h · ft^2 · °F) [turbulent free convection with a wall-to-room temperature difference of 10°F, from Eq. (2.19US)], and the outer surface coefficient is 3.7 Btu/(h · ft^2 · °F) [15 mi/h wind along a 15-ft-long wall, from Eq. (2.25US)]. Find the overall wall R value.

$\quad$ *Given:* $\quad h_{con,i} = 0.40$ Btu/(h · ft^2 · °F)

$h_{con,o} = 3.7$ Btu/(h · ft^2 · °F)

$\quad$ *Figure:* $\quad$ See Fig. 2.21.

$\quad$ *Find:* $\quad R_{tot}$

$\quad$ *Lookup values:* $\quad$ From Example 2.2 the R value of the solid wall $R_{th,wall}$ is 19.8 (h · ft^2 · °F)/Btu.

SOLUTION

From the equivalent circuit shown in Fig. 2.20, we see that the total wall U value is

$$R_{th,tot} = \frac{1}{h_{con,i}} + R_{th,wall} + \frac{1}{h_{con,o}}$$

Substituting numerical values gives

$$R_{th,tot} = \frac{1}{0.4} + 19.8 + \frac{1}{3.7} = 22.6 \ (h \cdot ft^2 \cdot °F)/Btu \ \left[4.0 \ (m^2 \cdot K)/W\right]$$

COMMENTS

Note that the inner and outer surface air films through which convection occurs account for 13 percent of the wall's thermal resistance.

$\quad$ For a more accurate calculation, the radiation losses from the walls must be added. If this is done, ASHRAE recommends the following values of total surface coefficients for peak load calculations:

$$h_{con+rad,i} = 1.46 \ Btu/(h \cdot ft^2 \cdot °F) \left[17.6 \ W/(m^2 \cdot K)\right]$$

$$h_{con+rad,o} = 6.0 \ Btu/(h \cdot ft^2 \cdot °F) \left[34 \ W/(m^2 \cdot K)\right] \quad \text{at 15 mi/h wind speed}$$

If these more realistic values are used, the percentage of this wall's resistance due to the surface effects drops to only 4 percent. The bulk of the insulating effect of a wall occurs in the insulation.

$\quad$ The Appendix CD-ROM contains surface conductance values for a number of other configurations and surface radiation properties. The values there show that surface emissivity can affect the combined radiation and convection surface conductance by a factor of 2.

Example 2.11: Radiant Barrier in Residential Wall

A building designer wishes to evaluate the R value of a 1-in-wide air gap in a wall for its insulating effect. The resistance to heat flow offered by convection is small, so she proposes lining the cavity's inner and outer surfaces with a highly reflecting aluminum foil film whose emissivity is 5 percent. Find the R value of this wall cavity, including both radiation and convection effects, if the temperatures of the walls facing the gap are 45 and 55°F.

The convection coefficient at the inner surface of both walls is 0.32 Btu/(h·ft²·°F) [from Eq. (2.19US) with 5°F temperature difference].

Given: $h_{con,i} = h_{con,o} = 0.32$ Btu/(h·ft²·°F)

$\epsilon_i = \epsilon_o = 0.05$

Figure: See Fig. 2.22.

Assumptions: The emittance of the cavity surfaces prior to installing the radiant barrier is 1.0. Consider a unit area so that $R = R_{th}$.

Find: R_{gap}

SOLUTION

As shown in Fig. 2.22, this problem involves both series and parallel resistances. The solution to this problem has two parts. We will find the thermal resistance of the wall cavity first *without* the barrier and then with the barrier.

The radiation coefficient from Eq. (2.45) for the standard cavity is

$$h_{rad} = 4(0.1714 \times 10^{-8}) \frac{(460 + 50)^3}{1/1.0 + 1/1.0 - 1} = 0.91 \text{ Btu}/(h \cdot ft^2 \cdot °F)$$

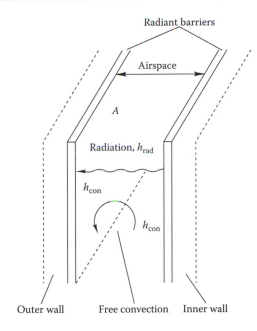

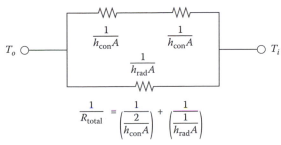

FIGURE 2.22
Wall radiant barrier equivalent circuit showing both convection and radiation resistances.

The radiation resistance without radiation shield is

$$R_{th,rad} = \frac{1}{h_{rad}} = \frac{1}{0.91} = 1.10$$

The total convection resistance is

$$R_{th,con} = \frac{1}{h_{con}} = \frac{1}{h_{con}} = \frac{2}{0.32} 6.3$$

Finally, these two parallel resistances are added reciprocally to find the total gap resistance

$$\frac{1}{R_{th,gap}} = \frac{1}{R_{th,rad}} + \frac{1}{R_{th,con}} = \frac{1}{1.10} + \frac{1}{6.3} = 1.07$$

from which

$$R_{th,gap} = 0.94 \ (h \cdot ft^2 \cdot {}^\circ F)/Btu \ [0.17 \ (m^2 \cdot K)/W]$$

The calculation is now repeated with the low-emittance radiant shield. The convection coefficient remains the same, but the radiation coefficient is different.

$$h_{rad} = 4(0.1714 \times 10^{-8}) \frac{(460 + 50)^3}{1/0.05 + 1/0.05 - 1} = 0.023 \ Btu/(h \cdot ft^2 \cdot {}^\circ F)$$

The overall wall cavity resistance *with radiant shield* is

$$\frac{1}{R_{th,gap}} = \frac{1}{R_{th,rad}} + \frac{1}{R_{th,con}} = \frac{1}{1/0.023} + \frac{1}{6.3} = 0.18$$

$$R_{th,gap} = 5.50 \ (h \cdot ft^2 \cdot {}^\circ F)/Btu \ [0.97 \ (m^2 \cdot K)/W]$$

COMMENTS

The thermal resistance of this gap is almost 6 times that of a standard wall gap. However, there are practical problems since the low radiant barrier emissivity is difficult to maintain. Dust accumulation on the surface can reduce emissivity markedly. For example, if the emissivity increases to 20 percent from the as-new value of 5 percent, the gap *R* value drops to 3.8, a 30 percent decrease.

The Appendix CD-ROM contains *R* value tabulations for other configurations of radiant barrier–air gap walls, ceilings, and floors. The table shows that *the width of the air gap has almost no effect on the thermal resistance.*

2.4.6.2 Thermal Bridges

In the context of this book, a *thermal bridge* is a local area of a building's envelope with relatively lower thermal resistance than exists in its surroundings. The wood stud considered in Example 2.2 is an example of a thermal bridge. The presence of the stud caused the wall heat flow to increase by 18 percent. Similar or greater increases are caused by structural members that penetrate walls (e.g., balcony supports in a high-rise building) or that support walls (e.g., the steel structure in a high rise).

Thermal bridges are unavoidable in conventional building practice and cause at least two significant difficulties:

- Heat losses and gains are increased.
- Lower temperature can cause condensation, leading to moisture problems in winter (material degradation, paint peeling, mold).

Figure 2.23 shows a more complex thermal bridge and its approximation by a thermal network. The construction shown represents a three-layer vertical wall supported by a concrete floor. The concrete floor is the thermal bridge. The thermal network is a combined

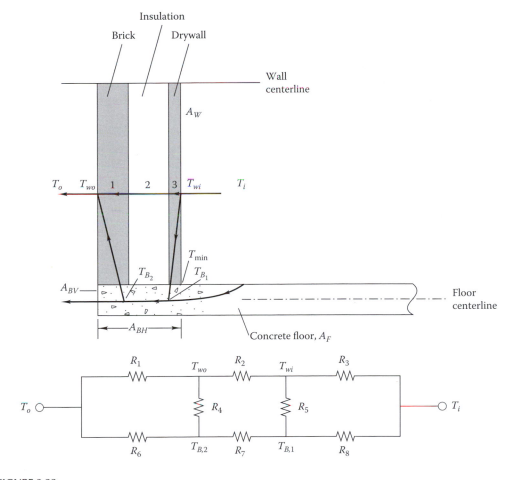

FIGURE 2.23

Example of a thermal bridge (concrete floor) with insulated wall above. The thermal network allows for two-dimensional calculations of the thermal bridge heat loss. Heavy lines with arrows show various heat flow paths. Each resistance can be found by using basic equations developed earlier for conduction and convection. Resistances R_1, R_6, R_3, and R_8 include surface film effects.

series-parallel arrangement that can be analyzed by using the rules for series and parallel circuits shown in Fig. 2.2. The resistances that are associated with the vertical heat flows include two series heat transfers—through the wall and through a portion of the floor (the subscript *B* in the figure denotes the thermal bridge area). According to Staelens (1988), this method is accurate to a few percent. Once the total heat flow has been found, the temperatures at each node can also be determined to ascertain if moisture condensation will occur under winter conditions. In the bridge shown, the critical temperature point where condensation may occur is at the inner corner of the wall at the floor.

The method used to analyze the stud in Example 2.2 underestimates the heat loss, since the lateral two-dimensional heat flows were not accounted for. However, according to ASHRAE (1999), the simple parallel-flow approach can be used if all materials involved in the bridge environment (the bridge and all building envelope components in contact with it) are nonmetals (wood, drywall, concrete). If any portion of the bridge has high conductivity (e.g., structural steel or building skin materials), the simple method cannot be used and that shown in Fig. 2.23 must be used.

2.5 Evaporation and Moisture Transfer

The evaporation of moisture is a primary mechanism for cooling of various surfaces. For example, the evaporation of water from surfaces in evaporative coolers is an important method for cooling buildings in arid parts of the world. In addition, the control of migration of moisture through building materials must be considered in building design to avoid structural damage. Although it is outside the scope of this text to deal with evaporation and moisture transfer in detail, we present the governing equations below.

There are two mechanisms of interest. First, the transfer of moisture within building materials is essentially a diffusion process governed by Fick's law (Bird, Stuart, and Lightfoot, 1960):

$$\dot{m}_w = -DA\frac{\partial C}{\partial x} \tag{2.46}$$

in which D is the *diffusivity*, A the area through which mass transfer takes place, and the derivative is the concentration gradient of water vapor in the material in which diffusion is taking place. This equation is analogous to Fourier's law of heat conduction, Eq. (2.1); and in cases for which D is constant, the solutions to the conduction equation are solutions to the diffusion equation if the same boundary conditions exist. This expression is used to determine the migration of moisture into and from building contents and structural members.

In winter, the interior of a building is more humid than the exterior. Hence, water vapor will tend to diffuse from the interior to the exterior. If the winter is cold, the migrating moisture may reach a location in the wall where the insulation (or other component of the wall such as a metal stud) is at the dew point (see Chap. 4) and condensation occurs. The condensed water can saturate insulation, rendering it useless as a heat loss barrier. If condensation levels are extreme, wooden structural members can deteriorate or steel ones can rust, compromising the structural integrity of a building.

Moisture migration is often controlled in cold climates by use of a *vapor barrier*—a material (most often plastic) essentially impervious to moisture. If the vapor barrier is placed on the heated side of building insulation, condensation will be avoided in insulation in winter. The situation in summer is not as clear-cut, but a higher vapor resistance at the building exterior, relative to the interior, is recommended. To be effective, the vapor barrier must be continuous without punctures or tears.

Following the heat-mass transfer analogy for another step, we see that the convective transport of mass is given by

$$\dot{m}_w = h_m A (C_s - C_f) \tag{2.47}$$

in which h_m is the convective *mass transfer coefficient*. Its value can be found from convective *heat transfer* equations by using the proper transformations as described in Bird, Stuart, and

TABLE 2.12

Chapter 2 Equation Summary

Heat Transfer Mode	Equations	Equation Number(s)	Comments
Conduction	$\dot{Q} = \dfrac{\Delta T}{R}$	(2.3)	Fourier's law
	$R = \dfrac{\Delta X}{kA}$	(2.4)	Thermal resistance
	$R_{\text{th}} = \dfrac{\Delta X}{k} = AR$	(2.5)	R value
	$U \equiv \dfrac{1}{R_{\text{th}}}$	(2.6)	U value
Parallel flow	$\dfrac{1}{R_{\text{tot}}} = \dfrac{1}{R_1} + \dfrac{1}{R_2} + \cdots + \dfrac{1}{R_n}$	Fig. 2.2	
Series flow	$R_{\text{tot}} = R_1 + R_2 + \cdots + R_n$	Fig. 2.2	
Ground coupling	$\dot{Q}_g = \left(\sum_{\text{depth}} \dfrac{\dot{Q}}{L\,\Delta T} \right) L (T_i - T_{g,\text{des}})$	(2.9)	$T_{g,\text{des}}$, winter ground temperature
Slab edge loss	$\dot{Q}_{\text{edge}} = F_2 P (T_i - T_o)$	(2.10)	
Convection	$\dot{Q} = h_{\text{con}} A\,\Delta T$	(2.13)	Definition of h_{con}
	$R = \dfrac{1}{h_{\text{con}} A}$	(2.15)	Thermal resistance
Natural, laminar	$h_{\text{con}} = \text{const} \left(\dfrac{\Delta T}{L} \right)^{1/4}$	(2.18) to (2.23)	Dimensional equations
Natural, turbulent	$h_{\text{con}} = \text{const}(\Delta T)^{1/3}$	(2.18) to (2.23)	Dimensional equations
Forced, turbulent	$h_{\text{con}} = \text{const} \left(\dfrac{v^4}{L} \right)^{1/5}$	(2.25) to (2.28)	Dimensional equations
Heat exchangers	$\dot{Q} = \epsilon (\dot{m}c)_{\text{min}} (T_{h,i} - T_{c,i})$	(2.32)	
Radiation	$\dot{q} = \sigma T^4$	(2.35)	Stefan–Boltzmann
	$\dot{Q}_{12} = \dfrac{A(\sigma T_1^4 - \sigma T_2^4)}{1/\epsilon_1 + 1/\epsilon_2 - 1}$	(2.43)	Parallel surfaces, closely spaced
	$h_{\text{rad}} = \dfrac{4\sigma T^3}{1/\epsilon_1 + 1/\epsilon_2 - 1}$	(2.45)	Parallel surfaces, closely spaced
	$\dot{Q}_{12} = \dfrac{A_1(\sigma T_1^4 - \sigma T_2^4)}{\rho_1/\epsilon_1 + 1/F_{12} + \rho_2 A_1/(\epsilon_2 + A_2)}$	(2.42)	Two-surface enclosure

Lightfoot (1960). And C_s and C_f are the surface and fluid concentrations of the species (water vapor in buildings) undergoing the mass transfer process. This equation governs the evaporation of moisture from surfaces. In later chapters, we will use results from this equation to analyze the performance of evaporative coolers and indirectly to find the relation of human comfort to the velocity of air moving over the skin.

2.6 Closure

In this chapter we have reviewed the key features of heat transfer that apply to buildings. Conduction and convection expressions will be used to find the heating and cooling requirements of buildings in Chaps. 7 and 8. In Chap. 9 we will discuss heat exchangers in greater detail, but the governing equations will be those presented here.

A great many equations have been developed in this chapter. Table 2.12 is a summary of the key ones.

Problems

The problems in this book are arranged by topic. The approximate degree of difficulty is indicated by a parenthetic italic number from 1 to 10 at the end of the problem. Problems are stated most often in USCS units; when similar problems are presented in SI units, it is done with approximately equivalent values in parentheses. The USCS and SI versions of a problem are not exactly equivalent numerically. Solutions should be organized in the same order as the examples in the text: given, figure or sketch, assumptions, find, lookup values, solution. For some problems, the Heating and Cooling of Buildings (HCB) software may be helpful. In some cases it is advisable to set up the solution as a spreadsheet, so the the design variations are easy to evaluate.

2.1 Oil is heated to 150°F (65°C) by a steam coil located in the bottom of a horizontal cylindrical tank. If the diameter is 5 ft (1.5 m) and the length 20 ft (6 m), find the heat loss if the tank is insulated with 2 in (5 cm) of polyisocyanurate and located in a room at 65°F (18°C). Compare the heat loss with that if the tank were insulated with fiberglass of the same thickness. Ignore the resistance to heat transfer offered by the metal tank and by convection at the outer surface of the insulation.   (5)

2.2 Calculate the thermal resistance of the typical wall shown schematically in Fig. P2.2. The inside surface is 0.5-in (1.2-cm) drywall, and the outside wall surface is 0.44-in-thick hardboard siding (see App. CD-ROM for thermal properties of these materials). The studs, headers, and floor plates in this wall are nominal 2-in by 6-in wood studs with an R value of 6.88 (°F · h · ft^2)/Btu [1.21 (K · m^2)/W] with true finished-wood dimensions of 1.5 by 5.5 in (3.8 by 13.5 cm). The wall cavities are insulated with fiberglass at $R = 19$ [3.35 (K · m^2)/W]. Ignore convection film resistances for now at the inner and outer wall surfaces (together they add less than one unit of R value in USCS units).   (8)

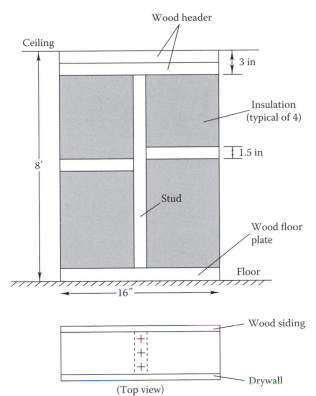

Wood header

Ceiling

3 in

Insulation
(typical of 4)

1.5 in

8'

Stud

Wood floor
plate

Floor

16"

Wood siding

Drywall

(Top view)

FIGURE P2.2

2.3 A roof is constructed as shown in Fig. 2.2b. The outer layer is plywood and shingles (total of $R = 3$ in USCS), and the center layer is 6 in (15 cm) of fiberglass. The innermost layer is 0.5-in (1.2-cm) drywall. On a day when the outer surface is 0°F (-18°C) and the indoor surface is 68°F (20°C), where in the wall is the temperature exactly equal to the freezing point of water? Based on this calculation, where should one place a vapor barrier designed to keep water vapor from diffusing into and freezing the insulation? (Since the R value of the wall assembly itself is so much larger than the thermal resistance offered by inner and outer surface air films, you may ignore them for this problem.)   (6)

2.4 An insulated 24×60 in (60×150 cm) rectangular duct carries 12,000 ft^3/min (5700 L) of air at 45°F (7°C) in a low-temperature air conditioning system at sea level. If the temperature rise in the cold air is not to exceed 1°F in every 100 ft of duct length (2°C in 100 m), what is the R value of insulation needed on the duct? Assume that the combined effect of the convection resistances on the inside of the duct and the outside of the insulation is $R = 1$ in USCS units (0.18 SI). This duct is located in an uncondi-tioned roof plenum that reaches 110°F (43°C) on a summer day.   (7)

2.5 Find the temperature in the unheated room, shown in Fig. P2.5. Some important properties of the space are shown in the table below. The unheated room is 8 ft (2.4 m) high. Ignore floor slab losses.   (7)

Building Component	USCS Units		SI Units	
	U Value, Btu/(h · ft² · F)	Adjacent Air Temp, °F	U Value, W/(m² · K)	Adjacent Air Temp., °C
Exterior wall	0.2	−5	1.13	−21
Ceiling + attic + roof	0.07	−5	0.40	−21
Interior walls	0.06	72	0.34	22

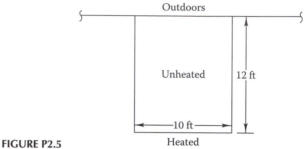

FIGURE P2.5

2.6 Repeat Prob. 2.5, but include the floor slab losses. Assume that the floor is a slab-on-grade design, and consider the full range of loss coefficients given in Table 2.6 (neither the details of wall construction nor the site details are given). Is the ignoring of slab losses in Prob. 2.5 justified?   (8)

2.7 In an energy audit of a building, you find that a 4-in-diameter (10-cm-diameter) pipe located in the 80°F (27°C) mechanical room is carrying superheated steam at 400°F (205°C). The pipe is insulated with 2 in (5 cm) of aluminum-jacketed fiberglass. Find the outer surface temperature of the insulation jacket. Ignore the heat transfer resistance offered by the insulation jacket. Comment on the safety of this installation.   (4)

2.8 The exterior wall of an old building constructed of face brick and common brick is shown in Fig. P2.8. If the outer surface is 10°F (−12°C) and the inner surface is 65°F (18°C), what is the heat flux through the wall? If the air gap is filled with 2.0-in (3.8-cm) polyisocyanurate board, by what percentage is the heat loss reduced?   (7)

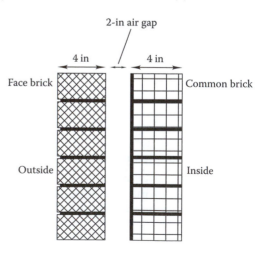

FIGURE P2.8

2.9 The inner and outer walls of a school building are constructed of 4-in (10-cm) concrete and brick, respectively, connected by 6-in-wide (15-cm-wide) grout stiffeners spaced every 4 ft (1.2 m). If the 4-in-deep (10-cm-deep) wall cavity is filled with extruded polystyrene except for the stiffeners, what is the wall thermal resistance? For convenience, just treat a unit height of wall. Grout is similar to concrete from a thermal viewpoint.   (6)

2.10 A buried pipe carries 45°F (7°C) chilled water in a 1-mi-long (1.6-km-long) loop serving a campus cooling system. To save initial cost, a contractor proposes to just bury the 20-in (50-cm) pipe 2 ft (0.6 m) below the surface, claiming that the ground itself serves as adequate insulation. Find the heat gained by the pipe on a summer day if the ground temperature is 100°F (38°C). Do you recommend insulation?   (7)

2.11 The floor of a 100 × 1000 ft (30 × 300 m) warehouse is an uninsulated slab on grade. Find the heat loss from the slab in December and April if the building is located in Denver, Colorado, using the Bahnfleth and Pederson approach.   (8)

2.12 A 32 × 100 ft (10 × 30 m), 7-ft-deep (2.1-m-deep) basement must be maintained at 72°F (22°C) during the winter in Denver, Colorado. The room above the basement is maintained at 72°F (22°C). How much heat is needed if the basement is uninsulated? If it is insulated to $R = 8.34$ ($R = 1.47$)?   (8)

2.13 If a wall consists of 1 in (2.5 cm) of plywood exterior siding, 1 in (2.5 cm) of polyisocyanurate, 3 in (7.5 cm) of fiberglass, and 0.5 (1.2 cm) of gypsum board, find the temperature at the interface between each material, and plot the temperature gradients in each material. In which material is the temperature gradient the steepest? Make your calculations for an outer wall surface temperature of 0°F (-17.3°C) and inner surface temperature of 70°F (21°C).   (10)

2.14 The difference between the air and the inside wall temperature in a room is 10°F (5.5°C). What is the free convection coefficient if the wall is 8 ft (2.4 m) high?   (3)

2.15 The surface temperature of an insulated horizontal 12-in-diameter (30-cm-diameter) steam pipe varies between 80 and 150°F (27 and 66°C) depending on steam usage. What effect does this variation have on the free convection heat transfer coefficient if the pipe is located in a room whose average temperature is 65°F (18°C)?   (5)

2.16 On a winter day, a nearly horizontal skylight in a building is 50°F (10°C). What is the heat loss from it if its area is 32 ft^2 (3 m^2) and if the room that it illuminates is 72°F (22°C)?   (4)

2.17 Air blows over the flat roof of a building at 30 mi/h (14 m/s) and 5°F (-15°C). What is the heat loss from the roof if its area is 320 ft^2 (30 m^2) and its temperature is 20°F (-6°C)?   (4)

2.18 Chilled water (45°F, 7°C) used to cool a building flows at 6 ft/s (2 m/s) in an 8-in- (20-cm-) outside-diameter pipe. If the pipe is uninsulated, what is the heat gain per 100 ft (100 m) of pipe if it is in a 70°F (21°C) room?   (5)

2.19 Repeat Prob. 2.18 if a 33% propylene glycol solution is used instead of pure water. (Glycol is used in some low-temperature cooling systems where chiller coil surface temperatures can drop below the freezing point of water.) [Use $\rho = 64$ lb/ft^2 and $c_p = 0.95$ Btu/(lb · °F).]   (6)

2.20 A double-pipe counterflow heat exchanger transfers heat between two water streams. What is the overall heat transfer conductance $U_o A_o$ if 200 gal/min (13 L/s) is heated in the tubes from 50 to 100°F (10 to 38°C) with 200 gal/min (13 L/s) of inlet water on the shell side at 115°F (46°C)? What is the effectiveness? (6)

2.21 Repeat Prob. 2.20, except use saturated steam at 230°F (110°C) as the heat source. Sufficient steam is supplied to avoid subcooling below 230°F (110°C). *Hint:* The flow rate of the condensing steam does not matter. (6)

2.22 Repeat Prob. 2.20 (with the same two fluid inlet temperatures and same volumetric flow rates), but assume that the fluid being heated is 33% propylene glycol. Account for the differences in specific heat and density as well as the difference in the heat transfer coefficient for glycol. Assume that 90 percent of the resistance to heat transfer in the heat exchanger of Prob. 2.20 is attributed to convection, equally split between the two water-to-tube surfaces. (9)

2.23 Suppose that the heat exchanger in Prob. 2.20 is connected by mistake as a parallel-flow heat exchanger instead of a counterflow exchanger. What will the effectiveness be if the value of $U_o A_o$ is unaffected by the change? (7)

2.24 Figure P2.24 shows a 10,000-ft^2 (1000-m^2) ceiling suspended 2 ft (60 cm) below the roof of a building. The bottom surface of the roof is 115°F (46°C), and the top of the ceiling is 80°F (27°C); both surfaces are gray ($\epsilon = 0.8$). Find the shape factor between the two surfaces and the radiation heat flux. Comment on the relative magnitude of the convection that also occurs in this situation. (4)

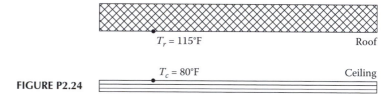

$T_r = 115°F$ Roof

$T_c = 80°F$ Ceiling

FIGURE P2.24

2.25 A radiation shield is to be considered for the attic shown in Fig. P2.24. It will be located 2 in (5 cm) above the ceiling and will have an emissivity of 0.05. *Considering only radiation* in the attic, by how much will the shield reduce the heat flow compared to the situation without the shield? (6)

2.26 The room shown in Fig. P2.26 is heated by a radiant floor maintained at 27°C. The other five surfaces of the space and the room air temperature are all at 20°C. The emissivity of all six surfaces can be taken to be 0.80. Find the total heat transfer from the floor to the other five surfaces by *radiation*. (5)

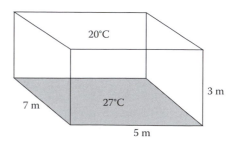

FIGURE P2.26

2.27 What is the view factor between the floor and the ceiling in Fig. P2.26? Between the floor and the 7- and 5-m-wide walls? (6)

2.28 If the room in Fig. P2.26 is modified by replacing one-half of one of the 5-m walls with a square pane of glass centered in the wall, how is the total floor-to-wall heat loss increased for this room if the glass is 10°C on a cool day? (8)

2.29 In dry climates, the effective sky temperature for radiation can be as much as 15°F (10°C) below the air temperature. Under such conditions, find the air temperature at which water could freeze on the top of a cold, sky-facing surface (the surface has no thermal connection to anything else) outdoors if heat is transferred from both sides of the surface by convection with $h_{con} = 1.8$ Btu/(h · ft² · °F) [10 W/(m² · K)]. (8)

2.30 A person stands in the room shown in Fig. P2.30. If she can be modeled as a 20-ft² (1.9-m²) cylinder with an emissivity of 0.80, what is the *radiation* heat transfer from her for the case of room surfaces, all at 70°F (22°C)? Make a reasonable assumption for the body's effective surface temperature, assuming summer clothing. (7)

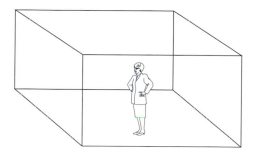

FIGURE P2.30

2.31 The effective surface temperature of the sun is 5760 K. At what wavelength does the sun emit the most energy? (3)

2.32 An incandescent lamp (assumed to be spherical) is housed in a protective plastic sphere for exterior building lighting. If the diameter of the sphere is twice the diameter of the lamp, what is the shape factor between the sphere and the lamp? What are the values of all other shape factors in this system (there are four total)? (5)

2.33 A sunspace on a residence is shown in Fig. P2.33. Find the shape factors between (a) the floor and the wall, (b) the floor and glass, and (c) the glass and the wall. (6)

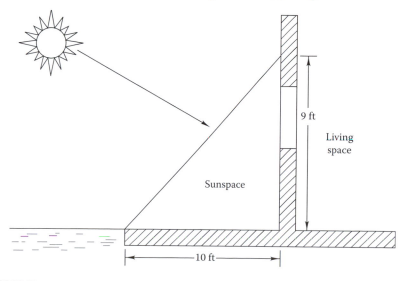

9 ft

Living space

Sunspace

10 ft

FIGURE P2.33

2.34 What is the heat flux between two gray orthogonal surfaces of equal size having a common edge if one is at 50°F and the other is at 80°F (10 and 27°C)? The surfaces are 8 ft (2.4 m) square and have emittances of 0.5 and 0.7. (6)

2.35 A vertical wall between the residential space and the adjacent garage of a home consists of 2×4 in (5×10 cm) framing on 16-in (40-cm) centers between two 0.5-in (12-mm) sheets of drywall (gypsum board). The stud space is uninsulated, and the still air temperatures on the two sides of the wall are 72 and 60°F (22 and 16°C). Find the resistance of the airspace between the drywall sheets as well as the overall U value of the wall, accounting for the stud thermal bridge. What is the wall heat flux? (5)

2.36 Find the wall U value if fiberglass insulation fills the stud space of Prob. 2.35. What is the percentage of reduction in wall heat flux due to the addition of this insulation? (5)

2.37 Figure P2.37 is a schematic diagram of a long, unventilated attic above a residence. What is the rate of heat transfer through the ceiling for the roof and ceiling properties given in the table (the "convection" coefficients include radiation as well)? What is the attic air temperature? (5)

	Ceiling	Roof
Area, m^2	20	23
R value, $(K \cdot m^2)/W$	5.0	0.5
Convection coefficient, upper, $W/(m^2 \cdot K)$	10	25
Convection coefficient, lower, $W/(m^2 \cdot K)$	10	8

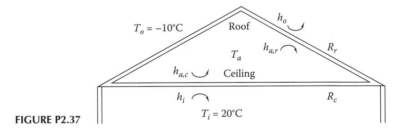

FIGURE P2.37

2.38 A roof assembly is shown schematically in Fig. P2.38. Draw the thermal analog circuit for the roof and its thermal bridge (the roof joist). There will be eight total resistances including the convection resistances. Describe each in words, and indicate which is likely to be the smallest and which the largest. (3)

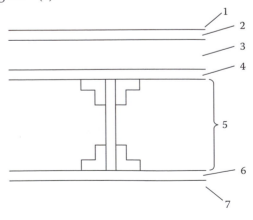

FIGURE P2.38

2.39 An uninsulated steam condensate return tank is to be insulated with 3 in (7.5 cm) of fiberglass. If the tank is cylindrical with height and radius both equal to 5 ft (1.5 m) and if the tank is at 250°F (121°C) and located in an 80°F (27°C) mechanical room, what will the steam energy savings be per year if the tank is used 8000 h/yr? If the steam is valued at $4.00 per million Btu ($3.80 per gigajoule), what is the cost saving due to the installation of insulation?   (7)

2.40 Figure 2.23 shows the thermal bridge produced by a concrete floor supporting a wall constructed of brick, insulation, and drywall. The minimum temperature on the building's internal surface is at the corner formed by the inside wall and the floor and is approximately equal to T_{b1}. For a wall constructed of 4-in (10-cm) brick, 4-in (10-cm) fiberglass insulation, and 0.5-in (1.2-cm) drywall and an 8-in (15-cm) concrete floor, find T_{b1} for a winter day on which the outdoor temperature is −10°F (−23°C) and the interior is 70°F (21°C). State any assumptions made.   (10)

2.41 Find the heat transfer from the floor to the *walls and room air* in Prob. 2.26 by both convection and radiation.   (9)

2.42 On a sunny day, 280 Btu/(h · ft²) (880 W/m²) strikes the dark, well-insulated horizontal roof of a building ($\alpha = 0.9$). There is no wind blowing, so heat is removed only by free convection and radiation from the roof (a negligible amount passes through the roof to the interior). What is the roof temperature under these conditions if the air temperature is 90°F (32°C)? Assume that radiation takes place to the atmosphere, also at 90°F (32°C).   (9)

2.43 The outside wall of a building absorbs 350 W/m² of solar flux on a 0°C day. If the wall surface temperature is 10°C and the convection coefficient is 35 W/(m² · K), how much heat flows through the wall?   (5)

References

ASHRAE (1989). *Handbook of Fundamentals*. American Society of Heating, Refrigeration and Air-Conditioning Engineers, Atlanta, GA.

ASHRAE (1999). *Standard 90.1-1999: Energy Efficient Design for Buildings except Low-Rise Residential Buildings*. American Society of Heating, Refrigerating and Air-Conditioning Engineers, Atlanta.

ASHRAE (2001). *Handbook of Fundamentals*, American Society of Heating, Refrigeration and Air-Conditioning Engineers, Atlanta; see also ASHRAE (1997), *Handbook of Fundamentals*.

Bahnfleth, W. P., and C. O. Pederson (1990). "A Three-Dimensional Numerical Study of Slab on Grade Heat Transfer." *ASHRAE Trans.* vol. 96, pt. 2.

Bird, R. B., W. E. Stuart, and E. N. Lightfoot (1960). *Transport Phenomena*. Wiley, New York.

Claridge, D. (1986). "Building to Ground Heat Transfer." *Proc. American Solar Energy Society Conf.*, Boulder, Colo., pp. 144–154.

Goswami, Y., F. Kreith, and J. F. Kreider (2000). *Principles of Solar Engineering*. Taylor and Francis, New York.

Holman, J. P. (1997). *Heat Transfer*, 8th ed. McGraw-Hill, New York.

Incropera, F. P., and D. P. Dewitt (1996). *Introduction to Heat Transfer*, 3d ed. Wiley, New York.

Karlekar B., and R. M. Desmond (1982). *Engineering Heat Transfer*. West Publishing, St. Paul, Minn.

Kays, W., and A. L. London (1984). *Compact Heat Exchangers*. McGraw-Hill, New York.

Kersten, M. S. (1949). "Thermal Properties of Soils." *University of Minnesota Institute of Technology, Engineering Experiment Station Bulletin*, vol. 28.

Krarti, M. (1987). "New Developments in Ground Coupling Heat Transfer." Ph.D. dissertation, University of Colorado, Boulder.

Krarti, M. (1999). "Building Foundation Heat Transfer," Chapter 6 of *Advances in Solar Energy*. Edited by Y. Goswami and K. Boer, ASES vol. 13, pp. 242–308.

Krarti, M., D. Claridge, and J. Kreider (1988a). "The ITPE Technique Applied to Steady-State Ground Coupling Problems." *Int. J. Heat and Mass Transfer*, vol. 31, pp. 1885–1898.

Krarti, M., D. Claridge, and J. Kreider (1988b). "The ITPE Technique Applied to Time-Varying, Two-Dimensional Ground Coupling Problems." *Int. J. Heat and Mass Transfer*, vol. 31, pp. 1899–1911.

Krarti, M., D. Claridge, and J. Kreider (1990). "The ITPE Technique Applied to Time-Varying, Three-Dimensional Ground Coupling Problems." *ASME J. Heat Transfer*, vol. 112, pp. 849–856.

Mathur, G. D. (1990). "Calculating Single Phase Heat Transfer Coefficients." *Heating, Piping and Air Conditioning*, vol. 62, no. 3, pp. 103–107.

Mitalas, G. P. (1982). "Basement Heat Loss Studies at DBR/NRC." Division of Building Research Paper No. 1045, National Research Council, Ottawa, Ont., Canada.

Nussbaum, O. J. (1990). "Using Glycol in a Closed Circuit System." *Heating, Piping and Air Conditioning*, vol. 62, no. 1, pp. 75–85.

Siegel, R., and J. R. Howell (1981). *Thermal Radiation Heat Transfer*. McGraw-Hill, New York.

Sparrow, E. M., and R. D. Cess (1978). *Radiation Heat Transfer*, augmented edition. Hemisphere, New York.

Staelens, P. G. (1988). "Thermal Bridges and Standardization." *ASHRAE Trans.*, vol. 94, pt. 2, pp. 1793–1801.

3

Review of Thermodynamic Processes in Buildings

3.1 Introduction to Thermodynamics

Thermodynamics is the study of energy in its various forms and transformations of energy from one form to another. In classical thermodynamics, one considers several forms of energy. Three of these—kinetic, potential, and internal—can be possessed by materials in building systems. Thermodynamics also considers two forms of energy in transport: heat and work. These differ in that they only exist as they are transferred across thermodynamic system boundaries. Heat and work are not properties of a material.

In thermodynamics, one also studies the restrictions that nature places on the transformations of energy. These restrictions are the first and second laws of thermodynamics. In addition to these fundamental, universally applicable laws, we will use the thermodynamic properties of materials to analyze their behavior in important thermodynamic processes.

The goals of this chapter are to

- Review thermodynamic essentials needed for building HVAC system design and analysis.
- Review thermodynamic cycles used in building equipment.
- Provide basic reference data for the loads and systems chapters, Chapters 6 to 11.

For those students who have already had a first course in engineering thermodynamics, we suggest that this chapter serve as a quick review of basic principles and as a source for thermal property data for HVAC equipment calculations, needed in Chaps. 9 and 10.

Thermodynamics is important in the design and analysis of buildings since the performance and efficiency of heating and cooling systems for buildings are determined to a large part by the restrictions which thermodynamics places on HVAC systems and equipment. In this chapter, we review those areas of thermodynamics relevant to commercial, institutional, and residential buildings. Figure 3.1 shows two examples of systems that are associated with buildings and which are analyzed by using the principles reviewed in this chapter—a power-producing system and a cooling system.

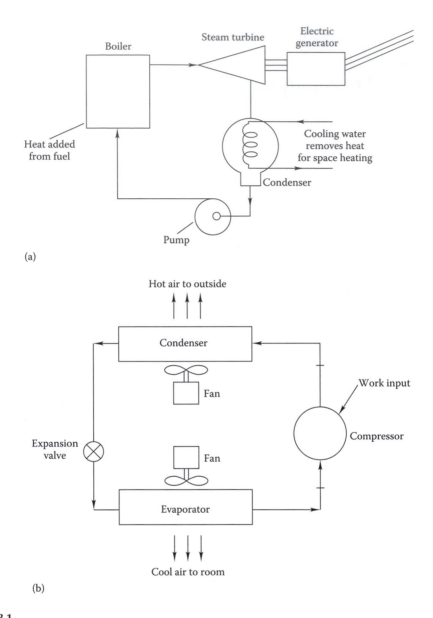

FIGURE 3.1
Schematic diagram of (a) cogeneration steam power plant for simultaneously producing electric power and building heating and (b) an air conditioning system for producing building cooling.

3.2 Thermodynamic Properties

The thermodynamic state of a single-substance system—e.g., air in a building, steam in a boiler, or refrigerant in an air conditioner—is defined by specifying *two independent, intensive* thermodynamic coordinates or properties. *Intensive* properties do not depend on the size or mass of the thermodynamic system, whereas *extensive* properties depend directly on the system size or mass. For example, pressure, temperature, and specific volume are intensive while volume and mass are extensive properties. There are six properties which are needed in building and HVAC system design:

- Pressure p
- Temperature T
- Specific volume v, or its inverse, density ρ ($\rho = 1/v$)
- Specific internal energy u
- Specific enthalpy ($h = u + pv$)
- Specific entropy s

These properties are all defined in standard thermodynamics texts (Cengel and Boles, 2000; Van Wylen and Sonntag, 1986), and we will not redefine them here.

We have two sources of the properties of materials—either simple equations of state, such as the ideal gas law, or tables computed from complex equations based on careful measurements. In this text, we will be able to use the ideal gas law for dry or moist air and all the gaseous constituents of air. We will use the tabular approach for properties of water and of refrigerants.

3.2.1 Ideal Gas

The concept of an ideal gas is embodied in the definition of the ideal gas constant. The ratio of pressure p times molar volume $\bar{v}$ divided by absolute temperature T is observed to approach a constant—the ideal gas constant—as the pressure of the gas is allowed to approach zero. That is,

$$\lim_{p \to 0} \frac{p\bar{v}}{T} = \mathcal{R} \tag{3.1}$$

in which $\mathcal{R}$ is the *universal gas constant*. Its value is 1545.35 ft·lb$_f$/(lb$_m$·mol·°R) or 8314.41 J/(kg·mol·K). One lb$_m$·mol is a pound-mole, i.e., the mass of 1 mol of a substance in pound-mass. (Recall by Avogadro's law that 1 g-mol of any substance contains 6.023×10^{23} molecules.) For example, 1 lb$_m$·mol of nitrogen contains 28 lb$_m$ of nitrogen.

Since we do not use molar volumes in HVAC design, we replace Eq. (3.1) with the following form, using the *gas constant* for a specific gas

$$pv = RT \tag{3.2}$$

or equivalently,

$$p = \rho RT \tag{3.3}$$

The gas constant R is calculated by dividing $\mathcal{R}$ by the molecular weight of the gas. Since the molecular weight of air is 28.97, the gas constant for air is 1545.45/28.97 = 53.35 ft·lb$_f$/(lb$_m$·°R). In SI units, the gas constant for air is 287 J/(kg·K).

Fortunately for the HVAC designer, the ideal gas law applies at pressures other than zero. In fact, the ideal gas law is very accurate for air and its principal constituents up to several atmospheres of pressure and at temperatures involved in buildings. Therefore, for HVAC engineering calculations involving air, the ideal gas law is entirely satisfactory. Note that neither steam nor refrigerants are ideal gases. We will treat their properties in Secs. 3.2.2 and 3.2.3.

The atmospheric pressure is needed in most calculations involving buildings. Of course, this pressure and the density of air both vary with altitude. An approximate expression, useful for design, relating density to altitude is

$$\rho = 39.8 \frac{e^{-H/27,000}}{T+460} \ \text{lb}_\text{m}/\text{ft}^3 \tag{3.4US}$$

in which T (°F) is the air temperature and the altitude above sea level is H (ft). In SI units, the equivalent expression is

$$\rho = 353 \frac{e^{-H/8230}}{T+273} \ \text{kg}/\text{m}^3 \tag{3.4SI}$$

in which T (°C) is the air temperature and the altitude above sea level is H (m). The atmospheric pressure can be found from Eq. (3.3).

Example 3.1: Ideal Gas Law

Find the mass of air enclosed in an office at 20°C if the dimensions of the room are 10 m × 10 m × 2.5 m. The room is located at sea level.

Given: $T = 20°C$, $H = 0$

Find: m_air

Lookup values: $p = 101.325$ kPa (standard atmospheric pressure at sea level)

SOLUTION

The ideal gas law applies to this problem. First we find the density (kg/m³), and then we multiply by the room volume to find the mass. From Eq. (3.3),

$$\rho = \frac{p}{RT}$$

Then substituting known values, we find that the density is

$$\rho = \frac{101,325 \ \text{Pa}}{[287 \ \text{J}/(\text{kg}\cdot\text{K})][(20+273) \ \text{K}]} = 1.20 \ \text{kg}/\text{m}^3$$

Finally, we can find the mass from the product of the density and room volume:

$$m_\text{air} = \rho V = (1.20 \ \text{kg}/\text{m}^3)(10 \ \text{m} \times 10 \ \text{m} \times 2.5 \ \text{m}) = 300 \ \text{kg}$$

COMMENTS

Relative to the other contents of a typical office, the mass of air is small. As we will see in later chapters on building dynamics, the mass of air can usually be ignored for such calculations.

TABLE 3.1

Properties of Common Gases

Gas	Molecular Weight	c_p Btu/(lb$_m$ · °F)	c_p kJ/(kg · C)	c_v Btu/(lb$_m$ · °F)	c_v kJ/(kg · °C)	R ft·lb/(lb$_m$ · °R)	R J/(kg · K)
Air	28.97	0.240	1.005	0.1715	0.718	53.35	287.1
Hydrogen (H$_2$)	2.016	3.42	14.32	2.43	10.17	767.0	4127
Helium (He)	4.003	1.25	5.234	0.75	3.14	386.3	2078
Methane (CH$_4$)	16.04	0.532	2.227	0.403	1.687	96.4	518.7
Water vapor (H$_2$O)	18.02	0.446	1.867	0.336	1.407	85.6	460.6
Acetylene (C$_2$H$_2$)	26.04	0.409	1.712	0.333	1.394	59.4	319.6
Carbon monoxide (CO)	28.01	0.249	1.043	0.178	0.745	55.13	296.6
Nitrogen (N$_2$)	28.02	0.248	1.038	0.177	0.741	55.12	296.6
Ethane (C$_2$H$_6$)	30.07	0.422	1.767	0.357	1.495	51.3	276
Oxygen (O$_2$)	32.00	0.219	0.917	0.156	0.653	48.24	259.6
Argon (A)	39.94	0.123	0.515	0.074	0.310	38.65	208
Carbon dioxide (CO$_2$)	44.01	0.202	0.846	0.156	0.653	35.1	188.9
Propane (C$_3$H$_8$)	44.09	0.404	1.692	0.360	1.507	35.0	188.3
Isobutane(C$_4$H$_{10}$)	58.12	0.420	1.758	0.387	1.62	26.6	143.1

Other properties of ideal gases can be found with assistance from the ideal gas law. For example, the specific heat at constant volume c_v and at constant pressure c_p can be used to find changes in the specific internal energy u and the specific enthalpy h. If the specific heats are constant,

$$u - u_o = c_v(T - T_o)$$ (3.5)

and

$$h - h_o = c_p(T - T_o)$$ (3.6)

where the subscripted variables indicate a specified but arbitrary reference state.

The final property of an ideal gas that is needed in second-law calculations is the entropy. The specific entropy of an ideal gas (relative to the reference state T_o, v_o, s_o) can be found from

$$s - s_o = c_v \ln \frac{T}{T_o} + R \ln \frac{v}{v_o}$$ (3.7)

Table 3.1 lists the specific heats and gas constants of a number of common gases. If extended temperature ranges are involved in a computation, tables of properties of air (Keenan and Kaye, 1948) rather than the preceding equations should be used to properly account for the variation of specific heat with temperature. This refinement is rarely needed in building-related calculations.

3.2.2 Water Vapor and Steam

In the majority of conditions where steam is used to transport energy in building systems, it cannot be treated as an ideal gas. In fact, there is no convenient equation for finding the properties of steam. As a result, one must resort to the steam tables (Keenan et al., 1978)

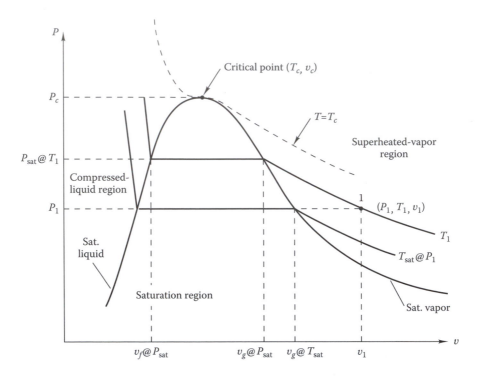

FIGURE 3.2
Pressure-volume diagram for water.

for properties of wet and superheated steam. However, when water vapor forms a constituent of air in very dilute proportions, the mixture can be treated as an ideal gas, as we shall see in Chap. 4.

Figure 3.2 shows a pressure-volume diagram (*p-v* diagram) for water including the region where liquid-vapor equilibrium conditions can exist. The key feature of this diagram is the shape of the saturation curve and the regions defined by it. To the left of the critical point and the saturation curve, only compressed liquid water exists; to the right of the saturation curve, only superheated vapor is present. In the liquid region, properties are primarily a function of temperature; pressure has a secondary effect. Conditions precisely on the saturation curve are called *saturated-liquid conditions* to the left of the critical point and *saturated-vapor* to the right.

Beneath the saturation curve, "wet steam" conditions exist. Wet steam is a mixture of saturated liquid and vapor. Since pressure and temperature are not independent properties beneath the saturation curve, a separate thermodynamic condition called the *quality x* is defined. The quality is the fraction of the total mass of the wet steam mixture that is vapor, i.e., not liquid. The quality is used frequently to find thermodynamic properties such as internal energy, specific volume, enthalpy, and entropy for wet steam mixtures. For example, one would use the quality to find the specific volume of wet steam as follows. By definition, the specific volume of steam is the volume of liquid water plus the volume of water vapor, all divided by the total mass. From the definition,

$$v = \frac{V}{m_{\text{tot}}} = \frac{m_f v_f + m_g v_g}{m_{\text{tot}}} \tag{3.8}$$

in which the subscript f refers to the liquid phase and g to the vapor (or gaseous) phase. From the definition of x one finds

$$x = \frac{m_g}{m_{tot}} \quad (3.9)$$

Since the mixture contains only liquid and vapor, the mass which is not vapor must be liquid. Hence,

$$\frac{m_f}{m_{tot}} = 1 - x \quad (3.10)$$

Using these results in Eq. (3.8), we have

$$v = v_f(1 - x) + x v_g \quad (3.11)$$

The phase-change specific volume $v_{fg} = v_g - v_f$ can be used to express this result differently:

$$v = v_f + x v_{fg} \quad (3.12)$$

Equations analogous to Eqs. (3.11) and (3.12) can be derived for internal energy, enthalpy, and entropy. These will be used later for analysis and design of chillers, boilers, and heat engines.

Complete steam tables need an entire book (Keenan et al., 1978). However, since we are interested in a relatively restricted portion of the full range of properties, simplified tables are adequate. These are included in the Appendix CD-ROM. Table 3.2 shows an excerpt of the saturated steam table in SI units that we will use to solve Example 3.2.

Example 3.2: Wet Steam Properties

Find the temperature, specific volume, and enthalpy of wet steam if the quality is 10 percent (very wet steam). The pressure is 150 kPa (0.15 MPa).

Given: $x = 0.10$,

$p = 0.15$ MPa

Assumptions: Thermodynamic equilibrium exists.

Find: T, v, h

Lookup values: From Table 3.2 we read the properties of saturated liquid and vapor at 0.15 MPa:

$T = 111.4°C$

$v_f = 0.001 \text{ m}^3/\text{kg}$

$v_g = 1.159 \text{ m}^3/\text{kg}$

$h_f = 467.11 \text{ kJ/kg}$

$h_g = 2693.6 \text{ kJ/kg}$

SOLUTION

Since Eq. (3.12) is the easiest approach to this problem, we will use it to find the volume and enthalpy. Finding the temperature is just a matter of looking up the saturation temperature corresponding to the stated pressure. We have already completed this portion of the problem.

TABLE 3.2

Saturated Water-Pressure Table

Press., kPa p	Sat. Temp., °C T_{sat}	Specific Volume, m³/kg		Internal Energy, kJ/kg			Enthalpy, kJ/kg			Entropy, kJ/(kg·K)		
		Sat. Liquid v_f	Sat. Vapor v_g	Sat. Liquid u_f	Evap. u_{fg}	Sat. Vapor u_g	Sat. Liquid h_f	Evap. h_{fg}	Sat. Vapor h_g	Sat. Liquid s_f	Evap. s_{fg}	Sat. Vapor s_g
40	75.87	0.001027	3.993	317.53	2159.5	2477.0	317.58	2319.2	2636.8	1.0259	6.6441	7.6700
50	81.33	0.001030	3.240	340.44	2143.4	2483.9	340.49	2305.4	2645.9	1.0910	6.5029	7.5939
75	91.78	0.001037	2.217	384.31	2112.4	2496.7	384.39	2278.6	2663.0	1.2130	6.2434	7.4564
MPa												
0.100	99.63	0.001043	1.6940	417.36	2088.7	2506.1	417.46	2258.0	2675.5	1.3026	6.0568	7.3594
0.125	105.99	0.001048	1.3749	444.19	2069.3	2513.5	444.32	2241.0	2685.4	1.3740	5.9104	7.2844
0.150	111.37	0.001053	1.1593	466.94	2052.7	2519.7	467.11	2226.5	2693.6	1.4336	5.7897	7.2233
0.175	116.06	0.001057	1.0036	486.80	2038.1	2524.9	486.99	2213.6	2700.6	1.4849	5.6868	7.1717
0.200	120.23	0.001061	0.8857	504.49	2025.0	2529.5	504.70	2201.9	2706.7	1.5301	5.5970	7.1271
0.225	124.00	0.001064	0.7933	520.47	2013.1	2533.6	520.72	2191.3	2712.1	1.5706	5.5173	7.0878
0.250	127.44	0.001067	0.7187	535.10	2002.1	2537.2	535.37	2181.5	2716.9	1.6072	5.4455	7.0527
0.275	130.60	0.001070	0.6573	548.59	1991.9	2540.5	548.89	2172.4	2721.3	1.6408	5.3801	7.0209
0.300	133.55	0.001073	0.6058	561.15	1982.4	2543.6	561.47	2163.8	2725.3	1.6718	5.3201	6.9919

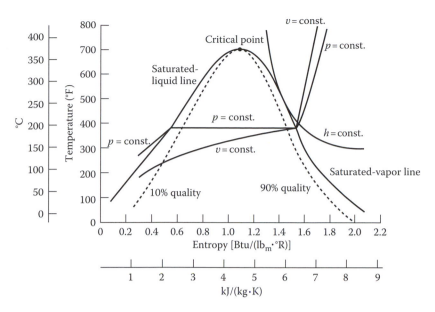

FIGURE 3.3
Temperature-entropy diagram for water.

For the specific volume we have

$$v = v_f + x v_{fg} = 0.001 + 0.10(1.159 - 0.001) = 0.117 \text{ m}^3/\text{kg}$$

Likewise, for the enthalpy,

$$h = h_f + x h_{fg} = 467.11 + 0.1(2693.6 - 467.1) = 689.8 \text{ kJ/kg}$$

When we discuss heat engines using water as the working fluid, we will find a *temperature-entropy (T-s)* diagram to be very useful since isentropic processes will appear as vertical lines on such a plot. Figure 3.3 is the *T-s* diagram for water. The shape of the saturation curve is similar to that in the *p-v* diagram. The corresponding regions—superheated, saturated, wet, and compressed liquid—are located similarly to those regions in the *p-v* diagram.

3.2.3 Refrigerants

Refrigerants are the working fluids in air conditioners, chillers, and refrigerators. The majority are chlorinated and fluorinated hydrocarbons (CFCs), although a number of other organic and inorganic compounds can be used. With the known deleterious effects of some conventional CFCs on the integrity of the earth's stratospheric ozone, the use of these compounds will be significantly curtailed in the next two decades. The two most widely used, harmful CFCs are refrigerant 11 (R11) and refrigerant 12 (R12). Both R11 and R12 are being phased out in favor of R123 and R134a, respectively. In this section we will use R134a as the example refrigerant, although the reader is cautioned that there are still many systems that use R22 and will continue to do so as long as the supply does not get too expensive.

Refrigerants are not ideal gases under the conditions present in cooling equipment for buildings. As a result, we cannot use the ideal gas approach to find thermodynamic state coordinates. Tables or the HCB software is to be used instead. We have included thermodynamic properties of R22 and R134a in both the Appendix CD-ROM and the book appendix, respectively. Table 3.3 is a portion of the R134a saturated properties table. It contains the same sort of data as the previous sample steam table. The HCB software contains thermodynamic property tables and charts (both p-h and T-s) for refrigerants 11, 12, 13, 14, 22, 114, 134a, 500, and 502.

A useful design tool for refrigeration cycle calculations is a pressure-enthalpy (p-h) diagram. In a later section of the chapter, the reason for this will become apparent.

TABLE 3.3

Refrigerant 134a Saturation Properties—USCS

Temp., °F	Pressure, psia	Density, lb_m/ft^3 Liquid	Volume, ft^3/lb_m Vapor	Enthalpy, Btu/lb_m Liquid	Enthalpy, Btu/lb_m Vapor	Entropy, $Btu/(lb_m \cdot °R)$ Liquid	Entropy, $Btu/(lb_m \cdot °R)$ Vapor
40	49.739	79.933	0.947	24.051	107.389	0.052154	0.218941
42	51.697	79.699	0.912	24.693	107.661	0.053426	0.218811
44	53.713	79.464	0.879	25.336	107.932	0.054697	0.218685
46	55.789	79.227	0.847	25.982	108.201	0.055967	0.218561
48	57.926	78.989	0.816	26.629	108.469	0.054235	0.218441
50	60.125	78.750	0.787	27.279	108.736	0.058502	0.218325
52	62.388	78.510	0.759	27.930	109.001	0.049768	0.218211
54	64.714	78.268	0.732	28.584	109.265	0.061032	0.218100
56	67.107	78.024	0.707	29.240	109.528	0.062295	0.217992
58	69.566	77.779	0.682	29.897	109.790	0.063556	0.217887
60	72.093	77.533	0.658	30.557	110.050	0.064817	0.217784
62	74.689	77.285	0.636	31.219	110.308	0.066076	0.217684
64	77.355	77.035	0.614	31.883	110.565	0.067335	0.217586
66	80.093	76.784	0.593	32.549	110.821	0.068592	0.217490
68	82.904	76.531	0.573	33.218	111.075	0.069848	0.217397
70	85.789	76.276	0.554	33.888	111.327	0.071103	0.217305
72	88.749	76.020	0.535	34.561	111.578	0.072357	0.217216
74	91.786	75.762	0.518	35.236	111.827	0.073611	0.217128
76	94.901	75.502	0.500	35.913	112.074	0.074863	0.217042
78	98.095	75.240	0.484	36.593	112.320	0.076115	0.216957
80	101.369	74.976	0.468	37.275	112.564	0.077366	0.216874
82	104.725	94.710	0.453	37.959	112.805	0.078616	0.216793
84	108.164	74.442	0.438	38.646	113.045	0.079865	0.216712
86	111.688	74.172	0.424	39.335	113.283	0.081115	0.216633
88	115.297	73.900	0.411	40.026	113.519	0.082363	0.216555
90	118.993	73.626	0.398	40.720	113.753	0.083611	0.216477
95	128.623	72.930	0.367	42.467	114.328	0.086730	0.216286
100	138.827	72.219	0.339	44.230	114.888	0.089848	0.216096
105	149.626	71.491	0.313	46.011	115.432	0.092967	0.215907
110	161.044	70.745	0.290	47.811	115.959	0.096087	0.215704

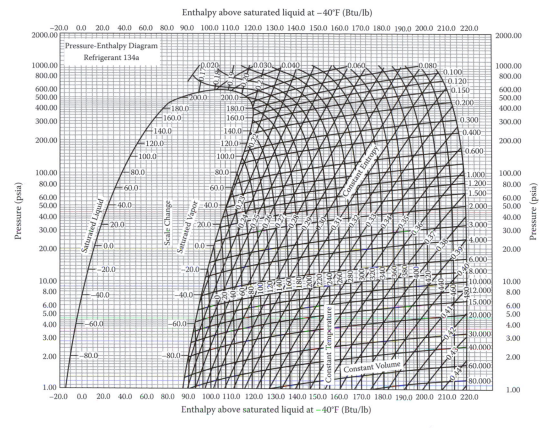

FIGURE 3.4

Pressure-enthalpy diagram for R134a. (Courtesy of ASHRAE, *Handbook of Fundamentals*, American Society of Heating, Refrigerating and Air-Conditioning Engineers, Atlanta, GA, 1989. With permission.)

The basic advantage of the *p-h* diagram is that three of the four steps in a basic refrigeration cycle can be plotted as straight lines. Of course, thermodynamic properties can be read directly from the chart instead of looking them up in a table. Figure 3.4 is the *p-h* diagram for R134a. The HCB software also plots *p-h* diagrams for all refrigerants listed in the previous paragraph. The reader is encouraged to try out the software to quickly compare the *p-h* and *T-s* diagrams for several refrigerants.

Example 3.3: R134a Properties

Find the enthalpy change in a process in which 38 lb_m of R134a at 80 percent quality is heated to 300°F in a constant-pressure process at 200 psia (lb_f/in^2 absolute).

    Given: $p_i = 200$ psia,

               $x_i = 0.80$

    Figure: See Fig. 3.4, the *p-h* diagram for R134a, or refer to the HCB software

    R134a *p-h* diagram.

    Assumptions: Thermodynamic equilibrium exists.

    Find: ΔH, the enthalpy change

SOLUTION

The solution to this problem is most quickly done by using the *p-h* diagram, since the process is just a horizontal line at 200 psia with the left endpoint at the 80 percent quality-line intersection. The right end of the process line is at the intersection of the 200-psia line and the 300°F line. The enthalpy at the left endpoint can be approximated by first finding the enthalpies at saturated liquid (53.5 Btu/lb$_m$) and saturated vapor (117.5 Btu/lb$_m$) and then interpolating. The enthalpy at the right endpoint can be read directly from th *p-h* diagram. Thus, the left and right endpoints are

$$h_1 = 53.5 + (117.5 - 53.5)0.8 = 104.7 \text{ Btu/lb}_m$$

and

$$h_2 = 165.0 \text{ Btu/lb}_m$$

The enthalpy change is the product of the enthalpy change per unit mass and the mass, 38 lb$_m$:

$$\Delta H = (38 \text{ lb}_m)[(165 - 104.7) \text{ Btu/lb}_m] = 2291 \text{ Btu}$$

COMMENTS

For a more accurate calculation, the reader may use the saturated and superheated R134a tables or the tables in the HCB software.

3.3 First Law of Thermodynamics

One of the constraints that nature places on processes is the first law of thermodynamics, commonly called the *law of conservation of energy*. In this section, we will summarize the two most useful forms of this law. The first law for closed thermodynamic systems, i.e., systems in which there is no mass crossing the boundary, will be discussed first. Then the first law for open systems in which mass flow does occur will be treated.

3.3.1 Closed-System First Law

A closed thermodynamic system is one in which mass neither enters nor leaves the system during the process under analysis. An example is the combustion of fuel and air confined in a cylinder of an internal combustion engine. The first law for a closed system relates heat added, work produced, and the change of internal energy of the material involved in the process, as shown in Fig. 3.5. The closed-system first law is less frequently used in HVAC design than the open-system first law. However, it does occur sufficiently often to warrant review.

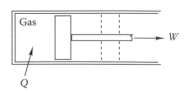

In equation form, the first law can be written in three ways. The first is expressed on an extensive basis, i.e., on a total-mass basis

$$Q - W = \Delta U \qquad (3.13)$$

FIGURE 3.5
Schematic diagram of closed thermo-dynamic system consisting of gas expanding in a piston-cylinder assembly with work output and heat input.

in which *Q* is the heat added (taken to be positive when an input to the system); *Q* has units of joules (or kilowatthours) or Btu (or foot-pound-force). The work W is taken to be positive when it is an output of the system; work has the same units as heat. Finally, ΔU is the change in internal energy of the substance undergoing the process. The internal energy change is positive if the internal energy at the final state is greater than that at the initial state. [In Eq. (3.13), kinetic and potential energy changes are ignored since they are small and not of interest in buildings.]

The second form of the first law is written on an intensive (per-unit-mass) basis as follows:

$$q - w = \Delta u \qquad (3.14)$$

In this expression, the symbols denote the intensive analogs of the quantities in Eq. (3.13).

It is sometimes useful to write the first law on a time-rate basis as follows:

$$\dot{Q} - \dot{W} = \frac{dU}{dt} \qquad (3.15)$$

The time rate of doing work $\dot{W}$ is called the *mechanical power*.

Example 3.4: First Law of Thermodynamics—Closed System

A closed tank of refrigerant 22 (R22) has a volume of 700 ft³ and is filled with 2800 lb$_m$ of a wet mixture of R22 at an initial pressure of 136.19 psia. The tank is moved out of doors and absorbs sufficient solar heat to cause saturated vapor alone to exist within the tank. What is the thermodynamic state (temperature, specific volume, and pressure) prior to the addition of heat? What are the final temperature and pressure after heating?

  Given: V, m, p_i

  Find: $v_i, p_f, T_f,$

  Lookup values: The R22 tables in the Appendix CD-ROM are used.

  Alternatively, one may use the HCB software.

SOLUTION

The initial (and final) specific volume is

$$v = \frac{700 \text{ ft}^3}{2800 \text{ lb}_m} = 0.25 \text{ ft}^3/\text{lb}_m$$

From the saturation table for R22 at 136.19 psia, we read that the initial temperature of the tank contents is

$$T_i = 70°F$$

As heat is added, the wet mixture of R22 becomes saturated vapor. Note that the process is a constant-volume process; hence one can find the final thermodynamic state by using the saturation tables again.

In the saturated vapor column of the table, we read the final temperature and pressure directly from the entries corresponding to a *saturated vapor* value of 0.25 ft³/lb$_m$. The values are (by interpolation):

$$T_f = 102°F \quad p_f = 216.45 \text{ psia}$$

3.3.2 Open-System First Law

The steady-state steady-flow energy equation is the form of the first law most often used in HVAC design. The applications are broad, including heating and cooling load calculations, chiller design, and heat distribution design among many others. The unsteady form of the first law in which flows, heat rates, and other terms in the energy equation vary with time is rarely used for design, but is needed for transient thermal analyses of buildings. However, in this section we will present only the steady form of the first law.

Figure 3.6 shows the generic open system for which we will write the first law in its three forms—total-mass basis, per-unit-mass basis, and rate basis. The figure shows an input stream and an output stream (of course, more than one input or output stream may be involved) with heat and work transfer across the system boundary. On a total-mass (extensive) basis, the open-system first law is

$$m_i\left(gz_i + \frac{v_i^2}{2} + h_i\right) + Q = m_o\left(gz_o + \frac{v_o^2}{2} + h_o\right) + W \qquad (3.16)$$

where
  m_i, m_o = inlet and outlet mass flow amounts, lb_m (kg)
  z_i, z_o = inlet and outlet system port elevations, ft (m)
  v_i, v_o = inlet and outlet fluid average velocities, ft/s (m/s)
  h_i, h_o = inlet and outlet specific enthalpies, Btu/lb_m (kJ/kg)

The sign convention on heat transfer Q is that heat added to a thermodynamic system is positive; work output by the system W is also positive; this is consistent with signs used in the closed-system first law.

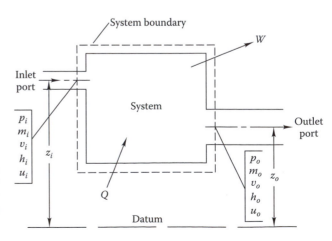

FIGURE 3.6
Schematic diagram of open thermodynamic system showing work output W and heat input Q along with mass inflow m_i and outflow m_o across the system boundary.

The first law on a per-unit-mass (intensive) basis is easily derived from Eq. (3.16) by dividing by the mass terms. The steady-state conservation of mass equation ensures that

$$m_i = m_o \tag{3.17}$$

Therefore, the first law on an intensive basis is

$$gz_i + \frac{v_i^2}{2} + h_i + q = gz_o + \frac{v_o^2}{2} + h_o + w \tag{3.18}$$

The heat and work transfer terms on an intensive basis are indicated by lowercase notation q and w. Finally, the rate-basis form of the steady first law is

$$\dot{m}_i \left(gz_i + \frac{v_i^2}{2} + h_i \right) + \dot{Q} = \dot{m}_o \left(gz_o + \frac{v_o^2}{2} + h_o \right) + \dot{W} \tag{3.19}$$

One selects the form to be used in a particular situation based on the type of problem. For example, in analyzing a refrigeration thermodynamic cycle, the mass flow rate may not be known if the required cooling rate is not known. In such a problem, one would use the intensive form, Eq. (3.18). On the other hand, if the cooling requirement (Btu/h or kW) of a building is to be found, Eq. (3.19) should be used. Note that the closed-system form of the first law can be derived from the open-system form by setting the kinetic, potential, and flow energy terms (that is, pv terms in the enthalpy $h = u + pv$) all to zero.

In USCS, units present a particular problem in the first law since different units are used for each term. The heat transfer term is most often expressed in Btu or Btu per hour whereas the work term may be in foot-pounds-force. The mass flow term may be in units of pound-mass per hour. The only consistent unit of these three is that used for work. Heat (in Btu) is converted to the consistent unit footpounds-force by multiplying by the mechanical equivalent of heat, 778 (ft·lb$_f$)/Btu. Kinetic and potential energy terms involving pound-mass are converted to consistent units by dividing by $g_c = 32.17$ (lb$_m$·ft)/(s^2·lb$_f$). In SI, no conversions are needed.

We will now illustrate the use of the first law in two examples related to buildings.

Example 3.5: Air Infiltration

Suppose that on a windy winter day at sea level it is determined that air leaks into a residential building at the rate of *one air change per hour*. In other words, all the warm air in the building is exchanged with outside air once each hour. This transfer of air occurs through small openings in the building, such as cracks around windows and loose-fitting building components. It is called *infiltration*.

Infiltration is balanced by an equal mass flow called *exfiltration*, since mass must be conserved and no change of air mass within the building occurs under steady conditions. If the outdoor air temperature is −10°C and the interior temperature is 22°C, how much heat must the heating system provide to heat the infiltrating air? The building has 150 m² of floor area and is two stories high with a net interior height of 7 m.

    Given: $T_i = 22°C$

$T_o = -10°C$

$\dot{V} = 1$ volume/h

$V = 150 \text{ m}^2 \times 7 \text{ m} = 1050 \text{ m}^3$

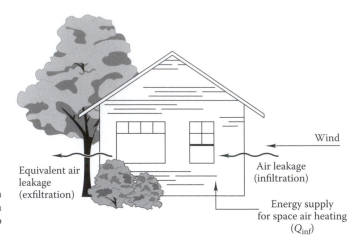

FIGURE 3.7

Sketch showing infiltration through building skin and heat input from building heating system needed to heat infiltrating air.

Figure: See Fig. 3.7.

Assumptions: Steady flow; air is an ideal gas, i.e., the enthalpy depends only on the temperature.

Find: $\dot{Q}_{inf}$

Lookup values: Specific heat of air $c_p = 1.01$ kJ/(kg·K). Density of air at 22°C and sea-level barometric pressure is 1.19 kg/m³ (it can be found from the ideal gas law).

SOLUTION

The solution uses the rate form of the open-system first law of thermodynamics. The kinetic and potential energy terms are identical on both sides of the equation if the small openings through which air infiltrates and leaves (exfiltrates) the building are uniformly distributed. No shaft work occurs in this process. The remaining terms in the first law are

$$\dot{m}h_i + \dot{Q}_{inf} = \dot{m}h_o$$

Rearranging, we have

$$\dot{Q}_{inf} = \dot{m}(h_o - h_i)$$

Equation (3.6) is used to find the enthalpy change as follows:

$$\dot{Q}_{inf} = \dot{m}c_p(T_o - T_i) \tag{3.20}$$

The mass flow rate can be found from the density and volumetric flow rate ($V = 150 \times 7 = 1050$ m³):

$$\dot{m} = \rho \dot{V} = \frac{(1.19 \text{ kg/m}^3)(1050 \text{ m}^3/\text{h})}{3600 \text{ s/h}} = 0.347 \text{ kg/s} \tag{3.21}$$

Now, all values on the right-hand side of Eq. (3.20) are known. The heat to be added is

$$\dot{Q}_{inf} = \dot{m}c_p(T_o - T_i)$$
$$= (0.347 \text{ kg/s})[1.01 \text{ kJ/(kg·K)}]\{[22 - (-10)]°C\} = 11.2 \text{ kW}$$

Example 3.6: Building Process—The Economizer

The interiors of commercial buildings often have large heat gains due to lighting, computers, and occupants. Therefore, even in cold and mild weather, many commercial buildings may require cooling. Under these conditions, outdoor air can be used for cooling buildings, thereby saving operating costs and energy that would otherwise be consumed by mechanical cooling systems. In the design of these "free cooling" systems, called *economizers,* it is necessary to calculate the temperature of air supplied to the building. For a constant cooling need in the building, one needs a constant supply flow rate $\dot{m}_s$ at a predictable temperature.

Derive the equation relating the amount of outdoor air needed as a function of the outdoor air temperature. Note that at low outdoor air temperatures, relatively little air is needed whereas at mild temperatures, a great deal of outdoor air is needed to meet the cooling load. Of course, at still higher temperatures, outdoor air can no longer be used for cooling, and the mechanical cooling system will operate. Also derive the equation for the outside temperature above which the economizer is inadequate for cooling.

Given: Constant T_s, $\dot{m}_s$

Figure: See Fig. 3.8. Note that part of the return air is reused. This is done to provide sufficient airflow circulation within building spaces. The amount of outdoor air needed, by itself, is insufficient to maintain comfortable air circulation under all conditions.

Assumptions: Steady state exists.

Find: Expression for $\dot{m}_o$, and T_o at the limiting condition called the *economizer cutoff point.*

SOLUTION

The thermodynamic system is the flowing air contained in the ductwork shown in the figure. Most of the terms in the open form of the energy balance equation are zero—heat added, work produced, and kinetic and potential energy change. The only nonzero terms are the enthalpy terms from which one can write the following enthalpy balance:

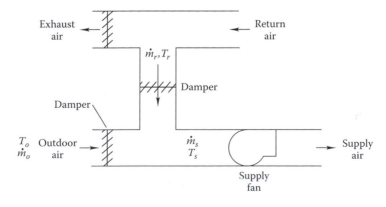

FIGURE 3.8
Airflow diagram for economizer cooling system.

$$\dot{m}_r c_{p,\,\text{air}} T_r + \dot{m}_o c_{p,\,\text{air}} T_o = \dot{m}_s c_{p,\,\text{air}} T_s$$

Canceling the specific heat of air from each side, we get

$$\dot{m}_r T_r + \dot{m}_o T_o = \dot{m}_s T_s \tag{3.22}$$

The conservation of mass equation is

$$\dot{m}_r + \dot{m}_o = \dot{m}_s \tag{3.23}$$

It is an easy matter to solve these two equations for the flow rate of outside air needed to maintain a given supply temperature with given outdoor and return air temperatures. The outdoor air is supplied at the outdoor temperature T_o whereas the air returning from the building at T_r is warmer, because of the heat added to room air by occupants and internal heat sources, such as lights and electronic equipment.

$$\dot{m}_o = \dot{m}_s \frac{T_r - T_s}{T_r - T_o} \tag{3.24}$$

The outdoor temperature capable of meeting design values of supply flow rate and temperature is obtained by rearranging Eq. (3.22) and is given by

$$T_o = \left(1 - \frac{\dot{m}_s}{\dot{m}_o}\right) T_r + \frac{\dot{m}_s}{\dot{m}_o} T_s \tag{3.25}$$

There are two limits to economizer performance. At the upper limit of economizer use (at the point where mechanical cooling is just about to be turned on), there is no return air (all air from the building is exhausted) and the supply temperature is essentially the same as the outdoor air. The air supply mass flow is

$$\dot{m}_o = \dot{m}_s = \dot{m}_{\text{max}} \tag{3.26}$$

At the other extreme, some outdoor air is needed whether space cooling is needed or not. This minimum flow $\dot{m}_{\text{min}}$ is set by the Uniform Building Code ventilation requirements, which will be discussed in Chap. 7 of this book. We can use Eqs. (3.25) and (3.26) to find the minimum outdoor temperature at which the economizer operates. Below this temperature, heat will need to be added to meet the supply air temperature requirement. The minimum outdoor temperature at which an economizer can be used is

$$T_{o,\,\text{min}} = \left(1 - \frac{\dot{m}_s}{\dot{m}_{\text{min}}}\right) T_r + \frac{\dot{m}_s}{\dot{m}_{\text{min}}} T_s \tag{3.27}$$

At outdoor temperatures less than this value, the minimum airflow required by ventilation $\dot{m}_{\text{min}}$ is used, and heat is added.

COMMENTS

Typical economizers operate up to about 55 to 60°F outdoor air temperature. We will see in Chap. 5 that the supply fan shown in the figure causes a modest temperature rise (1 to 2°F) in the air due to flow work and fan inefficiencies. The final example in Chap. 1 also addresses fan temperature rise.

3.4 Second Law of Thermodynamics and Thermodynamic Cycles

The second law of thermodynamics places a further restriction on energy conversion processes beyond that imposed by the first law. The second law recognizes that (1) only certain processes are allowed and (2) some forms of energy are of higher quality than others. For example, the first law for an isothermal closed system would have us believe that heat and work are equivalent forms of energy in transport across a system boundary. However, the Kelvin-Planck statement of the second law explicitly states that no process can exist which has as its sole effect the conversion of heat completely to work.

In the analysis of buildings, the second law is most important for placing limits on the efficiency of all involved thermodynamic cycles. These cycles include the heat pump and refrigeration cycles, which we discuss shortly. In this chapter, we will discuss the *idealized cycles* on which actual heating and cooling systems are based. The idealization that we make is that processes proceed very slowly (so that thermodynamic equilibrium exists at all times) and that all parts of the process are reversible.

Of course, practical heating and cooling systems used in buildings cannot use reversible processes since buildings and their occupants operate in finite time and cannot accommodate infinitely large heat exchangers, for example. However, the reversible paradigm offers considerable insight into important thermal cycles. *Real thermodynamic cycles* are discussed in Chaps. 9 and 10.

3.4.1 Carnot Cycle

The Carnot heat engine cycle is a connected set of four reversible processes, the outcome of which is the conversion of heat to work. Figure 3.9 shows the four processes. Beginning at the upper left-hand corner and moving clockwise,

1. Heat is added to the working fluid and work is produced isothermally.
2. Work is produced by an adiabatic expansion.
3. Heat is rejected to a low-temperature reservoir isothermally, and work is consumed in compressing the fluid.
4. Work is consumed in compressing the fluid adiabatically to its initial condition.

The first law would allow us to avoid the rejection of heat at the low temperature to maximize work output, but the second law requires the rejection of some heat. By use of the first law, it is shown in standard thermodynamics texts that the efficiency of the Carnot cycle is given by

$$\eta = \frac{W_{\text{net}}}{Q_h} = 1 - \frac{T_l}{T_h} \tag{3.28}$$

in which the temperatures are absolute. This expression places an upper limit on the efficiency at which a heat engine can convert heat to work, given the source and sink temperatures.

3.4.2 Carnot Refrigeration Cycle

A Carnot engine operated in the opposite direction to that shown in Fig. 3.9 has the net effect of removing heat from the low-temperature reservoir and rejecting it to

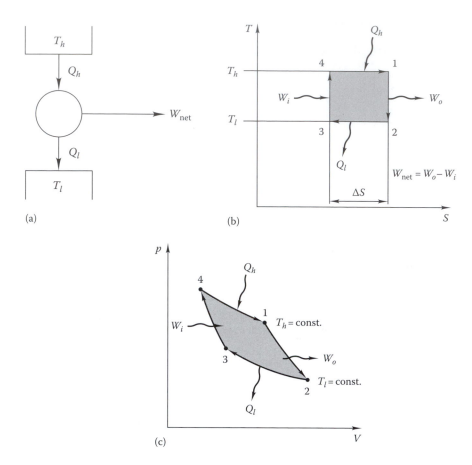

FIGURE 3.9
Schematic diagram of Carnot heat engine with cycle plotted on *T-S* and *p-V* coordinates.

the high-temperature reservoir. For example, heat can be extracted from the outdoors, can be "pumped up a temperature gradient," and can provide heat to the interior of a building. This "heat pump" will be described shortly.

If one applies the first law to each process in the refrigeration (or heat pump) cycle shown in Fig. 3.10, it is an easy matter to determine the efficiency of the process. Surprisingly, the efficiency is found to be greater than 1. In keeping with common engineering practice, we prefer to reserve the term *efficiency* for a ratio of effect to cause less than 1. As a result, for refrigeration and cooling cycles, we use the term *coefficient of performance* (COP) as the figure of merit. The cooling[1] COP is

$$\mathrm{COP}_c = \frac{Q_l}{W_{\mathrm{net}}} = \frac{T_l}{T_h - T_l} \tag{3.29}$$

---

[1] A dimensional number called the *energy efficiency ratio* (EER) is sometimes used instead of the dimensionless COP. See Chap. 9 for details.

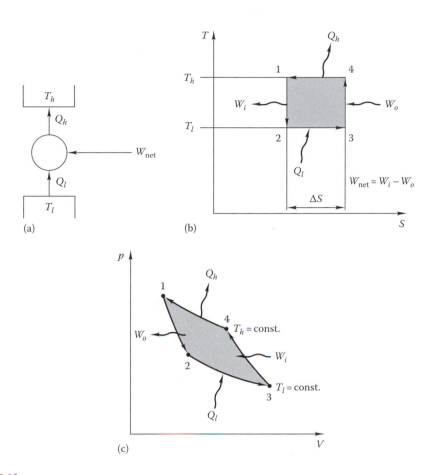

FIGURE 3.10
Schematic diagram of Carnot refrigerator with cycle plotted on *T-S* and *p-V* coordinates.

The cycle shown in Fig 3.10 is also the heat pump cycle, but the definition of the COP differs because the heat added to the building (the high-temperature reservoir) is the desired effect. Therefore, the numerator in Eq. (3.29) is replaced with Q_h, and the COP expression becomes

$$\text{COP}_{\text{hp}} = \frac{Q_h}{W_{\text{net}}} = \frac{Q_l + W_{\text{net}}}{W_{\text{net}}} = \text{COP}_c + 1 = \frac{T_h}{T_h - T_l} \tag{3.30}$$

Example 3.7: Carnot Heat Pump

If a Carnot heat pump operates between an outdoor temperature of $-10°C$ and a building interior temperature of 22°C, what is the COP?

Given: $T_l = 273 + (-10) = 263$ K

$T_h = 273 + 22 = 295$ K

Figure: See Fig. 3.10.

Assumptions: Reversible, Carnot cycle

Find: COP_{hp}

SOLUTION

The last part of Eq. (3.30) gives the solution directly:

$$\text{COP}_{hp} = \frac{295}{295 - 263} = 9.22$$

COMMENTS

If we were able to operate a heat pump of this efficiency each kilowatthour of electricity used to operate the heat pump (W_{net} in Fig. 3.9) would result in 9.22 kWh of heat being provided to the building's interior. Of course, practical heating equipment cannot achieve COPs of this level for reasons we shall discuss briefly next and in greater detail in Chaps. 9 and 11.

3.4.3 Practical Cycle Limitations Imposed by Thermodynamics

The cycles discussed in Sec. 3.4.2 cannot be realized in practice because many restrictions are placed on actual equipment due to thermal and materials limitations. For example, the working fluid of an actual refrigerator must be economical, have desired heat transfer properties, and be safe. In addition, heat exchangers, compressors, and turbines all have thermodynamic irreversibilities that cannot be avoided. As a result, the ideal cycles discussed above are replaced by similar, real cycles in actual equipment. In this section we will give two examples. In later chapters we will discuss these in detail.

A limit placed on all cycles has to do with the necessity for a temperature difference to exist in order for heat transfer to occur, as described in Chap. 2. According to the Carnot heat engine equation (3.28), the efficiency will be reduced because of the reduced temperature difference across the Carnot cycle itself. This occurs because some of the available temperature difference between the two fixed reservoirs is "lost" because of the requirements of real heat exchangers. A similar consideration reduces the COP of heat pump and cooling cycles. Fluid pressure drops due to friction further decrease the efficacy of Carnot-based cycles.

For practical purposes, the Carnot heat engine is replaced by the *Rankine cycle* based on it. Figure 3.11 shows the basic Rankine cycle plotted on a temperature-entropy diagram. The similarities to Fig. 3.9 are apparent. Two isothermal heat exchange processes are shown (2'-3, vapor condensation, and 5-1, water evaporation; point 5 occurs within the boiler and is not shown in the schematic diagram). An isentropic expansion for work production is also used. However, there are significant differences. If one were to operate the expansion, work-producing step 1-2 (most often by using a turbine) in the wet steam region, the turbine blade erosion and other problems would occur. Hence, wet steam from point 1 is superheated to point 1' so that the expansion can occur entirely in the dry steam region. Such practical modifications to the theoretical Carnot cycle cannot be avoided. The net result is a drop in efficiency from that predicted by Eq. (3.27).

The refrigeration cycle must also be modified for practical reasons. The most common result is a *reversed Rankine cycle,* as shown in Fig. 3.12. Again the resemblance to the Carnot refrigeration cycle is apparent. If we begin at the upper left-hand corner and proceed counterclockwise, the steps are

1. Throttling of high-pressure liquid to low-pressure wet mixture.
2. Heat addition to the cold fluid by evaporation to saturated vapor.

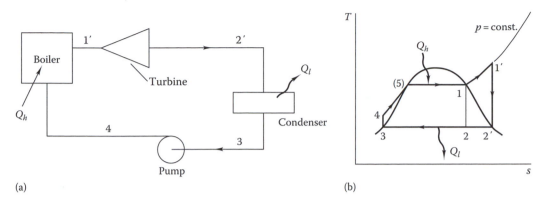

(a) (b)

FIGURE 3.11
Schematic diagram of the Rankine cycle and plot on *T-s* coordinates. Points are labeled to correspond with Fig. 3.9.

3. Dry compression of the saturated vapor to high pressure.
4. Heat rejection to the environment by condensing the hot vapor.

In addition to operation in the reverse direction relative to the Rankine power cycle, the refrigeration cycle differs in the fluid expansion step. No attempt is made to extract work in this step, since the equipment to do so is not economical. Instead, a simple throttling process is used. Since throttling is an isenthalpic process, it appears as a vertical line on the *p-h* diagram in Fig. 3.12b. Therefore, the *p-h* diagram is used more often than the *T-s* diagram to depict refrigeration cycles.

Although the compression process is shown as isentropic in Fig. 3.12, actual compressors are not isentropic. The actual compressor outlet condition is to the right (i.e., higher entropy) of point 4 by an amount depending on the compressor efficiency. Compressor efficiency is the ratio of the theoretical enthalpy rise between points 3 and 4 to the actual enthalpy rise in the physical equipment. The details of actual refrigeration equipment are discussed in Chap. 10.

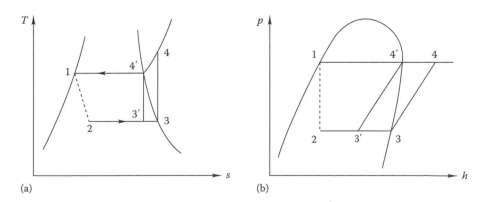

(a) (b)

FIGURE 3.12
Vapor compression cycle plotted on *T-s* and *p-h* coordinates. Points are labeled to correspond with Fig. 3.10.

TABLE 3.4

Chapter 3 Equation Summary

Topic	Equations	Equation Number	Notes
Thermodynamic properties	$pv = RT$	(3.2)	Ideal gas law
	$\rho = 39.8 \dfrac{e^{-H/27,000}}{460 + T}$	(3.3US)	Air density
	$u - u_o = c_v\,(T - T_o)$	(3.5)	Internal energy
	$h - h_o = c_p\,(T - T_o)$	(3.6)	Enthalpy
	$s - s_o = c_v \ln\dfrac{T}{T_o} + R \ln\dfrac{v}{v_o}$	(3.7)	Entropy
	$v = v_f(1 - x) + x v_g$	(3.11)	Wet mixtures, similar for h, u, and s
	$v = v_f + x v_{fg}$	(3.12)	Wet mixtures, similar for h, u, and s
First law	$Q - W = \Delta U$	(3.13)	Closed system
	$m_i\left(gz_i + \dfrac{v_i^2}{2} + h_i\right) + Q = m_o\left(gz_o + \dfrac{v_o^2}{2} + h_o\right) + W$	(3.16)	Open system
Second law	$\eta = 1 - \dfrac{T_l}{T_h}$	(3.28)	Carnot heat engine
	$COP_c = \dfrac{T_l}{T_h - T_l}$	From (3.29)	Carnot refrigerator

3.5 Summary

In this chapter we have discussed the key thermodynamic principles which govern the behavior of buildings and their heating and cooling systems. We will use the results of this chapter frequently in subsequent chapters. You should be familiar with the following as a result of your study of this review chapter:

1. Thermodynamic properties—ideal gases, real fluids.
2. The first law for open and closed systems.
3. The second law for power and refrigeration cycles.
4. Important thermal cycles—refrigeration and heat pump.
5. Essential differences between theoretical Carnot cycles and actual Rankine cycles used in HVAC equipment for buildings.

Table 3.4 summarizes the key equations used in this chapter.

Problems

The problems in this book are arranged by topic. The approximate degree of difficulty is indicated by a parenthetic italic number from 1 to 10 at the end of the problem. Problems are stated most often in USCS units; when similar problems are presented in SI units, it is done with approximately equivalent values in parentheses. The USCS and SI versions of a problem are not exactly equivalent numerically. Solutions should be organized in the same order as the examples in the text: given, figure or sketch, assumptions, find, lookup values, solution. For some problems, the Heating and Cooling of Buildings (HCB) software may be helpful. In some cases it is advisable to set up the solution as a spreadsheet, so that design variations are easy to evaluate.

3.1 Find the density of air at the elevation of Boulder, Colorado [5500 ft (1677 m)], and at the summit of Pike's Peak [14,110 ft (4302 m)] for an air temperature of 60°F (15°C). (3)

3.2 How much heat is added to a fixed volume of 2000 ft^3 (60 m^3) of air to raise it from 60°F (15°C) to 120°F (49°C)? (3)

3.3 The 30-m^3 compressed-air storage tank for a building pneumatic control system is at 25° C and 800 kPa. Enough air is allowed to escape to lower the pressure to 600 kPa; the temperature is 20° C. How much air was vented? (4)

3.4 If the pressure gauge on a 30-ft^3 (0.85-m^3) oxygen tank reads 30 psig (207 kPa), how much oxygen is in the tank when the temperature is 65°F (18°C) at sea level? (4)

3.5 How much air must be added to the tank in Prob. 3.4 to raise the pressure at the given temperature to 75 psig (500 kPa)? (4)

3.6 Complete the following table for water. (5)

T, °C	p, kPa	v, m$^3 \cdot$ kg	Phase
60		3.25	
	175		Saturated vapor
300	300		
100	10		
		0.001097	Saturated liquid
1000	10		

3.7 Complete the following table for water. (5)

T, °F	p, psia	v, ft^3/lb$_m$	Phase
170		30.0	
	60		Saturated vapor
400	20		
240	5.0		
		0.017209	Saturated liquid
220	14.7		

3.8 Complete the following table for R22. (5)

T, °F	p, psia	v, ft^3/lb$_m$	Phase
		0.6561	Saturated vapor
45	90.791		
10		0.5	
	83.280		Saturated liquid

3.9 Complete the following table for R134a.   (5)

T, °F	p, psia	v, ft^3/lb$_m$	Phase
		2.0528	Saturated vapor
150	50.0		
−10		0.090	
	1.00		Saturated liquid

3.10 Solve Probs. 3.8 and 3.9, using the HCB software.   (7)

3.11 A rigid tank with a volume of 5 m^3 contains 10 kg of saturated liquid-vapor mixture at 75° C. Heat is added to the water until all liquid is completely evaporated. What is the temperature?   (3)

3.12 Write a computer program to fit saturated water *vapor* enthalpy data as a function of temperature between 40 and 125°F (0 and 50°C). Use a third-degree polynomial.   (9)

3.13 Write a computer program or use a spreadsheet to fit saturated R134a *vapor* enthalpy data as a function of temperature between −20 and 40°F (−30 and 5°C). Use a third-degree polynomial.   (9)

3.14 Write a computer program or use a spreadsheet to find a linear equation for the specific heat of saturated liquid water as a function of temperature between 40 and 200°F (5 and 95°C). Recall that $c_p \equiv \Delta h / \Delta T$ and use CD-Rom enthalpy values, recalling that the effect of pressure on the enthalpy of liquid water is very small.   (9)

3.15 Write a computer program or use a spreadsheet to find a third-order equation for the saturation temperature as a function of pressure for water between 50 and 250°F (10 and 120°C).   (9)

3.16 Find the enthalpy, specific volume, and temperature of 70 percent and 90 percent quality steam at 40 psia (280 kPa).   (6)

3.17 Water is heated from 50 to 140°F (10 to 60°C) by mixing saturated steam at 25 psia (175 kPa) with it. If hot water is required at 200 gal/min (12 L/s), what is the steam rate in pounds per hour (kilograms per second)?   (4)

3.18 Refrigerant 134a is condensed from a saturated vapor to saturated liquid at 100°F by using water entering at 60°F. If the refrigerant flow rate is 500 lb/h and the water flow rate is 1500 lb/h, what are the heat rejection rate and the water outlet temperature?   (7)

3.19 How much electric power is needed to pump water from a 200-ft-deep (60-m-deep) well by using a 71 percent efficient pump powered by a 92 percent efficient motor? The friction pressure drop in the 200-ft pipe is 10 ft of water (30 kPa).   (5)

3.20 Refrigerant 22 is throttled from a saturated liquid at 136.2 psia (950 kPa) to 55 psia (380 kPa). What are the temperature drop, quality, and specific volume at the end of the throttling process?   (7)

3.21 Refrigerant 134a is throttled from a saturated liquid at 70°F (21°C) to a saturated liquid-vapor mixture at 36.1 psia (250 kPa). What are the temperature drop, quality, and specific volume at the end of the throttling process?   (7)

3.22 Air is heated in a duct with a 30-kW electric resistance coil. If the sea-level airflow is 300 m^3/min of 15°C air, what is the heating unit outlet temperature? We know that 300 W is lost from the surface of the heating coil housing.   (5)

3.23 Air at 90°F (27°C) and 14.7 psia (101.3 kPa) is heated by passing over a steam coil. If steam enters the coil at 30 psia (210 kPa), 400°F (205°C), and 30 lb/min (0.23 kg/s) and leaves at 25 psia (175 kPa) and 200°F (94°C), what is the air mass flow rate that produces a leaving air temperature of 140°F (60°C)? (7)

3.24 Standard atmospheric air enters the evaporator of a residential air conditioner at 350 ft^3/min (118 L/s) and 85°F (29°C). Refrigerant 22 at 69.6 psia (480 kPa) and 30 percent quality enters the coil at 3.5 lb/min (0.027 kg/s) and leaves as a saturated vapor. Calculate the exit air temperature from the evaporator and the rate of heat transfer from the air if the effect of water vapor in the air can be ignored due to low humidity. (8)

3.25 Standard atmospheric air enters the cooling coil of an air handling unit in a small commercial building at 10,000 ft^3/min (4700 L/s) and 83°F (28°C). Refrigerant 22 at 76 psia (500 kPa) and 35 percent quality enters the coil at 100 lb/min (0.75 kg/s) and leaves as a saturated vapor. Calculate the exit air temperature from the coil and the rate of heat transfer from the air if the effect of water vapor in the air can be ignored due to low humidity. (8)

3.26 A polytropic process law is given by $pv^n = k$, where k is constant. Derive an equation for the work done during a closed polytropic process beginning at (p_1, v_1) and ending at (p_2, v_2). (7)

3.27 Refrigerant 134a at 140°F (60°C) and 125 psia (840 kPa) enters the condenser of a chiller at 800 lb/h (0.10 kg/s) and exits as a saturated liquid at the same pressure. If the cooling water enters at 85°F (30°C) and leaves at 104°F (40°C), what is the mass flow rate of the cooling water? Assume that water is a compressed liquid at 30 psia (205 kPa). (7)

3.28 Water is pumped through the cooling system of a building at 300 gal/min (19 L/s). If the pressure drop is 18 psia (126 kPa), how much power does the pump exert on the flowing fluid? If the pump has an efficiency rate of 72 percent, what is the shaft power input to the pump? (*Hint:* When you are using the appropriate form of the first law, assume that water is incompressible.) (6)

3.29 Water is pumped through the cooling tower of a cooling plant with a 2-hp (1.5-kW) pump that is 72 percent efficient. The pressure drop in the piping is 15 psi (105 kPa). What is the water flow rate? Note that the motor imparts 2 hp (1.5 kW) of power to the pump input shaft. (6)

3.30 A building with a floor area of 100,000 ft^2 (10,000 m^2) and inside wall height of 9 ft (3 m) has an air infiltration (i.e., leakage) rate of 0.4 air change per hour. If the outdoor temperature is 0°F (−18°C) and the indoor temperature is 68°F (20°C), how much heat must be provided by the building heating system to warm the cold outside air for two possible building locations—the first at sea level and the second at Leadvile, CO [10,000 ft (3000 m)]? (6)

3.31 The ventilation air requirement of a building is 10,000 ft^3/min (4700 L/s), representing 18 percent of the total supply airflow to the HVAC system. What is the minimum outdoor temperature at which an economizer can be used without preheating the outdoor air if the building supply air temperature is 55°F (13°C) and the interior temperature of the building is 74°F (23°C)? (4)

3.32 As the energy manager of a building, you want to establish an energy balance on a boiler to periodically check its efficiency. The heat input is easy since you know the amount of natural gas consumed. The boiler produces z lb of steam per hour with 10 percent new feedwater (10 percent of the steam is lost due to poor steam trap performance and consumption for air humidification) supplied at 55°F (13°C) and 90 percent condensate return at y°F. On average, 5 percent of the steam is exhausted for the purpose of solids control. Describe how you would do an energy balance on the steam side of the boiler, and calculate the boiler efficiency.   (8)

3.33 An energy audit of a large building discovers that 20 billion Btu (20 billion kJ) of energy is being lost per year due to steam leaks. If the steam is saturated at 130 psia (900 kPa), how much water is lost due to the steam leaks? If water costs $1.00 per 1000 gal ($0.25 per 1000 L), what is the value of the lost water? How does this cost compare to the cost of lost steam, valued at $3.00 per 1000 lb ($6.50 per 1000 kg)?   (5)

3.34 How much heat is added to the R22 in Example 3.4?   (4)

3.35 A air conditioner removes heat from a residence at 1200 kJ/min while drawing power at the rate of 10 kW. What is the COP of this air conditioner, and how much heat is rejected to the outdoors?   (5)

3.36 The cooling system on a small office building maintained at 20°C removes heat from the building to offset summer heat gains. The heat gain from outside is 60 kW, and gains from people and computers within the office amount to 20 kW. If the COP of the cooling system is 2.4, what is the electric power consumption of the cooler?   (3)

3.37 A homeowner is considering replacing the electric resistance heating system in his residence with a heat pump. During one winter month, the resistance heaters (electric baseboards) consumed 2000 kWh, priced at 9¢/kWh. If a heat pump could operate during the same period with a COP of 2.4, what would be the savings in utility bills for that month?   (4)

3.38 A tightly sealed building is to be warmed from 41°F (5°C) to 73°F (23°C) with a heat pump having a COP of 3.0. If the power input to the heat pump is 17,750 Btu/h (5.2 kW), how long will it take to warm the interior? Assume that the heat capacity of the building interior and its furnishings is equivalent to 1300 lb (600 kg) of wood [specific heat 0.5 Btu/(lb · °F), 2.1 kJ/(kg · K)].   (4)

3.39 An air conditioning system on a house maintains the interior at 68°F (20°C) when the outside temperature is 95°F (35°C). The compressor motor is rated at 4.5 hp (3.4 kW). What is the maximum rate at which heat can be removed from the building interior?   (5)

3.40 What is the minimum power input required to maintain the interior of a building at 72°F (22°C) if heat gains from the 92°F (33°C) exterior are 8000 Btu/min (140 kW) and internal gains from people and lights are 2000 Btu/min (36 kW)?   (5)

3.41 If a building loses heat at the rate of 17,000 Btu/(h · °F) (9000 W/K) temperature difference from indoors (72°F, 22°C) to outdoors, what is the lowest outdoor temperature at which a Carnot heat pump can maintain the interior temperature of the building? This heat pump has a nominal power input, when it is running, of 80 kW. (Note that the heat pump controls the indoor temperature by cycling on and off; but when it is on, it always draws 80 kW.)   (5)

3.42 One of the factors that degrades the actual COP of heat pumps relative to the Carnot maximum is the need to transfer heat between heat sources and sinks and the Carnot cycle working fluid itself. Suppose that a building heat pump with low- and high-temperature reservoirs at 40°F (4.5°C) (outdoor air) and 72°F (22°C) (indoor air temperature) delivers 30,000 Btu/h (8.8 kW) of heat to the building interior. If the temperature differences needed to transfer heat from these two reservoirs to the Carnot fluid are both 10°F (5.5 K), what is the COP of this heat pump, accounting for heat transfer penalties? Compare it to the ideal Carnot COP without this penalty. (7)

3.43 Generalize the results of Prob. 3.42, and plot the COP of this heat pump for a range of outdoor temperatures between 0 and 60°F (−18 and 16°C). The heat transfer resistance between the two reservoirs and the heat pump working fluid is independent of the operating temperature. On the same graph, plot the ideal Carnot heat pump COP for comparative purposes. (9)

3.44 The temperature of the return-air plenum above the ceiling in a commercial building is determined by the heat input to this area by lights and the loss to the cooler space below as well as by the flow of air through the plenum. Figure P3.44 is a simplified diagram of such a plenum located above one of the floors in the interior of a building. Derive an equation for the plenum temperature T_p that includes the three effects just listed. Include the ceiling tile R value and area, the return airflow rate $\dot{m}_r$, the lighting fixture top heat loss $\dot{Q}_L$, and the zone temperature. How would your result be modified for the top floor of a building? (8)

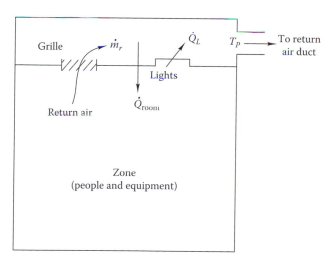

FIGURE P3.44

References

ASHRAE (1989). *Handbook of Fundamentals.* American Society of Heating, Refrigeration and Air-Conditioning Engineers, Atlanta, GA.

Cengel, Y. A., and M. A. Boles (2000). *Thermodynamics,* 3d ed. McGraw-Hill, New York.

Keenan, J. H., and J. Kaye (1948). *Gas Tables,* Wiley, New York.

Keenan, J. H., F. G. Keyes, P. G. Hill, and J. G. Moore (1978). *Steam Tables,* Wiley, New York.

Van Wylen, G. J., and R. E. Sonntag (1986). *Fundamentals of Classical Thermodynamics,* 3d ed. Wiley, New York.

4

Psychrometrics, Comfort, and Health

4.1 Introduction and Definitions

Psychrometry is the study of the measurement of the moisture content of atmospheric air. By extension, the term is commonly taken to mean the study of atmospheric moisture and its effect on buildings and building systems.

The moisture content of air is important since it has a major, direct effect on the comfort of building occupants. As we will learn later, comfort and health require that the moisture content of air be controlled within a relatively narrow range. In hot, humid weather, one must remove moisture by using a dehumidification process that is often a part of the air cooling process. In dry, cold weather, humidity is added to the air by humidifiers, which are often a part of the heating system. Since the latent heat of water is large (about 1050 Btu/lb or 2450 kJ/kg at room temperature), either the removal or the addition of moisture to air in a building can involve a significant amount of energy.

Moisture control in buildings is also needed to limit moisture collection in or on building components, such as insulation, window glass, and structural members. The goals of this chapter are to

- Introduce thermodynamic fundamentals having to do with psychrometrics.
- Analyze important building heating, cooling, and ventilation processes from a psychrometric point of view.
- Provide charts and tables needed to perform psychrometric calculations.

The majority of students taking this course have not been previously exposed to psychrometric analysis techniques. Therefore, you should have a good understanding of the material in this chapter before moving on to the loads and equipment chapters, Chaps. 6 to 11.

Several processes involving humidification, dehumidification, and mixing of streams with differing moist-air properties will be analyzed in detail in this chapter. All the analytical and design information in this chapter is based on application of the first law of thermodynamics and the conservation of mass along with properties of air and water vapor taken from basic thermodynamics (see, e.g., Chap. 3). We will assume that air–water vapor mixtures behave as ideal gases with constant specific heat. (The water vapor is in a superheated state in moist-air mixtures; even under saturated-air conditions, the assumption is valid for the water vapor since moist air is a very dilute mixture of water vapor and dry air.) The ideal gas assumption involves errors of less than 1 percent in building heating and cooling calculations (Kuehn et al., 1998).

4.2 Thermodynamic Fundamentals

There are a number of measures of the amount of moisture contained in atmospheric air. Before we begin the discussion of the thermodynamics of moist-air mixtures, we shall define their most important properties and indicate where each is used. However, first, we define the *dry-bulb temperature* T_d as the temperature of a moist-air mixture measured by a perfectly dry sensor, such as a thermocouple or thermometer. The measures of moist-air–water vapor content are as follows:

1. The *partial pressure* p_w of water vapor in the homogeneous air–water vapor mixture is used in the definition of relative humidity. Partial pressure is the pressure exerted by one gas component on a mixture of several gases. Dalton's law of additive pressures states that the sum of the partial pressures of all gases in a mixture is the total pressure exerted by a gas mixture.

2. The *relative humidity* ϕ is the ratio of the partial pressure of water vapor p_w to the saturation pressure of water vapor p_{sat} at the existing dry-bulb temperature. The saturation pressure is the pressure exerted by saturated water vapor at T_d; it can be read from steam tables, such as those provided on the CD-ROM. The relative humidity is the most common measure of moisture content—it is the fraction of moisture contained in an air–water vapor mixture relative to what a saturated-air–water vapor mixture at the same temperature would contain.[1]

3. The *humidity ratio* W is the ratio of the mass of water vapor to the mass of *dry* air in a moist-air mixture. The humidity ratio is used, e.g., to calculate the enthalpy of moist air. It has units of lb_w/lb_{da} or kg_w/kg_{da}, where the subscript "da" denotes dry air.

4. The *dew point temperature* T_{dew} is the temperature at which a moist-air mixture at a given humidity ratio becomes saturated; i.e., moisture begins to condense, as it is cooled at constant pressure. The dew point is used in design calculations to find the temperature at which condensation must be accounted for in equipment design.

5. The *wet-bulb temperature* T_{wet} is the temperature of a thermometer with a wetted wick over which air flows at a specified velocity. The wet-bulb temperature is an intermediate quantity used in many psychrometric calculations. Designers use it as a quick indication of the moisture content of air.

Using the ideal gas law, we will now find expressions relating the relative humidity and the humidity ratio to standard thermodynamic properties such as specific volume and pressure. To do this, consider the process 1-2 shown in Fig. 4.1 (the *T-s* diagram for water) in which superheated water vapor at T_d (point 1) is cooled at constant pressure p_w to a saturated-vapor state at the dew point T_{dew} (point 2). The mixture is assumed to be homogeneous and at thermodynamic equilibrium. In buildings, moist-air mixtures are usually between 0 and 60°C; therefore, the ideal gas laws will give accurate results for building design purposes. (Figure 4.1 also shows the saturation pressure p_{sat} located on the saturation curve at point 3, the temperature corresponding to the dry-bulb temperature of the moist air at condition 1.)

---

[1] Although the relative humidity is strictly defined as a ratio of mole fractions, the ideal gas assumption enables us to express the relative humidity more conveniently for working purposes as the ratio of pressures.

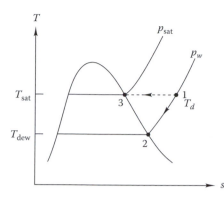

FIGURE 4.1
The T-s diagram for steam showing dry-bulb and dew point temperatures and saturation line.

We will begin with the working expression for ϕ:

$$\phi = \frac{P_w}{P_{\text{sat}}} \tag{4.1}$$

The ideal gas low can be used to replace this pressure ratio by a ratio of specific volumes

$$\phi = \frac{RT_d/v_d}{RT_{\text{sat}}/v_{\text{sat}}} = \frac{v_{\text{sat}}}{v_d} \tag{4.2}$$

If we consider equal volumes V of nonsaturated and saturated mixtures, the ratio of the masses of water vapor in the mixtures can be found from the definition of specific volume V/m. Using this Using this substitution in Eq. (4.2), we have

$$\phi = \frac{m_w}{m_{\text{sat}}} \tag{4.3}$$

Like the relative humidity, the humidity ratio can be expressed in a more useful fashion by using the ideal gas law to replace the mass fraction with a ratio of pressures. We will later show how the saturation pressure of water vapor can be calculated; of course, it can also be looked up in the steam tables in the appendix CD-ROM. The atmospheric pressure depends on the location for which the psychrometric calculations are being done.

The ideal gas law written on an extensive basis with the universal gas constant $\mathcal{R}$ and molecular weight M is

$$pV = m\left(\frac{\mathcal{R}}{M}\right)T \tag{4.4}$$

Equation (4.4) can be solved for the mass. For the masses of air and water vapor, we have

$$m_{\text{da}} = \frac{p_{\text{da}} V M_{\text{da}}}{\mathcal{R}T} \tag{4.5}$$

and

$$m_w = \frac{p_w V M_w}{\mathcal{R}T} \tag{4.6}$$

The ratio of m_w to m_{da} is the humidity ratio W by definition; therefore,

$$W = \frac{m_w}{m_{da}} = \frac{p_w M_w}{p_{da} M_{da}} \tag{4.7}$$

We now substitute the numerical values of the two molecular weights—28.96 for air and 18.02 for water—to get

$$W = \frac{p_w}{p_{da}} \left(\frac{18.02}{28.96} \right) = 0.622 \frac{p_w}{p_{da}} \tag{4.8}$$

Since the total of the partial pressure of water vapor and dry air is the atmospheric pressure p, by Dalton's law, we can replace p_{da} in Eq. (4.8) as follows:

$$W = 0.622 \frac{p_w}{p - p_w} \tag{4.9}$$

Because p_w is much less than p, the humidity ratio and partial pressure are related approximately linearly. We will use this observation later in construction of the psychrometic part.

An alternate form of Eq. (4.9) using the relative humidity is sometimes useful:

$$W = 0.622 \frac{\phi p_{sat}}{p - \phi p_{sat}} \tag{4.10}$$

The dry-air basis for the preceding quantities p_{da} and m_{da} is used since the composition of dry air is essentially invariant whereas the composition of moist air obviously varies with the moisture content. The use of the dry-air basis leads to less involved calculations than a moist-air basis, although the latter could be used equally well from a theoretical point of view. For altitude effects, see the tables in the appendices on the CD-ROM.

4.2.1 Adiabatic Saturation and the Wet-Bulb Temperature

In principle, the relative humidity or the humidity ratio could be used to specify the moisture content of atmospheric air. The problem is that neither can be measured directly with inexpensive and durable instruments. Hence, an indirect method using a straightforward temperature measurement is chosen. Consider the apparatus shown in Fig. 4.2 in

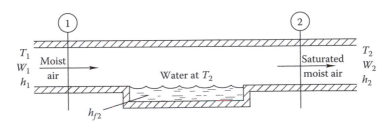

FIGURE 4.2
Schematic diagram showing adiabatic saturation process equipment.

which air at temperature T_1 enters an adiabatic (i.e., no heat transfer occurs *across* the surface bounding the apparatus) enclosure. The entering moist air becomes saturated as it passes over the liquid due to the evaporation that occurs. Heat is removed from the airstream to accomplish the evaporation, and the outlet temperature T_2 is lower than the inlet temperature as a result.

The adiabatic saturation outlet temperature T_2 is called the *thermodynamic wet-bulb temperature*. By applying the conservation of mass law to both the dry air and the water systems and by applying the open-system first law to the thermodynamic system consisting of both the water and the air, it is possible to find the unknown inlet humidity ratio W_1 (the details of this are contained in Sec. 4.4.1). This device is impractical for field use, so the psychrometer (a wet- and dry-bulb thermometer arrangement described below) is used instead. The wet-bulb temperature so measured is very close to the thermodynamic wet-bulb temperature and is substituted for it in common psychrometric calculations.

Figure 4.3 shows a device for finding the wet-bulb temperature approximately. The sling psychrometer is standard equipment in many laboratories and HVAC field engineering services. As air flows over the moistened wick surrounding the "wet bulb," water evaporates, extracting heat from the thermometer and lowering its temperature. At equilibrium, the thermometer bulb will be at a temperature at which the rate of evaporation times the heat of vaporization just balances the heat transferred to the bulb. The wet bulb's temperature is below the dry bulb's temperature because heat must be transferred to the wet bulb in order to evaporate moisture. Kuehn et al. (1998) offer a detailed heat and mass transfer analysis of the sling psychrometer.

The suggested velocity of air over the wet bulb is 7.5 to 15 ft/s (2.5 to 5 m/s). The wick over the bulb is made of fine cotton, which must be kept clean. Periodic replacement is needed depending on the water mineral content and air cleanliness. The service interval can be increased if distilled water is used. An alternative to the sling psychrometer is a

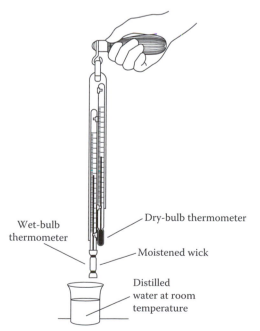

Wet-bulb thermometer

Dry-bulb thermometer

Moistened wick

Distilled water at room temperature

FIGURE 4.3
Sling psychrometer device for conveniently measuring wet- and dry-bulb temperatures.

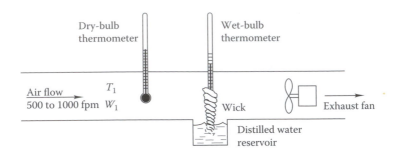

FIGURE 4.4
Psychrometry apparatus for measuring wet- and dry-bulb temperatures.

small duct through which air flows over a stationary wet-bulb thermometer. Figure 4.4 shows this device schematically.

The devices described measure a close approximation to the thermodynamic wet-bulb temperature. To find the relative humidity directly from the wet-bulb temperature (and the concurrently measured dry-bulb temperature), the following approach (Pallady, 1989) is recommended since it is more accurate than the expression given in ASHRAE (1989a, 2001).[2] The algorithm below is reported to be accurate to 0.3 percent on average (not accounting for errors in temperature measurement) and is based on the van der Waals equation. It uses the critical pressure p_c and temperature T_c of water to find the vapor pressure and humidity, as shown below. The relative humidity is calculated as follows from this alternative, working form of Eq. (4.1) in which the numerator p_w has been replaced by $p_{sat}(T_{wet}) - p_m$, defined as

$$\phi = \frac{p_{sat}(T_{wet}) - p_m}{p_{sat}(T_d)} \tag{4.11}$$

where

$p_{sat}(T_{wet})$ = saturation pressure at wet-bulb temperature T_{wet}; temperature units uses with equations in this section are *absolute*—°R or K

$p_{sat}(T_d)$ = saturation pressure at dry-bulb temperature T_d, °R (K)

p_m = partial pressure of water vapor due to *depression of wet-bulb temperature* below dry-bulb temperature; it does not express a physical process result but in just an intermediate quantity useful for relative humidity calculations. And p_m is given by

$$p_m = p\left(\frac{T_d - T_{wet}}{2725}\right)\left(1 + \frac{T_{wet} - 492}{1571}\right) \tag{4.12US}$$

or, in SI units,

$$p_m = p\left(\frac{T_d - T_{wet}}{1514}\right)\left(1 + \frac{T_{wet} - 273.2}{873}\right) \tag{4.12SI}$$

---

[2] Mathur (1989) reports the accuracy of the ASHRAE equation to be 1.2 to 1.5 percent between 50 and 150°F (35 and 90°C).

The two saturation partial pressures in Eq. (4.11) are found at dry- and wet-bulb temperatures from

$$p_{sat}(T) = p_c(10^{K(1-T_c/T)}) \tag{4.13}$$

where $p_{sat}(T)$ is the saturation pressure of water at temperature T. This expression for saturation pressure is to be applied at both the dry- and wet-bulb *absolute* temperatures. The critical pressure of water p_c is 3226 psia (22.1 MPa), and the critical temperature T_c is 1165.67°R (647.30 K).

The parameter K depends on the temperature and is given by

$$K = 4.39553 - 3.469\left(\frac{T}{1000}\right) + 3.072\left(\frac{T}{1000}\right)^2 - 0.8833\left(\frac{T}{1000}\right)^3 \tag{4.14US}$$

in which T is the dry- or wet-bulb temperature in degrees Rankine. In SI units, this dimensional expression is

$$K = 4.39553 - 6.2442\left(\frac{T}{1000}\right) + 9.953\left(\frac{T}{1000}\right)^2 - 5.151\left(\frac{T}{1000}\right)^3 \tag{4.14SI}$$

in which T is the dry- or wet-bulb temperature in kelvins.

In this method, the partial pressure of water vapor p_w needed to find ϕ from Eq. (4.1) is found indirectly from the difference between the saturation partial pressure *at the wet-bulb temperature* $p_{sat}(T_{wet})$ (note that this is different from p_{sat} used in the previous section) and the partial pressure of water vapor p_m due to the *depression of the wet-bulb temperature* below the dry-bulb temperature. These two partial pressures are not used directly in design but are convenient for finding the relative humidity (RH).

To summarize the calculation:

- Find both $p_{sat}(T_d)$ and $p_{sat}(T_{wet})$ from Eqs. (4.13) and (4.14), using the known wet-bulb and dry-bulb temperatures.
- Next, find p_m from Eq. (4.12).
- Finally, calculate the relative humidity, using Eq. (4.11).

The following example illustrates the method.

Example 4.1: Calculation of the Relative Humidity from the Dry- and Wet-Bulb Temperatures

Find the relative humidity and the humidity ratio of moist air at 204°F if the wet-bulb temperature is measured at 190°F by a sling psychrometer. The atmospheric pressure is that of a standard atmosphere, 14.696 psia.

Given: $T_d = 204°F = 664°R$ (369 K)

$T_{wet} = 190°F = 650°R$ (361 K)

$p = 14.696$ psia (101 kPa)

Find: ϕ, W

SOLUTION

The approach described in the previous paragraph will be used. From Eq. (4.14) the K values at the dry- and wet-bulb temperatures are found to be

$$K(T_d) = 3.18814 \quad K(T_{\text{wet}}) = 3.19611$$

The values of $p_{\text{sat}}(T_d)$ and $p_{\text{sat}}(T_{\text{wet}})$ are then found from Eq. (4.13):

$$p_{\text{sat}}(T_d) = 3226 \times 10^{3.18814(1-1165.67/664)} = 12.58 \text{ psia}$$

$$p_{\text{sat}}(T_{\text{wet}}) = 3226 \times 10^{3.19622(1-1165.67/660)} = 9.397 \text{ psia}$$

Finally, Eq. (4.12) is used to find p_m:

$$p_m = (14.696 \text{ psia})\left(\frac{664 - 650}{2725}\right)\left(1 + \frac{650 - 492}{1571}\right) = 0.0831 \text{ psia}$$

The relative humidity is found from Eq. (4.11):

$$\phi = \frac{9.397 - 0.0831}{12.58} = 0.74$$

The humidity ratio is readily found from Eq. (4.10), where p_{sat} was evaluated at T_d:

$$W = 0.622 \frac{\phi p_{\text{sat}}}{p - \phi p_{\text{sat}}} = 0.622\left(\frac{0.74 \times 12.58}{14.696 - 0.74 \times 12.58}\right) = 1.08 \text{ lb}_w/\text{lb}_{da}$$

COMMENTS

Observe that the capacity of hot air to absorb water vapor is significant. Both the relative humidity and the humidity ratio can also be found by using the HCB software[3] accompanying this book.

4.2.2 Dew Point Temperature

The dew point temperature is shown on the saturation curve in Fig. 4.1. It is the temperature at and below which moist air is saturated, since air at the dew point temperature contains the maximum possible amount of water vapor. The humidity ratio of air remains constant as it is cooled until the dew point is reached. Below the dew point, condensation occurs and the humidity ratio decreases.

The dew point is important in building design since condensation must be avoided within building elements such as walls or on HVAC system components such as chilled-water pipes. Of course, condensation is desirable in air conditioning systems since it is the mechanism by which humidity is removed from humid outdoor air before introducing such air into the interior of buildings for ventilation purposes.

The dew point temperature can be found from p_w by using the following expression (for temperatures above the freezing point) (ASHRAE, 1989a):

$$T_{\text{dew}} = 100.45 + 33.193\alpha + 2.19\alpha^2 + 0.1707\alpha^3 + 1.2063p_w^{0.1984} \, ^\circ\text{F} \qquad (4.15\text{US})$$

---

[3] The psychrometrics section of the HCB software can be used to solve nearly all the examples in this chapter along with many of the end-of-chapter problems.

in which $\alpha = \ln p_w$ and p_w has units of pounds per square inch absolute. In SI units, the dew point is

$$T_{\text{dew}} = 6.54 + 14.256\alpha + 0.7389\alpha^2 + 0.0948\alpha^2 + 0.4569p_w^{0.1984} \ \text{°C} \qquad (4.15\text{SI})$$

in which $\alpha = \ln p_w$ and p_w has units of kilopascals.

A convenient rule of thumb states that at 70 percent relative humidity a difference of 10°F (5.5°C) between the dew point and dry-bulb temperature exists. A difference of 20°F (11°C) between dew point and dry-bulb temperature corresponds roughly to a relative humidity of 50 percent. These and similar rules for other dew point–dry-bulb differences apply to within a few percent accuracy in the restricted range of dry-bulb temperatures involved in building design.

4.2.3 Moist-Air Enthalpy

The enthalpy of moist air is the sum of the enthalpy of the dry air and of the water vapor comprising the mixture. On an intensive basis (i.e., per unit mass of dry air), the enthalpy h of the mixture is

$$h = h_{\text{da}} + W h_g$$

where h_g is the enthalpy of water vapor.

From our discussion of the first law in Chap. 3, we recall that enthalpy changes, that is, Δh, are all that ever arise in steady first-law calculations in buildings. As a result, the enthalpy at the temperature on which we base the enthalpy calculation cancels out. For convenience, therefore, the standard base temperature for moist-air properties is 0°F or 0°C, depending upon which units are being used. In terms of temperature, we express the enthalpy of moist air as

$$h = c_{pa}T_d + W(h_{g,\text{ref}} + c_{pw}T_d) \qquad (4.16)$$

where
    h = enthalpy of moist air, Btu/lb (kJ/kg)
    c_{pa} = specific heat of dry air, Btu/(lb · °F) [kJ/(kg · °C]
    c_{pw} = specific heat of superheated water vapor, Btu/(lb · °F) [kJ/(kg · °C)]
    T_d = dry-bulb temperature, °F (°C)
    $h_{g,\text{ref}}$ = enthalpy of water vapor at appropriate reference temperature, Btu/lb (kJ/kg)
    W = humidity ratio, lb/lb$_{\text{da}}$ (kg/kg$_{\text{da}}$)

Substituting numerical values for the two specific heats and for the extrapolated value of enthalpy of saturated water vapor (at 0°F or 0°C), we have the following pair of dimensional equations ($h_{g,\text{ref}} = 1061.2$ Btu/lb is the enthalpy of saturated water vapor at 0°F)

$$h = 0.240T_d + W(1061.2 + 0.444T_d) \ \text{Btu/lb}_{\text{da}} \qquad (4.17\text{US})$$

in which T_d is in degrees Fahrenheit and

$$h = 1.0T_d + W(2501.3 + 1.86T_d) \ \text{kJ/kg}_{\text{da}} \qquad (4.17\text{SI})$$

in which T_d is in degrees Celsius; in SI units, $h_{g,ref} = 2501.3$ kJ/kg, the enthalpy of saturated water vapor at 0°C.

In Sec. 4.3.3 we discuss the use of tables of moist-air properties. These are to be used when maximum accuracy is required and the preceding equations for h are not sufficiently precise.

4.3 The Psychrometric Chart and Tables of Air Properties

4.3.1 Construction of the Chart

Although the equations that we have developed for the many properties of moist air are useful for computer calculations, it is convenient to have them plotted in chart form for easy reference during preliminary design of HVAC systems. The standard psychrometric chart serves this purpose. It is also of value in plotting psychrometric processes such as heating, cooling, mixing, drying, or humidification of airstreams.

The psychrometric chart is an xy plot with dry-bulb temperature T_d as the abscissa and the humidity ratio W as the ordinate.[4] Since these are two independent thermodynamic variables, all other properties of moist air can be expressed as functions of them at a given atmospheric pressure. On the standard psychrometric chart, the following moist-air properties are plotted:

- Relative humidity
- Wet-bulb temperature
- Vapor pressure
- Specific volume
- Enthalpy

Figure 4.5a to e shows how lines of constant properties listed appear on the psychrometric chart. Figure 4.6 is the psychrometric chart in USCS units, and Fig. 4.7 is presented in SI units. The charts shown are for the "normal" temperature range. Additional charts are available from ASHRAE for both low and elevated temperature ranges and for altitudes between sea level and 7500 ft (2260 m).

The expressions developed above depend on the local atmospheric pressure, which is assumed to be 1 atm if the elevation is not much different from sea level. However, at high altitude, the values of thermodynamic variables can be affected considerably by the lower local barometric pressure. As a result, high-altitude psychrometric charts must be used. The chart for 5000-ft elevation is included in the appendices on the CD-ROM as an example.

In the psychrometric chart, the relative humidity lines curve upward from the lower left to the upper right of the chart. The saturation curve (100 percent relative humidity) can be plotted from data in the steam tables at the desired atmospheric pressure. Since saturated-steam properties are given in steam tables as a function of pressure, one uses Eq. (4.9) to find the psychrometric chart ordinate, the humidity ratio, corresponding to the selected vapor pressure p_w. Lines of constant relative humidity less than 100 percent are plotted by

---

[4] Close inspection of the psychrometric charts shows that the coordinates are not quite orthogonal. The dry-bulb lines are nearly vertical but slope slightly to the left.

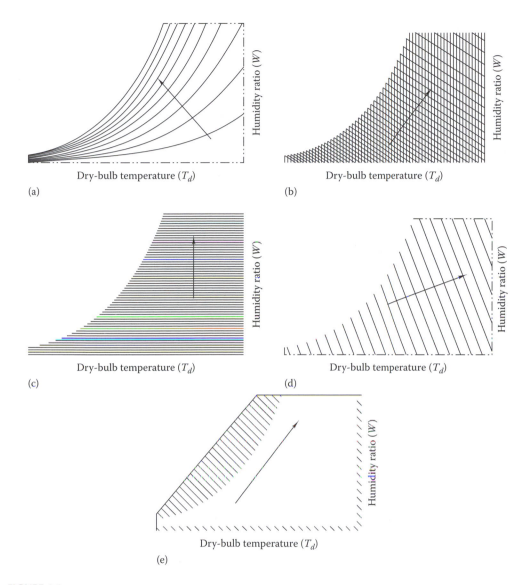

FIGURE 4.5
Skeleton psychrometric chart showing lines of (a) constant relative humidity, (b) constant wet- and dry-bulb temperatures, (c) constant humidity ratio, (d) constant specific volume, and (e) constant enthalpy.

linearly scaling vertical distances between the abscissa and the saturation curve by their respective values of relative humidity. The relative humidity values are written along the curves. The increment between lines is 10 percent in the two charts. On the 100 percent humidity line, the wet-bulb, dry-bulb, and dew point temperatures are identical by their definitions.

Wet-bulb temperature lines slope downward from left to right in the chart. By solving Eqs. (4.11) to (4.14) for dry-bulb temperatures as a function of the relative humidity for selected values of the wet-bulb temperature, one can plot lines of constant wet-bulb temperature, as shown in the figure; values are written along the lines. The increment

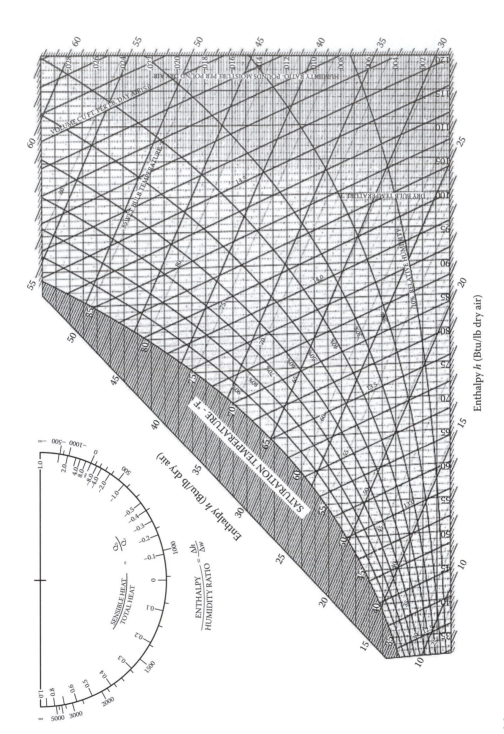

FIGURE 4.6

Psychrometric chart for sea level in USCS units. (Courtesy of ASHRAE, *Handbook of Fundamentals*, American Society of Heating, Refrigerating and Air-Conditioning Engineers, Atlanta, GA, 1989a. With permission.)

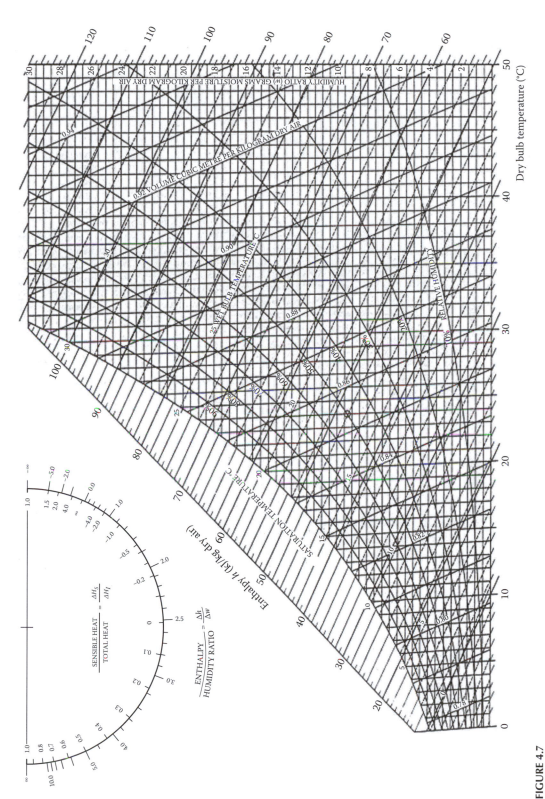

FIGURE 4.7

Psychrometric chart for sea level in SI units. (Courtesy of ASHRAE, *Handbook of Fundamentals*, American Society of Heating, Refrigerating and Air-Conditioning Engineers, Atlanta, GA, 1989a. With permission.)

between labeled lines in the two charts is 5°F (5°C). The T_{wet} lines are straight due to a fortuitous combination of heat and mass transfer properties for air and water vapor (Stoecker and Jones, 1982).

Enthalpy lines also slope downward from left to right in the chart. Values are noted at the left and right ends of the lines. The increment between marked lines in the two figures is 5 Btu/lb$_{da}$ (10 kJ/kg$_{da}$). Enthalpy lines and wet-bulb temperature lines are very nearly parallel, but lines of constant wet-bulb temperature are slightly steeper than lines of constant enthalpy. Therefore, it is best to use the auxiliary enthalpy scales (1) along the left diagonal border of the chart and (2) along the bottom and right side of the chart along with a straightedge for accurate readings. Enthalpy lines are plotted by selecting a value of enthalpy, inserting it into Eq. (4.17), and finding the temperature T_d for several values of the humidity ratio. The results, when plotted, provide the enthalpy lines on the psychrometric chart. For more accurate values, one can use the table of moist-air properties as described in the next section.

Specific-volume lines slope steeply downward from left to right of the chart at about 20° from the vertical. Lines of constant specific volume are found by solving Eq. (4.5) for the pressure $p_{da} = p - p_w$ as a function of the temperature and specific volume. Then, for a given specific volume, one finds $T - p_w$ pairs that are plotted. The specific-volume values are written along the lines. The increment between marked lines is 0.5 ft^3/lb$_{da}$ (0.02 m^3/kg$_{da}$) in the two charts.

Vapor pressure p_w lines are horizontal and parallel to lines of constant humidity ratio. Values are found by solving Eq. (4.9) for the vapor pressure as a function of the humidity ratio. In some psychrometric charts (not Figs. 4.6 and 4.7), these values are written along an auxiliary ordinate which is nearly linear. Likewise, dew point temperature lines are horizontal and parallel to lines of constant humidity ratio. Values are found from Eq. (4.15) and are optionally written along an auxiliary, nonlinear ordinate. However, the dew point is just as easily read along the saturation curve itself. Details of the geometric construction of the psychrometric chart are contained in Kuehn et al. (1998).

4.3.2 Use of the Chart

The psychrometric chart can be used in analyzing many processes involving moist atmospheric air. This section contains three examples of such calculations, and the following sections consider a number of *processes* which we will find to be important in the design of heating and cooling systems in buildings.

Example 4.2: Use of the Psychrometric Chart

Find the five listed properties for atmospheric air in a building maintained at 70°F (21°C). The wet-bulb temperature is measured to be 60°F (15.5°C), a typical interior condition in winter in heating climates.

Given: $T_d = 70°F$

$T_{wet} = 60°F$

Figure: See Fig. 4.6.

Assumptions: The atmospheric pressure is the standard atmosphere, 14.7 psia.

Find: ϕ, h, v, T_{dew}, p_w

SOLUTION

This problem can be solved easily by using the USCS psychrometric chart. Note that only one of the two independent variables (T_d) used to plot the chart is known; but two different *independent* properties are known. Hence, the problem can be solved.

First, find the intersection of the 70°F dry-bulb vertical line and the 60°F wet-bulb line. This intersection is approximately at the *57 percent relative humidity* point.

At the same intersection point, the following values can be read by interpolation. The *humidity ratio* is 0.0088 lb/lb$_{da}$. The enthalpy is read from the scales at the upper left and bottom of the chart. The *enthalpy* is 26.5 Btu/lb$_{da}$. By interpolation the *specific volume* is 13.55 ft^3/lb$_{da}$. A better approach than this rough interpolation is to calculate the specific volume from the partial pressure p_w and the dry-bulb temperature by using the ideal gas law:

$$v = \frac{R_{air} T_d}{p - p_w}$$

The partial pressure of water vapor p_w cannot be found from the chart but is easy to find from Eq. (4.9) since the humidity ratio W is already known:

$$p_w = \frac{p}{1 + 0.622/W} = \frac{14.696 \text{ psia}}{1 + 0.622/0.0088} = 0.21 \text{ psia}$$

An alternate method of finding p_w is to use the basic equation for the relative humidity, Eq. (4.1),

$$p_w = \phi p_{sat} = 0.57(0.3633 \text{ psia}) = 0.21 \text{ psia}$$

in which the saturation pressure has been read from the last column of Table 4.1 with appropriate units conversion. (Table 4.1 is discussed in detail shortly.)

The value of p_w can be used in the ideal gas law above to find a more accurate value of the specific volume.

The dew point is found at the intersection of the humidity ratio line (0.0088) and the saturation curve. The *dew point* is 54°F.

COMMENTS

The psychrometric chart can be read to about 1 to 2 percent accuracy. For better accuracy, the HCB software, tables, or psychrometric equations can be used.

Example 4.3: Sensible Cooling Process on Psychrometric Chart

Moist air is cooled from 40°C (104°F) and 30 percent relative humidity to 30°C (86°F). Does moisture condense? What are the values of the relative humidity and the humidity ratio at the process endpoint? (Latent and sensible loads are discussed in Chap. 7; the former involve moisture condensation and the latter, temperature changes only.)

Given: $T_{d1} = 40$°C

$\phi_1 = 0.30$

$T_{d2} = 30$°C

Figure: See Fig. 4.8, process 1-2.

Find: W_2, ϕ_2

TABLE 4.1

Thermodynamic Properties of Moist Air at Standard Barometric Pressure (14.696 psia)

Temp., °F	Saturated Humidity Ratio W_{sat} lb$_w$/lb$_{da}$	Volume, ft³/lb Dry Air			Enthalpy, Btu/lb Dry Air			Entropy, Btu/(lb Dry Air·°R)			Enthalph* h_f Btu/lb	Entropy* s_f Btu/(lb·°R)	Vapor Pressure† p_{sat} inHg
		v_{da}	$v_{d,s}$	v_{sat}	h_{da}	$h_{d,s}$	H_{sat}	s_{da}	$s_{d,s}$	s_{sat}			
10	0.0013158	11.832	0.025	11.857	2.402	1.402	3.804	0.00517	0.00315	0.00832	−154.13	−0.3141	0.062901
12	0.0014544	11.883	0.028	11.910	2.882	1.550	4.433	0.00619	0.00347	0.00966	−153.17	−0.3120	0.069511
14	0.0016062	11.933	0.031	11.964	3.363	1.714	5.077	0.00721	0.00381	0.01102	−152.20	−0.3100	0.076751
16	0.0017724	11.984	0.034	12.018	3.843	1.892	5.736	0.00822	0.00419	0.01241	−151.22	−0.3079	0.084673
18	0.0019543	12.035	0.038	12.072	4.324	2.088	6.412	0.00923	0.00460	0.01383	−150.25	−0.3059	0.093334
20	0.0021531	12.085	0.042	12.127	4.804	2.303	7.107	0.01023	0.00505	0.01528	−149.27	−0.3038	0.102798
22	0.0023703	12.136	0.046	12.182	5.285	2.537	7.822	0.01123	0.00554	0.01677	−148.28	−0.3018	0.113130
24	0.0026073	12.186	0.051	12.237	5.765	2.793	8.558	0.01223	0.00607	0.01830	−147.30	−0.2997	0.124396
26	0.0028660	12.237	0.056	12.293	6.246	3.073	9.318	0.01322	0.00665	0.01987	−146.30	−0.2977	0.136684
28	0.0031480	12.287	0.062	12.349	6.726	3.378	10.104	0.01420	0.00728	0.02148	−145.31	−0.2956	0.150066
30	0.0034552	12.338	0.068	12.406	7.206	3.711	10.917	0.01519	0.00796	0.02315	−144.31	−0.2936	0.164631
⋮													
55	0.009233	12.970	0.192	13.162	13.213	10.016	23.229	0.02715	0.02048	0.04763	23.11	0.0459	0.43592
56	0.009580	12.995	0.200	13.195	13.453	10.397	23.850	0.02762	0.02122	0.04884	24.11	0.0478	0.45205
57	0.009938	13.021	0.207	13.228	13.694	10.790	24.484	0.02808	0.02198	0.05006	25.11	0.0497	0.46870

58	0.010309	13.046	0.216	13.262	13.934	11.197	25.131	0.02855	0.02277	0.05132	26.11	0.0517	0.48589
59	0.010692	13.071	0.224	13.295	14.174	11.618	25.792	0.02901	0.02358	0.05259	27.11	0.0536	0.50363
60	0.011087	13.096	0.233	13.329	14.415	12.052	26.467	0.02947	0.02442	0.05389	28.11	0.0555	0.52193
61	0.011496	13.122	0.242	13.364	14.655	12.502	27.157	0.02994	0.02528	0.05522	29.12	0.0575	0.54082
62	0.011919	13.147	0.251	13.398	14.895	12.966	27.862	0.03040	0.02617	0.05657	30.11	0.0594	0.56032
63	0.012355	13.172	0.261	13.433	15.135	13.446	28.582	0.03086	0.02709	0.05795	31.11	0.0613	0.58041
64	0.012805	13.198	0.271	13.468	15.376	13.942	29.318	0.03132	0.02804	0.05936	32.11	0.0632	0.60113
65	0.013270	13.223	0.281	13.504	15.616	14.454	30.071	0.03178	0.02902	0.06080	33.11	0.0651	0.62252
66	0.013750	13.248	0.292	13.540	15.856	14.983	30.840	0.03223	0.03003	0.06226	34.11	0.0670	0.64454
67	0.014246	13.273	0.303	13.577	16.097	15.530	31.626	0.03269	0.03107	0.06376	35.11	0.0689	0.66725
68	0.014758	13.299	0.315	13.613	16.337	16.094	32.431	0.03315	0.03214	0.06529	36.11	0.0708	0.69065
69	0.015286	13.324	0.326	13.650	16.577	16.677	33.254	0.03360	0.03325	0.06685	37.11	0.0727	0.71479
70	0.015832	13.349	0.339	13.688	16.818	17.279	34.097	0.03406	0.03438	0.06844	38.11	0.0746	0.73966
71	0.016395	13.375	0.351	13.726	17.058	17.901	34.959	0.03451	0.03556	0.07007	39.11	0.0765	0.76528
72	0.016976	13.400	0.365	13.764	17.299	18.543	35.841	0.03496	0.03677	0.07173	40.11	0.0783	0.79167
73	0.017575	13.425	0.378	13.803	17.539	19.204	36.743	0.03541	0.03801	0.07343	41.11	0.0802	0.81882
74	0.018194	13.450	0.392	13.843	17.779	19.889	37.668	0.03586	0.03930	0.07516	42.11	0.0821	0.84684
75	0.018833	13.476	0.407	13.882	18.020	20.595	38.615	0.03631	0.04062	0.07694	43.11	0.0840	0.87567

Source: Courtesy of ASHRAE, *Handbook of Fundamentals*, American Society of Heating, Refrigerating and Air-Conditioning Engineers, Atlanta, GA, 1989a. With permission.)

* Saturated conditions.

† Multiply inHg by 0.49115 to convert to psia.

SOLUTION

First, we must determine whether moisture condenses. In terms of psychrometric variables, we need to determine if the final temperature is above or below the dew point. By referring to the SI psychrometric chart (Fig. 4.7), we observe that the dew point temperature $T_{dew} = 19°C$. Therefore, moisture does not condense, and the problem can be solved using the properties of moist air as plotted in Fig. 4.7 directly. This is a *sensible cooling* process, shown by points 1-2 in Fig. 4.8a; no latent heat removal is involved.

Second, we observe that the *humidity ratio* will not change since no moisture condensation has occurred. Therefore, the humidity ratio at the end of the process will be identical to that at the start, 0.014 kg/kg$_{da}$.

Third, the *relative humidity will* change since the relative ability of the cooler air to hold moisture is less than that of the warm air at the start of the process. Since this process is a constant-humidity-ratio process, it is plotted as a horizontal line on the psychrometric chart. One moves to the left from the initial point until the intersection with the $T_d = 30°C$ vertical line. At this point, we read a *relative humidity value* of 53 percent.

COMMENTS

The heat removed from the air during this process can be read from the psychrometric chart by finding the enthalpy at the two endpoints. Alternatively, Eq. (4.17) or the HCB software provides a more accurate result.

Example 4.4: Latent and Sensible Cooling Process on Psychrometric Chart

Moist air is cooled from 40°C (104°F) and 30 percent relative humidity to 15°C (50°F). Does moisture condense? What are the values of the relative humidity and the humidity ratio at the process endpoint?

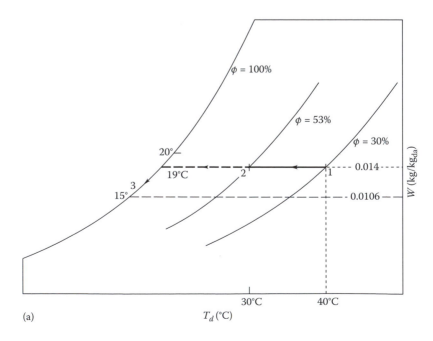

FIGURE 4.8
Cooling processes plotted on psychrometric chart: (a) Sensible cooling only (1–2); sensible and latent cooling (1–3).

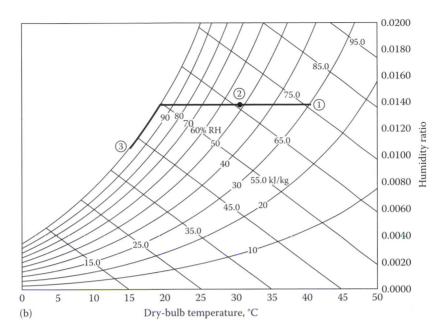

FIGURE 4.8 (continued)
(b) Sensible and latent process as calculated with HCB software.

Given: $T_{d1} = 40°C$

$\phi_1 = 0.30$

$T_{d2} = 15°C$

Figure: See Fig. 4.8a, process 1-3.

Find: W_3, ϕ_3

SOLUTION

From the previous example, the dew point is known to be 19°C for the moist air at the beginning of the process. Therefore, moisture condenses since the outlet temperature (15°C) for this case is below this dew point. This is both a sensible and a latent cooling process.

Inspection of process 1-3 in Fig. 4.8 shows that the *humidity ratio changes* since moisture condensation has occurred. The saturated air at the exit temperature of 15°C has a humidity ratio, according to Fig. 4.7, of 0.0106 kg/kg$_{da}$.

Finally, the *relative humidity changes* to the saturated value, 100 percent, as shown in Fig. 4.8a.

This process is a two-part process: (1) constant-humidity-ratio cooling until saturation and (2) constant-relative-humidity cooling with condensation of moisture. Figure 4.8b shows the process as it is displayed within the HCB software.

4.3.3 Tabulations of Moist Air Properties

Tables of thermodynamic properties of air (or computer software) are needed for calculations in which the accuracy of the psychrometric chart is inadequate. Table 4.1 is an excerpt of the complete moist-air table, which is included on the CD-ROM. This table contains air properties at 1 atm of pressure. It includes the following data from left to right: The second column is the saturated humidity ratio W_{sat} followed by columns of specific volume v, enthalpy h, and entropy s, each at three conditions: (1) dry air (denoted by "da"), (2) difference between dry air and saturated air (denoted by "d, s"), and (3) saturated air (denoted by "sat"). The reference state for enthalpy and entropy is 0°F and 1 atm pressure in Table 4.1.

For moist air that is neither saturated nor perfectly dry, the following equations are used to find the enthalpy and specific volume. Other properties are found similarly.

$$h = h_{da} + \frac{W}{W_{sat}} h_{d,s} \tag{4.18a}$$

$$v = v_{da} + \frac{W}{W_{sat}} v_{d,s} \tag{4.18b}$$

4.4 Psychrometric Processes for Buildings

In the next six sections, we will analyze a number of common psychrometric processes used in buildings. They will be referred to in subsequent chapters on heating and cooling systems.

The six processes described in the next sections include

- Adiabatic saturation
- Warming and humidification of dry, cold air
- Cooling and dehumidification of warm, moist air
- Evaporative cooler process
- Adiabatic mixing of airstreams
- Building air conditioning process

4.4.1 Adiabatic Saturation

In an earlier section, we described the concept of adiabatic saturation and its relation to the thermodynamic wet-bulb temperature. For practical purposes, this temperature is essentially the same as the wet-bulb temperature measured by a sling psychrometer. In this section, we will analyze adiabatic saturation and its relation to the evaporative cooling process.

Referring to Fig. 4.2, we can write the steady-state first law as

$$h_1 + (W_2 - W_1)h_{f2} = h_2 \tag{4.19}$$

where h_{f2} is the enthalpy of liquid water at T_2.

By definition of an adiabatic process, water is supplied to the basin at T_2 with enthalpy h_{f2} to maintain the water level.

The outlet temperature T_2, corresponding to a condition with T_1 at the inlet, is the wet-bulb temperature of the environmental inlet air. If we use Eq. (4.17) to find the enthalpy of moist air and use the following equation for liquid-water enthalpy (the reference temperature for liquid water is 32°F), we are able to solve for the humidity ratio W_1, recalling that the specific heat of water c_w is 1.0 Btu/(lb · °F).

$$h_{f2} = c_w(T_2 - 32) = T_2 - 32 \text{ Btu/lb} \tag{4.20SI}$$

Solving Eq. (4.19) for W_1 and using Eq. (4.17US) for enthalpy values, we find

$$W_1 = \frac{(1093 - 0.556T_2)W_2 - 0.240(T_1 - T_2)}{1093 + 0.444T_1 - T_2} \tag{4.21US}$$

Note that W_1 is nearly a linear function of the dry-bulb temperature since $1093 \gg 0.444T_1$ for a given wet-bulb temperature T_2. This explains why constant-wet-bulb-temperature lines look straight on the psychrometric chart.

In SI units, the expression for the inlet humidity ratio is

$$W_1 = \frac{(2501 - 2.391T_2)W_2 - 1.0(T_1 - T_2)}{2501 + 1.86T_1 - 4.186T_2} \tag{4.21SI}$$

Example 4.5 illustrates how the humidity ratio can be found by analyzing adiabatic saturator (essentially the same as a psychrometer) data.

Example 4.5: Adiabatic Saturation Process

If an adiabatic saturator (an idealization of an evaporative cooler) operates at 1 atm with an inlet dry-bulb temperature of 82°F (28°C) and an exit dry-bulb temperature of 65°F (18.5°C), what are the entering humidity ratio and relative humidity?

Given: T_{d1}, T_{d2}, p

Figure: See Figs. 4.2 and 4.9 (process 1-2).

Assumptions: Thermodynamic equilibrium exists. The local atmospheric pressure is 1 atm, or 14.696 psia.

Find: W_1, ϕ_1

Lookup values: At 65°F the saturation pressure is 0.62252 inHg from Table 4.1. This is equivalent to 0.306 psia according to the conversion factor in the footnote.

SOLUTION

As indicated in Fig. 4.9, this process is a constant-wet-bulb-temperature process, not a constant-enthalpy process, because water supplied to maintain the sump water level carries enthalpy with it. From Eq. (4.9) the humidity ratio at the exit can be found to be

$$W_2 = 0.622 \frac{0.306}{14.696 - 0.306} = 0.0132 \text{ lb/lb}_{da}$$

Equation (4.21US) can now be used to find the inlet humidity ratio

$$W_1 = \frac{(1093 - 0.556 \times 65)(0.0132) - 0.240(82 - 65)}{1093 + 0.444 \times 82 - 65} = 0.00927 \text{ lb/lb}_{da}$$

To find the inlet relative humidity, Eq. (4.9) can be solved for the water vapor pressure

$$W_1 = 0.00927 = 0.622 \frac{p_{w1}}{14.696 \text{ psia} - p_{w1}}$$

from which

$$p_{w1} = 0.216 \text{ psia}$$

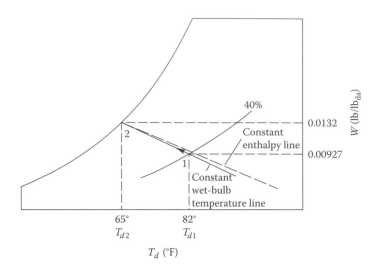

FIGURE 4.9
Adiabatic saturation process plot. The difference between the enthalpy and wet-bulb lines has been exaggerated. (The numerical values apply to Example 4.5.)

At 82°F the saturation pressure is 1.10252 inHg (0.542 psia) from the appendices in the CD-ROM. Finally from the definition of relative humidity we have

$$\phi_1 = \frac{p_{w1}}{p_{sat1}} \times 100\% = \frac{0.216}{0.542} \times 100 = 40\%$$

COMMENTS

Adiabatic saturation is the basic process involved in evaporative cooling of air-streams. It provides an economical method for reducing the dry-bulb temperature of air in climates with low ambient air humidity. The details of these systems will be discussed in Chap. 10.

4.4.2 Warming and Humidification of Cold, Dry Air

A generalization of the adiabatic saturation process just discussed is a process in which heat is added during evaporation. This method is used to simultaneously humidify and heat the air in buildings in winter, when low-humidity conditions could affect health. If no moisture is added, only *sensible* heating has been done; but if moisture and heat are added, both sensible and *latent* heating, respectively, have been accomplished. We will use the results of this section in Chap. 9 on heating equipment. The general results in this section also apply to the special case of only sensible heating when no moisture is added.

Application of the open-system first law of thermodynamics to the process shown in Fig. 4.10a results in the following energy balance

$$h_1 + \frac{\dot{Q}}{\dot{m}_{da}} + (W_2 - W_1)h_{f2} = h_2 \tag{4.22}$$

in which $\dot{Q}$ is the rate of heat addition to the moist air and $\dot{m}_{da}$ is the dry-air mass flow rate. Makeup water is supplied with enthalpy h_{f2} at T_{d2}.

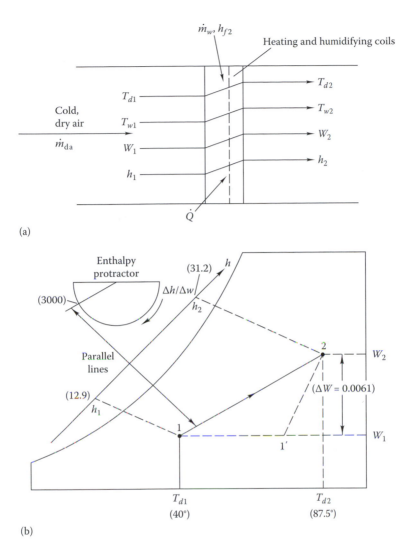

FIGURE 4.10
Heating and humidification process plot: (a) equipment schematic and air properties; (b) process plotted on psychrometric chart. (The numerical values in parentheses apply to Example 4.6.)

If one is to design heating and humidification systems, a graphical construct in the psychrometric chart can be useful. Figure 4.10b shows the warming and humidification process plotted on the chart. The definition of the humidity ratio and the conservation of mass can be used to relate water and mass flow rates. From this, the ratio of the enthalpy change to the humidity ratio change in Eq. (4.22) can be found:

$$\frac{\Delta h}{\Delta W} = \frac{\dot{Q} + \dot{m}_w h_{f2}}{\dot{m}_w} \tag{4.23}$$

To reach this result, we have used the conservation of mass

$$\dot{m}_w = \dot{m}_{da}(W_2 - W_1)$$

The enthalpy-humidity ratio is independent of the airflow rate, and one can construct a line with this slope on the psychrometric chart. The constructed line will have a slope related to the space heating load and the moisture added to the airstream. The principal utility of this ratio is that for a desired outlet (or inlet) state, the conditions of the inlet (or outlet) point must lie on a straight line with slope given by Eq. (4.23). Often psychrometric charts include a protractorlike construction which enables the user to establish the slope and then pass a straightedge through the known inlet point in order to find the outlet condition.

Figure 4.10b shows how the protractor is used for the air warming and humid-ification process 1-2. In actual practice, air is first warmed and then humidified in separate equipment, as shown by the two-step process 1-1' and 1'-2. Figures 4.6 and 4.7 include the protractor in the upper left-hand corner. Of course, if this graphical approach is not sufficiently accurate, the expression for enthalpy of moist air developed earlier or the tables can be used.

Example 4.6: Warming and Humidification of Cold, Dry Air

Outdoor air at 40°F (4.5°C) and 60 percent humidity is heated and humidified by saturated steam at 230°F (110°C). The airflow rate is 15,000 ft^3/min (7080 L/s), and heat is added to the air at the rate of 765,000 Btu/h (224 kW) while it absorbs 415 lb$_m$/h (0.0524 kg/s) of steam. What are the dry- and wet-bulb temperatures at the exit of this heater-humidifier if it is located at sea level?

Given: $T_{d1} = 40$°F

$\phi_1 = 60$ percent

$\dot{V}_{da} = 15,000$ ft^3/min $= 900,000$ ft^3/h

$\dot{m}_w = 415$ lb/h

$\dot{Q} = 765,000$ Btu/h

Figure: See Fig. 4.10b.

Find: T_{d2} and $T_{wet,2}$

Lookup values: Steam enthalpy at 230°F from the tables on the CD-ROM by interpolation, $h_g = 1157$ Btu/lb $= h_{f2}$.

The inlet condition moist-air enthalpy is read from the psychrometric chart (Fig. 4.6) as $h_1 = 12.9$ Btu/lb.

SOLUTION

This problem can be solved by using the graphical approach with the enthalpy protractor in Fig. 4.6. The enthalpy-humidity ratio needed to use the graphical construction is found from Eq. (4.23):

$$\frac{\Delta h}{\Delta W} = \frac{765,000 \text{ Btu/h} + 415 \text{ lb/h} \times 1157 \text{ Btu/lb}}{415 \text{ lb/h}} = 3000 \text{ Btu/lb}$$

This value is located on the enthalpy protractor, and a line is constructed in the chart with the same slope passing through the known process start point 1. To find the termination point 2 of the process, find the exit enthalpy as follows. First find the humidity ratio increase ΔW:

$$\Delta W = \frac{\dot{m}_w}{\dot{m}_{da}} = \frac{415 \text{ lb/h}}{(0.075 \text{ lb/ft}^3)(900,000 \text{ ft}^3/\text{h})} = 0.0061 \text{ lb/lb}_{da}$$

The end-state enthalpy is then found as follows:

$$h_2 = h_1 + \frac{\Delta h}{\Delta W}\Delta W = 12.9 + 3000(0.0061) = 31.2 \ \text{Btu/lb}$$

The outlet enthalpy is used to terminate the construction line at point 2, as shown in Fig. 4.10b. At point, we can read from the chart the following values of the outlet wet- and dry-bulb temperatures:

$$T_{\text{wet}} = 66.5°\text{F} \ (19.2°\text{C})$$

$$T_d = 87.5°\text{F} \ (30.8°\text{C})$$

COMMENTS

This example demonstrates how the protractor construction can save considerable time. If the chart had not been used, then complicated, iterative solutions involving a number of equations in Sec. 4.1 and 4.2 would have been needed.

4.4.3 Cooling and Dehumidification of Warm, Humid Air

Summertime occupant comfort in buildings in warm climates requires the removal of heat from the interior of buildings. Since occupants and outdoor ventilation air also introduce humidity into buildings, comfort requires control of humidity levels as described in Chaps. 7, 10, and 11. Air cooling and dehumidification are usually accomplished in the same piece of equipment, called the *air conditioner* in small buildings or the *air handler cooling coil* in large buildings. In this section, we will outline the joint dehumidification-cooling process. It is very simply the inverse of the humidification-heating process described in the previous section, and essentially the same analysis applies.

To achieve space cooling, one must supply air to the conditioned space at a dry-bulb temperature sufficiently below the space temperature to remove sensible heat gains. To achieve dehumidification, this same air must also be supplied at sufficiently low moisture content to be able to remove moisture gains in the space. Figure 4.11 depicts the general cooling-dehumidification process 1-2 schematically and on a psychrometric chart. This change of air conditions is most often achieved in a single piece of equipment—the cooling coil—installed in the air supply duct. A cold liquid circulates through the coil at a temperature sufficiently low to remove needed amounts of latent and sensible heat. If the coil surface temperature is below the dew point at inlet condition 1, moisture is condensed to liquid water. The liquid is drained from the coil as shown.

The steady-state first law of thermodynamics for the process shown in Fig. 4.11 is

$$\dot{Q} = \dot{m}_{\text{da}}\big[h_1 - h_2 - h_{fc}(W_1 - W_2)\big] \tag{4.24}$$

in which h_{fc} is the enthalpy of liquid water at the temperature at which it leaves the cooling coil.

Line 1-2' in Fig. 4.11b represents an impossible process since there is no *apparatus dew point* (ADP), i.e., no surface sufficiently cool for moisture to condense. Although such processes can be drawn on the chart, they cannot occur.

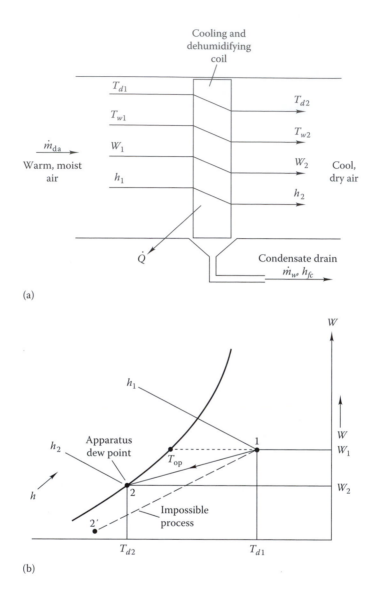

FIGURE 4.11
Cooling and dehumidification process plot: (a) equipment schematic and air properties; (b) process plotted on psychrometric chart. *Note that process 1-2' is impossible since the process line does not intersect the saturation curve; there is no apparatus dew point.*

Example 4.7: Cooling of Moist Air

Outdoor air at sea level enters an air conditioner's cooling coil at 2000 ft^3/min (944 L/s). The inlet dry-bulb condition is 100°F (38°C), and the wet-bulb condition is 75°F (24°C). If the air is cooled to 55°F (13°C) and 90 percent relative humidity, what is the coil heat transfer rate? Assume that the liquid water leaves the coil at 50°F.

Given: T_{d1}, T_{wet1}, T_{d2}, ϕ_2

Figure: See Fig. 4.11.

Find: $\dot{Q}$

Lookup values: From the sea-level psychrometric chart, read the following inlet and outlet values in USCS units. The enthalpy of liquid water at 50°F is 18.11 Btu/lb from the tables on the CD-ROM.

State	W, lb/lb$_{da}$	h, Btu/lb$_{da}$	v, ft^3/lb$_{da}$
1	0.013	38.4	14.4
2	0.0083	22.2	—

SOLUTION

The heat rate can be found from Eq. (4.24) since all values on the right-hand side are known. First we will find the mass flow rate

$$\dot{m}_{da} = \frac{(2000 \text{ ft}^3/\text{min})(60 \text{ min}/\text{h})}{14.4 \text{ ft}^3/\text{lb}} = 8333 \text{ lb}_{da}/\text{h}$$

$$\dot{Q} = \left(8333 \frac{\text{lb}_{da}}{\text{h}}\right)\left\{(38.4 - 22.2)\frac{\text{Btu}}{\text{lb}_{da}} - \left(18.11 \frac{\text{Btu}}{\text{lb}_w}\right)\left[(0.013 - 0.0083)\frac{\text{lb}_w}{\text{lb}_{da}}\right]\right\}$$

$$= (134.995 - 709) \text{ Btu/h} = 134 \text{ kBtu/h} (39.3 \text{ kW})(\sim 11.2 \text{ tons of cooling})$$

COMMENTS

This cooling effect must be provided by the chiller. We will discuss chillers in Chap. 10. Note that the amount of energy leaving the coil in the form of liquid condensate for this problem (709 Btu/h) is very small, about 0.5 percent of the total coil load.

The quantity of water removed is significant: $8333(0.013 - 0.0083) = 39.2$ lb$_w$/h (4.7 gal/h, 17.8 L/h). The designer must include a method of routing this condensed water to a drain for proper disposal.

4.4.4 Evaporative Cooler Process

It is a common human experience to have an evaporating liquid cool one's skin. This same phenomenon is used in mechanical equipment for buildings in several ways. For example, space cooling can be achieved in dry climates by passing air over or through a wet medium or spray. Sensible heat is removed from the airstream by evaporating some of the liquid. Cooling towers are used to reject heat from cooling equipment using the same principle.

Evaporative cooling is essentially an *incomplete adiabatic saturation process*, discussed earlier in this chapter. In actual practice, the exit airstream is often not entirely saturated or at the thermodynamic wet-bulb temperature. We will discuss the details of this cycle in Chap. 10 on cooling equipment. We can do a simple psychrometric calculation at this point, however, to illustrate how the cycle operates.

Example 4.8: Residential Evaporative Cooler

A residence in the Sonora Desert of Arizona is equipped with an evaporative cooler with an airflow rate of 3000 ft^3/min (1416 L/s) sized by the rule of thumb of 2 ft^3/min per square foot of residence area. If the outdoor conditions in summer are 110°F (43.5°C) and 20 percent relative humidity, what is the water flow rate if 80 percent of the available dry-bulb temperature depression is achieved?

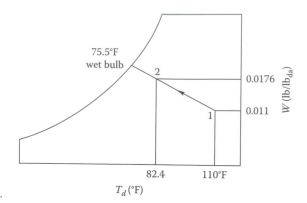

FIGURE 4.12
Evaporative cooler process plot (Example 4.8).

Given: T_{d1}, ϕ_1

Figure: See Fig. 4.12.

Assumptions: The site is at sea level. (The Sonora Desert is about 300 to 400 m above sea level; but to avoid the complications of using the basic psychrometric equations, we will assume a sea-level elevation and use the sea-level psychrometric chart.)

Find: $\dot{m}_w$

Lookup values: $W_1 = 0.011$ lb_w/lb_{da}, $T_{2,min} = 75.5°F$ (the wet-bulb temperature is the minimum temperature achievable in this process)

SOLUTION

The exit dry-bulb temperature is 82.4°F by the problem statement of 80 percent approach to the wet-bulb value [$82.4 = 110 - 0.80(110 - 75.5)$]. To find the exit humidity ratio, one moves along the 75.5°F wet-bulb line until the 82.4°F dry-bulb temperature is reached. At this point, read the exit humidity ratio $W_2 = 0.0176$ lb_w/lb_{da}.

From the chart we also read the inlet specific volume as 14.6 ft^3/lb_{da}.

Then the needed water supply rate is

$$\dot{m}_w = \dot{m}_{da}\Delta W = \frac{\dot{V}\Delta W}{v}$$

$$= \left(3000 \ \frac{ft^3}{min}\right)\frac{(0.0176 - 0.011) \ lb_w/lb_{da}}{14.6 \ ft^3/lb_{da}} = 1.36 \ lb/min$$

COMMENTS

This flow rate, which amounts to about 9.5 gal/h (0.01 L/s), is not excessive for a desert climate when compared to the amount of electric energy saved by using this approach instead of mechanical cooling. For example, if water costs $1 per 1000 gal (typical of an arid western state in the United States), the water consumed costs less than 1¢ per hour. The electricity to operate the fan in the evaporative cooler would cost another 2 to 3¢ per hour. On the other hand, an electric cooling system would cost about 3 to 8 times as much, depending on the weather and load as well as the electric system's COP.

4.4.5 Adiabatic Mixing of Airstreams

An important process in buildings is the adiabatic mixing of airstreams, i.e., mixing without addition or extraction of heat. For example, outdoor air is often mixed with a portion of the air returned from a conditioned space upstream of a cooling coil prior to delivery to the building's occupied space. This mixing is undertaken because the airflow should be larger than the minimum just required to meet loads to ensure proper air motion and comfort in enclosed spaces. Whenever two moist airstreams are mixed, both energy and mass of each species—dry air and water vapor—must be conserved. By using these principles, it is straightforward to find the mixed exit condition and to plot the entire mixing process on a psychrometric chart. We will show how this is done in this section.

Figure 4.13a depicts the mixing of two airstreams of different flow rates, temperatures, and moisture contents. Using the nomenclature of this figure, one can write the conservation of energy and conservation of mass (for both dry air and water), respectively, as follows:

$$\dot{m}_{da1}h_1 + \dot{m}_{da2}h_2 = \dot{m}_{da3}h_3 \tag{4.25}$$

$$\dot{m}_{da1} + \dot{m}_{da2} = \dot{m}_{da3} \tag{4.26}$$

$$\dot{m}_{da1}W_1 + \dot{m}_{da2}W_2 = \dot{m}_{da3}W_3 \tag{4.27}$$

The outlet airflow can be eliminated in Eqs. (4.25) and (4.27) by using Eq. (4.26) to establish the following ratios:

$$\frac{\dot{m}_{da1}}{\dot{m}_{da2}} = \frac{h_2 - h_3}{h_3 - h_1} = \frac{W_2 - W_3}{W_3 - W_1} \tag{4.28}$$

If we consider the enthalpy and humidity ratio coordinates on the psychrometric chart in Fig. 4.13b, it is easy to see that this expression represents a straight line on the chart, as shown. The mixed conditions are placed between the two inlet conditions at distances from these conditions inversely proportional to the ratios of mass flow (e.g., line segment 23 is inversely proportional to airflow $\dot{m}_{da2}$ at point 2). Since enthalpy is a linear function of

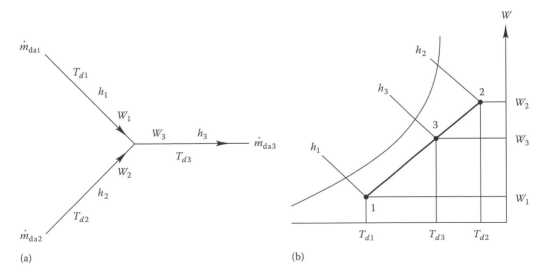

(a)   (b)

FIGURE 4.13
Adiabatic mixing: (a) schematic diagram of flow streams; (b) plot of process on psychrometric chart.

the dry-bulb temperature according to Eq. (4.17), the temperature of the mixture is also proportionately located as shown along the abscissa.

The exit conditions can also be solved for analytically with this result:

$$T_{d3} = \frac{\dot{m}_{da1} T_{d1} + \dot{m}_{da2} T_{d2}}{\dot{m}_{da} + \dot{m}_{da2}} \tag{4.29}$$

$$W_3 = \frac{\dot{m}_{da1} W_1 + \dot{m}_{da2} W_2}{\dot{m}_{da1} + \dot{m}_{da2}} \tag{4.30}$$

$$h_3 = \frac{\dot{m}_{da1} h_1 + \dot{m}_{da2} h_2}{\dot{m}_{da1} + \dot{m}_{da2}} \tag{4.31}$$

It is seen that the outlet properties are just the weighted averages of the inlet properties. Note that the outlet wet-bulb temperature must be found from psychrometric calculations; it is not necessarily the weighted average of the inlet wet-bulb temperatures.

If the inlet stream specific volumes are not too different, one can divide Eq. (4.29) by the average specific volume of the two inlet streams to replace the mass flow rate terms by corresponding volumetric flow rate terms (units of liters per second or cubic feet per minute are typical). This form of the equation is useful for preliminary design since volumetric flow rates are most often the basis of design calculations.

4.4.6 The Building Air Conditioning Process

The building air conditioning process consists of three steps (refer to Fig. 4.14):

1. Room air at condition 3 (for this discussion the fan temperature rise is ignored, so that states 3 and 4 are the same) is mixed with outdoor air required for ventilation at condition 0 to form air at state 5. Note that the mixed-air condition 5 is found by using the rule for mixed streams embodied in Eqs. (4.28) to (4.31).

2. This mixed air (state 5) enters the cooling coil, where it is cooled and dehumidified to condition 1.

3. The cooled dehumidified air is supplied to the space by way of the supply fan at state 2.

Figure 4.14a shows the process schematically, and Fig. 4.14b is a plot of the process on the psychrometric chart. In winter, the process operates similarly with heat addition instead of extraction from the mixed airstream.

The *sensible heat ratio* (SHR), or sensible heat factor, is used to analyze air conditioning processes. It is defined as the ratio of the sensible heat removed from an airstream to the total (latent + sensible) heat removed:

$$\text{SHR} \equiv \frac{\dot{Q}_{sens}}{\dot{Q}_{sens} + \dot{Q}_{lat}} \tag{4.32}$$

The protractor included with the psychrometric charts has an extra scale useful for cooling system calculations. The *inner* scale is the SHR scale. Its use will be illustrated by Example 4.9.

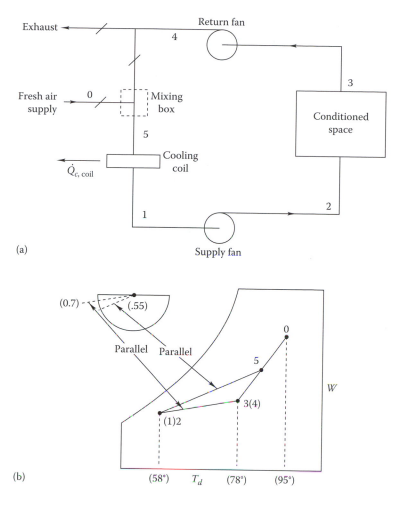

FIGURE 4.14

Building air conditioning process: (a) system schematic diagram showing state points, exhaust, and ventilation flows; (b) process plot on psychrometric chart (ignoring duct losses and fan air temperature rise) showing use of SHR protractor. (The numerical values in parentheses apply to Example 4.9.)

Example 4.9: Building Air Conditioning

A space is to be maintained at 78°F (25.5°C) and 50 percent relative humidity. The total cooling load (heat to be removed from the space to maintain comfort) is 120,000 Btu/h (35 kW) of which 70 percent is sensible heat. Ventilation air at 1000 ft³/min (472 L/s) is required on a day when the conditions are 95°F (35°C) and 55 percent relative humidity. What are the space air supply rate and the cooling coil rating?

Figure: See Fig. 4.14.

Assumptions: The location is at sea level. The supply air temperature to the space is 58°F (14°C), a typical value for commercial buildings in the United States. The duct heat transfer and the fan air temperature rise are ignored for simplicity.

Given: SHR = 0.70

$\dot{Q}_{c,space} = 120{,}000$ Btu/h

$\dot{V} = 1000$ ft^3/min $= 60{,}000$ ft^3/h

$\dot{T}_{d2} = 58°$F

Outdoor and indoor air conditions

Find: $\dot{Q}_{c,coil}, \dot{m}_{da2}$

Lookup values: $V_0 = 14.4$ ft^3/lb

SOLUTION

Space conditions: Line 2-3. The psychrometric chart (Fig. 4.6) will be used for a graphical solution. The conditioned space conditions lie on line 2-3 in Fig. 4.14. Condition 3 (78°F, 50 percent humidity) is already known from the problem statement. The slope of line 2-3 is determined from the inner scale of the protractor, as shown at the given SHR value of 0.7. The location of point 2 is determined by the fact that it lies on line 2-3 and has a known dry-bulb temperature of 58°F. This is sufficient information to find the space supply airflow rate.

The first law for this process is

$$\dot{Q}_{c,space} = \dot{m}_{da2}(h_3 - h_2)$$

Solving for the mass flow rate gives

$$\dot{m}_{da2} = \frac{\dot{Q}_{c,space}}{h_3 - h_2}$$

The enthalpies can be read from Fig. 4.6 at points 2 and 3. The values are

$$h_2 = 23 \text{ Btu/lb} \quad h_3 = 30 \text{ Btu/lb}$$

The flow rate is

$$\dot{m}_{da2} = \frac{120{,}000 \text{ Btu/h}}{(30 - 23) \text{ Btu/lb}} = 17{,}140 \text{ lb/h } (3770 \text{ ft}^3/\text{min}, 1780 \text{ L/s})$$

Cooling coil: Line 1-5. State 5 must be found to specify this line segment. Since the endpoints of the line are known, the inverse relation between mass flows and line segment lengths [Eq. (4.28)] is used to locate 5. The outdoor mass flow is

$$\dot{m}_{da0} = \frac{\dot{V}_0}{v_0} = \frac{60{,}000 \text{ ft}^3/\text{h}}{14.4 \text{ ft}^3/\text{lb}} = 4170 \text{ lb}_{da}/\text{h}$$

The ratio of air mass flows is $4170/17{,}140 = 0.243$. Therefore, point 5 is located 24.3 percent of the distance from point 3 along line segment 3-0. The properties at point 5 can be read from the psychrometric chart. They are 82°F dry-bulb and 69°F wet-bulb temperatures as well as 33.4 Btu/lb$_m$ moist-air enthalpy. The dry-bulb temperature and enthalpy could have been calculated by using the weighted average rule; e.g., the mixed-air temperature is $0.243 \times 95°$F $+ (1 - 0.243)(78°$F$) = 82°$F, as read from the chart.

Line 1-5 can now be constructed by connecting points 5 and 1. The slope of the resulting line is transposed to the protractor, and the *coil's sensible heat ratio* SHR$_{coil}$ can be read off as 0.55. This quantity is of importance in cooling coil selection.

Finally, the coil heat removal rate ("coil cooling load") can be found.

$$\dot{Q}_{c,coil} = \dot{m}_{da2}(h_5 - h_1)$$
$$= (17{,}140) \text{ lb/h}[(33.4 - 23) \text{ Btu/lb}] = 178{,}300 \text{ Btu/h (52.2 kW)}$$

COMMENTS

Peak-condition coil information (flow, SHR, and heat rate) is supplied to manufacturers for coil costing, selection, and fabrication purposes. Note that the coil and zone cooling loads are quite different because the coil must remove zone heat and humidity gains along with the latent and sensible loads imposed by the need for outdoor air for ventilation.

4.5 Thermal Comfort

The next two sections deal with comfort and health. Comfort and health depend on many complex and interrelated phenomena, involving objective conditions as well as subjective perception. However, the HVAC engineer, being neither psychologist nor medical doctor, needs simple objective design criteria that will ensure acceptability by the vast majority of occupants. Here the primary concerns of the HVAC engineer are thermal comfort and air quality, and we will present the appropriate design criteria. This is an important subject: The trend of the past suggests that people will demand increasingly higher standards of comfort and health.

4.5.1 Thermal Balance of Body and Effective Temperature

The physical basis of comfort lies in the thermal balance of the body. The heat produced by the body's metabolism must be dissipated to the environment; otherwise, the body would overheat. Roughly speaking, if the rate of heat transfer is higher than the rate of heat production, the body cools down and we feel cold; if the rate is lower, we feel hot.

This is a complex problem in transient heat transfer, involving radiation, convection, conduction and evaporation, and many variables, from the wetness of the skin to the composition of the clothing. In this book we discuss only the simplest aspects of the problem to provide a basis for understanding the ASHRAE comfort recommendations.

The total energy production rate of the body is the sum of the production rates of heat $\dot{Q}$ and of work $\dot{W}$ and can be written in the form

$$\dot{Q} + \dot{W} = MA_{sk} \tag{4.33}$$

where
  A_{sk} = total surface area of skin
  M = rate of metabolic energy production per surface area, customarily expressed in units of

$$1 \text{ met} = 18.4 \text{ Btu}/(\text{h} \cdot \text{ft}^2)(58.2 \text{ W/m}^2) \tag{4.34}$$

Metabolic rates for various activities are shown in Table 4.2. Metabolic rate M is defined to include the production of work, whereas only the heat production rate matters for

TABLE 4.2

Metabolic Rate M for Various Activities

Activity	met	W/m^2	Btu/(h · ft^2)
Reclining	0.8	46.6	14.8
Seated, quiet	1.0	58.2	18.4
Standing, relaxed	1.2	69.8	22.1
Sedentary activity (office, dwelling, lab, school)	1.2	69.8	22.1
Light activity, standing (shopping, lab, light industry)	1.6	93.1	29.5
Medium activity, standing (shop assistant, domestic work, machine work)	2.0	114.4	36.9
Heavy activity, if sustained (heavy machine work, garage work)	3.0	174.6	55.3

Source: Courtesy of ASHRAE, *Standard 55-1992: Thermal Environmental Conditions for Human Occupancy*, American Society of Heating, Refrigerating and Air-Conditioning Engineers, Atlanta, GA, 1992. With permission.

thermal comfort. In practice, the difference is entirely negligible compared to overall uncertainties, since work represents at most a few percent of M for normal indoor activities. As the area A_{sk} is on the order of 16 to 22 ft^2 (1.5 to 2 m^2) for an adult, this implies heat production rates on the order of 340 Btu/h (100 W) for typical indoor activities.

To analyze the dissipation of the heat $\dot{Q}$ to the environment, we neglect storage effects because the temperature of the interior of the body is remarkably constant, between 36 and 37°C under normal conditions, maintained by controlling the perspiration rate and the flow of blood to the outer regions of the body. Thus $\dot{Q}$ can be set equal to the instantaneous heat flow to the environment. It is convenient to distinguish several major heat transfer modes by writing

$$\dot{Q} = \dot{Q}_{con} + \dot{Q}_{rad} + \dot{Q}_{evap} + \dot{Q}_{res,sen} + \dot{Q}_{res,lat} \tag{4.35}$$

where the first three terms refer to the skin (convection, radiation, and evaporation) and the last two terms to respiration (sensible and latent), as indicated by the subscripts.

Let us begin with the convective transfer $\dot{Q}_{con}$ from the skin, expressing it in terms of convection coefficient h_{con} and surface area A_{cl} and surface temperature T_{cl} of the clothed body

$$\dot{Q}_{con} = A_{cl}h_{con}(T_{cl} - T_a) \tag{4.36}$$

In this equation h_{con} and T_{cl} are averaged over the body surface (skin or clothing) that is in contact with the air at dry-bulb temperature T_a.

The radiative heat loss can be a bit more complicated because different surfaces of the environment may have significantly different temperatures, a situation that occurs, for instance, during cold weather in a room with single-glazed exterior windows and warm interior walls. To simplify the analysis, it is convenient to define a *mean radiant temperature T_r* of the environment as the temperature of an imaginary isothermal enclosure with which a human body would exchange the same radiation as with the actual environment. The emissivities of most indoor surfaces are around 0.9, sufficiently high that one can calculate T_r as if all surfaces were black. In that case the radiative heat loss per unit area of the body is

$$\sigma(T_{cl}^4 - T_r^4) = \sigma \sum_n F_{cl-n}(T_{cl}^4 - T_n^4) \tag{4.37}$$

where the sum runs over all surfaces with which the body can exchange direct radiation, F_{cl-n} is the radiation shape factor from the body to the nth surrounding surface, and σ is the Stefan-Boltzmann constant. Since the sum of the shape factors over an enclosure is unity, the T_{cl}^4 terms drop out of this equation and it follows that T_r is given by

$$T_r^4 = \sum_n F_{cl-n} T_n^4 \tag{4.38}$$

Inside buildings the temperature differences are often small enough that T_r can be approximated by

$$T_r \approx \sum_n F_{cl-n} T_n \tag{4.39}$$

For example, if one-half of the surroundings is at 30°C and the other half at 20°C, the difference between these two expressions is only 0.1°C. In terms of the mean radiant temperature T_r, the radiative heat loss of the body can then be written as

$$\dot{Q}_{rad} = A_{cl} h_{rad} (T_{cl} - T_r) \tag{4.40}$$

where h_{rad} is the radiative heat transfer coefficient.

If one defines a total heat transfer coefficient

$$h_{c+r} = h_{con} + h_{rad} \tag{4.41}$$

and a so-called *operative temperature* T_{op}

$$T_{op} = \frac{h_{con} T_a + h_{rad} T_r}{h_{con} + h_{rad}} \tag{4.42}$$

then one can write the sum of Eqs. (4.36) and (4.40) in the form

$$\dot{Q}_{con} + \dot{Q}_{rad} = A_{cl} h_{c+r} (T_{cl} - T_{op}) \tag{4.43}$$

Often T_{op} is close to the simple average of T_a and T_r since h_{con} and h_{rad} are not very different indoors. The point of this exercise is to obtain a single temperature index T_{op} for characterizing an environment that really depends on two separate temperatures T_a and T_r: any combination of T_a and T_r with the same T_{op} has the same heat transfer $\dot{Q}_{con} + \dot{Q}_{red}$.

For a more accurate analysis, taking into account the dependence of the convective heat transfer coefficient h_{con} on the velocity v of the air surrounding a person, one can use the occupant convective heat transfer coefficients in the appendices which list h_{con} as a function of v.

Example 4.10

Consider a room 3 m × 3 m × 3 m ($\approx$10 ft × 10 ft × 10 ft) all but one of whose surfaces are at 20°C (68°F) while the remaining 3 m × 3 m surface is a window at 10°C; the dry-bulb temperature is

$T_a = 21°C$. Find the mean radiant temperature and the operative temperature for a person at the center of the room.

Given: Radiation temperatures of surfaces and T_a

Find: T_{op}

Assumption: $h_{rad} = h_{con}$

Lookup value: We need the shape factor F from the center of the room to the window. It could be determined from formulas or graphs in books on radiative heat transfer (see references of Chap. 2). For the determination of mean radiant temperatures in buildings, it will usually be sufficient to evaluate the solid angle subtended by the surface in question as seen from the center of the body, and to approximate the shape factor by this solid angle after dividing by 4π, the solid angle of a complete enclosure. In the present case the solid angle is easy to determine because of the symmetry of the room: seen from the center the window fills one-sixth of the total field of view, and therefore $F = \frac{1}{6} = 0.167$.

SOLUTION

In this case Eq. (4.39) for the average radiation temperature T_r has only two terms because there are only two different surface temperatures, with corresponding shape factors F and $1 - F$. The result is

$$T_r = F \times 10°C + (1 - F) \times 20°C = 18.3°C$$

Next the operative temperature is easy to find from Eq. (4.42). It is the simple average of T_a and T_r because of $h_{rad} = h_{con}$.

$$T_{op} = \frac{21°C + 18.3°C}{2} = 19.7°C$$

In like manner one can include the evaporative skin loss $\dot{Q}_{evap}$, to obtain a temperature index that accounts for humidity. We will not discuss the details here, and we mention only that one can define a quantity, called *adiabatic equivalent temperature*, which is a linear combination of T_{op} and of the vapor pressure of the air and which depends on skin wetness and on clothing permeability. This quantity achieves the goal of combining the three characteristics of the environment (radiation temperature, dry-bulb air temperature, and humidity) in a single temperature index that completely determines the total heat loss from the skin. It is used as basis of the ASHRAE comfort chart because any combination of humidity, dry-bulb temperature, and surrounding surface temperatures will cause the same heat loss from the body if the adiabatic equivalent temperature is the same.

In passing we mention another temperature index, the *effective temperature* ET*. Like the adiabatic equivalent temperature, it is a linear combination of T_{op} and of the vapor pressure, and it is often used in the analysis of thermal comfort. Effective temperature ET* is the temperature of an isothermal black enclosure with 50 percent relative humidity where the body surface would experience the same heat loss as in the actual space. The 50 percent level is chosen as the reference because it is the most common and the most desirable level of humidity.

4.5.2 The Perception of Comfort

The ASHRAE comfort recommendations attempt to define objective conditions that will be found satisfactory by most people most of the time. There are, however, subjective factors to the perception of comfort that can complicate matters considerably. For best success the designer should be aware of these factors.

It is interesting to look at the distribution of responses of a large group of individuals when exposed to the same thermal environment. They can be asked to vote according the ASHRAE thermal sensation scale:

+3   hot
+2   warm
+1   slightly warm
 0   neutral
−1   slightly cool
−2   cool
−3   cold.

The distribution of votes will always show considerable scatter, no matter what the thermostat setting. A useful index of the acceptability of an environment is the *percent of people dissatisfied* (PPD), defined as people voting outside the range −1 to +1. When PPD is plotted versus the mean vote of a large group, one typically finds a distribution like Fig. 4.15. This graph shows that even under optimal conditions (i.e., mean vote = 0) approximately 5 percent are dissatisfied with the thermal environment.

Sometimes there are objective reasons for different preferences. For example, some dependence of temperature preferences on sex and age has been observed: Women tend to prefer somewhat higher temperatures, and so do adults over 60. Whether that is due to biological differences or to differences in clothing or activity is of little concern to the building designer because he or she has no influence over either of these factors. It is best to design the building and the HVAC system to allow the occupants to obtain their preferred conditions.

Often the situation is further complicated by the subtle relationships between objective conditions and subjective perception of comfort. The recent literature highlights the importance of subjective variables such as satisfaction with one's work, aesthetic aspects of one's work environment, the quality of the lighting, colors, views, noise, and odors.

For instance, the degree of control over one's environment is crucial for understanding the difference between comfort perceptions at home and at work. At home most people can freely adjust thermostat, windows, and lighting levels as they please. Being in control,

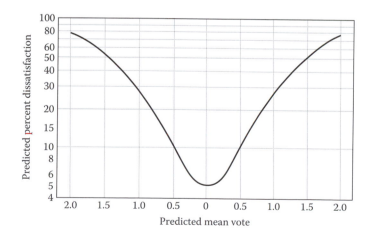

FIGURE 4.15

Percentage of people dissatisfied as a function of mean vote. (Courtesy of ASHRAE, *Standard 55-1992: Thermal Environmental Conditions for Human Occupancy*, American Society of Heating, Refrigerating and Air-Conditioning Engineers, Atlanta, GA, 1992. With permission.)

they are likely to be satisfied, even at times when objective conditions are outside the conventional comfort zone. Owners of a passive solar home are a good example. By and large, people who choose such a home want to live in tune with nature and minimize the consumption of nonrenewable energy. Their willingness to accept indoor temperature fluctuations may appear unbelievable to designers who are accustomed to the stringent temperature tolerances of most office buildings.

At work, by contrast, many people do not have much control over their physical environment: The thermostat settings are determined by others (the boss or coworkers), and ventilation cannot be adjusted independently. This is particularly true with the open office space design. Under the same thermal environment, people at work are more likely to be dissatisfied than people at home.

An interesting case in point is the possibility of opening the windows, something that is usually missing in commercial buildings in the United States. Some people feel a certain amount of claustrophobia if they cannot open their windows; they are more likely to find the air stuffy than people who can open their windows—even if the windows actually remain shut and the physical conditions are the same. In fact, people often keep the windows shut anyway, as protection from street noise, dust, or draft. More important than the actual opening is the freedom to choose.

These considerations suggest that it may be worth designing a building in a way that allows the occupants a certain amount of individual control over their environment, even if the initial cost is increased by extra thermostats, adjustable vents, light switches, and operable windows. Comfort and health are essential for productivity in the modern work environment.

Even though we do not discuss acoustic comfort, we emphasize its importance because the HVAC engineer can make an important contribution by choosing quiet equipment and adding sufficient sound insulation. Otherwise the hapless occupants may suffer buzzing air vents, humming motors, or droning compressors.

4.5.3 The ASHRAE Comfort Chart

The purpose of ASHRAE Standard 55-1992 is to specify a thermal environment that is acceptable to at least 80 percent of the occupants. Table 4.3 shows the optimal operative temperature and the acceptable range for light sedentary activity at 50 percent relative humidity and at mean airspeed ≤ 0.15 m/s (30 ft/min), for the indicated levels of clothing insulation. A graphic presentation is shown in Fig. 4.16. Clothing insulation is customarily measured in units of *clo*, defined as

TABLE 4.3

Optimal Operative Temperature and Acceptable Range for Light Sedentary Activity at 50% Relative Humidity and at Mean Airspeed ≤ 0.15 m/s (30 ft/min)

Season	Typical Clothing	I_{cl}[clo]	Optimum Operative Temperature	Acceptable Range
Winter	Heavy slacks, long-sleeve shirt, and sweater	0.9	22°C	20–23.5°C
			71°F	68–75°F
Summer	Light slacks and short-sleeve shirt	0.5	24.5°C	23–26°C
			76°F	73–79°F
	Minimal	0.05	27°C	26–29°C
			81°F	79–84°F

Source: Courtesy of ASHRAE, *Standard 55-1992: Thermal Environmental Conditions for Human Occupancy*, American Society of Heating, Refrigerating and Air-Conditioning Engineers, Atlanta, GA, 1992. With permission.

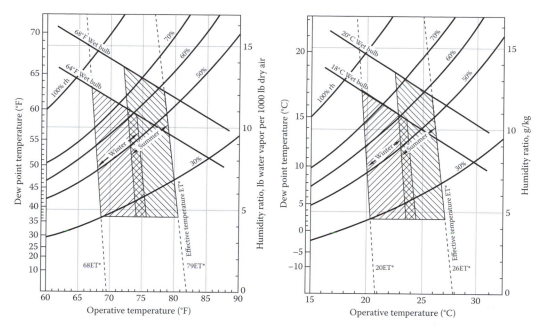

FIGURE 4.16

Acceptable ranges of operative temperature and humidity, for sedentary activity and typical summer and winter clothing. (Courtesy of ASHRAE, *Standard 55-1992: Thermal Environmental Conditions for Human Occupancy*, American Society of Heating, Refrigerating and Air-Conditioning Engineers, Atlanta, GA, 1992. With permission.)

$$1 \text{ clo} = 0.88 \ (\text{ft}^2 \cdot \text{h} \cdot \text{F})/\text{Btu} \ \left[0.155 \ (\text{m}^2 \cdot \text{K})/\text{W}\right] \tag{4.44}$$

Some examples of clo values are listed in Table 4.4. The effect of clothing insulation on the ASHRAE comfort recommendations is shown in Fig. 4.17. However, at lower temperatures the perception of comfort depends also on maintaining sufficiently uniform insulation over the body, in particular the hands and feet. For sedentary occupancy of more than an hour, the minimum operative temperature should not be below 18°C (65°F).

When the activity level is above 1.2 met, ASHRAE Standard 55-1992 recommends that the criteria for operative temperature be reduced according to

$$T_{\text{op, active}} = T_{\text{op, sedentary}} - (1.0 + \text{clo})(M - 1.2) \times 5.4°\text{F} \tag{4.45US}$$

$$T_{\text{op, active}} = T_{\text{op, sedentary}} - (1.0 + \text{clo})(M - 1.2) \times 3.0 \text{ K} \tag{4.45SI}$$

TABLE 4.4

Clothing Insulation, clo

Clothing	Insulation, clo
Underwear (briefs, or bra plus panties)	0.05
Summer ensemble (lightweight slacks or skirt, short-sleeve shirt or blouse, plus accessories)	0.35–0.6
Winter ensemble (heavy slacks or skirt, long-sleeve shirt or blouse, warm sweater or jacket, plus accessories	0.8–1.2

1 clo = 0.88 ft^2 · h · °F/Btu = 0.155 m^2 · K/W.

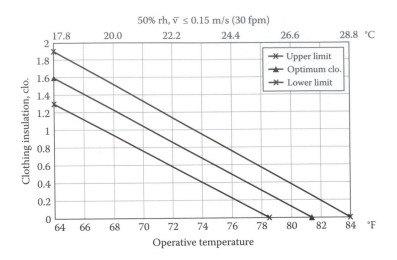

FIGURE 4.17
Clothing insulation necessary to be within the ASHRAE 80 percent acceptability limits as a function of operative temperature, during light sedentary activity (≤1.2 met). (Courtesy of ASHRAE, *Standard 55-1992: Thermal Environmental Conditions for Human Occupancy*, American Society of Heating, Refrigerating and Air-Conditioning Engineers, Atlanta, GA, 1992. With permission.)

where clo = clothing insulation, clo; M = activity level, met; and $T_{op,sedentary}$ designates the value at 1.2 met. Recommended operative temperatures for active people are shown in Fig. 4.18 as a function of metabolic rate.

The ASHRAE comfort chart in Fig. 4.16 indicates the acceptable ranges of operative temperature and humidity during light sedentary activity, assuming typical summer or winter clothing, respectively. This chart is similar to the psychrometric chart, but the abscissa is the operative temperature T_{op} rather than the dry-bulb air temperature T_a. The ordinate is the humidity ratio W, and the dew point temperature is indicated on the left-hand scale. The relative-humidity lines of the psychrometric chart that have been superimposed on Fig. 4.16 are, strictly speaking, correct only if the radiation temperature is equal to the dry-bulb air temperature (in which case $T_{op} = T_a$). The acceptable values of

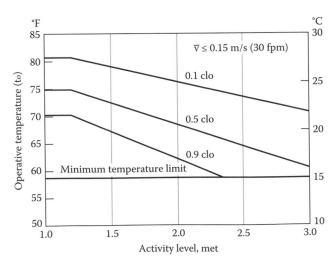

FIGURE 4.18
Recommended operative temperatures for active people, as a function of metabolic rate. (Courtesy of ASHRAE, *Standard 55-1992: Thermal Environmental Conditions for Human Occupancy*, American Society of Heating, Refrigerating and Air-Conditioning Engineers, Atlanta, GA, 1992. With permission.)

T_a and W are indicated by the shaded zones. The chart shows different zones for winter and for summer because comfort depends on the insulation value of the clothing.

The left and right boundaries of the shaded zones are lines of constant adiabatic equivalent temperature, hence lines of constant heat loss. They are sloped from upper left to lower right because the evaporative heat loss from the body decreases as the humidity ratio W of the air increases. This agrees with the common experience that hot, humid weather is less comfortable than hot, dry weather at the same T_a.

The upper and lower humidity limits in Fig. 4.16 have been chosen on the basis of issues such as dry skin, eye irritation, respiratory health, and microbial growth. One should also avoid the condensation of moisture on building surfaces.

Example 4.11

Suppose the room of Example 4.10 has a humidity ratio $W=0.005$. Do the conditions at the center of the room satisfy the ASHRAE comfort criteria?

Given: $T_{op}=19.7°C$ and $W=0.005$; comfort chart of Fig. 4.16

Find: Are these conditions within comfort zone?

SOLUTION

The humidity ratio does satisfy the ASHRAE criterion, even though only barely. At $W=0.005$ the minimum T_{op} of the comfort chart is 20.3°C; the conditions of this example are slightly too cold, but the difference is not significant since the boundaries of the comfort zone are in any case not absolutely sharp.

COMMENT

To improve the comfort, one could obviously increase one or several of the temperature components of T_{op}. But an interesting alternative is to raise the humidity ratio W to 0.011.

The boundaries of the ASHRAE comfort zone were chosen to ensure that at least 80 percent of all occupants will find the environment thermally acceptable. The further one strays from these values, the greater the PPD (percent of people dissatisfied). Expectations and recommendations for comfortable temperatures have evolved considerably since the beginning of the century, and only during the last two decades do they seem to have approached saturation (in the sense of preference if cost were no concern).

There are several additional physical variables that can affect comfort. Air movement plays a role because the convective heat transfer from the body depends on air velocity. An excess may be perceived as *draft*, and a lack as stuffiness. The recommendations for airspeed depend on temperature and humidity. As long as the latter are within the comfort zone of Fig. 4.16, there is no minimum requirement for airspeed. In hot weather the upper range of temperatures could be extended if the airspeed is increased. Figure 4.19 indicates how much the air temperature can be increased above the summer comfort zone by increasing the airspeed. However, the acceptability of increased airspeed depends on the control the occupants have; also, one should keep in mind that beyond 160 ft/min (0.8 m/s) loose paper and other light objects may be blown away.

In a cold environment, high air movement would cause draft. Draft is a complex phenomenon that depends, among other factors, on turbulence (i.e., rms variation of airspeed). Sensitivity to draft is greatest where the skin is exposed at the head and ankles. Figure 4.20

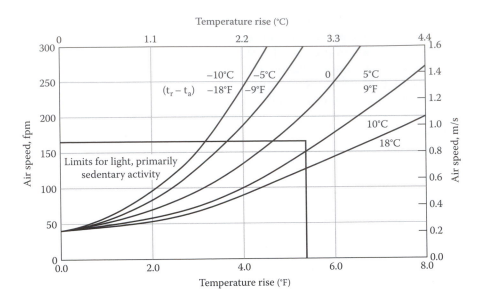

FIGURE 4.19
Airspeed required to increase the air temperature above the summer comfort zone. (Courtesy of ASHRAE, *Standard 55-1992: Thermal Environmental Conditions for Human Occupancy*, American Society of Heating, Refrigerating and Air-Conditioning Engineers, Atlanta, GA, 1992. With permission.)

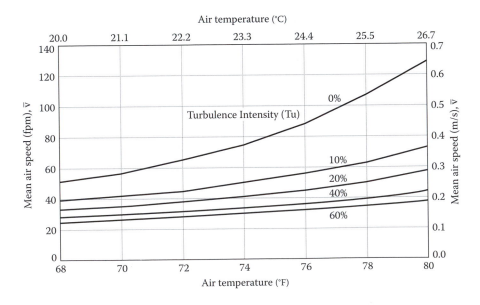

FIGURE 4.20
Recommended limits for mean airspeed to reduce the risk of draft, as a function of air temperature and of turbulence intensity. (Courtesy of ASHRAE, *Standard 55-1992: Thermal Environmental Conditions for Human Occupancy*, American Society of Heating, Refrigerating and Air-Conditioning Engineers, Atlanta, GA, 1992. With permission.)

shows the recommended limits for mean airspeed to reduce the risk of draft, as a function of air temperature and of turbulence intensity. The latter is defined as the ratio of the standard deviation to the average of the airspeed. In conventionally ventilated spaces the turbulence intensity is usually in the range of 30 to 60 percent. In rooms with displacement ventilation or without ventilation it may be lower.

Inhomogeneities of temperature should not be too large. For instance, the vertical air temperature difference should not exceed 5°F (3 K) between the levels of 4 in (0.1 m) and 67 in (1.7 m) above the floor. Likewise, large temperature drifts should be avoided. Specific recommendations for that and other details can be found in ASHRAE Standard 55-1992.

4.6 Air Quality and Ventilation

4.6.1 Basic Considerations

Air quality is essential, not just for comfort but also for health and productivity. Since the 1970s concern over air quality has risen, in part because buildings have become tighter and ventilation rates have been reduced to conserve energy. Also there has been recognition of the importance of indoor pollution, both natural (radon) and artificial (e.g., formaldehyde emanating from furnishings and carpets). *Sick building* has entered our vocabulary.

The causes of the sick building syndrome are often difficult to pin down completely. Simple objective parameters such as air exchange rates or formaldehyde concentrations are relatively easy to measure, but in some cases there may be complex problems involving odors or dust with bacteria.

To help ensure that indoor environments are healthy, ASHRAE has devised Standard 62-1999, *Ventilation for Acceptable Indoor Air Quality.* Ventilation with outdoor air is of course essential for supplying "fresh air," and usually a well-designed ventilation system, in conjunction with filtering, will be sufficient. But an understanding of a few basic issues is advisable, to forestall problems that would be expensive to correct by a retrofit.

ASHRAE Standard 62-1999 defines indoor air quality in terms of upper limits for the concentrations of air pollutants, called *contaminants* by ASHRAE. For the most important contaminants these limits are listed in Table 4.5. The limits in part *a* of this table are equal to the outdoor limits prescribed by the Environmental Protection Agency. In addition, ASHRAE has established guidelines for several indoor pollutants, as shown in part *b* of Table 4.5.

The indoor concentration of a contaminant is equal to the one outdoors if a building has no sources or sinks. One type of sink is, of course, a filter or cleaning device that removes a contaminant before it enters a building. For some contaminants absorption by the surfaces in a building can also be a significant sink. Here is a brief description of the most important contaminants.

Carbon Dioxide (CO_2) Carbon dioxide is exhaled as a by-product of metabolism, making the concentration in buildings higher than that outdoors. It has no harmful effects on the human body, and the concentrations in buildings rarely reach levels high enough to have an appreciable impact on the availability of oxygen needed for breathing. But since elevated concentrations of CO_2 are correlated with human bioeffluents and odors, CO_2 is a convenient proxy indicator for the latter. Standard 62-1999 recommends that the indoor concentration of CO_2 be not more than 700 ppm above the outdoor concentration, the latter being in the range of 300 to 500 ppm.

TABLE 4.5

Guidelines for Concentrations of Air Contaminants. From ASHRAE Standard 62-1999

Contaminant	Long Term Concentration			Short Term Concentration		
	$\mu g/m^3$	ppm	Averaging Period	$\mu g/m^3$	ppm	Averaging Period
a. National primary ambient air quality standards for outdoor air as set by the U.S. Environmental Protection Agency						
Sulfur dioxide	80	0.03	1 yr	365	0.14	24 h
Total particles	50		1 yr	150		24 h
Carbon monoxide				10,000	9	8 h
Oxidants (ozone)				235	0.12	1 h
Nitrogen dioxide	100	0.055	1 yr			

b. Contaminants of indoor origin

Contaminant	Concentration	Exposure Time
Carbon dioxide	Not more than 700 ppm above outdoor concentration (typically 300 to 500 ppm)	Continuous
Ozone	100 $\mu g/m^3$ (0.05 ppm)	Continuous
Radon	4 pCi/L (148 Bq/m^3)	Annual average

Carbon Monoxide (CO)   Carbon monoxide, the result of incomplete combustion, is toxic because it binds preferentially to hemoglobin, thereby blocking the uptake of oxygen by the blood. Significant quantities can be emitted by motor vehicles, especially those without properly functioning catalytic converter, and by poorly regulated boilers and furnaces. Obviously, buildings should be designed in such a way that exhaust from such sources cannot enter; e.g., a loading dock must never be placed next to an air intake of the ventilation system.

Sulfur Dioxide (SO$_2$)   Sulfur dioxide is a gas, produced by the combustion of fuels that contain sulfur, in particular, coal and oil. It is an irritant that can cause or aggravate various respiratory problems, both directly and after transformation to sulfates, especially sulfuric acid (the main constituent of acid rain), by chemical reactions in the atmosphere. There are no significant sources of SO$_2$ in buildings.

Nitrous Oxides (NO$_x$)   Nitrous oxides NO$_x$ designate an unspecified mixture of NO and NO$_2$. Most high-temperature combustion processes produce some NO which is rapidly oxidized to NO$_2$ in the atmosphere. Both NO and NO$_2$ are quite reactive, with complicated chemical reactions that lead to the production of HNO$_3$ (another constituent of acid rain) and of O$_3$. Both NO$_x$ and the secondary pollutants created by it are widely believed to cause or aggravate respiratory problems. There are few sources of NO$_x$ in buildings, except open flames, for instance, in gas stoves.

Ozone (O$_3$)   Ozone is a strong oxidant, and exposure to ozone is harmful to most living organisms (however, ozone in the stratosphere is beneficial because it protects us from harmful UV radiation). Ozone air pollution results from the combination of volatile organic compounds (VOCs) and NO$_x$ in the presence of light, often called photochemical smog. Whereas this process is negligible inside buildings, there is some direct production of ozone by copiers, laser printers, and some deodorizers.

VOCs   Most VOCs have no direct health impacts, but they are undesirable as a precursor of O$_3$. Among VOCs with direct health impacts there is formaldehyde as well as certain aromatic compounds (e.g., benzene) which are carcinogenic. Formaldehyde irritates eyes and mucous membranes and can cause respiratory problems, in addition to being a carcinogen. It is used in the manufacture of many materials for buildings, e.g., carpets, insulation, pressed board, and paper products.

Particulate Matter   Particulate matter (PM) often has a subscript indicating the maximum diameter, in micrometers, of the particles that have been counted. Suspended particles in the air, also known as dust, have been increasingly recognized as a health hazard. Dust is a complex mixture of particles of inorganic and organic origin, often contaminated by unhealthy bacteria. There are many different sources, including natural soil, sea salt, pollen, bacteria, and fragments of dead skin. The most harmful particles are emitted by combustion processes, including tobacco smoke. In addition, there are secondary particles, in particular sulfates and nitrates, that are created by oxidation of SO_2 and NO_x in the atmosphere. Characteristics of particulate matter are indicated in Fig. 4.21, together with cleaning equipment suitable for removing it. Particles larger than 10 μm are of little

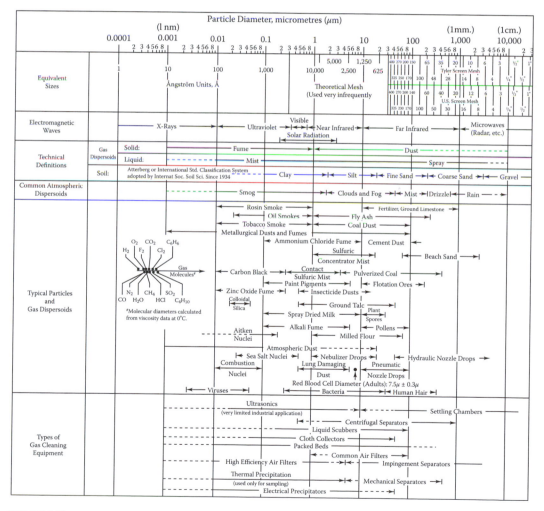

FIGURE 4.21

Characteristics of particulate matter and cleaning equipment suitable for removing it. (Courtesy of ASHRAE, *Standard 62-1999. Ventilation for Acceptable Indoor Air Quality*, American Society of Heating, Refrigerating and Air-Conditioning Engineers, Atlanta, GA, 1999. With permission.)

concern for health because they get filtered by the upper respiratory tract. Air quality regulations are concerned with PM_{10} (*inhalable particles*) and more recently with $PM_{2.5}$ because the latter are small enough to enter the deepest recesses of the lungs.

Radon   Radon is a noble gas, i.e., chemically inert, but it can cause lung cancer because it is radioactive, with a half-life of about 4 days. It is continuously emitted by the decay of naturally occurring radioactive elements in certain geologic formations. If it enters through the foundations into a building, the concentration can reach levels that are considered undesirable. In regions with high levels of radon, it is advisable to reduce the penetration of radon into a building by, e.g., making the foundations sufficiently airtight.

Tobacco Smoke   Smoking is a singularly effective means of delivering a host of dangerous contaminants (carbonaceous particles, CO, benzene, nicotine, etc.) into the lungs not only of smokers but also of nonsmokers who share the same space. A pack of cigarettes a day reduces life expectancy by about 8 years, to say nothing of the degraded quality of life due to the associated cancers and cardiopulmonary diseases (Doll et al., 1994). The health impacts of passive smoking are also significant. For the comfort of nonsmokers and for the health of everybody, there is a trend, at least in North America and the EU, to restrict smoking in buildings or confine it to limited areas.

There are four approaches to controlling the air quality in buildings:

- Use of outdoor air
- Control of contaminants at the source
- Air distribution in the space
- Air cleaning

If control of contaminants at the source is feasible, it is preferable because it avoids the energy consumption for conditioning extra outdoor air. There are many examples. One can choose building materials with minimal outgassing of contaminants. Air from smoking areas should be exhausted directly to the outside rather than being returned to the air distribution system. Local exhaust fans can remove local contaminants before they get mixed with the return air, for instance, from stoves or from rooms where toxic chemicals are stored such as paints, cleaners, or insecticides. However, care should be taken to make sure the contaminants are actually removed by such exhaust fans: Motion of air is not easily controlled by suction alone.

4.6.2 Ventilation

The effect of ventilation depends on the airflow patterns inside the building. The situation is simple if one can assume the indoor air to be uniformly mixed at all times. In most buildings there is a fair amount of natural and/or forced convection that mixes the air, and the well-mixed model is often an acceptable approximation. It is also intuitive and simple to analyze. Consider what happens when there is a source of air pollution in a well-mixed building which is ventilated with outdoor air at a rate $\dot{V}_o$. The source can be characterized in terms of the rate N_{pol} at which it adds contaminants to the space. If the outdoor concentration C_o of the contaminant were zero, the indoor concentration C_i would be the ratio $N_{pol}/\dot{V}_o$. If C_o is not zero, the term $N_{pol}/\dot{V}_o$ represents the increase $C_i - C_o$ due to the source. Therefore the *well-mixed model* can be written as

$$C_i = C_o + \frac{N_{pol}}{\dot{V}_o} \tag{4.46}$$

The concentrations have units of mass per volume, e.g., $\mu g/m^3$, if N_{pol} is in units of g/s; and the concentrations are dimensionless, e.g., ppm, if N_{pol} is in units of volume per second. Conversely, if source strength N_{pol} and outdoor concentration C_o are given, one can solve this equation for the ventilation rate $\dot{V}_o$ that is necessary to keep the indoor concentration from exceeding a level C_i.

Example 4.12

The CO_2 production at an activity level of 1.2 met is $N_{CO_2} = 0.005$ L/s. What outdoor airflow $\dot{V}_o$ per occupant is needed at this activity level if the CO_2 concentration is not to exceed $C_i = 1050$ ppm? The outdoor concentration is $C_o = 350$ ppm.

Given: $N_{CO_2} = 0.005$ L/s.

$C_i = 1050$ ppm

$C_o = 350$ ppm

$C_i - C_o = 700$ ppm as per ASHRAE Standard 62-1999

Find: $\dot{V}_o$

Assumptions: Perfect mixing

SOLUTION

Insert $\dot{V}_{CO_2}$ for N_{pol} in Eq. (4.46) and solve for $\dot{V}_o$. Inserting numbers, we find

$$\dot{V}_o = \frac{0.005 \text{ L/s}}{0.00105 - 0.00035} = 7.1 \text{ L/s } (15.1 \text{ ft}^3/\text{min})$$

Typical ventilation systems in commercial buildings recirculate air at a high rate, the recirculation rate being several times higher than the rate at which fresh air is taken in. Also the air coming out of the air diffusers has sufficiently high velocities to cause much local mixing. As a result, the air in buildings with mechanical ventilation systems tends to be fairly well mixed. Nonetheless there may be zones with relatively stagnant air, for instance, between partitions in open-plan offices. In such zones the local airflow rate is, in effect, lower than the building average, and the resulting local concentration of contaminants is higher. Insufficient mixing can cause local air quality problems.

Before proceeding to guidelines for ventilation, we mention one other flow situation that is easy to analyze. It is called plug flow because the fluid moves like a plug through the space.

Example 4.13

What is the average air velocity if ventilation air is made to flow like a plug from floor to ceiling? Assume that the height of the room is 3.0 m ($\approx$10 ft) and the ventilation rate (total of outdoor air and recirculated air) can be expressed as $\dot{V}/V = 3.0$ air changes per hour.

Given: $h = 3.0$ m and $\dot{V}/V = 3.0$ h^{-1}

Find: v of air

Assumption: Plug flow, from bottom to top

SOLUTION

Consider a room of floor area A. The flow rate $\dot{V}$ is related to velocity v and area A by

$$\dot{V} = vA$$

The volume is

$$V = Ah.$$

Combining the last two equations and solving for v, we find

$$v = \frac{h\dot{V}}{V} = 3.0 \text{ m} \times 3.0 \text{ h}^{-1} = 9.0 \text{ m/h} = 0.0025 \text{ m/s } (0.0082 \text{ ft/s})$$

in the direction from bottom to top.

COMMENTS

1. Conservation of mass implies that this is the lowest possible airspeed that achieves the specified ventilation rate.
2. There are ventilation systems (called displacement ventilation) that achieve this kind of plug flow. They are used in some buildings, especially in Sweden.
3. To achieve well-mixed flow, the velocity must be much larger than this minimum (of course, to maintain comfort it should not exceed the upper limits mentioned in Sec. 4.5.3). Data for air diffusers and the resulting air velocities in rooms will be presented in Chap. 11.

ASHRAE Standard 62-1999, *Ventilation for Acceptable Indoor Air Quality*, offers two paths toward compliance: the ventilation rate procedure and the indoor air quality procedure. The ventilation rate procedure is simpler and will usually be chosen for conventional designs. It prescribes the minimal outdoor airflow rate that must be delivered to a space, for a large number of building types and uses. We have reproduced in Table 4.6 the minimal outdoor airflow rates for the most common applications. For sites where the outdoor air is not of sufficient quality, treatment of the outdoor air is also prescribed by Standard 62-1999.

The second path involves verifying that the air in the building will actually have the required air quality. It offers the designer much greater flexibility, because any method can be used for treating the air, and the flow of outdoor air can tailored precisely to the actual requirements.

It is instructive to consider the relation between Table 4.5 and Table 4.6 in view of the outdoor airflow rate per occupant that we have calculated in Example 4.12. Based on the level of CO_2 prescribed in Table 4.5b, we found an outdoor airflow rate of 15.1 ft^3/min per occupant, very close to the lowest outdoor air requirement per occupant of 15 ft^3/min in Table 4.6. The choice of the CO_2 level in Table 4.5b is in fact based on the observation that CO_2 production is closely correlated with metabolic activity and associated body odors. The minimum outdoor air rates in Table 4.6 have been set to avoid stuffiness from body odors, rather than being determined by oxygen demand. In fact, at these outdoor airflow rates, the oxygen content changes only a fraction of a percent as a result of metabolism.

There are several approaches to providing fresh outdoor air in a building. The simplest is to do nothing and rely on natural infiltration through cracks and leaks in the building envelope. In parts of a building where this would not be sufficient, for instance, kitchens and bathrooms, one can enhance the air supply by adding air vents with dampers; for better control one can add exhaust fans. (Air vents are also used in basements and attics to prevent moisture problems.) That is still the customary design for the residential sector in the United States. Unfortunately with natural infiltration the air supply will, most of the time, be either too high (causing energy waste) or too low (causing air quality problems).

TABLE 4.6

Outdoor Air Requirements for Ventilation

Application	Estimated Maximum Occupancy, Persons per 1000 ft² or 100 m² Net Occupiable Space	Outdoor Air Requirements				Comments
		cfm per Person	l/s per Person	cfm/ft²	l/(s · m²)	
a. Commercial Facilities (offices, stores, shops, hotels, sports facilities)						
Dry cleaners, laundries						Dry-cleaning processes may require more air.
Commercial laundry	10	25	13			
Commercial dry cleaner	30	30	15			
Storage, pickup	30	35	18			
Coin-operated laundries	20	15	8			
Coin-operated dry cleaner	20	15	8			
Food and beverage service						
Dining rooms	70	20	10			
Cafeteria, fast food	100	20	10			
Bars, cocktail lounges	100	30	15			Supplementary smoke removal equipment may be required.
Kitchens (cooking)	20	15	8			Makeup air for hood exhaust may require more ventilating air. The sum of the outdoor air and transfer air of acceptable quality from adjacent spaces shall be sufficient to provide an exhaust rate of not less than 1.5 cfm/ft² [7.5 l/(s · m²)].
Garages, repair, service stations						
Enclosed parking garage				1.50	7.5	
Auto repair rooms				1.50	7.5	Distribution among people must consider worker location and concentration of running engines; stands where engines are run must incorporate systems for positive engine exhaust withdrawal. Contaminant sensors may be used to control ventilation.

(continued)

TABLE 4.6 (continued)

Outdoor Air Requirements for Ventilation

Application	Estimated Maximum Occupancy, Persons per 1000 ft² or 100 m² Net Occupiable Space	Outdoor Air Requirements				Comments
		cfm per Person	l/s per Person	cfm/ft²	l/(s · m²)	
Hotels, motels, resorts, dormitories				cfm/ room	l/s · room	
Bedrooms				30	15	Independent of room size.
Living rooms				30	15	
Baths				35	18	Installed capacity for intermittent use.
Lobbies	30	15	8			
Conference rooms	50	20	10			
Assembly rooms	120	15	8			
Dormitory sleeping areas	20	15	8			
Gambling casinos	120	30	15			See also food and beverage services, merchandising, barber and beauty shops, garages. Supplementary smoke removal equipment may be required.
Offices						
Office space	7	20	10			Some office equipment may require local exhaust.
Reception areas	60	15	8			
Telecommunication centers and data entry areas	60	20	10			
Conference rooms	50	20	10			
Public spaces						
Corridors and utilities				0.05	0.25	
Public restrooms, cfm/WC, or cfm/urinal		50	25			Normally supplied by transfer air.

Locker and dressing rooms				0.5	2.5	Local mechanical exhaust with no recirculation recommended.
Smoking lounge	70	60	30			
Elevators				1.00	5.0	Normally supplied by transfer air.
Retail stores, sales floors, and show room floors						
Basement and street	30			0.30	1.50	
Upper floors	20			0.20	1.00	
Storage rooms	15			0.15	0.75	
Dressing rooms				0.20	1.00	
Malls and arcades	20			0.20	1.00	
Shipping and receiving	10			0.15	0.75	
Warehouses	5			0.05	0.25	
Smoking lounge	70	60	30			Normally supplied by transfer air, local mechanical exhaust; exhaust with no recirculation recommended.
Specialty shops						
Barber	25	15	8			
Beauty	25	25	13			
Reducing salons	20	15	8			
Florists	8	15	8			Ventilation to optimize plant growth may dictate requirements.
Clothiers, furniture				0.30	1.50	
Hardware, drugs, fabric	8	15	8			
Supermarkets	8	15	8			
Pet shops				1.00	5.00	
Sports and amusement						
Spectator areas	150	15	8			
Game room	70	25	13			
Ice arenas (playing areas)				0.50	2.50	When internal combustion engines are operated for maintenance of playing surfaces, increased ventilation rates may be required.
Swimming pools (pool and deck area)				0.50	2.50	Higher values may be required for humidity control.

(continued)

TABLE 4.6 (continued)

Outdoor Air Requirements for Ventilation

Application	Estimated Maximum Occupancy, Persons per 1000 ft² or 100 m² Net Occupiable Space	Outdoor Air Requirements				Comments
		cfm per Person	l/s per Person	cfm/ft²	l/(s · m²)	
Playing floors (gymnasium)	30	20	10			
Ballrooms and discos	100	25	13			
Bowling alleys (seating areas)	70	25	13			
Theaters						
Ticket booths	60	20	10			Special ventilation will be needed to eliminate special stage effects (e.g., dry ice vapors, mists).
Lobbies	150	20	10			
Auditorium	150	15	8			
Stages, studios	70	15	8			
Transportation						
Waiting rooms	100	15	8			
Platforms	100	15	8			
Vehicles	150	15	8			Ventilation within vehicles may require special considerations.
Workrooms						
Meat processing	10	15	8			Spaces maintained at low temperatures (−10 to +50°F, or −23 +10°C) are not covered by these requirements unless the occupancy is continuous. Ventilation from adjoining spaces is permissible. When the occupancy is intermittent, infiltration will normally exceed the ventilation requirement.
Photo studios	10	15	8			
Darkrooms	10		8	0.50	2.50	
Pharmacy	20	15	8			

Application						Comments
Bank vaults	5	15	8			
Duplicating, printing				0.50	2.50	Installed equipment must incorporate positive exhaust and control (as required) of undesirable contaminants (toxic or otherwise).
b. Institutional Facilities						
Education						Special contaminant control systems may be required for processes or functions including laboratory animal occupancy.
Classroom	50	15	8			
Laboratories	30	20	10			
Training shop	30	20	10			
Music rooms	50	15	8			
Libraries	20	15	8			
Locker rooms				0.50	2.50	
Corridors				0.10	0.50	
Auditoriums	150	15	8			
Smoking lounges	70	60	30			Normally supplied by transfer air. Local mechanical exhaust with no recirculation recommended.
Hospitals, nursing and convales centhomes						Special requirements or codes and pressure relationships may determine minimum ventilation rates and filter efficiency. Procedures generating contaminants may require higher rates.
Patient rooms	10	25	13			
Medical procedure	20	15	8			
Operating rooms	20	30	15			
Recovery and ICU	20	15	8			
Autopsy rooms				0.50	2.50	Air shall not be recirculated into other spaces.
Physical therapy	20	15	8			
Correctional facilities						
Cells	20	20	10			
Dining halls	100	15	8			
Guard stations	40	15	8			

This table prescribes supply rates of acceptable outdoor air required for acceptable indoor air quality. These values have been chosen to dilute human bioeffluents and other contaminants with an adequate margin of safety and to account for health variations among people and varied activity levels.

The best control over air supply can be provided by mechanical ventilation systems; one simply specifies the capacity of the fan(s) and ducts, and one obtains the required outdoor air flow. Mechanical ventilation is the predominant approach for commercial buildings in the United States.

Quite generally for the design of ventilation systems it is important to keep in mind the following rules:

- Avoid air intake near loading docks, furnace vents, and other local sources of air pollution.
- Avoid short-circuiting between exhaust and intake.
- Place exhaust vents or exhaust fans near indoor sources of odor or pollution, e.g., bathrooms, kitchens, and open flames.

Example 4.14

What outdoor airflow $\dot{V}_o$ is needed for an office building of 1500-m^2 floor area if the expected density of occupants is 1 per 15 m^2?

Given: 1500-m^2 floor area and 15 m^2 per occupant

Find: $\dot{V}_o$

Assumptions: The outdoor air has acceptable quality (i.e., it meets the EPA requirements for long-term exposure, Table 4.5).

SOLUTION

The expected number of occupants is

$$N = \frac{1500}{15} = 100$$

Table 4.6 lists the ASHRAE-recommended outdoor airflow rate for offices as 10 L/s per occupant. Multiplying this number by the number of occupants, we obtain

$$\dot{V}_o = 100 \times 10 \text{ L/s} = 1000 \text{ L/s} = 1.0 \text{ m}^3/\text{s (2000 cfm)}$$

For the design of ventilation systems, one should pay attention to the fact that in most buildings perfect mixing, as assumed above in Eq. (4.46), is not achieved. Frequently a fraction S of the supply air does not reach the zones of a building where the occupants actually are, namely, the space between the floor and a height of approximately 1.80 m (72 in), and more than 0.6 m (2 ft) from the walls or the air conditioning equipment. This implies that some of the outdoor air that the ventilation system tries to introduce into the conditioned space is exhausted to the outside without being used; this air cannot be counted as contributing toward meeting the ventilation requirement. To quantify this effect, let us consider the flow rates in a typical air distribution system if a fraction S of the supply air does not reach the occupied zones, as shown in Fig. 4.22, where

$\dot{V}_o$ = flow rate of outdoor air into building

$\dot{V}_s$ = flow rate of supply air into building

$\dot{V}'_e$ = flow rate at which air is exhausted from building via central exhaust duct

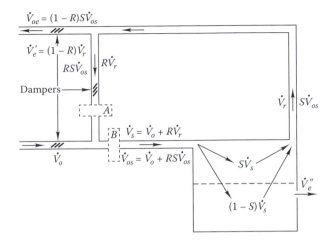

FIGURE 4.22
Flow rates in typical air distribution system (highly schematic). In italics this figure also indicates the outdoor air portion of the various flow rates. A and B designate possible filter locations.

$\dot{V}_e''$ = flow rate at which air is lost from building by exfiltration or local exhaust fans
$\dot{V}_r$ = flow rate of return air from conditioned space
$R = 1 - \dot{V}_e'/\dot{V}_r$ = fraction of return air that is recirculated
$\dot{V}_{os}$ = flow rate at which outdoor air is supplied to conditioned space
$\dot{V}_{oe}$ = flow rate at which unutilized outdoor air is exhausted from building

It is natural to define ventilation effectiveness E_v as the fraction of $\dot{V}_o$ that is really utilized

$$E_v = 1 - \dot{V}_{oe}/\dot{V}_o \tag{4.47}$$

The mass balance for the supply duct is

$$\dot{V}_s = \dot{V}_o + R\dot{V}_r \tag{4.48}$$

as indicated in Fig. 4.22. In italics this figure also indicates the outdoor air portion of the various flow rates. By analogy to Eq. (4.48), $\dot{V}_{os}$ is given as

$$\dot{V}_{os} = \dot{V}_o + RS\dot{V}_{os} \tag{4.49}$$

since $S\dot{V}_{os}$ is the flow of unutilized outdoor air in the return duct. And $\dot{V}_{oe}$ is given by

$$\dot{V}_{oe} = (1 - R)S\dot{V}_{os} \tag{4.50}$$

Combining Eqs. (4.47) to (4.50), one readily finds

$$E_v = \frac{1 - S}{1 - RS} \tag{4.51}$$

Example 4.15

An office building has been designed to just meet the ASHRAE ventilation standard, but due to the placement of the diffusers, 40 percent of the supply air does not reach the occupied space.

A fraction $R = 0.7$ of the return air is recirculated. If one does not want to change the diffusers, by what factor should one increase the outdoor airflow rate to meet the standard?

Given: $S = 0.4$

$R = 0.7$

Find: Inverse of ventilation effectiveness E_v

SOLUTION

From Eq. (4.51) the fraction of the outdoor air utilized is $E_v = (1 - 0.4)/(1 - 0.4 \cdot 0.7) = 0.83$. The outdoor airflow rate would have to be increased by a factor $1/0.83 = 1.2$ to really meet the standard.

4.6.3 Cleaning of Air

A variety of methods are available for cleaning the air in buildings. Most widely used, by far, are filters, and they can be quite effective against dust. For judging the performance of air filters, the following criteria are the most important:

- Efficiency (measures ability to remove particles from an airstream)
- Pressure drop across the filter (affects fan power and energy consumption)
- Dust-holding capacity (determines how often the filter should be cleaned or replaced, to avoid excessive flow resistance)

There are three broad categories of air filters: *fibrous media unit filter* (to be replaced or cleaned when full); *renewable media filter* (where new media are continually introduced into the airstream and old media are removed); and *electronic air cleaners* (which apply a high electric potential to sweep dust out of the airstream). Whereas the performance of fibrous media unit filters changes as they fill up with dust, the other two filter categories perform with essentially constant pressure drop and efficiency. To avoid the overloading of unit filters, they should be inspected periodically; or even better, a pressure gauge should be installed to indicate when they are full.

Roughly speaking, the efficiency of a filter is the percentage of the particles that it removes from an airstream. However, determining the efficiency of a filter is not a simple matter because the efficiency can vary strongly with type and size of particles. This is illustrated in Fig. 4.23 where the percentage of particles removed is plotted versus particle size. One sees that smaller particles are more difficult to filter out, which incidentally explains why $MP_{2.5}$ penetrate more deeply into the lungs than larger particles.

Several different test methods have been developed, each one appropriate for particular applications. The weight arrestance test measures the weight fraction of a standardized dust that the filter removes from the airstream; e.g., if this test yields an efficiency figure of 80 percent, the filter removes 80 percent by weight of this type of dust in the airstream. The dust spot efficiency test compares the discoloration effect of an airstream with and without the filter. Yet another test is the DOP (di-octyl-phthalate) penetration test that uses smoke produced by condensation of DOP vapor and measures the smoke concentration upstream and downstream; this latter test is used especially for high-performance filters. Even though the efficiency figures determined by these tests can be completely different (see Table 4.7), they are valuable for ranking and selecting filters for a given application.

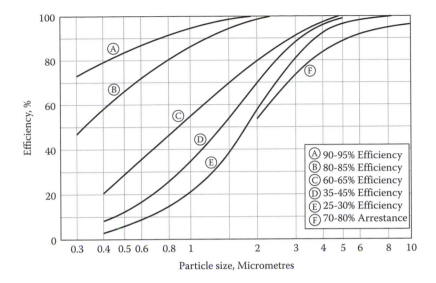

FIGURE 4.23
Efficiency of removing particles versus particle size, for several filter types. These curves are approximations based on manufacturers' data. They do not correspond to results of tests recognized by ASHRAE and should not be used for HVAC design. (Courtesy of ASHRAE, *Standard 55-1992: Thermal Environmental Conditions for Human Occupancy,* American Society of Heating, Refrigerating and Air-Conditioning Engineers, Atlanta, GA, 1992. With permission.)

TABLE 4.7

Some Characteristics of Filters

Type	Weight Arrestance, Percent	Dust Spot Efficiency, Percent	Pressure Drop, Clean, inWG	Pressure Drop, Max., inWG	Comments
Viscous impingement	50–80	<20		0.5	Low cost, good for lint, poor for dust
Dry-type extended surface					
Textile nonwoven media	Up to 80	Up to 60	0.05–0.25	0.5–0.7	Average offices, laboratories, etc.
Fine glass fibers, electret fibers, etc.	Up to 90				Above average offices, laboratories, etc.
HEPAandULPA	>98	>80	0.5–2.0		Clean rooms, hospitals, etc.

There are a wide variety of filters, and these values are very approximate and should not be used for design; rather the manufacturer's data should be consulted.

There are a wide variety of filter types. To provide a brief overview, we present in Table 4.7 some information extracted from Chapter 10 of ASHRAE (1988). One type of filter, called *viscous impingement filter,* is usually a flat panel made of coarse fibers (e.g., glass fibers or metallic wool) with high porosity, coated with a viscous substance to act as adhesive for dust particles. This type of filter has low pressure drop, low cost, high efficiency for lint but low efficiency for dust. Another type is the dry media filter, where fibers are much more closely spaced to form a dense mat, without adhesive on the fibers. Their efficiency can be significantly higher than that of viscous impingement filters.

Where especially clean air is demanded, one uses HEPA (high-efficiency particulate air) or ULPA (ultralow-penetration air) filters, made of submicrometer-size glass fiber paper in an extended surface configuration of deep space folds. They are used in applications such as hospitals and clean rooms. It may be advantageous to use combinations of different filter types, e.g., an inexpensive low-efficiency filter upstream of a more costly high-efficiency filter, thereby extending the life of the latter.

The filter efficiency required in a given application depends on the quality of the outdoor air, on the rate at which contaminants are produced in the space, on the ratio of the air that is recirculated, and of course on the indoor air quality that one would like to achieve. The pertinent design equations are generalizations of Eq. (4.46), and they can be found in Appendix E of the ventilation standard ASHRAE (1999). Here we show the equations for VAV systems with constant outdoor airflow; they also apply to constant-volume systems if the VAV flow reduction factor F_r is set to unity. The equation for filter location A in Fig. 4.22 is

$$\dot{V}_o = \frac{N - E_v F_r R E_f C_s \dot{V}_r}{E_v (C_s - C_o)} \tag{4.52a}$$

and the one for filter location B is

$$\dot{V}_o = \frac{N - E_v F_r R E_f C_s \dot{V}_r}{E_v \left[C_s - (1 - E_f) C_o\right]} \tag{4.52b}$$

with
$\dot{V}_o$ = flow rate of outdoor air into building
C_s = concentration of contaminant in supply air
C_o = concentration of contaminant in outdoor air
N = rate at which contaminant is added to space
$\dot{V}_r$ = flow rate of return air from conditioned space
$R = 1 - \dot{V}'_e / \dot{V}_r$ = fraction of return air that is recirculated
E_v = ventilation effectiveness
E_f = filter efficiency
F_r = flow reduction factor of VAV system

Example 4.16

An office building with constant-volume system has a filter in location A with filter efficiency of 70 percent for environmental tobacco smoke (ETS). There are 10 occupants, 3 of whom are smokers. Data of NRC (1986) indicate that smoking one cigarette per hour produces between 5 and 10 μg/s of $PM_{2.5}$. Taking the average, one can assume a production rate of ETS of $3 \cdot 7.5 = 22.5$ μg/s. The outdoor airflow rate is 20 cfm per person, the supply airflow rate 50 cfm per person, and the return airflow rate 40 cfm per person. The ventilation effectiveness is 0.83. Assume that there is no ETS outdoors. What is the concentration of ETS in the supply air, and how does it compare with EPA standards for air quality?

   Given: $\dot{V}_o = 200$ cfm
   $\dot{V}_s = 500$ cfm

$\dot{V}_r = 400$ cfm

$E_f = 0.7$

$E_v = 0.83$

$N = 22.5$ $\mu g/s = 1350$ $\mu g/min$ of $PM_{2.5}$ from 3 smokers each smoking 1 cigarette per hour

$C_o = 0$

Find: C_s

SOLUTION

First one needs the recirculation factor; by solving Eq. (4.48),

$$R = \frac{\dot{V}_s - \dot{V}_o}{\dot{V}_r} = \frac{500 - 200}{400} = 0.75$$

Now apply Eq. (4.52a) with $F_r = 1$ because it is a constant-volume system, and solve it for C_s, with the result

$$C_s = \frac{N + E_v C_o V_o}{E_v(RE_f V_r + V_o)}$$

$$= \frac{1350 \ \mu g/min}{0.83(0.75 \times 0.7 \times 400 \ \text{cfm} + 200 \ \text{cfm})}$$

$$= 3.97 \ \mu g/ft^3 = 140 \ \mu g/m^3 \text{ of } PM_{2.5} \tag{4.53}$$

There is no EPA standard for ETS, but since much of ETS is particulate matter, a relevant comparison is with the EPA standard for outdoor concentration of PM_{10}; see Table 4.5. The concentration limit is 50 $\mu g/m^3$ for long-term exposures. If the toxicity of ETS is the same as that of PM_{10}, the EPA limit is largely exceeded (note that the EPA standards in Table 4.5 are based on consideration of human health). While current epidemiological evidence is not sufficient to quantify the toxicity of each component of PM_{10} in outdoor air and of ETS, it certainly incriminates carbonaceous particles from combustion which are common to both. Furthermore, note that the ratio of $PM_{2.5}/PM_{10}$ for outdoor air in the United States is typically 0.6, which implies that 50 $\mu g/m^3$ of PM_{10} is equivalent to about 33 $\mu g/m^3$ of $PM_{2.5}$. Thus it may be prudent to reduce the concentration in this case significantly, perhaps by a factor of $140/33 \approx 4$. Looking at the numbers in Eq. (4.53), one sees that not enough can be achieved by increasing E_f or E_v. Even with $E_f = 1 = E_v$ one still finds $C_s = 95 \ \mu g/m^3$. Filtering has relatively little effect because it acts on the concentration only after the smoke has been emitted into the occupied space. For a major reduction of the concentration one needs either a very large increase in the airflow rates (which is costly) or a reduction in the production of ETS.

COMMENT

The details of the formula for the concentration depend on characteristics of the air distribution system. But the key feature is the same for all, namely, that the concentration decreases in inverse proportion with airflow—as expected intuitively. These numbers highlight the wisdom of avoiding the production of tobacco smoke in buildings.

One of the considerations in designing a filter system is the pressure drop. Obviously high-pressure drops should be avoided in the interest of energy efficiency. Data for the pressure drop are furnished by the manufacturer, but only at design flow rates. In practice, the flow rates will usually be different, and it is necessary to extrapolate to other flow rates.

For this purpose we anticipate a result from Chap. 5, namely, that the pressure drop across an air filter is proportional to the square of the flow rate $\dot{V}$

$$\Delta p = \text{constant} \times \dot{V}^2 \tag{4.54}$$

This is illustrated in Example 4.17.

Example 4.17

Design a filter system for the office of Example 4.14. As will be explained in later chapters, typical air distribution systems recirculate much of the return air, so that the total supply airflow rate is about 3 to 10 times larger than the outdoor rate. For the building of Example 4.14 the latter is $\dot{V}_o = 1.0$ m^3/s. Suppose that the total supply flow rate is $\dot{V}_s = 5.0$ m^3/s. One of the filters under consideration comes in units of area 0.6 m × 0.6 m (24 in × 24 in) and 0.2-m (8-in) depth, and the manufacturer's data specify a design flow rate of 0.81 m^3/s (1725 cfm) with a pressure drop of 87 Pa (0.35 inWG).

Given: $\dot{V}_s = 5.0$ m^3/s

$A_{\text{filter}} = 0.6$ m × 0.6 m per unit $= 0.36$ m^2

$\dot{V}_{\text{design}} = 0.81$ m^3/s with $\Delta p_{\text{design}} = 87$ Pa

Find: Number of units, and pressure drop in this application

SOLUTION

Dividing the total flow by the design flow per filter unit, we find that we would need

$$\frac{5.0 \text{ m}^3/\text{s}}{0.81 \text{ m}^3/\text{s per unit}} = 6.17 \text{ units}$$

Since we can only install an integral number of units, we must choose either 6 or 7. With 6 units the flow per unit is (5.0 m^3/s)/6 $= 0.833$ m^3/s. To find the corresponding pressure drop, we apply Eq. (4.54), with the result

$$\Delta p = \Delta p_{\text{design}} \left(\frac{\dot{V}}{\dot{V}_{\text{design}}} \right)^2 = 87 \text{ Pa} \left(\frac{0.833 \text{ m}^3/\text{s}}{0.81 \text{ m}^3/\text{s}} \right)^2 = 92.1 \text{ Pa (0.37 inWG)}$$

With 7 units the corresponding result is 67.7 Pa (0.27 inWG), significantly lower.

COMMENT

The velocity of the airstream is obtained by dividing the flow rate by the area. For the solution with 6 units we find

$$v = \frac{\dot{V}}{A_{\text{filter}}} = \frac{0.833 \text{ m}^3/\text{s}}{0.36 \text{ m}^2} = 2.3 \text{ m/s (456 ft/min)}$$

As for other methods of air cleaning, we mention washing with water spray (which controls humidity and cleans air at the same time); adsorption (e.g., by charcoal or zeolite); and chemical cleaning (apparatus similar to air washer, but with chemicals instead of plain water). These methods are necessary if one wants to remove gases and vapors, for which filters are not very effective.

TABLE 4.8

Chapter 4 Equation Summary

Topic	Equations	Equation Number	Notes
Psychrometric properties	$\phi = \dfrac{p_w}{p_{\text{sat}}}$	(4.1)	Relative humidity
	$W = 0.622 \dfrac{p_w}{p - p_w}$	(4.9)	Humidity ratio
	$\phi = \dfrac{p_{\text{sat}}(T_{\text{wet}}) - p_m}{p_{\text{sat}}(T_d)}$	(4.11)	ϕ from T_d and T_{wet}
	$p_{\text{sat}}(T) = p_c 10^{K(1 - T_c/T)}$	(4.13)	Auxiliary equation used with (4.11) for ϕ from T_d and T_{wet}
	$T_{\text{dew}} = 100.45 + 33.193\alpha + 2.319\alpha^2$ $+ 0.17074\alpha^3 + 1.2063 p_w^{0.1984}$	(4.15US)	Dew point temperature
	$h = c_{\text{pa}}T_d + W(h_{g,\text{ref}} + c_{pw}T_d)$	(4.16)	Moist-air enthalpy
	$h = h_{da} + \dfrac{W}{W_{\text{sat}}} h_{d,s}$	(4.18a)	Tabular air properties interpolation
Psychrometric processes	$W_1 = \dfrac{(1093 - 0.556T_2)W_2 - 0.240(T_1 - T_2)}{1093 + 0.444T_1 - T_2}$	(4.21US)	Adiabatic saturation
	$\dfrac{\Delta h}{\Delta W} = \dfrac{\dot{Q} + \dot{m}_w h_{f2}}{\dot{m}_w}$	(4.23)	Warming (cooling and humidifying (dehumidifying) moist air
	$\dfrac{\dot{m}_{\text{da1}}}{\dot{m}_{\text{da2}}} = \dfrac{h_2 - h_3}{h_3 - h_1} = \dfrac{W_2 - W_3}{W_3 - W_1}$	(4.28)	Adiabatic mixing
	$\text{SHR} \equiv \dfrac{\dot{Q}_{\text{sens}}}{\dot{Q}_{\text{sens}} + \dot{Q}_{\text{lat}}}$	(4.32)	Sensible heat ratio

4.7 Summary

In this chapter the basics of psychrometrics and a number of applications have been developed. The key assumption made was that air–water vapor mixtures behave as ideal gases. We found that the psychrometric chart was a convenient method for plotting processes involving heating, cooling, mixing, humidifying, and dehumidifying of moist airstreams. When accuracy better than that possible with a chart is needed, the equations presented in Section 4.2 should be used.

The comfort and health of occupants depend on the amount of clothing as well as the thermal and chemical contents of the building air. Filtration, thermal and humidity control subsystems, and adequate ventilation are essential. The designer has as one of his or her most important assignments the assurance of proper comfort and health conditions in occupied spaces in buildings.

Table 4.8 summarizes the important equations developed in this chapter.

Problems

The problems in this book are arranged by topic. The approximate degree of difficulty is indicated by a parenthetic italic number from 1 to 10 at the end of the problem. Problems are stated most often in USCS units; when similar problems are presented in SI units, it is

done with approximately equivalent values in parentheses. The USCS and SI versions of a problem are not exactly equivalent numerically. Solutions should be organized in the same order as the examples in the text: given, figure or sketch, assumptions, find, lookup values, solution. For some problems, the Heating and Cooling of Buildings (HCB) software included in this book may be helpful. In some cases it is advisable to set up the solution as a spreadsheet, so that design variations are easy to evaluate.

4.1 Calculate the humidity ratio, enthalpy, and specific volume of saturated air at 14.696 psia (101.3 kPa), using the ideal gas law and table of moist-air properties at 20°F (−6.5°C) and at 70°F (21°C). (3)

4.2 The air in a room at sea level[5] is 68°F (20°C) and 50 percent relative humidity (RH). Will moisture condense on a window whose surface is at 45°F (7°C)? If the room is 15 ft (4.5 m) square and 8 ft (2.5 m) high, how much water is contained in the room? (4)

4.3 For a site where the atmospheric pressure is 13.5 psia (93 kPa), find the relative humidity, humidity ratio, dew point, and enthalpy for a condition where the dry-bulb temperature is 95°F (35°C) and the wet-bulb temperature is 60°F (15.5°C). (5)

4.4 For a site at sea level, find the relative humidity, humidity ratio, dew point, and enthalpy for a condition where the dry-bulb temperature is 100°F (38°C) and the wet-bulb temperature is 55°F (13°C). (5)

4.5 At 5000 ft (1500 m) where the atmospheric pressure is 12.2 psia (84 kPa), find the relative humidity, humidity ratio, dew point, and enthalpy for a condition where the dry-bulb temperature is 90°F (32°C) and the wet-bulb temperature is 55°F (13°C). (5)

4.6 Write a computer program or use a spreadsheet to derive an empirical third-order polynomial equation for the enthalpy of saturated sea-level air as a function of the dry-bulb temperature in the range 10 to 130°F (−12 to 55°C). (8)

4.7 Calculate the dew point temperature for a sea-level site where the dry- and wet-bulb temperatures are 75 and 55°F (24 and 13°C), respectively. (5)

4.8 A chilled-water line carries chilled water at 45°F (7°C) through a room at 70°F (21°C) and 60 percent relative humidity (at sea level). How much fiberglass insulation is needed on the pipe to avoid condensation? (8)

4.9 During very cold weather at a 5000-ft-high (1500-m-high) mountain site in Colorado, the interior surface of a single-glazed window can reach 40°F (4°C). If the room is at 68°F (20°C), what is the maximum relative humidity that can exist in the room without condensation's occurring on the window? (4)

4.10 If the relative humidity at a sea-level site is 50 percent for a dry-bulb temperature of 80°F (26.5°C), calculate the moist-air enthalpy. (5)

4.11 Use the tables of moist-air properties to find the enthalpy and specific volume for air at 70°F (21°C) if the humidity ratio is 0.008. (4)

4.12 Use the tables of moist-air properties to find the enthalpy and specific volume for air at 125°F (51°C) if the humidity ratio is 0.040. (4)

4.13 A glass of ice water at 40°F (4.5°C) condenses moisture on its exterior outdoors on an 85°F (29°C) day. What is the minimum relative humidity needed for the condensation to occur if the location is at 5000 ft (1500 m)? (4)

---

[5] In the problems, the term *sea level* means that the atmospheric pressure is 14.696 psia (101.325 kPa). The effect of altitude on density can be found from Eq. (3.3) or from the respective table in the CD-ROM.

4.14 Complete the following table, using the psychrometric chart or tables and equations for moist-air properties: (6)

p_{atm}	T_d	W	h	ϕ	T_{wet}	T_{dew}
14.696 psia	95°F			50 percent		
29.92 inHg	70°F				50°F	
101.325 kPa		0.0070	55 kJ/kg			
101.325 kPa	20°C	0.010				
12 psia	55°F					45°F
101.325 kPa	30°C				30°C	
101.325 kPa	−10°C			90 percent		

4.15 Solve Prob. 4.14 using the HCB software. (5)

4.16 Calculate and plot the dew point and enthalpy of air at 70°F (21°C) dry-bulb temperature for relative humidities of 10, 30, 50, 70, and 90 percent. (7)

4.17 Calculate the humidity ratio, enthalpy, and specific volume for saturated air at 30°F (−1°C) and 80°F (27°C) at sea level. (4)

4.18 Calculate the humidity ratio, enthalpy, and specific volume for saturated air at 40°F (4°C) and 80°F (27°C) at a 5000-ft (1500-m) elevation. (4)

4.19 Complete the following table, using either the psychrometric chart or the HCB software:

Point	p_{atm}	T_d	W	h	ϕ	T_{wet}	T_{dew}
A	14.696 psia	95°F			45 percent		
B	14.696 psia	70°F				56°F	

What is the sensible heat ratio for the cooling process *AB*? (5)

4.20 Complete the following table, using either the psychrometric chart or the HCB software:

Point	p_{atm}	T_d	W	h	ϕ	T_{wet}	T_{dew}
A	101.325 kPa	35°C			50 percent		
B	101.325 kPa	20°C				14°C	

What is the sensible heat ratio for the cooling process *AB*? (5)

4.21 An automotive air conditioner cools 250 ft³/min (120 L/s) of 95°F (35°C) and 35 percent relative humidity sea-level air to saturation at 45°F (7°C). How much moisture must be drained from the evaporator per hour? If the COP of this air conditioner is 2.2 and the air conditioning compressor belt drive has an efficiency of 75 percent, how much power is extracted from the automobile engine to operate the air conditioner? (6)

4.22 Describe the steps in the algorithm used to calculate the humidity ratio from measured wet- and dry-bulb temperatures.   (5)

4.23 Construct the relative humidity curves on a psychrometric chart for 2500-ft altitude in USCS units, using the graphics routines in a commercial spreadsheet software package. Check your solution by comparison with Fig. 4.6 and the electronic appendices. Optionally you could add enthalpy and wet-bulb temperature lines.   (9)

4.24 Solve Example 4.3 for a 5000-ft (1500-m) altitude.   (4)

4.25 Air at 100°F (38°C) and 30 percent relative humidity is cooled to 70°F (21°C). How much moisture condenses per pound (kilogram) of air?   (4)

4.26 Air at 100°F (38°C) and 30 percent relative humidity is cooled to 55°F (13°C). How much moisture condenses per pound (kilogram) of air?   (4)

4.27 One hundred ft^3/min (47 L/s) of air is to be humidified by saturated steam at 212°F (100°C). Sea-level air enters the steam humidifier at 55°F (13°C) dry-bulb temperature and 39°F (4°C) wet-bulb temperature. What is the steam flow rate if the air is humidified to 80 percent relative humidity? What is the temperature of the humidified air?   (5)

4.28 Air enters a steam humidifier at 50°F (10°C) dry-bulb temperature and 35°F (2°C) wet-bulb temperature at a flow rate of 1000 ft^3/min (470 L/s). The air is humidified by saturated steam at 230°F (110°C). What is the steam flow rate if the air is humidified to 90 percent relative humidity? What is the temperature of the humidified air? The location of the humidifier is at 5000-ft (1500-m) altitude.   (5)

4.29 Suppose that an HVAC design engineer used the sea-level psychrometric chart instead of the 5000-ft chart to solve Prob. 4.28. What would the error in the steam flow rate be?   (8)

4.30 On a winter day, outdoor air at 10°F (−12°C) and 70 percent relative humidity is heated to 70°F (21°C) in a residence. If the occupants require that the indoor humidity be 60 percent for comfort reasons, how much moisture per unit of dry air must be added to the outdoor air if the atmospheric pressure is 14.696 psia (101.325 kPa)?   (4)

4.31 Air at 34°F dry-bulb temperature and 33°F wet-bulb temperature must be heated to 68°F, the interior temperature of a residence at sea level. The airflow rate is 800 ft^3/min. What is the heat rate? How much of is due to the presence of moisture?   (5)

4.32 Humidity is added to the outdoor air in Prob. 4.30 by evaporating water sprayed into the airstream. How much energy must be added to the air to increase the humidity level to the required 60 percent?   (6)

4.33 Moist air at 70°F (21°C) dry-bulb temperature and 45°F (7°C) wet-bulb temperature is humidified to a final dew point temperature of 55°F (13°C) by addition of saturated 230°F (110°C) steam. If the airflow rate is 200 lb/min (1.5 kg/s), what is the required steam flow rate? What is the final dry-bulb temperature?   (5)

4.34 Outdoor air at 40°F (4.5°C) and 60 percent humidity is heated and humidified by steam at 230°F (110°C). The airflow rate is 30,000 ft^3/min (14,000 L/s), and heat is added to the air at the rate of 1,500,000 Btu/h (440 kW) while it absorbs 800 lb_m/h (0.10 kg/s) of steam. What are the dry- and wet-bulb temperatures at the exit of this heater-humidifier if it is located at sea level?   (5)

4.35 Figure P4.35 shows the typical arrangement for mixing two airstreams at the air handler of a commercial building HVAC system at sea level. Outdoor air at

35°C dry-bulb temperature and 25°C wet-bulb temperature mixes with 25°C and 50 percent relative humidity return indoor air from the building in the mass flow ratio of 1:5. What are the enthalpy, relative humidity, humidity ratio, and dry-bulb temperature of the mixed air? If the fan flow rate is 12 kg/s, how much energy and moisture must be removed from the mixed airstream to provide 25°C, 50 percent relative humidity supply air to the building?   (7)

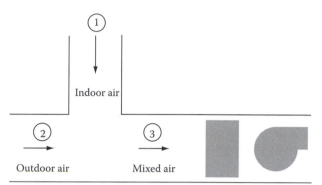

FIGURE P4.35

4.36 Cold air at 10°C dry-bulb temperature and 5°C wet-bulb temperature is mixed with warm, moist air at 25°C dry-bulb temperature and 20°C wet-bulb temperature in a ratio of 1 : 2, respectively. Use the sea-level psychrometric chart to find the mixed-air condition: dry-bulb and wet-bulb temperatures, relative humidity, humidity ratio, and specific volume.   (5)

4.37 Repeat Example 4.7, using moist-air property tables instead of the psychrometric chart.   (7)

4.38 An evaporative cooler is able to cool air by 85 percent of the difference between the entering air dry- and wet-bulb temperatures. If inlet air is at 100°F (38°C) and 25 percent relative humidity, what is the outlet condition (T_d, ϕ, and W)? How much water is evaporated if the airflow is 10,000 ft³/min (4700 L/s) at 3000-ft (1000-m) elevation [$p_{atm} = 13.2$ psia (91 kPa)]?   (8)

4.39 Repeat Example 4.8, using the moist-air property tables and psychrometric equations rather than the psychrometric chart. How close is the agreement between the two?   (7)

4.40 Because the 2000 ft³/min (950 L/s) exhaust fan in a chemistry laboratory has no on/off switch, it operates continuously. How much energy is wasted because this fan operates needlessly for 14 h/day for the 150-day cooling season if the interior air condition is 70°F and 50 percent relative humidity? On average, the outdoor condition at this sea-level site is 80°F (26°C) and 50 percent relative humidity. The air conditioner's COP is 2.4. If electric power costs 8¢/kWh, what is the value of this wasted energy?   (7)

4.41 At a site at 5000 ft (1500 m), air is cooled from 80°F (27°C) dry-bulb temperature and 75°F (24°C) wet-bulb temperature to saturation at the 55°F (13°C) outlet of a cooling coil. How much water is removed? How much latent heat is removed, and how much sensible heat? What is the SHR?   (7)

4.42 At a sea-level site, air is cooled from 80°F (27°C) dry-bulb temperature and 75°F (24°C) wet-bulb temperature to saturation at the 55°F (13°C) outlet of a cooling coil.

How much water is removed? How much latent heat is removed, and how much sensible heat? What is the SHR?   (7)

4.43 An office is occupied by 30 persons who each produce 200 Btu/h (58 W) of sensible heat and 0.25 lb/h (0.1 kg/h) of moisture. The office is to be maintained at 72°F (22°C) and 50 percent relative humidity. Conditioned air is supplied at 60°F (15°C) to meet the sensible and latent loads. What is the SHR? To meet the loads, what must the humidity ratio and mass flow rate of the supply air be? The office is located at sea level.   (8)

4.44 A high school classroom is occupied by 20 students who each produce 180 Btu/h (53 W) of sensible heat and 0.20 lb/h (0.09 kg/h) of moisture. The class is to be maintained at 68°F (20°C) and 55 percent relative humidity. Conditioned air is supplied at 60°F (15°C) to meet the sensible and latent loads. What is the SHR? To meet the loads, what must the humidity ratio and mass flow rate of the supply air be? The office is located at 5000-ft (1500-m) altitude.   (8)

4.45 Find the cooling coil load for the following entering and required leaving air conditions for an airflow of 10,000 ft^3/min (4700 L/s) at sea level.   (7)

Case	Inlet State		Outlet State	
	Dry Bulb, °F	Wet Bulb, °F	Dry Bulb, °F	Relative Humidity, Percent
1	80	67	50	90
2	80	60	50	90
3	45	44	50	90

4.46 Complete the following table of cooling coil performance that compares the loads at various coil outlet temperature control conditions.   (7)

Point	$T_{d,in}$, °F	W_{in}	$T_{d,out}$, °F	$T_{dew,out}$, °F	W_{out}	Latent Load, Btu/h	Load, Load, Btu/h	SHR
1	78	0.010386	65	58.2				
2	78	0.010386	63	57.5				
3	78	0.010386	61	56.3				
4	78	0.010386	59	55.0				
5	78	0.010386	57	53.8				
6	78	0.010386	55	52.0				

4.47 An economizer must deliver 10,000 ft^3/min (4700 L/s) at 56°F (13.5°C) dry-bulb temperature. The return air from the building is at 75°F (24°C) and 50 percent relative humidity, and the outside air condition is 35°F (2°C) and 40 percent relative humidity. How much outside air is needed, and what is the mixed-air humidity ratio?   (9)

4.48 An economizer must deliver 50,000 ft^3/min (23,500 L/s) at 55°F (13°C) dry-bulb temperature. The return air from the building is at 72°F (22°C) and 45 percent relative humidity, and the outside air condition is 32°F (0°F) and 50 percent relative humidity. How much outside air is needed, and what is the mixed-air humidity ratio?   (7)

4.49 What is the inlet humidity ratio to an ideal evaporative cooler (i.e., one that cools inlet air to the inlet air wet-bulb temperature) that produces saturated outlet air at 63°F (17°C) from 85°F (29°C) dry-bulb temperature inlet air? The cooler is at sea level. (4)

4.50 Work Prob. 4.49 for a site where the atmospheric pressure is 12.8 psia (88 kPa). (4)

4.51 An adiabatic saturator (Sec. 4.2.1) operating at sea level with entering air of 80°F (26°C) has a leaving air temperature of 65°F (18°C). Compute the entering-air humidity ratio and relative humidity. (5)

4.52 What effects are taken in account by (a) the mean radiant temperature, (b) the operative temperature, and (c) the adiabatic equivalent temperature? (3)

4.53 Why are the temperature limits of the ASHRAE comfort chart inclined rather than parallel to the y axis? (3)

4.54 Determine whether the following conditions are expected to be comfortable for light office work.

(a) Summer: air temperature = radiant temperature = 23°C (73.4°F), relative humidity = 60 percent.

(b) Summer: air temperature = 22°C (71.6°F) and radiant temperature = 28°C (82.4°F) (assume that the convective and radiative heat transfer coefficients are equal), relative humidity = 30 percent.

(c) Winter: air temperature = radiant temperature = 20°C (68°F), relative humidity = 30 percent. (5)

4.55 Consider a room with the following conditions:

- Room dimensions are 2.5 m × 4 m × 5 m (8.2 ft × 13.12 ft × 16.41 ft).
- One side has dimensions 2.5 m × 4 m (8.2 ft × 13.12 ft) and is entirely glazed with interior surface temperature 10°C (50°F).
- The other surfaces are at 20°C (68°F).
- The air is at 22°C (71.6°F) dry-bulb temperature and 30 percent relative humidity.

(a) What is the mean radiant temperature?

(b) What is the operative temperature?

(c) Are these conditions within the comfort limits of ASHRAE? (6)

4.56 Find the highest permissible T_{op} and W for the summer zone of the ASHRAE comfort chart if the relative humidity is 50 percent. Assume $T_{op} = T_a$. (4)

4.57 Estimate the contribution of thermal storage in the body relative to the steady-state heat loss if the temperature of the body changes by 0.5 K (0.28°F) over 10 h (typical day-night swing). Assume constant steady-state heat loss corresponding to an activity level of 1 met and a surface area of 1.8 m² (19.37 ft²). Treat the body as isothermal with a heat capacity of 250 kJ/K (130 Btu/°F). (5)

4.58 To estimate the effect of hot or cold food on the thermal balance of the body, consider the cooling power of drinking 0.1 L/h (0.0264 gal/h) of cold drinks at 5°C (41°F) (assuming no change in evaporative transfer from the body). Is it significant? (4)

4.59 An air handler in a hospital is to supply a zone that contains rooms for 50 patients. Find the required minimum outdoor airflow rate for this zone. (4)

4.60 Suppose there is a source of NO_x in a building that produces 100 $\mu g/s$ of NO_x. If the air inside the building is always well mixed, and if the outdoor air has already an NO_x concentration of 50 μ/m^3, what outdoor airflow is needed to satisfy the conditions of Table 4.5 in the building? (4)

4.61 An office building 100 ft × 100 ft × 13 ft (30.48 m × 30.48 m × 3.95 m) has an infiltration rate of 0.6 air change per hour during the heating season. Is this sufficient to meet the fresh air requirements of 75 occupants, or is mechanical ventilation necessary if the windows are to remain closed? (4)

4.62 Two airstreams, both at 5000 cfm, are well mixed inside an air handling unit. One airstream is at 80°F and 80 percent relative humidity, and the other is at 50°F and 80 percent relative humidity. What are the resulting mixed airstream temperature and relative humidity? (3)

4.63 Which sample of moist air has the higher density: 50°F at 10 percent relative humidity or 50°F at 90 percent relative humidity? (3)

4.64 Which sample of moist air has the higher density: 50°F at 50 percent relative humidity or 80°F at 50 percent relative humidity? (3)

4.65 An economizer mode attempts to mix outside air and building return air to minimize the amount of energy needed to condition the resulting mixed airstream to match the desired supply air conditions. Suppose the conditions are outside air at 90°F and 40 percent relative humidity, the return air at 80° F and 70 percent relative humidity, and the supply air set point of 55°F at 80 percent relative humidity. Should the economizer control use mostly outside air or mostly building return air? (4)

4.66 The World Trade Center in New York contains approximately 100 million ft^3 of air at 70°F and 60 percent relative humidity. Assuming a ventilation rate of 0.5 air change per hour and ambient air design conditions of 92°F dry-bulb and 76°F wet-bulb temperature, how much water is removed each hour from the outdoor air entering the building? (4)

4.67 You exhale air at about 80°F and 50 percent relative humidity. What outdoor air conditions must be met before you start to see your breath? (3)

4.68 Air leaves a cooling coil at 55°F and a humidity ratio of 0.008 lb water/lb air. Does this supply air meet the requirement of a minimum of 85 percent relative humidity? This air then passes through a fan that heats up the airstream by 2°F before being supplied to the building. Is the 85 percent relative humidity requirement met? (2)

4.69 Air at 90°F and 50 percent relative humidity is cooled to 70°F. How much moisture condenses out of the air? (2)

4.70 Air at 90°F and 50 percent relative humidity is cooled to 60°F. How much moisture condenses out of the air? (2)

4.71 Data from a coastal weather station record a daytime high temperature of 90°F and a relative humidity of 37 percent. At night the temperature drops to 60°F, and the relative humidity reaches 100 percent. How much has the humidity ratio varied throughout the day? (3)

4.72 Some buildings use *night purging*, where cool night air is used to cool the building mass in the evening to reduce cooling energy used the following day. However, bringing in the cool night air can also cause the building and its contents to absorb a lot of moisture that can actually increase the latent load. If the building return air is a constant 78°F at 70 percent relative humidity without night purging and 76°F at 77 percent relative humidity, does night purging make sense? Explain your answer. (6)

4.73 A small office building has a volume of about 50,000 ft^3. If the forced ventilation rate if 0.5 air change per hour (ACH) and there are 30 occupants who give off 0.01 ft^3 of CO_2 per minute, how does the CO_2 concentration vary throughout the day? Draw a graph from 8 a.m. to 5 p.m. showing the concentration, assuming an ambient concentration of 350 ppm of CO_2 and an initial building concentration of the same amount. Ignore any infiltration effects, and assume that the building air is thoroughly mixed. (8)

4.74 A party is held in a house during the winter with all the doors and windows closed. When the guests all leave at midnight, the concentration of CO_2 in the house is 1400 ppm. If the house has a volume of 25,000 ft^3 and an infiltration rate of 0.3 ACH, how long before the CO_2 concentration in the house goes below 500 ppm (assume an ambient concentration of 350 ppm and that no one remains in the house)? (7)

4.75 In Prob. 4.74, how long would it take if the doors and windows were opened, increasing the effective ventilation rate to 1 ACH? (7)

4.76 Most relative humidity sensors have an accuracy of about ±3 percent. If an airstream is measured to have a temperature of 55°F and a relative humidity of 90 percent, assuming the stated error of the humidity sensor, what is the possible range of air enthalpy for this airstream? (3)

4.77 A cooling coil at sea level removes 100,000 Btu/h from an airstream. The air flows into the coil at 10,000 cfm. What is the leaving air temperature from the coil if the entering air is at 80°F and 50 percent relative humidity? (6)

4.78 In Prob. 4.77, what is the leaving air temperature if the coil is located in Leadville, Colorado, at 10,000 ft above sea level? (6)

4.79 The *zone ventilation effectiveness* η_{vent} of a conditioned space is used to quantify the "short-circuiting" of air from the supply diffusers to the return grill. It is defined such that the true amount of supply air delivered to the occupied zone is $Q_{supply} \cdot \eta_{vent}$. If a clean room requires two air changes per hour and has a zone ventilation effectiveness of 0.8 then what is the required Q_{supply}?

4.80 A smoker produces approximately 7.5 $\mu g/s$ of environmental tobacco smoke (ETS). What is the minimum required ventilation rate if the goal is to keep the ETS concentration below 50 $\mu g/m^3$? Assume steady-state conditions and that the air is well mixed in the zone. Note that the production rate is given here in mass per time and the concentration in mass per volume, by contrast to Eq. (4.46) and Example 4.12 where the production rate is in volume per time (as appropriate for a gaseous pollutant) and the concentration in ppm. (3)

4.81 A smoker producing ETS at a rate of 7.5 $\mu g/s$ lives in a house of volume $V = 500$ m^3 with an average outdoor airflow rate corresponding to an air exchange rate $\dot{V}_o = 1.0$ air change per hour. What is the resulting steady-state concentration of ETS if the air is well mixed in the house? Compare with the EPA standard for exposure to particulate matter in Table 4.5.

4.82 You are designing the ventilation system (see Fig. P4.82) for a smoking lounge for 10 persons. The outdoor air intake rate $\dot{V}_o$ is 30 L/s per person as recommended by ASHRAE Standard 62-1989. Smokers introduce environmental tobacco smoke at an average rate of 7.5 $\mu g/s$. What is the resulting steady-state indoor air concentration of

ETS in $\mu g/m^3$ if there is no filter and if all 10 occupants are smokers? Neglect any air exchange through the building envelope, and assume perfect mixing in the building. Compare with the EPA standard for exposure to particulate matter in Table 4.5.   (3)

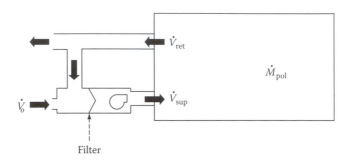

FIGURE P4.82
Typical ventilation system

4.83 Using the parameters from Prob. 4.82, what is the resulting ETS concentration if the filter has a weight arrestance efficiency of 80 percent? Assume a supply airflow rate $\dot{V}_{sup}$ of 4 times the outdoor air intake rate.   (10)

References

ASHRAE (1981). *Standard 55-1981: Thermal Environmental Conditions for Human Occupancy.* American Society of Heating, Refrigerating and Air-Conditioning Engineers, Atlanta.

ASHRAE (1988). *Handbook of Equipment.* American Society of Heating, Refrigerating and Air-Conditioning Engineers, Atlanta.

ASHRAE (2001, 1989a). *Handbook of Fundamentals.* American Society of Heating, Refrigerating and Air-Conditioning Engineers, Atlanta.

ASHRAE (1989b). *Standard 62-1989: Ventilation for Acceptable Indoor Air Quality.* American Society of Heating, Refrigerating and Air-Conditioning Engineers, Atlanta.

ASHRAE (1992). *Standard 55-1992: Thermal Environmental Conditions for Human Occupancy.* American Society of Heating, Refrigerating and Air-Conditioning Engineers, Atlanta.

ASHRAE (1996). *Handbook of Systems and Equipment.* American Society of Heating, Refrigerating and Air-Conditioning Engineers, Atlanta.

ASHRAE (1997). *Handbook of Fundamentals.* American Society of Heating Refrigerating and Air-Conditioning Engineers, Atlanta.

ASHRAE (1999). *Standard 62-1999. Ventilation for Acceptable Indoor Air Quality.* American Society of Heating, Refrigerating and Air-Conditioning Engineers, Atlanta.

Doll, R., R. Peto, K. Wheatley, R. Gray, and I. Sutherland (1994). "Mortality in Relation to Smoking: 40 Years' Observations on Male British Doctors." *British Medical J.*, vol. 309, pp. 901–911.

Janssen, J. E. (1989). "Ventilation for Acceptable Indoor Air Quality." *ASHRAE J.*, October 1989, pp. 40–48.

Kuehn, T. H., J. L. Threlkeld, and J. W. Ramsey (1998). *Thermal Environmental Engineering*, 3d ed. Prentice-Hall, Englewood Cliffs, N.J.

Mathur, G. D. (1989). "Predicting Wet Vapor Saturation Pressure." *Heating, Piping and Air Conditioning*, pp. 103–104.

National Research Council (1986). *Environmental Tobacco Smoke: Measuring Exposures and Assessing Health Effects*. Committee on Passive Smoking. Board on Environmental Studies and Toxicology. National Academy Press, Washington. Available from http://www.ulib.org/webRoot/ Books /National_Academy_Press_Books/env_tobacco_smoke/0000001.htm

Pallady, P. H. (1989). ''Computing Relative Humidity Quickly.'' *Chem. Eng.*, vol. 96, pp. 255–257, November; see also, by the same author, ''Evaluating Moist Air Properties,'' *Chem. Eng.*, October 24, 1984, vol. 91, pp. 117–118.

Stoecker, W. F., and J. W Jones (1982). *Refrigeration and Air Conditioning*. McGraw-Hill, New York.

5

Fundamentals of Fluid Mechanics in Building Systems

5.1 Introduction

Liquids and gases comprise the fluids used for the delivery of heating or cooling in building mechanical systems. The most common liquids are water (and aqueous solutions) and refrigerants whereas the most common gas is atmospheric air. In this chapter we review the key features of fluid mechanics that pertain to buildings. The goals of this chapter are to

- Review the essentials of incompressible fluid mechanics as they apply to building mechanical systems.
- Provide basic data needed to calculate flow rates and pressure drops in building ductwork and piping systems.

After an introduction of basic concepts such as pressure, viscosity, and the equations of fluid motion, a summary of pressure loss calculation methods is given. The prime movers in HVAC systems—pumps and fans—are then described, followed by a short section on flow measurement. Chapter 11, Secondary Systems for Heating and Cooling, will apply the fundamental concepts summarized in this chapter to the design of complete air and liquid HVAC systems. For additional detail on fluid mechanics, the reader is referred to Kreider (1985), White (1998), or Streeter et al. (1997).

5.2 Basic Concepts

This section reviews key, basic concepts in fluid mechanics including pressure, fluid properties, and the mechanical energy equation.

5.2.1 Pressure

Static pressure is defined as the stress (force per unit area) normal to a surface at any point on any plane in a fluid at rest. In equation form, it is given by

$$p \equiv \lim_{A \to 0} \frac{F}{A} \qquad (5.1)$$

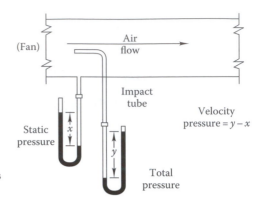

FIGURE 5.1
Pressure measurement arrangements
for airflow in a duct.

The consistent units of pressure are pound-force per square foot in the USCS and the pascal[1] in SI. An inconsistent unit of pressure, pound-force per square inch (abbreviated "psi"), is often used in the USCS. One psi (1 psi) is equivalent to 6895 Pa; 1 atm is equivalent to 14.696 psi or 101,325 Pa.

In practice, pressures are often referenced to the local atmospheric pressure. Pressures so expressed are called *gauge pressures*. To convert from gauge pressure to absolute pressure, one simply adds the atmospheric pressure to the gauge pressure:

$$p_{abs} = p_g + p_{atm} \tag{5.2}$$

where
p_{abs} = absolute pressure
p_g = gauge pressure
p_{atm} = local atmospheric pressure

In thermodynamic calculations, one must always use the absolute pressure, as described in Chap. 3. Gauge pressures can be used whenever pressure differences are involved.

Figure 5.1 shows how pressure measurements are made in an air duct. A similar approach is used for liquid measurements. The *static* gauge pressure is the pressurization of the duct air above the local atmospheric pressure. The *total* pressure is the pressure measured by a tube facing into the flow. The difference between the total and static pressures is due to the velocity of the fluid and is called the *velocity* pressure.

This measurement technique for air systems leads to another pressure unit in common use in the HVAC industry. Pressures in air systems are often expressed in terms of *water column* or *water gauge* (abbreviated WG). In USCS units, the unit is inches of water gauge (inWG); in SI units, the water column height can be measured in centimeters or millimeters.

5.2.2 Fluid Properties

Viscosity μ is a measure of the resistance of fluids to flow. For common fluids such as water and air, the viscosity is the proportionality constant in the expression relating shear in the fluid to the velocity gradient:

---

[1] To get a feeling for the magnitude of this unit, it may be helpful to recall the famous apocryphal story about the discovery of the law of gravitation and note that an apple weighs about 1 N (newton). When one slices the apple thinly enough to cover 1 m², the pressure is 1 N/m² = 1 Pa.

$$\tau = \mu \frac{du}{dy} \tag{5.3}$$

The common units of viscosity are pound-mass per foot per second (an inconsistent unit derived from the consistent unit pound-force–second per square foot) in USCS and pascal-seconds in SI. Another common viscosity unit is the poise (abbreviated P) from the old cgs system of units (1 Pa·s = 10 P). The viscosity of water at 70°F is approximately 1 cP, 0.001 Pa·s, or 0.00067 $lb_m/(ft·s)$. The conversion factors among various viscosity units are given in the conversion factor table inside the cover of the book.

Density is also a frequently used property of fluids. It was discussed in Chap. 3.

Shortly we will see that the quotient of viscosity and density is important in fluid dynamics calculations. This combination of properties is a property itself, called the *kinematic viscosity*:

$$v = \frac{\mu}{\rho} \tag{5.4}$$

The viscosity is a strong function of temperature for both air and water. For air, the viscosity is expressed well by a power law proportionality over a limited range of temperatures (for example, 50 to 250°F):

$$\mu \propto T^{0.67} \tag{5.5}$$

in which the proportionality constant will depend on the units used.

The viscosity [$lb_m/(ft·s)$] of pure water is given by Perry et al. (1999):

$$\mu = \frac{1}{32.0(D + \sqrt{8078.4 + D^2}) - 1786} \tag{5.6US}$$

where

$$D \equiv (0.556T - 26.21)°F \tag{5.7US}$$

In SI units of pascal-seconds, the viscosity of water is

$$\mu = \frac{0.1}{2.1482(D + \sqrt{8078.4 + D^2}) - 120} \tag{5.6SI}$$

where

$$D \equiv (T - 8.435)°C \tag{5.7SI}$$

A common liquid used in HVAC systems where water could freeze is an aqueous solution of ethylene or propylene glycol. Glycol solutions are denser and more viscous and provide poorer heat transfer (lower thermal conductivity and lower specific heat) than pure water in the same application. Table 5.1 is an example of some properties of a 30% aqueous solution of propylene glycol. Thirty percent propylene glycol freezes at approximately 6°F (−15°C). The appendix CD-ROM contains tables of fluid properties for air, water, glycol solutions, and other liquids used in building HVAC systems.

TABLE 5.1

Properties of 30% (by Weight) Aqueous Solution of Propylene Glycol

Temperature, °F	Viscosity, cP	Density, lb/ft³	Thermal Conductivity, Btu/(h · ft · °F)	Specific Heat, Btu/(lb · °F)
20	9.08	64.95	0.242	0.926
40	5.44	64.69	0.249	0.930
60	3.44	64.40	0.255	0.935
80	2.32	64.08	0.261	0.939
100	1.65	63.72	0.266	0.943
120	1.24	63.34	0.270	0.948
140	0.97	62.92	0.274	0.952
160	0.80	62.48	0.278	0.956
180	0.67	62.00	0.281	0.961
200	0.59	61.49	0.283	0.965

Multiply	By	To Obtain
cP	0.001	Pa · s
Lb/ft³	16.01	kg/m³
Btu/(h · ft · °F)	1.73	W/(m · °C)
Btu/(lb · °F)	4.19	kJ/(kg · °C)
cP	0.000672	$lb_m/(ft · s)$

Source: Courtesy of Born, D.W., *ASHRAE Trans.*, 95, pt. 2, 969, 1989. With permission.

5.2.3 Equation of Motion

The first law of thermodynamics [Eq. (3.19)] governs fluid motion in pipes and ducts in buildings. It is most useful when expressed on a rate basis in steady state as

$$\dot{Q} - \dot{W}_s = \dot{m}_o \left(gz_o + \frac{v_o^2}{2} + h_o \right) - \dot{m}_i \left(gz_i + \frac{v_i^2}{2} + h_i \right) + \dot{W}_f \qquad (5.8)$$

where
$\dot{Q}$ = rate of heat addition
$\dot{m}$ = mass flow rate
v = average velocity
z = elevation above datum
$\dot{W}_f$ = power expended in overcoming friction
$\dot{W}_s$ = shaft power
h = enthalpy
o, i = outflow and inflow points

as defined in Chap. 3. In steady state, the conservation of mass law requires that the two mass flow rates be the same.

HVAC designers often express the pressure terms in Eq. (5.8) in terms of "head," i.e., the physical height of a column of the *flowing fluid* in a manometer such as that shown in Figure 5.1. (If air were flowing, the column of air would be very tall indeed; therefore, more dense water is used as the manometer fluid.) Equation (5.8) is converted to the *head* basis by dividing it by the mass flow rate and acceleration of gravity g:

$$h_q - h_s - \left(z_o + \frac{v_o^2}{2g} + \frac{u_o}{g} + \frac{p_o}{\rho_o g}\right) - \left(z_i + \frac{v_i^2}{2g} + \frac{u_i}{g} + \frac{p_i}{\rho_i g}\right) + h_f \tag{5.9}$$

in which u represents the internal energy and the head terms h_q, h_s, and h_f correspond to the heat and work terms in Eq. (5.8); do not confuse the head terms with the enthalpy, both of which customarily use the same symbol h. The frictional head loss h_f is usually expressed as a multiple of the *velocity head* $v^2/(2g)$ as follows:

$$h_f \equiv \frac{\Delta p_f}{\rho g} = K\frac{v^2}{2g} \tag{5.10}$$

The dimensionless proportionality constant K depends on the flow situation, type of fluid, and size and shape of the fluid conduit. This is discussed in the next section.

If the heat flow, friction, and shaft work terms are zero in Eq. (5.8), the resulting expression is the *Bernoulli* equation for *incompressible isothermal* flow:

$$gz_o + \frac{v_o^2}{2} + \frac{p_o}{\rho} = gz_i + \frac{v_i^2}{2} + \frac{p_i}{\rho} \tag{5.11}$$

Customarily the units used for density in USCS applications of the Bernoulli equation are pound-mass per cubic foot. For dimensional consistency, the first and second terms on both sides of Eq. (5.11) must be divided by the conversion factor g_c [32.2 $\text{lb}_m \cdot \text{ft}/(\text{lb}_f \cdot \text{s}^2)$]. No correction of this type is needed with SI units.

The Bernoulli equation can be generalized to the case of flow with friction by adding a friction term to the right-hand side. This is the equation most often used in HVAC fluid flow design:

$$gz_o + \frac{v_o^2}{2} + \frac{p_o}{\rho} = gz_i + \frac{v_i^2}{2} + \frac{p_i}{\rho} + K\frac{v^2}{2} \tag{5.12}$$

The velocity v used in the friction term depends on the reference velocity for K, as described below. It is not necessarily either v_i or v_o.

5.3 Pressure Losses in Liquid and Air Systems

Flow of fluids through conduits (ducts and pipes) is subject to the parasitic loss due to friction. In this section, methods for finding the pressure drop in pipes and ducts are summarized. Since the flow in HVAC systems is turbulent,[2] only turbulent pressure drop equations will be included.

---

[2] So-called low-temperature cooling systems may involve laminar flow at low loads. See Kreider (1985) for laminar-flow friction factor expressions.

5.3.1 Pressure Losses in Piping and Pipe Fittings

The pressure drop in straight pipe is given by the D'arcy-Weisbach equation

$$\Delta p_f = f \left(\frac{L}{D_h} \right) \left(\rho \frac{v^2}{2} \right) \tag{5.13}$$

where
   Δp_f = pipe pressure drop
   L = pipe length
   v = average fluid velocity
   D_h = *hydraulic diameter*, defined as 4 times the ratio of flow area A (i.e., conduit cross section) to wetted perimeter P (e.g., the circumference of fully filled pipe or duct):

$$D_h \equiv \frac{4A}{P} \tag{5.14}$$

Note that the proportionality constant K in Eq. (5.10) is related to the friction factor f by $K = fL/D_h$ in Eq. (5.13) for straight pipe.

   The friction factor f is found from an independent equation that involves fluid and pipe properties along with the average velocity v. There are a number of expressions for f. A simple, explicit equation is

$$f = \frac{1.325}{\left\{ \ln \left[\epsilon/(3.7D_h) + 5.74/\mathrm{Re}^{0.9} \right] \right\}^2} \tag{5.15}$$

where
   ϵ/D_h = roughness-to-diameter ratio, shown in Figure 5.2 for common types of pipe
   Re = *Reynolds number*, defined as
   $\mathrm{Re} \equiv vD_h/v$
   v = kinematic viscosity from Eq. (5.4)

Equation (5.15) is equal in accuracy to the more commonly used, implicit Colebrook equation for f given by

$$\frac{1}{\sqrt{f}} = -0.87 \ln \left(\frac{\epsilon}{3.7D_h} + \frac{2.52}{\mathrm{Re}\sqrt{f}} \right) \tag{5.16}$$

Figure 5.3 is a plot of the friction factor as a function of the Reynolds number with the pipe roughness ratio as a parameter. For rapid calculations during preliminary design, it is adequate for estimating the friction factor. As shown on this chart, called the *Moody diagram*, turbulent flow exists for Reynolds numbers above about 3000. Also note that the friction factor is independent of the Reynolds number for sufficiently large values of Re for rough pipe. This is the area of the Moody diagram to the right of the dashed line.

   Another method of estimating the pipe pressure drop is to use a dimensional chart such as that given in Figure 5.4a and b. Pressure drop (in feet of fluid flowing per 100 ft of pipe length) is plotted versus flow (in gallons per minute) with actual pipe diameter (in inches)

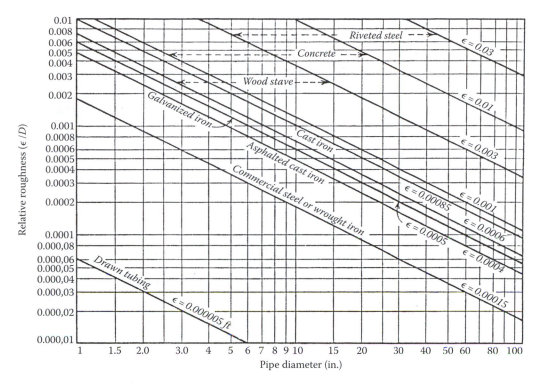

FIGURE 5.2
Pipe internal roughness ratio ϵ/D versus pipe diameter D (in). The absolute roughness ϵ is shown in feet. (Courtesy of Tuve, G., *Mechanical Engineering Experimentation*, McGraw-Hill, New York, 1961. With permission.)

as a parameter. Lines sloping downward from the left are used to find the fluid velocity. Figure 5.4a and b apply for pure water in steel pipe that has been in place for several years. For new copper or smooth plastic pipe, the pressure drops read from this chart should be multiplied by the constant 0.62. Glycols are often used for freeze protection in HVAC system piping exposed to cold air in heating coils or elsewhere. Figure 5.4c is a plot of the correction which must be applied to values read from Figure 5.4a and b when ethylene glycol is used for freeze protection. In addition to increased pressure drops, glycol solutions have poorer heat transfer properties than pure water, as discussed in Chap. 2. Proper attention to the corrosive possibilities of glycols is a necessity, as described in Chap. 11.

Pipe sizes are stated in terms of the *nominal* diameter, which is roughly related to the actual inside diameter of the pipe. Table 5.2 is an excerpt from the detailed table of steel pipe properties in the appendix CD-ROM. For standard pipe (schedule 40), nominal 1-in pipe is seen to have an outside diameter of 1.315 in and an inside diameter of 1.049 in with a resulting wall thickness of 0.133 in. Extra-strong pipe (schedule 80) of the same nominal size has the same outside diameter but a smaller inside diameter, as a consequence of the greater wall thickness of schedule 80 pipe.

Plastic pipe is manufactured in dimensions based on standard steel dimensions but with greater wall thickness for the same operating pressure.

Nominal copper piping sizes are all $\frac{1}{8}$ in less than the physical outside diameter. Table 5.3 shows sizes available in type L copper pipe. Type L is the standard weight of drawn

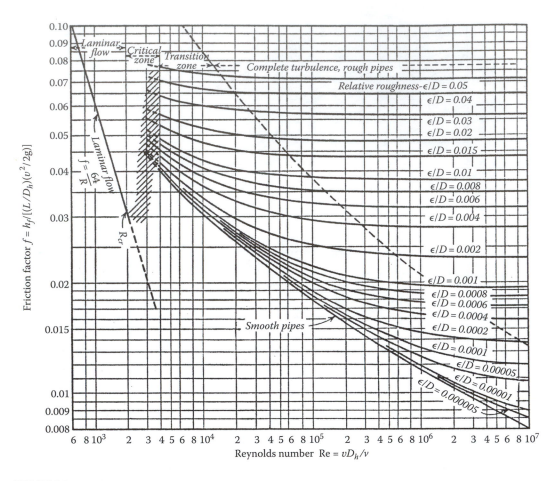

FIGURE 5.3
The Moody diagram with friction factor as a function of Reynolds number and relative roughness as a parameter. The hydraulic diameter is D_H. (Courtesy of Tuve, G., *Mechanical Engineering Experimentation*, McGraw-Hill, New York, 1961. With permission.)

copper piping for pressurized piping applications. The appendix CD-ROM contains additional data on type K piping (a thicker, soft tempered tubing suitable for field bending) and type M tubing (a thinner-walled material that cannot be bent and is intended for lower-pressure use).

The selection of *pipe size* involves a tradeoff between noise, material cost, and pumping power. As we will see in Example 5.1, the pressure drop (and pumping power) decreases with the fifth power of pipe diameter whereas pipe cost increases roughly linearly with diameter. One manufacturer recommends the sizes of piping for specific flow rates shown in Table 5.4. According to Waller (1990), noise is produced by rapid flow transitions; in straight pipe, water without entrained air produces little noise at velocities below 30 ft/s (10 m/s). Erosion (in copper piping) is another consideration that limits water velocities to no more than 6 ft/s (2 m/s). ASHRAE (1989, 2001) recommends that velocities be limited to 5 ft/s (1.7 m/s) in plastic pipe.

Example 5.1: Friction Factor

Find the flow, given the pressure drop and pipe size. Water at 38°C flows through 400-mm cast-iron pipe with a pressure drop of 225 Pa/m. What is the volumetric flow rate?

Given: $\Delta P/L = 225$ Pa/m (typical design rule of thumb)

$D = 400$ mm

Fluid: water at 38°C

Assumptions: Fully developed, turbulent flow

Find: $\dot{V}$

Lookup values: Kinematic viscosity at 38°C: $v = 0.6831 \times 10^{-6}$ m²/s

roughness $\epsilon = 0.00085$ ft (Figure 5.2) from which the roughness ratio is

$$\frac{\epsilon}{D} = \frac{0.00085 \text{ ft} \times 304.8 \text{ mm/ft}}{400 \text{ mm}} = 0.00064$$

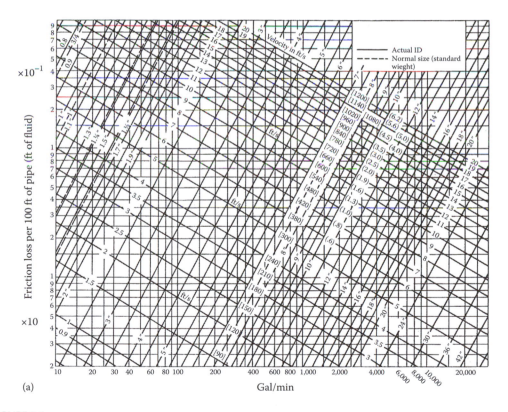

(a)

FIGURE 5.4

(a) Pipe friction chart for seasoned steel pipe in USCS units in the range of 10 to 30,000 gal/min with friction factor ranging from 0.2 to 100 ftWG.

(*continued*)

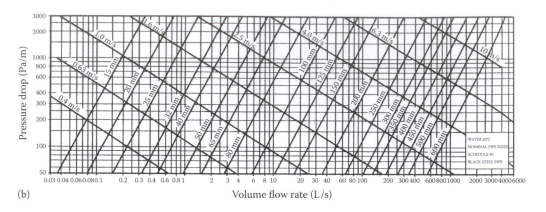

(b)

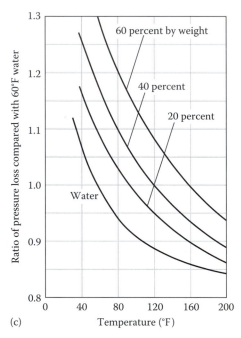

(c)

FIGURE 5.4 (continued)

(b) Pipe friction chart for seasoned steel pipe in SI units. (c) Correction factor for ethylene glycol flowing in steel pipe. Multiplier is applied to values read from the pipe friction chart. (Courtesy of Nussbaum, O.J., *Heat. Piping Air Cond.*, 62, 75, January 1990. With permission.)

SOLUTION

Equation (5.15) cannot be used to solve this problem directly, since neither the friction factor *f* nor the velocity *v* in the Reynolds number are known. However, a rearrangement of Eq. (5.15) is possible that will lead to a solution. The D'arcy-Weisbach equation (5.13) can be expressed in terms of volumetric flow rate as follows since $\Delta p = \rho g h_f$ and $\dot{V} = Av$:

$$\frac{h_f}{L} = \frac{f(L/D)(v^2/2)}{gL} = f\frac{\dot{V}^2/(2D)}{g[(\pi/4)D^2]^2} \tag{5.17}$$

TABLE 5.2

Standard Dimensions of Selected Small Steel Pipes

Nominal Pipe Size, in	Nominal Metric Size, mm	Schedule No.	Outside Diameter, in	Inside Diameter, in	Wall Thickness, in	Flow Area, in^2
$\frac{1}{2}$	13	40 (S)	0.840	0.622	0.109	0.3040
		80 (X)	0.840	0.546	0.147	0.2340
$\frac{3}{4}$	19	40 (S)	1.050	0.824	0.113	0.5330
		80 (X)	1.050	0.742	0.154	0.4330
1	25	40 (S)	1.315	1.049	0.133	0.8640
		80 (X)	1.315	0.957	0.179	0.7190
$1\frac{1}{4}$	32	40 (S)	1.660	1.380	0.140	1.495
		80 (X)	1.660	1.278	0.191	1.283
$1\frac{1}{2}$	38	40 (S)	1.900	1.610	0.145	2.036
		80 (X)	1.900	1.500	0.200	1.767
2	50	40 (S)	2.375	2.067	0.154	3.355
		80 (X)	2.375	1.939	0.218	2.953

S = standard pipe; X = extra-strong pipe. 1 in = 2.540 cm; 1 in^2 = 6.452 cm^2. See also the appendix in the CD-ROM.

TABLE 5.3

Standard Dimensions of Copper Pipe

Classification	Nominal Tube Size, in	Outside Diameter, in	Stubbs Gauge	Wall Thickness, in	Inside Diameter, in	Flow Area, in^2
Type L	$\frac{3}{8}$	$\frac{1}{2}$	19	0.035	0.430	0.146
	$\frac{1}{2}$	$\frac{5}{8}$	—	0.040	0.545	0.233
	$\frac{3}{4}$	$\frac{7}{8}$	—	0.045	0.785	0.484
	1	$1\frac{1}{8}$	—	0.050	1.025	0.825
	$1\frac{1}{4}$	$1\frac{3}{8}$	—	0.055	1.265	1.256
	$1\frac{1}{2}$	$1\frac{5}{8}$	—	0.060	1.505	1.78
	2	$2\frac{1}{8}$	—	0.070	1.985	3.094
	$2\frac{1}{2}$	$2\frac{5}{8}$	—	0.080	2.465	4.77
	3	$3\frac{1}{8}$	—	0.090	2.945	6.812
	$3\frac{1}{2}$	$3\frac{5}{8}$	—	0.100	3.425	9.213

1 in = 2.540 cm; 1 in^2 = 6.452 cm^2. See also the appendix in the CD-ROM.

Taking the square root of Eq. (5.17) and solving for $f^{-1/2}$, we have

$$f^{-1/2} = \frac{\dot{V}\sqrt{8}}{\pi\sqrt{g(h_f/L)D^5}} \qquad (5.18)$$

The Reynolds number can also be expressed as a function of volumetric flow rate as follows:

$$Re \equiv \frac{vD}{v} = \frac{\dot{V}}{(\pi/4)D^2}\frac{D}{v} = \frac{4\dot{V}}{\pi Dv} \qquad (5.19)$$

TABLE 5.4

Recommended Pipe Sizes, Flows, and Velocities

Nominal Pipe or Tubing Sizes, in	Steel Pipe				Copper Tubing			
	Minimum gal/min	Velocity, ft/s	Maximum gal/min	Velocity, ft/s	Minimum gal/min	Velocity, ft/s	Maximum gal/min	Velocity, ft/s
$\frac{1}{2}$	—	—	1.8	1.9	—	—	1.5	2.1
$\frac{3}{4}$	1.8	1.1	4.0	2.4	1.5	1.0	3.5	2.3
1	4.0	1.5	7.2	2.7	3.5	1.4	7.5	3.0
$1\frac{1}{4}$	7.2	1.6	16	3.5	7	1.9	13	3.5
$1\frac{1}{2}$	14	2.2	23	3.7	12	2.1	20	3.6
2	23	2.3	45	4.6	20	2.1	40	4.1
$2\frac{1}{2}$	40	2.7	70	4.7	40	2.7	75	5.1
3	70	3.0	120	5.2	65	3.0	110	5.2
$3\frac{1}{2}$	100	3.2	170	5.4	90	3.1	150	5.2
4	140	3.5	230	5.8	130	3.5	210	5.6
5	230	3.7	400	6.4	—	—	—	—
6	350	3.8	610	6.6	—	—	—	—
8	600	3.8	1200	7.6	—	—	—	—
10	1000	4.1	1800	7.4	—	—	—	—
12	1500	4.3	2800	8.1	—	—	—	—

Source: Courtesy of Taco, Inc., Cranston, RI. With permission.

When Eqs. (5.18) and (5.19) are inserted into Eq. (5.16), this is the result:

$$\dot{V} = -0.966\sqrt{gD^5(h_f/L)}\ln\left[\frac{\epsilon}{3.7D} + \frac{1.782vD}{\sqrt{gD^5(h_f/L)}}\right] \tag{5.20}$$

For this example, the quantity under the radical has the value [recall that $h_f = \Delta p/(\rho g)$]

$$\sqrt{gD^5\left(\frac{h_f}{L}\right)} = \left(9.81 \times 0.4^5 \times \frac{225}{9.81 \times 993}\right)^{1/2} = 0.0481 \text{ m}^3/\text{s} \tag{5.21}$$

From Eq. (5.20) the volumetric flow rate is

$$\dot{V} = -0.966 \times 0.0481 \ln\left(\frac{0.00064}{3.7} + \frac{1.782 \times 0.6831 \times 10^{-6} \times 0.4}{0.0481}\right)$$

$$= 0.400 \text{ m}^3/\text{s} \ (400 \text{ L/s}, 6350 \text{ gal/min})$$

COMMENTS

Designers often use rules of thumb such as 3 to 6 ftWG (mWG) of pressure drop per 100 ft (m) of pipe length (or Pa/m) for preliminary design. This example illustrates how to accurately find the flow in such a case. Of course, a rough solution can be found from Figure 5.4b, which gives a flow of 400 L/s, too.

Straight lengths of pipe are connected by elbows and tees, are interrupted by valves to control flow, and must include ancillary equipment such as filters. Pipe fittings such as valves, elbows, tees, and other equipment have pressure drops that are of the same order of magnitude as pressure losses through straight pipe. As noted earlier, the pressure drop through fittings is given by Eq. (5.10). We denote the pressure loss coefficient for fittings by K_f. Therefore, the fitting pressure loss is

$$\Delta p_f = K_f \rho \frac{v^2}{2} \tag{5.22}$$

Or in terms of head of flowing fluid, the pressure loss is

$$h_f = K_f \frac{v^2}{2g} \tag{5.23}$$

The total loss in a single loop of uniformly sized pipe consisting of lengths of straight pipe and fittings is

$$h_L = \frac{v^2}{2g} \left(\frac{fL}{D} + \sum_{\text{fittings}} K_f \right) \tag{5.24}$$

Note that the pipe length L includes the length of both the straight pipe and the fittings. The K_f factors account for the *excess* pressure drop above that which would occur if the physical length of the fitting were replaced by straight pipe of the same length. Figure 5.5 shows sketches of common fittings used in HVAC systems. The K_f value for many fittings is a strong function of pipe size. This information is contained in Figure 5.6. The left column lists the fitting type, and the left scale is the internal pipe diameter in inches. The right scale shows the K_f values. Each large dot in the center of the figure has a number associated with it corresponding to an entry in the list to the left. To find the value of K_f for a specific fitting, pass a straightedge through the pipe diameter and the dot corresponding to the fitting. The extension of the straightedge to the right scale gives the value of K_f. Fittings for which K_f is independent of the pipe size have their coded dots located on the K_f axis in Figure 5.6. If a size change occurs in a fitting (e.g., items 46 to 55 in Figure 5.6), the velocity to be used in Eqs. (5.23) and (5.24) is the velocity in the smaller-diameter connection.

Example 5.2: Pumped-Fluid Pipe Circuit

Figure 5.7 shows a simple circuit for pumping pure ethylene glycol from an outdoor holding tank to an indoor tank at the same elevation. Plastic pipe is used; its actual inside diameter is 5.05 in (nominal 5-in pipe), and the roughness height is 0.00002 ft. If the flow rate is 500 gal/min on a day when the fluid temperature is 77°F, what is the pump shaft input power requirement if the pump efficiency is 67 percent?

Given: $D = 5.05$ in $= 0.42$ ft

$\epsilon = 0.00002$ ft

$\dot{V} = 500$ gal/min $= 1.12$ ft³/s

$\eta_p = 0.67$

Figure: See Figure 5.7.

Find: $\dot{W}_p$

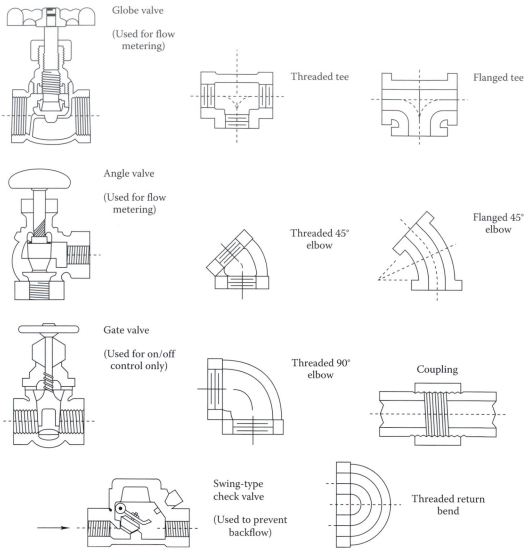

FIGURE 5.5
Typical pipe fittings used in HVAC systems.

Lookup values: Glycol viscosity $\mu = 109 \times 10^{-4}$ lb$_m$/(ft · s)
Glycol density $\rho = 68.47$ lb$_m$/ft^3
Both are from the appendix in the CD-ROM.

SOLUTION

The first step is to determine the K_f values for each fitting by using Figure 5.6. For the square-edged exit from the tank (fitting 23), $K_f = 0.5$. Likewise for the pipe exit to the receiving tank, $K_f = 1.0$ (one velocity head is lost). The flanged gate valve (assumed to be fully open) has $K_f = 0.04$ according to the nomograph (fitting 35). Finally, the flanged swing check valve has $K_f = 2.0$ (fitting 32). The sum of the K_f values is 3.54.

No.   Fitting
1. Coupling or union
2. Elbow, 90°, std., threaded
3. Elbow, 90°, std., flanged
4. Elbow, 90°, std., welded
5. Elbow, 90°, long-radius, threaded
6. Elbow, 90°, long-radius, flanged
7. Elbow, 90°, long-radius, welded
8. Elbow, 45°, std., threaded
9. Elbow, 45°, std., flanged
10. Elbow, 45°, std., welded
11. Elbow, 45°, long radius, flanged
12. Elbow, 45°, long radius, welded
13. Return bend, std., threaded
14. Return bend, std., flanged
15. Return bend, std., welded
16. Return bend, long-radius flanged
17. Return bend, long-radius welded
18. Tee, line flow, threaded
19. Tee, branch flow, threaded
20. Tee, line flow, flanged
21. Tee, branch flow, flanged
22. Bell-mouth tank entrance
23. Square-edged tank exit
24. Inward-projecting ripe
25. Circular miter, 25°
26. Circular miter, 30°
27. Circular miter, 35°
28. Circular miter, 40°
29. Foot valve
30. Angle valve, threaded
31. Angle valve, flanged, > 4 in.
32. Swing check-valve, flanged
33. Lift check-valve
34. Gate valve, threaded
35. Gate valve, flanged
36. Globe valve, flanged, > 5 in.
37. Plug valve, threaded
38. Plug valve, flanged
39. Butterfly valve, 5°
40. Butterfly valve, 10°
41. Butterfly valve, 20°
42. Butterfly valve, 40°
43. Plug-cock valve, 5°
44. Plug-cock valve, 10°
45. Plug-cock valve, 20°
46. Sudden enlargement, $d/D = 5$
47. Sudden enlargement, $d/D = 2.5$
48. Sudden enlargement, $d/D = 1.67$
49. Sudden enlargement, $d/D = 1.25$
50. Sudden contraction, $d/D = 0.8$
51. Sudden contraction, $d/D = 0.6$
52. Sudden contraction, $d/D = 0.4$
53. Sudden contraction, $d/D = 0.2$
54. Orifice, $d/D = 0.6$
55. Orifice, $d/D = 0.8$

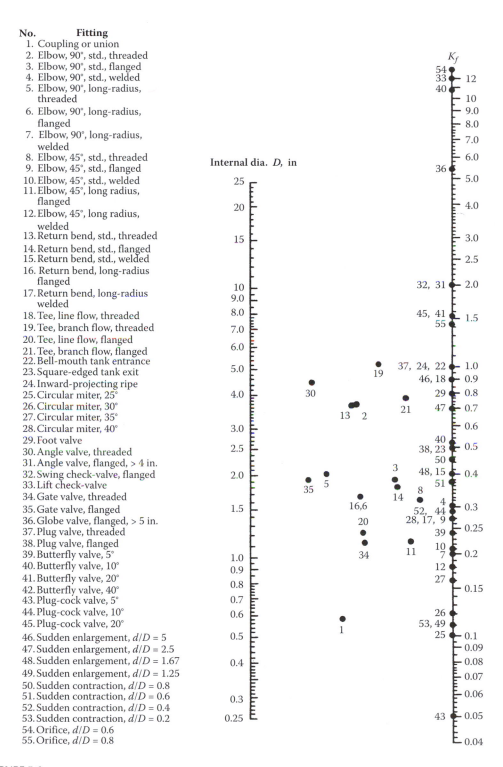

FIGURE 5.6

Nomograph for pipe fitting friction factors. (Courtesy of Rao, K.V.K., *Chem. Eng.*, 89, 1982. With permission.)

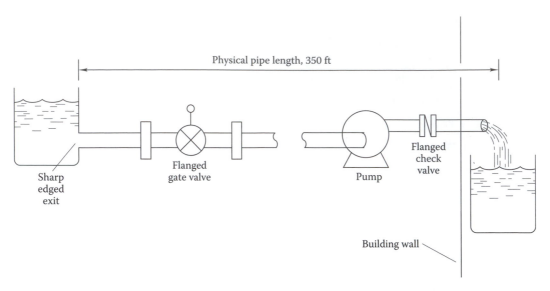

FIGURE 5.7
Piping diagram for Example 5.2.

The friction drop in the 350 ft of pipe is found next. The mean velocity is

$$v = \frac{\dot{V}}{A} = \frac{1.12 \text{ ft}^3/\text{s}}{(\pi/4)(5.05/12)^2 \text{ ft}^2} = 8.0 \text{ ft/s} \tag{5.25}$$

Next the Reynolds number is calculated:

$$Re = \frac{(8.0 \text{ ft/s})[(5.05/12) \text{ ft}](68.47 \text{ lb}_m/\text{ft}^3)}{109 \times 10^{-4} \text{ lb}_m/(\text{ft} \cdot \text{s})} = 21{,}150 \tag{5.26}$$

Since the Reynolds number is above the laminar limit (~3000), the flow is turbulent. Equation (5.15) is used to find the friction factor f:

$$f = \frac{1.325}{\{\ln[0.00002/(3.7 \times 0.42) + 5.74/21{,}150^{0.9}]\}^2} = 0.0256 \tag{5.27}$$

The reader can confirm this value approximately by referring to Figure 5.3. The pump head is found from Eq. (5.24):

$$h_f = \frac{8.0^2}{2g}\left[0.0256\left(\frac{350}{0.42}\right) + 3.54\right] = 24.8 \text{ ft} \tag{5.28}$$

Finally the power imparted to the fluid is found from

$$\dot{W} = \dot{V}\Delta p = \dot{V}\rho g h_f \tag{5.29}$$

$$= (1.12 \text{ ft}^3/\text{s})\left[\frac{68.47 \text{ lb}_m/\text{ft}^3}{32.2 \text{ lb}_m \cdot \text{ft}/(\text{lb}_f \cdot \text{s}^2)}(32.2 \text{ ft/s}^2)(24.8 \text{ ft})\right]$$

$$= 1902 \text{ (ft} \cdot \text{lb}_f)/\text{s} = 3.46 \text{ hp} \tag{5.30}$$

Since the pump is 67% efficient, the pump shaft input power is

$$\dot{W}_{act} = \frac{3.46}{0.67} = 5.2 \text{ hp} \tag{5.31}$$

COMMENTS

Where USCS units are used, pumps are rated in terms of gallons per minute of flow and feet of head. The solution of this problem is given in *feet of glycol*, not feet of water, the usual fluid used to specify the pump pressure rise.

During preliminary design, piping layouts are not made in sufficient detail to allow a proper calculation of fitting pressure drops. At this stage of design, it is appropriate to estimate fitting losses as 50% of the straight-pipe loss (which can be determined during early design). To this must be added the pressure drops of control valves and equipment such as coils, boilers, and chillers.

Fitting pressure losses are sometimes presented in terms of the *equivalent length* of straight pipe that would have the same pressure loss as the fitting. If a fitting is to be replaced by an equivalent length L_{eq} of pipe, then the following equality must hold, according to Eqs. (5.13) and (5.23):

$$f\left(\frac{L_{eq}}{D}\right)\left(\rho\frac{v^2}{2}\right) = K_f \rho \frac{v^2}{2} \tag{5.32}$$

From this we can solve for the equivalent length

$$L_{eq} = \frac{K_f}{f} D \tag{5.33}$$

This relationship shows the fundamental shortcoming with the equivalent-length approach. Even though K_f and D are constant for a given pipe fitting under various flow conditions, the friction factor is *not*, unless one is operating in the fully turbulent region where f is independent of Re (the region to the right of the dashed line in Figure 5.3). Since this is not always the case, the equivalent-length method must be used with caution. Later, when we discuss the flow of air in ducts where the Reynolds number is lower, the equivalent-length method is even a poorer approximation. The equivalent-length method is *not recommended* for use in HVAC design.

The pressure drop for *steam flow* in steel pipe can be found from Figure 5.8 in a two-step process. Figure 5.8a is used to find the pressure drop and velocity for atmospheric-pressure steam. Then Figure 5.8b is used to correct the velocity for the actual steam pressure. A correction for the pressure drop is not made. In Figure 5.8a, an example is shown by the dashed line. Steam at 100 psig and 6700 lb/h is flowing through a pipe with a pressure drop of 11 psi/100 ft. The intersection of the specified pressure drop line and the vertical steam flow line (about 2500 lb/h at 0 psig) indicates that a 2.5-in line should be used and that a 32,000 ft/min velocity of 0-psig steam would result. The fact of entering Figure 5.8b at 32,000 ft/min and dropping down and to the right indicates that the actual velocity of 100-psig steam would be about 13,000 ft/min. Typical steam velocities should not exceed

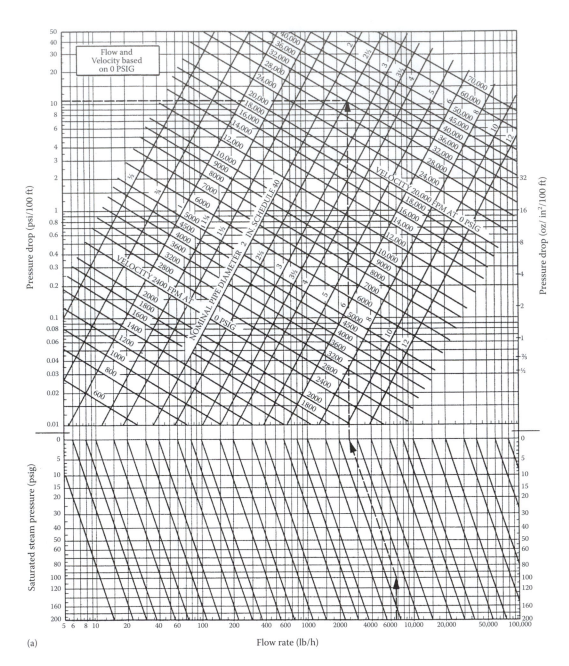

(a)

FIGURE 5.8

(a) Steam flow friction charts. After you find the values for 0-psig steam, use the second part of the figure to correct for the actual steam pressure. Dashed lines represent the text example. (Courtesy of ASHRAE, *Handbook of Fundamentals*, American Society of Heating, Refrigerating and Air-Conditioning Engineers, Atlanta, GA, 1989, 2001. With permission.)

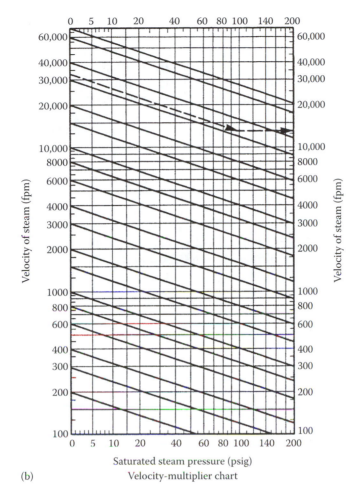

(b)

Saturated steam pressure (psig)

Velocity-multiplier chart

FIGURE 5.8 (continued)
(b) Chart used to adjust 0-psig velocities from Fig. 5.8a.

15,000 ft/min. Additional information on steam piping is contained in Grimm and Rosaler (1990) and ASHRAE (1989, 2001).

5.3.2 Pressure Losses in Ducts and Duct Fittings

The dimensionless equations governing the pressure loss in airflow confined in circular or rectangular ducts are the same as those that apply for the flow of water, Eqs. (5.13) and (5.15). Likewise, the pressure drop through duct fittings is given by an expression similar to Eq. (5.22).

   In dimensional form, however, one cannot use the pipe friction chart in Figure 5.4 for airflow. Instead Figure 5.9, prepared for standard air (ASHRAE, 2001), must be used. This particular chart applies for round, galvanized-steel ducts with joints every 48 in (equivalent to an average roughness $\epsilon = 0.0003$ ft). The chart may be used without correction (to ±5% accuracy) for altitudes up to 1500 ft, for temperatures between 40 and 100°F, for other duct materials with medium smoothness, and duct pressurizations of ±20 inWG. For materials or environments outside this range, the friction factor expression, Eq. (5.15), must be used. Note that common fiberglass ductwork is not classified as medium smooth, and Figure 5.9 should not be used for it. Wright (1945) describes the construction of the duct friction chart.

   Since ducts consume a relatively large building volume compared to pipes, it is sometimes necessary to use rectangular ducts to fit within confined spaces between floors or between the ceiling and the floor above in commercial buildings. Rectangular ducts of the same cross-sectional area as round ducts have a higher frictional loss, but Figure 5.9a can be used for rectangular ducts if an *equivalent diameter for friction* is used. The equivalent diameter for friction can be calculated from (USCS units)

$$D_{eq,f} = 1.30 \frac{(WH)^{0.625}}{(W + H)^{0.25}} \tag{5.34US}$$

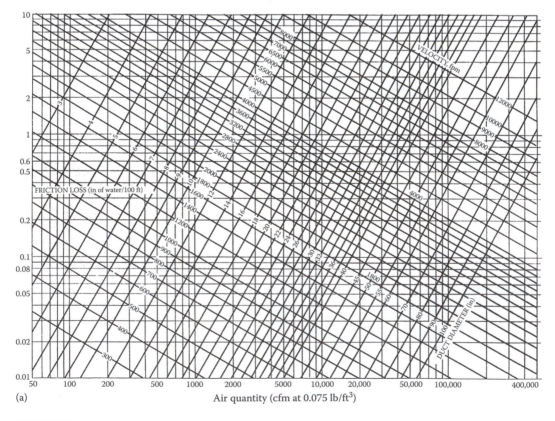

(a)   Air quantity (cfm at 0.075 lb/ft³)

FIGURE 5.9

(a) Duct friction chart in USCS units based on average roughness of $\epsilon = 0.0003$ ft and standard air.

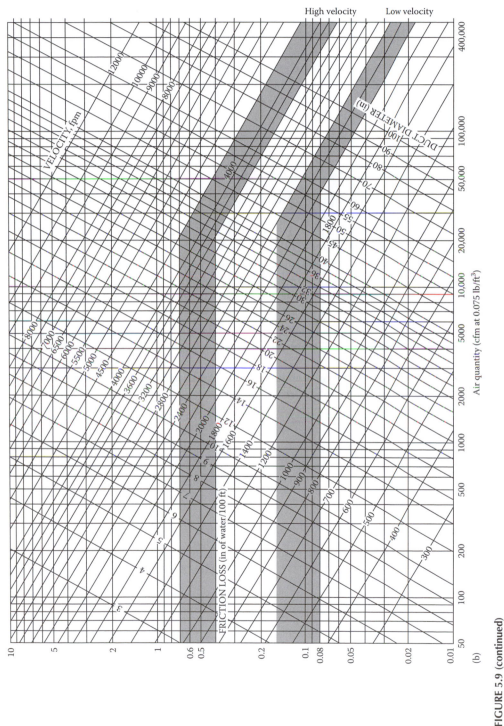

FIGURE 5.9 (continued)

(b) Duct friction chart in SI units based on average roughness of $\epsilon = 0.091$ mm and standard air. (Courtesy of ASHRAE, *Handbook of Fundamentals*, American Society of Heating, Refrigerating and Air-Conditioning Engineers, Atlanta, GA, 1989, 2001. With permission.)

where

$D_{eq,f}$ = diameter of circular duct with volumetric flow rate and friction drop equivalent to specified rectangular duct, in (Note that a circular duct with this diameter will *not* have the same average velocity as the rectangular duct.)

W = width of rectangular duct, in

H = height of rectangular duct, in

An additional complication present in airflows but not in water flows is the variation of the pressure drop with density. According to Eq. (5.13), the pressure drop at a given velocity (not a given mass flow rate) is proportional to the density. Therefore, if one knows the pressure drop at a given density (condition 1), the pressure drop at a different density (condition 2) is given by

$$h_{f,2} = h_{f,1}\frac{\rho_2}{\rho_1} \tag{5.35}$$

This linear dependence ignores the slight effect of density on the friction factor itself by way of its Reynolds number dependence.

In practical design situations, it is necessary to maintain a constant mass flow, not a constant volumetric flow, because the heating or cooling capacity of an airstream depends ultimately on the mass flow. Under these conditions, the designer first determines the pressure drop for a volume of *standard* air equal to the needed volume of nonstandard air. Then the correction from Eq. (5.35) is applied to the result. The variation of air density with altitude and temperature is given in the appendix in the CD-ROM.

Example 5.3: Duct Friction

What is the pressure drop in 250 ft of 30-in-diameter standard round duct for an actual airflow volume of 14,000 ft^3/min at an altitude where the density is 0.064 lb_m/ft^3?

Given: $L = 250$ ft

$\rho_{act} = 0.064$ lb_m/ft^3

$\dot{V} = 14,000$ ft^3/min

$D = 30$ in

Figure: See Figure 5.9a.

Assumptions: Steady, isothermal flow

Find: h_f

Lookup values: $\rho_{std} = 0.075$ lb_m/ft^3 (from the ideal gas law)

SOLUTION

The problem is solved in two steps. First, the friction loss for the actual volume of air is read from Fig. 5.9a. From the figure at a duct size of 30 in, we read 0.25 inWG/100 ft. Therefore,

$$h_{f,std} = \frac{0.25 \text{ inWG}}{100 \text{ ft}} \times 250 \text{ ft} = 0.63 \text{ in WG} \tag{5.36}$$

Next the density correction is applied according to Eq. (5.35):

$$h_{f,act} = 0.63 \text{ inWG}\left(\frac{0.064}{0.075}\right) = 0.54 \text{ in WG} \tag{5.37}$$

> **COMMENTS**
>
> If the mass flow, not the volumetric flow, were specified at nonstandard conditions, there would be two effects. The velocity in the duct would vary inversely as the air density, and the pressure drop would vary linearly with the density. The net effect on the pressure drop is calculated by the two-step procedure above if the duct friction chart is used.

Pressure losses in duct fittings are given by the general expression

$$\Delta p_f = C\rho \frac{v^2}{2} \tag{5.38}$$

where C is the dimensionless fitting (or local loss) coefficient; it is the airflow analog of K_f used in pipe fitting calculations. (As described in the previous section, the equivalent-length approach is not appropriate for duct pressure drop calculations.)

Table 5.5 is an example of a fitting loss tabulation. This table applies for smooth elbows in round ducts with fully developed entering flow. Depending on the elbow radius r, the C value can be read from the second line of the table. This value is denoted by C_{90} since it applies for a 90° elbow. For other elbow angles between 0° and 180°, a correction K_θ is needed. It is read from the fourth line of Table 5.5. The appendix in the CD-ROM contains selected values of loss coefficients for many round duct fittings. The fittings listed in the appendix include round

- Converging duct entries
- Smooth elbows
- Three-, four-, and five-piece elbows
- Rectangular elbows
- Elbows with turning vanes
- Transitions
- Converging tees
- Diverging tees

For rectangular duct and combined rectangular-round duct C values, the reader is referred to extensive tables in ASHRAE (2001) or Idelchik et al. (1986). Example 11.2 illustrates the use of the C tables in an air transport system. Example 5.4 illustrates the use of Table 5.5 for a single fitting.

TABLE 5.5

Friction Coefficients for Smooth-Radius Duct Elbows

Coefficients for 90° elbows

r/D	0.5	0.75	1.0	1.5	2.0	2.5
C_{90}	0.71	0.33	0.22	0.15	0.13	0.12

Angle correction factors K_θ

θ	0	20	30	45	60	75	90	110	130	150	180
K_θ	0	0.31	0.45	0.60	0.78	0.90	1.00	1.13	1.20	1.28	1.40

Source: Courtesy of ASHRAE, *Handbook of Fundamentals*, American Society of Heating, Refrigerating and Air-Conditioning Engineers, Atlanta, GA, 1989, 2001. With permission.

Example 5.4: Pressure Loss in an Elbow

What is the pressure drop in an 8-in-diameter smooth 90° elbow through which 250 ft³/min of standard air is flowing? The radius of the elbow is 8 in.

Given: $\dot{V} = 250$ ft³/min

$r/D = 8/8 = 1.0$

Figure: See Table 5.5.

Find: Δp_f

Lookup values: From Table 5.5 the C value is found to be

$C = 0.22$

SOLUTION

The average air velocity is

$$v = \frac{\dot{V}}{A} = \frac{250 \text{ ft}^3/\text{min}}{(\pi/4)\left(\frac{8}{12}\right)^2 \text{ ft}^2} = 716 \text{ ft/min} \tag{5.39}$$

The pressure drop is found from Eq. (5.38):

$$\Delta p_f = 0.22 \times \frac{0.075 \text{ lb}_m/\text{ft}^3}{32.2 \text{ (lb}_m \cdot \text{ft)}/(\text{lb}_f \cdot \text{s}^2)} \times \frac{(716/60)^2 \text{ ft}^2/\text{s}^2}{2} = 0.036 \text{ lb}_f/\text{ft}^2 \tag{5.40}$$

This consistent unit for pressure drop, denoted by Δp_f, is not used in practice. Instead, inches water gauge, the "head" denoted by h_f and described in the first section of this chapter, is used. The pressure drop is converted to these units as follows:

$$h_f = 0.036 \, \frac{\text{lb}_f}{\text{ft}^2} \times \frac{\text{ft}^2}{144 \text{ in}^2} \times \frac{27.7 \text{ inWG}}{\text{lb}_f/\text{in}^2} = 0.0069 \text{ inWG}$$

COMMENTS

The pressure loss at this typical flow is small. Elbows in round ducts with moderate turning radii do not have excessive losses.

   The lengthy unit conversion in this example can be avoided by using a *dimensional equation* for *standard air*. In USCS units, if velocity is in feet per minute, the pressure drop in a fitting is given by

$$h_f = C\left(\frac{v}{4005}\right)^2 \text{ inWG} \tag{5.41US}$$

In SI units with the velocity in meters per second, the fitting pressure loss in pascals for standard air is

$$\Delta p_f = C\left(\frac{v}{1.29}\right)^2 \text{ Pa} \tag{5.41SI}$$

These equations can be used for nonstandard air at the same velocity by multiplying the calculated pressure loss for standard air by the ratio of the actual air density to that of standard air.

5.4 Prime Movers

The pressure losses that occur in liquid and air systems are balanced by pressure increases in pumps and fans, respectively. The application of these components is described in greater detail in Chap. 11, Secondary Systems for Heating and Cooling. In this chapter we give an overview of these prime movers as they are used in HVAC systems.

The most common method for producing pressure in HVAC fluid loops is by use of centrifugal fans and pumps. A photograph of a fan is shown in Fig. 5.10. Within the housing of centrifugal machinery, the shaft torque is converted to a force that propels fluid by increasing the pressure forces acting on it. Figure 5.11 shows a single blade of the type used in either a fan or a pump. The blade rotates counterclockwise (in the *forward-curved* design) with outer tip tangential velocity u_o. The fluid leaving the blading has a velocity V_o with a useful tangential component V_{to}. As shown in Kreider (1985), the theoretical head produced by the centrifugal device is given by

$$h_{\text{ideal}} = \frac{u_o V_{to}}{g} \tag{5.42}$$

The actual performance of pumps and fans must account for losses and nonideal flow within the blading. Manufacturers present data for real machinery in terms of a head-capacity curve, as shown in Fig. 5.12. The data are measured by using standard procedures. Details of the use of such curves are given in Chap. 11. Their principal application is to select the *operating point* of a system. The operating point is the point at which the system pressure loss [given by Eq. (5.24) for liquid or air systems] is just balanced by the pressure developed by the pump or fan. If Eq. (5.24) is plotted on Fig. 5.12, it is approximately a parabola[3]

FIGURE 5.10
Photograph of a typical centrifugal fan used for building exhaust.

---

[3] The curve is not exactly parabolic since f and Kf can vary weakly with v.

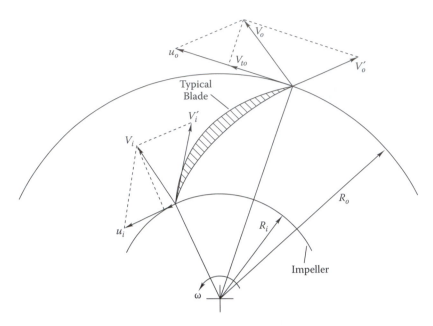

FIGURE 5.11
Velocity diagram for centrifugal machinery. Here V' is the fluid velocity relative to the blade, u is the blade tip velocity and V is the absolute fluid velocity.

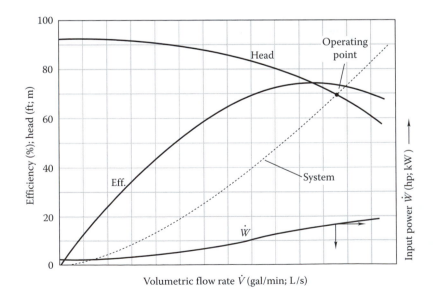

FIGURE 5.12
Typical pump or fan performance curve (solid curve). System curve is approximately parabolic (dashed). Intersection point is the system operating point.

passing through the origin. The operating point is the intersection of the two curves, as shown in the figure. Fan and pump curves provided by manufacturers often contain additional data such as efficiency, input power, and other design information. These complete curves are discussed in Chap. 11.

Since flow requirements in HVAC systems vary considerably in response to building loads that depend on the weather, time of day, and season of the year, fans and pumps may need to function at a number of different operating points, determined, e.g., by different fan or pump rotational speeds. As the speed, head, and flow rate vary, the power requirements also vary. The relationships among these key variables are called the *fan laws* for fans and the *affinity laws* for pumps. Table 5.6 summarizes the six fan laws (derived in standard fluid mechanics texts). The only requirement for the fan laws to apply is that the *efficiency remain constant* and that *geometric similarity in blading and housing be retained* as sizes change. This table gives the relationships among the following parameters that can vary during the operation of a fan or of one fan geometrically similar to another fan:

- Rotational speed N
- Volumetric flow $\dot{V}$
- Shaft power $\dot{W}$
- Pressure rise h
- Air density ρ
- Rotor diameter D

For example, the first line of the table shows that an increase in the speed of a fan of fixed size with constant air density increases the flow rate linearly, the head quadratically, and the power cubically.

TABLE 5.6

Summary of Fan Laws

	Rotational Speed N	Flow Rate $\dot{V}$	Fan Power $\dot{W}$	Produced Head h	Air Density ρ	Rotor Diameter D
Speed change	Varies	$\dot{V}_2 = \dot{V}_1 \dfrac{N_2}{N_1}$	$\dot{W}_2 = \dot{W}_1 \left(\dfrac{N_2}{N_1}\right)^3$	$h_2 = h_1 \left(\dfrac{N_2}{N_1}\right)^2$	Fixed	Fixed
Diameter change (fixed tip speed)*	$N_2 = N_1 \dfrac{D_1}{D_2}$	$\dot{V}_2 = \dot{V}_1 \left(\dfrac{D_2}{D_1}\right)^2$	$\dot{W}_2 = \dot{W}_1 \left(\dfrac{D_2}{D_1}\right)^2$	Fixed	Fixed	Varies
Diameter change*	Fixed	$\dot{V}_2 = \dot{V}_1 \left(\dfrac{D_2}{D_1}\right)^3$	$\dot{W}_2 = \dot{W}_1 \left(\dfrac{D_2}{D_1}\right)^5$	$h_2 = h_1 \left(\dfrac{D_2}{D_1}\right)^2$	Fixed	Varies
Density change	Fixed	Fixed	$\dot{W}_2 = \dot{W}_1 \dfrac{\rho_2}{\rho_1}$	$h_2 = h_1 \dfrac{\rho_2}{\rho_1}$	Varies	Fixed
Density change	$N_2 = N_1 \left(\dfrac{\rho_1}{\rho_2}\right)^{1/2}$	$\dot{V}_2 = \dot{V}_1 \left(\dfrac{\rho_1}{\rho_2}\right)^{1/2}$	$\dot{W}_2 = \dot{W}_1 \left(\dfrac{\rho_1}{\rho_2}\right)^{1/2}$	Fixed	Varies	Fixed
Density change (fixed mass flow)	$N_2 = N_1 \dfrac{\rho_1}{\rho_2}$	$\dot{V}_2 = \dot{V}_1 \dfrac{\rho_1}{\rho_2}$	$\dot{W}_2 = \dot{W}_1 \left(\dfrac{\rho_1}{\rho_2}\right)^2$	$h_2 = h_1 \dfrac{\rho_1}{\rho_2}$	Varies	Fixed

* At the same rating point, e.g., at the maximum efficiency point.

Example 5.5: Fan Laws—Fan Speed Increase

If a fan delivers 20,000 ft^3/min of standard air at a static pressure of 3.0 inWG at a fan speed of 500 r/min and a power input of 18 hp, what speed, input power, and static pressure are required if the flow rate is increased to 28,000 ft^3/min by increasing the rotational speed?

Given: $\dot{V}_1 = 20,000 \ ft^3/min$

$N_1 = 500 \ r/min$

$h = 3.0 \ inWG$

$\dot{W}_1 = 18 \ hp$

$\dot{V}_2 = 28,000 \ ft^3/min$

Table: See Table 5.6.

Assumptions: The system to which the fan is connected is unchanged, and the fan efficiency is unchanged as well.

Find: $N_2, h_2, \dot{W}_2$

SOLUTION

The first line of Table 5.6 is used, since it applies to speed changes with fixed air density and fan size. The flow rate, input power, and produced head are found by using the equations in the third, fourth, and fifth columns of the table. To find the head and power, the new fan speed must be known. It is found by solving the flow speed equation for rotational speed, since the flow at both conditions is known:

$$N_2 = N_1 \frac{\dot{V}_2}{\dot{V}_1} = (500 \ r/min) \left(\frac{28,000 \ ft^3/min}{20,000 \ ft^3/min} \right) = 700 \ r/min \qquad (5.43)$$

The new head is now found from the fifth column of the table:

$$h_2 = h_1 \left(\frac{N_2}{N_1} \right)^2 = 3.0 \ inWG \left(\frac{700}{500} \right)^2 = 5.88 \ inWG \qquad (5.44)$$

Finally, the power input is found from the equation in the fourth column of Table 5.6:

$$\dot{W}_2 = \dot{W}_1 \left(\frac{N_2}{N_1} \right)^3 = 18 \ hp \left(\frac{700}{500} \right)^3 = 49.4 \ hp \qquad (5.45)$$

COMMENTS

The fan laws are among the most important tools for the air system designer. It is important to understand their applications while keeping in mind the limitations described earlier.

The performance of pumps operating at fixed efficiency is also governed by a set of very useful rules known as the *pump affinity laws*. Table 5.7 summarizes the affinity laws. The affinity laws are essentially the same as the fan laws in Table 5.6, except that the number of equations is smaller since liquids are essentially incompressible and density effects do not occur.

The power required to produce a fluid pressure rise across a pump or fan can be found from the first law of thermodynamics, Eq. (3.19). Kinetic and potential energy difference

TABLE 5.7

Summary of Pump Affinity Laws

	Flow Rate $\dot{V}$	Pump Power $\dot{W}$	Produced Head h
Speed Change (fixed rotor size)	$\dot{V}_2 = \dot{V}_1 \dfrac{N_2}{N_1}$	$\dot{W}_2 = \dot{W}_1 \left(\dfrac{N_2}{N_1}\right)^3$	$h_2 = h_1 \left(\dfrac{N_2}{N_1}\right)^2$
Diameter change (fixed rotor speed)*	$\dot{V}_2 = \dot{V}_1 \left(\dfrac{D_2}{D_1}\right)^3$	$\dot{W}_2 = \dot{W}_1 \left(\dfrac{D_2}{D_1}\right)^5$	$h_2 = h_1 \left(\dfrac{D_2}{D_1}\right)^2$
Density change	Fixed	$\dot{W}_2 = \dot{W}_1 \dfrac{\rho_2}{\rho_1}$	Fixed

* At the same rating point, e.g., at the maximum efficiency point.

terms are small, if not zero; heat transfer is small; and internal energy changes are negligible. The remaining terms in the first law are

$$\dot{W}_{\text{fluid}} = \dot{m}\frac{p_i - p_o}{\rho} \tag{5.46}$$

where
$\dot{W}_{\text{fluid}} = $ power input to fluid
$\dot{m} = $ mass flow rate
$p_i - p_o = $ inlet to outlet pressure rise

Alternatively, the shaft power can be expressed in terms of volumetric flow and pressure rise by using the equivalence $\dot{m} = \rho\dot{V}$:

$$\dot{W}_{\text{fluid}} = \dot{V}(p_i - p_o) \tag{5.47}$$

The power given by these expressions is seen to be negative, indicating that work is done on the fluid. The shaft power input is greater than the power input to the fluid because of turbulence and other inefficiencies within pumping machinery. The ratio of the fluid power to the shaft power is *thepump efficiency* η_p:

$$\eta_p = \frac{\dot{W}_{\text{fluid}}}{\dot{W}_{\text{shaft}}} \tag{5.48}$$

The fan efficiency is defined in a similar manner.

Sometimes it is useful to express the shaft power input in dimensional equations that employ the units used by practitioners. For *pumps*, the dimensional equations are as follows (units of each term are shown in parentheses):

$$\dot{W}_{\text{shaft}} \ (\text{hp}) = \frac{\dot{V} \ (\text{gal/min})\,H \ (\text{ft})}{3960\eta_{\text{pump}}} \tag{5.49US}$$

In SI units, the equivalent expression is

$$\dot{W}_{\text{shaft}} \ (\text{kW}) = \frac{\dot{V} \ (\text{m}^3/\text{h})\,H \ (\text{m})}{369\eta_{\text{pump}}} = \frac{\dot{V} \ (\text{L/s})\,H \ (\text{m})}{102\eta_{\text{pump}}} \tag{5.49SI}$$

In both equations, the head H is in feet (meters) of water. For other fluids, the shaft work expressions should be multiplied by the actual pumped-fluid specific gravity.

The dimensional equations for *fans* are similar, but the constants in the denominator differ since the customary units used for fan calculations differ from those used with pumps. In USCS units,

$$\dot{W}_{\text{shaft}} \ (\text{hp}) = \frac{\dot{V} \ (\text{ft}^3/\text{min}) \, H \ (\text{inWG})}{6356 \eta_{\text{fan}}} \tag{5.50US}$$

In SI units,

$$\dot{W}_{\text{shaft}} \ (\text{kW}) = \frac{\dot{V} \ (\text{m}^3/\text{s}) \, H \ (\text{cmWG})}{10.06 \eta_{\text{fan}}} = \frac{\dot{V} \ (\text{L/s}) \, \Delta p \ (\text{kPa})}{985.5 \eta_{\text{fan}}} \tag{5.50SI}$$

In determining the capacity of cooling coils, it is necessary to account for the *temperature rise* that occurs along with the pressure rise across a fan, because it can impose an additional cooling requirement. From the first law we can write (replacing the enthalpy difference by using specific heat)

$$\frac{\dot{W}_{\text{fluid}}}{\eta_f} = \dot{m} c_p \Delta T \tag{5.51}$$

Equation (5.46) can be used to simplify this expression, with the result

$$\Delta T = \frac{\Delta p}{\eta_f c_p \rho} \tag{5.52}$$

For a pressure rise in a fan of 1.0 inWG, the temperature rise with a 100% efficient fan is 0.37°F. For a 70% efficient fan, the temperature rise is 0.53°F. In commercial building HVAC systems, a pressure rise of several inches water gauge is common. The resulting temperature rise can be 2 to 3°F and *must be considered* when cooling coils and central cooling plants are sized. Note that "fan heat" occurs at the location of the air supply fan; therefore, if the fan is *downstream* of a cooling coil, the cooling load appears as a zone load whereas for a fan located *upstream* of the coil, the load appears directly as an additional load on the coil. Additional fan heat is also produced by return fans, as discussed in Chap. 11. If HVAC fan motors are located in the airstream, the air temperature rise is increased due to motor inefficiencies. This effect is accounted for by dividing the right side of Eq. (5.52) by the motor efficiency.

A similar thermal effect occurs when pressure increases in water flow through pumps. However, the density and specific heat of water are much larger than those for air, and according to Eq. (5.52), the temperature rise will be small. For example, the temperature rise in water with a pressure increase across a typical pump operating at 100 ft of head is 0.05°F. The dissipation of pressure energy by viscosity in piping also produces heat. For this example, the temperature rise is 0.13°F. Hence the total temperature increase is 0.18°F—about a 1% effect ignored by most designers. [A similar fluid heating in insulated ducts does not occur since adiabatic flow with friction in a compressible fluid such as air results in a minute temperature drop (<0.01°F) within the ducting system.]

5.5 Flow Measurement

Flow measurement in liquid streams and airstreams in buildings is needed for control and energy management purposes. It is not possible to measure velocity directly, however, unless expensive, Doppler-type devices are used. Instead, we use equipment that produces a pressure drop which can be converted to velocity by an equation specific to the equipment. Several devices are in common use—the venturi, orifice plate, and Pitot tube. The first and second produce an average flow value whereas the third produces a point measurement.

Figure 5.13 shows a venturi meter with two pressure taps. The *average throat velocity* is given by

$$v = CE\sqrt{\frac{2\Delta p}{\rho}} \tag{5.53}$$

where E depends on β, the throat-to-pipe-diameter ratio:

$$E = (1 - \beta^4)^{-1/2} \tag{5.54}$$

The venturi coefficient C corrects for differences in velocity between the uniform streamline flow [assumed in the Bernoulli equation from which Eq. (5.53) is derived] and the actual flow through a venturi including viscous effects. For the universal venturi tube

$$C = 0.9797 \tag{5.55}$$

When USCS units of density (lb_m/ft^3) are used in Eq. (5.53), the density value must be divided by g_c [32.2 ($lb_m \cdot ft)/(s^2 \cdot lb_f)$] for dimensional consistency.

Figure 5.14 shows an orifice plate used to produce a pressure difference related to the fluid velocity in the pipe (or duct) in which it is inserted. Equation (5.53) is used to find the average velocity, but the orifice coefficient for the pressure tap locations shown in the figure is given by

$$C = 0.5959 + 0.0312\beta^{2.1} - 0.184\beta^8 + 0.039\frac{\beta^4}{1 - \beta^4} - 0.0158\beta^3 + 91.71\frac{\beta^{2.5}}{Re_{D_{pipe}}^{0.75}} \tag{5.56}$$

The Reynolds number is based on the pipe diameter. Since the velocity v is not known a priori, all terms but the last are used to estimate v. This value is then used to find the Reynolds

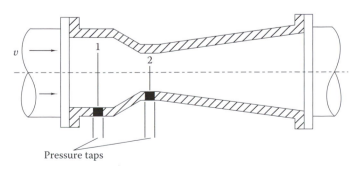

Pressure taps

FIGURE 5.13
Cross section of venturi meter.

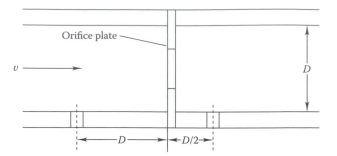

FIGURE 5.14
Orifice plate cross section with D and $D/2$ pressure taps.

number in the final term, and the final value of v is calculated. A different expression for C is used if different pressure tap locations are used (Kreider, 1985; Miller, 1983).

Orifice plates cause a significant *parasitic pressure drop* in any fluid stream in which they are located, but they have a low first cost. Venturi meters are more expensive initially but have much smaller pressure penalties than orifice plates. Life cycle power costs must be considered in the selection of flow measurement equipment that will be permanently installed.

Incidentally, Eq. (5.53) also applies for small orifices which occur in the envelopes of buildings. Of course, there one is not measuring flows in ducts, but rather is attempting to find air leakage rates given the pressure difference between the interior and exterior of a building. Instead of the square root, however, building leakage appears to depend on the pressure difference raised to the 0.65 power (Palmiter, 1991). We will use this result in Chap. 7 when calculating air infiltration rates.

Point rather than average velocities are sometimes needed in fluid streams in buildings for sensor calibration or equipment control. The Pitot tube shown in Figure 5.15 is used for this purpose. The local velocity is given by

$$v = \sqrt{\frac{2\Delta p}{\rho}} \tag{5.57}$$

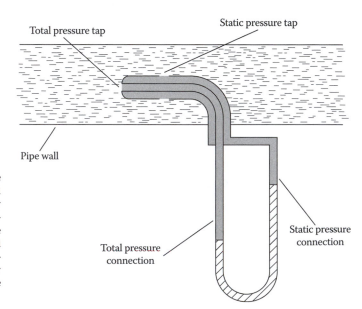

FIGURE 5.15
Pitot tube. When the densities of the flowing fluid and the manometer fluid are different, the pressure exerted by the column of flowing fluid (cross-hatched) must be accounted for in the Pitot equation. This is not required when a water column is used to measure airflow, however, since the density of air is much less than that of the manometer fluid, water.

Some manufacturers provide a multiplying coefficient similar to C in Eq. (5.53). However, if none is provided, Eq. (5.57) is used directly without adjustment. Pitot tubes can also be used to find the average velocity in a duct by averaging several point velocity measurements.

Example 5.6: Pitot Tube

A Pitot tube in an airstream registers a pressure reading of 2.1 inWG. What is the velocity of this airstream?

Given: $\Delta p = 2.1 \text{ inWG} \times [(144 \text{ lb}_f/\text{ft}^2)/27.7 \text{ inWG}] = 10.9 \text{ lb}_f/\text{ft}^2$

Figure: See Fig. 5.15.

Assumptions: Standard air in steady flow; $\rho = 0.075 \text{ lb}_m/\text{ft}^3$

Find: v

SOLUTION

Equation (5.57) is used to find the velocity:

$$v = \sqrt{\frac{2 \times 10.9 \text{ lb}_f/\text{ft}^2}{(0.075 \text{ lb}_m/\text{ft}^3)/\left[32.2 (\text{lb}_m \cdot \text{ft})/(\text{lb}_f \cdot \text{s}^2)\right]}} = 97 \text{ ft/s}$$

Turbine flow meters use a different approach to measure average velocity (or, equivalently, volumetric flow) in pipes. Figure 5.16 shows how a rotor is caused to move by the passage of fluid. Turbines are capable of 1% accuracy over long periods if properly maintained and if the measured fluid is kept clean by continuous filtration. Although expensive, one turbine meter in each fluid loop—e.g., the building's chilled-water loop—can serve as the measurement standard against which other flow devices can be periodically calibrated. Venturi meters are a good compromise between the permanent pressure penalty of

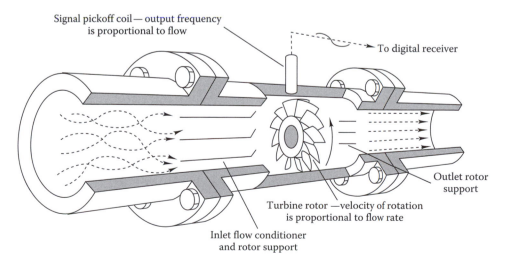

FIGURE 5.16
Cutaway drawing of turbine flow meter.

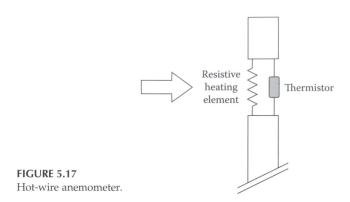

FIGURE 5.17
Hot-wire anemometer.

orifice plates and the initial cost of turbine meters when only average velocity (or, equivalently, the volumetric flow) is needed. Benedict (1984) discusses details of fluid flow measurement.

5.5.1 Hot-Wire Anemometers

Hot-wire anemometers rely on heating an airstream to determine the airflow rate. Most hot-wire anemometers operate by varying the amount of current passing through a resistive heating element in order to maintain a constant temperature (usually around 200°F) at a downstream thermistor. See Fig. 5.17. Other types keep the current constant and measure the change of downstream temperature. Since the response of the thermistor will also depend on the air temperature, it is necessary to measure this value as well.

The current draw of a constant temperature hot-wire anemometer is

$$i = \sqrt{(A + B\sqrt{\dot{M}})(T_H - T_A)} \tag{5.58}$$

where A and B are the calibration constants, $\dot{M}$ is the mass flow rate, and T_H and T_A are the hot-wire and air temperatures, respectively. Note that this equation identifies the mass flow rate, not the volumetric flow rate. The mass flow rate is, of course, independent of density and of thermal conductivity. The density is used only to find the resulting air velocity in the vicinity of the sensor.

5.6 Summary

In this chapter we have reviewed the principles that govern the flow of fluid in ducts and pipes. Methods for finding the pressure losses in straight ducts and in fittings were summarized. Fans and pumps used to overcome frictional losses in conduits were described generally from a component performance viewpoint. The details of integrating prime movers into complete fluid systems will be discussed in Chap. 11.

Table 5.8 summarizes the important equations developed in this chapter having to do with the fluid mechanics of H VAC systems.

TABLE 5.8

Chapter 5 Equation Summary

Topic	Equation	Equation Number	Notes
Pressure drop	$\Delta p_f = f\left(\dfrac{L}{D_h}\right)\left(\rho\dfrac{v^2}{2}\right)$	(5.13)	D'arcy-Weisbach equation
	$f = \dfrac{1.325}{\{\ln[\epsilon/3.7D + 5.74/\mathrm{Re}^{0.9}]\}^2}$	(5.15)	Pipe turbulent friction factor
	$h_f = K_f\dfrac{v^2}{2g}$	(5.23)	Pipe fitting pressure drop
	$h_f = C\dfrac{v^2}{2g}$	(5.38)	Duct fitting pressure drop
Pumps and fans	$\dot{W}_{\text{fluid}} = \dot{V}(p_i - p_o)$	(5.47)	Pump or fan power input to fluid
	$\dot{W}_{\text{shaft}} \text{ (hp)} = \dfrac{\dot{V} \text{ (gal/min) } H \text{ (ft)}}{3960\eta_{\text{pump}}}$	(5.49US)	Dimensional pump input power equation
	$\dot{W}_{\text{shaft}} \text{ (hp)} = \dfrac{\dot{V} \text{ (ft}^3/\text{min) } H \text{ (inWG)}}{6356\eta_{\text{fan}}}$	(5.50US)	Dimensional fan input power equation
	$\Delta T = \dfrac{\Delta p}{\eta_f c_p \rho}$	(5.52)	Fan air temperature rise
Flow measurement	$v = \sqrt{\dfrac{2\Delta p}{\rho}}$	(5.57)	Pitot tube
	$v = CE\sqrt{\dfrac{2\Delta p}{\rho}}$	(5.53)	Venturi meter and orifice plate

Problems

The problems in this book are arranged by topic. The approximate degree of difficulty is indicated by a parenthetic italic number from 1 to 10 at the end of the problem. Problems are stated most often in USCS units; when similar problems are presented in SI units, it is done with approximately equivalent values in parentheses. The USCS and SI versions of a problem are not exactly equivalent numerically. Solutions should be organized in the same order as the examples in the text: given, figure or sketch, assumptions, find, lookup values, solution. For some problems, the Heating and Cooling of Buildings (HCB) software may be helpful. In some cases it is advisable to set up the solution as a spreadsheet, so that design variations are easy to evaluate.

5.1 Calculate the viscosity of pure water at 80 and 110°F (27 and 43°C). (*3*)

5.2 Convert Eq. (5.5) to an equality $\mu = kT^{0.67}$ by using the viscosity of air at 50, 100, and 150°F (10, 38, and 66°C). How sensitive is the proportionality constant k to the temperature? Does the use of absolute temperature improve the accuracy? (*7*)

5.3 What is the velocity pressure in standard air flowing at 200 ft/s (6 m/s)? Express the answer in inches water gauge (pascals). (*3*)

5.4 What is the velocity pressure in standard air flowing at 200 ft/min (1.0 m/s)? Express the answer in inches water gauge (pascals). (*3*)

5.5 Figure P5.5 shows a tank drain. At what point does the water stream strike the ground as a function of z_1 and z_2? Ignore friction effects. (5)

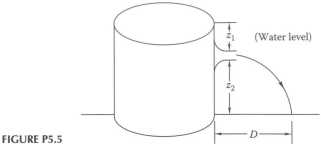

FIGURE P5.5

5.6 What are the pressure differences $p_2 - p_1$ and $p_3 - p_1$ in the tee shown in Fig. P5.6 if $D_2 = D_1/2 = 2D_3/3 = 4$ in (10 cm)? The inlet water flow rate of 3 ft^3/s (0.085 m^3/s) is equally split between the two exit streams. Ignore friction losses. (6)

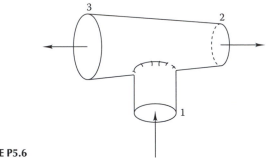

FIGURE P5.6

5.7 Water flows at 6 ft/s (2 m/s) through a straight 500-ft (150-m) pipe up a grade to a tank such that the pipe outlet is 30 ft (10 m) above the inlet where the pump outlet pressure is 150 psig (1 MPa). What is the pressure just prior to the pipe exit if friction losses are ignored for this 2-in-diameter (5-cm-diameter) pipe? How much power is provided to the fluid by the pumping system if the pump inlet pressure is 30 psig? (4)

5.8 What is the pressure p_2 in Fig. P5.8 if $y_1 = 60$ cm, $y_2 = 1$ m, $y_3 = 20$ m, and $\dot{V} = 0.006$ m^3/s? Neglect the friction losses in the 3-cm pipe used throughout. The nozzle diameter is 1 cm. (3)

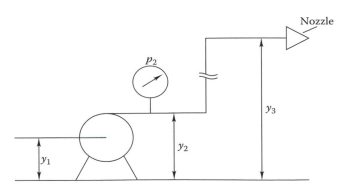

FIGURE P5.8

5.9 How much shaft work is required to pump 10 gal/min (6 L/s) of octane from one tank into another 50 ft (15 m) above? Friction losses are 6 percent of the static head, and the pipe size is uniform.   (4)

5.10 Calculate the velocity head in water flowing at 6 ft/s (2 m/s) in feet of water gauge (pascals).   (3)

5.11 Calculate the friction factor, using Eq. (5.15), in schedule 40, nominal, 4-in commercial steel pipe in which 60°F (15°C) water flows at 7 ft/s (2 m/s). What would the friction factor be if the fluid were 40 percent ethylene glycol?   (6)

5.12 Calculate the friction factor, using Eq. (5.15), in schedule 40, nominal, 2-in commercial steel pipe in which 180°F (82°C) water flows at 5 ft/s (1.5 m/s).   (4)

5.13 Water at 100°F (38°C) flows through nominal 5-in-diameter, schedule 40 commercial steel pipe with a typically used pressure drop of 4 ftWG/100 ft (4 m/100 m) of pipe. What is the volumetric flow rate?   (5)

5.14 Chilled water at 55°F (13°C) flows through nominal 5-in-diameter, schedule 40 commercial steel pipe with a pressure drop of 5 ftWG/100 ft (5 m/100 m) of pipe. What is the volumetric flow rate?   (5)

5.15 The schematic diagram in Fig. P5.15 represents the hot water distribution system in a residential baseboard heating system. The physical length of the pipe is 300 ft (90 m) of nominal $\frac{3}{4}$-in type L copper pipe (baseboard units are essentially straight lengths of pipe and are included in the 300 ft). The fluid flow rate is 4 gal/min (0.25/L/s) of 180°F (82°C) water. The boiler heat exchanger has the same pressure drop as 30 ft (10 m) of straight pipe; the valves are threaded, and the elbows are soldered. What is the pressure drop that the pump must overcome if all valves are wide open?   (7)

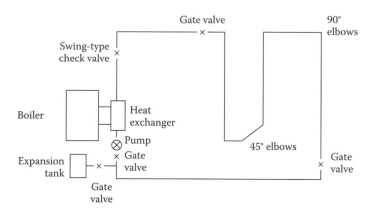

FIGURE P5.15

5.16 What would the pump motor energy savings be in Prob. 5.15 if nominal 1-in type L copper pipe and fittings were used instead of $\frac{3}{4}$-in and if the heating system operated 3500 h/yr? The combined pump motor efficiency is 0.70.   (9)

5.17 Figure P5.17 is the schematic diagram of a water piping loop. What is the pressure drop if the flow rate is 200 gal/min through 4.5-in nominal schedule 40 steel pipe if all fittings are threaded? What is the water velocity in the pipe?   (4)

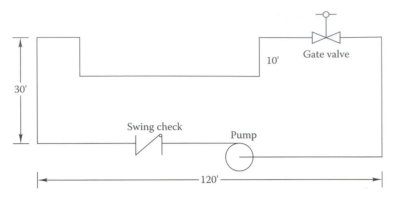

5.18 Figure P5.17 is the schematic diagram of a water piping loop. What is the pressure drop if the flow rate is 10 L/s through 100-mm nominal schedule 40 steel pipe? What is the water velocity in the pipe? (4)

5.19 A chilled-water loop carrying 200 gal/min (13 L/s) consists of 200 ft (60 m) of straight nominal 3-in schedule 40 steel pipe with one branch tee, one line flow tee, eight standard 90° elbows, and four gate valves. All fittings are flanged. Additional pressure drop occurs in the chiller (equivalent to 20 ft of water) and in the air handler cooling coil (equivalent to 30 ft of water). Comment on the water velocity and pressure drop of the straight pipe. (6)

5.20 Work Prob. 5.19 with nominal 3.5-in schedule 40 steel pipe, and compare to the same nominal size of type L copper pipe. (8)

5.21 What is the equivalent diameter (for pressure drop) of a duct 20 in × 48 in (50 cm × 120 cm)? (2)

5.22 If 0.25 inWG (60 Pa) is available to overcome friction in 300 ft (90 m) of 25-in-diameter (60-cm-diameter) ductwork, what is the volumetric flow rate at sea level?[4] What is the air velocity? (4)

5.23 If 0.25 inWG (60 Pa) is available to overcome friction in 300 ft (90 m) of 20-in-diameter (50-cm-diameter) ductwork, what is the volumetric flow rate at 5000-ft (1500-m) elevation? What is the air velocity? (4)

5.24 The velocity in a duct is 1800 ft/min (9 m/s). If 5000 ft³/min (2530 L/s) is to flow in this duct, what are the diameter and pressure drop in 150 ft (45 m). (4)

5.25 The pressure drop in the duct system of an HVAC system carrying 5000 ft³/min (2360 L/s) of 55°F (13°C) air is 2.0 inWG (500 Pa) at sea level. If the HVAC system is duplicated at an altitude of 5000 ft (1500 m), what will the pressure drop be if the volumetric flow rate remains the same? (3)

5.26 Repeat Prob. 5.25 except that the mass flow must remain constant, not the volumetric flow rate. (4)

5.27 If the sea-level system in Prob. 5.25 carries air at 120°F (49°C) rather than at 55°F (13°C), what will the pressure drop be for the same volumetric flow? (3)

---

[4] In the problems, *sea level* means that the atmospheric pressure is 14.696 psia (101.325 kPa). The effect of altitude on density can be found from Eq. (3.3).

5.28 Standard air [$\rho = 0.075$ lb/ft^3 (1.2 kg/m^3)] flowing through a heating coil (turned off) and downstream ductwork has a pressure drop of 1.0 inWG (250 Pa); the coil pressure drop is 0.3 inWG. If sufficient heat is added to the air to decrease the density to 0.062 lb/ft^3 (1.0 kg/m^3), what is the system pressure drop? What is the design point for the fan for this system?   (3)

5.29 A duct 4 ft × 4 ft carries 25,000 ft^3/min. What is the pressure drop per 100 ft?   (4)

5.30 An air distribution system carrying 10,000 ft^3/min (4700 L/s) consists of 750 ft (230 m) of 24-in-diameter (60-cm-diameter) ductwork, six $r/D = 2$, 90°, smooth elbows, one filter [0.2 inWG (50 Pa)], and a full-diameter butterfly damper. Find the system pressure drop if the damper is wide open ($\theta = 0°$) at sea level.   (5)

5.31 Consider the air system described in Prob. 5.30. As the butterfly damper is closed in 10° increments from fully open to 60°, how does the system pressure drop increase, assuming that sufficient fan capacity exists to force the required air amount through the system for all damper positions? Plot the system pressure drop versus damper position.   (7)

5.32 Work Prob. 5.30 at an elevation of 6000 ft (1800 m).   (5)

5.33 Figure P5.33 shows an air distribution system schematically. Find the pressure drop between the fan outlet and point *B* if the duct velocity is 1800 ft/min (9 m/s) everywhere. Ignore the pressure drop for (1) the tees (straight-through flow) and (2) duct-size changes. (See Chap. 11 for the method for treating these—they are *not* to be neglected in duct design.) The diffuser pressure drop is 0.1 inWG (25 Pa).   (9)

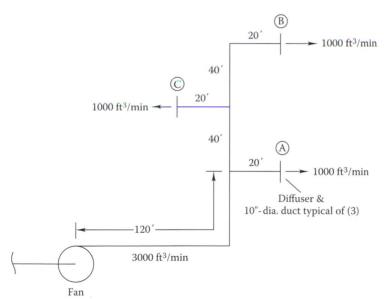

FIGURE P5.33

5.34 Rework Prob. 5.33 if the duct velocity is reduced to 1500 ft/min (7.6 m/s).   (9)

5.35 What is the shaft power input to a 68 percent efficient fan that moves 5000 ft^3/min (2400 L/s) against a 2.0-inWG (500-Pa) pressure drop at sea level? What would the power input be at 5000-ft (1500-m) altitude?   (3)

5.36 What is the shaft power input to an 85 percent efficient pump that moves 1000 gal/min (64 L/s) against a 3-psi (21-kPa) pressure drop? What is the electric power consumption if an 87 Percent efficient motor is used to drive the pump?   (3)

5.37 A duct system is designed to carry 9000 ft³/min (4200 L/s) with a pressure drop of 0.80 inWG (200 Pa). If the duct is connected to the oversized fan whose fan curve is shown by the down- and right-sloping solid line in Fig. P5.37, what will be the flow rate and pressure drop in the system? Curve *B* in the figure is the duct resistance curve for this case.   (3)

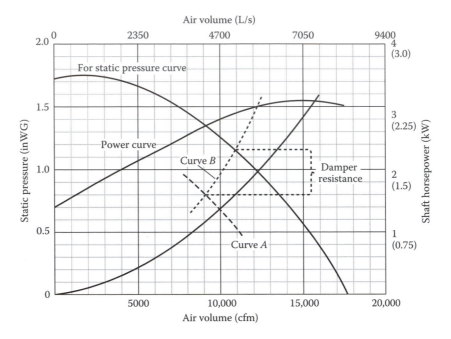

FIGURE P5.37

5.38 The flow rate determined in Prob. 5.37 is approximately 2000 ft³/min larger than the design. If a flow damper is inserted in the system, how much pressure drop must a damper produce to reduce the flow to the design level of 9000 ft³/min (4200 L/s)?   (4)

5.39 The flow rate determined in Prob. 5.37 is approximately 2000 ft³/min larger than the design. If the fan speed is reduced to provide flow equal to the design level of 9000 ft³/min (4200 L/s), what will the new fan speed be? Curve *A* in the figure is the fan curve for the reduced-speed case.   (3)

5.40 The system curve shown in Fig. P5.37 by a solid line has a system operating point of 12,500 ft³/min (5900 L/s) and 0.97 inWG (240 Pa). If a damper that has a 0.4-inWG (100-Pa) pressure drop is inserted in the system, what will the new flow rate be?   (4)

5.41 The system curve shown in Fig. P5.37 by a solid line has a system operating point of 12,500 ft³/min (5900 L/s) and 0.97 inWG (240 Pa). If the fan speed is reduced from 575 to 475 r/min, what will the system flow rate and fan power be? State any assumptions.   (4)

5.42 The system curve shown in Fig. P5.37 by a solid line has a system operating point of 12,500 ft³/min (5900 L/s) and 0.97 inWG (240 Pa). If the fan speed is increased from 575 to 675 r/min, what will the system flow rate and fan power be? State any assumptions.   (4)

5.43 Plot the fan efficiency curve for the fan whose characteristics are shown in Fig. P5.37.   (7)

5.44 If the fan whose sea-level characteristics are shown in Fig. P5.37 is moved to a 5000-ft (1500-m) altitude, how will the system, fan, and power curves change?   (5)

5.45 A chilled-water system requires a 100 gal/min (63 L/s) pump with 40 ftWG (115 kPa) of head. Select the most appropriate pump from the pump curves in Fig. P5.45. State the efficiency, motor power input, impeller size, and NPSH requirement for your selection. Note that the pump may not be a precise match to the specified flow condition.   (3)

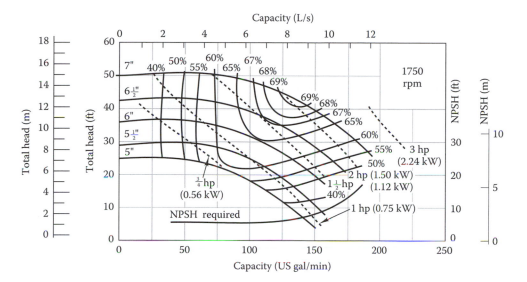

FIGURE P5.45

5.46 Suppose that you selected a 1750 r/min pump with a 7.0-in impeller in Prob. 5.45. What will the actual flow be? Plot the system and pump curves to solve this problem. Since the flow will be higher than that specified in Prob. 5.45, a flow control valve in series with the pump is partly closed. What will the efficiency and pump power be if the control valve is closed sufficiently to produce the desired 100 gal/min (6.3 L/s)? Plot the system curve for the partially closed valve situation also. Is there a more energy-efficient way to achieve design point operation than by use of this valve?   (8)

5.47 One more energy-efficient approach to controlling the flow in Prob. 5.46 is to slow the pump down with a variable-speed drive until the flow is at the desired level of 100 gal/min. Plot the new pump curve and the original system curve with the control valve wide open. Use the pump affinity laws to find the new pump curve and the power input. What is the pump efficiency?   (7)

5.48 Describe analytically the effect of pipe size on pump power consumption. If the cost of the pipe depends on the amount of material in the pipe (recall that larger pipes have thicker walls), how would you go about selecting pipe sizes based on the joint minimization of pipe material cost and pump initial and operating costs? The costs of pump and motor assemblies vary according to the 0.6 power of their size.   (9)

5.49 The wind speed is measured by a Pitot tube facing upwind. If the reading on a manometer connected to the tube is 1.23 in (3.1 cm) gauge oil (specific gravity of 0.826), what is the wind speed in feet per second (meters per second) at sea level and at 2500 ft (750 m) above sea level?   (4)

5.50 The velocity of water at the centerline of a pipe is measured by a Pitot tube facing upstream. If the reading on a manometer connected to the tube is 6 in (15 cm) of water, what is the water velocity in feet per second (meters per second)?   (4)

5.51 Size a venturi meter through which water flows to produce a pressure differential of 6.0 psia (41 kPa). The water flows in an 8-in (20-cm) pipe at 6 ft/s (2 m/s). What is the throat diameter of a universal venturi tube for this application?   (6)

5.52 Water at 40°C flows through a 120-mm-diameter pipe. It is proposed to measure the flow with a 30-mm, sharp-edged orifice with D and $D/2$ taps. If the measured pressure difference is 30 kPa, what is the flow rate?   (7)

5.53 Select the appropriate orifice to measure 50°F (10°C) water flowing at 700 lb/s (320 kg/s) in an 18-in-diameter (45-cm-diameter) pipe. The value of Δp must not exceed 80 inWG (20 kPa). An iterative solution will be required.   (8)

References

ASHRAE (1989, 2001). *Handbook of Fundamentals*. American Society of Heating, Refrigerating and Air-Conditioning Engineers, Atlanta.

Benedict, R. P. (1984). *Fundamentals of Temperature, Pressure and Flow Measurements*. Wiley, New York.

Born, D. W. (1989). "Inhibited Glycols for Corrosion and Freeze Protection in Water-Based Heating and Cooling Systems." *ASHRAE Trans.*, vol. 95, pt. 2, pp. 969–975.

Grimm, N. R., and R. C. Rosaler (1990). *Handbook of HVAC Design*. McGraw-Hill, New York.

Idelchik, I. E., G. R. Malyavskaya, and E. Fried (1986). *Handbook of Hydraulic Resistance*. Hemisphere, New York.

Kreider, J. F. (1985). *Principles of Fluid Mechanics*. Allyn and Bacon, Boston.

Miller, R. W. (1983). *Flow Measurement Engineering Handbook*. McGraw-Hill, New York.

Moody, L. F. (1944). "Friction Factors for Pipe Flow." *ASME Trans.*, vol. 66, p. 671.

Nussbaum, O. J. (1990). "Using Glycol in a Closed Circuit System." *Heating, Piping and Air Conditioning*, vol. 62, January, pp. 75–85.

Palmiter, L. (1991). Personal conversation.

Perry, R. H., D. W. Green, and J. O. Malrnay, eds. (1999). *Perry's Chemical Engineers' Handbook,* 6th ed. McGraw-Hill, New York.

Rao, K. V. K. (1982). "Nomograph for Pipe Fitting Friction Factors." *Chem. Eng.*, vol. 89.

Streeter, V. L., E. B. Wylie, and D. Bedford (1997). *Fluid Mechanics*. McGraw-Hill, New York.

Tuve, G. (1961). *Mechanical Engineering Experimentation*. McGraw-Hill, New York.

Waller, B. (1990). "Piping from the Beginning." *Heating, Piping and Air Conditioning,* vol. 62, October, pp. 51–71.

White, F. (1998). *Fluid Mechanics*, 4th ed. McGraw-Hill, New York.

Wright, D. K. (1945). "New Friction Factor for Round Ducts." *ASHVE J.*, pp. 577–584.

6

Solar Radiation and Windows

Solar radiation is an important term in the energy balance of a building, and one must account for it in a calculation of loads. This is particularly true for perimeter zones and for peak cooling loads. Apart from thermal effects, an understanding of solar radiation is necessary for the design of daylighting.

For peak loads, one needs the characteristics of solar radiation on a short time scale: hourly to daily. Peak heating loads occur during cold winter nights, and in most cases the contribution of solar radiation can be neglected (with the possible exception of buildings with significant thermal storage). For cooling loads, on the other hand, the solar contribution can be very important and should be evaluated with care. Usually the peak cooling load occurs on sunny days.

For the solar contribution to *annual energy consumption,* much longer time scales are relevant. Since one is interested in the consumption over the life of a building, the solar radiation data should be representative of the average over many years. Certainly the calculation should take the entire year into account. But fine time resolution of the solar radiation is rarely necessary for this purpose. For hand calculations, one often works in monthly steps, for the building may be sensitive to the month-to-month variations of solar radiation. Hourly variations have little influence on annual consumption: They tend to average out (especially when damped by the thermal inertia of buildings).

Thus, the insolation data needed by the designer are of two types: data that represent sunny days and data that represent long-term averages. If the design is done by means of an hour-by-hour computer simulation, the weather data should be representative of both the long-term average and days with peak loads. Much work has been done to assemble data sets that satisfy these requirements, and now such data are furnished routinely with the purchase of one of the standard simulation programs. For hand calculations, such an approach would be unmanageable, and simple models are used instead. Such models remain interesting despite the advance of the computer; they are, in effect, a concise summary of the essential characteristics of solar radiation and thus are a valuable guide for one's intuition. Being easier to manipulate than large data sets, they are also useful as research tools. Therefore, we present two solar radiation models, one for clear days and one for long-term averages.

The radiation received by a surface depends on the incidence angle. Accordingly, we begin, in Sec. 6.1, with a discussion of the required geometric relationships; the shading by obstacles or overhangs is also addressed.We proceed to a discussion of extraterrestrial insolation, in Sec. 6.2, because it is needed as the basis for most insolation models. In Sec. 6.3 we proceed to terrestrial solar radiation, including the separation of direct and diffuse components that must be made to calculate the radiation on tilted surfaces. Here we also present two models for solar radiation: one model for clear days and one for long-term averages. In Secs. 6.4 and 6.5, we address the calculation of solar heat gains in buildings. For opaque surfaces, it is convenient to use the concept of sol-air temperature, introduced in Sec. 6.4. In the final section, we treat the important subject of windows, presenting the calculation of the U value and shading coefficient as well as a brief survey of the technologies of high-performance glazings.

6.1 Solar Geometry

The earth moves about the sun in an elliptic orbit, completing one revolution per year (= 365.25 days). In addition, the earth rotates about its polar axis once per day. The polar axis is inclined at an angle of 23.45°. from the normal of the plane of the orbit (= ecliptic), as shown in Fig. 6.1. These rotations and the inclination between the polar axis and the ecliptic are responsible for the day-night cycle and the seasons. Some of the relevant geometric relationships are not intuitively obvious; they can be derived most simply by means of vector algebra. Here we state only the results, referring to Rabl (1985) for proofs.

For solar energy calculations, it is convenient to use coordinates fixed in the earth, as if the sun moved around the earth. For this point of view, one needs to define two important concepts: solar time and declination.

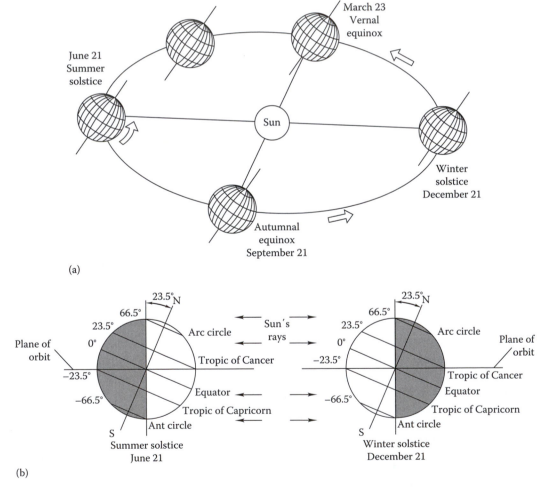

FIGURE 6.1
Geometry of the earth's orbit and inclination of polar axis: (a) entire orbit; (b) enlarged detail, solstices.

6.1.1 Solar Time

A brief discussion is needed about the specification of the time of day. *Universal time, or Greenwich civil time* $t_{\text{civ Gre}}$, is the time along the Greenwich meridian, the meridian of zero longitude. For other longitudes, *local civil time* $t_{\text{loc civ}}$ is retarded relative to $t_{\text{civ Gre}}$ by $\frac{1}{15}$ h for each degree of longitude west of Greenwich, since one full cycle corresponds to 360° longitude in 24 h. In most parts of the world, clocks are set to the same time within a time zone covering approximately 15° of longitude (although the boundaries may be quite irregular). That is the *standard time* t_{std} of the time zone, defined by the reference value of the longitude. For instance, in the contiguous United States, the reference meridians for the time zones are 75°W for Eastern, 90°W for Central, 105°W for Mountain, and 120°W for Pacific standard time. The *local civil time* $t_{\text{loc civ}}$ differs from standard time by $\frac{1}{15}$ h = 4 min for each degree of difference in longitude from the reference meridian.

Currently, most parts of western Europe and North America have instituted an advancement of the clock by 1 h during the summer half of the year (the "spring ahead, fall back" rule). This is called *daylight saving time* (DST). It reduces overall energy consumption by improving the match between human activities and the availability of daylight.

The time dependence of the incidence angles is simpler to express in terms of solar time than standard time. Solar time is based on the apparent motion of the sun as seen from a point on the surface of the earth, the deviation from local civil time being due to the nature of the orbit. Solar noon is the time when the sun reaches the highest point in the sky; it can differ from noon of local civil time by as much as one-quarter hour. The difference between solar noon and noon of local civil time is called the *equation of time* E_t. It is a function of the time of year and can be approximated by

$$E_t = 9.87 \sin 2B - 7.53 \cos B - 1.5 \sin B \text{ min} \tag{6.1}$$

with

$$B = 360° \times \frac{n - 81}{364} \quad \text{for } n\text{th day of year}$$

It is plotted in Fig. 6.2. *Solar time* t_{sol} (in hours) is defined, for each day, by

$$t_{\text{sol}} = t_{\text{loc civ}} + \frac{E_t}{60 \text{ min/h}} \tag{6.2}$$

Its relation to standard time t_{std} (in hours) is

$$t_{\text{sol}} = t_{\text{std}} + \frac{L_{\text{std}} - L_{\text{loc}}}{15°/\text{h}} + \frac{E_t}{60 \text{ min/h}} \tag{6.3}$$

where L_{std} and L_{loc} designate the longitudes (in degrees) of the time zone and the location, respectively. In regions with daylight saving time, one has to subtract 1 h from daylight saving time to obtain t_{std} during the summer half of the year.

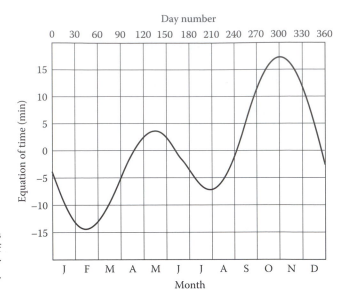

FIGURE 6.2
Equation of time E_t [Eq. (6.1)] (months are rounded to 30 days). (Courtesy of Duffie, J.A. and Beckman, W.A., *Solar Engineering of Thermal Processes*, Wiley, New York, 1980.)

Example 6.1

At what time t_{DST} (central European daylight saving time) is the sun due south in Paris, on July 21?

Given: $t_{sol} = 12{:}00$ at $L_{loc} = -2.48°$ (2.48°E), with $L_{std} = -15°$ (15°E)

Find: t_{std}

Lookup values: $E_t = -6$ min, from Fig. 6.2 or Eq. (6.1)

SOLUTION

Solve Eq. (6.3) for standard time:

$$t_{std} = 12{:}00 - 4[-15 - (-2.48)] - (-6 \text{ min}) = 12{:}56$$

The corresponding daylight saving time is

$$t_{DST} = 12{:}56 + 1{:}00 = 13{:}56$$

The relation between solar and daylight saving time is

$$t_{sol} = t_{DST} - 1{:}56$$

COMMENTS

The difference between solar and standard time is large because Paris lies quite far from the meridian of its time zone. Daylight saving time t_{DST} is almost 2 h ahead of solar time t_{sol} in this case.

6.1.2 Declination

Considering the geometry of Fig. 6.1 from the point of view of the earth, one can say that the sun traverses, each day and in solar time, one circular orbit around the earth. In general, this orbit does not lie in the plane of the equator; rather the line from sun to

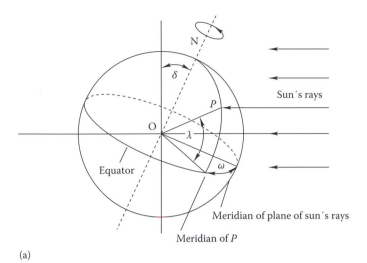

(a)

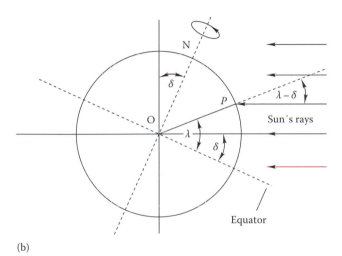

(b)

FIGURE 6.3
Latitude λ, hour angle ω, and declination δ. O = center of earth, N = north pole, P = point on earth's surface. (a) Three-dimensional view. (b) Cross-sectional view at solar noon.

earth makes an angle δ relative to the equatorial plane, as shown in Fig. 6.3. This angle is called the *declination*, and it is given by

$$\sin \delta = - \sin 23.45° \cos \frac{360° \times (n + 10)}{365.25} \tag{6.4}$$

where n = day of the year (with $n = 1$ for January 1). Numerical values can be found in Fig. 6.4.

The declination is a crucial quantity for calculating incidence angles. Consider a point P at latitude λ, as shown in Fig. 6.3. The incidence angle of the sun on the earth's surface at P is the angle between the normal of the surface at P and the line from P to the sun. It is called the *zenith angle* θ_s of the sun (see also Fig. 6.5). At solar noon, the ray from the sun to P is coplanar with the cross section of the earth through P and the poles; and since all are in a plane, the angles are easy to determine. From Fig. 6.3b one sees immediately that the zenith

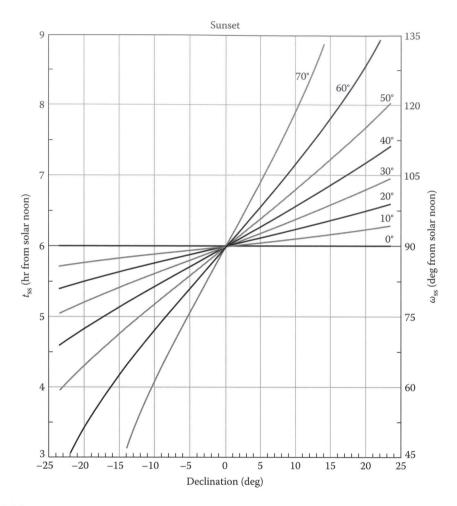

FIGURE 6.4
Sunset time t_{ss} and sunset hour angle ω_{ss} as functions of declination δ, the curves being labeled by latitude $\lambda(0°, 10°, \ldots, 70°N)$.

angle of the sun at noon is equal to the difference between the latitude and the declination. Incidence angles at other times of the day are far less obvious because they involve three-dimensional geometry. It can be shown that the the zenith angle θ_s for any latitude λ and any time of day and year is given by

$$\cos\theta_s = \cos\lambda \cos\delta \cos\omega + \sin\lambda \sin\delta \tag{6.5}$$

where

$$\omega = \frac{(t_{sol} - 12\,\text{h}) \times 360°}{24\,\text{h}} \tag{6.6}$$

is the *solar hour angle*.

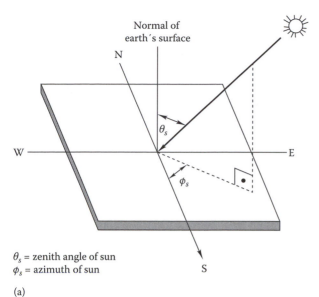

θ_s = zenith angle of sun
ϕ_s = azimuth of sun

(a)

θ_p = tilt of surface
ϕ_p = azimuth of surface
θ_i = incidence angle of sun on surface

(b)

FIGURE 6.5
Zenith angle and azimuth. (a) Zenith angle θ_s and azimuth ϕ_s of sun. (b) Zenith angle θ_p and azimuth ϕ_p of a plane and angle of incidence θ_i of sun on this plane.

Sunset occurs when the zenith angle reaches 90°. Setting $\cos \theta_s = 0$ in Eq. (6.5), we find that the sunset hour angle ω_{ss} is given by

$$\cos \omega_{ss} = -\tan \lambda \tan \delta \qquad (6.7)$$

The corresponding sunset time t_{ss} is plotted in Fig. 6.4. The lengths of day and of night are equal when ω_{ss} is 90°, which happens for $\delta = 0$. From Eq. 6.4 for δ, the corresponding days of the year follow as

$$n = -10 + 90° \times \frac{365.25}{360°} \qquad (= \text{March 23})$$

and

$$n = -10 + 270° \times \frac{365.25}{360°} \quad (=\text{September 21})$$

These dates are called *equinoxes* (whose Latin root implies "equal night"), and they mark the official beginning of spring and of fall. Declination and day length reach their extreme values at the *solstices*, June 21 and December 21.

For latitudes beyond $90° - 23.45° = 66.55°$ (the arctic circles), the absolute value of $\tan \lambda$ $\tan \delta$ can exceed unity at certain times of the year, implying that there is no real solution for ω_{ss}. This is the region of midnight sun and of winter days without daylight. The zone between 23.45°S and 23.45°N is called the *tropics*. Here the sun appears directly overhead at certain times of the year.

Example 6.2

What is the solar zenith angle at 16:00 (4:00 p.m.) daylight saving time in Paris, on July 21?

> *Given:*   Conditions of Example 6.1.
>
> Latitude $\lambda = 48.82°$
>
> $t_{sol} = t_{DST} - 1{:}56 \text{ h} = 14.07 \text{ h}$ from Example 6.1
>
> Day of year $n = 181 + 21 = 202$
>
> *Find:* θ_s

SOLUTION

First use Eq. (6.4) with $n = 202$ to find the declination

$$\sin \delta = -\sin 23.45° \cos \frac{360° \times (202 + 10)}{365.25} = 0.3482$$

Hence $\delta = 20.38°$; and the hour angle $\omega = 31.05°$ from Eq. (6.6). Finally find the zenith angle θ_s from Eq. (6.5):

$$\cos \theta_s = \cos 48.82° \cos 20.38° \cos 31.05° + \sin 48.82° \sin 20.38° = 0.7909$$

Hence $\theta_s = 37.73°$.

To indicate the position of the sun uniquely, one needs an additional quantity, the azimuth ϕ_s shown in Fig. 6.5a (in terms of a rotation about the vertical, ϕ_s is the angle from due south). One can show that it is related to the hour angle ω, declination δ, and zenith angle θ_s by

$$\sin \phi_s = \frac{\cos \delta \sin \omega}{\sin \theta_s} \tag{6.8}$$

It is positive in the afternoon, with ϕ_s pointing west. (Zenith and azimuth are the standard angles of the spherical coordinate system.) Instead of the zenith angle, many people employ the complement, called the *solar altitude angle*:

$$\text{Solar altitude angle} = 90° - \theta_s \tag{6.9}$$

For the analysis of shading problems, a plot of the solar altitude angle versus ϕ_s will be helpful, as in Fig. 6.6, with one curve for each month of the year, indicating the hour of the day along each curve.

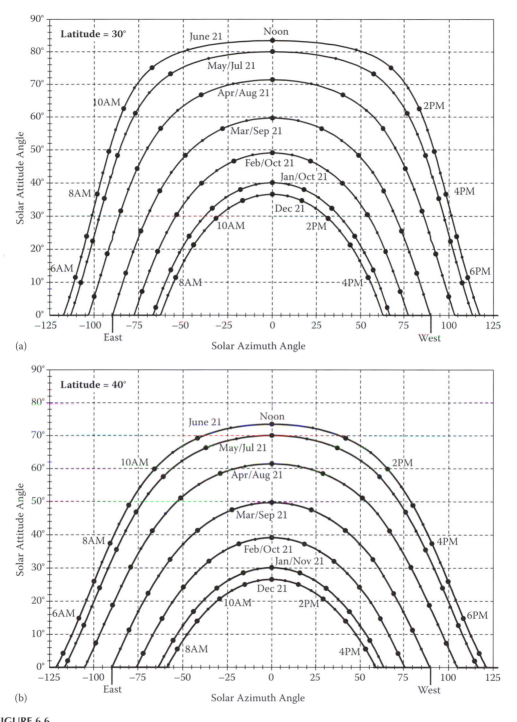

FIGURE 6.6
(a) Sun path diagram: solar altitude angle ($=90° - \theta_s$) versus azimuth ϕ_s. Time in legends is solar time. Latitude $\lambda = 30°$. (b) Latitude $\lambda = 40°$.

(continued)

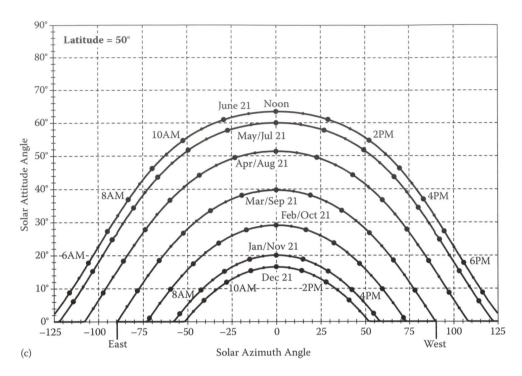

(c)

FIGURE 6.6 (continued)
(c) Latitude $\lambda = 50°$.

For the calculation of incidence angles on arbitrary planes, it is convenient to specify the orientation of the plane in terms of the tilt angle θ_p and azimuth ϕ_p of the surface normal (positive for orientations west of south), as indicated in Fig. 6.5b. The zenith angle θ_p of a plane is its tilt from the horizontal. In terms of these quantities, the incidence angle θ_i of the sun on the plane (= angle between normal of plane and line to sun) can be written in the form

$$\cos \theta_i = \sin \theta_s \sin \theta_p \cos (\phi_s - \phi_p) + \cos \theta_s \cos \theta_p \qquad (6.10)$$

This simplifies considerably for the important case of vertical planes ($\theta_p = 90°$):

$$\cos \theta_i|_{\text{vert}} = \sin \theta_s \cos (\phi_s - \phi_p) \qquad (6.11)$$

Example 6.3

Calculate the incidence angle of the sun on a west-facing vertical wall in Paris, at 16:00 daylight saving time on July 21.

Given: $\delta = 20.38°$

$\omega = 31.05°$

$\theta_s = 37.73°$ (conditions of Example 6.2)

$\theta_p = 90°$

$\phi_p = 90°$ for west-facing vertical wall

Find: $\cos \theta_i$

SOLUTION

From Eq. (6.8)

$$\sin \phi_s = \frac{\cos 20.38° \sin 31.05°}{\sin 37.73°} = 0.7901$$

Hence $\phi_s = 52.20°$. From Eq. (6.11)

$$\cos \theta_i = \sin 37.73° \cos(52.20° − 90°) = 0.4835$$

Hence $\theta = 61.08°$.

6.1.3 Shading and Overhangs

Rays from the sun to a point of a building are frequently blocked by other buildings, by trees, or by overhangs. The design of overhangs is important because it permits control of solar heat gains. Detailed analysis of the geometric relationships can be tedious.

Nowadays computer programs offer welcome relief. But even in the age of computers, it is useful to develop a certain intuition about some general features of shading.

Fig. 6.7 shows how the sun path diagram of Fig. 6.6 can be used to determine the times of day and year when an object will cast a shade at a point. One superimposes the outline

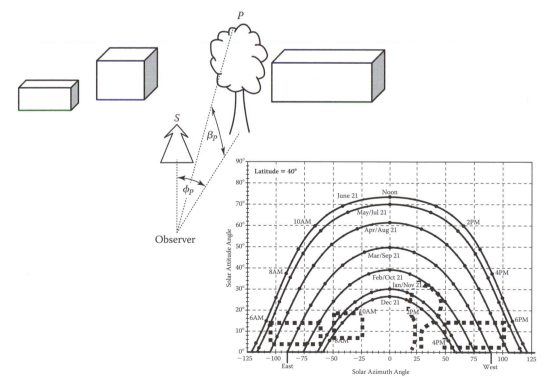

FIGURE 6.7

Outline of horizon superimposed on sun path diagram, to determine incidence of shading. β_P and ϕ_P are the altitude and azimuth angles of point P as seen from the position of the observer.

of the horizon, as seen from the point in question, on the sun path diagram. The point is shaded when the sun path passes below this horizon outline (the heavy solid line in Fig. 6.7).

Example 6.4

Suppose one wants to build a solar house with a ground-mounted collector at the position of the observer in Fig. 6.7. Will the tree cause shading?

    Given: The tree is indicated by the dotted outline in the sun path diagram of Fig. 6.7.

    Find: Times when the collector will be shaded by the tree

SOLUTION

There is no shade at all from the middle of February to the end of October. During the winter months, some shade is cast, but for at most 2 h, between 1 p.m. and 3 p.m.

COMMENTS

Shading is, of course, undesirable for the performance of a solar collector, but in practice it cannot always be completely avoided. In the present case it might be acceptable if the tree is deciduous.

    The method of Fig. 6.7 requires a separate horizon outline for each point where shading is to be evaluated. This method is practical when the distance to the shading object is large compared to the dimensions of the surface whose solar exposure is to be studied. But for cases such as an overhang above a window, the dimensions of the shaded surface are not small compared to the shading object, and different points of the window would see different horizon outlines. To analyze shading by nearby objects, we present a formula for calculating the location of the shadow cast by a point obstacle.

    Consider the shading of a rectangular window that is set back a distance z into a vertical wall, as shown in Fig. 6.8. Suppose the position of the sun is given in terms of the azimuth ϕ_s and zenith θ_s, while the azimuth of the wall is ϕ_p (defined, analogous to ϕ_s, as the angle

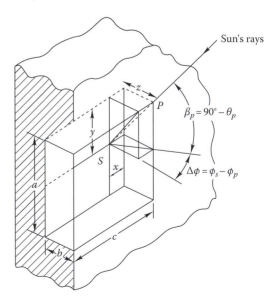

FIGURE 6.8
Coordinates of the shade $S = (x, y, z)$ cast by a point P (in this example P is corner of a rectangular window recess).

of the surface normal from due south, in terms of a rotation about the vertical; see Fig. 6.5b). Then the azimuth of the sun relative to the wall is

$$\Delta\phi = \phi_s - \phi_p \tag{6.12}$$

Let x, y, and z be the coordinates of point S, the shadow of the outer corner P of the window. These coordinates are relative to P: x is the width of the shadow cast by the vertical edge, analogously for the height y and the horizontal width z, as shown in Fig. 6.8. It is easy to see that x is given by

$$x = z \tan \Delta\phi \tag{6.13}$$

Using vector algebra, one can prove that

$$y = z\frac{\cot\theta_s}{\cos\Delta\phi} \quad \text{with } \Delta\phi = \phi_s - \phi_p \tag{6.14}$$

Having the formula for the shadow of any point, one can analyze any shading problem point by point.

Of particular importance in practice are straight shading devices such as an overhang above a window. The analysis is straightforward if the shading device is long enough that end effects can be neglected. One calculates the zenith and azimuth angles of the sun and evaluates the depth y of the shade according to Eq. (6.14).

Example 6.5

A southwest-facing window of height $a = 2$ m (6.56 ft) has right at its top a horizontal overhang of width $b = 0.5$ m (1.64 ft). The overhang extends far enough to either side that the window is shaded uniformly along its entire length. The incidence angles of the sun are as in Example 6.3. What fraction of the window is shaded?

Given: $a = 2$ m

$b = 0.5$ m

$\phi_s = 52.085°$

$\theta_s = 37.69°$

No end effects

Sketch: Fig. 6.8

Find: y/a

SOLUTION

$$\phi_p = 45° \quad \text{for southwest-facing vertical surface}$$

$$\Delta\phi = 52.085° - 45° = 7.085°$$

$$y = b\frac{\cot\theta_s}{\cos\Delta\phi} = 0.5\,\text{m} \times \frac{\cot 37.69°}{\cos 7.085°} = 0.65\,\text{m}$$

$$\frac{y}{a} = \frac{0.65}{2.00} = 0.33$$

One-third of the window is shaded at this time.

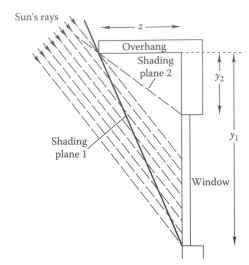

FIGURE 6.9
Shading of south-facing window
by horizontal overhang.

Sometimes a designer may want a device that will block direct solar radiation from entering a window during certain times of the day or the year. That may be desirable for two reasons: to reduce the cooling loads and/or to avoid uncomfortable lighting in perimeter offices due to excessive contrast. With fixed devices, it is generally not possible to satisfy all the requirements. Relatively good results are possible for south-facing windows, and the analysis is particularly easy. Consider a south-facing window with a horizontal overhang of width z, as in Fig. 6.9.

For this case one can show that the depth y of the shadow attains its extreme value for the day at solar noon (see, e.g., Rabl, 1985). In winter, the extreme value is a minimum, given by

$$y_{min} = z \cot(\lambda - \delta) \quad \text{for south-facing overhangs when } \delta < 0 \text{ (winter)} \qquad (6.15a)$$

while in summer the value at noon is a maximum

$$y_{max} = z \cot(\lambda - \delta) \quad \text{for south-facing overhangs when } \delta > 0 \text{ (summer)} \qquad (6.15b)$$

Thus, if one chooses z such that the shadow falls on the bottom edge of the window according to Eq. (6.15b) with a certain declination $\delta > 0$ (summer half of the year), then the window will be completely shaded for all days with declination greater than this value of δ. For passive solar heating, one would like the overhang to allow direct insolation in winter while blocking it in summer. Example 6.6 shows to what extent that is possible.

Example 6.6

A south-facing window of height $\Delta y = y_1 - y_2 = 2$ m (6.56 ft) is to receive full sunshine until February 15 while being completely shaded beginning May 15 (see Fig. 6.9). What is the required width z of the overhang and at what height y_2 should it be above the top of the window? The site is Denver, Colorado, at latitude 40°.

  Given: $\lambda = 40°$

  $n = 46$ for February 15

$n = 135$ for May 15

Sketch: Fig. 6.9

Find: z

SOLUTION

First we need the declination for these dates, from Eq. (6.4):

$$\delta_{Feb} = -13.13° \quad \text{for Feb. 15} \quad \delta_{May} = 18.51° \quad \text{for May 15}$$

Then we find the position of the shade ($=$ distance below overhang) from Eq. (6.15):

$$y_{Feb} = z \cot(\lambda - \delta_{Feb}) = y_2 \quad \text{on Feb. 15}$$
$$y_{May} = z \cot(\lambda - \delta_{May}) = y_1 \quad \text{on May 15}$$

The difference $\Delta y = y_{May} - y_{Feb}$ should be equal to the height 2 m (6.56 ft) of the window. Combine these equations and solve for z. The answer is

$$z = \frac{\Delta y}{\cot(\lambda - \delta_{May}) - \cot(\lambda - \delta_{Feb})} = 1.12 \text{ m (3.67 ft)}$$

Then $y_{Feb} = 0.84$ m (2.76 ft) $= y_2 =$ height of overhang above window.

COMMENTS

The dimensions are not unreasonable for a house.

The constraints of passive solar heating cannot be satisfied to perfection. The shading effects are necessarily determined by the declination, while the heating and cooling loads reach their extreme values about a month later than the solar declination. The situation is particularly awkward around the equinoxes, yielding either too little heat in spring or too much in the fall. Furthermore, the better one wants to separate the shaded and unshaded phases, the larger the shading device needs to be relative to the window. The problem is much simpler in tropical locations where solar heating is of no interest and the sun is so high that most of it can be blocked from a south- or north-facing window by an overhang of reasonable dimensions.

If one wants to achieve better control of solar gains, one must vary the transmission. Adjustable exterior blinds do allow good control, but they are bulky mechanical devices, prone to malfunction. Internal blinds can be lighter and cheaper than external blinds, but with lower thermal performance to the extent that some of the radiation is absorbed inside the conditioned space. Switchable coatings of the future, such as electrochromic coatings, promise a more elegant solution.

The prospects of technology should not make us forget a natural alternative. Deciduous trees or vines provide shade and sunlight in a seasonal pattern well matched to cooling and heating loads. Besides, the beauty of plants has universal appeal. Of course, there are certain expenses for the upkeep, and the control is not perfect. It takes years to grow plants to the desired size—not a negligible problem since buildings should be comfortable right from the start.

A major cause of cooling loads is solar radiation on east- or west-facing windows. In summer there are long periods when the sun can reach these facades with fairly small angles of incidence. The range of incidence angles is so wide as to make it impossible to block all this radiation, short of eliminating the windows completely. But partial blocking can be achieved by combining horizontal overhangs with vertical shades to the south of each window.

6.2 Extraterrestrial Insolation

The solar irradiance outside the earth's atmosphere at normal incidence and at the mean sun-earth distance is called the *solar constant;* its value is 435.2 Btu/(h · ft^2) (1373 W/m^2). Due to the slight eccentricity of the orbit, the actual value of the extraterrestrial irradiance I_0 varies by ±3.3 percent. A good fit for I_0 is

$$I_0 = \left(1 + 0.033\cos\frac{360° \times n}{365.25}\right) \times 435.2 \text{ Btu/(h · ft}^2) \qquad (6.16\text{US})$$

$$I_0 = \left(1 + 0.033\cos\frac{360° \times n}{365.25}\right) \times 1373 \text{ W/m}^2 \qquad (6.16\text{SI})$$

with $n =$ day of year (=1 for January 1). Since its peak occurs in winter (of the northern hemisphere), the seasonal variations in the northern hemisphere are somewhat smaller than they would be if the earth's orbit were circular.

The extraterrestrial insolation is a useful quantity because many solar radiation models are based on it, e.g., the clear-day model presented in Sec. 6.3.

Example 6.7

Calculate the extraterrestrial irradiance that would be incident on the west-facing vertical wall of Example 6.3 if there were no atmosphere.

  Given: $n = 202$
  $\theta_i = 61.08°$
  Find: $I_0 \cos \theta_i$

SOLUTION

First we evaluate the factor in Eq. (6.16):

$$\cos\frac{360° \times n}{365.25} = -0.94497$$

The extraterrestrial irradiance at normal incidence is

$$I_0 = [1 + 0.033(-0.94497)](1373 \text{ W/m}^2) = 1330 \text{ W/m}^2 \quad \text{from Eq. (6.16)}$$

Accounting for the incidence angle, we find

$$I_0 \cos 61.08° = (1330 \text{ W/m}^2)(0.4836) = 643.2 \text{ W/m}^2 \text{ [204 Btu/(h · ft}^2)]}$$

Also of interest is the total daily extraterrestrial radiation H_0 on a horizontal surface. It is obtained by integrating the cosine of the incidence angle θ_s of Eq. (6.5) from sunrise to sunset, multiplied by I_0. When the sunset hour angle ω_{ss} of Eq. (6.7) has a real value, the result is

$$H_0 = \frac{\tau_{\text{day}}}{\pi} I_0 \cos\lambda \cos\delta \left(\sin\omega_{ss} - \frac{\pi\omega_{ss}}{180°} \cos\omega_{ss} \right) \qquad (6.17a)$$

with $\tau_{\text{day}} = 24\text{ h} = 86{,}400$ s. Otherwise we have

$$H_0 = 0 \quad \text{when } \tan\lambda \tan\delta < -1 \qquad (6.17b)$$

for days without sun, or for days with midnight sun,

$$H_0 = \frac{\tau_{\text{day}}}{\pi} I_0 \sin\lambda \sin\delta \quad \text{when } \tan\lambda \tan\delta > 1 \qquad (6.17c)$$

For the northern hemisphere, H_0 is plotted in Fig. 6.10. If the extraterrestrial radiation I_0 were independent of the day of year n, this plot would be exact for southern latitudes as well, after substituting $\lambda \rightarrow -\lambda$ and $n \rightarrow n + 183$. The variation of I_0 disturbs this symmetry slightly, by at most ±3 percent on January 1 and July 1. One feature of this graph may come as a surprise: Under the midnight sun, the polar regions would receive more solar radiation per day than the tropics, if there were no atmosphere. Of course, the total amount of radiation that reaches the surface during this relatively brief period is quite limited, and the polar regions stay cold.

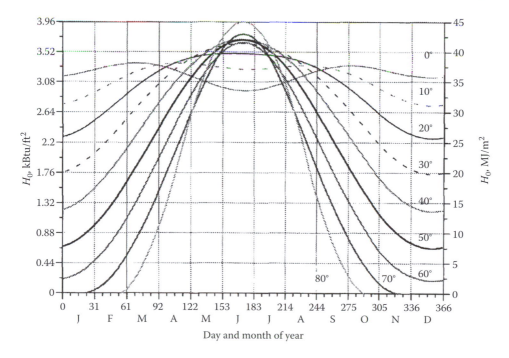

FIGURE 6.10
Extraterrestrial daily horizontal irradiation H_0 as function of day of year, for latitudes 0°, 10°,..., 80° north.

6.3 Insolation Data and Models

6.3.1 Components of Solar Radiation

A clear terminology is needed to keep track of the various types of solar radiation that need to be distinguished. We will use the term *irradiance* and the symbol I for radiative power in W/m^2 [(Btu/(h · ft^2)]. The radiative energy during a certain time interval such as an hour or a day is designated by *irradiation* and H in J/m^2 (Btu/ft^2). (We have already used this distinction in Sec. 6.2.) Subscripts are added to show where and how the radiation is incident. *Direct* radiation from the solar disk, also called *beam radiation*, is indicated by the subscript "dir," and *diffuse* radiation by a subscript "dif." The sum of direct and diffuse radiation incident on a surface is usually called *global radiation* (with subscript "glo"), even though the term *hemispherical* would be more logical since one is talking about radiation incident on one side of a surface only. An additional subscript may indicate the surface on which the radiation is received. For instance, the direct irradiance on the horizontal is designated by $I_{\text{dir,hor}}$. The vague word *insolation* can serve when distinctions are unnecessary.

Insolation data have been collected in many places (Hulstrom, 1989). The type and quality of data span a wide range. Long-term data collection is costly, all the more if one wants hourly rather than daily time resolution or anything beyond global horizontal radiation. Most weather stations do not even measure solar radiation but provide only some indirect estimate such as cloud cover or hours of sunshine. Such information is nonetheless useful because it improves the interpolation between stations where radiation has actually been measured. In the United States, the SOLMET network (NOAA, 1978) has been the major source of measured hourly data. Until the late 1970s it contained only 26 stations with hourly global horizontal data; it was expanded during the peak years of solar energy research, but fell into neglect as funds were cut. Data are also available from many private networks.

Measured direct or diffuse data are exceptional. For instance, the direct radiation "data" on the tapes for the 26 SOLMET stations were actually calculated based on an algorithm developed from a few special sites with measured data for direct radiation. Specialty data, e.g., measured radiation on tilted surfaces, are extremely rare.

To bridge the gap between what is needed and what has actually been measured, various correlations and models have been developed. For example, if only the global horizontal irradiance $I_{\text{glo,hor}}$ is given, one can estimate the diffuse irradiance I_{dif} (also on the horizontal surface) from the equation (Erbs et al., 1982)

$$\frac{I_{\text{dif}}}{I_{\text{glo}}} = \begin{cases} 1.0 - 0.09k_T & \text{for } 0 \leq k_T \leq 0.22 \\ 0.9511 - 0.1604k_T & \\ \quad +4.388k_T^2 - 16.638k_T^3 + 12.336k_T^4 & \text{for } 0.22 \leq k_T \leq 0.80 \\ 0.165 & \text{for } 0.80 \leq k_T \end{cases} \tag{6.18}$$

where

$$k_T = \frac{I_{\text{glo}}}{I_0 \cos \theta_s} \tag{6.19}$$

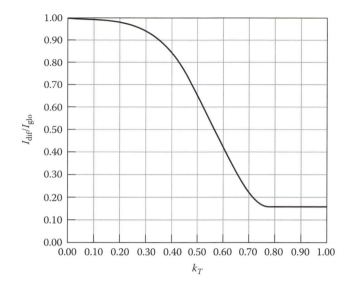

FIGURE 6.11
Correlation Eq. (6.18) between diffuse irradiance I_{dif} and global irradiance I_{glo} (hourly averages, on horizontal) and hourly clearness index k_T.

called the hourly *clearness index*,[1] is the ratio of terrestrial and extraterrestrial irradiance, with θ_s being the zenith angle (= incidence angle on the horizontal). This correlation is plotted in Fig. 6.11. The general shape of the curve is very plausible. Low k_T corresponds to overcast skies when all the radiation is diffuse; as k_T increases, the diffuse component, as a fraction of the total, decreases, down to a minimum of 0.165. The exact form of the correlation between diffuse and global insolation depends on the time interval over which the insolation is averaged (hourly, daily, or monthly). In Fig. 6.14 we will encounter an analogous correlation for monthly averages.

The effects of solar radiation depend on not only the intensity but also its *spectrum*. The spectrum varies somewhat with atmospheric conditions. The spectral distribution (intensity as a function of wavelength) is shown in Fig. 6.12 for air mass zero (= extraterrestrial) and for air mass two, with the air mass defined as $1/\cos\theta_s$, which is the length of the path through the atmosphere in units of the length at normal incidence. A little less than one-half of the energy is in the visible range, 0.4 to 0.7 μm. Most of the remainder is in the near infrared, up to 2.0 μm. The spectrum becomes important for buildings when the transmissivity of the glazing varies with the wavelength. Of special interest are glazings that transmit daylight while blocking the infrared portion to minimize cooling loads.

6.3.2 Radiation on Tilted Surfaces

Solar radiation on tilted surfaces contains three components: direct radiation, diffuse sky radiation, and radiation reflected from the ground. The direct component is the product of the direct normal irradiance I_{dir} and the cosine of the incidence angle. For diffuse radiation, from sky and from ground, one also needs to consider the angular distribution. The simplest hypothesis is isotropy. More accurate anisotropic models have been developed (e.g., Hay, 1979), but for buildings the accuracy of the isotropic model will usually be sufficient.

---

[1] This is now the customary terminology in solar energy (Duffie and Beckman, 1980). Clearness index k_T [and its daily equivalent K_T, defined in (Eq. 6.27) later on] has nothing to do with the atmospheric clearness number that is used in the clear-day insolation model of the ASHRAE Handbook.

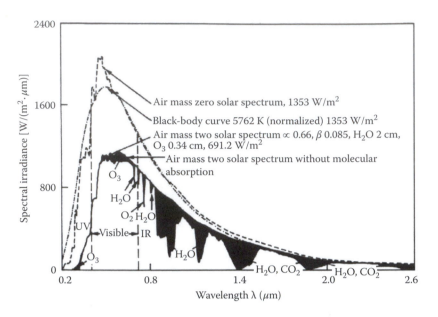

FIGURE 6.12

Solar spectrum for air mass zero (= extraterrestrial) and for air mass two, air mass being defined as $1/\cos\theta_s$. The black portions indicate molecular absorption. (Absolute values of irradiance are based on an old value of solar constant and should be rescaled by 1373/1353.) (Courtesy of Kreith, F. and Kreider, J.F., *Principles of Solar Engineering*, Hemisphere/McGraw-Hill, New York, 1978.)

If an unshaded flat surface is tilted at an angle θ_p, it sees a fraction

$$F_{sky} = \frac{1 + \cos\theta_p}{2} \qquad (6.20)$$

of isotropic radiation from the sky and a fraction

$$F_{grd} = \frac{1 - \cos\theta_p}{2} \qquad (6.21)$$

of isotropic radiation reflected by the ground (these factors are radiation shape factors; see Sec. 2.4.4). Therefore the global irradiance on the tilted plane is

$$I_{glo_p} = I_{dir}\cos\theta_i + I_{dif}\frac{1 + \cos\theta_p}{2} + I_{glo,hor}\rho_g\frac{1 - \cos\theta_p}{2} \qquad (6.22)$$

where

$$I_{glo,hor} = I_{dir}\cos\theta_s + I_{dif} \qquad (6.23)$$

is the global irradiance on the horizontal and ρ_g is the reflectivity of the ground. For the latter, one usually assumes 0.2 without snow cover and 0.7 with snow cover, unless better information is available. Typical data for the most common surfaces are listed in Table 6.1. The reflectivity depends strongly on the condition of the surface. Snow cover, in particular, can greatly increase the reflectivity. There can also be some variation with the angle of

TABLE 6.1

Reflectivity of Common Exterior Surfaces

Surface	Reflectivity ρ_g
Natural surfaces (no vegetation)	
Snow (fresh)	0.75
Soils (clay, loam, etc.)	0.14
Water (relatively large incidence angles)	0.07
Artificial surfaces	
Bituminous and gravel roof	0.13
Blacktop, old	0.10
Building surfaces, dark (red brick, dark paints, etc.)	0.27
Building surfaces, light (light brick, light paints, etc.)	0.60
Concrete, new	0.35
Concrete, old	0.25
Crushed rock surface	0.20
Earth roads	0.04
Vegetation	
Coniferous forest (winter)	0.07
Leaves, dead	0.30
Forests in autumn, ripe field crops, plants	0.26
Grass, dry	0.20
Grass, green	0.26

Sources: Courtesy of Hunn, B.D. and Calafell, D.O., *Sol. Energy*, 19, 87, 1977; Threlkeld, J.L., *Thermal Environmental Engineering*, 2nd edn, Prentice-Hall, Englewood Cliffs, NJ, 1970.

incidence, although it is usually not significant in view of the overall uncertainties, as shown by the data in Fig. 14.5 of Threlkeld (1970).

For the important case of vertical surfaces, the angle $\theta_p = 90°$ and Eq. (6.22) becomes simply

$$I_{\text{glo,vert}} = I_{\text{dir}} \cos \theta_i + \frac{I_{\text{dif}}}{2} + \frac{I_{\text{glo,hor}} \rho_g}{2} \tag{6.24}$$

Strictly speaking, these equations are correct only when the ground in front of the surface is not shaded. But with partially shaded ground, such as for a west-facing surface in the morning, a correct calculation becomes so complicated that the gain in accuracy has usually not been deemed worth the effort.

6.3.3 Clear Sky Radiation

Solar radiation under clear skies can be represented with fairly good accuracy by simple models because the transparency of clear atmospheres does not vary all that much with time or location. The model used by ASHRAE (1989) is based on research done more than 40 years ago by Threlkeld and Jordan (1958). Since then Hottel (1976) has developed a model that is only slightly more complicated but is more flexible and, we believe, more accurate because it had the benefit of better computers and more extensive data. It includes the effect of altitude (above sea level) and offers a choice between two visibility levels, important features not found in the ASHRAE model.[2]

According to Hottel, the direct irradiance I_{dir} at normal incidence can be calculated from the extraterrestrial irradiance I_0 of Eq. (6.16) and the zenith angle θ_s of Eq. (6.5) by a correlation with three coefficients:

$$I_{dir} = I_0 \left[a_0 + a_1 \exp \left(-\frac{k}{\cos \theta_s} \right) \right] \tag{6.25}$$

This model treats the atmosphere as a superposition of three types of gas; the first term represents a black and a clear gas, and the second term represents a gray gas. The coefficients depend on the state of the atmosphere; they are listed in Table 6.2 for 23- and 5-km visibility, with correction factors r_0, r_1, and r_k that depend on the time of year and climate. There is also a strong dependence on the altitude above sea level A; in Table 6.2 it has units of kilometers and must be less than 2.5 km.

Figure 6.13 shows the clear day irradiance as function of zenith angle θ_s, for midlatitude summer conditions at $A = 0$ and 1 km (0.622 mi) above sea level.

One also needs the diffuse insolation. For clear skies, the diffuse irradiance on a horizontal surface can be estimated from a relation due to Liu and Jordan (1960):

$$I_{dif} = (0.271 I_0 - 0.2939 I_{dir})(\cos \theta_s) \tag{6.26}$$

TABLE 6.2

Coefficients of Clear Day Model

	23-km (14.3-mi) Visibility	5-km (3.1-mi) Visibility

a. Coefficients a_0, a_1, and k as a function of altitude A above sea level (in kilometers), for two levels of visibility

a_0	$r_0[0.4237 - 0.00821 \times (6.0 - A)^2]$	$r_0[0.2538 - 0.0063 \times (6.0 - A)^2]$
a_1	$r_1[0.5055 + 0.00595 \times (6.5 - A)^2]$	$r_1[0.7678 + 0.0010 \times (6.5 - A)^2]$
k	$r_k[0.2711 + 0.01858 \times (2.5 - A)^2]$	$r_k[0.2490 + 0.0810 \times (2.5 - A)^2]$

b. Correction factors r_0, r_1, and r_k

	r_0			
	Visibility			
Climate Type	**23 km**	**5 km**	r_1	r_k
Tropical	0.95	0.92	0.98	1.02
Midlatitude summer	0.97	0.96	0.99	1.02
Subarctic summer	0.99	0.98	0.99	1.01
Midlatitude winter	1.03	1.04	1.01	1.00

Source: Courtesy of Hottel, H.C., *Solar Energy*, 18, 129, 1976.

---

[2] The reader might wonder about the consistency of using Hottel's model with the cooling load factors and transfer functions of the ASHRAE Handbook (presented in the following chapters), since the latter were calculated on the basis of the ASHRAE insolation model. There is really no problem, because the cooling load factors and transfer functions depend only on the time-of-day profile, not on the magnitude of the insolation. The profiles are sufficiently similar between the models (and even between different latitudes)—so similar in fact that a single profile has been assumed for the calculation.

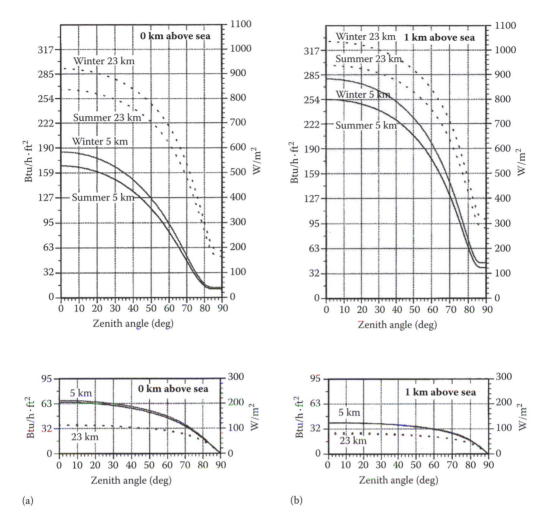

FIGURE 6.13
Clear day irradiance as function of zenith angle θ_s, under midlatitude conditions for two levels of visibility (5 km = 3.1 mi and 23 km = 14.3 mi) and two times of year (winter = January 21 and summer = July 21), for northern hemisphere. Direct irradiance I_{dir} at normal incidence and diffuse irradiance I_{dif} on horizontal. (a) At 0 km above sea level. (b) At 1 km (0.622 mi) above sea level.

Example 6.8

Calculate the clear day insolation for the conditions of Example 6.7, assuming 5-km visibility and $A = 0.1$ km above sea level. Also $I_0 = 1330$ W/m² from Example 6.7.

Find: I_{dir}, I_{dif}

Lookup values: From Table 6.2b, $r_0 = 0.96$, $r_1 = 0.99$, and $r_k = 1.02$.

SOLUTION

First we evaluate the coefficients in Table 6.2a. They turn out to be

$$a_0 = 0.0331 \quad a_1 = 0.8007 \quad k = 0.7299$$

From Eq. (6.25) the direct irradiance at normal incidence is

$$I_{dir} = (1330 \text{ W/m}^2)\left[0.0331 + 0.8007 \exp\left(-\frac{0.7299}{\cos 37.73°}\right)\right]$$

$$= 467.2 \text{ W/m}^2 \ [148 \text{ Btu/(h} \cdot \text{ft}^2)]$$

Equation (6.26) then yields the diffuse irradiance on the horizontal as

$$I_{dif} = [(0.271 \times 1330 - 0.2939 \times 467.2) \text{ W/m}^2](\cos 37.73°)$$

$$= 176.5 \text{ W/m}^2 \ [(55.9 \text{ Btu/(h} \cdot \text{ft}^2)]$$

COMMENTS

This I_{dir} is very much lower than the value of 839 W/m^2 for July 21 at 14:00 at latitude 48°, from table 9 of chap. 27 of ASHRAE (1989). However, with the 23-km atmosphere, we would have obtained $I_{dir} = 782$ W/m^2 (and $I_{dif} = 103$ W/m^2), much closer to the ASHRAE value. The absorption and scattering of the 5-km atmosphere greatly reduce the direct insolation.

Example 6.9

Calculate the clear day irradiance on a vertical west-facing wall for the conditions of Example 6.8, if the reflectivity of the ground is $\rho_g = 0.2$.

  Given: $\theta_p = 90°$ for vertical wall

  From Example 6.3, $\theta_i = 61.08°$ and $\theta_s = 37.73°$

  From Example 6.8, $I_{dir} = 467.2$ W/m^2 and $I_{dif} = 176.5$ W/m^2

  Find: $I_{glo,vert}$

SOLUTION

First find the global horizontal irradiance $I_{glo,hor} = 467.2 \cos 37.73° + 176.5 = 546.0$ W/m^2, from Eq. (6.23). Then insert into Eq. (6.24) for the irradiance on the vertical surface

$$I_{glo, vert} = 467.2 \cos 61.08° + \frac{176.5}{2} + 546.0 \times \frac{0.2}{2}$$

$$= 225.9 + 88.3 + 54.6 = 369 \text{ W/m}^2 \ [117 \text{ Btu/(h} \cdot \text{ft}^2)]$$

6.3.4 Long-Term Average Insolation

In this section we present a method that needs only a minimum of input: the daily *clearness index* K_T, defined as the ratio of terrestrial and extraterrestrial radiation

$$K_T = \frac{H_{glo, hor}}{H_0} \tag{6.27}$$

Here $H_{glo,hor}$ is the daily global irradiation at the earth's surface, and H_0 is the extraterrestrial daily irradiation on the same surface.

   By defining the clearness index, one separates two independent causes for the variability of terrestrial solar radiation: the atmosphere and the geometry of the earth's motion. On heavily overcast days, K_T may be as low as 0.05 to 0.1 while on clear days it is around 0.7 to 0.75 (but not more than 0.8, at least near sea level). Monthly averages, designated by $\bar{K}_T$, range from 0.3 for very cloudy climates such as upstate New York to 0.75 for the peak of the Sunbelt. Monthly average data for $\bar{K}_T$ are widely available. The appendices on the CD-ROM contain contour maps for $\bar{K}_T$.

Using only this piece of information, one can calculate the long-term average irradiance on any surface, and the accuracy is sufficient to reproduce the monthly average radiation incident on solar collectors within a few percent (Collares-Pereira and Rabl, 1979). The starting point is thus either $\bar{K}_T$ or equivalently $\bar{H}_{glo, hor}$; the overbar indicates long-term averages.

The first step is to estimate the average daily diffuse irradiation on the horizontal from the correlation

$$\frac{\bar{H}_{dif}}{\bar{H}_{glo, hor}} = 0.775 + 0.347 \frac{(\omega_{ss} - 90°)\pi}{180°}$$
$$- \left[0.505 + 0.261 \frac{(\omega_{ss} - 90°)\pi}{180°} \right] \cos \frac{360°(\bar{K}_T - 0.9)}{\pi} \qquad (6.28)$$

which is plotted in Fig. 6.14.

As the second step, one converts the daily irradiation to long-term average irradiance at any moment of the day, by the correlations

$$\bar{I}_{glo, hor} = r_{glo}(\omega_{ss}, \omega)\bar{H}_{glo, hor} \qquad (6.29)$$

and

$$\bar{I}_{dif} = r_{dif}(\omega_{ss}, \omega)\bar{H}_{dif, hor} \qquad (6.30)$$

where ω_{ss} and ω are the hour angles corresponding to sunset time [Eq. (6.7) and Fig. 6.4] and to time of day [Eq. (6.6)]. The correlation functions are given by

$$r_{dif}(\omega_{ss}, \omega) = \frac{\pi}{\tau_{day}} \frac{\cos \omega - \cos \omega_{ss}}{\sin \omega_{ss} - (\pi\omega_{ss}/180°)(\cos \omega_{ss})} \qquad (6.31)$$

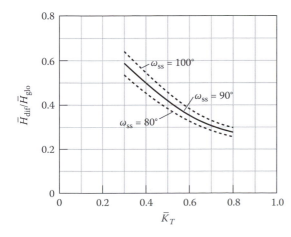

FIGURE 6.14
Correlation of Collares-Pereira and Rabl (1979) for ratio of long-term averages of daily diffuse and global solar irradiation, as function of clearness index $\bar{K}_T$ and sunset hour angle ω_{ss}.

and

$$r_{\text{glo}}(\omega_{\text{ss}}, \omega) = (a + b \cos \omega) r_{\text{dif}}(\omega_{\text{ss}}, \omega) \qquad (6.32)$$

with $\tau_{\text{day}} = 24\text{ h} = 86{,}400\text{ s}$,

$$a = 0.4090 + 0.5016 \sin(\omega_{\text{ss}} - 60°) \qquad (6.33)$$

and

$$b = 0.6609 - 0.4767 \sin(\omega_{\text{ss}} - 60°) \qquad (6.34)$$

They are plotted in Fig. 6.15. The units are h^{-1} for the left-hand scale and 10^{-6} s^{-1} for the right-hand scale.

The direct normal irradiance follows as

$$\bar{I}_{\text{dir}} = \frac{\bar{I}_{\text{glo, hor}} - \bar{I}_{\text{dif}}}{\cos \theta_s} \qquad (6.35)$$

With the isotropy assumption for the diffuse component, one can thus compute the average irradiance on any surface at any time. Integrating from sunrise to sunset, one obtains the daily average.[3]

Example 6.10

Find the long-term average irradiance on the vertical west-facing wall of Examples 6.2, 6.3, and 6.9 (west-facing vertical wall in Paris, 16:00 daylight saving time on July 21) if $\bar{K}_T = 0.47$.

  Given: $\omega = 31.05°$
  $\cos \theta_s = 0.7909$
  Find: $\bar{I}_{\text{glo, vert}}$

SOLUTION

The extraterrestrial irradiation is $H_0 = 39.70\text{ MJ}/(\text{m}^2 \cdot \text{day})$, from Eq. (6.17a), from which the terrestrial irradiation follows as

$$\bar{H}_{\text{glo, hor}} = \bar{K}_T H_0 = 0.47 \times 39.70 = 18.66\text{ MJ}/(\text{m}^2 \cdot \text{day})$$

Then we need the sunset hour angle; it is found from Eq. (6.7) after we insert δ and λ of Example 6.2:

$$\omega_{\text{ss}} = 115.12°$$

---

[3] The equations of this section provide an alternative model for calculating clear day radiation if one inserts $K_T = 0.75$. Numerically the results are close to the model of Hottel. The latter has, however, several advantages: It has been explicitly validated against clear day data, and it can account for detailed conditions of the atmosphere (elevation above sea level, visibility, type of climate).

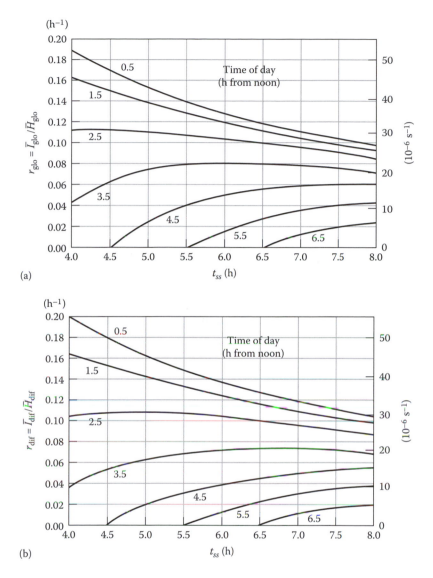

FIGURE 6.15

Correlation between long-term averages of daily total irradiation and instantaneous irradiance versus sunset hour t_{ss}: (a) for diffuse insolation; (b) for global insolation.

For the conversion from daily to instantaneous insolation, we evaluate the following quantities, using Eq. (6.31) to (6.34):

$$a = 0.8205 \quad b = 0.2698 \quad r_{dif} = 26.51 \times 10^{-6} \text{ s}^{-1} \quad r_{glo} = 27.88 \times 10^{-6} \text{ s}^{-1}$$

The diffuse component is obtained from Eq. (6.28) or Fig. 6.14:

$$\frac{\bar{H}_{dif}}{\bar{H}_{glo, hor}} = 0.523$$

Hence the daily diffuse irradiation is $\bar{H}_{dif} = 9.76$ MJ/m^2. Now we can multiply by the conversion factors r_{glo} and r_{dif} with the result

$$\bar{I}_{glo,hor} = r_{glo}\bar{H}_{glo,hor} = (27.88 \times 10^{-6}\ s^{-1})(18.66\ MJ/m^2) = 520.4\ W/m^2$$

$$\bar{I}_{dif} = r_{dif}\bar{H}_{dif} = (26.51 \times 10^{-6}\ s^{-1})(9.76\ MJ/m^2) = 258.7\ W/m^2$$

Then the beam normal irradiance is found from Eq. (6.35) with $\cos\theta_s = 0.7909$:

$$\bar{I}_{dir} = \frac{520.4 - 258.7}{0.7909} = 330.9\ W/m^2$$

Finally, the global irradiance on the vertical surface, from Eq. (6.22), is

$$\bar{I}_{glo,vert} = 330.9 \times \cos 61.08° + \frac{258.7}{2} + 520.4 \times \frac{0.2}{2}$$
$$= 160.0 + 129.4 + 52.0 = 341.4\ W/m^2\ [108\ Btu/(h \cdot ft^2)]$$

Even though the calculations involve fairly complicated combinations of trigonometric functions, it turns out that the results for daily totals can be approximated by a simple correlation, as shown by Potter et al. (1989). Using the data of SERI (1980), they have calculated the monthly average irradiation $\bar{H}_{glo,vert}$ on vertical surfaces of the principal orientations for latitudes from 30° to 45°. They found that the relation between $\bar{H}_{glo,vert}$ and $\bar{K}_T$ can be represented by the simple linear model

$$\bar{H}_{glo,vert} = a\bar{K}_T + b \tag{6.36}$$

with coefficients a and b that depend on the latitude, time of year, and surface orientation, as listed in Table 6.3. The accuracy is excellent, the average error being less than 1 percent. The results for January, April/October, and July are shown in Fig. 6.16, for north-, east/west-, and south-facing surfaces.

TABLE 6.3

Coefficients of $\bar{H}_{glo,vert} = a\bar{K}_T + b$ of Eq. (6.36), btu/ft^2 per Day

	North				East/West				South			
January												
$\lambda =$	30°	35°	40°	45°	30°	35°	40°	45°	30°	35°	40°	45°
$a =$	326	217	209	210	1288	1171	1066	927	2783	2925	2861	2651
$b =$	113	122	86	49	−66	−63	−63	−52	−283	−352	−327	−258
April/October												
$\lambda =$	30°	35°	40°	45°	30°	35°	40°	45°	30°	35°	40°	45°
$a =$	459	425	430	448	1849	1872	1891	1904	1310	1544	1784	2018
$b =$	298	304	285	260	−31	−49	−67	−84	137	85	17	−51
July												
$\lambda =$	30°	35°	40°	45°	30°	35°	40°	45°	30°	35°	40°	45°
$a =$	848	817	830	854	1917	2010	2082	2189	599	897	1136	1430
$b =$	251	269	268	266	3	−24	−40	−77	367	291	243	164

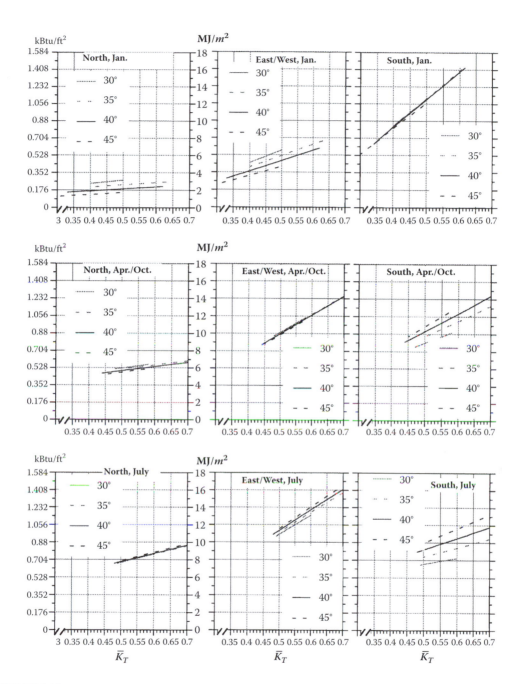

FIGURE 6.16

Monthly average daily total irradiation $\bar{H}_{glo,vert}$ on vertical surfaces in January, April, and July as function of monthly average clearness index $\bar{K}_T$ and latitude λ (in degrees). (Courtesy of Potter, R. et al., Simplification of monthly solar radiation calculations for a vertical surface, in *Proceedings of 1989 Conference of the American Section of the International Solar Energy Society*, Denver, CO, p. 450, 1989.)

6.4 Sol-Air Temperature

This section develops a concept that is convenient for analyzing solar heat gains due to absorption by opaque exterior surfaces of a building.

If the surrounding environment can be characterized by a single temperature T_o for radiation and for convection, the resulting heat flow per unit area $q = \dot{Q}/A$, from environment to surface, is given by

$$q = h_o(T_o - T_s) \tag{6.37}$$

where
h_o = surface heat transfer coefficient (radiation plus convection, as described in Chap. 2)
T_s = surface temperature

Now if in addition a radiative flux αI is absorbed on the surface, the total heat flow at the surface becomes

$$q_{\text{tot}} = h_o(T_o - T_s) + \alpha I \tag{6.38}$$

To simplify the load calculations, it is preferable to work with an equation of the form of Eq. (6.37) rather than specify both a temperature and a flux. That can be accomplished if one replaces the outdoor air temperature T_o by an equivalent temperature

$$T_o + \frac{\alpha I}{h_o}$$

With that replacement Eq. (6.37) yields the same result as Eq. (6.38).

Not only the solar flux but also infrared exchanges can be treated in this manner— in particular those with the sky when the radiation temperature of the sky is below the air temperature. Both effects are included in what is called the *sol-air temperature*, defined as

$$T_{os} = T_o + \frac{\alpha I}{h_o} - \frac{\Delta q_{\text{ir}}}{h_o} \tag{6.39}$$

where
h_o = surface heat transfer coefficient for radiation and convection, W/(m$^2 \cdot$K) [Btu/(h $\cdot$ ft$^2 \cdot$ °F)]
I = global solar irradiance on surface, W/m^2 [Btu/(h $\cdot$ ft^2)]
Δq_{ir} = correction to infrared radiation transfer between surface and environment if sky temperature is different from T_o, W/m^2 [Btu/(h $\cdot$ ft^2)]

In practice, one assumes that $\Delta q_{\text{ir}}/h_o$ varies from zero for vertical surfaces to 7°F (3.9 K) for upward-facing surfaces (the sky overhead is colder than the rest of the environment). In terms of the sol-air temperature, the total heat transfer at the surface is therefore

$$q_{\text{tot}} = h_o(T_{os} - T_s) \tag{6.40}$$

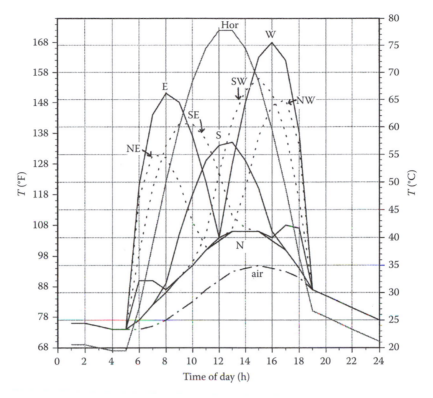

FIGURE 6.17
Sol-air temperature for horizontal and vertical surfaces as a function of time of day for summer design conditions, July 21 at 40° latitude, assuming $\alpha/h_o = 0.30$ (h · ft^2 · °F)/Btu [0.052 (m^2 · K)/W]. The curves overlap when there is no direct radiation on a surface. (Courtesy of ASHRAE, *Handbook of Fundamentals*, American Society of Heating, Refrigerating and Air-Conditioning Engineers, Atlanta, GA, 1989, Table 26.1.)

Outdoor temperatures for design conditions can be stated in terms of sol-air temperatures. This is convenient because solar radiation on opaque surfaces will then be automatically included in the resulting load calculation. In Fig. 6.17 we have plotted the sol-air temperature for horizontal and vertical surfaces versus time of day, based on the ASHRAE summer design conditions for dark surfaces with $\alpha/h_o = 0.30$ (h · ft^2 · °F)/Btu [0.052 (m^2 · K)/W]. A list of values can also be found in the appendices on the CD-ROM. The sol-air temperature at other values of α/h_o can be obtained from this figure by scaling the difference between the curves for the air temperature T_o and the sol-air temperature for the surface in question (for the horizontal this is not quite correct because of Δq_{ir}).

As far as the associated heat gains through an opaque wall or roof are concerned, the outdoor temperature appears to be augmented by an amount $\Delta T_{os} = T_{os} - T_o$. If the total heat transfer coefficient (including h_o) is U, the extra heat flow is $U\Delta T_{os}$. Keeping only the solar contribution in ΔT_{os}, we find that the solar heat gain $\Delta \dot{Q}_{sol}$ through an opaque surface is

$$\Delta \dot{Q}_{sol} = \frac{UI\alpha}{h_o} \tag{6.41}$$

Example 6.11

Find the solar heat gain through a wall.

Given: $U = 0.24$ W/(m^2 · K) [0.042 Btu/(h · ft^2 · °F)] also called R23.8 $\alpha/h_o = 0.30$ (h · ft^2 · °F)/ Btu [0.052 (m^2 · K)/W], summer design conditions for dark surface

Solar irradiance on wall $I = 369$ W/m^2 [117 Btu/(h · ft^2)] = conditions of Example 6.9

Find: $\Delta \dot{Q}_{sol}$

SOLUTION

From Eq. (6.41) we find

$$\Delta \dot{Q}_{sol} = [0.24\,\text{W}/(\text{m}^2 \cdot \text{K})](369\,\text{W}/\text{m}^2)[0.052(\text{m}^2 \cdot \text{K})/\text{W}]$$
$$= 4.6\,\text{W}/\text{m}^2\,[1.46\,\text{Btu}/(\text{h} \cdot \text{ft}^2)]$$

COMMENTS

The ratio of solar heat gain to incident solar irradiance is $4.6/369 = 0.012$, hardly more than 1 percent. Quite generally, solar heat gains *per unit area* through opaque surfaces are one to two orders of magnitude smaller than solar heat gains through glazing.

It is interesting to reflect upon the relation between solar radiation and outdoor temperature. The sun provides the major term in the energy balance of the earth, orders of magnitude larger than the geothermal heat production. If it were not for the sun, our environment would not be much above the temperature of outer space, about 3 K. Obviously solar radiation is the dominant driving force for outdoor temperature. But there are time lags due to the thermal inertia of the earth and its atmosphere. As explained in Chap. 8, the effective inertia depends on how far a heat pulse can penetrate from the surface. As it turns out, the diurnal heat pulse can warm up only a thin surface layer (on the order of a few centimeters) plus the atmosphere. The resulting inertia delays the peak of T_{os} by a few hours relative to the sun: The warmest hours of the day are reached a few hours after noon. The seasonal heat pulse penetrates on the order of 1 m into the soil, and the resulting phase lag is about 1 month. Outside the tropics, the coldest (hottest) days of the year tend to occur about 1 month after the winter (summer) solstice.

6.5 Windows

From the point of view of energy, windows are the most important element of the envelope of a building, providing at once insulation and solar radiation. With conventional glazing, most of the energy consumption of a building goes quite literally out of the windows. In typical houses, roughly one-third of the total heat loss passes through the glazings. Also most of the cooling load is due to solar heat gains through glazing. Furthermore, much of the air infiltration occurs around the edges of windows. On the other hand, there is the potential of energy savings from solar light and heat. Clearly, windows merit attention.

6.5.1 *U* Values of Windows

The energy balance of a single-pane window is shown in Fig. 6.18 (assuming one-dimensional heat flow). To begin, let us leave out solar radiation; it will be included in Sec. 6.5.3.

The heat flow from interior to exterior passes through three resistances in series: the resistance at the interior surface, the resistance of the glass, and the resistance at the exterior surface. Therefore, the total resistance R_{th} of the glazing is the sum of the three resistances in Fig. 6.18b, and it is related to the U value of the glazing by $UA = 1/R_{th}$. Canceling the area A, one finds that the U value is given by

$$\frac{1}{U} = \frac{1}{h_i} + \frac{\Delta x}{k} + \frac{1}{h_o} \tag{6.42}$$

where

h_i = heat transfer coefficient (combined for radiation and conduction/convection) at interior surface

h_o = heat transfer coefficient (combined for radiation and conduction/convection) at exterior surface

Δx = thickness of glass

k = conductivity of glass

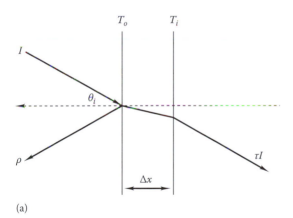

(a)

(b)

(c)

FIGURE 6.18
Energy balance of a single-pane window (neglecting heat capacity of glass), (a) Physical configuration. (b) Thermal network without solar radiation. (c) Thermal network with solar radiation (and neglecting resistance in glass).

The surface heat transfer coefficients can vary with ambient conditions; for highest accuracy, one would have to resort to detailed formulas for the radiative and convective heat transfers. In most building applications, one can use lumped parameters h_i and h_o and standard design values—a tremendous simplification. The ASHRAE 1989 design values are (see appendices on the CD for additional standard values for other cases)

$$h_i = 1.46 \text{ Btu}/(\text{h} \cdot \text{ft}^2 \cdot {}^\circ\text{F}) \ [8.29 \text{ W}/(\text{m}^2 \cdot \text{K})] \quad \text{year-round}$$

and

$$h_o = 6.0 \text{ Btu}/(\text{h} \cdot \text{ft}^2 \cdot {}^\circ\text{F}) \ [34.0 \text{ W}/(\text{m}^2 \cdot \text{K})] \quad \text{winter}$$
$$= 4.0 \text{ Btu}/(\text{h} \cdot \text{ft}^2 \cdot {}^\circ\text{F}) \ [22.7 \text{ W}/(\text{m}^2 \cdot \text{K})] \quad \text{summer}$$

The sensitivity of the total U value to variations in h_o is greatest for single glazing; the more insulating the glazing, the lower the variability of the U value.

Inserting the conductivity of glass $k \approx 0.6 \text{ Btu}/(\text{h} \cdot \text{ft} \cdot {}^\circ \text{F}) \ [\approx 1.0 \text{ W}/(\text{m} \cdot \text{K})]$ and $\Delta x = 5$ mm (0.20 in) as a typical glazing thickness into Eq. (6.42) for U, one finds that the resistance of the glass

$$\frac{\Delta x}{k} \approx 0.005 \ (\text{m}^2 \cdot \text{K})/\text{W}$$

is negligible compared to the surface resistances, with the contribution of the interior resistance $1/h_i \approx 1/[8.29 \text{ W}/(\text{m}^2 \cdot \text{K})] = 0.12 \ (\text{m}^2 \cdot \text{K})/\text{W}$ being about 24 times larger by itself. Even with acrylic or polycarbonate, whose conductivity is 4 times lower than that of glass, the contribution is small. Henceforth, we omit the $\Delta x/k$ term from the equations.

The thermal resistance of windows can be enhanced by adding further panes. With double glazing, neglecting the resistance of the glass itself, one obtains

$$\frac{1}{U} = \frac{1}{h_i} + \frac{1}{h_s} + \frac{1}{h_o} \tag{6.43}$$

where $h_s =$ heat transfer coefficient of the space between the panes. The latter is, of course, the sum of the term h_{rad} for radiation and h_{con} for conduction/convection across one airspace (see Sec. 2.4.6):

$$h_s = h_{\text{rad}} + h_{\text{con}} \tag{6.44}$$

Each additional pane reduces the U value of the glazing by adding another $1/h_s$ to the right-hand side of Eq. (6.43).

The variation of the heat transfer coefficient h_s with the separation d between the panes is of interest because the cost of the frame increases with d. When d is small (less than about 0.3 in ≈ 8 mm for air under normal conditions), there is no convection, and h_{con} varies as k_{air}/d, where k_{air} is the conductivity of air. As d is increased beyond the threshold for convection, h_{con} becomes approximately constant. This effect can be seen in Fig. 6.19, where the U value of several window types is plotted as function of the width of the separation between panes. There is no point in making the width larger than about 10 mm (0.40 in).

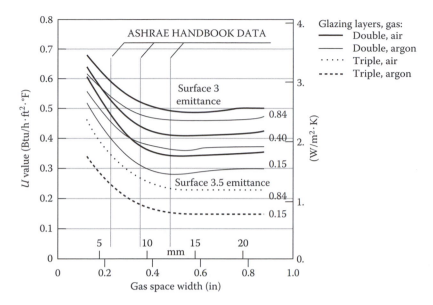

FIGURE 6.19

Center-of-glass U value for double- and triple-pane glass at ASHRAE winter design conditions, as function of separation between panes. (Courtesy of McCabe, M.E., *ASHRAE J.*, June, 56, 1989.)

The above formulas do not take into account edge effects. In most windows, the conductance of the frame is so different from that of the glazing that it should not be neglected. Furthermore, many multipane windows use continuous aluminum spacers around the edge of the glazing to keep the panes apart, thus degrading the insulating value. Therefore, ASHRAE (1989) recommends that the overall U value U_{tot} of a window be calculated as an area-weighted average

$$U_{av} = \frac{U_{cg}A_{cg} + U_{eg}A_{eg} + U_fA_f}{A_{cg} + A_{eg} + A_f} \tag{6.45}$$

where the subscripts cg, eg, and f refer to center of glass, edge of glass, and frame, respectively. The difference between the center-of-glass and edge-of-glass values can be seen in Fig. 6.20 for a variety of spacer materials. ASHRAE values for heat transfer coefficients of typical frames are listed in Table 6.4.

The importance of frame and spacer materials has been emphasized by a number of authors. It is particularly true in small windows. Traditionally, aluminum has been the favorite material for the frame and spacer, by virtue of its durability and low cost. Unfortunately its conductivity is very high. While an aluminum frame does not render single glazing any worse, its use in insulating glazing defeats the purpose.

Example 6.12

Calculate the U value of a double-glazed air-filled (air gap $= 9$ mm $= 0.35$ in) window of gross area 0.5 m $\times$ 1.0 m with low-emissivity coating of emissivity 0.15 on surface 3, with aluminum spacer and aluminum frame. Frame width $= 0.05$ m and edge width $=$ spacer width $= 0.01$ m.

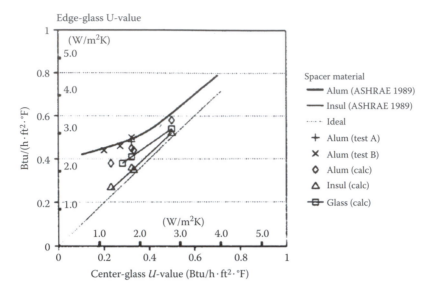

FIGURE 6.20

Center-of-glass and edge-of-glass U values for a variety of spacer materials ("ideal" corresponds to a spacer with the same U value as the glazing). (Courtesy of McCabe, M.E., *ASHRAE J.*, June, 56, 1989.)

TABLE 6.4

Heat Transfer Coefficients U_f of Typical Frames

Material	Btu/(h · ft² · °F)	W/(m² · K)
Aluminum without thermal break	1.9	10.8
Aluminum with thermal break	1.0	5.68
Wood or vinyl	0.4	2.27

Source: Courtesy of McCabe, M.E., *ASHRAE J.*, June, 56, 1989.

Given: Center-of-glass U value of Fig. 6.19 (double-glazed, airspace width 9 mm, $\epsilon = 0.15$) and edge-of-glass U value of Fig. 6.20 for aluminum spacer

Find: U_{av}

Sketch: Fig. 6.21

Lookup values: $U_{cg} = 1.9$ W/(m² · K) [0.335 Btu/(h · ft² · °F)] from Fig. 6.19

$U_{eg} = 2.8$ W/(m² · K) [0.493 Btu/(h · ft² · °F)] from Fig. 6.20

$U_f = 10.8$ W/(m² · K) [1.902 Btu/(h · ft² · °F)] from Table 6.4

SOLUTION

We list the areas and U values in this table:

	Length	Width	A	U	AU
Frame	1.00	0.50	$1.00 \times 0.50 - 0.90 \times 0.40 = 0.140$	10.800	1.512
Edge of glass	0.90	0.40	$0.90 \times 0.40 - 0.88 \times 0.38 = 0.026$	2.800	0.072
Center	0.88	0.38		0.334	1.900	0.635
Sum			0.500		2.219

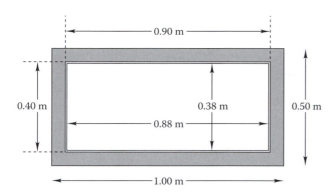

FIGURE 6.21
Dimensions of frame, edge of glass (spacer), and center of glass for Example 6.12, drawn to scale.

Dividing the sum of the UA values by the sum of the A values according to Eq. (6.45), we find

$$U_{av} = \frac{2.219}{0.500} = 4.44 \text{ W/(m}^2 \cdot \text{K)} \text{ [0.782 Btu/(h} \cdot \text{ft}^2 \cdot °\text{F)]}$$

COMMENTS

This is more than twice as large as the U value of the glazing itself. A bad frame can dramatically degrade the performance of a good window.

6.5.2 Surface Temperature of Glazing

Since the perception of comfort depends not only on air temperature but also on radiation temperature, it is of interest to calculate the temperature T_s of the interior glazing surface as a function of outdoor temperature. Let us designate by R' the resistance of the glazing without the surface heat transfer coefficient h_i to the indoor air. The total resistance $R = 1/(UA)$ of the glazing is, of course, the sum

$$R = R' + \frac{1}{h_i A} \tag{6.46}$$

since these two resistances are in series. The heat flow $h_i A(T_i - T_s)$ from air to surface must be equal to the flow from the surface to the outside $(T_s - T_o)/R'$:

$$h_i A(T_i - T_s) = \frac{T_s - T_o}{R'} \tag{6.47}$$

Straightforward algebra yields the surface temperature as the following weighted average of indoor and outdoor temperatures:

$$T_s = \left(1 - \frac{U}{h_i}\right) T_i + \frac{U}{h_i} T_o \tag{6.48}$$

This is plotted in Fig. 6.22 as function of T_o and for several values of U, assuming $T_i = 20°$C and $h_i = 8.29 \text{ W/(m}^2 \cdot \text{K)}$.

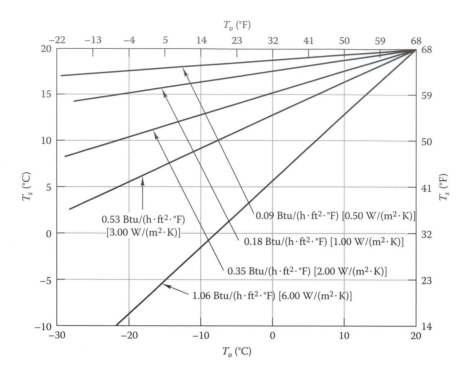

FIGURE 6.22
Interior glass surface temperature, Eq. (6.48), for $T_i = 20°C$ and for several U values.

6.5.3 Shading Coefficient

Of the solar irradiance I incident on glazing, a portion ρI is reflected, a portion αI is absorbed in the glazing, and a portion τI is transmitted to the interior. By energy conservation, the sum of reflectivity ρ, absorptivity α, and transmissivity τ equals unity (see Chap. 2):

$$\rho + \alpha + \tau = 1 \tag{6.49}$$

The variation of these three quantities with the incidence angle θ_i is shown in Fig. 6.23 for three types of glass. The radiation transmitted to the interior of the building is assumed to be entirely absorbed, by virtue of the cavity effect, which causes the number of reflections to be so large that only a negligible amount of radiation can reemerge through the opening.

The radiation absorbed in the glazing raises the temperature of the latter, thus changing the heat flow. While this term could be treated in terms of a sol-air temperature, it is more convenient to combine it with the radiation absorbed in the interior. Consider the energy terms indicated in Fig. 6.18 for heat flow through a single pane of glass. In the absence of solar radiation, the heat gain $\dot{Q}$ of the interior per unit window A area would be only the conductive term

$$\frac{\dot{Q}}{A} = U(T_o - T_i) \quad \text{if no solar gain} \tag{6.50}$$

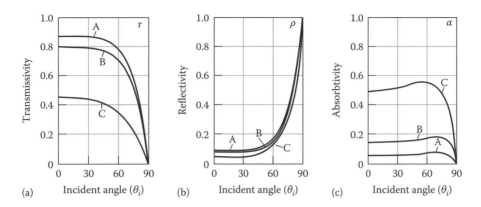

FIGURE 6.23
Solar transmissivity, reflectivity, and absorptivity, as function of incidence angle θ_i, for three types of glass: A = DSA (double-strength sheet); B = 6-mm (0.25-in) clear glass; C = 6-mm (0.25-in) gray, bronze, or green tinted heat-absorbing glass. (Courtesy of ASHRAE, *Handbook of Fundamentals*, American Society of Heating, Refrigerating and Air-Conditioning Engineers, Atlanta, GA, 1989. With permission.)

corresponding to the U value of the glazing. The sign is chosen to give a positive gain when the outside temperature T_o is above the inside temperature T_i. In the presence of solar radiation, the total heat gain becomes

$$\frac{\dot{Q}}{A} = U(T_o - T_i) + \tau I + h_i \Delta T_{\text{sol}} \tag{6.51}$$

where the terms on the right represent conduction (without solar), transmitted solar radiation, and extra heat gain due to solar radiation absorbed by the glass, with ΔT_{sol} being the temperature rise of the glass from absorption of solar radiation.

To render this equation more useful, one eliminates ΔT_{sol} in favor of I. Since one can assume linear approximations for all the relevant heat transfer equations, ΔT_{sol} is independent of T_i and T_o. It can therefore be determined from the energy balance of the glass when T_i and T_o are zero, by setting the absorbed solar flux equal to the sum of the heat flows to inside and outside:

$$\alpha I = h_i \Delta T + h_o \Delta T_{\text{sol}} \tag{6.52}$$

After solving for ΔT, one obtains

$$\frac{\dot{Q}}{A} = U(T_o - T_i) + \tau I + \frac{\alpha I h_i}{h_i + h_o} \tag{6.53}$$

The two rightmost terms can be combined as the total solar heat gain $\dot{Q}_{\text{sol}}$ through glazing

$$\frac{\dot{Q}_{\text{sol}}}{A} = F I \tag{6.54}$$

with a constant of proportionality called the *solar heat gain coefficient F*, given by

$$F = \tau + \alpha \frac{U}{h_o} \quad \text{with } U = \frac{h_i h_o}{h_i + h_o} \tag{6.55}$$

for single-pane glass. For double glazing an analogous calculation yields

$$F = \tau + \alpha_o \frac{U}{h_o} + \alpha_i U \left(\frac{1}{h_s} + \frac{1}{h_o} \right) \quad \text{with } \frac{1}{U} = \frac{1}{h_i} + \frac{1}{h_s} + \frac{1}{h_o} \tag{6.56}$$

where h_s = heat transfer coefficient of the space between the two panes and α_i and α_o are the absorptivities of the inner and outer panes, respectively. As a check, note that in the limit $1/h_s \to 0$ this formula correctly reproduces the single-pane result with $\alpha_o + \alpha_i = \alpha$.

In addition to the properties of the glass, F depends on the incidence angle and on the surface heat transfer coefficients. A list of all the relevant quantities for all glazing types of interest would be too long for most practical purposes. To simplify the presentation, one defines a shading coefficient as the ratio of F for the glazing in question to F for reference glazing, taking double-strength sheet (called DSA) glass with optical properties of Fig. 6.23 for the latter and assuming standard summer conditions. Then the reference glazing has $F = 0.87$, and the *shading coefficient (SC)* becomes

$$SC = \frac{F}{0.87} = 1.15F \tag{6.57}$$

Values for several glazing types are listed in Table 6.5. (The term *shading coefficient* may appear strange since high values correspond to high solar gains, but it has become the

TABLE 6.5

Shading Coefficient SC and Solar Transmissivity τ for Several Conventional Glazing Types, for Winter and Summer Conditions*

| Type of Glazing | Nominal Thickness of Each Pane | | | SC | |
	mm	in	τ	Winter	Summer
Single glazing					
Clear	3	$\frac{1}{8}$	0.86	1.00	1.00
	6	$\frac{1}{4}$	0.78	0.94	0.95
	10	$\frac{3}{8}$	0.72	0.90	0.92
	13	$\frac{1}{2}$	0.67	0.87	0.88
Heat-absorbing	3	$\frac{1}{8}$	0.64	0.83	0.85
	6	$\frac{1}{4}$	0.46	0.69	0.73
	10	$\frac{3}{8}$	0.33	0.60	0.64
	13	$\frac{1}{2}$	0.24	0.53	0.58
Double glazing					
Clear out, clear in	3	$\frac{1}{8}$	0.71	0.88	0.88
	6	$\frac{1}{4}$	0.61	0.81	0.82
Heat-absorbing out, clear in	6	$\frac{1}{4}$	0.36	0.55	0.58

Source: Courtesy of ASHRAE, *Handbook of Fundamentals*, American Society of Heating, Refrigerating and Air-Conditioning Engineers, Atlanta, GA, 1989. With permission.

* See also Table 6.6 for high-performance glazing.

accepted terminology.) Shading coefficients for drapes and blinds can be found in Chap. 27 of ASHRAE (1989).

Within the framework of the ASHRAE procedure for cooling load calculations (described in Secs. 7.6 and 7.7), the solar heat gain through DSA glass is designated as solar heat gain factor (SHGF)

$$\text{SHGF} = FI \quad \text{with } F = 0.87 \tag{6.58}$$

(its daily maximum is listed in the SHGF tables on the CD-ROM). Thus the *instantaneous solar heat gain* through glazing is

$$\dot{Q}_{\text{sol}} = A \times \text{SC} \times \text{SHGF} \tag{6.59}$$

These SHGF values are based on a ground reflectivity $\rho_g = 0.2$; for other ground surfaces, they can be corrected by the procedure on page 27.25 of ASHRAE (1989).

Example 6.13

Find the instantaneous heat gain *FI* for the conditions of Example 6.9 for DSA glass with $F = 0.87$. Compare with the SC $\times$ SHGF value of ASHRAE (1989) chap. 27, table 9.

Given: $I = 369$ W/m^2 [117 Btu/(h $\cdot$ ft^2)] irradiance on west-facing vertical wall at $t_{\text{sol}} = 14.07$ h on July 21, latitude $\lambda = 48.82°$

Find: *FI*; compare with SC $\times$ SHGF

SOLUTION

$$FI = 0.87 \times 369 \text{ W/m}^2 = 321 \text{ W/m}^2$$

For comparison with ASHRAE (1989), note that SC $= 1$ for this glass and SHGF $= 453$ W/m^2 [143.6 Btu/(h $\cdot$ ft^2)] on west-facing wall at 14:00 and $\lambda = 48°$, from chap. 27, table 9.

COMMENTS

1. The results are quite different because we have assumed the 5-km visibility atmosphere for Hottel's model. With 23-km visibility, we would have obtained $I = 501$ W/m^2 and $FI = 0.87 \times 501$ W/m$^2 = 436$ W/m^2 [138.2 Btu/(h $\cdot$ ft^2)], very close to the ASHRAE value.
2. It is interesting to compare this heat gain with the heat gain through an opaque wall for the same solar flux, as calculated in Example 6.11. The latter is only 4.6 W/m^2, almost two orders of magnitude smaller. For buildings with typical ratios of transparent to opaque surfaces, the solar gains through transparent surfaces tend to dominate by far.

6.5.4 High-Performance Glazing

The ideal window is a perfect insulator, with variable transmissivities for the visible and the infrared portions of the solar spectrum, to permit as much control of daylight and heat gains as possible. Recent years have seen vigorous progress in the development of transparent insulators and of coatings with variable transmissivity, and there is good hope that windows of the future will come quite close to the ideal.

There are several methods for reducing the heat loss of windows. Obviously one can add further glass panes, but beyond three panes, one faces diminishing returns. Thin plastic films avoid the cost and weight of glass panes. For greatest effect, they should be coated with a heat mirror. Such coatings do not have to interfere with the transmission of solar radiation because the wavelengths of thermal infrared are quite different from those of the infrared and visible portions of the solar spectrum. While uncoated glass or plastic surfaces have an emissivity of 0.84, quite a few durable coatings are now available with emissivities in the range of 0.2 to 0.4. Heat mirror coatings are a powerful means for reducing the U value of windows, as can be seen from Fig. 6.19.

To combat conductive/convective heat transfer, one can replace the air between the panes by argon. Figure 6.19 shows that the improvement over air is appreciable. Such windows are now in commercial production.

The technology of windows is in rapid evolution, and it is difficult to foresee which products will be the best choice for future buildings. To give an indication of what is available today, we list in Table 6.6 a variety of glazing types; they are representative of generic rather than specific products. The most important performance characteristics are the thermal resistance $R_{th} = 1/U$, the shading coefficient SC, and the transmissivity τ_v, for visible solar radiation.

Recently Arasteh et al. (1988) published the achievement of center-of-glass U values around 0.11 Btu/(h · ft² · °F) [0.7 W/(m² · K)] in a triple-glazed krypton-filled window with two low-emissivity coatings. Its overall thickness is 1 in (25 mm) or less, compatible with the sash and frame requirements of most manufacturers. Further work is needed to develop frames and spacers of comparable performance.

Evacuation would be even better, of course. While the atmospheric pressure would be prohibitive for large unsupported panes, recent work at the Solar Energy Research Institute (Benson et al., 1988) has demonstrated the feasibility of producing flat evacuated windows where the glass panes are separated by tiny glass beads, with about 0.039-in (1-mm)

TABLE 6.6

Data of Generic Glazing Types

Type	1/U		ϵ	SC	τ_v	K_e
	(m² · K)/W	(h · ft² · °F)/Btu				
1. Reflective double glazing (bronze)	0.44	2.5	0.40	0.20	0.10	0.5
2. Tinted double glazing (bronze)	0.35	2.0		0.57	0.47	0.8
3. Clear double glazing	0.35	2.0		0.82	0.80	1.0
4. Low-ϵ double glazing, bronze	0.53	3.0	0.15	0.42	0.41	1.0
5. Low-ϵ double glazing, clear	0.53	3.0	0.15	0.66	0.72	1.1
6. Tinted double glazing, green	0.35	2.0		0.56	0.67	1.2
7. Low-ϵ single glazed, green	0.16	0.9	0.35	0.53	0.65	1.2
8. Triple glazing, green: two panes with low-ϵ coated polyester film in airspace	0.63	3.6	0.15	0.47	0.58	1.2
9. Low-ϵ double glazing, green	0.53	3.0	0.15	0.41	0.61	1.5

Source: Courtesy of Sweitzer, G. et al., *ASHRAE Trans.*, 93, pt. 1, 1987.

Notes: ASHRAE winter conditions; ϵ = emissivity on coated surface (= surface number 2, counted from outside to inside), all other surfaces uncoated (emissivity = 0.84); SC = shading coefficient; τ_v = transmissivity for visible radiation; $K_e = \tau_v/SC$ is proportional to the luminous efficacy of transmitted daylight. The glazings are numbered in order of increasing K_e.

diameter. The U value would be impressive: The center-of-glass value is expected to be 0.062 Btu/(h · ft^2 · °F) [0.35 W/(m^2 · K)], and with a reasonable frame, the area-averaged value would be around 0.088 Btu/(h · ft^2 · °F) [0.5 W/(m^2 · K)]. The commercial promise is not yet clear. It will depend on the visual appearance of the glass beads and, of course, on the cost.

Aerogels are another interesting approach. They are basically a glass foam whose structure is smaller than the wavelength of light. Thus they are quite clear, even at a thickness of several centimeters. Their density is about one-tenth that of glass, while the thermal conductivity is comparable to still air. The conductivity can be cut in half by evacuating the aerogel. Aerogels are quite fragile, and in practical windows they would by sandwiched between glass panes for protection. Fabrication cost and optical clarity need further improvement, but some companies are already close to commercializing the product.

While visual appearance is, of course, a crucial concern for vision windows, there are numerous applications where transparent insulators can be used even if they lack clarity. For many skylights, diffuse glazings are preferable in any case, and for transom windows they may be acceptable. For flat-plate solar collectors, the specularity of the cover is irrelevant. For all such cases, an interesting option is the use of plastic honeycombs made of acrylic or polycarbonate enclosed between flat panes, with cell walls normal to the panes to minimize optical losses. The U value is about 0.176 Btu/(h · ft^2 · °F) [1.0 W/(m^2 · K)] for a thickness of 4 in (0.1 m), while offering a transmissivity for diffuse solar radiation of 0.78. When ordinary glass panes are added on both sides for protection, the transmissivity decreases to 0.60 and the U value improves slightly.

In recent years, several types of transparent insulating panels have become, or are about to become, available at reasonable cost. This is opening interesting possibilities for energy-efficient design, and research activity in this field is vigorous. One of the consequences is that Trombe's old idea of solar storage walls has gained new impetus (Jesch, 1989).

Among other possibilities, we might mention windows that use ventilation air to recover heat or to reduce heat losses. For instance, Fig. 6.24 shows an air curtain window where the heat of the exhaust air compensates, in effect, the heat loss through the window, instead of being vented directly to the outside. As a variation, the flow could be reversed, drawing the supply air through such a window; the heat lost through the inner pane would be brought back into the building. Such windows appear to be expensive, and they are not widely used at present (which in turn prevents the cost reductions achievable by mass production).

For the control of solar radiation, the traditional tools have been mechanical: overhangs and blinds, fixed or movable. Fixed shading devices lack flexibility for good control of heat gains. Movable devices on the outside of a building are expensive because they must be sturdy enough to withstand wind. For the inside of windows, venetian blinds are classic, but their thermal performance is worse than that of exterior blinds because venetian blinds cause much solar radiation to be absorbed inside the building. Automatic adjustment of mechanical shading devices is possible, but in practice there have been many problems. Far more elegant would be a coating whose transmissivity could be varied by an electric signal. Several solutions have been developed during the past decade. Electrochromic coatings can be applied to glass, and they allow the solar transmissivity to vary continuously by applying a voltage. Samples have been produced whose transmissivity varies from 0.2 to 0.8, and research is continuing. Liquid crystals, enclosed between glass panes, allow the switching between two levels of transmissivity. Finally, there are thermochromic materials whose transmissivity varies with temperature.

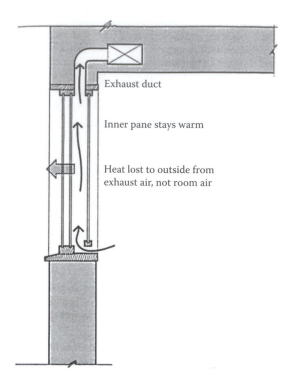

Exhaust duct

Inner pane stays warm

Heat lost to outside from
exhaust air, not room air

FIGURE 6.24
Air curtain window. (Courtesy of Nisson, J.D.N.
and Dutt, G., *The Superinsulated Home Book*, Wiley,
New York, 1985.)

Problems

The problems in this book are arranged by topic. The approximate degree of difficulty is indicated by a parenthetic italic number from 1 to 10 at the end of the problem. Problems are stated most often in USCS units; when similar problems are presented in SI units, it is done with approximately equivalent values in parentheses. The USCS and SI versions of a problem are not exactly equivalent numerically. Solutions should be organized in the same order as the examples in the text; given, figure or sketch, assumptions, find, lookup values, solution. For some problems, the Heating and Cooling of Buildings (HCB) software may be helpful. In some cases it is advisable to set up the solution as a spreadsheet, so that design variations are easy to evaluate.

6.1 What is the purpose of the equation of time?   (2)

6.2 What is the value of the solar constant?   (1)

6.3 How much does the extraterrestrial normal solar irradiance vary during the course of the day and year?   (5)

6.4 What, approximately, is the highest value of the solar irradiance that might be incident on the surface of a building?   (3)

6.5 What characteristics of a window are relevant for the thermal analysis of a building?   (3)

6.6 A window has a shading coefficient of 0.64 and a transmittance for solar radiation of 0.33. What is the instantaneous solar heat gain when the solar irradiance incident on the window is 200 Btu/(h · ft^2) (631 W/m^2)? What is the SHGF?   (4)

6.7 Find the sol-air temperature for a vertical surface when the incident solar radiation is 200 Btu/(h · ft^2) (631 W/m^2) and the air temperature 80°F (26.7°C).

(a) Assume $\alpha/h_o = 0.15$ (h · ft^2 · °F)/Btu [0.026 (m^2 · K)/W], summer design conditions for light surface.

(b) $\alpha/h_o = 0.3$ (h · ft^2 · °F)/Btu [0.052 (m^2 · K)/W], summer design conditions for dark surface. (4)

6.8 Find the global horizontal irradiance when the direct normal irradiance is 700 W/m^2 [221.9 Btu/(h · ft^2)], the diffuse horizontal irradiance 150 W/m^2 [47.6 Btu/(h · ft^2)], and the angle of incidence is 30°. (4)

6.9 Use Fig. 6.16 to find the average daily solar heat gain through a west-facing vertical window with shading coefficient SC = 0.9 in Albuquerque, New Mexico ($\lambda = 35$°N), in January. (3)

6.10 Find the length of the day (sunrise to sunset) at the summer and winter solstices in

(a) Honolulu ($\lambda = 21.03$°N)

(b) Stockholm, Sweden ($\lambda = 59.35$°N) (5)

6.11 Estimate the azimuth angle (from due south) of the chalk board in your classroom. What is the angle of incidence of the sun on this surface at solar noon, equinox? (5)

6.12 (a) Write an equation for the number of hours per day when direct solar radiation can reach an unshaded fixed surface at arbitrary tilt and zero azimuth.

(b) Evaluate this equation for the case of tilt = latitude at summer solstice and at winter solstice. (6)

6.13 Consider a sundial built as a vertical rod of 1-m (3.281-ft) length that casts a shadow on a flat horizontal surface. The location is Princeton, New Jersey, with approximate latitude 40°N and longitude 75°W.

(a) What time of day (solar time) and time of year is it when the shadow is 0.50 m (1.641 ft) long, pointing due north? How many solutions are there?

(b) What time of day (solar time) and time of year is it when the shadow is 0.50 m (1.641 ft) long, pointing 45° (in the horizontal plane) east of north? (By contrast to part *a*, this requires two equations in two unknowns; but they can be solved in closed form.)

(c) What are the corresponding standard times? (10)

6.14 Consider a south-facing unshaded vertical window at 1:30 p.m. solar time on January 21. The latitude is 45°N.

(a) Find the zenith and azimuth angles of the sun.

(b) Find the incidence angle on the window.

(c) Suppose the direct normal irradiance is 700 W/m^2 [221.9 Btu/(h · ft^2)] and the diffuse horizontal irradiance is 100 W/m^2 [31.70 Btu/(h · ft^2)]. Find the global irradiance on the window if the ground is a diffuse reflector with reflectivity 0.2.

(d) How does the answer in part *c* change if the reflectivity is increased to 0.7, a value typical of snow?

(e) How does the answer in part *d* change if the ground is specular instead of diffuse? (10)

6.15 Two buildings in a town at a latitude 40°N are arranged like Fig. P6.15.

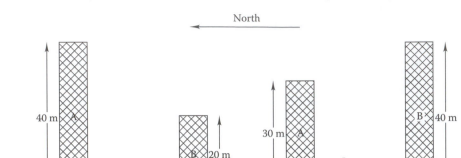

FIGURE P6.15

(a) Find the portion of building A that is shaded by building B at 11:00 a.m. on August 21.

(b) Superimpose the outline of building B as seen from the corner P on the sun path diagram of Fig. 6.6. (For simplicity, take only the corners and join them by straight lines, even though, strictly speaking, nonvertical straight lines look curved when viewed in a sun path diagram.)

(c) Use this diagram to estimate how many hours a window located at the corner P can receive direct sunlight on August 21.   (9)

6.16 Set up a spreadsheet to calculate and plot, as a function of time of day, the incidence angle of the sun on the horizontal and vertical surfaces facing the four cardinal directions. Add comments for documentation, and define names for the key variables (declination, latitude, etc.) to facilitate future reuse and expansion of the spreadsheet.

(a) Produce the plots for equinox and for the solstices, at latitude 40°N.

(b) Add to this spreadsheet the direct and diffuse irradiance according to Hottel's clear day model, for a midlatitude summer atmosphere with 23-km visibility at sea level. Plot the direct normal and the diffuse horizontal irradiance versus time of day for July 21, at latitude 40°N.

(c) Add to the spreadsheet the global irradiance on vertical surfaces facing the four cardinal directions, and plot the result. Assume a ground reflectivity of 0.2.

(d) Add to the spreadsheet the global irradiance on a south-facing roof at tilt = latitude, and plot the result, at latitude 40°N. Assume a ground reflectivity of 0.2.   (10)

6.17 You want to measure the height of a telephone pole without climbing it. The latitude is 40°, and the length of the shadow at solar noon on the summer solstice is 5 m. How tall is the telephone pole?   (2)

6.18 You want to measure the height of a telephone pole without climbing it and without waiting for a solstice. The location is Boulder, Colorado (40.00°N and 105.27°W). You find that the shadow is 10 m at 10 a.m. MDT on May 15. How tall is the telephone pole?   (6)

6.19 You are designing a sanctuary in Truth or Consequences, New Mexico (33.23°N and 107.27°W), in which you want the sun to illuminate a golden sphere at sunrise on the summer solstice as the sunlight passes through a notch between two rocks. The ground is flat at that location. At what direction (i.e., angle from due east) should the sphere be placed relative to the notch? (4)

6.20 At what latitude can you collect the greatest amount of energy on a horizontal surface during the day of June 21 if the sky is clear? At what latitude is the lowest amount collected? (2)

6.21 Use the clear day model of Hottel to calculate total direct and diffuse irradiance on a horizontal surface at solar noon on the summer solstice at the top of Mt. Whitney (4418 m) and at the beach in Los Angeles, California. (5)

6.22 Calculate the U value for the following double-glazed windows:

(a) Ordinary glass

(b) Ordinary glass with vacuum between the panes

(c) Low-emittance coating with $\varepsilon = 0.05$ on both surfaces facing the gap

(d) Low-emittance coatings as in part c but with a vacuum between the panes For simplicity, assume the heat transfer coefficients as given in Example 2.11. (6)

6.23 Calculate the shading coefficient for double glazing with $\alpha_o = 0.5$ and $\alpha_i = 0.05$, assuming $U = 0.5 \text{ Btu}/(\text{h} \cdot \text{ft}^2 \cdot {}^\circ\text{F})$. (4)

6.24 Frost can form on a surface when the humidity is sufficiently high and the surface temperature falls below freezing. What is the highest outdoor temperature at which frost can form on the inside of a single-glazed window if it is 60°F indoors? Assume a U value of 1 $\text{Btu}/(\text{h} \cdot \text{ft}^2 \cdot {}^\circ\text{F})$. (3)

6.25 You are designing an atrium with single-glazed fenestration, and you are worried about the possibility of the glass cracking when it is heated by the sun and then suddenly hit by cold water from a nearby sprinkler. Estimate the surface temperature of the glazing under the following conditions:

- Outside air temperature is 40°C.
- Solar radiation incident on the glazing is 1000 W/m^2.
- The inside air temperature is 30°C.
- The U value of the glazing is 6.0 W/(m$^2 \cdot$ K).
- The solar absorptance of the glazing is 0.5.
- Inside surface convection coefficient is $h_i = 10$ W/(m$^2 \cdot$ K).
- Outside surface convection coefficient is $h_o = 20$ W/(m$^2 \cdot$ K). (7)

6.26 Assuming a solar heat gain factor of 200 $\text{Btu}/(\text{h} \cdot \text{ft}^2)$, what is the highest instantaneous solar heat gain among the nine glazing types listed in Table 6.6? What is the lowest? (2)

References

Arasteh, D., S. Selkowitz, and J. Wolfe (1989). "The Design and Testing of a Highly Insulating Glazing System for Use with Conventional Window Systems." Report LBL-24903 TA-257 (November). Lawrence Berkeley Laboratory, Berkeley, Calif. *J. Solar Energy Eng.*, vol. 111, No. 1, pp. 44–53.

ASHRAE (1989). *Handbook of Fundamentals*. American Society of Heating, Refrigerating and Air-Conditioning Engineers, Atlanta.

Benson, D. K., T. F. Potter, and C. B. Christensen (1988). "Vacuum Insulating Window R&D: An Update." *Proc. American Council for Energy Efficient Economy 1988 Summer Study*, vol. 3, p. 3.21.

Cinquemani, V., J. R. Owenby, Jr., and R. G. Baldwin (1978). "Input Data for Solar Systems." Report prepared by the National Climatic Data Center, Asheville, N.C., under contract E(49–26)-1041 between the U.S. Department of Commerce and the U.S. Department of Energy.

Collares-Pereira, M., and A. Rabl (1979). "The Average Distribution of Solar Radiation: Correlations between Diffuse and Hemispherical and between Hourly and Daily Insolation Values." *Solar Energy*, vol. 22, pp. 155–164.

Duffie, J. A., and W. A. Beckman (1980). *Solar Engineering of Thermal Processes*. Wiley, New York.

Erbs, D. G., S. A. Klein, and J. A. Duffie (1982). "Estimation of the Diffuse Radiation Fraction for Hourly, Daily and Monthly-Average Global Radiation." *Solar Energy*, vol. 28, p. 293.

Hay, J. E. (1979). "Calculation of Monthly Mean Solar Radiation for Horizontal and Inclined Surfaces." *Solar Energy*, vol. 23, p. 301.

Hottel, H. C. (1976). "A Simple Model for Estimating the Transmissivity of Direct Solar Radiation through Clear Atmospheres." *Solar Energy*, vol. 18, p. 129.

Hulstrom, R. L., ed. (1989). *Solar Resources*. M.I.T. Press, Cambridge, Mass.

Hunn, B. D., and D. O. Calafell (1977). "Determination of Average Ground Reflectivity for Solar Collectors." *Solar Energy*, vol. 19, p. 87.

Jesch, L. F., ed. (1989). *Transparent Insulation Technol. 3d Int. Workshop Proc.* The Franklin Company Consultants, Birmingham, U.K.

Kreith, F., and J. F. Kreider (1978). *Principles of Solar Engineering*. Hemisphere/McGraw-Hill, New York.

Liu, B. Y. H., and R. C. Jordan (1960). "The Interrelationship and Characteristic Distribution of Direct, Diffuse and Total Solar Radiation." *Solar Energy*, vol. 4, p. 34.

Mazria, E. (1979). *The Passive Solar Energy Book*. Rodale, Emmaus, Pa.

McCabe, M. E. (1989). "Window *U*-Values: Revisions for the 1989 ASHRAE *Handbook of Fundamentals*." *ASHRAE J.*, June, p. 56.

Nisson, J. D. N. (1985). *Windows and Energy-Efficiency: Principles, Practice and Available Products*. Cutter Information Corp., Arlington, Mass.

Nisson, J. D. N., and G. Dutt (1985). *The Superinsulated Home Book*. Wiley, New York.

NOAA (1978). "SOLMET: Hourly Solar Radiation—Surface Meteorological Observations." National Climatic Data Center of the National Oceanic and Atmospheric Administration, Asheville, N.C.

Potter, R., N. Karkamaz, and J. F. Kreider (1989). "Simplification of Monthly Solar Radiation Calculations for a Vertical Surface." *Proc. 1989 Conf. American Section, Int. Solar Energy Soc.*, p. 450.

Rabl, A. (1985). *Active Solar Collectors and Their Applications*. Oxford University Press, New York.

SERI (1980). *Insolation Data Manual*. Report SERI/SP-755–789, Solar Energy Research Institute, Golden, Colo.

Sweitzer, G., D. Arasteh, and S. Selkowitz (1987). "Effects of Low-Emissivity Glazings on Energy Use Patterns in Nonresidential Daylighted Buildings." *ASHRAE Trans.*, vol. 93, pt. 1.

Threlkeld, J. L. (1970). *Thermal Environmental Engineering*, 2d ed. Prentice-Hall, Englewood Cliffs, N.J.

Threlkeld, J. L., and R. C. Jordan (1958). "Direct Solar Radiation Available on Clear Days." *ASHRAE Trans.*, vol. 64, p. 65.

7

Heating and Cooling Loads

Heating and cooling loads are the thermal energy that must be supplied to or removed from the interior of a building in order to maintain the desired comfort conditions. That is the demand side of the building, addressed in this chapter and Chap. 8. Once the loads have been established, one can proceed to the supply side and determine the performance of the required heating and cooling equipment; this is discussed in Chaps. 9 to 12.

Of primary concern to the designer are the maximum or peak loads, because they determine the capacity of the equipment. They correspond to the extremes of hot and cold weather, called *design conditions*. But while in the past it was common practice to limit oneself to the consideration of peak loads, examination of annual performance has now become part of the designer's job. The oil crises have sharpened our awareness of energy, and the computer revolution has given us the tools to compute the cost of energy and to optimize the design of the building. In this chapter we address the calculation of peak loads. Methods for the determination of annual energy requirements are presented in Chap. 8.

A load calculation consists of a careful accounting of all the thermal energy terms in a building. While the basic principle is simple, a serious complication can arise from storage of heat in the mass of the building. In practice, this is very important for peak cooling loads, even in lightweight buildings typical in the United States. For peak heating loads, the heat capacity can be neglected unless one insists on setting the thermostat back even during the coldest periods. For annual energy consumption, the effect of heat capacity depends on the control of the thermostat: It is negligible if the indoor temperature is constant but can be quite significant with thermostat setback or setup.

This chapter begins with models for air exchange, in Sec. 7.1. In Sec. 7.2 we discuss the design conditions, heat loss coefficient, and thermal balance of a building. Then we address the need for zoning, i.e., the separate treatment of different parts of a building where the loads are too dissimilar to be lumped together. In Sec. 7.4 we examine the limitations of a steady-state analysis. A steady-state method for peak heating loads is presented in Sec. 7.5. Peak cooling loads are calculated in Sec. 7.6, by using a modified steady-state method, the CLF-CLTD method. To provide an algorithm for dynamic load calculations, the transfer function method is presented in Sec. 7.7; using this method has become relatively simple, thanks to computers with spreadsheets. Recent updates to the energy calculation methods are presented in Sec. 7.8.

Some further topics related to loads are discussed in Chap. 8, in particular, the calculation of annual energy consumption, the use of thermal networks for dynamic modeling, the concept of the time constant, and the relation between thermal networks and the transfer function method.

The calculation of loads presented here does not take into account the losses in the distribution system. These losses can be quite significant, especially in the case of uninsulated ducts, and they should be taken into account in the analysis of the HVAC system. Distribution systems are the province of Chap. 11.[1]

7.1 Air Exchange

Fresh air in buildings is essential for comfort and health, and the energy for conditioning this air is an important term. The designer must pay attention: Not enough air, and one risks the sick building syndrome; too much air, and one wastes energy. The supply of fresh air, or air exchange, is stated as the flow rate $\dot{V}$ of the outdoor air that crosses the building boundary and needs to be conditioned [ft^3/min (m^3/s or L/s)]. Often it is convenient to divide it by the building volume, as $\dot{V}/V$, expressing it in units of air changes per hour. Even though it is customary to state the airflow as the volumetric rate, the mass flow $\dot{m} = \rho\dot{V}$ would be more relevant for most applications in buildings. The relation between mass flow and volume flow depends on the density ρ, which varies quite significantly with temperature and pressure.

To estimate the air exchange rate, the designer has two sources of information: data from similar buildings and models. The underlying phenomena are complicated, and a simple comparison with other buildings may not be reliable. The modeling approach can be far more precise, but a fair amount of effort may be required.

It is helpful to distinguish two mechanisms that contribute to the total air exchange:

- *Infiltration* (uncontrolled airflow through all the little cracks and openings in a real building)
- *Ventilation* (natural ventilation through open windows or doors and mechanical ventilation by fans)

7.1.1 Data for Air Exchange

Air exchange rates can be measured directly by means of a tracer gas. Sulfur hexafluoride (SF$_6$) has been a favorite because it is inert and harmless, and it can be detected at concentrations above 1 part per billion (ppb). The equipment is relatively expensive but allows determination at hourly or even shorter intervals (Sherman et al., 1980; Grimsrud et al., 1980). More recently a low-cost alternative has been developed that uses passive perfluorocarbon sources and passive samplers (Dietz et al., 1985). Each source can cover up to several hundred cubic meters of building volume, at a cost of about $50, but the method yields only averages over sampling periods of several weeks (in fact, it averages the inverse of the air exchange rate). Carbon dioxide is interesting as a tracer gas because it is produced by the occupants and can be used for monitoring indoor air quality; as a measure of air

---

[1] For future reference we note that the loads of each zone, as calculated in this chapter, *include* the contribution of the outdoor air change. However, for the analysis of air-based central distribution systems, it is convenient to *exclude* the contribution of ventilation air from the zone loads and to count it instead as load at the air handler. Keeping the load due to ventilation air apart is straightforward because this load is instantaneous. This is illustrated in Chap. 11.

exchange, it is uncertain to the extent that the number of occupants and their metabolism are not known.

An entirely different method of determining the airtightness of a building is pressurization with a blower door (a special instrumented fan that is mounted in the frame of a door for the duration of the test). To obtain accurate data, one needs fairly high pressures, around 0.2 to 0.3 inWG (50 to 75 Pa), which are higher than natural conditions in most buildings. The extrapolation to lower values requires assumptions about the exponent in the flow-pressure relation, and it is not without problems (see Chap. 23 of ASHRAE, 1989a). In buildings with mechanical ventilation, one could bypass the need for a blower door by using the ventilation system itself, if the pressure is sufficiently high.

In the past, not much attention was paid to airtight construction, and older buildings tend to have rather high infiltration rates, in the range of 1 to 2 air changes per hour. With current conventional construction in the United States, one finds lower values, around 0.3 to 0.7. These values are seasonal averages; instantaneous values vary with wind and the indoor-outdoor temperature difference. When infiltration is insufficient to guarantee adequate indoor air quality, forced ventilation becomes necessary. The required air exchange rate depends, of course, on the density of occupants. In residential buildings, the density is relatively low, and with conventional U.S. construction, infiltration is likely to be sufficient. But it is certainly possible to make buildings much tighter than 0.3 air change per hour of infiltration. That is, in fact, standard practice in Swedish houses, which are built to such high standards that uncontrolled infiltration rates are around 0.1 air change per hour; mechanical ventilation supplies just the right amount of outdoor air, and an air-to-air heat exchanger minimizes the energy consumption. In the United States for buildings with forced ventilation, ASHRAE ventilation Standard 62–87 applies. Good sites for mechanical exhaust are the kitchen and bathrooms, to remove indoor air pollution and excessive humidity. In France, building codes require exhaust fans with minimum continuous flow in kitchens and bathrooms of all new residential buildings.

Uncontrolled air exchange is highly dependent on wind and on temperatures. Even with closed windows, it can vary by a factor of 2 or more, being lower in summer than in winter. The variability of air exchange is indicated schematically in Fig. 7.1 as the relative frequency of occurrence for three types of house: a leaky house and a moderately tight house, both with natural infiltration, and a very tight house with mechanical ventilation. The last guarantees adequate supply of fresh air at all times, without the energy waste of conditioning unnecessary air.

With open windows, the air exchange rate is difficult to predict accurately. It varies with the wind, and it is highly dependent on the aerodynamics of the building and its surroundings. The designer needs data on ventilation rates with open windows to assess comfort conditions in buildings with operable windows during the transition season between heating and cooling. In Fig. 7.2, we present data for natural ventilation that have been measured in a two-story house. Depending on which windows are open and where the ventilation is measured, the air change rate is seen to vary between 1 and 20 per hour.

7.1.2 Models for Air Leakage

For a more precise model of air exchange, recall from Chap. 5 that the flow through an opening is proportional to the area and to some power of the pressure difference:

$$\dot{V} = Ac\, \Delta p^n \qquad (7.1)$$

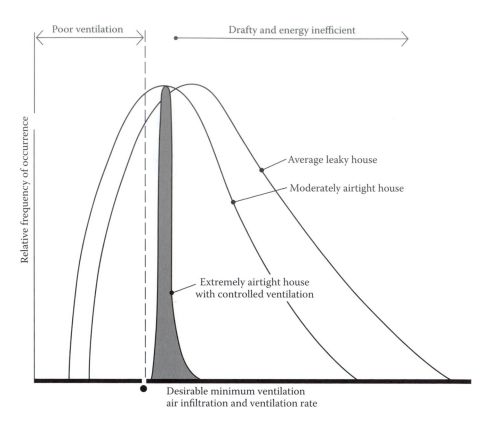

FIGURE 7.1
Variability of air exchange for three types of house, plotted as relative frequency of occurrence versus air change rate. (Courtesy of Nisson, J.D.N. and Dutt, G., *The Superinsulated Home Book*, Wiley, New York, 1985.)

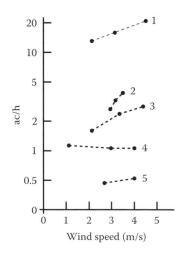

	Open Windows	**Measuring Point**
1	All upper floor	Upper floor
2	Upper floor windward	Upper floor
3	All upper floor	Lower floor
4	Upper floor leeward	Upper floor
5	None	Whole house

FIGURE 7.2
Measured ventilation rates, as a function of wind speed, in a two-story house with windows open on lower floor. The curves are labeled according to the location of open windows and measuring point as above. (Courtesy of Achard, P. and Gicquel, R., *European Passive Solar Handbook*, Commission of the European Communities, Directorate General XII for Science, Research and Development, Brussels, Belgium, 1986.)

where
    A = area of opening, $ft^2(m^2)$
    $\Delta p = p_o - p_i$ = pressure difference between outside and inside, inWG (Pa)
    c = flow coefficient, $ft/(min \cdot inWG^n)$ $[m/(s \cdot Pa^n)]$
    n = exponent, between 0.4 and 1.0 and usually around 0.65 for buildings

In general, different openings may have different coefficients and exponents. This equation is an approximation, valid only for a certain range of pressures and flows; different n and C values may have to be used for other ranges. There is another problem in applying this equation to buildings: The width of an opening can change with pressure. Blower door tests have shown that the apparent leakage area can be significantly higher for over-pressurization than for underpressurization; external pressure tends to compress the cracks of a building (Lydberg and Honarbakhsh, 1989). In buildings without mechanical ventilation, the pressure differences under natural conditions are positive over part of the building and negative over the rest; here it is appropriate to average the blower door data over positive and negative pressure differences. In buildings where over- or underpressure is maintained, the leakage area data should correspond to those conditions.
    The total flow is obtained by summing over all openings k as

$$\dot{V} = \sum_k A_k c_k \, \Delta p_k^{n_k} \quad \text{(include only terms with } \Delta p_k > 0) \tag{7.2}$$

where
    A_k = leakage area
    c_k = flow coefficient
    n_k = exponent
    $\Delta p_k = p_o - p_i$ = local pressure difference

If one sums naively over all openings of a building, the result (averaged over momentary fluctuations) is zero, because the quantity of air in a building does not change. The flow into the building must equal the flow out; the former corresponds to positive terms in the sum, the latter to negative terms. What interests us here is the energy needed for conditioning the air that flows into the building. Therefore, the sum includes only the terms with $p_o > p_i$.
    As indicated by the subscript k, all the terms can vary from one point to another in the building. Therefore, a fairly detailed calculation may be required. Typical air leakage sites in houses are shown in Fig. 7.3, and data for leakage areas can be found in Table 7.1 for a wide variety of building components.
    Actually, in many applications, one need not worry about this. For relatively small buildings *without* mechanical ventilation, the LBL model of Sec. 7.1.4 can be used; this model bypasses Eq. (7.2) by correlating the flow directly with the wind speed, temperature difference, and total leakage area. In many (if not most) buildings *with* mechanical ventilation, one maintains a significant pressure difference between interior and exterior.[2] If this pressure difference is larger than the pressures induced by wind and temperature, the latter can be neglected and all terms in Eq. (7.2) have the same sign. Before proceeding to these applications, we have to discuss the origin of the pressure differences.

---

[2] Overpressure in the building allows better control and comfort. Underpressure can be maintained with smaller ducts and lower cost, but at the risk of condensation, freezing, and possibly draft.

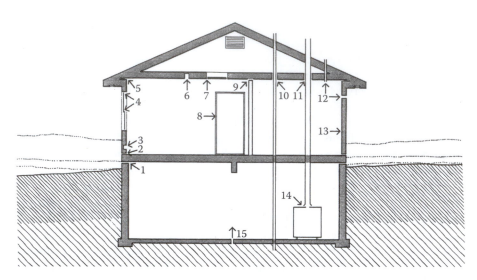

FIGURE 7.3

Typical air leakage sites in a house. (1) Joints between joists and foundation; (2) joints between sill and floor; (3) electrical boxes; (4) joints at windows; (5) joints between wall and ceiling; (6) ceiling light fixtures; (7) joints at attic hatch; (8) cracks at doors; (9) joints at interior partitions; (10) plumbing-stack penetration of ceiling; (11) chimney penetration of ceiling; (12) bathroom and kitchen ventilation fans; (13) air-vapor barrier tears; (14) chimney draft air leaks; (15) floor drain (air enters through drain tile). (Courtesy of Nisson, J.D.N. and Dutt, G., *The Superinsulated Home Book*, Wiley, New York, 1985.)

TABLE 7.1

Data for Effective Leakage Areas of Building Components at 0.016 inWG (4 Pa)

Component	Best Estimate	Maximum	Minimum
Sill foundation—Wall			
Caulked, in^2/ft of perimeter	0.04	0.06	0.02
Not caulked, in^2/ft of perimeter	0.19	0.19	0.05
Joints between ceiling and walls			
Joints, in^2/ft of wall (only if not taped or plastered and no vapor barrier)	0.07	0.12	0.02
Windows			
Casement			
Weatherstripped, in^2/ft^2 of window	0.011	0.017	0.006
Not weatherstripped, in^2/ft^2 of window	0.023	0.034	0.011
Awning			
Weatherstripped, in^2/ft^2 of window	0.011	0.017	0.006
Not weatherstripped, in^2/ft^2 of window	0.023	0.034	0.011
Single-hung			
Weatherstripped, in^2/ft^2 of window	0.032	0.042	0.026
Not weatherstripped, in^2/ft^2 of window	0.063	0.083	0.052
Double-hung			
Weatherstripped, in^2/ft^2 of window	0.043	0.063	0.023
Not weatherstripped, in^2/ft^2 of window	0.086	0.126	0.046

TABLE 7.1 (continued)

Data for Effective Leakage Areas of Building Components at 0.016 inWG (4 Pa)

Component	Best Estimate	Maximum	Minimum
Single-slider			
Weatherstripped, in^2/ft^2 of window	0.026	0.039	0.013
Not weatherstripped, in^2/ft^2 of window	0.052	0.077	0.026
Double-slider			
Weatherstripped, in^2/ft^2 of window	0.037	0.054	0.02
Not weatherstripped, in^2/ft^2 of window	0.074	0.110	0.04
Doors			
Single door			
Weatherstripped, in^2/ft^2 of door	0.114	0.215	0.043
Not weatherstripped, in^2/ft^2 of door	0.157	0.243	0.086
Double door			
Weatherstripped, in^2/ft^2 of door	0.114	0.215	0.043
Not weatherstripped, in^2/ft^2 of door	0.16	0.32	0.1
Access to attic or crawl space			
Weatherstripped, in^2 per access	2.8	2.8	1.2
Not weatherstripped, in^2 per access	4.6	4.6	1.6
Wall—Window frame			
Wood frame wall			
Caulked, in^2/ft^2 of window	0.004	0.007	0.004
No caulking, in^2/ft^2 of window	0.024	0.038	0.022
Masonry wall			
Caulked, in^2/ft^2 of window	0.019	0.03	0.016
No caulking, in^2/ft^2 of window	0.093	0.15	0.082
Wall—Door frame			
Wood wall			
Caulked, in^2/ft^2 of door	0.004	0.004	0.001
No caulking, in^2/ft^2 of door	0.024	0.024	0.009
Masonry wall			
Caulked, in^2/ft^2 of door	0.0143	0.0143	0.004
No caulking, in^2/ft^2 of door	0.072	0.072	0.024
Domestic hot water systems			
Gas water heater (only if in conditioned space), in^2	3.1	3.9	2.325
Electric outlets and light fixtures			
Electric outlets and switches			
Gasketed, in^2 per outlet and switch	0	0	0
Not gasketed, in^2 per outlet and switch	0.076	0.16	0
Recessed light fixtures, in^2 per fixture	1.6	3.10	1.6
Pipe and duct penetrations through envelope			
Pipes			
Caulked or sealed, in^2 per pipe	0.155	0.31	0
Not caulked or sealed, in^2 per pipe	9.30	1.55	0.31

(continued)

TABLE 7.1 (continued)

Data for Effective Leakage Areas of Building Components at 0.016 inWG (4 Pa)

Component	Best Estimate	Maximum	Minimum
Ducts			
Sealed or with continous vapor barrier, in^2 per duct	0.25	0.25	0
Unsealed and without vapor barrier, in^2 per duct	3.7	3.7	2.2
Fireplace			
Without insert			
Damper closed, in^2 per fireplace	10.7	13.0	8.4
Damper open, in^2 per fireplace	54.0	59.0	50.0
With insert			
Damper closed, in^2 per fireplace	5.6	7.1	4.03
Damper open or absent, in^2 per fireplace	10.0	14	6.2
Exhaust fans			
Kitchen fan			
Damper closed, in^2 per fan	0.775	1.1	0.47
Damper open, in^2 per fan	6.0	6.5	5.6
Bathroom fan			
Damper closed, in^2 per fan	1.7	1.9	1.6
Damper open, in^2 per fan	3.1	3.4	2.8
Dryer vent			
Damper closed, in^2 per vent	0.47	0.9	0
Heating ducts and furnace—Forced-air systems			
Ductwork (only if in unconditioned space)			
Joints taped or caulked, in^2 per house	11	11	5
Joints not taped or caulked, in^2 per house	22	22	11
Furnace (only if in conditioned space)			
Sealed combustion furnace, in^2 per furnace	0	0	0
Retention head burner furnace, in^2 per furnace	5	6.2	3.1
Retention head plus stack damper, in^2 per furnace	3.7	4.6	2.8
Furnace with stack damper, in^2 per furnace	4.6	6.2	3.1
Air conditioner			
Wall or window unit, in^2 per unit	3.7	5.6	0

Source: Courtesy of ASHRAE, *Handbook of Fundamentals*, American Society of Heating, Refrigerating and Air-Conditioning Engineers, Atlanta, GA, 1989a.

For conversion to SI units: $1\ in^2 = 6.45\ cm^2$, $1\ ft^2 = 0.0929\ m^2$, and $1\ in^2/ft^2 = 69\ cm^2/m^2$.

7.1.3 Pressure Terms

The pressure difference $\Delta p = p_o - p_i$ is the sum of three terms:

$$\Delta p = \Delta p_{\text{wind}} + \Delta p_{\text{stack}} + \Delta p_{\text{vent}} \qquad (7.3)$$

The first is due to wind, the second to the stack effect (like the flow induced in a heated smokestack), and the third to forced ventilation, if any. We take the pressure differences to be positive when they cause air to flow toward the interior. The flow depends only on the total Δp, not on the individual terms. The relative contribution of the wind, stack, and ventilation terms varies across the envelope, and because of the nonlinearity, one cannot calculate separate airflows for each of these effects and add them at the end.

The wind pressure is given by Bernoulli's equation (in SI units)

$$p_{\text{wind}} = \frac{\rho}{2}(v^2 - v_f^2) \text{ Pa} \tag{7.4SI}$$

where
    $v =$ wind speed (undisturbed by building), m/s
    $v_f =$ final speed of air at building boundary
    $\rho =$ air density, kg/m^3

In USCS units, we have

$$p_{\text{wind}} = \frac{\rho}{2g_c}(v^2 - v_f^2) \text{ lb}_f/\text{ft}^2 \tag{7.4US}$$

with $g_c = 32.17$ (lb$_m \cdot$ ft)/(lb$_f \cdot$ s^2), the wind speed being in feet per second, and the air density in pound-mass per cubic foot.

Under *standard conditions* of 14.7 psi (101.3 kPa) and 68°F (20°C), the density is

$$\rho = 0.075 \text{ lb}_m/\text{ft}^3 \quad (\rho = 1.20 \text{ kg/m}^3)$$

but one should note that the density of outdoor air can deviate more than 20 percent above (winter at sea level) or below (summer in the mountains). In USCS units, the ratio of ρ to g_c has the value

$$\frac{\rho}{g_c} = 0.00964 \text{ inWG/(mi/h)}^2$$

under standard conditions if pressure is in inches water gauge and wind speed in miles per hour.

The wind speed is strongly modified by terrain and obstacles, being significantly higher far above the ground (see Chap. 14 of ASHRAE, 1989a). Since the final speed v_f is awkward to determine, a convenient shortcut is to use Eq. (7.4) with $v_f = 0$, multiplying it instead by a pressure coefficient C_p:

$$p_{\text{wind}} = C_p \frac{\rho}{2} v^2 \tag{7.5}$$

The quantity $p_{\text{wind}}/C_p = (\rho/2)v^2$ is plotted versus wind speed v in Fig. 7.4a. Numerical values for C_p can be gleaned from Fig. 7.4b, where this coefficient is plotted as a function of the angle between the wind and the surface normal. Typical values are in the range from approximately -0.6 to 0.6, depending on the direction of the wind.

Actually we are interested in the pressure difference between the interior and exterior of a building. If the interior of an entire floor offers no significant flow resistance, one can find

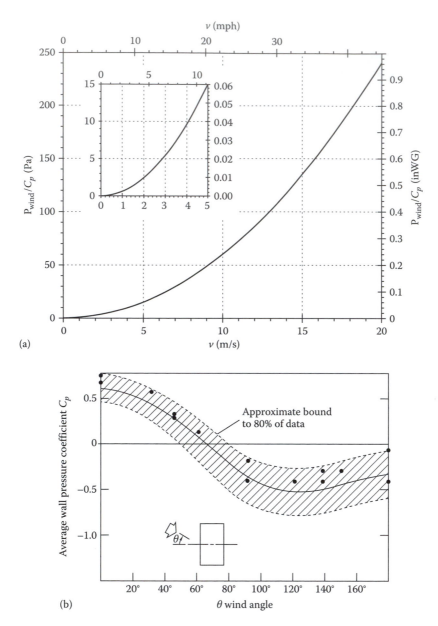

FIGURE 7.4
(a) Correlations for wind pressure. Wind pressure plotted as $p_{wind}/C_p = (\rho/2)v^2$ versus wind speed v. (b) Correlations for wind pressure. Typical values of pressure coefficient C_p of Eq. (7.5) for a rectangular building as a function of wind direction. (The dots indicate the values from fig. 14.6 of that reference.) (Courtesy of ASHRAE, *Handbook of Fundamentals*, American Society of Heating, Refrigerating and Air-Conditioning Engineers, Atlanta, GA, 1989a.)

the indoor pressure due to wind by averaging the flow coefficient over all orientations of the surrounding wall. Since that average is approximately −0.2, the local pressure difference $p_o - p_i$ at a point of the wall is, in that case,

$$\Delta p_{wind} = \Delta C_p \frac{\rho}{2} v^2 \qquad (7.6)$$

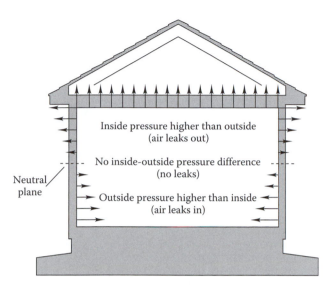

FIGURE 7.5
Air leakage due to stack effect during heating season. (Courtesy of Nisson, J.D.N. and Dutt, G., *The Superinsulated Home Book*, Wiley, New York, 1985.)

with $\Delta C_p = C_p - (-0.2)$ being the difference between the local pressure coefficient and the average.

The *stack effect* is the result of density differences between air inside and outside the building. In winter, the air inside the building is warmer and hence less dense than the air outside. Therefore, the indoor pressure difference (bottom versus top) is less than the outdoor pressure difference between the same heights. Consequently there is an indoor-outdoor pressure difference. It varies linearly with height, and the level of neutral pressure is at the midheight of the building, as suggested by Fig. 7.5, if the leaks are uniformly distributed. During the cooling season when indoor air is colder than the outside, the effect is reversed.

The pressure difference is given by

$$\Delta p_{\text{stack}} = -C_d \rho_i g \, \Delta h \, \frac{T_i - T_o}{T_o} \tag{7.7SI}$$

$$\Delta p_{\text{stack}} = -C_d \frac{\rho_i g}{g_c} \, \Delta h \, \frac{T_i - T_o}{T_o} \tag{7.7US}$$

where
  ρ_i = density of air in building = 0.075 lb_m/ft^3 (1.20 kg/m^3)
  Δh = vertical distance from neutral pressure level, up being positive, ft (m)
  g = 32.17 ft/s^2 (9.80 m/s^2) = acceleration due to gravity
    $[g_c = 32.17 \ (\text{lb}_m \cdot \text{ft})/(\text{lb}_f \cdot \text{s}^2)]$
  T_i and T_o = indoor and outdoor absolute temperatures, °R (K)
    C_d = draft coefficient, a dimensionless number to account for resistance to airflow between floors

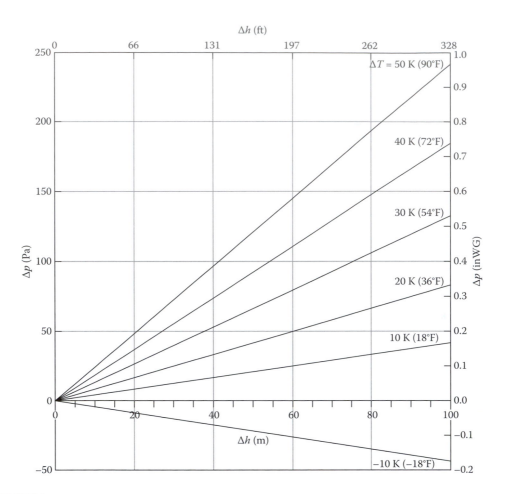

FIGURE 7.6
Pressure difference due to stack effect.

The draft coefficient ranges from about 0.65 for typical modern office buildings to 1.0 if there is no resistance at all. Equation (7.7) is plotted in Fig. 7.6 as a function of $\Delta T = T_i - T_o$ and Δh, assuming air at 75°F (24°C). Since the relation is linear in Δh, this figure can be read outside the range shown by simply changing the scales of the axes. For a brief summary one can say that the stack pressure amounts to

$$\frac{\Delta p_{\text{stack}}}{C_d \, \Delta h \, \Delta T} = 0.04 \ \text{Pa}/(\text{m} \cdot \text{K}) \tag{7.8SI}$$

$$\frac{\Delta p_{\text{stack}}}{C_d \, \Delta h \, \Delta T} = 0.00014 \ \text{lb}_f/(\text{ft}^2 \cdot \text{ft} \cdot {}^\circ\text{R}) \tag{7.8US}$$

The stack effect tends to be relatively small in low-rise buildings, up to about five floors, but in high-rise buildings it can dominate and should be given close attention.

Finally, in buildings with *mechanical ventilation*, there is the pressure difference Δp_{vent} if the intake and exhaust flow are not equal. The resulting pressure difference depends on the design and operation of the ventilation system and on the tightness of the building. In addition, there is some coupling to the wind and stack terms. Thus the determination of Δp_{vent} may be somewhat difficult. However, as we now show, the situation is simple when Δp_{vent} is larger in magnitude than the wind and stack terms. This is an important case since many designers aim for slight overpressurization of commercial buildings by making the outdoor air intake larger than the exhaust flow.

Consider how the pressures and flows are related in a fairly tight building where mechanical ventilation maintains overpressure or underpressure Δp_{vent} relative to the outside; for simplicity, we assume it uniform in the entire building. The law of conservation of mass implies that the net airflow provided by the ventilation system equals the net leakage $\dot{V}$ across the envelope, as calculated, according to Eq. (7.2), by summing over all leakage sites k of the building envelope:

$$\dot{V} = \sum A_k C_k \, \Delta p_k^{n_k} \tag{7.9}$$

with

$$\Delta p_k = \Delta p_{\text{wind},k} + \Delta p_{\text{stack},k} + \Delta p_{\text{vent}}$$

For simplicity, let us assume at this point one single value n for the exponent. Then the pressure term on the right-hand side of Eq. (7.9) can be rewritten in the form

$$\Delta p_k^n = \Delta p_{\text{vent}}^n \, (1 + x_k)^n$$

with

$$x_k = \frac{\Delta p_{\text{wind},k} + \Delta p_{\text{stack},k}}{\Delta p_{\text{vent}}} \tag{7.10}$$

As long as $|x_k| < 1$, the binomial expansion can be used, with the result[3]

$$\dot{V}_{\text{vent}} = \Delta p_{\text{vent}}^n \sum A_k c_k \left[1 + n x_k + \frac{n(n-1)}{2} x_k^2 + \cdots \right]$$

The quantity x_k is positive in some parts of the building, negative in others. In fact, if the distribution of cracks is approximately symmetric (top-bottom and windward-leeward), then for each term with positive x_k there will also be one with approximately the same coefficient but negative x_k. Thus the linear terms in the expansion tend to cancel. The higher-order terms are small, beginning with x_k^2 which is multiplied by $n(n-1)/2$, a factor that is always less than $\frac{1}{8}$ in absolute value since $0 < n < 1$. Thus the contributions of the x_k-dependent terms are much smaller than that of the leading term. Therefore, *if a building*

---

[3] The *i*th power of x_k in this series is multiplied by the binomial coefficient

$$\binom{n}{i} = \frac{n!}{i!(n-i)!}$$

is pressurized to Δp_{vent} *by mechanical ventilation and if wind and stack pressures are smaller than* Δp_{vent}, *then it is indeed a fair approximation to neglect them altogether and write*

$$\dot{V} \approx \Delta p^n \sum A_k c_k \quad \text{with } \Delta p = \Delta p_{vent} \tag{7.11}$$

the sum covering the entire envelope of the building. Had we allowed for different exponents n_k in Eq. (7.9), Δp with its exponent would remain inside the sum, but the conclusion about the negligibility of stack and wind terms continues to hold.

Example 7.1

The mechanical ventilation system of a three-story office building maintains a constant over-pressure Δp_{vent}. Can the wind and stack terms be neglected under the following conditions?

Given: $\Delta h = 0.5 \times$ total building height $= 0.5 \times (3 \times 3 \text{ m}) = 4.5 \text{ m (15 ft)}$

$\Delta p_{vent} = -25$ Pa (-0.1 inWG)

$T_i = 20°C \ (68°F)$

$T_o = -10°C \ (14°F)$

$v = 6.7$ m/s (15 mi/h) (winter design conditions, per Sec. 7.2.1)

Find: Δp_{wind} and Δp_{stack}, and compare with Δp_{vent}

SOLUTION

For the wind term, $|\Delta C_P| \lesssim 0.5$ from Fig. 7.4b. From Fig. 7.4a at 6.7 m/s, we find $|\Delta p_{wind}|/\Delta C_p = 28$ Pa; hence $|\Delta p_{wind}| \lesssim 0.5 \times 28 \text{ Pa} = 14 \text{ Pa}$. For the stack term $\Delta T = 30$ K (54°F), from Fig. 7.6 we find $|\Delta p_{stack}| = 6$ Pa. Hence

$$|\Delta p_{stack} + \Delta p_{wind}| = 20 \text{ Pa} < |\Delta p_{vent}|$$

During most of the year, v and ΔT are less than the design conditions, and thus the wind and stack terms are smaller than the values calculated here. Their contribution is smaller than that of the ventilation terms and can, indeed, be neglected.

7.1.4 LBL Model for Air Leakage

To apply Eq. (7.2), one needs data for leakage areas and flow coefficients of all the components of a building. Much research has been done to obtain such data, both for components and for complete buildings, e.g., by pressurizing a building with a blower door. Data were presented in Table 7.1.

The total leakage area is obtained by adding all the leakage areas of the components, as illustrated in Example 7.2. Once the total leakage area has been found, either by such a calculation or by a pressurization test, the airflow $\dot{V}$ can be estimated by the following model, developed at Lawrence Berkeley Laboratory (LBL) as reported by ASHRAE (1989a):

$$\dot{V} = A_{leak} \sqrt{a_s \Delta T + a_w v^2} \text{ L/s} \tag{7.12}$$

TABLE 7.2

Stack Coefficient a_s

	Number of Stories		
	One	Two	Three
Stack coefficient a_s, $(ft^3/min)^2/(in^4 \cdot °F)$	0.0156	0.0313	0.0471
Stack coefficient a_s, $(L/s)^2/(cm^4 \cdot K)$	0.000145	0.000	0.000435

Source: Courtesy of ASHRAE, *Handbook of Fundamentals*, American Society of Heating, Refrigerating and Air-Conditioning Engineers, Atlanta, GA, 1989a.

TABLE 7.3

Wind Coefficient a_w

Shielding Class	Description	Wind Coefficient a_w, $(L/s)^2/[cm^4 \cdot (m/s)^2]$ Number of Stories			Wind Coefficient a_w, $(ft^3/min)^2/[in^4 \cdot (mi/h)^2]$ Number of Stories		
		One	Two	Three	One	Two	Three
1	No obstructions or local shielding	0.000319	0.000420	0.000494	0.0119	0.0157	0.0184
2	Light local shielding; few obstructions, a few trees or small shed	0.000246	0.000325	0.000382	0.0092	0.0121	0.0143
3	Moderate local shielding; some obstructions within two house heights, thick hedge, solid fence, or one neighboring house	0.000174	0.000231	0.000271	0.0065	0.0086	0.0101
4	Heavy shielding; obstructions around most of perimeter, buildings, or trees within 10 m in most directions; typical suburban shielding	0.000104	0.000137	0.000161	0.0039	0.0051	0.0060
5	Very heavy shielding; large obstructions surrounding perimeter within two house heights; typical downtown shielding	0.000032	0.000042	0.000049	0.0012	0.0016	0.0018

Source: Courtesy of ASHRAE, *Handbook of Fundamentals*, American Society of Heating, Refrigerating and Air-Conditioning Engineers, Atlanta, GA, 1989a.

where

A_{leak} = total effective leakage area of building, cm^2

a_s = stack coefficient of Table 7.2, $(L/s)^2/(cm^4 \cdot K)$

$\Delta T = T_i - T_o$, K

a_w = wind coefficient of Table 7.3, $(L/s)^2/[cm^4 \cdot (m/s)^2]$

v = average wind speed, m/s

This model is applicable to single-zone buildings *without* mechanical ventilation.

Example 7.2

Estimate the effective leakage area for a simple wood-frame building. It is built as a rectangular box 12 m × 12 m × 2.5 m (39.4 ft × 39.4 ft × 8.2 ft) with a flat roof. The windows cover 20 percent of the sides. This basic building will be used for many examples in this book, with further details

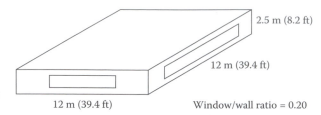

FIGURE 7.7
Sketch of the building used for many
of the examples in this chapter.

being specified as needed. Here we need to spell out characteristics of the airtightness only; they
are listed in the following table, which also leads up to the answer.

Given: Single-story wood-frame construction in shape of a rectangular box 12 m ×
12 m × 2.5 m (39.4 ft × 39.4 ft × 8.2 ft)

Window/wall ratio $= 0.20$

Figure: Fig. 7.7 (this basic structure will be used by many examples in this chapter)

SOLUTION

Taking numbers from Table 7.1 for each of the components and adding, we obtain the following:

Component	Area, m², or Perimeter, m (given)	Leakage Area per Area or Perimeter (from Table 7.1)	Leakage Area, cm²
Walls at sill (sill uncaulked)	48 m	4 cm²/m	192
Walls at roof not taped or plastered, no vapor barrier	48 m	1.5 cm²/m	72
Windows: double-hung, not weatherstripped	24 m²	6 cm²/m²	144
Window frames: no caulking	24 m²	1.7 cm²/m²	41
Total			449

COMMENTS

This is a caricature of a building, kept as simple as possible with just enough detail to illustrate the
principles. For a real building one would have to include further details such as doors, electric
outlets, and fireplace.

Example 7.3

Find the air exchange rate for the building of Example 7.2 when $\Delta T = T_i - T_o = 20$ K (36°F) and
wind speed $v = 6.7$ m/s (15 mi/h).

Given: Building of Example 7.2 in suburban neighborhood

$\Delta T = T_i - T_o = 20$ K

$v = 6.7$ m/s

$A_{leak} = 449$ cm²

Intermediate quantities: Stack coefficient $a_s = 0.000145$ (L/s)²/(cm⁴·K) from Table 7.2.
Wind coefficient $a_w = 0.000104$ (L/s)²/[cm⁴·(m/s)²] from Table 7.3

SOLUTION

From Eq. (7.12) one obtains the infiltration rate

$$\dot{V} = 449 \text{ cm}^2$$

$$\times \sqrt{0.000145 \,(\text{L/s})^2(\text{cm}^{-4} \cdot \text{K}) \times 20 \text{ K} + 0.000104(\text{L/s})^2/[\text{cm}^4 \cdot (\text{m/s})^2] \times (6.7 \text{ m/s})^2}$$

$$= 39.1 \text{ L/s} = 140.6 \text{ m}^3/\text{h}$$

Since the volume is $V = 360 \text{ m}^3$, the air change rate is 0.39 per hour.

7.1.5 Further Correlations for Building Components

Most commercial buildings have features that are not included in the model of the previous section:

- Mechanical ventilation
- Revolving or swinging doors
- Curtain walls (i.e., the non-load-bearing wall construction commonly employed in commercial buildings)

These features can be analyzed by using the method presented in the *Cooling and Heating Load Calculation Manual* (1979, with new edition in 1992), published by ASHRAE. Here we give a brief summary. The airflows are determined from the following correlations, where Δp is the total pressure difference at each point of the building, calculated as described above.

For residential-type doors and windows, the flow per length l_p of perimeter is given by an equation of the form

$$\frac{\dot{V}}{l_p} = k(\Delta p)^n \quad \dot{V}, \text{ ft}^3/\text{min}; \; l_p, \text{ ft}; \; \Delta p, \text{ inWG} \tag{7.13US}$$

Numerical values can be found in Fig. 7.8 for windows and residential-type doors, and in Fig. 7.9 for swinging doors when they are closed, as function of the pressure difference Δp, for several values of k corresponding to different types of construction. The exponent n is 0.5 for residential-type doors and windows and 0.65 for swinging doors. The larger values for n and for k for the latter account for the larger cracks of swinging doors. Since the equations are dimensional, all quantities must be used with the specified units. We have added dual scales to the graphs, so Figs. 7.8 to 7.11 can be read directly in both systems of units.

Obviously the airflow increases markedly when doors are opened. Fig. 7.10 permits an estimate of the airflow through swinging doors, both single-bank and vestibule-type, as a function of traffic rate. The coefficient $C \; [(\text{ft}^3/\text{min})/(\text{inWG})^{-0.5}]$ for Fig. 7.10a is found from Fig. 7.10b for the number of people passing the door per hour. The equation for the flow in part a is

$$\dot{V} = C(\Delta p)^{0.5} \quad \dot{V}, \text{ ft}^3/\text{min}; \; \Delta p, \text{ inWG} \tag{7.14US}$$

| | Windows | | Doors |
Coefficient	Wood Double-Hung (Locked)	Other Types	(Residential Type)
$k = 1.0$ "tight"	Weatherstripped, small gap width 0.4 mm ($\frac{1}{64}$ in)	Weatherstripped: wood casement and awning windows, metal casement windows	Very small perimeter gap and perfect-fit weatherstripping—often characteristic of new doors
$k = 2.0$ "average"	Nonweatherstripped, small gap width 0.4 mm ($\frac{1}{64}$ in); or weather-stripped, large gap width 2.4 mm ($\frac{3}{32}$ in)	All types of sliding windows, weatherstripped (if gap width is 0.4 mm, this could be "tight"; or nonweatherstripped metal casement windows (if gap width is 2.4 mm, this could be "loose")	Small perimeter gap with stop trim, good fit around door, and weather-stripping
$k = 6.0$ "loose"	Nonweatherstripped, large gap width 2.4 mm ($\frac{3}{32}$ in)	Nonweatherstripped vertical and horizontal sliding windows	Large perimeter gap with poor-fitting stop trim and weatherstripping, or small perimeter gap without weatherstripping

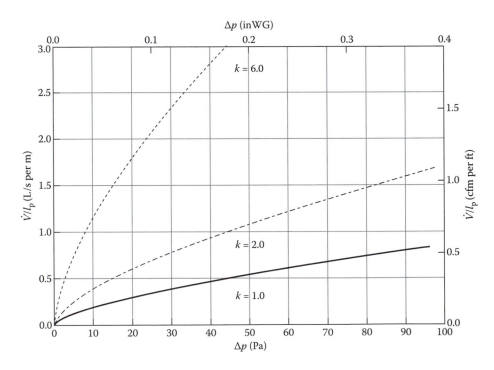

FIGURE 7.8
Window and residential-type door air infiltration $\dot{V}$ per perimeter length l_p. The curves correspond to Eq. (7.13), with $n = 0.65$ and coefficient k [(ft^3/min)/(inWG)$^{0.65}$] according to construction type, as shown in the above table.

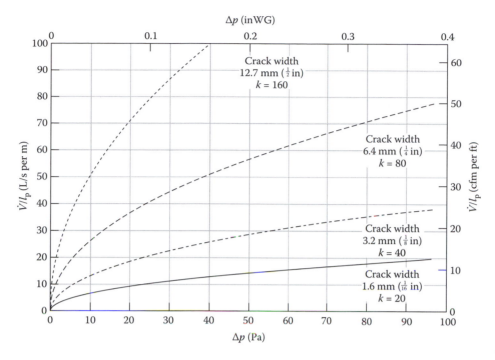

FIGURE 7.9
Infiltration through closed swinging door cracks, $\dot{V}$ per perimeter length l_p. The curves correspond to Eq. (7.13), with $n = 0.5$ and coefficient k [(ft³/min)/(inWG)$^{0.5}$].

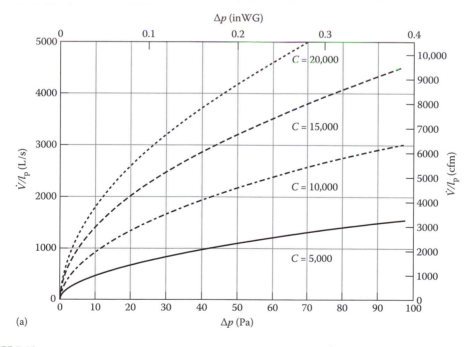

(a)

FIGURE 7.10
Infiltration due to door openings as a function of traffic rate: (a) infiltration (with $n = 0.5$).

(continued)

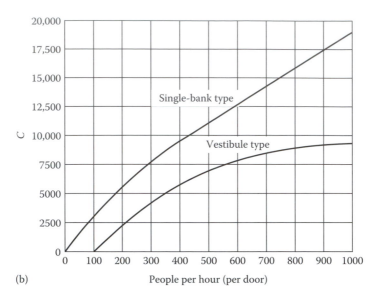

(b)   People per hour (per door)

FIGURE 7.10 (continued)
(b) Coefficient C [(ft^3/min)/(inWG)$^{0.5}$].

Analogous information on flow per unit area of curtain wall can be determined by the equation

$$\frac{\dot{V}}{A} = k(\Delta p)^{0.65} \quad \dot{V}, \text{ ft}^3/\text{min}; A, \text{ ft}^2; \Delta p \text{ inWG} \tag{7.15US}$$

It is presented in Fig. 7.11 for three construction types, corresponding to the indicated values of the coefficient K (ft^3/min)/(ft$^2 \cdot$ inWG$^{0.65}$).

Example 7.4

Estimate how much air leaks out of a single-story office building of average construction when it is pressurized to $\Delta p = 50$ Pa (0.2 inWG) [assume that wind and stack pressures can be neglected, per Eq. (7.11)].

Given: One-story building 30 m × 30 m × 2.5 m (98.4 ft × 98.4 ft × 8.2 ft); curtain wall, alternating with floor-to-ceiling windows 3 m × 2.5 m each (9.84 ft × 8.2 ft); windows cover 50 percent of the entire wall area; four swinging doors 2 m × 1 m (6.56 ft × 3.28 ft), single-bank type, crack width 3.2 mm ($\frac{1}{8}$ in); floor area 15 m^2 (161 ft^2) per occupant.

Estimate the average traffic over 10 h by assuming occupants make two entries and two exits per day; increase the result by 25 percent to account for visitors.

Assumptions: For simplicity, neglect infiltration through the roof and through cracks at the edges of the roof and the floor.

Find: Air exchange $\dot{V}$

Coefficient	Construction
$K = 0.22$ "tight"	Close supervision of workmanship; joints are redone when they appear inadequate
$K = 0.66$ "average"	Conventional
$K = 1.30$ "loose"	Poor quality control, or older building where joints have not been redone

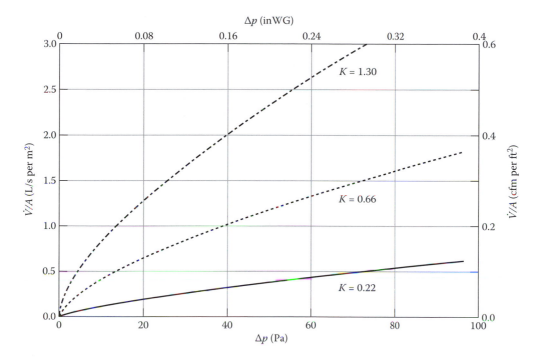

FIGURE 7.11
Infiltration per area of curtain wall for one room or one floor. The curves correspond to Eq. (7.15), with coefficient $K\ [(\mathrm{ft}^3/\mathrm{min})/(\mathrm{ft}^2 \cdot \mathrm{inWG})^{0.65}]$ according to construction type.

SOLUTION

$$\text{Perimeter of windows} = 20 \times (6\ \mathrm{m} + 5\ \mathrm{m}) = 220\ \mathrm{m} \quad k = 2.0\ (\text{"average"})$$

$$\text{Perimeter of doors} = 24\ \mathrm{m} \quad k = 40$$

$$\text{Curtain wall area} = 4 \times 30\ \mathrm{m} \times 2.5\ \mathrm{m} = 300\ \mathrm{m}^2 \quad K = 0.66$$

$$\text{Number of occupants} = \frac{900\ \mathrm{m}^2}{15\ \mathrm{m}^2} = 60$$

$$\text{Traffic per door} = \frac{60 \times 4 \times 1.25\ \text{people}}{4\ \text{door} \times 10\ \mathrm{h}} = 7.5\ \text{passages/h}$$

At $\Delta p = 50$ Pa, we read from Fig. 7.8,

$$\frac{\dot{V}}{l_p} = 1.1\ \mathrm{L/(s \cdot m)} \quad \text{perimeter for windows}$$

hence

$$\dot{V} = 1.1 \text{ (L/s)/m} \times 220 \text{ m} = 242 \text{ L/s}$$

From Fig. 7.9,

$$\frac{\dot{V}}{l_p} = 28 \text{ L/(s} \cdot \text{m) perimeter for doors when closed}$$

hence

$$\dot{V} = 28 \text{ (L/s)/m} \times 24 \text{ m} = 672 \text{ L/s}$$

From Fig. 7.10b, coefficient C is around 250, and from Fig. 7.10a the extra flow through the doors due to traffic, per door, is 50 L/s, to be multiplied by 4 for four doors, to give $\dot{V} = 200$ L/s. From Fig. 7.11, $\dot{V}/A = 1.2$ L/(s $\cdot$ m^2) for the curtain wall; hence $\dot{V} = 1.2$ (L/s)/m$^2 \times 300$ m$^2 = 360$ L/s.
   The total flow is

$$\dot{V} = 242 + 672 + 200 + 360 = 1474 \text{ L/s}$$
$$= 1.474 \text{ m}^3/\text{s} \times 3600 \text{ s/h} = 5306 \text{ m}^3/\text{h} \ (3123 \text{ ft}^3/\text{min})$$

Since the building volume is $V = 2250$ m^3, this corresponds to (5069 m^3/h)/(2250 m^3) = 2.36 air changes per hour.

COMMENTS

The number of occupants is (900 m^2)(1 occupant/15 m^2) = 60; hence the airflow per occupant is (1474 L/s)/60 = 24.7 L/s. Since ASHRAE Standard 62–87 specifies only 7.5 L/s per person as minimum, the air exchange is about 3 times larger than necessary. The dominant contribution comes from the doors, 672 L/s when they are closed and 200 L/s from traffic. In the interest of energy conservation, tighter doors should be considered and perhaps the replacement by vestibule-type doors.

7.2 Principles of Load Calculations

7.2.1 Design Conditions

Loads depend on the indoor conditions that one wants to maintain and on the weather. The latter is not known in advance. If the HVAC equipment is to guarantee comfort at all times, it must be designed for peak conditions. What are the extremes? For most buildings it would not be practical to aim for total protection, by choosing the most extreme weather on record and adding a safety margin. Such oversizing of the HVAC equipment would be excessive, not just in first cost but also in operating cost: Most of the time, the equipment would run with poor part-load efficiency. Therefore compromise is called for, reducing the cost of the HVAC equipment significantly while accepting the risk of slight discomfort under rare extremes of weather. The greater the extreme, the rarer the occurrence.
   To help with the choice of design conditions, ASHRAE has published weather statistics corresponding to several levels of probability. They are the conditions that are exceeded at

the site in question during a specified percentage of time of an average season. For warm conditions the ASHRAE *Handbook of Fundamentals* lists design conditions for the 0.4, 1.0, and 2 percent levels. For cooling these percentage probabilities refer to 12 months (8760 h). To see what these statistics imply, consider Washington, D.C., where the 1 percent level for the dry-bulb temperature is 92°F (33°C). Here the temperature is above 92°F during 0.010×8760 h $= 87.6$ h of an average summer. Since the hottest hours are concentrated during afternoons rather than spread over the entire day, the corresponding number of days can be considerably higher than 1 percent of the year.

The cold weather design conditions are listed as 99.0 and 99.6 percent conditions because that is the percentage of the typical year when the temperature is *above* these levels.

A certain amount of judgment is needed in the choice of design conditions. For ordinary buildings, it is customary to base the design on the level of 1.0 percent in summer and 99.0 percent in winter. For critical applications such as hospitals or sensitive industrial processes or for lightweight buildings, one may prefer the more stringent level of 0.4 percent in summer (99.6 percent in winter). Thermal inertia can help reduce the risk of discomfort: It delays and attenuates the peak loads, as will be explained in the following sections. Therefore, one may move to a less stringent level for a given application if the building is very massive.[4]

For peak heating loads, one need not bother with solar radiation because the extremes occur during winter nights. For cooling loads, solar radiation is crucial, but its peak values are essentially a function of latitude alone. For opaque surfaces, the effect of solar radiation is treated by means of the sol-air temperature; for glazing, by means of the solar heat gain factor. Design values for the solar heat gain factor for a set of surface orientations and latitudes can be found in ASHRAE (2001).

As for humidity and latent loads, the ASHRAE tables include design wet-bulb temperatures, also at the 0.4, 1.0, and 2 percent levels, along with the coincident dry-bulb temperature. Alternatively, the tables also show the mean coincident wet-bulb temperatures, defined as the average wet-bulb temperature at the corresponding dry-bulb values (also at the 0.4, 1.0, and 2 percent levels). For winter no wet-bulb temperature data are given. Usually this poses no serious problem because latent loads during the heating season are zero if one does not humidify. If one does humidify, uncertainties in the value of the outdoor humidity have little effect on the latent load because the absolute humidity of outdoor air in winter is very low.

Wind speed is another weather-dependent variable that has a bearing on loads. Traditionally the ASHRAE (2001) value

$$v_{\text{win}} = 15 \text{ mi/h} (6.7 \text{ m/s}) \tag{7.16a}$$

has been recommended for heating loads, if there is nothing to imply extreme conditions (such as an exposed hilltop location). For cooling loads, a value one-half as large is recommended

$$v_{\text{sum}} = 7.5 \text{ mi/h} (3.4 \text{ m/s}) \tag{7.16b}$$

---

[4] As a guide for the assessment of the relation between persistence of cold weather and thermal inertia, we note that according to studies at several stations, temperatures below the design conditions can persist for up to a week ASHRAE (2001).

because wind tends to be less strong in summer than in winter. Of particular interest is the surface heat transfer coefficient (radiation plus convection) h_o for which ASHRAE (2001) recommends the design values.

$$h_{o,\text{win}} = 6.0 \text{ Btu}/(\text{h} \cdot \text{ft}^2 \cdot {}^\circ\text{F}) \ [34.0 \text{ W}/(\text{m}^2 \cdot \text{K})] \tag{7.17a}$$

$$h_{o,\text{sum}} = 4.0 \text{ Btu}/(\text{h} \cdot \text{ft}^2 \cdot {}^\circ\text{F}) \ [22.7 \text{ W}/(\text{m}^2 \cdot \text{K})] \tag{7.17b}$$

This coefficient is only one of several components of the calculation of thermal loads, and it enters only through the building heat transmission coefficient defined in the next section. The better a building is insulated and tightened, the less its heat transmission coefficient K_{tot} depends on wind. With current practice for new construction in the United States, typical wind speed variations may change the heat transmission coefficient by about 10 percent relative to the value at design conditions. Temperature and humidity for normal indoor activities should be within the comfort region delineated in Chap. 2. The comfort chart indicates higher indoor temperatures in summer than in winter because of the difference in clothing.

7.2.2 Building Heat Transmission Coefficient

One of the most important terms in the heat balance of a building is the heat flow across the envelope. As discussed in Chap. 2, heat flow can be assumed to be linear in the temperature difference when the range of temperatures is sufficiently small; this is usually a good approximation for heat flow across the envelope. Thus one can calculate the heat flow through each component of the building envelope as the product of its area A, its conductance U, and the difference $T_i - T_o$ between the interior and outdoor temperatures. The calculation of U (or its inverse, the R_{th} value) is described in Chap. 2. Here we combine the results for the components to obtain the total heat flow.

The total conductive heat flow from interior to exterior is

$$\dot{Q}_{\text{cond}} = \sum_k U_k A_k (T_i - T_o) \tag{7.18}$$

with the sum running over all parts of the envelope that have a different composition. It is convenient to define a total *conductive heat transmission coefficient* K_{cond}, or UA value, as

$$K_{\text{cond}} = \sum_k U_k A_k \tag{7.19}$$

so that the conductive heat flow for the typical case of a single interior temperature T_i can be written as

$$\dot{Q}_{\text{cond}} = K_{\text{cond}}(T_i - T_o) \tag{7.20}$$

In most buildings, the envelope consists of a large number of different parts; the greater the desired accuracy, the greater the amount of detail to be taken into account.

As a simplification, one can consider a few major groups and use effective values for each. The three main groups are glazing, opaque walls, and roof. The reason for distinguishing the wall and the roof lies in the thickness of the insulation: Roofs tend to be better insulated because it is easier and less costly to add extra insulation there than in the walls. With these three groups one can write

$$K_{\text{cond}} = U_{\text{glaz}}A_{\text{glaz}} + U_{\text{wall}}A_{\text{wall}} + U_{\text{roof}}A_{\text{roof}} \tag{7.21}$$

if one takes for each the appropriate effective value. For instance, the value for glazing must be the average over glass and framing, as described in Sec. 6.5.1. Results for aggregate U values for walls and roofs of typical construction can be found in the appendices of the CD-ROM.

In the energy balance of a building, there is one other term that is proportional to $T_i - T_o$. It is the flow of sensible heat [W (Btu/h)] due to air exchange:

$$\dot{Q}_{\text{air}} = \rho c_p \dot{V}(T_i - T_o) \tag{7.22}$$

where
$\rho =$ density of air
$c_p =$ specific heat of air
$\dot{V} =$ air exchange rate, ft^3/h (m^3/s)

At standard conditions, 14.7 psia (101.3 kPa) and 68°F (20°C), the factor ρc_p has the value

$$\rho c_p = 0.018 \text{ Btu}/(\text{ft}^3 \cdot {}^{\circ}\text{F}) \text{ [1.2 kJ}/(\text{m}^3 \cdot \text{K)]} \tag{7.23}$$

In USCS units, if $\dot{V}$ is in cubic feet per minute, it must be converted to cubic feet per hour by multiplying by 60 (ft^3/h)/(ft^3/min). It is convenient to combine the terms proportional to $T_i - T_o$ by defining the total heat transmission coefficient K_{tot} of the building as the sum of conductive and air change terms:

$$K_{\text{tot}} = K_{\text{cond}} + \rho c_p \dot{V} \tag{7.24}$$

Example 7.5 illustrates the calculation of K_{tot} with typical numbers for a house.

Example 7.5

Calculate the heat transmission coefficient K_{tot} for the building of Example 7.2 under the following assumptions.

Given: Building of Example 7.2, rectangular box 12 m × 12 m × 2.5 m (39.4 ft × 39.4 ft × 8.2 ft) with flat roof (see Fig. 7.7). The insulation is fiberglass with $k = 0.06$ W/(m·K) [0.10 Btu/ (h·ft·°F)], thickness $\Delta x = 0.25$ m (10 in) in the roof and $\Delta x = 0.15$ m (6 in) in the walls. The windows are double-glazed with

$$U_{\text{glaz}} = 3.0 \text{ W}/(\text{m}^2 \cdot \text{K)[0.53 Btu}/(\text{h} \cdot \text{ft}^2 \cdot {}^{\circ}\text{F)]}$$

and they cover 20 percent of the sides. The air exchange rate is 0.5 per hour.

Assumptions: For simplicity, treat the opaque surfaces as if they consisted only of the fiberglass, without covering and without the indoor and outdoor surface film coefficients. Neglect thermal bridges due to studs. Assume that all values are independent of temperature and wind.

Find: Heat transmission coefficient K_{tot}

SOLUTION

$$K_{\text{tot}} = 205 \text{ W/K } [389 \text{ Btu/(h} \cdot {}^\circ\text{F)}]$$

as can be seen from this table:

Component	Area A m²	Thickness Δx m	Conductivity k W/(m·K)	U ($k/\Delta x$ for Roof and Walls) W/(m²·K)	k (UA for Conduction, $\rho c_p \dot{V}$ for Air Exchange) W/K
Roof 12 m × 12 m, flat	144	0.250	0.060	0.24	34.6
Walls (opaque) (height 2.5 m)	96	0.150	0.060	0.40	38.4
Glazing (20% of sides)	24			3.00	72.0
Air exchanges (0.5/h)					60.0
Total					205.0

COMMENTS

This example is oversimplified. For greater realism, one can use the U values from the appendices in the CD-ROM. Thermal bridges due to studs in the walls should be included per Example 2.2.

A more refined calculation would take surface heat transfer coefficients into account, as well as details of the construction, using the data from the appendices in the CD-ROM. In practice, such details can take up most of the effort. Example 7.6 shows how the heat loss coefficients vary if the effects of wind on surface heat transfer and infiltration are included.

Example 7.6

For the building of Example 7.5, model the infiltration according to the LBL model of Eq. (7.12) and add an outdoor surface heat transfer coefficient (fit to the data in Fig. 2.10)

$$h_o = 1.8 + 0.19v \quad h_o, \text{Btu/(h} \cdot \text{ft}^2 \cdot {}^\circ\text{F)}; v, \text{ft/s} \tag{7.25US}$$

$$h_o = 10.0 + 3.5v \quad h_o, \text{W/(m}^2 \cdot \text{K)}; v, \text{m/s} \tag{7.25SI}$$

on all exterior surfaces. For simplicity, still neglect h_i for the wall and roof (or assume that it is included in the k/t values of Example 7.5—in any case it does not vary with v, the variable of interest here).

    Given: Leakage area $A \doteq 449 \text{ cm}^2$ from Example 7.2; wind and stack coefficients of Example 7.3

SOLUTION

Take the LBL model of Eq. (7.12) for airflow as in Example 7.3, and calculate the U values of the wall and roof according to

$$\frac{1}{U} = \frac{\Delta x}{k} + \frac{1}{h_o}$$

using for $\Delta x/k$ the inverse of the values listed for the roof and wall under the $k/\Delta x$ column of the table in Example 7.5. For instance, for the walls we have

$$\frac{1}{U} = \frac{1}{0.40} + \frac{1}{10.0 + 3.5v} \; (m^2 \cdot K)/W$$

For windows the U value of 3.0 W/(m² · K) already includes the design surface heat transfer coefficient of $h_o = 34$ W/(m² · K). Therefore, we calculate the wind dependence of U by first subtracting this contribution and then adding $1/h_o$ at arbitrary v:

$$\frac{1}{U} = \frac{1}{3.0 \; W/(m^2 \cdot K)} - \frac{1}{34 \; W/(m^2 \cdot K)} + \frac{1}{10.0 + 3.5v}$$

At $h_o = 34$ W/(m² · K), this reproduces, of course, $U = 3.0$ W/(m² · K).

The result is shown in Fig. 7.12. The total heat loss coefficient K_{tot} equals 192 W/K [364 Btu/(h · °F)] at the winter and 176 W/K [334 Btu/(h · °F)] at the summer design values of wind speed. Most of the variation comes from air infiltration. The wall and roof are sufficiently well insulated that h_o has almost no effect.

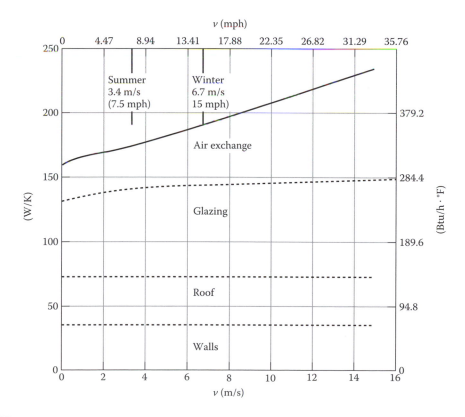

FIGURE 7.12
Variation with wind speed of the heat loss coefficient K_{tot} and its components (indicated by dashed lines) for the house of Example 7.6.

7.2.3 Heat Gains

Heat gains affect both heating and cooling loads. In addition to solar gains, already discussed in Chap. 6, there are heat gains from occupants, lights, and equipment such as appliances, motors, computers, and copiers. Power densities for lights in office buildings are around 20 to 30 W/m^2. For lights and for resistive heaters, the nominal power rating (i.e., the rating on the label) is usually close to the power drawn in actual use. But for office equipment, that would be quite misleading; the actual power has been measured to be much lower, often by a factor of 2 to 4 (Norford et al., 1989). Some typical values are indicated in Table 7.4. In recent years, the computer revolution has brought a rapid increase in electronic office equipment, and the impact on loads has become quite important, comparable to lighting. The energy consumption for office equipment is uncertain: Will the occupants turn off the computers between uses or keep them running nights and weekends?

For special equipment such as laboratories or kitchens, it is advisable to estimate the heat gains by taking a close look at the inventory of the equipment to be installed, paying attention to the possibility that much of the heat may be drawn directly to the outside by exhaust fans.

Heat gain from occupants depends on the level of physical activity. Nominal values are listed in Table 7.5. It is instructive to reflect on the origin of this heat gain. The total heat

TABLE 7.4

Typical Heat Gain Rates for Several Kinds of Equipment.

Equipment	Heat Gain		Comments
	Btu/h	**W**	**Comments**
Television set	170–340	50–100	
Refrigerator	340–680	100–200	Recent models more efficient
Personal computer (desktop)	170–680	50–200	Almost independent of use while turned on
Impact printer	34–100	10–30 standby	Increases about twofold during printing
Laser printer	510	150 standby	Increases about twofold during printing
Copier	500–1000	150–300 standby	Increases about twofold during printing

Sources: Courtesy of ASHRAE, *Handbook of Fundamentals*, American Society of Heating, Refrigerating and Air-Conditioning Engineers, Atlanta, GA, 1989a; Norford, L.K. et al., Electronic office equipment: The impact of market trends and technology on end use demand, in *Electricity: Efficient End Use and New Generation Technologies, and Their Planning Implications*, Johansson, T.B. et al. (eds), Lund University Press, Lund, Sweden, 1989, 427–460.
Note that measured values are often less than half of the nameplate rating.

TABLE 7.5

Nominal Heat Gain Values from Occupants.

Activity	Total		Sensible		Latent	
	Btu/h	**W**	**Btu/h**	**W**	**Btu/h**	**W**
Seated at rest	340	100	240	70	100	30
Seated, light office work	410	120	255	75	150	45
Standing or walking slowly	495	145	255	75	240	70
Light physical work	850	250	310	90	545	160
Heavy physical work	1600	470	630	185	970	285

Source: Courtesy of ASHRAE, *Handbook of Fundamentals*, American Society of Heating, Refrigerating and Air-Conditioning Engineers, Atlanta, GA, 1989a.

gain must be close to the caloric food intake, since most of the energy is dissipated from the body as heat. An average of 100 W corresponds to

$$100\ \text{W} = 0.1\ \frac{\text{kJ}}{\text{s}} \times \frac{1\ \text{kcal}}{4.186\ \text{kJ}} \times \left(24 \times 3600\ \frac{\text{s}}{\text{day}} \right) = 2064\ \frac{\text{kcal}}{\text{day}}$$

indeed a reasonable value compared to the typical food intake (note that the dietician's calorie is really a kilocalorie). The latent heat gain must be equal to the heat of vaporization of the water that is exhaled or transpired. Dividing 30 W by the heat of vaporization of water, we find a water quantity of $30\ \text{W}/(2450\ \text{kJ/kg}) = 12.2 \times 10^{-6}\ \text{kg/s}$, or about 1.1 kg/24 h. That also appears quite reasonable.

The latent heat gain due to the air exchange is

$$\dot{Q}_{air,lat} = \dot{V}\rho h_{fg}(W_o - W_i) \tag{7.26}$$

where

$\dot{V}$ = volumetric air exchange rate, ft³/min (m³/s or L/s)
ρ = density, $\text{lb}_\text{m}/\text{ft}^3$ (kg/m³)
ρh_{fg} = 4840 Btu/(h · ft³/min) [3010 W/(L/s)] at standard conditions
W_i, W_o = humidity ratios of indoor and outdoor air

7.2.4 Heat Balance

Loads are the heat that must be supplied or removed by the HVAC equipment to maintain a space at the desired conditions. The calculations are like accounting. One considers all the heat that is generated in the space or flows across the envelope; the total energy, including the thermal energy stored in the space, must be conserved according to the first law of thermodynamics. The principal terms are indicated in Fig. 7.13. Outdoor air, occupants, and possibly certain kinds of equipment contribute both sensible and latent heat terms.

Load calculations are straightforward in the static limit, i.e., if all input is constant. As discussed in the following section, that is usually an acceptable approximation for the calculation of peak heating loads. But for cooling loads, dynamic effects (i.e., heat storage) must be taken into account because some of the heat gains are absorbed by the mass of the

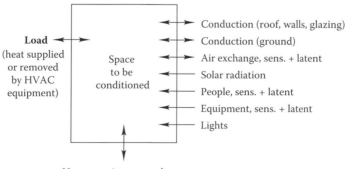

FIGURE 7.13
The terms in a load calculation.

building and do not contribute to the loads until several hours later. Dynamic effects are also important whenever the indoor temperature is allowed to float.

Sometimes it is appropriate to distinguish several aspects of the load. If the indoor temperature is not constant, the instantaneous load of the space may differ from the rate at which heat is being supplied or removed by the HVAC equipment. The load for the heating or cooling plant is different from the space load if there are significant losses from the distribution system or if part of the air is exhausted to the outside rather than returned to the heating or cooling coil.

It is convenient to classify the terms of the static energy balance according to the following groups. The sensible energy terms are

1. Conduction through building envelope other than ground

$$\dot{Q}_{cond} = K_{cond}(T_i - T_o) \tag{7.27}$$

2. Conduction through floor $\dot{Q}_{floor}$

3. Heat due to air exchange (infiltration and/or ventilation), at rate $\dot{V}$,

$$\dot{Q}_{air} = \dot{V}\rho c_p(T_i - T_o) \tag{7.28}$$

4. Heat gains from solar radiation, from lights, from equipment (appliances, computers, fans, etc.), and from occupants

$$\dot{Q}_{gain} = \dot{Q}_{sol} + \dot{Q}_{lit} + \dot{Q}_{equ} + \dot{Q}_{occ} \tag{7.29}$$

Combining the heat loss terms and subtracting the heat gains, one obtains the total *sensible load*

$$\dot{Q} = \dot{Q}_{cond} + \dot{Q}_{air} + \dot{Q}_{floor} - \dot{Q}_{gain} \pm \dot{Q}_{stor} \tag{7.30}$$

where we have added a term $\dot{Q}_{stor}$ on the right to account for storage of heat in the heat capacity of the building (the terms *thermal mass* and *thermal inertia* are also used to designate this effect). A dynamic analysis includes this term; a static analysis neglects it.

We have kept $\dot{Q}_{floor}$ as a separate item because it should not be taken proportional to $T_i - T_o$ except in cases like a crawl space, where the floor is in fairly direct contact with outside air. More typical is conduction through massive soil, for which the methods of Sec. 2.2.4 are appropriate. In traditional construction, the floor term has usually been small, and often it has been neglected altogether. But in superinsulated buildings it can be relatively important.

Using the total heat transmission coefficient K_{tot},

$$K_{tot} = K_{cond} + \dot{V}\rho c_p \tag{7.31}$$

one can write the sensible load in the form

$$\dot{Q} = K_{tot}(T_i - T_o) + \dot{Q}_{floor} - \dot{Q}_{gain} \pm \dot{Q}_{stor} \tag{7.32}$$

For signs we take the convention that $\dot{Q}$ is positive when there is a heating load and negative when there is a cooling load. Sometimes, however, we will prefer a plus sign for cooling loads. In that case, we will add subscripts c and h with the understanding that

$$\dot{Q}_c = -\dot{Q} \quad \text{and} \quad \dot{Q}_h = \dot{Q} \tag{7.33}$$

The *latent heat gains* are mainly due to air exchange, equipment (such as in the kitchen and bathroom), and occupants. Their sum is

$$\dot{Q}_{lat} = \dot{Q}_{lat,air} + \dot{Q}_{lat,occ} + \dot{Q}_{lat,equ} \tag{7.34}$$

The total load is the sum of the sensible and the latent loads.

During the heating season, the latent gain from air exchange is usually negative [with the signs of Eq. (7.26)] because the outdoor air is relatively dry. A negative $\dot{Q}_{lat}$ implies that the total heating load is greater than the sensible heating load alone—but this is relevant only if there is humidification to maintain the specified humidity ratio W_i. For buildings without humidification, one has no control over W_i, and there is not much point in calculating the latent contribution to the heating load at a fictitious value of W_i.

Example 7.7

Find the sensible peak heating load for a house under the following conditions.

Given: Site in Washington, D.C.
$K_{tot} = 186$ W/K [353 Btu/(h·°F)]
$\dot{Q}_{gain} = 1.0$ kW (3.4 kBtu/h)
$T_i = 22$°C (71.6°F)
Assumptions: Steady state. Neglect heat transfer to floor.
Lookup values: Winter design condition (97.5 percent) for Washington, D.C.;
$T_o = -10$°C (14°F), from tables in the appendices of the CD-ROM.

SOLUTION

From Eq. (7.32)

$$\dot{Q}_h = K_{tot}(T_i - T_o) - \dot{Q}_{gain}$$
$$= (186 \text{ W/K})\{[22 - (-10)] \text{ K}\} - 1.0 \text{ kW} = 5.95 - 1.0$$
$$= 4.95 \text{ kW} (16.9 \text{ kBtu/h})$$

Example 7.8

Estimate the latent load of humidification for the house of Example 7.7 if the air exchange rate is 0.5 per hour and the indoor relative humidity is to be 30 percent (usually sufficient in winter). What is the uncertainty due to not knowing the humidity of the outdoor air?

Given: Volume of 360 m³
$T_o = -10$°C (humidity not known)
$T_i = 22$°C with 30 percent relative humidity

Air change rate $= 0.5$ per hour (h^{-1})

Lookup values: Absolute humidities from psychrometric chart (Fig. 4.7)

Indoor humidity $W_i = 0.0048$ kg/kg$_{da}$

The outdoor humidity W_o cannot be larger than 0.0016 kg/kg$_{da}$ (at 100 percent relative humidity)

SOLUTION

Airflow

$$\dot{V} = (0.5\,\text{h}^{-1})(360\,\text{m}^3) = 180\,\text{m}^3/\text{h} = 50\,\text{L/s}$$

The humidity difference $W_i - W_o$ is bounded by $0.0032 < W_i - W_o < 0.0048$; the mean of these values is 0.0040; and the relative error is 20 percent at most. The latent gain at the mean is [see Eq. (7.26)]

$$\dot{Q}_{\text{lat}} = \dot{V}\rho h_{fg}(W_o - W_i)$$
$$= (50\,\text{L/s})[3010\,\text{W/(L/s)}](-0.0040) = -0.60\,\text{kW}$$

Being negative, it makes a positive contribution to the heating load if one decides to humidify. The uncertainty due to not knowing the outdoor humidity is only 20 percent, negligible in this context, because the latent load is only about one-tenth of the sensible load.

7.3 Storage Effects and Limits of Static Analysis

The storage term $\dot{Q}_{\text{stor}}$ in Eq. (7.32) is the rate of heat flow into or out of the mass of the building, including its furnishings and even the air itself. The details of the heat transfer depend on the nature of the building, and they can be quite complex.

One of the difficulties can be illustrated by considering an extreme example: a building that contains in its interior a large block of solid concrete several meters thick. The conductivity of concrete is relatively low, and diurnal temperature variations do not penetrate deeply into the block; only in the outer layer, to a depth of roughly 0.20 m, are they appreciable. Thus the bulk of the block does not contribute any storage effects on a diurnal time scale. The static heat capacity, defined as the product of the mass and the specific heat, would overestimate the storage potential because it does not take into account the temperature distribution under varying conditions.

To deal with this effect, some people (e.g., Sonderegger, 1978) have used the concept of *effective heat capacity* C_{eff}, defined as the periodic heat flow into and out of a body divided by the temperature swing at the surface. The effective heat capacity depends on the rate of heat transfer and on the frequency. The effective heat capacity is smaller than the static heat capacity, approaching it in the limit of infinite conductivity or infinitely long charging and discharging periods. As a rule of thumb, for diurnal temperature variations, the effective heat capacity of walls, floors, and ceilings is roughly 40 to 80 percent of the static heat capacity, assuming typical construction of buildings in the United States (wood, plaster, or concrete 3 to 10 cm thick). For items such as furniture that are thin relative to the depth of temperature variations, the effective heat capacity approaches the full static value.

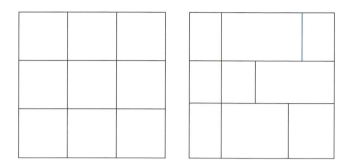

FIGURE 7.14
Variation of schematic floor plan to show that surface area of interior walls is independent of arrangement.

Further complications arise from the fact that the temperatures of different parts of the building are almost never perfectly uniform. For instance, sunlight entering a building is absorbed by the floor, walls, and furniture and raises their temperatures. The air itself does not absorb any appreciable solar radiation and is warmed only indirectly. Thus the absorbed radiation can cause heat to flow from the building mass to the air, even if the air is maintained thermostatically at uniform and constant temperature.

To show that these storage effects can be significant, let us return to the house of Example 7.5 and try to estimate its effective heat capacity. Most of the effective heat capacity lies in the plywood and drywall of the inner surfaces. Glass wool, like most insulation materials, has negligible heat capacity. The outer skin, be it aluminum siding or even brick veneer, does not contribute much because it is only weakly coupled to the interior air. The furniture might be equivalent to one plywood floor. For the interior wall surfaces, let us suppose a floor plan as in Fig. 7.14 with nine rectangular rooms (actually they need not be the same size, because the wall surface area depends only on the number of rooms, not on their arrangement). With nine rooms, the number of interior surfaces is 4 for the outer walls and $2 \times 4 = 8$ for the inner partitions, each contributing 12 m $\times$ 2.5 m, for a total of 360 m². Suppose the plywood and the plasterboard are each 1.3 cm thick. Then the static heat capacity adds up to about 9000 kJ/K, as shown in Table 7.6.

Since these components are fairly thin compared to the penetration depth of diurnal heat pulses, their effective heat capacity is almost as large as the static one. We have neglected

TABLE 7.6

Estimate of Static Heat Capacity for the House of Example 7.5, Not Counting the Wooden Support Beams in Walls, Floor, and Roof. Material Properties from Appendices in CD-ROM

Component	Area A m²	Thickness ΔX m	Volume $V = At$ m³	Specific Heat C_p kJ/(kg·K)	Density ρ kg/m³	ρc_p kJ/(m³·K)	Heat Capacity $C = V\rho c_p$ kJ/K
Drywall, roof	144	0.013	1.87	1.20	800	960.0	1797
Drywall, walls	360	0.013	4.68	1.20	800	960.0	4493
Plywood, floor	144	0.013	1.87	1.09	545	594.1	1112
Furniture (same as floor)							1112
Air	144	2.50	360.0	1.00	1.20	1.20	432
Total (without brick floor)							8946

the wooden support beams in the walls, floor, and roof; their static capacity is about 4500 kJ/K, but they are less directly coupled to the interior, and they might contribute about 2000 kJ/K of dynamic capacity. Thus the effective heat capacity of this type of house is on the order of 11,000 kJ/K. The air accounts for less than 5 percent and is usually ignored.

A brick or concrete floor would increase the heat capacity significantly. Concrete has a ρc_p value around 1500 kJ/(m$^3 \cdot$ K). If the concrete floor is 0.1 m thick, its static heat capacity is 144 m$^2 \times 0.1$ m $\times 1500$ kJ/(m$^3 \cdot$ K) $= 21,600$ kJ/K. Since it is fairly thick, its effective heat capacity is around one-half that value. Still, a concrete floor can easily double the effective heat capacity of the house.

To get an idea of the importance of the heat capacity, suppose that during the course of a summer afternoon the average temperature of the mass increases by 2 K in 5 h, implying a rate of temperature rise of

$$\dot{T} = \frac{2 \text{ K}}{5 \text{ h}} = 0.4 \text{ K/h}$$

This can happen even when the thermostat is set to maintain the air at constant temperature: Under dynamic conditions, air and mass can have different temperatures. Assuming an effective heat capacity $C_{\text{eff}} = 11,000$ kJ/K, the storage term in the heat balance, Eq. (7.32), is

$$\dot{Q}_{\text{stor}} = C_{\text{eff}}\dot{T} = (11,000 \text{ kJ/K})(0.4 \text{ K/h}) \approx 1.2 \text{ kW}$$

For summer design conditions the temperature difference $T_i - T_o$ is around 25°C $-$ 35°C $= -10$°C, and the corresponding steady-state conductive heat loss (with $K_{\text{tot}} = 205$ W/K from Example 7.5) is

$$K_{\text{tot}}(T_i - T_o) = -2.05 \text{ kW}$$

with a minus sign because it a heat gain of the building. If there were no storage, the cooling load would be 2.05 kW (for simplicity, we assume there are no other terms in the thermal balance). But 1.2 kW of this heat gain is soaked up by the storage, and thus the effective cooling load at this moment is only $2.05 - 1.2 = 0.85$ kW. Under these conditions, the storage effect reduces the cooling load by one-half. The stored heat gain does not appear as load until later: There is a time lag. One can speak of the *thermal inertia* (or *thermal mass*).

This example explains how the heat capacity can cause a time lag in the cooling load and can reduce the peak load. That is particularly important in commercial buildings, where concrete floor slabs are quite common and where a time lag of several hours can shift much of the load past the hours of occupancy, to a time when temperature control is no longer critical.

Heat capacity tends to be more important for cooling than for heating loads, for a number of reasons. Summer heat flows are more peaked than those in winter. Peak heating loads correspond to times without sun, and the diurnal variation of $T_i - T_o$ is small compared to its maximum in most climates. By contrast, for peak cooling loads, the diurnal variation of $T_i - T_o$ is comparable to its maximum, and solar gains are crucial. Also, in climates with cold winters, heating loads are larger than cooling loads, and the storage terms, for typical temperature excursions, are relatively less important in winter than in summer.

Consequently, the traditional steady-state calculation of peak heating loads was well justified for buildings with constant thermostat setpoint. But since the oil crisis, thermostat setback has become common practice for energy conservation. Thermostat setback can have a sizable impact on peak heating loads because setback recovery occurs during the early morning hours on top of the peak heat loss; in Sec. 7.5 we discuss this point in greater detail. Peak cooling loads, by contrast, are usually not affected by thermostat setup because recovery is not coincident with the peak gains.

Storage effects for latent loads are difficult to analyze (see, e.g., Fairey and Kerestecioglu, 1985, and Kerestecioglu and Gu, 1990), and most of the current computer programs for building simulation, such as DOE2.1 and BLAST (see Sec. 7.7.1), do not account for moisture exchange with the building mass. In practice, this neglect of moisture storage is usually not a serious problem. Precise humidity control is not very important in most buildings. Where it is important, e.g., in hospitals, temperature and humidity are maintained constant around the clock. When the air is at constant conditions, the moisture in the materials does not change much and storage effects can be neglected; but in buildings with intermittent operation, these effects can be large, as shown by Fairey and Kerestecioglu (1985) and by Wong and Wang (1990).

From this discussion emerge the following *recommendations for the importance of dynamic effects*:

1. They can significantly reduce the peak cooling loads, with or without thermostat setup.

2. They can be neglected for peak heating loads, except if thermostat setback recovery is to be applied even during the coldest periods of the year.

3. For the calculation of annual consumption, they can have an appreciable effect if the indoor temperature is not kept constant.

4. Storage of *latent* heat is neglected for most applications.

Thus a simple static analysis is sufficient for some of the problems the designer is faced with, but not for the peak cooling load. To preserve much of the simplicity of the static approach in a method for peak cooling loads, ASHRAE has developed the CLF-CLTD method which modifies the terms of a static calculation to account for thermal inertia. This method is presented in Sec. 7.6; it can be used for standard construction if the thermostat setpoint is constant.

7.4 Zones

So far we have considered the interior as a single zone at uniform temperature—a fair approximation for simple houses, for certain buildings without windows (such as warehouses), or for buildings that are dominated by ventilation. But in large or complex buildings, one usually has to calculate the loads separately for a number of different zones. There may be several reasons. An obvious case is a building where different rooms are maintained at different temperatures, e.g., a house with an attached sunspace. Here the heat balance equation is written for each zone, in the form of Eq. (7.32) but with an additional term

$$\dot{Q}_{j-k} = U_{j-k}A_{j-k}(T_j - T_k) \qquad (7.35)$$

for the heat flow between zones j and k.

However, even when the entire building is kept at the same temperature, multizone analysis becomes necessary if the spatial distribution of heat gains is too nonuniform. Consider, e.g., a building with large windows on the north and south sides, during a sunny winter day when the gains just balance the total heat loss. Then neither heating nor cooling would be required, according to a one-zone analysis. But how can the heat from the south get to the north?

Heat flow is the product of the heat transfer coefficient and the temperature difference, as in Eq. (7.35). Temperature differences between occupied zones are small, usually not more than a few kelvins; otherwise there would be complaints about comfort. The heat transfer coefficients between zones are often not sufficiently large for effective redistribution of heat, especially if there are walls or partitions. This is demonstrated by Example 7.9.

Example 7.9

Consider the loads in adjacent south and north zones of a building on a sunny winter day, assuming for simplicity that the loads are due only to the windows (10 m²) on each facade. How much heat could flow between these two zones if they are separated by a thin wall with a U value of 4.0 W/(m² · K) and area of 20 m² if the temperature difference is $\Delta T = 1$ K? How does this heat flow compare to the zone loads?

Given: Heat transmission coefficient to outside $UA = 3$ W/(m² · K) $\times 10$ m² $= 30$ W/K for each of the facades (north and south):

$$T_o = -10°C$$

Indoor temperatures are uniform within each zone, with $T_{i,N} = 20°C$ in north zone and $T_{i,S} = 21°C$ in south zone:

$$\dot{Q}_{sol,S} = (10 \text{ m}^2)(200 \text{ W/m}^2) = 2000 \text{ W on south window}$$
$$\dot{Q}_{sol,N} = (10 \text{ m}^2)(10 \text{ W/m}^2) = 100 \text{ W on north window}$$

Interzone heat transfer coefficient

$$K_{NS} = [4.0 \text{ W/(m}^2 \cdot \text{K})](20 \text{ m}^2) = 80 \text{ W/K}$$

SOLUTION

The zone loads are

$$\dot{Q}_N = UA(T_i - T_o)_N - \dot{Q}_{sol,N} = (30 \text{ W/K})(30 \text{ K}) - 100 \text{ W} = 800 \text{ W}$$
$$= \text{north zone } (>0, \text{ heating load})$$

$$\dot{Q}_S = UA(T_i - T_o)_S - \dot{Q}_{sol,S} = (30 \text{ W/K})(31 \text{ K}) - 2000 \text{ W} = -1070 \text{ W}$$
$$= \text{south zone } (<0, \text{ cooling load})$$

The heat flow between the zones is $\Delta T \times K_{NS} = 1$ K $\times 80$ W/K $= 80$ W. This is small compared to the zone loads, 800 and -1070 W, respectively. The HVAC system must control the two zones separately, if unacceptably large temperature differences between the zones are to be avoided.

If there could be perfect heat flow between the zones, the total load would be only $800 - 1070 = -270$ W, a small cooling load (which in practice would be reduced to zero by raising T_i a few degrees).

COMMENTS

In this example, the zones are defined by an obvious physical boundary. But even without the partition there would be a problem of unacceptable temperature gradients. Such gradients within a room are more difficult to calculate (and in this example we have bypassed the complication by assuming uniformity within each zone). In buildings with air distribution systems, the problem is alleviated to the extent that the system mixes the air between the zones—in effect, enhancing the interzone heat transfer.

To make a point, we have chosen the numbers in this example a bit on the extreme side, but the basic phenomenon is very common: Often the interzone heat transfer is so small that excess heat gains in one zone bring little or no benefit for heating loads in another zone—hence the thermodynamically perverse fact that many large buildings require simultaneous heating and cooling.

The problem of divergent zone loads is one of the prime targets for energy conservation in large buildings (and we will take up the topic again in Chap. 14). The first step is to reduce the loads through the envelope, by improved insulation and control of solar radiation: The smaller the loads, the smaller the differences between the loads. Careful attention must be paid to the design of the HVAC system and the choice of its zones. Finally, there is the possibility of heat recovery and redistribution between zones by heat pumps.

The basic criterion for zoning is the ability to control the comfort conditions; the control is limited by the number of zones one is willing to consider. *To guarantee comfort, the HVAC plant and distribution system must be designed with sufficient capacity to meet the load of each zone.* In choosing the zones for a multi-zone analysis, the designer should try to match the distribution of heat gains and losses. A common and important division is between interior and perimeter zones, because the interior is not exposed to the changing environment. Different facades of the perimeter should be considered separately for cooling load calculations, as suggested in Fig. 7.15. Corner rooms should be assigned to the facade with which they have the most in common; usually this will be the facade where a corner room has the largest windows. Corner rooms are often the critical rooms in a zone, requiring more heating or cooling (per unit floor area) than single-facade rooms of the same zone.

Actually there are different levels to a zoning analysis, corresponding to different levels of the HVAC system. To be specific, we anticipate some topics from Chap. 11. In an air system, there are major zones corresponding to each air handler. Within each air handler

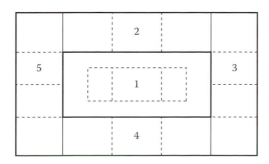

FIGURE 7.15

Example of recommended zoning. Thick lines represent zones, labeled 1 through 5. Dashed lines represent subzones.

zone, the air ducts, air outlets, and heating or cooling coils must have sufficient capacity and sufficient controllability to satisfy the loads of each subzone; the design flow rates for each room are scaled according to the design loads of the room. For best comfort (and if cost were no constraint), each zone should have its own air handler and each room its own thermostat. There is a tradeoff between equipment cost and achievable comfort, and the best choice depends on the circumstances. If temperature control is critical, one installs separate air handlers for each of the five zones in Fig. 7.15 and separate thermostats for each room. To save equipment cost, one often assigns several zones to one air handler and several rooms to one thermostat; but the more divergent the loads, the more problematic the control. For the building of Fig. 7.15, a single air handler and five thermostats may be adequate if the distribution of heat gains is fairly uniform and if the envelope is well insulated with good control of solar gains.

Another example is a house whose air distribution system has a single fan (typical of all but the largest houses). Even though there is only one major zone, the detailed design of the distribution system demands some attention to subzones. Within each room, the peak heating capacity should match the peak heat loss. Also, it is advisable to place heat sources close to points with large heat loss, i.e., under windows (unless they are highly insulating).

The choice of zones is not always clear-cut, and the design process may be iterative. Depending on the distribution of gains and losses, one may want to assign several rooms to a zone, one room to a zone, or even several zones to a room (if it is very large). With finer zonal detail one improves the control of comfort, but at the price of greater calculation effort and higher HVAC system cost. In an open office space, there is no obvious boundary between interior and perimeter; here a good rule is to make the perimeter zone as deep as the penetration depth of direct solar radiation, typically a few meters. Spaces connected by open doors, e.g., offices and adjacent hallways, can sometimes be treated as a single zone. Separate zones are advisable for rooms with large computers or energy-intensive equipment. In multistory buildings, one may want to treat the top floor apart from the rest.

The calculation of peak *heating* loads and capacities can often be done without considering separate perimeter zones, because peak heating loads occur when there is no sun; with uniform internal gains, the corresponding thermal balance is uniform around the perimeter. But while the calculation can be carried out for a single zone, the operation requires multiple zones: The heating system must allow separate control of different facades to compensate for the variability of solar gains during the day. For *cooling* loads, a multizone analysis is essential, even at the calculation stage, because the loads occur when the sun is shining.

As discussed in Sec. 7.3, peak cooling loads require a dynamic analysis whereas peak heating loads can be estimated quite well by static models (at least in the absence of thermostat setback). Compared to heating loads, the calculation of cooling loads of large buildings is thus doubly complicated: It requires multiple zones and dynamic analysis if one wants reasonable accuracy.

A related issue is the coincidence between peak loads of different zones. To determine the capacity of the central plant, one needs to know the peak load of the totality of zones served by the plant. This is usually less than the simple sum of the individual peak loads because of noncoincidence. The term *diversity* is used to designate the ratio of the actual system peak to the sum of the individual peak loads. In practice, one often finds a diversity around 0.6 to 0.8 for large buildings or groups of buildings (e.g., university campuses); for better estimates at the design stage, computer simulations are recommended (see Sec. 7.7.1).

7.5 Heating Loads

Since the coldest weather may occur during periods without solar radiation, it is advisable not to rely on the benefit of solar heat gains when calculating peak heating loads (unless the building contains long-term storage). If the indoor temperature T_i is constant, a static analysis is sufficient, and the calculation of the peak heating load $\dot{Q}_{h,max}$ is very simple: Find the design heat loss coefficient K_{tot}, multiply by the design temperature difference $T_i - T_o$, and subtract the internal heat gains on which one can count during the coldest weather:

$$\dot{Q}_{h,\,max} = K_{tot}(T_i - T_o) - \dot{Q}_{gain} \qquad (7.36)$$

In fact we have demonstrated this procedure already with the house of Example 7.7, which we recall at this point. It is a building with

$$K_{tot} = 186 \text{ W/K } [353 \text{ Btu/(h} \cdot {}^\circ\text{F)]}$$
$$\dot{Q}_{gain} = 1.0 \text{ kW } (3.4 \text{ kBtu/h})$$

design conditions

$$T_o = -10^\circ\text{C } (14^\circ\text{F}) \quad \text{and} \quad T_i = 22^\circ\text{C } (71.6^\circ\text{F})$$

resulting in a design heat load of

$$\dot{Q}_{h,\,max} = 4.95 \text{ kW } (16.9 \text{ kBtu/h})$$

Example 7.8 has shown that the latent load of humidification (if a humidifier is used) would increase the load by about 10 percent. The intervening discussion has provided the justification for having treated the house as a single zone and for having neglected dynamic effects if T_i remains constant.

What would happen if the thermostat were set back at night? For a rough indication, recall from Sec. 7.3 that this house (of lightweight construction typical in the United States) would require heat input at the rate of 1.2 kW if its temperature were to be increased by 2 K in 5 h. For setback recovery after winter nights, one might want rates that are several times faster, say, 4 K in 2.5 h. Assuming that the heat input is proportional to the warmup rate,[5] the extra load for setback recovery (also known as the *pickup load*) would be 4×1.2 kW $= 4.8$ kW, comparable to the static design heat load. In this case, the capacity of the heating system would have to be doubled relative to the case without setback.

In a given situation, the required extra capacity depends on the amount of setback $T_i - T_o$, the acceptable recovery time, and building construction. For reasonable accuracy a dynamic analysis is recommended. Optimizing the capacity of the heating system involves a tradeoff between energy savings and capacity savings, with due attention to part-load efficiency. As a general rule for residences, ASHRAE (1989a) recommends over-sizing by about 40 percent for a night setback of 10°F (5.6 K), to be increased to 60 percent oversizing if there is additional setback during the day. In any case, some flexibility can be

---

[5] Actually, at faster rates the effective heat capacity is smaller (the heat pulse takes longer than 1 h to penetrate the entire mass), and so the real increment for setback recovery is less. We do not know how much, without doing a detailed dynamic analysis, but in any case this example confirms the warning about the incompatibility of static analysis and setback recovery.

provided by adapting the operation of the building: If the capacity turns out insufficient, one can reduce the depth and duration of the setback during the coldest periods.

In commercial buildings with mechanical ventilation, the demand for extra capacity during setback recovery is reduced if the outdoor air intake is closed during unoccupied periods. In winter that should always be done for energy conservation (unless air quality problems demand high air exchange at night).

7.6 CLTD/CLF Method for Cooling Loads

Because of thermal inertia, it is advisable to distinguish several heat flow rates. The *heat gain* is the rate at which heat is transferred to or generated in a space. The *cooling load* is the rate at which the cooling equipment would have to remove thermal energy from the air in the space in order to maintain constant temperature and humidity. Finally, the *heat extraction rate* is the rate at which the cooling equipment actually does remove thermal energy from the space.[6]

Conductive heat gains and radiative heat gains do not enter the indoor air directly; rather they pass through the mass of the building, increasing its temperature relative to the air. Only gradually are they transferred to the air. Thus their contribution to the cooling load is delayed, and there is a difference between heat gain and cooling load. Averaged over time, these rates are, of course, equal, by virtue of the first law of thermodynamics.

The heat extraction rate is equal to the cooling load only if the temperature of the indoor air is constant (as assumed in this section). Otherwise the heat flow to and from the building mass causes the heat extraction rate to differ from the cooling load (a feature to be analyzed in Sec. 7.7.4).

To account for transient effects without having to resort to a full-fledged dynamic analysis, a special shorthand method has been developed that uses the *cooling load temperature difference (CLTD)* and *cooling load factor (CLF)*. To explain the principles, note that the cooling load due to conduction across an envelope element of area A and conductance U would be simply

$$\dot{Q}_{c,\text{cond}} = UA(T_o - T_i) \tag{7.37}$$

under static conditions, i.e., if indoor temperature T_i and outdoor temperature T_o were both constant. When the temperatures vary, this is no longer the case because of thermal inertia. But if the temperatures follow a periodic pattern, day after day, $\dot{Q}_{c,\text{cond}}$ will also follow a periodic pattern. Once $\dot{Q}_{c,\text{cond}}$ has been calculated, one can define a CLTD as the temperature difference that gives the same cooling load when multiplied by UA. If such temperature differences are tabulated for typical construction and typical temperature patterns, they can be looked up for quick determination of the load. Thus the conductive cooling load is

$$\dot{Q}_{c,\text{cond},t} = UA(\text{CLTD}_t) \tag{7.38}$$

where the subscript t indicates the hour t of the day.

Likewise, if there is a constant radiative heat gain in a zone, the corresponding cooling load is simply equal to that heat gain. If the heat gain follows a periodic pattern, the cooling

---

[6] In later chapters on HVAC systems, we will encounter yet another rate, the *coil load*; it is the rate at which the cooling coil removes heat from the air, and it can be different from the heat extraction rate due to losses in the distribution system.

load also follows a periodic pattern. The cooling load factor (CLF) is defined such that it yields the cooling load at hour t when multiplied by the daily maximum $\dot{Q}_{max}$ of the heat gain:

$$\dot{Q}_{c,\mathrm{rad},t} = \dot{Q}_{max}(\mathrm{CLF}_t) \tag{7.39}$$

The CLFs account for the fact that radiative gains (solar, lights, etc.) are first absorbed by the mass of the building, becoming a cooling load only as they are being transferred to the air. Only convective gains can be counted as cooling load without delay. Some heat gains, e.g., from occupants, are partly convective and partly radiative; the corresponding CLFs take care of that.

The CLTDs and CLFs of ASHRAE have been calculated by means of the transfer functions discussed in Sec. 7.7. To keep the bulk of numerical data within reasonable limits, only a limited set of standard construction types and operating conditions has been considered. Some correction factors are provided to extend the applicability, however, without escaping the constraint that the indoor temperature T_i be constant.

It is also important to note that most energy calculation methods are not static— they are constantly undergoing revision and improvement. The CLTD/CLF method has likewise experienced modifications. The remainder of this section discusses the original method as presented by ASHRAE, and that is still used by a large number of designers and engineers. For an overview of the updated CLTD method (now called the CLTD/SCL/CLF method), see Sec. 7.8.1. It is stressed that the student should understand the original method so that the evolution of the calculations can be better appreciated.

Excerpts of the CLTDs and CLFs are incorporated in the software included with this book. The software offers a user-friendly interface for entering the building description and for plotting the results. The reader is urged to try it out. Varying a few parameters and watching the effects on the computer screen are the best way to develop your intuition about the subject, without the tedium of lengthy hand calculations.[7]

One of the incidental advantages of such software is that it guides the user systematically through all the load components that must be considered. If one has to do the job by hand, it is advisable to use a worksheet such as the one reproduced in Fig. 7.16 to make sure that nothing is overlooked. The calculation needs to be done for the hour when the peak occurs. That hour can be guessed if a single load dominates, because in that case it is the hour with the largest value of CLTD or CLF. If several loads with noncoincident peaks are of comparable importance, the hour of the combined peak may not be entirely obvious, and the calculation may have to be repeated several times. *In most buildings, peak cooling loads occur in the afternoon or early evening.* Figures 7.17 to 7.21 give an indication when the components of the cooling load are likely to reach their peak.

The steps of the calculation are summarized in the worksheet of Fig. 7.16. We now proceed to discuss these steps, illustrating them by filling out the worksheet for a zone of an office building. The procedure has to be carried out for each zone of the building.

For *walls* and *roofs*, the conductive cooling load at time t is calculated by inserting the appropriate CLTD into Eq. (7.38). Numerical values can be found in the HCB software for a variety of construction types and materials. For roofs there are 13 types, listed as roof no. 1, 2, . . . , 13. For walls there are seven construction types, designated group A, B, . . . , G. The thermal properties of the materials can also be found on the CD-ROM.

---

[7] But you should not succumb to the temptation of blindly accepting the results. Even when the software is free of bugs, the input might not represent what you intended. An independent check is always advisable when you are doing a calculation for the first time or when the result is important.

Job ID	Date			Initials	
Site	Latitude			Longitude	
Design conditions	Indoor temp.		Rel. humid.	Outdoor temp.	Rel. humid.
Room	Identification			Dimensions	

Latent loads						Instantaneous
		$\dot{V}$	W_o	W_i	$\Delta W = W_o - W_i$	$\dot{Q}_{lat} = \rho \times h_{fg} \times \dot{V} \times \Delta W$
Air exchange						
			N = number		$\dot{Q}_{lat/unit}$	$\dot{Q}_{lat} = N \times \dot{Q}_{lat/unit}$
Appliances						
People						
TOTAL LATENT						

Sensible loads					hour t		hour t	
Component and orientation	Construction type		U	A	CLTD$_t$		$\dot{Q}_t = U \times A \times CLTD_t$	
Walts								
Roof								
Glazing conduction								
Glazing solar		A	SC	SHGF$_{max}$	CLF$_t$		$\dot{Q}_t = A \times SC \times SHGF_{max} \times CLF_t$	
Air exchange		V	$\dot{V}$	T_i	T_o		$\dot{Q}_t = \rho \times c_p \times \dot{V} \times (T_o - T_i)$ (instantaneous)	
Internal patitions			U	A	ΔT across partition		$\dot{Q} = U \times A \times \Delta T$ (instantaneous)	
Ceiling								
Floor								
Sides								
Ducts								
Internal gains		number	gain /unit	$\dot{Q}$	CLF$_t$		$\dot{Q}_t = \dot{Q} \times CLF_t$	
Appliances								
Fans								
Lights								
Motors								
People								
TOTAL SENSIBLE								

FIGURE 7.16

Worksheet for CLTD/CLF method for a specific zone. At sea level $\rho c_p = 1.08$ [Btu/(h · °F)]/(ft^3/min) [1.2 (W/K)/(L/s)] and $\rho h_{fg} = 4840$ (Btu/h)(ft^3/min) [3010 W/(L/s)].

Example 7.10

What is the peak conductive cooling load for a sunlit west-facing curtain wall under the following conditions? (Examples 7.10 to 7.16 carry out a CLTD/CLF calculation for a room in an office building.)

Given: Metal curtain wall construction type G with 3-in insulation, sunlit, dark color

$A_{wall} = 18 \text{ m}^2$

$U_{wall} = 0.52 \text{ W/(K} \cdot \text{m}^2)$

Design conditions same as for the CLTDs of the lookup tables, in particular: July 21, $T_{o,max} = 35.0°C$ (95°F), daily range = 11.7 K (21°F), $T_i = 25.5°C$ (78°F), latitude 40°N. (Urbana, Illinois, e.g., has these values as 1 percent design conditions.)

Lookup values: CLTDs for group G walls, orientation W

SOLUTION

The maximum CLTD occurs at 17:00. Filling out the worksheet (Fig. 7.16), we have the following:

Sensible loads					hour t			hour t		
					16:00	17:00	18:00	16:00	17:00	18:00
Component and orientation	Construc-tion type		U $[\text{W/m}^2 \cdot \text{K}]$	A $[\text{m}^2]$		$CLTD_t$ $[K]$			$\dot{Q}_t = U \times A \times CLTD_t$ $[W]$	
Walls										
Wests	*typeG*		*0.52*	*18*		*40*			*374.4*	

COMMENTS

1. In this manner, the conductive loads are calculated for all components of the envelope of the room or zone.
2. A steady-state calculation would have given us $0.52 \times 18 \times [(75.5 - 25.5) \text{ W}] = 468 \text{ W}$ at 16:00, since the peak of the sol-air temperature for a dark, west-facing surface is 75.5°C at 16:00. The real peak is 20 percent smaller and occurs 1 h later. The thermal inertia reduces and retards the peak. The effect is significant even for the lightweight wall of this example!

We have plotted CLTD versus time for three roof types in Fig. 7.17. The heavier the construction, the smaller the amplitude and the later the peak. Fig. 7.18 shows analogous results for sunlit walls having the four cardinal orientations. For these CLTDs, the following conditions have been assumed:

- High absorptivity for solar radiation ("dark")
- Solar radiation for 40°N on July 21
- $T_i = 25.5°C$ (78°F)
- T_o has a mean of $T_{o,av} = (T_{o,max} + T_{o,min})/2 = 29.4°C$ (85.0°F) and a *daily range* = $T_{o,max} - T_{o,min} = 11.7°C$ (21.0°F), with $T_{o,max} = 35.0°C$ (95.0°F) being the design temperature
- Outdoor convective heat transfer coefficient $h_o = 17 \text{ W/(m}^2 \cdot \text{K)}$ [3.0 Btu/(h $\cdot$ ft^2 $\cdot$ °F)]

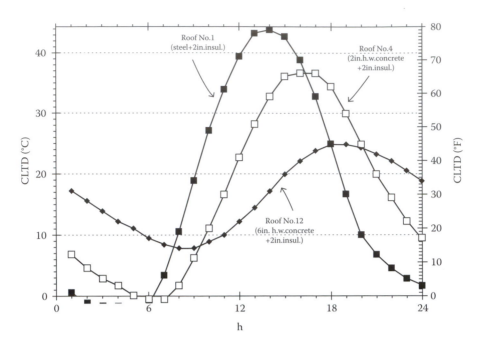

FIGURE 7.17
CLTDs for three roof types.

- Indoor convective heat transfer coefficient $h_i = 8.3 \, \text{W}/(\text{m}^2 \cdot \text{K})$ $[1.46 \, \text{Btu}/(\text{h} \cdot \text{ft}^2 \cdot \text{°F})]$
- No forced ventilation or air ducts in the ceiling space

When conditions are different, one should correct the CLTDs according to the formulas

$$\text{CLTD}_{\text{cor}} = (\text{CLTD} + \text{LM})K + (25.5\text{°C} - T_i) + (T_{o,\text{av}} - 29.4\text{°C}) \qquad (7.40\text{SI})$$

$$\text{CLTD}_{\text{cor}} = (\text{CLTD} + \text{LM})K + (78\text{°F} - T_i) + (T_{o,\text{av}} - 85\text{°F}) \qquad (7.40\text{US})$$

where
  LM = correction factor for latitude and month, from CD-ROM
  K = color adjustment factor
$T_i, T_{o,\text{av}}$ = actual values for application

And $T_{o,\text{av}}$ is obtained by subtracting $0.50 \times$ daily range from $T_{o,\text{max}}$, the design temperature of the site, which is listed in the lookup tables along with the daily range. The color correction K is 1.0 for dark and 0.5 for light surfaces; values less than 1.0 should be used only when one is confident that the surface will permanently maintain low absorptivity.

How about other construction types? Two factors are affected: the U value and the CLTD. One should always use the correct U value for the actual construction in Eq. (7.38). As for the CLTD, one should select the construction type that is closest in terms of mass and heat capacity. If there is additional insulation in a wall, one should pass, for each additional $1.23 \, (\text{m}^2 \cdot \text{K})/\text{W}$ $[7.0 \, (\text{ft}^2 \cdot \text{F} \cdot \text{h})/\text{Btu}]$ of resistance, to the wall group with the preceding letter of the alphabet (if the insulation is on the indoor side of the mass; if it is on the outside,

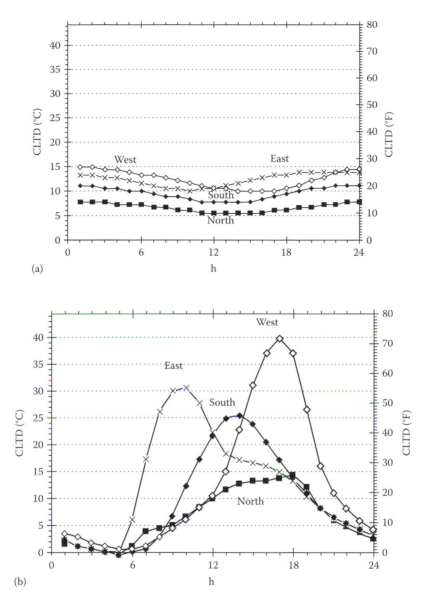

FIGURE 7.18
CLTDs for sunlit walls. Four orientations are shown, for two construction types: (a) group A walls [heavy, for example, 8-in (200-mm) concrete with insulation]; (b) group G walls (light, e.g., frame wall or curtain wall).

advance two letters in the alphabet). If this is not possible because one is already at letter A, use the following CLTDs [before applying any correction according to Eq. (7.40)]:

N	NE or NW	E or W	SE or SW	S
6.1 K (11°F)	9.4 K (17°F)	12.2 K (22°F)	11.6 K (21°F)	9.4 K (17°F)

For each additional 1.23 (m² · K)/W [7.0 (ft² · °F · h)/Btu] of resistance in a roof, select a construction type whose mass and heat capacity are approximately the same but whose

CLTD peaks 2 h later. If this is not possible because one has already selected the roof with the longest time lag, use CLTD $= 16$ K (29°F).

Example 7.11

How does the result of Example 7.10 change if one adds insulation to increase the R_{th} value by 1.23 (m$^2 \cdot$ K)/W, and if the site is Miami, Florida, with 2.5 percent design conditions?

   Given: Site in Miami, latitude 25.5°N

$T_{o,max} = 32.2$°C (90°F)

Daily range $= 8.3$ K (15°F)

$T_i = 25.5$°C (78°F), all from design conditions in the lookup tables

$U_{wall} = 0.32$ W/(K $\cdot$ m^2)

Lookup values: Latitude and month correction LM $= 0.0$, from the lookup tables. Color adjustment factor $K = 1.0$. To account for the extra insulation, we use wall group F instead of G.

We find that the peak is now CLTD $= 33$ K (uncorrected) and occurs at 19:00 from the lookup tables.

SOLUTION

The daily average temperature is $T_{o,av} = 32.2$°C $- (8.3$ K$)/2 = 28.05$°C. Inserting these values into Eq. (7.40), we obtain

$$\text{CLTD}_{cor} = (33 - 0.0)(1.0) + [(28.05 - 29.4)\text{ K}] = 31.65\text{ K}$$

The contribution to the cooling load is 18 m$^2 \times$ [0.32 W/(K $\cdot$ m^2)] $\times$ 31.65 K $= 182$ W at 19:00.

For *windows*, one treats conductive and solar heat gains separately, according to the decomposition

Heat gain through glass $=$ conduction due to $T_i - T_o$

   $+$ heat gain due to solar radiation transmitted through or absorbed by glass   (7.41)

The conductive part is calculated as in Eq. (7.38)

$$\dot{Q}_{c,cond,glaz,t} = UA(\text{CLTD}_{glaz,t}) \tag{7.42}$$

with the corresponding CLTDs in the appendices on the CD-ROM. It is plotted in Fig. 7.19. No orientation is indicated because they exclude solar radiation; thus they are applicable even for skylights.

Solar gains through windows are treated by means of the solar heat gain factor SHGF. It is defined as the instantaneous heat gain [Btu/(h $\cdot$ ft^2) (W/m^2)] due to solar radiation through reference glazing, as discussed in Eq. (6.58). There are two components in this solar gain: the radiation absorbed in the glass and the radiation transmitted through the glass. The latter is assumed to be totally absorbed in the interior of the building, a reasonable assumption in view of the cavity effect. The radiation absorbed in the glass raises its temperature, thereby changing the conductive heat flow. The SHGF combines this latter contribution with the radiation transmitted to the interior. For glazing types other than the reference glazing, one multiplies by the shading coefficient SC, defined in Sec. 6.5 and listed in Tables 6.5 and 6.6.

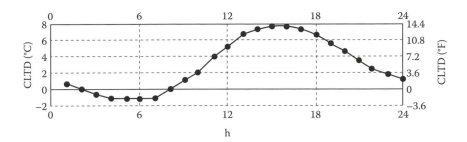

FIGURE 7.19
CLTDs for glass.

To calculate the contribution to the cooling load, the daily maximum of the solar heat gain is multiplied by the cooling load factor. Thus the actual cooling load at time t due to solar radiation is given by the formula

$$\dot{Q}_{c,\text{sol},t} = A \times SC \times SHGF_{\text{max}} \times CLF_t \qquad (7.43)$$

where

$A =$ area, ft^2 (m^2)

$SC =$ shading coefficient, Tables 6.5 and 6.6

$SHGF_{\text{max}} =$ maximum solar heat gain factor, Btu/(h · ft^2) (W/m^2), from the HCB lookup tables

$CLF_t =$ cooling load factor for time t, from the HCB lookup tables

$SHGF_{\text{max}}$ is the value of SHGF at the hour when the radiation attains its maximum for a particular month, orientation, and latitude. The CLF takes into account the variation of the solar radiation during the day, as well as the dynamics of its absorption in the mass of the building and the gradual release of this heat. A separate set of CLFs is given for each orientation and for each of three construction types, characterized in terms of the mass of building material per floor area: light $= 30$ lb/ft^2 (146 kg/m^2), medium $= 70$ lb/ft^2 (341 kg/m^2), and heavy $= 130$ lb/ft^2 (635 kg/m^2). Each set comprises all hours from 1 to 24. A subset is plotted in Fig. 7.20.

Example 7.12

Find the cooling load at 17:00 due to an unshaded west-facing window under the conditions of Example 7.10.

Given: West-facing unshaded double-glazed window with

$A = 6$ m^2

$U = 3.0$ W/(m^2 · K)

$SC = 0.80$

Light construction

Lookup values: $CLTD_t = 7.0$ K at $t = 17$:00 for glass, from the HCB tables

$SHGF_{\text{max}} = 681$ W/m^2 from the HCB tables

$CLF_t = 0.64$ at $t = 17$:00 for west-facing window from Fig. 7.20, or the HCB tables. Light construction

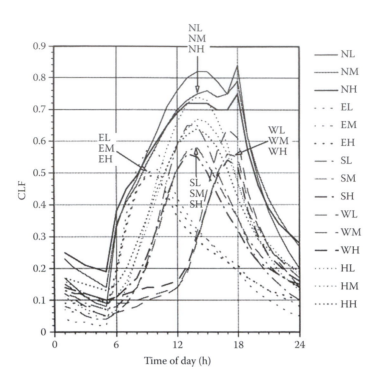

FIGURE 7.20

Cooling load factors for glass without interior shading for five orientations (E = east, S = south, W = west, N = north, H = horizontal) and three construction types (L = light, M = medium, H = heavy).

SOLUTION

Filling in the worksheet, we obtain

Sensible loads					hour t			hour t		
					16:00	*17:00*	*18:00*	*16:00*	*17:00*	*18:00*
Component and orientation	Construction type		U [W/m²·K]	A [m²]		CLTD$_t$ [K]			$\dot{Q}_t$=U×A×CLTD$_t$ [W]	
Glazing conduction										
West			3.00	6.00		7			126.0	
Glazing solar		A [m²]	SC	SHGF$_{max}$ [W/m²]		CLF$_t$			$\dot{Q}_t$=A×SC×SHGF$_{max}$×CLF$_t$ [W]	
West		6.00	0.80	681		0.64			2092.0	

The sum of conductive and radiative gain through the window is 126.0 + 2092.0 = 2218 W.

In analogous fashion, CLFs have been computed for heat gains from *internal heat sources*. There are different factors for each of the three major categories: *occupants* (see Table 7.5 for $\dot{Q}_{occ}$ and the HCB lookup tables for CLF$_{occ,t}$)

$$\dot{Q}_{occ,t} = \dot{Q}_{occ}(\text{CLF}_{occ,t}) \tag{7.44}$$

lights (see the HCB lookup tables for CLF_t)

$$\dot{Q}_{lit,t} = \dot{Q}_{lit}(CLF_{lit,t}) \tag{7.45}$$

and equipment such as *appliances* (see Table 7.4 for $\dot{Q}_{app}$ and the HCB lookup tables for CLF_t)

$$\dot{Q}_{app,t} = \dot{Q}_{app}(CLF_{app,t}) \tag{7.46}$$

In these equations, the $\dot{Q}$ Btu/h (W) on the right-hand side is the rate of heat production, assumed constant for a certain number of hours and zero the rest of the time. The $\dot{Q}_t$ on the left-hand side is the resulting cooling load at hour t, for t from 1 to 24. For each load profile there is a different set of CLFs.

In Fig. 7.21 we have plotted a set of CLFs for lights. The lights are left on for 10 h/day, and the time axis shows the number of hours after the lights have been turned on. Two construction types are shown, representing the highest and the lowest thermal inertia in the tables. Once again, the CLF curves show how the actual loads are attenuated and delayed by the thermal inertia.

It is interesting to check the consistency of the CLFs with the first law of thermodynamics, which says in this case that the daily total of the cooling load must be equal to the daily total of the heat input. This check is easy to perform for lights, equipment, and occupants because the assumed input profile is simply on/off (unlike the solar radiation profile, which varies with the sun-earth geometry). Therefore, the sum of the CLFs over 24 h equals the number of hours during which there is a constant nonzero input. For example, in Fig. 7.21 the areas under the curves for light and for heavy are the same: 10 h.[8]

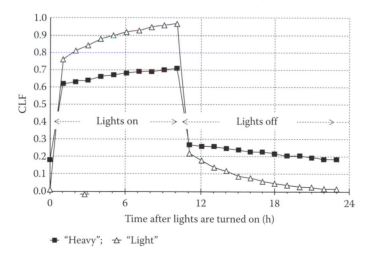

FIGURE 7.21
Cooling load factors for lights that are on 10 h/day, for two extreme construction types: light and heavy (*a* coefficient and *b* classification, as per the HCB lookup tables, are 0.45D for heavy and 0.75A for light).

---

[8] CLF data are provided only for a limited number of standard scenarios, and in certain situations one might like to interpolate between CLFs for different operating times, e.g., between the CLFs for 8 and 10 h if lights are on 9 h/day. The result would show two peaks, one at 8 h and one at 10 h, rather than a single peak at 9 h, because averaging between the CLFs for 8 and 10 h is equivalent to assuming that one-half of the lights are on for 8 h and the other half for 10 h.

Example 7.13

Find the cooling load due to internal gains, for the conditions of Example 7.10.

Given: Conditions of Example 7.10. Power for lighting is 25 W/m^2 average over a floor area of 64 m^2 from 8:00 to 18:00, and 0 the rest of the time.

Lookup values: CLFs for lights from Fig. 7.20, or the HCB lookup tables

SOLUTION

Continuing with the worksheet, we obtain

Sensible loads				hour t 16:00	hour t 17:00	hour t 18:00	hour t 16:00	hour t 17:00	hour t 18:00
Internal gains	num-ber	gain /unit	$\dot{Q}$ [W]		CLF$_t$			$\dot{Q}_t = \dot{Q} \times CLF_t$ [W]	
Lights	64 m^2	25 W/m^2	1600		0.84			1344.0	

Example 7.14

What is contribution of air exchange for the conditions of Example 7.10?

Given: 1.0 air change per hour, corresponding to $\dot{V} = 53.3$ L/s

Lookup values: $T_o = 33.8°C$ at 17:00, from the appendices on the CD-ROM.

SOLUTION

Filling out the worksheet of Fig. 7.16, we obtain

Sensible loads				hour t 16:00	hour t 17:00	hour t 18:00	hour t 16:00	hour t 17:00	hour t 18:00
Air exchange	V [m^3]	$\dot{V}$ [1/s]	T_i [°C]		T_o [°C]			$\dot{Q} = 1.2W/(K \cdot 1/s) \times \dot{V} \times (T_o - T_i)$ [W]	
	192	53.33	25.5		33.8			531.2	

Equations (7.38) to (7.46), together with the corresponding tables, as summarized in Fig. 7.16, are what is known as the *ASHRAE CLTD/CLF method for cooling load calculations*. Note the assumptions that have been made. In particular, the *indoor temperature T_i is assumed to be constant*. Implicit in all the tables is the assumption of *periodic conditions*, corresponding to a series of identical design days. Such limitations are the price of simplicity. If one wants to analyze features such as variable occupancy (weekday or weekend) or thermostat setup, one must resort to a dynamic analysis.

Since thermostat setup and reduced weekend heat gains are frequently encountered in commercial buildings, one may wonder about the applicability of the CLF/CLTD method for such cases. If the building is kept at constant T_i, the method can indeed be used (for identical weather there is a relatively small increase in cooling load from Monday to Friday, and since it is preceded by four days with identical conditions, the prediction of the peak is reliable). Thermostat setup, on the other hand, necessitates a dynamic analysis, as presented in Sec. 7.7.

Example 7.15

What is the peak sensible cooling load for a room of an office building for the following conditions?

Given: Most of these elements have already been presented in Examples 7.10 to 7.14.

Floor area, 8 m × 8 m; height, 3 m

Curtain wall of construction type G on west and south facades

Roof type 4 with 2-in insulation $U = 0.693$ W/(m^3 · K), without suspended ceiling

Window of area 6 m^2 on west side

No heat transfer across other surfaces

Air exchange rate 1.0 h^{-1}

Lights and occupants from 8:00 to 18:00

Average density of lights 25 W/m^2 of floor area

5 occupants seated

Design conditions of CLTD/CLF tables

SOLUTION

Most of the components of the cooling load have already been calculated in earlier examples. The complete calculation is listed in this worksheet. In view of the west-facing facade and its window, the peak load is likely to occur late in the afternoon. The exact time will be found by considering several hours.

Sensible loads					hour t			hour t		
					16:00	17:00	18:00	16:00	17:00	18:00
Component and orientation	Construction type		U [W/m^2 ·K]	A [m^2]		CLTD$_t$ [K]			$\dot{Q}_t = U \times A \times CLTD_t$ [W]	
Walls										
South	typeG		0.52	24	21	17	14	262.1	212.2	174.7
West	typeG		0.52	18	37	40	37	346.3	374.4	346.3
Roof	type4		0.693	64	36	37	37	1596.7	1641.0	1641.0
Glazing conduction										
West			3.00	6.00	8	7	7	144.0	126.0	126.0
Glazing solar		A [m^2]	SC	SHGF$_{max}$ [W/m^2]		CLF$_t$			$\dot{Q}_t = A \times SC \times SHGF_{max} \times CLF_t$ [W]	
West		6.00	0.80	681	0.57	0.64	0.61	1863.2	2092.0	1994.0
Air exchange		V [m^3]	$\dot{V}$ [1/s]	T_i [°C]		T_o [°C]			$\dot{Q} = 1.2 W/(K \cdot 1/s) \times \dot{V} \times (T_o - T_i)$ [W]	
		192	53.33	25.5	34.4	33.8	32.7	569.6	531.2	460.8
Internal gains		number	gain /unit	$\dot{Q}$ [W]		CLF$_t$			$\dot{Q}_t = \dot{Q} \times CLF_t$ [W]	
Lights		64 m^2 8:00 to 18:00	25 W/m^2	1600	0.83	0.84	0.85	1328.0	1344.0	1360.0
People		5 8:00 to 18:00	70 W	350	0.85	0.87	0.89	297.5	304.5	311.5
TOTAL SENSIBLE								6407	6625	6414

COMMENTS

1. The peak is 6625 W (22.6 kBtu/h) at 17:00, for a floor area of 64 m² (710 ft²).
2. The design of this room is anything but optimized. A reduction of the peak load would bring a double benefit, by reducing both the cost of the cooling equipment and the energy cost. The largest contribution comes from the radiative gain through the window; this can be reduced (easily to less than one-half) by selecting glass with a lower shading coefficient and/or by using shades. The second largest term is the roof; 2 in of additional insulation would reduce it to about one-half. Lights make up the third largest term; here one should consider more efficient luminaires and/or dimming in response to availability of daylight (see Chap. 13).

Example 7.16

What is the latent load for the room of Examples 7.10 to 7.15?

Given: Absolute humidities

$W_i = 0.0104$ indoor

$W_o = 0.0159$ outdoor

1.0 air change per hour, corresponding to $\dot{V} = 53.33$ L/s

5 occupants with 30-W latent load each

SOLUTION
Filling out the worksheet of Fig. 7.16, we obtain

Latent loads					Instantaneous
	$\dot{V}$ [1/s]	W_o	W_i	$\Delta W = W_o - W_i$	$\dot{Q}_{lat} = 3010\ \mathrm{W}/(1/s) \times \dot{V} \times \Delta W$ [W]
Air exchange	53.33	0.0159	0.0104	0.0055	883
	N = number			$\dot{Q}_{lat/unit}$	$\dot{Q}_{lat} = N \times \dot{Q}_{lat/unit}$
People	5			30	150
TOTAL LATENT					1033

COMMENTS
The total cooling load is the sum of the sensible and latent loads:

$$6625 + 1033 = 7658\ \mathrm{W}$$

The sensible portion is relatively large in this case (but should be reduced by better design, as explained above).

7.7 Transfer Functions for Dynamic Load Calculations

7.7.1 Basis of the Method

The load $\dot{Q}$ can be considered the *response* of the building or room to the *driving terms* ($T_i, T_o, \dot{Q}_{sol}$, etc.) that act on it. The transfer function method calculates the response of a system by making the following basic assumptions:

- *Discrete time steps* (all functions of time are represented as series of values at regular time steps, *hourly* in the present case)
- *Linearity* (the response of a system is a linear function of the driving terms and of the state of the system)
- *Causality* (the response at time t can depend only on the past, not on the future)

As an example, suppose there is a single driving term $u(t)$ and the response is $y(t)$. To make the expressions more readable, let us indicate the time dependence as a subscript, in the form $y(t) = y_t$, $u(t) = u_t$, and so on. Then according to the transfer function model, the relation between the response and the driving term is of the form

$$y_t = -(a_1 y_{t-1\Delta t} + a_2 y_{t-2\Delta t} + \ldots + a_n y_{t-n\Delta t})$$
$$+ (b_0 u_t + b_1 u_{t-1\Delta t} + b_2 u_{t-2\Delta t} + \cdots + b_m u_{t-m\Delta t}) \tag{7.47}$$

with time step

$$\Delta t = 1 \text{ h} \tag{7.48}$$

where a_1 to a_n and b_0 to b_m are coefficients that characterize the system; they are independent of the driving term or response. Eq. (7.47) is obviously linear. It satisfies causality because y_t depends only on the past values of the response ($y_{t-1\Delta t}$ to $y_{t-n\Delta t}$) and on present and past values of the driving terms[9] (u_t to $u_{t-m\Delta t}$).

The past state of the system enters because of the coefficients a_1 to a_n and b_1 to b_m; this is how thermal inertia is taken into account. The response is instantaneous only if these coefficients are zero. The greater their number and magnitude, the greater the weight of the past.

The accuracy of the model increases as the number of coefficients is enlarged and as the time step is reduced. For load calculations, hourly time resolution and a handful of coefficients per driving term will suffice. The coefficients are called *transfer function coefficients*.

Incidentally, the relation between u and y could be written in symmetric form

$$a_0 y_t + a_1 y_{t-1\Delta t} + \cdots + a_n y_{t-n\Delta t} = b_0 u_t + b_1 u_{t-1\Delta t} + \cdots + b_m u_{t-m\Delta t} \tag{7.49}$$

which is equivalent because one can divide both sides of the equation by a_0. Since the roles of u and y are symmetric, one can use the same model to find, e.g., the load (i.e., the heat $\dot{Q}$ to be supplied or removed) as a function of T_i, or T_i as a function of $\dot{Q}$.

Equation (7.49) can be readily generalized to the case where there are several driving terms. For instance, if the response T_i is determined by two driving terms, heat input $\dot{Q}$, and outdoor temperature T_o, then one can write the transfer function model in the form

$$a_{i,0} T_{i,t} + a_{i,1} T_{i,t-1\Delta t} + \cdots + a_{i,n} T_{i,t-n\Delta t}$$
$$= a_{o,0} T_{o,t} + a_{o,1} T_{o,t-1\Delta t} + \cdots + a_{o,m} T_{o,t-m\Delta t}$$
$$+ a_{Q,0} \dot{Q}_t + a_{Q,1} \dot{Q}_{t-1\Delta t} + a_{Q,2} \dot{Q}_{t-2\Delta t} + \cdots + a_{Q,r} \dot{Q}_{t-r\Delta t} \tag{7.50}$$

with the three sets of transfer function coefficients $a_{i,0}$ to $a_{i,n}$, $a_{o,0}$ to $a_{o,m}$, and $a_{Q,0}$ to $a_{Q,r}$. This equation can be considered an algorithm for calculating $T_{i,t}$, hour by hour, given the previous values of T_i and the driving terms T_o and $\dot{Q}$. Likewise, if T_i and T_o were given as driving terms, one could calculate $\dot{Q}$ as response.

---

[9] A series such as Eq. (7.47) is also known as a *time series*.

Any set of response and driving terms can be handled in this manner. Thus loads can be calculated hour by hour, for any driving terms (meteorological data, building occupancy, heat gain schedules, etc.), and it is, in fact, the method used by the computer simulation program DOE2.1 (Birdsall et al., 1990). In addition to the loads, DOE2.1 simulates the HVAC equipment and calculates operating costs.

Once the necessary numerical values of the transfer function coefficients have been obtained, the calculation of peak loads is simple enough for a spreadsheet. One specifies the driving terms for the peak day and iterates an equation like Eq. (7.50) until the result converges to a steady daily pattern. Transfer function coefficients have been calculated and listed for a wide variety of standard construction types (ASHRAE, 1989a), and some excerpts will be presented here. They are also included in the HCB software supplied with this book [some of the numbers may be slightly different because they are taken from a more recent computerized version (PREP, 1990)].

In the remainder of this section, we discuss the ASHRAE transfer function method in detail; it is also included in the software. The method involves three steps:

1. Calculation of the conductive heat gain (or loss) for each distinct component of the envelope, by Eq. (7.51)
2. Calculation of the load of the room at constant temperature, based on this conductive heat gain (or loss) as well as any other heat source in the room, by Eq. (7.56)
3. Calculation of the heat extraction (or addition) rate for the cooling (or heating) device and thermostat setpoints of the room, by Eq. (7.61)

7.7.2 Conductive Heat Gain

The conductive heat gain (or loss) $\dot{Q}_{cond,t}$ at time t through the roof and walls is calculated according to the formula

$$\dot{Q}_{cond,t} = -\sum_{n\geq1} d_n \dot{Q}_{cond,\,t-n\Delta t} + A\left(\sum_{n\geq0} b_n T_{os,\,t-n\Delta t} - T_i \sum_{n\geq0} c_n\right) \tag{7.51}$$

where

$\quad A$ = area of roof or wall, m^2 (ft^2)
$\quad \Delta t$ = time step = 1 h
$\quad T_{os,t}$ = sol-air temperature of outside surface at time t [defined in Eq. (6.39) with data in Fig. 6.17 and the appendices on the CD-ROM]
b_n, c_n, d_n = coefficients of conduction transfer function

The indoor temperature T_i is multiplied by the sum of the c_n values, so the individual c_n coefficients are not needed (this is because T_i is assumed constant at this point; the extension to arbitrary T_i comes in Sec. 7.7.4). In general, the initial value $\dot{Q}_{cond,t}=0$ is not known; its value does not matter if the calculation is repeated over a sufficient number of time steps until the resulting pattern becomes periodic within the desired accuracy. Usually a few days to a week will be sufficient.

Numerical values of the coefficients of the conduction transfer function are listed in Table 7.7: roofs in Table 7.7a and walls in Table 7.7b. If the room in question is adjacent to rooms at a different temperature, the heat gain across the partitions is also calculated according to Eq. (7.51).

TABLE 7.7

Coefficients of Conduction Transfer Function*

(Layer Sequence Left to Right = Inside to Outside)		$n=0$	$n=1$	$n=2$	$n=3$	$n=4$	$n=5$	$n=6$	Σc_n	U	δ	λ
a. Roofs												
Layers E0 A3 B25 E3 E2 A0	b_n	0.000487	0.03474	0.01365	0.00036	0.00000	0.00000	0.00000	0.05362	0.080	1.63	0.97
Steel deck with 3.33-in insulation	d_n	1.00000	−0.35451	0.02267	−0.00005	0.00000	0.00000	0.00000				
Layers E0 A3 B14 E3 E2 A0	b_n	0.00056	0.01202	0.01282	0.00143	0.00001	0.00000	0.00000	0.02684	0.055	2.43	0.94
Steel deck with 5-in insulation	d_n	1.00000	−0.60064	0.08602	−0.00135	0.00000	0.00000	0.00000				
Layers E0 E1 B15 E4 B7 A0	b_n	0.00000	0.00065	0.00339	0.00240	0.00029	0.00000	0.00000	0.00673	0.043	4.85	0.82
Attic roof with 6-in insulation	d_n	1.00000	−1.34658	0.59384	−0.09295	0.00296	−0.00001	0.00000				
Layers E0 B22 C12 E3 E2 C12 A0	b_n	0.00059	0.00867	0.00688	0.00037	0.00000	0.00000	0.00000	0.01652	0.138	5.00	0.56
1.67-in insulation with 2-in h.w. concrete RTS	d_n	1.00000	−1.11766	0.23731	−0.00008	0.00000	0.00000	0.00000				
Layers E0 E5 E4 B12 C14 E3 E2 A0	b_n	0.00000	0.00024	0.00217	0.00251	0.00055	0.00002	0.00000	0.00550	0.057	6.32	0.60
3-in insulation w/4-in l.w. conc. deck and susp. clg.	d_n	1.00000	−1.40605	0.58814	−0.09034	0.00444	−0.00006	0.00000				
Layers E0 E5 E4 C5 B6 E3 E2 A0	b_n	0.00001	0.00066	0.00163	0.00049	0.00002	0.00000	0.00000	0.01477	0.090	7.16	0.16
1-in insul. w/4-in h.w. conc. deck and susp. clg.	d_n	1.00000	−1.24348	0.28742	−0.01274	0.00009	0.00000	0.00000				
Layers E0 E5 E4 C13 B20 E3 E2 A0	b_n	0.00001	0.00060	0.00197	0.00086	0.00005	0.00000	0.00000	0.00349	0.140	7.54	0.15
6-in h.w. deck w/0.76-in insul. and susp. clg.	d_n	1.00000	−1.39181	0.46337	−0.04714	0.00058	0.00000	0.00000				
Layers E0 E5 E4 B15 C15 E3 E2 A0	b_n	0.00000	0.00000	0.00002	0.00014	0.00024	0.00011	0.00002	0.00053	0.034	10.44	0.30
6-in insul. w/6-in l.w. conc. deck and susp. clg.	d_n	1.00000	−2.29459	1.93694	−0.75741	0.14252	−0.01251	0.00046				
Layers E0 C13 B15 E3 E2 C12 A0	b_n	0.00000	0.00000	0.00007	0.00024	0.00016	0.00003	0.00000	0.00050	0.045	10.48	0.24
6-in h.w. deck w/6-in ins. and 2-in h.w. RTS	d_n	1.00000	−2.27813	1.82162	−0.60696	0.07696	−0.00246	0.00001				
b. Walls												
Layers E0 A3 B1 B13 A3 A0	b_n	0.00768	0.03498	0.00719	0.00006	0.00000	0.00000	0.00000	0.04990	0.066	1.30	0.98
Steel siding with 4-in insulation	d_n	1.00000	−0.24072	0.00168	0.00000	0.00000	0.00000	0.00000				
Layers E0 E1 B14 A1 A0 A0	b_n	0.00016	0.00545	0.00961	0.00215	0.00005	0.00000	0.00000	0.01743	0.055	3.21	0.91
Frame wall with 5-in insulation	d_n	1.00000	−0.93389	0.27396	−0.02561	0.00014	0.00000	0.00000				
Layers E0 C3 B5 A6 A0 A0	b_n	0.00411	0.03230	0.00474	0.00047	0.00000	0.00000	0.00000	0.05162	0.191	3.33	0.78
4-in h.w. concrete block with 1-in insulation	d_n	1.00000	−0.76963	0.04014	−0.00042	0.00000	0.00000	0.00000				

(continued)

TABLE 7.7 (continued)

Coefficients of Conduction Transfer Function*

(Layer Sequence Left to Right = Inside to Outside)		$n=0$	$n=1$	$n=2$	$n=3$	$n=4$	$n=5$	$n=6$	Σc_n	U	δ	λ
Layers E0 A6 C5 B3 A3 A0	b_n	0.00099	0.00836	0.00361	0.00007	0.00000	0.00000	0.00000	0.01303	0.122	5.14	0.41
4-in h.w. concrete with 2-in insulation	d_n	1.00000	−0.93970	0.04664	0.00000	0.00000	0.00000	0.00000				
Layers E0 E1 C8 B6 A1 A0	b_n	0.00000	0.00061	0.00289	0.00183	0.00018	0.00000	0.00000	0.00552	0.109	7.11	0.37
8-in h.w. concrete block with 2-in insulation	d_n	1.00000	−1.52480	0.67146	−0.09844	0.00239	0.00000	0.00000				
Layers E0 A2 C2 B15 A0 A0	b_n	0.00000	0.00000	0.00013	0.00044	0.00030	0.00005	0.00000	0.00093	0.043	9.36	0.30
Face brick and 4-in l.w. conc. block with 6-in insulation	d_n	1.00000	−2.00875	1.37120	−0.37897	0.03962	−0.00165	0.00002				
Layers E0 C9 B6 A6 A0 A0	b_n	0.00000	0.00005	0.00064	0.00099	0.00030	0.00002	0.00002	0.00200	0.106	8.97	0.20
8-in common brick with 2-in insulation	d_n	1.00000	−1.78165	0.96017	−0.16904	0.00958	−0.00016	0.00000				
Layers E0 C11 B6 A1 A0 A0	b_n	0.00000	0.00001	0.00019	0.00045	0.00022	0.00002	0.00000	0.00089	0.112	10.20	0.13
12-in h.w. concrete with 2-in insulation	d_n	1.00000	−2.12812	1.53974	−0.45512	0.05298	−0.00158	0.00002				

Source: Courtesy of ASHRAE, *Handbook of Fundamentals*, American Society of Heating, Refrigerating and Air-Conditioning Engineers, Atlanta, GA, 1989a. With permission.

* Also shown are the decrement factor λ and time lag δ for periodic heat flow (see Sec. 8.3.3); U, b_n and c_n are in Btu/(h · ft^2 · °F); d_n and λ are dimensionless; and δ is in hours [1 Btu/(h · ft^2 · °F) = 5.678 W/(m^2 · K)]. For definition of layer codes and thermal properties, see the tables on the CD-ROM.

It is instructive to establish the connection of the transfer function coefficients with the U value. In the steady-state limit, i.e., when $\dot{Q}_{cond}$, T_{os} and T_i are all constant, Eq. (7.51) becomes

$$\dot{Q}_{cond} \sum_{n\geq0} d_n = A\left(T_{os}\sum_{n\geq0} b_n - T_i\sum_{n\geq0} c_n\right) \quad \text{where } d_0 = 1 \tag{7.52}$$

Since in that limit we also have

$$\dot{Q}_{cond} = AU(T_{os} - T_i) \tag{7.53}$$

the coefficients of T_{os} and T_i must be equal

$$\sum_{n\geq0} b_n = \sum_{n\geq0} c_n \tag{7.54}$$

and the U value is given by

$$U = \frac{\displaystyle\sum_{n\geq0} c_n}{\displaystyle\sum_{n\geq0} d_n} \tag{7.55}$$

Example 7.17

Calculate the conductive heat gain per square meter of an exterior vertical wall, of dark color and facing west, for summer design conditions (July 21) if $T_i = 25°C$.

Given: Exterior wall of 0.10-m (4-in) concrete with 0.05-m (2-in) insulation on the outside[10]

$T_i = 25°C$

Sol-air temperature of Fig. 6.17 or the appendices on the CD-ROM

Find: $\dot{Q}_{cond}$

Lookup values: The transfer function coefficients for walls and roofs are listed in Table 7.7. But, as explained in the footnote, we have used ASHRAE (1981) where the following values were given:

$$
\begin{aligned}
b_0 &= 0.00312 & \sum c_n &= 0.0734 & d_0 &= 1.0000 \\
b_1 &= 0.04173 & & & d_1 &= -0.94420 \\
b_2 &= 0.02736 & & & d_2 &= 0.05025 \\
b_3 &= 0.00119 & & & d_3 &= -0.00008
\end{aligned}
$$

The d's are dimensionless, and the b's and c's are in $W/(m^2 \cdot K)$. All other coefficients are zero, and the U value is 0.693 $W/(m^2 \cdot K)$ [as the reader can verify by inserting the coefficients into Eq. (7.55)].

---

[10] This example had been calculated with values from the 1981 edition of the ASHRAE *Handbook of Fundamentals*. The precise wall (type 32 in that edition) is no longer listed in the newer editions, but it is close to the fourth wall in Table 7.7b. Since the educational content of the example is not affected, we have left the numbers as they were, in all the examples of this section.

SOLUTION

Arranging a spreadsheet with three long columns (one for t, one for $T_{os,t}$, and one for $\dot{Q}_{cond,t}$, per unit wall area A, according to Eq. (7.51), and taking $\dot{Q}_{cond,t} = 0$ for $t < 0$), we obtained the results in the following table (printed here side by side for successive days). For the first $\dot{Q}_{cond,t}/A$ entry at $t = 1$, the detail of the calculation is

$$\frac{\dot{Q}_{cond,1}}{A} = -\frac{d_1 \dot{Q}_{cond,1-1}}{A} - \frac{d_2 \dot{Q}_{cond,1-2}}{A} - \frac{d_3 \dot{Q}_{cond,1-3}}{A}$$
$$+ b_0 T_{os,1-0} + b_1 T_{os,1-1} + b_2 T_{os,1-2} + b_3 T_{os,1-3} - T_i \sum_{n \geq 0} c_n$$
$$= -(-0.94420)(0.00) - 0.05025(0.00) - (-0.00008)(0.00)$$
$$+ 0.00312(24.4) + 0.04173(25.0) + 0.02736(26.1)$$
$$+ 0.00119(27.2) - 25(0.0734)$$
$$= 0.03$$

t	$T_{os,t}$	$\dfrac{\dot{Q}_{cond,t}}{A}$	$\dfrac{\dot{Q}_{cond,t+24}}{A}$	$\dfrac{\dot{Q}_{cond,t+48}}{A}$	$\dfrac{\dot{Q}_{cond,t+72}}{A}$
h	°C			W/m²	
−2	27.2	0.00			
−1	26.1	0.00			
0	25.0	0.00			
1	24.4	0.03	9.29	9.82	9.85
2	24.4	0.00	8.22	8.69	8.72
3	23.8	−0.04	7.25	7.67	7.70
4	23.3	−0.11	6.36	6.73	6.76
5	23.3	−0.21	5.53	5.86	5.88
6	25.0	−0.32	4.79	5.08	5.10
7	27.7	−0.33	4.20	4.46	4.48
8	30.0	−0.17	3.85	4.08	4.10
9	32.7	0.16	3.74	3.94	3.95
10	35.0	0.66	3.83	4.01	4.02
11	37.7	1.29	4.10	4.26	4.27
12	40.0	2.04	4.54	4.68	4.69
13	53.3	2.94	5.15	5.28	5.29
14	64.4	4.40	6.37	6.48	6.49
15	72.7	6.59	8.34	8.44	8.44
16	75.5	9.26	10.81	10.90	10.91
17	72.2	12.02	13.40	13.48	13.48
18	58.8	14.40	15.62	15.69	15.69
19	30.5	15.77	16.86	16.92	16.92
20	29.4	15.39	16.36	16.41	16.41
21	28.3	14.13	14.98	15.03	15.03
22	27.2	12.84	13.60	13.64	13.64
23	26.1	11.60	12.28	12.32	12.32
24	25.0	10.42	11.02	11.05	11.06
Average	37.95				8.97

At the start of the fourth day, the pattern has stabilized within 0.5 percent of the true value, certainly good enough for this purpose.

COMMENTS

The conductive gain reaches its maximum at 19:00, which is 3 h after the peak of the sol-air temperature.

As a check, it is strongly recommended that you always verify that the solution satisfies the *steady-state limit*. Multiplying the average temperature difference $T_{os} - T_i = (37.95 - 25)$ K by $U = 0.693$ W/(m$^2 \cdot$K), we find 8.97 W/m^2 while the 24-h average of $\dot{Q}_{cond}$ is 8.97 W/m^2; they are indeed equal within the accuracy of the calculation.

7.7.3 The Load at Constant Temperature

The above calculation of the conductive heat gain (or loss) is to be repeated for each portion of the room envelope that has a distinct composition. The relation between these conductive gains and the total load depends on the construction of the entire room. For example, a concrete floor can store a significant fraction of the heat radiated by lights or by a warm ceiling, thus postponing its contribution to the cooling load of the room.

For each heat gain component $\dot{Q}_{gain}$, the corresponding cooling load $\dot{Q}_c$ (or reduction of the heating load) at constant T_i is calculated by using another set of coefficients, the coefficients v_n and w_n, of the *room transfer function*

$$
\begin{aligned}
\dot{Q}_{c,t} = v_0 \dot{Q}_{gain,t} + v_1 \dot{Q}_{gain,t-\Delta t} + v_2 \dot{Q}_{gain,t-2\Delta t} + \cdots \\
- w_1 \dot{Q}_{c,t-\Delta t} - w_2 \dot{Q}_{c,t-2\Delta t} - \cdots
\end{aligned}
\tag{7.56}
$$

with the subscript t indicating time, as before. The coefficient w_0 of $\dot{Q}_{c,t}$ is not shown because it is set equal to unity. The coefficients for a variety of room construction types are listed in Tables 7.8 and 7.9. In these tables, all coefficients with index 2 or higher are zero. Since w_0 is unity, Table 7.8 shows only a single coefficient w_1. Again, it is instructive to take the steady-state limit and check the consistency with the first law of thermodynamics. It requires that the sum of the v_n values equal the sum of the w_n values:

$$
\sum_{n \geq 0} v_n = \sum_{n \geq 0} w_n \quad \text{where } w_0 = 1
\tag{7.57}
$$

The entries of Tables 7.8 and 7.9 do indeed satisfy this condition.

Therefore, Eq. (7.56) has to be applied separately to each of the heat gain types in Table 7.9, and the resulting cooling load components $\dot{Q}_{c,t}$ are added to obtain the total cooling load of the room at time t. The heat gain types are as follows:

- Solar gain (through glass without interior shade) and the radiative component of heat from occupants and equipment
- Conduction through envelope and solar radiation absorbed by interior shade
- Lights
- Convective gains (from air exchange, occupants, equipment)

For lights the coefficients depend on the arrangement of the lighting fixture and the ventilation system.

TABLE 7.8

The w_1 Coefficient of the Room Transfer Function ($w_0 = 1.0$ and Higher Terms are Zero)

	Room Envelope Construction[†]				
	2-in (51-mm) Wood Floor	3-in (76-mm) Concrete Floor	6-in (152-mm) Concrete Floor	8-in (203-mm) Concrete Floor	12-in (305-mm) Concrete Floor
Room Air* Circulation and S/R Type	Specific Mass per Unit Floor Area, lb/ft²				
	10	40	75	120	160
Low	−0.88	−0.92	−0.95	−0.97	−0.98
Medium	−0.84	−0.90	−0.94	−0.96	−0.97
High	−0.81	−0.88	−0.93	−0.95	−0.97
Very high	−0.77	−0.85	−0.92	−0.95	−0.97
	−0.73	−0.83	−0.91	−0.94	−0.96

Source: Courtesy of ASHRAE, *Handbook of Fundamentals*, American Society of Heating, Refrigerating and Air-Conditioning Engineers, Atlanta, GA, 1989a. With permission.

* Circulation rate:

Low: Minimum required to cope with cooling load from lights and occupants in interior zone. Supply through floor, wall, or ceiling diffuser. Ceiling space not used for return air, and $h = 0.4$ Btu/(h · ft² · °F) [2.27 W/(m² · K)], where $h =$ inside surface convection coefficient used in calculation of w_1 value.

Medium: Supplied through floor, wall, or ceiling diffuser. Ceiling space not used for return air, and $h = 0.6$ Btu/(h · ft² · °F) [3.41 W/(m² · K)].

High: Room air circulation induced by primary air of induction unit, or by room fan and coil unit. Ceiling space used for return air, and $h = 0.8$ Btu/(h · ft² · °F) [4.54 W/(m² · K)].

Very high: Used to minimize temperature gradients in a room. Ceiling space used for return air, and $h = 1.2$ Btu/(h · ft² · °F) [6.81 W/(m² · K)].

† Floor covered with carpet and rubber pad; for a bare floor or if covered with floor tile, take next w_1 value down the column.

Example 7.18

Find the cooling load for a room that has only three types of heat gain:

(a) Conduction through 10 m² of the exterior wall of Example 7.17
(b) 300 W of electric lights that are turned on from 9:00 to 18:00
(c) Air exchange (ventilation plus infiltration) at a constant rate of 5.208 L/s (corresponds to 0.5 air change per hour for a volume of 15 m² × 2.5 m)

(This could be a windowless office area surrounded on all sides but one by rooms at the same temperature; in practice, the ventilation may be turned off when the office is unoccupied, but here we assume constant air exchange for simplicity.) In addition to the conditions of Example 7.17, assume the following: envelope construction type of medium, 150-mm (6-in) concrete floor, room air circulation type medium, ordinary furniture, and vented light fixtures.

Given: $A = 10$ m² exterior wall, with $\dot{Q}_{cond}/A$ from Example 7.17, 300-W lights from 9:00 to 18:00, air exchange at a constant rate of $\dot{V} = 5.208$ L/s, hence

$$\dot{V}\rho c_p = 6.25 \text{ W/K}$$

envelope construction type medium, 150-mm (6-in) concrete floor, room air circulation type medium, ordinary furniture, and vented light fixtures.

Find: $\dot{Q}_c$ for $T_i = 25°C$

TABLE 7.9

The v Coefficients of the Room Transfer Function* (Only v_0 and v_1 are Nonzero)

Heat Gain Component	Room Envelope Construction[†]	Dimensionless	
		v_0	v_1
Solar heat gain through glass[‡] with	Light	0.224	$1 + w_1 - v_0$
no interior shade; radiant heat from	Medium	0.197	$1 + w_1 - v_0$
equipment and people	Heavy	0.187	$1 + w_1 - v_0$
Conduction heat gain through exterior	Light	0.703	$1 + w_1 - v_0$
walls, roofs, partitions, doors,	Medium	0.681	$1 + w_1 - v_0$
windows with blinds, or drapes	Heavy	0.676	$1 + w_1 - v_0$
Convective heat generated by equipment	Light	1.000	0.0
and people, and from ventilation	Medium	1.000	0.0
and infiltration air	Heavy	1.000	0.0

Heat Gain From Lights[§]

Furnishings	Air Supply and Return	Type of Light Fixture	v_0	v_1
Heavyweight simple furnishings, no carpet	Low rate; supply and return below ceiling $(V \le 0.5)$[¶]	Recessed, not vented	0.450	$1 + w_1 - v_0$
Ordinary furnishings, no carpet	Medium to high rate, supply and return below or ceiling $(V \ge 0.5)$	Recessed not vented	0.550	$1 + w_1 - v_0$
Ordinary furnishings, with or without carpet on floor	Medium to high rate, or induction unit or fan and coil, supply and return below, or through ceiling, return air plenum $(V \ge 0.5)$	Vented	0.650	$1 + w_1 - v_0$
Any type of furniture, with or without carpet	Ducted returns through light fixtures	Vented or free-hanging in air-stream with ducted returns	0.750	$1 + w_1 - v_0$

Source: Courtesy of ASHRAE, *Handbook of Fundamentals*, American Society of Heating, Refrigerating and Air-Conditioning Engineers, Atlanta, GA, 1989a. With permission.

* The transfer functions in this table were calculated by procedures outlined in Mitalas and Stephenson (1967) and are acceptable for cases where all heat gain energy eventually appears as cooling load. The computer program used was developed at the National Research Council of Canada, Division of Building Research.

[†] The construction designations denote the following:
Light construction: such as frame exterior wall, 2-in (51-mm) concrete floor slab, approximately 30 lb of material/ft^2 (146 kg/m^2) of floor area.
Medium construction: such as 4-in (102-mm) concrete exterior wall, 4-in (102-mm) concrete floor slab, approximately 70 lb of building material/ft^2 (341 kg/m^2) of floor area.
Heavy construction: such as 6-in (152-mm) concrete exterior wall, 6-in (152-mm) concrete floor slab, approximately 130 lb of building material/ft^2 (635 kg/m^2) of floor area.

[‡] The coefficients of the transfer function that relate room cooling load to solar heat gain through glass depend on where the solar energy is absorbed. If the window is shaded by an inside blind or curtain, most of the solar energy is absorbed by the shade and is transferred to the room by convection and long-wave radiation in about the same proportion as the heat gain through walls and roofs; thus the same transfer coefficients apply.

[§] If room supply air is exhausted through the space above the ceiling and lights are recessed, such air removes some heat from the lights that would otherwise have entered the room. This removed light heat is still a load on the cooling plant if the air is recirculated, even though it is not a part of the room heat gain as such. The percent of heat gain appearing in the room depends on the type of lighting fixture, its mounting, and the exhaust airflow.

[¶] V is room air supply rate in (ft^3/min)/ft^2 of floor area

Lookup values: From Table 7.8 we find $w_1 = -0.94$, the same for all heat gain types. Table 7.9 yields (with $a = 0.65$ from the appendices on the CD-ROM) $v_0 = 0.681$, $v_1 = 1 + w_1 - v_0 = -0.621$, and $v_2 = 0$ for the conductive component, and $v_0 = 0$, $v_1 = 0.65$, and $v_2 = 1 + w_1 - v_1$ for the lights. (These are the values from the 1981 edition. Since 1989 there has been a shift in the v coefficients for lights by one time steprelative to the 1981 edition of the ASHRAE *Handbook of Fundamentals*; the newer values would read $v_0 = 0.65$, $v_1 = 1 + w_1 - v_0$, and $v_2 = 0$.)

SOLUTION

We have set up a spreadsheet with two sets of columns, one for conduction and one for lights; for each of these, the cooling load is programmed according to Eq. (7.56) with the appropriate coefficients. The heat gain $\dot{Q}_{cond}$ is the result of Example 7.17, multiplied by $A = 10$ m². Not knowing the initial conditions, we repeat the calculation until the convergence is acceptable after 4 days, printed here side by side, (a) for conduction, (b) for lights, and (c) for ventilation. In part c we also show the total cooling load in the last column.

(a) *Conductive* component $\dot{Q}_{c,t}$ of cooling load at time t. The second column is $A = 10$ m² times the last ($\dot{Q}_{cond}/A$) column of Example 7.17. For the first $\dot{Q}_{c,t}$ entry at $t = 1$, the detail of the calculation is

$$\dot{Q}_{c,1} = v_0 \dot{Q}_{cond,1} + v_1 \dot{Q}_{cond,1-1} - W_1 \dot{Q}_{c,1-1}$$
$$= 0.681(98.5) + [1 + (-0.94) - 0.681](110.6) - (-0.94)(0.0)$$
$$= -1.6$$

t	$\dot{Q}_{cond}$	$\dot{Q}_{c,t}$	$\dot{Q}_{c,t+24}$	$\dot{Q}_{c,t+48}$	$\dot{Q}_{c,t+72}$
h	W	W	W	W	W
−2	136.4	0.0			
−1	123.2	0.0			
0	110.6	0.0			
1	98.5	−1.6	76.8	94.6	98.6
2	87.2	−3.3	70.5	87.2	90.9
3	77.0	−4.8	64.5	80.2	83.7
4	67.6	−6.3	58.8	73.6	76.9
5	58.8	−7.8	53.4	67.3	70.4
6	51.0	−9.2	48.4	61.4	64.4
7	44.8	−9.8	44.3	56.6	59.3
8	41.0	−9.1	41.8	53.3	55.9
9	39.5	−7.1	40.7	51.5	54.0
10	40.2	−3.8	41.1	51.3	53.6
11	42.7	0.5	42.7	52.3	54.5
12	46.9	5.9	45.6	54.6	56.6
13	52.9	12.4	49.8	58.2	60.1
14	64.9	23.0	58.1	66.1	67.9
15	84.4	38.9	71.8	79.3	81.0
16	109.1	58.4	89.4	96.4	98.0
17	134.8	78.9	108.1	114.7	116.2
18	156.9	97.4	124.8	131.0	132.4
19	169.2	109.3	135.1	140.9	142.2
20	164.1	109.4	133.6	139.1	140.4
21	150.3	103.3	126.1	131.2	132.4
22	136.4	96.7	118.1	122.9	124.0
23	123.2	90.0	110.1	114.7	115.7
24	110.6	83.4	102.3	106.6	107.6
Average	89.66				89.03

(b) Cooling load $\dot{Q}_{c,t}$ due to heat gain $\dot{Q}_{\text{lights}}$ from *lights*. For the first $\dot{Q}_{c,t}$ entry at $t = 10$, the detail of the calculation is

$$\dot{Q}_{c,10} = v_0 \dot{Q}_{\text{lights},10} + v_1 \dot{Q}_{\text{lights},10-1} + v_2 \dot{Q}_{\text{lights},10-2} - w_1 \dot{Q}_{c,10-1}$$
$$= 0(300) + 0.65(300) + [1 + (-0.94) - 0.65](0) - (-0.94)(0.0)$$
$$= 195$$

t	$\dot{Q}_{\text{lights}}$	$\dot{Q}_{c,t}$	$\dot{Q}_{c,t+24}$	$\dot{Q}_{c,t+48}$	$\dot{Q}_{c,t+72}$
h	W	W	W	W	W
−2	0	0			
−1	0	0			
0	0	0			
1	0	0	30.9	37.9	39.5
2	0	0	29.1	35.7	37.2
3	0	0	27.3	33.5	34.9
4	0	0	25.7	31.5	32.8
5	0	0	24.1	29.6	30.9
6	0	0	22.7	27.8	29.0
7	0	0	21.3	26.2	27.3
8	0	0	20.1	24.6	25.6
9	300	0	18.9	23.1	24.1
10	300	195	212.7	216.7	217.6
11	300	201	218.0	221.7	222.6
12	300	207	222.9	226.4	227.2
13	300	213	227.5	230.8	231.6
14	300	218	231.9	235.0	235.7
15	300	223	235.9	238.9	239.6
16	300	228	239.8	242.6	243.2
17	300	232	243.4	246.0	246.6
18	0	236	246.8	249.2	249.8
19	0	45	55.0	57.3	57.8
20	0	42	51.7	53.9	54.3
21	0	40	48.6	50.6	51.1
22	0	37	45.7	47.6	48.0
23	0	35	42.9	44.7	45.1
24	0	33	40.4	42.0	42.4
Average	112.5				112.25

(c) *Ventilation*. Since ventilative heat gain is instantaneous, its transfer function coefficients are $v_0 = 1$, $v_1 = 0 = v_2$, and $w_1 = 0$; hence Eq. (7.56) is trivial.

$$\dot{Q}_{c,\text{vent},t} = (T_{o,t} - T_{i,t})\dot{V}_t \rho c_p \quad \text{(for simplicity we consider only sensible load)}$$

For the first $\dot{Q}_{c,\text{vent}}$ entry at $t = 1$, the detail of the calculation is

$$\dot{Q}_{c,\text{vent}} = (24.4°C - 25.0°C)(6.25 \text{ W/K}) = -3.9 \text{ W}$$

The last column shows the *total* load $\dot{Q}_{c,tot}$ = conduction + lights + ventilation = sum of loads from parts a, b, and c.

t	T_o	$\dot{Q}_{c,vent}$	$\dot{Q}_{c,tot}$
h	°C	W	W
−2	27.2	14.1	
−1	26.1	7.1	
0	25.0	0.0	
1	24.4	−3.9	134
2	24.4	−3.9	124
3	23.8	−7.7	111
4	23.3	−10.9	99
5	23.3	−10.9	90
6	23.3	−10.9	82
7	23.8	−7.7	79
8	25.0	0.0	82
9	26.6	10.3	88
10	28.3	21.2	292
11	30.5	35.3	312
12	32.2	46.2	330
13	33.8	56.5	348
14	34.4	60.3	364
15	35.0	64.2	385
16	34.4	60.3	401
17	33.8	56.5	419
18	32.7	49.4	432
19	30.5	35.3	235
20	29.4	28.2	223
21	28.3	21.2	205
22	27.2	14.1	186
23	26.1	7.1	168
24	25.0	0.0	150
Average	28.31	21.26	222.50

COMMENTS

The peak load is 432 W at 18:00. To check the consistency of the results, note that the average of $\dot{Q}_{c,tot}$ in the last column of part a is 89.03 W, in acceptable agreement with $UA(T_{os} − T_i)_{av} = 10 \text{ m}^2 \times$ [0.693 W/(m² · K)] × [(37.95 − 25) K] = 89.74 W (with $T_{os,av}$ from Example 7.17).

Since the lights are on 9 h/day, their average power is 300 W × $\frac{9}{24}$ = 112.5 W, in good agreement with the 24-h average of the $\dot{Q}_{c,t+72}$ column of part b.

The sum of the averages for conduction, lights, and ventilation

$$89.03 + 112.25 + 21.26 = 222.54 \text{ W}$$

agrees with the average 222.50 W of the total load, last column of part c.

It is interesting to compare these results with a steady-state calculation, adding simply the instantaneous heat gains. The instantaneous contribution from the lights is 300 W, constant from 9:00 to 17:00. From Fig. 6.17 we find that the sol-air temperature for a dark, west-facing wall reaches its peak

$$T_{os,\,max} = 75.5°C \quad \text{at } 16:00$$

and the corresponding conduction is

$$UA(T_{os} - T_i)_{max} = (10 \text{ m}^2)[0.693 \text{ W}/(\text{m}^2 \cdot \text{K})][(75.5 - 25)\text{K}] = 350 \text{ W}$$

The ventilation heat gain, listed in part c of Example 7.18, is seen to reach its peak of 64.2 W at 15:00, and at 16:00 it is 60.3 W. Therefore, the peak load of the steady-state calculation occurs between 15:00 and 16:00. Noting that at 15:00 the values for the conductive term are $T_{os,max} = 72.7°C$ and 330.6 W, we see that from 15:00 to 16:00 the decrease in ventilation load is small compared to the increase in conduction. Therefore, we can say that the steady-state peak is very close to 16:00, with 350 W for conduction, 60.3 W for ventilation, and 300 W for lights, giving a total of 710 W. The real peak reaches only 60 percent of that, and it occurs a little later, at 18:00.

While specific numbers vary a great deal with the circumstances, the general pattern is common to all peak cooling loads: *Thermal inertia attenuates and delays the peak contributions of individual load components.* The total peak is usually less than the result of a steady-state calculation, although it could be more if the time delays act in the sense of making the loads coincide.

Daily average loads, by contrast to peak loads, can be determined by a static calculation, if the average indoor temperature is known; that follows from the first law of thermodynamics. But if the thermostat allows floating temperatures, the indoor temperature is, in general, not known without a dynamic analysis.

With the transfer functions described so far, one can calculate peak loads when the indoor temperature T_i is constant. That is how the cooling load factors and cooling load temperature differences of Sec. 7.6 have been determined. We now address the generalization to variable T_i.

7.7.4 Variable Indoor Temperature and Heat Extraction Rate

The indoor temperature T_i may vary, not only because of variable thermostat setpoints but also because of limitations of the HVAC equipment (capacity, throttling range, imperfect control). The extension to variable T_i requires one additional transfer function.

Recall that the behavior of a room can be described by a relation like Eq. (7.50) which links the output (room temperature T_i) to all the relevant input variables (outdoor temperature T_o, heat input or extraction by the HVAC system $\dot{Q}$, solar heat gains, etc.)

$$a_{i,0}T_{i,k} + a_{i,1}T_{i,k-1} + \cdots + a_{i,l}T_{i,k-l}$$
$$= a_{o,0}T_{o,k} + a_{o,1}T_{o,k-1} + \cdots + a_{o,m}T_{o,k-m}$$
$$+ a_{Q,0}\dot{Q}_k + a_{Q,1}\dot{Q}_{k-1} + \cdots + a_{Q,n}\dot{Q}_{k-n} + \cdots \qquad (7.58)$$

A separate set of transfer function coefficients is needed for each input variable with different time delay characteristics; here we have indicated only T_o and $\dot{Q}$ explicitly. Now consider two different control modes, mode 1 with the constant value $T_{i,\text{ref}}$ assumed in Secs. 7.7.2 and 7.7.3 and mode 2 with arbitrary T_i, all input being the same except for $\dot{Q}$. Let

$$\delta T_i = T_{i,\text{ref}} - T_i \tag{7.59}$$

and

$$\delta \dot{Q} = \dot{Q}_{\text{ref}} - \dot{Q} \tag{7.60}$$

designate the differences in T_i and $\dot{Q}$ between these two control modes. Taking the difference between Eq. (7.50) for mode 1 and for mode 2, we see that all variables other than δT_i and $\delta \dot{Q}$ drop out. The transfer function between δT_i and $\delta \dot{Q}$ is called the *space air transfer function*, and following ASHRAE practice, its coefficients are designated by p_n $(= a_{Q,n})$ and g_n $(= a_{i,n})$

$$\sum_{n \geq 0} p_n \delta \dot{Q}_{t-n\Delta t} = \sum_{n \geq 0} g_{n,t}\, \delta T_{i,t-n\Delta t} \tag{7.61}$$

A subscript t has been added to g_n to allow the transfer function to vary with time if the air exchange rate varies. Numerical values as per ASHRAE can be obtained from Table 7.10. While p_n is listed directly, g_n is given in terms of g_n^* from which g_n is calculated according to

$$\begin{aligned}
g_{0,t} &= g_0^* A + p_0 K_{\text{tot},t} \\
g_{1,t} &= g_1^* A + p_1 K_{\text{tot},t-\Delta t} \\
g_{2,t} &= g_2^* A
\end{aligned} \tag{7.62}$$

where $A = $ floor area and $K_{\text{tot},t}$ W/K [Btu/(h · °F)] is the total heat transmission coefficient of the room. The latter is the sum of conductive and air change terms according to Eq.(7.24):

$$K_{\text{tot},t} = K_{\text{cond}} + \rho c_p \dot{V}_t \tag{7.24}$$

and a subscript t for time dependence has been added to allow for the possibility of variable air change. Of course, K_{cond} is the sum of the conductance-area products for the envelope of the room.

TABLE 7.10

Normalized Coefficients of Space air Transfer Function.

Room Envelope Construction	p_0	p_1	g_0^*	g_1^*	g_2^*	g_0^*	g_1^*	g_2^*
	Dimensionless		Btu/(h · ft² · °F)			W/(m² · K)		
Light	1.00	−0.82	1.68	−1.73	0.05	9.54	−9.82	0.28
Medium	1.00	−0.87	1.81	−1.89	0.08	10.28	−10.73	0.45
Heavy	1.00	−0.93	1.85	−1.95	0.10	10.50	−11.07	0.57

Source: Courtesy of ASHRAE, *Handbook of Fundamentals*, American Society of Heating, Refrigerating and Air-Conditioning Engineers, Atlanta, GA, 1989a. With permission.

To verify the consistency of these coefficients with the first law of thermodynamics, let us take the steady-state limit where $\delta\dot{Q}$, δT_i, and $K_{tot,t}$ are constant and can be pulled outside the sum. Replacing the g_n by Eq. (7.62), we find

$$\delta\dot{Q}\sum p_n = \delta T_i\left(A\sum g_n^* + K_{tot}\sum p_n\right) = \text{steady-state limit} \tag{7.63}$$

A look at the numerical values of g_n^* in Table 7.10 shows that their sum vanishes. Thus the equation reduces to

$$\delta\dot{Q} = \delta T_i K_{tot} \tag{7.64}$$

as it should. Since K_{tot} is positive, $\delta\dot{Q}$ and δT_i have the same sign; this means that $\delta\dot{Q}$ is positive for positive heat input to the room. If we want to state cooling loads $\dot{Q}_c$ as positive quantities, we should therefore take $\dot{Q}_c = -\dot{Q}$ and $\delta\dot{Q}_c = -\delta\dot{Q}$. In particular, if we call the cooling load at temperature T_i the *heat extraction rate* $\dot{Q}_x$ (because that is the rate at which the HVAC equipment must extract heat to obtain the temperature T_i), we can write Eq. (7.61) in the form

$$\sum_{n\geq0} p_n(\dot{Q}_{x,t-n\Delta t} - \dot{Q}_{c,\text{ref}}) = \sum_{n\geq0} g_{n,t}(T_{i,\text{ref}} - T_{i,t-n\Delta t}) \tag{7.65}$$

where $\dot{Q}_{c,\text{ref}}$ is the cooling load at the constant temperature $T_{i,\text{ref}}$.

Using Eq. (7.61), one can calculate $\delta\dot{Q}$ for any δT_i, or δT_i for any $\delta\dot{Q}$. It can also be used for a mixed regime where δT_i is specified for certain hours and $\delta\dot{Q}$ for others. The calculation proceeds from one hour to the next, solving for δT_i or $\delta\dot{Q}$ as appropriate. The daily cycle is iterated until the result converges to a stable pattern.

Example 7.19

Find T_i and $\dot{Q}_c$ for the room of Example 7.18 if the thermostat is set at $T_{i,\text{ref}} = 25°C$ from 8:00 until 18:00 and if the air conditioner is turned off from 18:00 until 8:00. The air conditioner has sufficient capacity to satisfy the peak load.

Given: Room of Example 7.18, with control schedule thermostat set at 25°C from 8:00 to 18:00, free-float otherwise.

$A_{\text{floor}} = 15$ m^2

$K_{\text{cond}} = UA = 6.93$ W/K from Example 7.17

$\dot{V}\rho c_p = 6.25$ W/K from Example 7.18 and still constant

$\dot{Q}_{c,\text{ref}} = \dot{Q}_{c,\text{tot}}$ in last column of output of Example 7.18c

Find: T_i, $\dot{Q}_x$

Assumptions: Same as in Example 7.18, but $T_i = 25°C$ only from 8:00 until 18:00, floats freely otherwise.

Lookup values: Looking up Table 7.10 for medium construction and using Eq. (7.62), we find the following coefficients for the room transfer function, where g_n is obtained from g_n^* by using Eq. (7.62) with $A_{\text{floor}} = 15$ m^2 and $K_{tot} = K_{\text{cond}} + \dot{V}\rho C_p = 13.18$ W/K.

n	p_n	g_n^*	g_n
0	1.0	10.28	167.55
1	−0.87	−10.73	−172.56
2	0	0.45	6.75

SOLUTION

We have set up a spreadsheet with three columns (t, $\delta T_{i,t} = T_{i,\text{ref},t} - T_{i,t}$ and $\delta \dot{Q}_{c,t} = \dot{Q}_{x,t} - \dot{Q}_{c,\text{ref},t}$). From $t = 8$ to 17 (as well as for $t < 0$), we set $\delta T_i = 0$ and solve for $\delta \dot{Q}_c$ according to Eq. (7.61); for the remaining hours we set $\delta \dot{Q}_c = 0$ and solve for δT_i according to Eq. (7.61). The cycle is repeated for 4 days, and the results are printed side by side. The last columns show the actual values of temperature $T_i = T_{i,\text{ref}} - \delta T_i$ and the heat extraction rate $\dot{Q}_x = \dot{Q}_{c,\text{ref}} + \delta \dot{Q}_c$.

For the first $\delta \dot{Q}_{c,t}$ entry at $t = 1$, the detail of the calculation is

$$\delta \dot{Q}_1 = -p_1 \delta \dot{Q}_{1-1} + g_{n,0} \delta T_{i,1-0} + g_{n,1} \delta T_{i,1-1} + g_{n,2} \delta T_{i,1-2}$$
$$= -(-0.87)(150) + 167.55(-0.02) + (-172.56)(0.00) + 6.75(0.00)$$
$$= 134$$

t	δT_i	$\delta \dot{Q}_c$	δT_i	$\delta \dot{Q}_c$	δT_i	$\delta \dot{Q}_c$	δT_i	$\delta \dot{Q}_c$	$T_{i,t}$	$\dot{Q}_{x,t}$
	t		$t+24$		$t+48$		$t+72$			
h	°C	W	°C	W	°C	W	°C	W	°C	W
−2	0.00	186								
−1	0.00	168								
0	0.00	150								
1	−0.02	134	−1.92	134	−2.32	134	−2.41	134	27.4	0
2	−0.07	124	−1.94	124	−2.34	124	−2.43	124	27.4	0
3	−0.09	111	−1.94	111	−2.34	111	−2.42	111	27.4	0
4	−0.10	99	−1.94	99	−2.33	99	−2.41	99	27.4	0
5	−0.12	90	−1.94	90	−2.33	90	−2.41	90	27.4	0
6	−0.15	82	−1.95	82	−2.33	82	−2.41	82	27.4	0
7	−0.19	79	−1.97	79	−2.35	79	−2.43	79	27.4	0
8	0.00	37	0.00	−258	0.00	−320	0.00	−334	25.0	415
9	0.00	33	0.00	−211	0.00	−263	0.00	−274	25.0	362
10	0.00	29	0.00	−184	0.00	−229	0.00	−238	25.0	531
11	0.00	25	0.00	−160	0.00	−199	0.00	−207	25.0	520
12	0.00	22	0.00	−139	0.00	−173	0.00	−180	25.0	510
13	0.00	19	0.00	−121	0.00	−151	0.00	−157	25.0	505
14	0.00	17	0.00	−105	0.00	−131	0.00	−137	25.0	500
15	0.00	14	0.00	−92	0.00	−114	0.00	−119	25.0	503
16	0.00	13	0.00	−80	0.00	−99	0.00	−103	25.0	505
17	0.00	11	0.00	−69	0.00	−86	0.00	−90	25.0	509
18	−2.52	432	−2.94	432	−3.02	432	−3.04	432	28.0	0
19	−1.76	235	−2.19	235	−2.28	235	−2.30	235	27.3	0
20	−1.82	223	−2.24	223	−2.33	223	−2.35	223	27.4	0
21	−1.87	205	−2.29	205	−2.37	205	−2.39	205	27.4	0
22	−1.90	186	−2.31	186	−2.40	186	−2.42	186	27.4	0
23	−1.91	168	−2.32	168	−2.41	168	−2.43	168	27.4	0
24	−1.92	150	−2.32	150	−2.41	150	−2.43	150	27.4	0
Average									26.4	203

In Fig. 7.22, the result is plotted and compared with the constant-set-point case of Example 7.18.

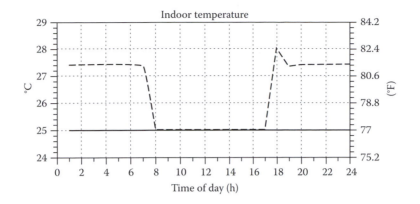

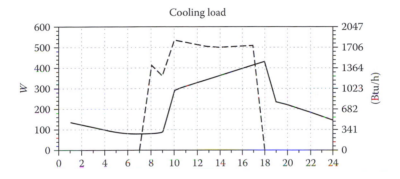

FIGURE 7.22
Cooling load and indoor temperature for room of Examples 7.18 and 7.19. Solid line = constant setpoint at 25°C; dashed line = setpoint at 25°C from 8:00 to 18:00, floating the rest of the time.

COMMENTS

1. To check the consistency with the steady-state limit, evaluate the conduction and ventilation loads at the new average of T_i of 26.4°C. The load from lights remains, of course, unaffected. The loads at average conditions are

$$\text{Conduction} = UA(T_{os} - T_i)_{av} = 10 \text{ m}^2 \times [0.693 \text{ W}/(\text{m}^2 \cdot \text{K})]$$
$$\times [(37.95 - 26.4)\text{K}] = 80.04 \text{ W}$$

$$\text{Ventilation} = \rho c_p (T_o - T_i)_{av} = 6.250[(28.31 - 26.4) \text{ K}] = 11.94 \text{ W}$$
$$\text{Lights} = 112.25 \text{ W}$$

Their sum is 204 W, in very good agreement with the average 203 W of $\dot{Q}_x$.

2. Compared to the fixed thermostat setpoint in Example 7.18, the setup reduces the average load from 222.5 to 203 W, a saving of about 10 percent.

3. Savings from thermostat setup depend on the relation between indoor and outdoor temperatures and on the relative importance of conduction, ventilation, and internal gains. In this example, internal gains dominate and setup recovery increases the peak load (a situation that is unlikely when skin loads dominate).

4. In climates with cool, dry nights, significant *savings* may be achieved *by increasing the ventilation at night* (in this example, we have assumed constant ventilation; its load contribution becomes negative at night, but the magnitude is relatively small). If outdoor air at night is cool but moist, absorption of humidity in the building may increase the latent load during the following day.

Capacity limitations of the HVAC equipment can also be included: One checks, at each hour with thermostatic control, whether the actual load (at T_i = thermostat setpoint) exceeds the capacity. If it does, one solves for T_i instead, setting $\dot{Q}_x$ equal to the capacity for this hour.

Likewise, one can account for the throttling range of a control system which modulates the heat extraction rate according to the control law shown in Fig. 7.23. Stated as an equation, this means that the heat extraction rate $\dot{Q}_x$ is determined by the room temperature T_i according to

$$\dot{Q}_{x,t} = \begin{cases} \dot{Q}_{max} & \text{for } T_{i,t} > T_{max} \\ \dot{Q}_{set} + \dot{Q}'(T_{i,t} - T_{set}) & \text{for } T_{min} < T_{i,t} < T_{max} \\ \dot{Q}_{min} & \text{for } T_{i,t} < T_{min} \end{cases} \qquad (7.66)$$

where

$$\dot{Q}' = \frac{\dot{Q}_{max} - \dot{Q}_{min}}{T_{max} - T_{min}}$$

and T_{set} is the thermostat setpoint; we have added the subscript t to indicate that this equation applies instantaneously at each hour t. At each new hour t, Eqs. (7.65) and (7.66) can be considered as a system of two equations for two unknowns: $\dot{Q}_{x,t}$ and $T_{i,t}$. After finding the solution, one repeats the process for the next hour.

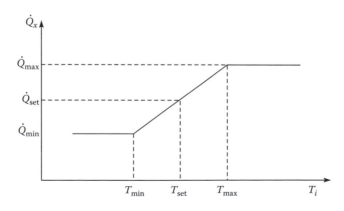

FIGURE 7.23
Control law of Eq. (7.66) for heat extraction rate $\dot{Q}_x$ (solid line) as function of room temperature T_i.

7.8 New Methods for Load Calculations

In the ever-present quest for more accurate load prediction methods, new procedures are being developed all the time. In the past few years, a couple of new calculation procedures have been created that are relatively similar to the existing methods but implement enough differences to warrant mention. The CLTD/CLF method has been updated to include a new term, the *solar cooling load (SCL)*, and is now called the CLTD/SCL/CLF method. The *radiant transfer series (RTS)* method is a noniterative modification of the transfer function method. An overview of both of these new methods is presented in this section.

7.8.1 CLTD/SCL/CLF Method for Calculating Cooling Loads

The CLTD/CLF method presented in Sec. 7.6 has been modified in a number of important ways. The selection of roof and wall CLTD values requires a number of lookup tables but allows the use of essentially any arbitrary wall construction. In addition, the solar heat gain factors have been replaced with a term called the solar cooling load. While the overall methodology of the CLTD/CLF method has been preserved, a full description of the necessary computations of the new method is presented here to avoid confusion with the original method.

Roof CLTD Value Selection   The CLTD/SCL/CLF method uses 10 types of roofs. The roof types are numbered 1, 2, 3, 4, 5, 8, 9, 10, 13, and 14. The roof type chosen depends on the principal roof material, the location of the mass in the roof, the overall R value of the roof, and whether or not there is a suspended ceiling. Table 7.11 shows the cross-reference chart used to select a roof type.

TABLE 7.11

Cross-Reference Table Used to Determine Roof Type

Mass Location	Principal Roof Material Description	ASHRAE Code	Susp. Ceiling	R Value (ft²·h·°F)/Btu 0–5	5–10	10–15	15–20	20–25	25–30
Inside insulation	2-in HW concrete	C12	No	2	2	4	4	5	
			Yes	5	8	13	13	14	
	1-in wood	B7	No	1	2	2	4	4	
			Yes		4	5	9	10	10
Evenly spaced	2-in HW concrete	C12	No	2					
			Yes	3					
	Steel deck	A3	No	1	1	1	2	2	
			Yes	1	1	2	2	4	
	Attic-ceiling combination	n/a	No	1	2	2	2	4	
Outside insulation	2-in HW concrete	C12	No	2	3	4	5	5	
			Yes	3	3	4	5		

The tables of new roof CLTD values are calculated based on an indoor temperature of 78°F, maximum and mean outdoor temperatures of 95 and 85°F, respectively, and a daily range of 21°F. Once the 24 CLTD values are selected, they are each adjusted by

$$\text{Corrected CLTD} = \text{CLTD} + (78 - T_i) + (T_{om} - 85) \tag{7.67}$$

Where T_i is the actual inside design dry-bulb temperature and T_{om} is the mean outside design dry bulb temperature,

$$T_{om} = \text{outside design dry-bulb temperature} - \frac{\text{daily range}}{2} \tag{7.68}$$

No adjustments to the CLTD are recommended for color or ventilation. The CLTD charts are usually published for several different latitudes; interpolation between the latitudes for an exact site is acceptable.

Wall CLTD Value Selection   The CLTD/SCL/CLF uses 15 wall types numbered sequentially 1 through 16 with no wall 8. The wall type is chosen based on the principal wall material, the secondary wall material, the location of the mass in the wall, and the overall wall R value. Table 7.12 shows an example cross-reference chart used to select a wall type. The tables of wall CLTD values are broken down by latitude. The wall CLTDs were calculated using the same conditions as the roof CLTD values and may require adjustments based on the actual inside and ambient conditions. Interpolation between the tables may be necessary to obtain the correct values for a given site.

Once the roof and wall CLTD values have been selected and adjusted as necessary, the conductive heat flow through the roof and walls is calculated for each hour (h) as in the original CLTD/CLF method

$$q(h) = UA[\text{CLTD}(h)] \tag{7.69}$$

where
$\quad U = $ overall heat transfer coefficient for surface, Btu/(h · ft² · °F)
$\quad A = $ area of surface
CLTD = cooling load temperature difference

Glass CLTD Value Selection   The glass CLTD values remain the same as they were in the original method. As with the roof and wall CLTDs, the fenestration CLTD values may need to be corrected based on Eqs. (7.67) and (7.68). The conductive load calculation from the glass uses the same method as for the roof and walls. The CLTD values for the glass are given in Table 7.13.

Solar Cooling Load   The new method replaces the maximum solar heat gain factor with the solar cooling load. This new value is used to calculate the radiative (solar) heat gain through any glass surface in the building. The radiative solar gains are then given by

$$q(h) = A(\text{SC})(\text{SCL}) \tag{7.70}$$

where A is the area of the glass surface, SC is the shading coefficient, and SCL is the solar cooling load factor. The shading coefficient is the ratio of the actual solar heat gain to that from the reference window used to calculate the SCL.

TABLE 7.12

Example Wall Type Selection Table. Values are Shown for Mass Located Inside Insulation

Secondary Material	R Value, (ft² · h · °F)/Btu	Principal Wall Material (ASHRAE Material Code)										
		A2	C1	C2	C3	C4	C5	C6	C7	C8	C17	C18
Stucco and/or plaster	2.0–2.5	5					5					
	2.5–3.0	5	3		2	5	6			5		
	3.0–3.5	5	4	2	2	5	6			6		
	3.5–4.0	5	4	2	3	6	6	10	4	6		5
	4.0–4.75	6	5	2	4	6	6	11	5	10		10
	4.75–5.5	6	5	2	4	6	6	11	5	10		10
	5.5–6.5	6	5	2	5	10	7	12	5	11		10
	6.5–7.75	6	5	4	5	11	7	16	10	11		11
	7.75–9.0	6	5	4	5	11	7		10	11		11
	9.0–10.75	6	5	4	5	11	7		10	11	4	11
	10.75–12.75	6	5	4	5	11	11		10	11	4	11
	12.75–15.0	10	10	4	5	11	11		10	11	9	12
	15.0–17.5	10	10	5	5	11	11		11	12	10	16
	17.5–20.0	11	10	5	9	11	11		15	16	10	16
	20.0–23.0	11	10	9	9	16	11		15	16	10	16
	23.0–27.0								16		15	
Steel or other lightweight siding	2.0–2.5	3			2	3	5					
	2.5–3.0	5	2		2	5	3			5		
	3.0–3.5	5	3	1	2	5	5			5		
	3.5–4.0	5	3	2	2	5	5	6	3	5		5
	4.0–4.75	6	4	2	2	5	5	10	4	6		5
	4.75–5.5	6	5	2	2	6	6	11	5	6		6
	5.5–6.5	6	5	2	3	6	6	11	5	6		6
	6.5–7.75	6	5	2	3	6	6	11	5	6		10
	7.75–9.0	6	5	2	3	6	6	12	5	6		11
	9.0–10.75	6	5	2	3	6	6	12	5	6	4	11
	10.75–12.75	6	5	2	3	6	7	12	6	11	4	11
	12.75–15.0	6	5	2	4	6	7	12	10	11	5	11
	15.0–17.5	10	6	4	4	10	7		10	11	9	11
	17.5–20.0	10	10	4	4	10	11		10	11	10	11
	20.0–23.0	11	10	4	5	11	11		10	11	10	16
	23.0–27.0								10		11	16

Using the SCL value tables requires that you know the number of walls, floor covering, inside shading, and a number of other variables for the zone. The tables are also broken down by building type, with different tables for zones in

- Single-story buildings
- Top floor of multistory buildings
- Middle floors of multistory buildings
- First floor of multistory buildings

TABLE 7.13

CLTD Values for Fenestration

Hour	CLTD	Hour	CLTD	Hour	CLTD	Hour	CLTD
1	1	7	−2	13	12	19	10
2	0	8	0	14	13	20	8
3	−1	9	2	15	14	21	6
4	−2	10	4	16	14	22	4
5	−2	11	7	17	13	23	3
6	−2	12	9	18	12	24	2

Table 7.14 gives the zone types for the SCL for the first story of multistory buildings. The zone type listed here is for the SCL *and is not necessarily the same zone type used for the CLF tables*. Once the zone type has been determined, the SCL can be found from tables such as Table 7.15.

TABLE 7.14

Zone Type for Solar Cooling Load for First Story of Multistory Buildings

Number of Walls	Midfloor Type	Ceiling Type	Floor Covering	Partition Type	Inside Shade	Zone Type
1 or 2	2.5-in concrete	With	Carper	Gypsum	Full	A
					Half to none	B
				Concrete block	Full	B
					Half to none	C
			Vinyl	Gypsum	Full	C
					Half to none	C
				Concrete block	Full	D
					Half to none	D
		Without	Carpet	Gypsum	—	B
				Concrete block	Full	C
					Half to none	C
			Vinyl	Gypsum	Full	C
					Half to none	D
				Concrete block	Full	C
					Half to none	D
	1-in wood	—	Carpet	Gypsum	Full	A
					Half to none	B
				Concrete block	Full	B
					Half to none	C
			Vinyl	Gypsum	Full	B
					Half to none	C
				Concrete block	Full	C
					Half to none	D

(*continued*)

TABLE 7.14 (continued)

Zone Type for Solar Cooling Load for First Story of Multistory Buildings

Number of Walls	Midfloor Type	Ceiling Type	Floor Covering	Partition Type	Inside Shade	Zone Type
			Carpet	Gypsum	Full	A
			Carpet	Gypsum	Half to none	B
		With	Carpet	Concrete block	—	B
			Vinyl	Gypsum	Full	C
			Vinyl	—	Half to none	C
	2.5-in concrete		Vinyl	Concrete block	Full	C
			Carpet	Gypsum	—	B
			Carpet	Concrete block	Full	B
3		Without	Carpet	Concrete block	Half to none	C
			Vinyl	Gypsum	Full	C
			Vinyl		Half to none	C
			Vinyl	Concrete block	Full	C
				Gypsum	Full	A
			Carpet		Half to none	B
				Concrete block	—	B
	1-in wood	—		Gypsum	Full	B
			Vinyl		Half to none	C
				Concrete block	Full	C
					Half to none	C
			Carpet	Gypsum	Full	A
		With	Carpet	Gypsum	Half to none	B
			Vinyl	Gypsum		C
	2.5-in concrete		Carpet	Gypsum	—	B
4		Without	Vinyl	Gypsum		B
			Carpet	Gypsum	Full	A
	1-in wood	—			Half to none	A
			Vinyl	Gypsum	Full	B
					Half to none	C

Accounting for Adjacent Zones The CLTD/SCL/CLF method treats the conductive heating load from any adjacent spaces through internal partitions, ceilings, and floors as a simple steady-state energy flow

$$q \text{ (h)} = UA(T_a - T_r) \tag{7.71}$$

where T_a is the temperature in the adjacent space and T_r is the temperature of the room in question.

Occupant Loads People within a space add both sensible and latent loads to the space. The heating load at any given hour due to the occupants is given as

$$q \text{ (h)} = NF_d[q_s \cdot \text{CLF (h)} + q_l] \tag{7.72}$$

TABLE 7.15

Solar Cooling Load for Sunlit Glass at 36° North Latitude for July (Units: $Btu/h - ft_2$)

Zone Type/Orientation		1	2	3	4	5	6	7	8	9	10	11	12	13	14	15	16	17	18	19	20	21	22	23	24
A	N	0	0	0	0	0	25	29	28	32	36	39	40	41	39	36	32	33	36	12	6	3	1	1	0
	NE	0	0	0	0	0	79	129	139	120	84	58	50	45	41	37	32	26	17	7	3	2	1	0	0
	E	0	0	0	0	0	86	153	184	182	155	107	67	54	45	39	33	26	17	7	3	2	1	0	0
	SE	0	0	0	0	0	42	90	125	142	140	119	86	58	48	40	34	27	17	7	3	2	1	0	0
	S	0	0	0	0	0	8	17	24	36	53	70	80	79	68	52	38	29	18	7	3	2	1	0	0
	SW	0	0	0	0	0	8	17	24	30	35	38	57	90	122	141	144	127	85	32	15	8	4	2	1
	W	1	0	0	0	0	8	17	24	30	35	38	40	66	115	159	188	191	149	53	25	12	6	3	2
	NW	1	0	0	0	0	8	17	24	30	35	38	40	40	56	93	129	148	127	43	21	10	5	2	1
	Hor	0	0	0	0	0	20	66	120	171	215	246	263	265	251	221	178	124	66	28	13	7	3	2	1
B	N	2	2	1	1	1	21	25	25	29	33	36	38	38	38	35	32	33	35	15	10	7	5	4	3
	NE	2	1	1	1	1	68	109	120	108	81	61	54	50	46	42	37	30	22	12	9	6	5	3	3
	E	2	2	1	1	1	73	130	158	161	143	106	75	63	55	48	41	34	25	14	10	7	5	4	3
	SE	2	2	1	1	1	36	77	107	124	125	111	85	64	55	48	41	33	24	14	10	7	5	4	3
	S	2	2	1	1	1	7	14	21	31	47	61	71	72	65	52	41	33	24	13	9	7	5	4	3
	SW	6	4	3	3	2	8	15	21	27	31	35	51	80	108	126	131	119	86	43	29	20	14	11	8
	W	8	6	5	4	3	9	16	22	27	32	35	37	60	101	140	166	172	141	63	42	29	20	15	11

	Dir																								
	NW	8	11	16	22	32	49	117	132	115	84	52	38	37	35	31	27	21	15	8	2	3	4	5	6
	Hor	11	14	19	26	37	53	88	137	182	215	237	244	237	218	188	148	103	57	19	3	4	5	6	8
C	N	6	6	7	8	10	14	34	31	29	32	34	35	35	33	31	27	24	25	24	3	4	4	5	5
	NE	8	9	10	11	13	16	24	32	37	41	44	46	48	51	68	95	111	106	71	5	5	6	6	7
	E	11	12	13	15	17	20	29	37	43	47	52	56	62	89	124	145	148	128	77	6	7	8	8	9
	SE	9	11	12	13	15	18	27	35	40	45	49	55	73	97	112	114	102	77	40	5	6	7	8	8
	S	7	8	9	10	11	14	22	30	35	45	57	65	65	58	45	31	22	17	10	4	4	5	6	6
	SW	14	16	18	21	26	34	74	105	118	116	102	77	50	35	32	29	25	20	14	8	9	10	12	13
	W	18	21	24	28	34	48	122	155	154	132	98	59	37	36	34	31	27	22	16	11	12	13	15	16
	NW	14	15	18	21	26	37	104	122	108	80	50	36	36	35	32	29	25	20	14	8	9	10	11	12
	Hor	27	30	34	38	44	53	81	122	161	192	212	220	217	203	178	145	107	66	31	16	17	19	22	24
D	N	9	10	11	12	14	17	32	30	29	31	32	32	31	30	27	25	21	22	21	5	6	6	7	8
	NE	12	14	15	17	19	22	29	35	40	43	46	47	49	51	63	82	93	87	59	7	8	9	10	11
	E	16	18	20	22	25	28	36	43	48	53	57	60	65	84	110	124	123	105	65	10	11	12	13	15
	SE	15	16	18	20	22	25	33	40	45	49	52	56	70	87	97	96	85	65	36	9	10	11	12	13
	S	11	12	13	15	16	19	26	31	36	43	52	57	56	49	39	27	20	16	11	6	7	8	9	9
	SW	22	24	27	30	35	41	72	95	103	100	87	67	45	33	31	28	25	21	17	13	14	16	18	20
	W	28	31	34	39	45	55	111	134	131	112	84	53	36	34	33	30	27	24	20	16	18	20	22	25
	NW	21	23	26	29	34	42	92	104	92	69	45	34	34	32	30	27	24	20	16	12	14	15	17	18
	Hor	41	46	51	56	63	72	95	128	157	179	191	195	189	174	153	125	94	62	35	24	27	30	33	37

where N is the number of people in the space and F_d is the diversity factor. As implied by Eq. (7.72), the latent load is assumed to immediately translate into a cooling load on the system while the sensible load is subject to some time delay as dictated by the mass of the room, i.e., its capability to absorb heat and release it at a later time. The diversity factor F_d takes into account the variability of the actual number of occupants in the space and has typical values as given in Table 7.16.

The CLF values come from tables. To find the CLF, it is first necessary to determine the zone type. This is done in a similar fashion as for the solar cooling loads. That is, the building type, room location, and floor coverings must be known before the zone type can be found. Table 7.17 gives the zone types for people, equipment, and lights for interior (nonperimeter) zones. Note that the zone type for occupants and equipment is not the same as for the lighting. The same holds true for the solar cooling load: The zone type for occupants is not the same as the zone type for the SCL.

Once the zone type has been determined, the occupant CLF is found from the lookup tables, such as shown in Table 7.18. This table shows values for type A zones only; the zones get progressively more massive for types B, C, and D. Figure 7.24 shows the cooling load factors for type A and D zones that are occupied for 12 h.

Note that the occupant CLF will be 1.0 for all hours in a building with high occupant density (greater than 1 person per 10 ft²) such as auditoriums and theaters. The CLF will also be 1.0 in buildings where there is 24-h/day occupancy.

TABLE 7.16

Typical Diversity Factors for Occupants in Large Buildings

Building Type	F_d
Apartment	0.40–0.60
Industrial	0.85–0.95
Hotel	0.40–0.60
Office	0.75–0.90
Retail	0.80–0.90

TABLE 7.17

Zone Types for Used in Determining the CLF. This Table is Used for Interior (i.e., Nonperimeter) Zones Only

Zone Parameters				Zone Type	
Room Location	Middle Floor	Ceiling Type	Floor Covering	People and Equipment	Lights
Single story	N/A	N/A	Carpet	C	B
	N/A	N/A	Vinyl	D	C
Top floor	2.5-in concrete	With	Carpet	D	C
	2.5-in concrete	With	Vinyl	D	D
	2.5-in concrete	Without	*	D	B
	1-in wood	*	*	D	B
Bottom floor	2.5-in concrete	With	Carpet	D	C
	2.5-in concrete	*	Vinyl	D	D
	2.5-in concrete	Without	Carpet	D	D
	1-in wood	*	Carpet	D	C
	1-in wood	*	Vinyl	D	D
Midfloor	2.5-in concrete	N/A	Carpet	D	C
	2.5-in concrete	N/A	Vinyl	D	D
	1-in wood	N/A	*	C	B

* The effect of this parameter is negligible in this case.

TABLE 7.18

Occupant Cooling Load Factors for Type A Zones

		Total Hours that Space is Occupied								
		2	4	6	8	10	12	14	16	18
	1	0.75	0.75	0.75	0.75	0.75	0.75	0.76	0.76	0.77
	2	0.88	0.88	0.88	0.88	0.88	0.88	0.88	0.89	0.89
	3	0.18	0.93	0.93	0.93	0.93	0.93	0.93	0.94	0.94
	4	0.08	0.95	0.95	0.95	0.95	0.96	0.96	0.96	0.96
	5	0.04	0.22	0.97	0.97	0.97	0.97	0.97	0.97	0.97
	6	0.02	0.10	0.97	0.97	0.97	0.98	0.98	0.98	0.98
Hour after occupants enter space	7	0.01	0.05	0.33	0.98	0.98	0.98	0.98	0.98	0.98
	8	0.01	0.03	0.11	0.98	0.98	0.98	0.99	0.99	0.99
	9	0.01	0.02	0.06	0.24	0.99	0.99	0.99	0.99	0.99
	10	0.01	0.02	0.04	0.11	0.99	0.99	0.99	0.99	0.99
	11	0.00	0.01	0.03	0.06	0.24	0.99	0.99	0.99	0.99
	12	0.00	0.01	0.02	0.04	0.12	0.99	0.99	0.99	1.00
	13	0.00	0.01	0.02	0.03	0.07	0.25	1.00	1.00	1.00
	14	0.00	0.01	0.01	0.02	0.04	0.12	1.00	1.00	1.00
	15	0.00	0.00	0.01	0.02	0.03	0.07	0.25	1.00	1.00
	16	0.00	0.00	0.01	0.01	0.02	0.04	0.12	1.00	1.00
	17	0.00	0.00	0.01	0.01	0.02	0.03	0.07	0.25	1.00
	18	0.00	0.00	0.00	0.01	0.01	0.02	0.05	0.12	1.00
	19	0.00	0.00	0.00	0.01	0.01	0.02	0.03	0.07	0.25
	20	0.00	0.00	0.00	0.01	0.01	0.02	0.03	0.05	0.12
	21	0.00	0.00	0.00	0.00	0.01	0.01	0.02	0.03	0.07
	22	0.00	0.00	0.00	0.00	0.01	0.01	0.02	0.03	0.05
	23	0.00	0.00	0.00	0.00	0.00	0.01	0.01	0.02	0.03
	24	0.00	0.00	0.00	0.00	0.00	0.01	0.01	0.02	0.03

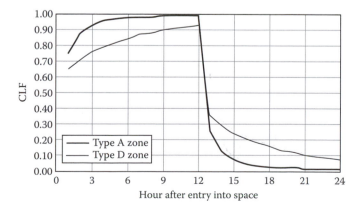

FIGURE 7.24

Occupant cooling load factors for type A and type D zones for a space that is occupied for 12 h.

Lighting Loads   At any given hour the load due to the lighting are approximated as

$$q\,(h) = \text{watts} \cdot F_d F_{sa}[\text{CLF}\,(h)] \tag{7.73}$$

where watts is the total lamp wattage in the space, F_d is the diversity factor, and F_{sa} is a ballast special allowance factor. The diversity factor F_d takes into account the variability of the actual wattage of lights on at any given time and has typical values given in Table 7.19.

TABLE 7.19

Typical Diversity Factors for Lighting in Large Buildings

Building Type	F_d
Apartment	0.30–0.50
Industrial	0.80–0.90
Hotel	0.30–0.50
Office	0.70–0.85
Retail	0.90–1.00

The lighting CLF values come from tables and are found in a fashion similar to that for the occupants. Remember that the zone types for lighting are not necessarily the same zone types for the solar cooling load or the occupants. Note that the lighting CLF will be 1.0 for buildings in which the lights are on 24 h/day or where the cooling system is shut off at night or on the weekends.

If the calculations are done in inch-pound units, then the result from Eq. (7.73) is multiplied by 3.41 to convert watts to Btu per hour.

Appliance and Equipment Loads   Equipment can add heat either through resistive heating or from electrical motors operating in the equipment. The CLTD/SCL/CLF method accounts for both types of equipment heat separately. In addition, the equipment loads are further broken down into sensible or latent components. The latent components are assumed to become immediate loads on the cooling system. The latent loads are found in tables devoted to hospital equipment, restaurant equipment, and office equipment. Latent loads are cited only for the hospital and restaurant equipment. An example of these kinds of loads is given in Table 7.20. The sensible component of the loads is adjusted by

$$q\,(h) = q_{sa}[\text{CLF}\,(h)] \tag{7.74}$$

where q_{sa} is the sensible heat gain per appliance as found from the tables. The cooling load factor is found by first determining the zone type and then looking up the CLF in a table appropriate for that zone type, as was done for the occupants and lighting. While the zone type is similar for occupants and equipment, it may not be the same as for lighting.

The total cooling load in the space is then found from the sum of the sensible and latent loads. If there is a cooling load due to equipment with electrical motors that run equipment in the space, then the space cooling load is incremented by

$$q\,(h) = 2545 \cdot \frac{\text{HP}}{\eta} \cdot F_l F_u[\text{CLF}\,(h)] \tag{7.75a}$$

where HP is the rated horsepower of the motor, η is the efficiency, F_l is the load factor average (power used divided by rated horsepower, typically around 12), and F_u is the motor use factor (accounting for intermittent use). The term 2545 converts from hp to Btu per hour. Equation (7.75a) assumes that both the equipment and the motor are located within the space. If the equipment is in the space but the motor is located outside the space, then this equation is derated by the motor efficiency:

$$q\,(h) = 2545 \cdot \text{HP} \cdot F_l F_u[\text{CLF}\,(h)] \tag{7.75b}$$

TABLE 7.20

Partial List of Recommended Heat Gain for Restaurant Equipment

| Appliance Type | Size | Maximum Input Rating | | Heat Gain Rate, Btu/h | | |
| | | | | Unhooded | | Hooded |
		W	Btu/h	Sensible	Latent	Sensible
Barbeque (pit), per 5 lb of food capacity	80–300 lb	200	680	440	240	210
Barbeque (pressurized), per 5 lb of food capacity	45 lb	470	1,600	550	270	260
Blender, per gallon of capacity	0.25–1.0 gal	1,800	6,140	4,060	2,080	1,980
Braising pan, per gallon of capacity	27–35 gal	400	1,360	720	380	510
Cabinet (large hot holding)	16.3–17.3 ft^3	2,080	7,100	610	340	290
Cabinet (large hot serving)	37.6–40.5 ft^3	2,000	6,820	610	310	280
Cabinet (large proofing)	16.0–17.0 ft^3	2,030	6,930	610	310	280
Cabinet (small hot holding)	3.3–6.5 ft^3	900	3,070	270	140	130
Cabinet (very hot holding)	17.3 ft^3	6,150	20,980	1,880	960	850
Can opener		170	580	580		0
Coffee brewer	12 cups/2 burners	1,660	5,660	3,750	1,910	1,810
Coffee heater, per boiling burner	1–2 burners	670	2,290	1,500	790	720
Coffee heater, per warming burner	1–2 burners	100	340	230	110	110
Coffee/hot-water holding urn, per gallon of capacity	3.0 gal	460	1,570	580	200	260
Coffee urn (large), per gallon of capacity	6.0–10.0 gal	2,500	8,530	2,830	1,430	1.36
Coffee urn (small), per gallon of capacity	3.0 gal	1,580	5,390	1,770	920	850
Cutter (large)	18-in bowl	750	2,560	2,560		820
Cutter (small)	14-in bowl	370	1,260	1,260		410
Cutter and mixer (large)	7.5–11.3 gal	3,730	12,730	12,730		4,060

Conversely, if the motor is inside the space but it acts on equipment outside the space, the cooling load is incremented by

$$q\,(h) = 2545 \cdot HP \cdot \frac{1-\eta}{\eta} \cdot F_l F_u [CLF\,(h)] \qquad (7.75c)$$

As with the lighting, the CLF is always 1.0 when the cooling system does not operate 24 h/day.

Air Infiltration The sensible and latent cooling loads introduced by infiltration are treated the same way in the CLTD/SCL/CLF method as they were in the original CLTD/CLF method. Specifically, the infiltrating air is assumed to immediately become a load on the cooling system.

7.8.2 Radiant Time-Series Method for Hourly Cooling Load Calculations

The radiant time-series (RTS) method is a new method currently under development by ASHRAE. This method is similar to the transfer function method except that the iterative computations for the conductive heat flows and room radiative transfer functions have been replaced by a set of 24 response factors that are used for the calculations at each hour. The principal differences between the transfer function method and the RTS method are outlined in this section.

Surface Conduction Heat Transfer   The transfer function method uses an iterative process to calculate the conductive heat flow across the roof and wall surfaces of a building. Depending on the driving forces and the wall material, this may require several repetitions of each day's values before the iteration converges. The RTS method replaces this iteration with a simple summation

$$q_{\text{cond}}\ (h) = A \sum_{j=0}^{23} Y_{pj}(T_{e,\theta-j\delta} - T_r) \tag{7.76}$$

where A is the surface area, Y_{pj} is the jth response factor, $T_{e,\theta-j\delta}$ is the sol-air temperature from j hours ago, and T_r is the space temperature, assumed to be constant. The response factors for the walls and roof can be found from lookup tables (such as Table 7.21) similar to those created for the transfer function coefficients. Once the conductive loads have been calculated, the transmitted solar heat and window conductive heat gains through each window, the absorbed solar gain, and the internal gains are calculated the same as with the transfer function method.

Radiative and Convective Fractions   When all the heat flows into the building have been calculated, the loads must be further broken down into the radiative and convective components. Table 7.22 shows recommended values for the radiative fraction; the convective fraction is simple:

$$\text{Load convective fraction} = 1 - \text{load radiative fraction} \tag{7.77}$$

The convective fraction immediately becomes a cooling load on the building HVAC system. The radiative portion is absorbed by the building materials, furniture, etc., and is convected into the space as a time-lagged and attenuated cooling load, as described next.

Conversion of Radiant Loads   The radiant loads are converted to hourly cooling loads through the use of radiant time factors. Similar to the response factors, the time factors estimate the cooling load based on past and present heat gains.

$$q_{\text{cool}}\ (h) = \sum_{j=0}^{23} r_j q_{\theta-j\delta} \tag{7.78}$$

where r_o is the fraction of the load convected to the space at the current time, r_1 is the fraction at the previous hour, and so forth. This step replaces the zone transfer function of the transfer function method.

Two sets of radiative time factors must be determined for each zone: one for the transmitted solar heat gain and one for all other types of heat gain. The difference between

TABLE 7.21

Period Response Factors for Representative Roof Types 1 through 8, Btu/(h · ft² · °F)

	Roof Type							
	1	**2**	**3**	**4**	**5**	**6**	**7**	**8**
Y_{P0}	0.004870	0.000556	0.006192	0.000004	0.000105	0.003675	0.001003	0.003468
Y_{P1}	0.036463	0.012356	0.044510	0.000658	0.002655	0.034908	0.009678	0.022622
Y_{P2}	0.026468	0.020191	0.047321	0.004270	0.007678	0.054823	0.017455	0.045052
Y_{P3}	0.008915	0.012498	0.035390	0.007757	0.008783	0.050193	0.017588	0.047168
Y_{P4}	0.002562	0.005800	0.026082	0.008259	0.007720	0.041867	0.015516	0.042727
Y_{P5}	0.000708	0.002436	0.019215	0.006915	0.006261	0.034391	0.013169	0.037442
Y_{P6}	0.000193	0.000981	0.014156	0.005116	0.004933	0.028178	0.011038	0.032544
Y_{P7}	0.000053	0.000388	0.010429	0.003527	0.003844	0.023078	0.009213	0.028228
Y_{P8}	0.000014	0.000152	0.007684	0.002330	0.002982	0.018900	0.007678	0.024472
Y_{P9}	0.000004	0.000059	0.005661	0.001498	0.002309	0.015478	0.006397	0.021212
Y_{P10}	0.000001	0.000023	0.004170	0.000946	0.001787	0.012675	0.005328	0.018386
Y_{P11}	0.000000	0.000009	0.003072	0.000591	0.001383	0.010380	0.004437	0.015937
Y_{P12}	0.000000	0.000003	0.002264	0.000366	0.001070	0.008501	0.003696	0.013814
Y_{P13}	0.000000	0.000001	0.001668	0.000225	0.000827	0.006962	0.003078	0.011973
Y_{P14}	0.000000	0.000001	0.001229	0.000138	0.000640	0.005701	0.002563	0.010378
Y_{P15}	0.000000	0.000000	0.000905	0.000085	0.000495	0.004669	0.002135	0.008995
Y_{P16}	0.000000	0.000000	0.000667	0.000052	0.000383	0.003824	0.001778	0.007797
Y_{P17}	0.000000	0.000000	0.000491	0.000032	0.000296	0.003131	0.001481	0.006758
Y_{P18}	0.000000	0.000000	0.000362	0.000019	0.000229	0.002564	0.001233	0.005858
Y_{P19}	0.000000	0.000000	0.000267	0.000012	0.000177	0.002100	0.001027	0.005077
Y_{P20}	0.000000	0.000000	0.000196	0.000007	0.000137	0.001720	0.000855	0.004401
Y_{P21}	0.000000	0.000000	0.000145	0.000004	0.000106	0.001408	0.000712	0.003815
Y_{P22}	0.000000	0.000000	0.000107	0.000003	0.000082	0.001153	0.000593	0.003306
Y_{P23}	0.000000	0.000000	0.000079	0.000002	0.000063	0.000945	0.000494	0.002866

Source: Courtesy of Spitler, J.D. and Fisher, D.E., *Int. J. HVAC&R Res.*, 5(2), 125, 1999.

TABLE 7.22

Typical Radiative Fraction of Building Heat Gains

Heat Gain Type	Typical Radiative Fraction
Occupants	0.7
Suspended fluorescent lighting, unvented	0.67
Recessed fluorescent lighting, vented to return air	0.59
Recessed fluorescent lighting, vented to supply and return air	0.19
Incandescent lighting	0.71
Equipment	0.2–0.8
Conductive heat gain through walls	0.63
Conductive heat gain through roofs	0.84
Transmitted solar radiation	1.0
Solar radiation absorbed by window glass	0.63

TABLE 7.23

Summary of Load Calculations

Item	Method and Comments	Section in Book
Zones	Define zones. Zone-part of building that can be assumed to have uniform loads. For each zone carry out the steps below	7.4
Design conditions	Determine appropriate values of temperatures (T_i and T_o) and humidities (W_i and W_o) for peak conditions at site in question	7.2
Conduction	$K_{cond} = \sum_K U_k A_k$	7.2.2
Air change	$\dot{V}\rho c_p$	7.1
	For relatively simple buildings, use LBL model, Eq. (7.12); otherwise use correlations of Sec. 7.1.5	
Heat gains	Solar, lights, equipment, occupants	7.2.3
Heating load, sensible	$\dot{Q}_{h,max} = K_{tot}(T_i - T_o) - \dot{Q}_{gain}$, with $K_{tot} = K_{cond} + \dot{V}\rho c_p$	7.5
Cooling load, "static," sensible	$\dot{Q}_{c,cond,t} = UA(\mathrm{CLTD}_t)$ $\dot{Q}_{c,rad,t} = \dot{Q}_{max}(\mathrm{CLF}_t)$	7.6
Cooling load, dynamic, sensible	Transfer function method Eqs. (7.51) to (7.66)	7.7
Latent loads	Latent gain from air change $\dot{Q}_{air,lat} = \dot{V}\rho h_{fg}(W_o - W_i)$, also latent gains from occupants and from equipment	7.2.3

the two is that the former is assumed to be absorbed by the floor only while the latter is assumed to be evenly distributed throughout the space. The radiant time factors are determined through a zone heat balance model as described by Spitler et al. (1997).

7.9 Summary

We have described the tools for the calculation of heating and cooling loads. The focus in this chapter has been on peak loads; annual loads and energy consumption are addressed in Chap. 8. The procedure begins with the definition of the zones and the choice of the design conditions.

   This is followed by a careful accounting of all thermal energy terms, including conduction, air change, and heat gains. The formulas for the load calculation depend on whether the thermostat setpoint is constant or variable. The first case is much simpler, allowing a static calculation (for heating loads) or a quasi-static calculation (for cooling loads). Correct analysis of loads for variable setpoints requires a dynamic method; in such a case, the transfer function method can be used both for heating and for cooling. For latent loads, a static calculation is usually considered sufficient. The key points are summarized in Table 7.23.

Problems

The problems in this book are arranged by topic. The approximate degree of difficulty is indicated by a parenthetic italic number from 1 to 10 at the end of the problem. Problems are stated most often in USCS units; when similar problems are presented in SI units, it is

done with approximately equivalent values in parentheses. The USCS and SI versions of a problem are not exactly equivalent numerically. Solutions should be organized in the same order as the examples in the text: given, figure or sketch, assumptions, find, lookup values, solution. For some problems, the Heating and Cooling Buildings (HCB) software included may be helpful. In some cases it is advisable to set up the solution as a spreadsheet, so that design variations are easy to evaluate.

7.1 Find the infiltration rate through a curtain wall of area 300 ft² (27.88 m²) if the indoor-outdoor pressure difference is 0.2 inWG (49.8 Pa). (4)

7.2 Find the indoor-outdoor pressure difference at the base of a one-story building of height 3 m (9.84 ft) without mechanical ventilation if the pressure coefficient $\Delta C_p = 0.3$, wind speed $v = 6.7$ m/s (15 mi/h), and $T_i - T_o = 15$ K (27.0°F). (5)

7.3 A one-story building with dimensions 100 ft × 100 ft × 13 ft (30.48 m × 30.48 m × 3.96 m) has an infiltration rate of 0.2 air change per hour during unoccupied periods. During the day, it is occupied by 80 people, each requiring 20 ft³/min (10 L/s) of outdoor air. Find the contribution of the air change to the total heat loss coefficient during unoccupied and occupied periods. Assume that during occupied periods there is only controlled ventilation, no infiltration. (5)

7.4 A 10-story office building with floor dimensions 50 ft × 100 ft (15.24 m × 30.48 m) and a height of 130 ft (39.6 m) has curtain walls with windows that are fixed and airtight. The window/wall ratio is 0.5. The draft coefficient for airflow between floors is $C_d = 0.65$. There are two vestibule-type doors on each of the two 100-ft (30.48-m) facades. The traffic rate corresponds to each of the occupants [one per 150 ft² (13.94 m²) of gross floor area] making on average five entrances or exits per 10 h. The indoor and outdoor temperatures are 70°F (21.1°C) and 20°F (−6.7°C), and the wind is incident parallel to the 50-ft (15.24-m) facade at 15 mi/h (6.7 m/s). Assume that infiltration through the roof is negligible. In other words, the only infiltration occurs through the curtain walls and through the doors.

  (a) Compute the pressure differences for each wall due to the stack effect and wind for floors 1, 5, and 10, with due attention to signs (infiltration or exfiltration).

  (b) Compute the total infiltration rates for these floors if the ventilation system is balanced for neutral pressure.

  (c) Compute the total infiltration rates for these floors if the ventilation system creates an overpressure of 0.1 inWG (24.9 Pa). (10)

7.5 There is an odor in your room that you would like to get rid of before your mother visits later this afternoon. The concentration of the odor-causing substance needs to be reduced by a factor of 10 in order to become unnoticeable. Using Fig. 7.2, try to estimate the number of hours that the windows should be open before your mother's visit. (7)

7.6 Suppose you live in a two-story house in a residential neighborhood that has 10 double-hung 1 m × 1.5 m windows that are not weather-stripped. The indoor temperature is constant at 20°C.

  (a) Estimate the leakage area from just the windows.

  (b) Assuming an average winter wind speed of 2 m/s and an average outdoor temperature of 5°C, what is the infiltration rate?

  (c) Estimate the associated heating load with this infiltration.

(d) If energy costs are $5/GJ, what is the cost of heating this air?

(e) How much money could you save per year if the windows had weather stripping? (6)

7.7 Find the design heat load for a building with total heat loss coefficient $K_{tot} = 5000$ Btu/(h · °F) (2640 W/K) if $T_i = 70°F$ (21.1°C) and $T_o = 0°F$ (−17.8°C) if one can count on 10 kW of heat gain from equipment that will be left on around the clock. (4)

7.8 Consider a single-family detached residence. It is a single-story building with an attached heated garage. (Assume that the garage is insulated the same and maintained at the same level of comfort as the house living space.) The house has the following characteristics.

General	Gross floor area (including garage)	2972 ft²
	Living gross floor area	2516 ft²
	Perimeter (including garage)	360 ft
	Wall height	8 ft
	Gross wall area	2880 ft²
	Window area (openings)	332 ft²
	Door area (openings)	70 ft²
	Garage door area	120 ft²
	Roof area (including garage)	3420 ft²
Wall construction	Wood siding, bevel, 0.5 in × 8 in lapped	
	Plywood sheathing, 0.5 in	
	Framing, nominal 2 ×4, 16-in center, with fiberglass batt insulation	
	Gypsum, 0.5 in	
Roof construction	Wood shingles, cedar shake	
	Felt building paper	
	Plywood sheathing, 0.5 in	
	Framing, nominal 2 × 6 trusses, 24-in centers	
	23° roof pitch	
Ceiling construction	Cellulose insulation, blown in, 8 in	
	Framing, nominal 2 × 6 trusses, 24-in centers	
	Gypsum, 0.5 in	
Window construction	Insulating glass, double, 0.5-in airspace, wood sash and frame	
	Loosely drawn interior drapery	
Door construction	Solid-core wood doors, 1.75 in, no glazing, no storm doors	
Garage door construction	$U = 0.55$ Btu/(h · ft² · °F)	

(a) Calculate overall *UA* values for each element of the building envelope, i.e., wall, roof, window, door.

(b) What is the heating load contribution due to conduction if the outdoor temperature is constant at 5°F and the indoor temperature is maintained at 70°F?

(c) How much would the conduction portion of the building heat loss be reduced by replacing the 0.5-in plywood sheathing in the walls with 0.75-in styrofoam (expanded, extruded polystyrene)?

(d) If you needed to reduce conduction through the envelope by 20 percent, how would you do it?

7.9 This problem uses the house description from Prob. 7.8. This house is to be built in suburban Denver. It will be built on a 0.3-acre lot, which is large compared to most suburban densities, and will have several small trees and shrubs near the house. It is estimated that the house will have an air change rate of 0.5 air change per hour under typical winter conditions of 10 mi/h wind speed and an outdoor temperature of 30°F. (Assume the house is maintained at 70°F.)

(a) Estimate the effective infiltration leakage area of the house, using the LBL model.

(b) Calculate the heating load due to infiltration under design winter conditions of 15 mi/h wind speed and 2°F outdoor temperature.

7.10 This problem uses the house description from Prob. 7.9. The house has a heated full basement under all but the garage. (The garage measures 22.5 ft × 19.5 ft.) The basement foundation wall is insulated on the outside with R4 insulation. The wall is 7.5 ft high with 1.5 ft above grade. The garage is built on a slab-on-grade floor. The slab foundation wall is also insulated below grade with the same insulation. Assuming that the average winter air temperature is 25°F and that the soil conductivity is 0.8 Btu/(h · ft · °F), estimate the design heat loss from the basement and garage slab. Clearly state your assumptions.

7.11 Find the conductive heat loss coefficient of a one-story building with the following characteristics: dimensions 100 ft × 100 ft × 13 ft (30.48 m × 30.48 m × 3.96 m), steel deck roof with 3.33-in (0.08-m) insulation (see Table 7.7a), steel siding wall with 4-in (0.1-m) insulation (see Table 7.7b), window/wall ratio of 0.4, and double-glazed windows. Neglect heat exchange with the ground.   (5)

7.12 A wall consists of the following components (order from inside to outside):

- 0.75-in (1.8-cm) gypsum plaster
- 4.0-in (0.1-m) glass wool insulation
- 4.0-in (0.1-m) facebrick

  (a) Calculate the *U* value for wind speed of 0 and 15 mi/h (6.7 m/s).

  (b) What is the inside surface temperature of the wall if $T_i = 70°F$ (21.1°C) and $T_o = 0°F$ (−17.8°C)?

  (c) Consider qualitively what would happen if the order of insulation and brick were interchanged. Would the *U* value change? How would the peak conductive cooling load change?   (7)

7.13 Wall consists of the following components (order from inside to outside): 0.75 in of gypsum plaster, 2 in of foam insulation, 4 in of lightweight concrete.

(a) Calculate the U value for wind speeds of 0 and 15 mi/h.

(b) What is the inside surface temperature of the wall if $T_i = 70°F$ and $T_o = 0°F$?

(c) Consider qualitatively what would happen if the order of the insulation and concrete were interchanged. Would the U value change? How would the peak conductive cooling load change?   (7)

7.14 Find the design heat load for the following building. The design outdoor temperature is $T_o = 0°F$ ($-17.8°C$) at night and until 9:00 a.m. At night the building is unoccupied, at indoor temperature $T_i = 50°F$ ($10°C$) and without internal gains, and the total heat loss coefficient K_{tot} is 30 kBtu/(h · °F) (15.8 kW/K). Starting at 8:00, the building is occupied by 500 occupants, with 250 kW (853 kBtu/h) of gains from lights and equipment, and $T_i = 70°F$; the ventilation system is also turned on at 8:00, increasing the total heat loss coefficient K_{tot} to 45 kBtu/(h · °F) (23.7 kW/K). Assume steady-state conditions; i.e., neglect pickup loads. Does the peak occur during occupied or unoccupied periods?   (6)

7.15 Consider a single-family, detached, single-story house located in suburban Long Island. The location has a design outdoor temperature of 13°F, and the indoor temperature is typically maintained at 70°F during the heating season. The house is built on a slab-on-grade and has no garage. The house has the following characteristics:

General		
	Length	56 ft
	Width	32 ft
	Wall height	8 ft
	Perimeter	176 ft
	Floor area	1,792 ft^2
	Gross wall area	1,408 ft^2
	Net wall area	1,114 ft^2
	Window area (openings)	252 ft^2
	Door area (openings)	42 ft^2
	Roof area	2,060 ft^2
	Gross volume (living area)	14,336 ft^3
Foundation	Slab-on-grade with 8-in foundation walls and vertical board insulation on outside of foundation wall	
	R value of insulation	5.4 (h · ft^2 · °F)/Btu
Construction net U values	Walls	0.084 Btu/(h · ft^2 · °F)
	Windows	0.420 Btu/(h · ft^2 · °F)
	Doors	0.350 Btu/(h · ft^2 · °F)
	Roof/ceiling (based on ceiling area)	0.030 Btu/(h · ft^2 · °F)

Using the information given above, perform the following calculations to determine the design heating load for the house.

(a) What is the design heat loss due to transmission heat transfer through the combined above-grade, building envelope (walls, windows, door, and ceiling)?

(b) What is the design heat loss due to heat transfer through the slab-on-grade floor?

(c) The infiltration leakage area has been determined to be 91.6 in^2 (590 cm^2). If the design winter wind velocity is 15 mi/h (6.7 m/s), what is the volumetric flow rate of infiltration air?

(d) What is the design heat load for the house? Ignore the effects of any moisture requirements, and make the conservative assumption that there are no heat gains due to solar, lights, equipment, or people. Assume that the air density is 0.075 lb/ft^3. (Note that 1 ft^3/h $=0.00787$ L/s.)

(e) If this house were located in Madison, Wisconsin, what would be the design outdoor temperature used to calculated the design heating load? (10)

7.16 Two people own a house with a total conductive heat loss coefficient $K_{cond}=120$ W/K [227.4 Btu/(h·°F)] and with a ventilation system that is controlled by an occupancy sensor to provide 7.5 L/s (15 ft^3/min) of outdoor air per person for the number of people actually present. They are giving a party for 30 guests. Everyone is so pleased that they are staying until steady-state conditions are established. Assume that the heat gains from other than occupants remain constant at 1 kW (3412 Btu/h).

(a) Estimate the latent and sensible heat gains as well as the outdoor air requirements, before and during the party.

(b) If the outdoor and indoor temperatures are $T_o=0$°C (32°F) and $T_i=20$°C (68°F), how much does the heating load change relative to steady-state conditions before the party?

(c) Does the latent heat gain have any effect on the heating load if there is no humidity control in the house? (8)

7.17 Estimate the temperature of the air in an unheated attached garage for these steady-state conditions:

- Dimensions are 5 m × 5 m × 3 m (16.41 ft × 16.41 × 9.85 ft).
- One wall is adjacent to house with $U=0.5$ W/(m^2·K) [0.088 Btu/(h·ft^2·°F)].
- The other walls (including door) and the roof are in contact with the outside and have $U=5$ W/(m^2·K) [0.88 Btu/(h·ft^2·°F)].
- The floor consists of 0.1 m (4 in) of concrete, and the ground underneath is assumed to have a uniform temperature of 12°C (53.6°F).
- The air exchange rate is 0.1 per hour.
- $T_o=-10$°C (14°F). (7)

7.18 If the house of Example 7.5 (Fig. 7.7) has a floor consisting of 4-in (0.1-m) concrete and 4-in (0.1-m) of Styrofoam, in direct contact with the ground, how do the peak load and annual heating load change? (6)

7.19 Find the latent cooling load per occupant for sedentary activity if the outdoor airflow rate is 20 ft^3/min per occupant and the outdoor air is 92°F (33.3°C) dry-bulb temperature with 74°F (23.3°C) wet-bulb temperature. Indoor conditions are 70°F (21°C) and 50 percent relative humidity. (6)

7.20 The installed lighting power in a building with $A_{floor} = 10,000$ ft^2 (929 m^2) is 1.8 W/ft^2 (19.4 W/m^2). The construction type is heavy. Find the contribution to the cooling load at 17:00 if the lights are left on from 9:00 to 17:00.   (5)

7.21 Suppose you work in an office with 1000 W of lighting. The lights are turned on at 8 a.m. Calculate the cooling load due to the lights if the lights are on 10 and 12 h/day. Assume that coefficient A of the room is 55 and the B classification is C. Now try to estimate the cooling load if the lights are on 11 h/day by interpolating between the 10 and 12 h/day CLFs. Plot all three results (load versus hour of day) on the same graph. To what extent does the 11 h/day calculation make sense?   (7)

7.22 Find the contribution to the cooling load at 17:00 in July at 32°N due to a window of area 100 ft^2 (9.29 m^2) and the following configurations:

(a) The window faces north.

(b) The window faces south and is unshaded.

(c) The window faces south and is shaded.   (6)

7.23 Calculate and compare the peak solar cooling load due to a south-facing vertical window of area 100 ft^2 (9.29 m^2) at latitude 32°N in July, if the construction type is medium, for several window designs:

(a) Unshaded window with clear double glazing (SC = 0.88)

(b) Same as (a) but completely shaded externally from direct solar radiation.

(c) Consider the glazing types in Tables 6.5 and 6.6. Which one would you select to minimize cooling loads?   (6)

7.24 The following three conductive heat load profiles were calculated for south-facing walls in Denver in July. The walls are 4-, 8-, and 12-in concrete. Without doing any calculations, identify which wall is which and justify your answer.   (2)

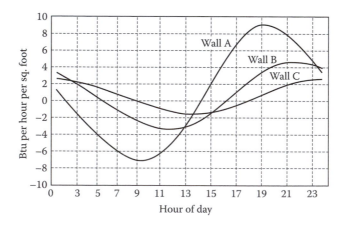

FIGURE P7.24

7.25 Calculate and plot the incremental cooling load for a house if an extra 500 W of lights is left on from 16:00 to 24:00. Assume that the house is of lightweight construction and is conditioned to a constant indoor temperature. Is the daily average load increase equal to the daily average heat output?   (7)

7.26 Modify the default values of the building construction in the HCB software, and observe the effect on the peak cooling load. In particular, find the modifications that maximize the cooling load and those that minimize it. Consider, one by one, modifications of (a) the roof, (b) the walls, (c) the floor, and (d) the windows (choosing among the glazing types in Tables 6.5 and 6.6).   (6)

7.27 Consider the default commercial building described in the HCB computer program. Use the program to calculate the peak cooling load for the default case and for each of the eight individual modifications. (Consider each individual modification to the default building, not cumulatively.)

Modifications

(a) Change the roof to 2-in heavyweight concrete with 1-in insulation.

(b) Change the floor to 4-in concrete slab.

(c) Change all walls to 4-in common brick with 2-in insulation.

(d) Change infiltration to 1 air change per hour.

(e) Move all north windows to the south side.

(f) Double the installed light wattage.

(g) Double the number of people.

(h) Double the installed equipment wattage.   (7)

7.28 Use the CLTD method to calculate and plot the cooling load per square meter (square foot) due to conduction through a flat roof of construction type 5 in the HCB tables on the CD-ROM.

(a) Assume a light surface [$K = 0.5$ in Eq. (7.40)].

(b) Assume a dark surface [$K = 1.0$ in Eq. (7.40)].

(c) How large is the corresponding difference in daily total cooling load for a roof of area 500 m^2 (5380 ft^2)?   (6)

7.29 Consider the design of a small office building with the following specifications:

- Site Denver, Colorado
- T_i constant 24 h/day at 78°F (25.6°C) in summer
- Size 30 ft × 30 ft × 8 ft (9.14 m × 9.14 m × 2.44 m), orientation due south
- Open floor plan to allow treating it as one zone
- Equal windows on all facades, with window/wall ratio $= 0.25$
- 0.5 air change per hour from 8:00 a.m. to 6:00 p.m., 0.2 air change per hour the rest of the time
- Double-pane glass with $U = 0.5$ Btu/(h · ft^2) [2.84 W/(m^2 · K)] and SC $= 0.82$
- Lights and office equipment 3.0 W/ft^2 (32.28 W/m^2) from 8:00 a.m. to 6:00 p.m., 0.5 W/ft^2 (5.38 W/m^2) the rest of the time
- Occupants 0.01 per square foot (0.108 per square meter) from 8:00 a.m. to 6:00 p.m.
- Roof type 3 in the HCB tables on the CD-ROM, but with 4-in (0.1-m) glass wool insulation, color dark, without suspended ceiling
- Wall group D in the HCB tables on the CD-ROM with 2-in insulation

Neglect heat exchange with the ground.

(a) Calculate the peak and the daily total cooling load.

(b) Try to reduce the peak cooling load by selecting, in the HCB tables, the roof type that yields the lowest peak CLTD.

(c) Try to reduce the peak cooling load further by selecting, in the HCB tables, the wall type that yields the lowest peak CLTD.

(d) Try to reduce the peak cooling load still further by means of exterior shading devices that keep direct solar radiation from all the windows.   (10)

7.30 In Example 7.18, the cooling load of a simple zone was calculated with the transfer function method. Since T_i is constant, the CLTD/CLF method could also be used. Compare the results of these two methods for the loads due to (a) conduction, (b) lights, and (c) ventilation.   (6)

7.31 How much do personal computers increase the indoor temperature in a small office building without air conditioning, if they are left on continuously? Assume a floor area of 500 m^2 (5380 ft^2), total heat transmission coefficient $K_{tot} = 1200$ W/K [2274 Btu/(h · °F)], and heat production per floor area 10 W/m^2 (0.93 W/ft^2) continuous.   (6)

7.32 Use the transfer function method to estimate how much the heat from personal computers increases the indoor temperature in a small office building without air conditioning, if they are left on 8 h/day. Does the answer depend on the temperature that the building would have in the absence of computers? Assume

- Floor area of 500 m^2 (5380 ft^2)
- Total heat transmission coefficient $K_{tot} = 1200$ W/K [2274 Btu/(h · °F)]
- Heat production per floor area 10 W/m^2 (107.6 W/ft^2) while turned on
- Construction type medium   (10)

7.33 Redo Example 7.19 if the cooling equipment is limited to a peak thermal output of 400 W.   (10)

7.34 Using the default building in the HCB loads menus, compare the effect of a dark roof versus a light roof on the total cooling load for Phoenix in July. Now change the roof type to include more insulation. How does this affect the total cooling load?   (3)

7.35 Suppose a fire breaks out in the dormitory room next to yours. For simplicity, assume the air temperature in your room remains constant at 70°F and that the average air temperature in the adjacent room is 600°F which is achieved immediately after the fire starts and then remains constant. The wall between the room consists of 6 in of heavyweight concrete. Unfortunately, you have taped a picture of your sweetheart to this wall. If the paper ignites when it reaches 451°F, how long after the fire starts will the picture catch fire?   (10)

References

Achard, P., and R. Gicquel (1986). *European Passive Solar Handbook*. Commission of the European Communities, Directorate General XII for Science, Research and Development, Brussels, Belgium.

ASHRAE (1979). *Cooling and Heating Load Calculation Manual*. GRP 158. American Society of Heating, Refrigerating and Air-Conditioning Engineers, Atlanta.

ASHRAE (1987). *Standard 62-87: Ventilation for Acceptable Indoor Air Quality*. American Society of Heating, Refrigerating and Air-Conditioning Engineers, Atlanta.

ASHRAE (1989a). *Handbook of Fundamentals*. American Society of Heating, Refrigerating and Air-Conditioning Engineers, Atlanta.

ASHRAE (1989b). *Standard 90.1-1989: Energy Efficient Design of New Buildings, except Low-Rise Residential Buildings*. American Society of Heating, Refrigerating and Air-Conditioning Engineers, Atlanta.

ASHRAE (2001). *Handbook of Fundamentals*. American Society of Heating, Refrigerating, and Air-Conditioning Engineers, Atlanta.

Birdsall, B., W. F. Buhl, K. L. Ellington, A. E. Erdem, and F. C. Winkelmann (1990). "Overview of the DOE2.1 Building Energy Analysis Program." Report LBL-19735, rev. 1. Lawrence Berkeley Laboratory, Berkeley, Calif.

Dietz, R. N., T. W. Ottavio, and C. C. Cappiello (1985). "Multizone Infiltration Measurements in Homes and Buildings Using Passive Perfluorocarbon Tracer Method." *ASHRAE Trans.*, vol. 91, pt. 2.

Fairey, P. W., and A. A. Kerestecioglu (1985). "Dynamic Modelling of Combined Thermal and Moisture Transport in Buildings: Effects on Cooling Loads and Space Conditions." *ASHRAE Trans.*, vol. 91, pt. 2A, p. 461.

Grimsrud, D. T., M. H. Sherman, J. E. Janssen, A. N. Pearman, and D. T. Harrje (1980). "An Intercomparison of Tracer Gases Used for Air Infiltration Measurements." *ASHRAE Trans.*, vol. 86, pt. 1.

Kerestecioglu, A. A., and L. Gu (1990). "Theoretical and Computational Investigation of Simultaneous Heat and Moisture Transfer in Buildings: 'Evaporation and Condensation' Theory." *ASHRAE Trans.*, vol. 96, pt. 1.

Lydberg, M., and A. Honarbakhsh (1989). "Determination of Air Leakiness of Building Envelopes Using Pressurization at Low Pressures." Swedish Council for Building Research. Document D19:1989, Gävle, Sweden.

Mitalas, G. P., and D. G. Stephenson (1967). Room Thermal Response Factors. *ASHRAE Trans.*, vol. 73, pt. 2.

Nisson, J. D. N., and G. Dutt (1985). *The Superinsulated Home Book*. Wiley, New York.

Norford, L. K., A. Rabl, J. P. Harris, and J. Roturier (1989). "Electronic Office Equipment: The Impact of Market Trends and Technology on End Use Demand." In T. B. Johansson et al., eds. *Electricity: Efficient End Use and New Generation Technologies, and Their Planning Implications*. Lund University Press, Lund, Sweden, pp. 427–460.

Pederson, C. O., D. E. Fisher, J. D. Spitler, and R. J. Liesen (1998). *Cooling and Heating Load Calculation Principles*. American Society of Heating, Refrigerating and Air-Conditioning Engineers, Atlanta.

PREP (1990). Included in "TRNSYS—A Transient System Simulation Program." Solar Energy Laboratory, Engineering Experiment Station Report 38-12, University of Wisconsin, Madison.

Sherman, M. H., D. T. Grimsrud, P. E. Condon, and B. V. Smith (1980). "Air Infiltration Measurement Techniques." *Proc. 1st IEA Symp.*, Air Infiltration Centre, London. Also Lawrence Berkeley Laboratory Report LBL 10705, Berkeley, Calif.

Sonderegger, R. C. (1978). "Diagnostic Tests Determining the Thermal Response of a House." *ASHRAE Trans.*, vol. 84, pt. 1, p. 691.

Spitler, J. D., and D. E. Fisher (1999). "On the Relationship Between the Radiant Time Series and Transfer Function Methods for Design Cooling Load Calculations." *Int. J. of HVAC&R Research*, vol. 5, no. 2, pp. 125–138.

Spitler J. D., D. E. Fisher, and C. O. Pedersen (1997). "The Radiant Time Series Cooling Load Calculation Procedure." *ASHRAE Trans.*, vol. 103, no. 2, pp. 503–515.

Wong, S. P. W., and S. K. Wang (1990). "Fundamentals of Simultaneous Heat and Moisture Transfer between the Building Envelope and the Conditioned Space Air." *ASHRAE Trans.*, vol. 96, pt. 2.

8

Annual Energy Consumption and Special Topics

In the first section of this chapter, we present methods for the calculation of annual energy consumption, a crucial topic for every building designer. Both steady-state and dynamic methods are described.

In the rest of the chapter, we work to provide a better understanding of the modeling process. In Sec. 8.2 we present thermal networks, since they provide a framework that is general, yet conceptually simple for the modeling of buildings. The *RC* network is presented and solved as the simplest possible dynamic model. It serves to explain the important concept of the time constant and to estimate the warm-up and cooldown times of a building.

In Sec. 8.3 we address the extension to more complicated networks, and we explain the connection with transfer function models and with the transient heat conduction equation. We also treat the response to periodic driving terms.

In Sec. 8.4, we introduce the topic of inverse models (i.e., models derived from data); they are useful for diagnostics (e.g., weather correction of consumption data) and for optimal control. Finally, in Sec. 8.5, we discuss the comparison between models and measured data because the modeler should understand the limitations of his or her predictions. We close with a brief summary of the principal methods.

8.1 Annual Consumption

The yearly energy consumption Q_{yr} is needed to evaluate the operating cost—an essential step if one wants to come close to an optimal design, in the sense of minimizing the life cycle cost of a building. Consumption Q_{yr} is the time integral of the instantaneous consumption over the heating or cooling season; the instantaneous consumption is the instantaneous load divided by the efficiency of the heating or cooling equipment. Several methods are available, depending on the complexity of the case and the amount of detail one wants to take into account. A major distinction is between steady-state methods (based on degree-days or temperature bins) and dynamic methods (e.g., based on transfer functions).

Degree-day methods are the simplest. They are appropriate if the utilization of the building and the efficiency of the HVAC equipment can be considered constant. For situations where efficiency or conditions of utilization vary with outdoor temperature, one can calculate the consumption for different values of the outdoor temperature and multiply it by the corresponding number of hours; this approach is used in the bin methods. As a starting point for these steady-state methods, one needs the value of T_o below which heating becomes necessary; that is called the *balance-point temperature*.

When the indoor temperature is allowed to fluctuate, one leaves the domain of simple steady-state models. Various correction terms have been developed to permit an approximate treatment of some dynamic effects with steady-state methods. For greatest accuracy, the use of full-fledged dynamic models is recommended, a choice that is becoming ever more natural with the advance of computer technology. We do not describe dynamic correction terms for steady-state methods because we believe that interest in them is waning in favor of dynamic simulation programs.

But even in an age when computers can calculate the energy consumption of a building at the touch of a key, the concepts of degree-day and balance-point temperature remain valuable tools for intuition. The severity of a climate can be characterized concisely in terms of degree-days. Also the degree-day method and its generalizations can provide a simple estimate of annual loads, which can be quite accurate if the indoor temperature is constant. For these reasons, we present the basic steady-state methods in detail.

8.1.1 Balance-Point Temperature and Degree-Days

The *balance-point temperature* T_{bal} of a building is defined as that value of the outdoor temperature T_o where, for the specified value of T_i, the total heat loss is equal to the free heat gain (from the sun, occupants, lights, etc.):

$$K_{\text{tot}}(T_i - T_{\text{bal}}) = \dot{Q}_{\text{gain}} \tag{8.1}$$

where K_{tot} [Btu/(h · °F) (W/K)] is the total heat loss coefficient of the building. Hence the balance-point temperature is

$$T_{\text{bal}} = T_i - \frac{\dot{Q}_{\text{gain}}}{K_{\text{tot}}} \tag{8.2}$$

Heating is needed only when T_o drops below T_{bal}. The rate of energy consumption of the heating system is

$$\dot{Q}_h = \frac{K_{\text{tot}}}{\eta_h}[T_{\text{bal}} - T_o(t)] \quad \text{when } T_o < T_{\text{bal}} \tag{8.3}$$

where η_h = efficiency of the heating system (also designated AFUE, for annual fuel-use efficiency, details to be discussed in Chapter 9). Detailed models for η_h are presented in Chapter 9. For now we make the simplifying assumption that η_h, T_{bal}, and K_{tot} are constant. Then the annual heating consumption can be written as an integral:

$$Q_{h,\text{yr}} = \frac{K_{\text{tot}}}{\eta_h} \int [T_{\text{bal}} - T_o(t)]_+ \, dt \tag{8.4}$$

where the plus sign subscript on the bracket indicates that only positive values are to be counted. This integral of the temperature difference is a convenient summary of the effect of outdoor temperatures on a building. In practice, it is approximated by a sum of averages over short time intervals (daily or hourly), and the result is called *degree-days* or degree-hours.

If daily average values of outdoor temperature T_o are used for evaluating the integral, one obtains the degree-days for heating $D_h(T_{bal})$ as[1]

$$D_h(T_{bal}) = 1 \text{ day} \times \sum_{days} (T_{bal} - T_o)_+ \tag{8.5}$$

with dimensions of °F · days (K · days). It is a function of T_{bal}, reflecting the roles of T_i, heat gain, and the loss coefficient. Also T_{bal} is called the *base* of the degree-days. In terms of degree-days, the annual heating consumption is

$$Q_{h,yr} = \frac{K_{tot}}{\eta_h} D_h(T_{bal}) \tag{8.6}$$

Degree-days or degree-hours for a balance-point temperature of 18°C in Europe or 65°F (18.3°C) in the United States have been widely tabulated, based on the observation that this has represented average conditions in typical buildings (that the appropriate value has been changing with improving insulation is a point which we discuss next). This value is understood whenever T_{bal} is not indicated explicitly. A contour map of heating degree-days for the United States is shown in Fig. 8.1. The extension of degree-day data to different bases is discussed next.

Example 8.1

Find the annual heating bill for a house in New York under the following conditions.
  Given: Heat transmission coefficient $K_{tot} = 205$ W/K [388.5 Btu/(h · °F)]
Heat gain $\dot{Q}_{gain} = 569$ W (1941 Btu/h)
$T_i = 21.1°C$ (70°F)
Efficiency of heating system $\eta = 0.75$
Fuel price $8/GJ ($8.44/MBtu)
Assumptions: K_{tot}, $\dot{Q}_{gain}$, T_i, and η_h are constant.
Lookup values: Degree-days $D_h = 2800$ K · days (5040°F · days) from Fig. 8.1 or Fig. 8.3.

SOLUTION

First we must evaluate the balance-point temperature

$$T_{bal} = T_i - \frac{\dot{Q}_{gain}}{K_{tot}} = 21.1°C - \frac{569 \text{ W}}{205 \text{ W/K}} = 18.3°C$$

---

[1] The question arises whether shorter time intervals should be used for calculating degree-days. With a summation over hourly values of T_o, one would obtain the degree-days in the form

$$D_h(T_{bal}) = \frac{1 \text{ day}}{24 \text{ h}} \times 1 \text{ h} \times \sum_{hours} (T_{bal} - T_o)_+$$

Because of the nature of short-term fluctuations of T_o, the result increases somewhat as the averaging interval is made finer. One could argue that the daily interval is more appropriate because the time constant (see Sec. 8.2.2) of most buildings is closer to a day than to an hour, and roughly speaking, the thermal inertia of a building averages the effect of fluctuations over the time constant. In any case, most degree-day tabulations do not indicate the details of the calculation (which explains some of the discrepancies between different sources).

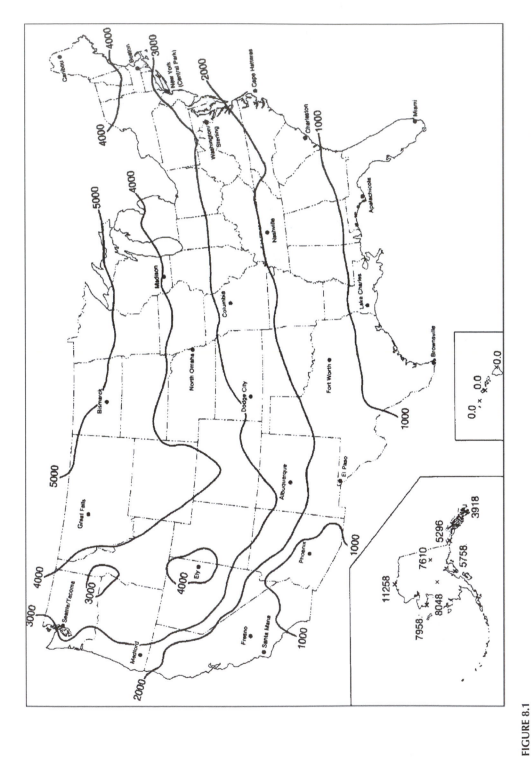

FIGURE 8.1
Annual heating degree-days (K · days) for the United States, for base of 65°F (18.3°C). Inserts show Alaska and Hawaii. To convert to °F · days, multiply by $\frac{9}{5}$. (Courtesy of SERI, Solar Radiation Energy Resource Atlas of the United States, Solar Energy Research Institute, Golden, CO, Report SERI/SP-642-1037, 1981.)

Now T_{bal} happens to have the standard value, and we can proceed directly with the value $D_h = 2800$ K · days. Inserting this into Eq. (8.6) (remember the conversion of units from days to seconds), we obtain the annual heating energy

$$Q_{h,yr} = \frac{K_{tot}D_h}{\eta_h} = 205 \text{ W/K} \times 2800 \text{ K} \cdot \text{days} \times 24 \text{ h/day} \times \frac{3600 \text{ s/h}}{0.75}$$

$$= \frac{49.58 \text{ GJ}}{0.75} = 66.11 \text{ GJ (62.67 MBtu)}$$

The cost is 66.11 GJ × \$8/GJ = \$529.

Cooling degree-days can be defined in an analogous expression as

$$D_c(T_{bal}) = 1 \text{ day} \times \sum_{days} (T_o - T_{bal})_+ \tag{8.7}$$

While the definition of the balance-point temperature is the same for cooling as for heating, in a given building its numerical value for cooling is, in general, different from that for heating because T_i, Q_{gain}, and K_{tot} can be different. Note that heating and cooling degree-days, if based on the same T_{bal}, are related by

$$D_h(T_{bal}) - D_c(T_{bal}) = (365 \text{ days})(T_{bal} - T_{o,av}) \tag{8.8}$$

where $T_{o,av}$ = annual average of T_o. This follows from the fact that the terms of Eq. (8.5) are nonzero if and only if the terms of Eq. (8.7) vanish, and vice versa. Standard cooling degree-day data are widely available for a base of 18°C or 65°F; for example, Fig. 8.2 shows the data for the United States.

In some cases, it is desirable to have heating and cooling degree-days for shorter time periods than a year, and listings of monthly values are indeed quite common. The relation of Eq. (8.8) continues to hold, of course, with the appropriate number of days on the right side. Monthly data can be found in the HCB software.

While we postpone the presentation of a general model for the variation of degree-days with the base temperature, we show in Fig. 8.3a how the heating degree-days vary with T_{bal} for a particular site, New York. We have obtained this plot by evaluating Eq. (8.5) with data for the number of hours per year during which T_o is within 5°F temperature intervals centered at 72, 67, 62, ..., −8°F (from the weather data tables on the CD-ROM). The data for the number of hours in each interval ("bin") are included as labels in this plot. Analogous curves, without these labels, are shown in Fig. 8.3b for three further sites: Houston, Washington, D.C., and Denver.

Example 8.2

Estimate the annual cooling degree-days in New York for a base of $T_{bal} = 23.1$°C (73.6°F) and for a base of $T_{bal} = 18.1$°C (64.7°F).

Given: $D_h(T_{bal})$ of Fig. 8.3a

Find: $D_c(T_{bal})$

Lookup values: $T_{o,av} = 12.5$°C, from the weather data tables on the CD-ROM.

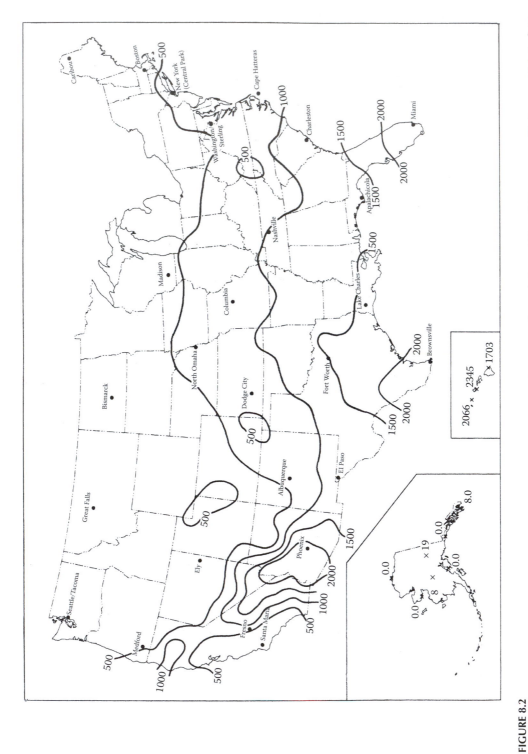

FIGURE 8.2
Annual cooling degree-days (K · days) for the United States, for base of 65°F (18.3°C). Inserts show Alaska and Hawaii. To convert to °F · days, multiply by $\frac{9}{5}$. (Courtesy of SERI, Solar Radiation Energy Resource Atlas of the United States, Solar Energy Research Institute, Golden, CO, Report SERI/SP-642-1037, 1981.)

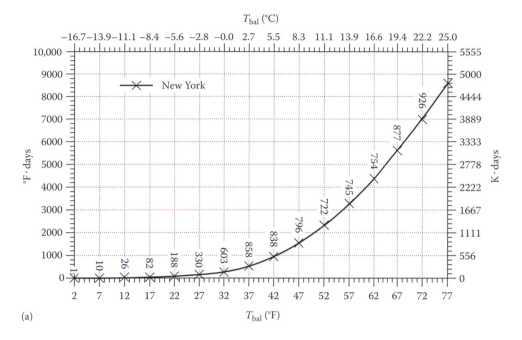

(a)

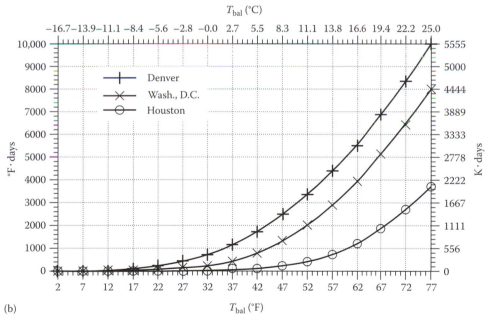

(b)

FIGURE 8.3

Annual heating degree-days $D_h(T_{bal})$ as function of T_{bal}. Based on the weather data tables on the CD-ROM. (a) For New York, with labels = h/yr when T_o is in 5°F (2.7 K) bin centered at the point (interpreting x axis as T_o values). (b) For Denver, Houston, and Washington, D.C., degree-days only.

SOLUTION

Reading Fig. 8.3a, we find

$$D_h(23.1°C) \approx 4080 \text{ K} \cdot \text{days (7344°F} \cdot \text{days)}$$

From Eq. (8.8) we have

$$D_c(23.1°C) = D_h(23.1°C) - (365 \text{ days}) [(23.1 - 12.5) \text{ K}]$$
$$\approx 208 \text{ K} \cdot \text{days (375°F} \cdot \text{days)}$$

Likewise for 18.1°C,

$$D_h(18.1°C) \approx 2778 \text{ K} \cdot \text{days (5000°F} \cdot \text{days)}$$

and

$$D_c(18.1°C) = D_h(18.1°C) - 365 \text{ days}[(18.1 - 12.5) \text{ K}]$$
$$\approx 734 \text{ K} \cdot \text{days (1321°F} \cdot \text{days)}$$

A calculation of cooling energy consumption is more difficult than one for heating. Here are some of the reasons. For cooling, the equation analogous to Eq. (8.6) would be

$$Q_c = \frac{K_{tot}}{\eta_c} D_c(T_{bal}) \tag{8.9}$$

for a building whose K_{tot} value never changes. That assumption is acceptable during the heating season when windows are kept shut and the air exchange rate is fairly constant. But during the cooling season, one can get rid of heat gains and postpone the onset of cooling by opening the windows or increasing the ventilation (in buildings with mechanical ventilation, this is called the *economizer mode*, discussed in Example 3.6 and Sec. 11.1). The air conditioner is needed only when the outdoor temperature goes beyond the threshold T_{max}. This threshold is given by an equation analogous to Eq. (8.2), with the replacement of the closed-window heat transmission coefficient K_{tot} by its value K_{max} for open windows

$$T_{max} = T_i - \frac{\dot{Q}_{gain}}{K_{max}} \tag{8.10}$$

Now K_{max} can vary quite a bit with wind, but to simplify the discussion, let us assume a constant value. The resulting cooling load is shown schematically in Fig. 8.4 as a function of T_o. The solid line is the load with open windows or increased ventilation; the dashed line is the load if K_{tot} is kept constant. One can show (by breaking the area under the solid line into a rectangle and a triangle) that the annual cooling load for this mode is

$$Q_c = K_{tot}[D_c(T_{max}) + (T_{max} - T_{bal})N_{max}] \tag{8.11}$$

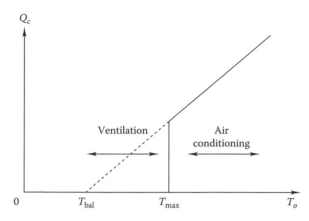

FIGURE 8.4
Cooling load as a function of outdoor temperature T_o. Below T_{max}, cooling can be avoided by using ventilation.

where $D_c(T_{max})$ is the cooling degree-days for base T_{max}, and N_{max} is the number of days during the season when T_o rises above T_{max}. This model of air conditioning is very schematic, of course. In practice, heat gains and ventilation rates are variable, to say nothing of erratic occupant behavior in utilizing the windows and air conditioner. Also in commercial buildings with the economizer mode, the extra fan energy for the increased ventilation should not be overlooked.

Example 8.3

Estimate the annual cooling load, with and without a window opening, for a house in New York under the following conditions:

Given: Heat transmission coefficient $K_{tot} = 205$ W/K [388.5 Btu/(h · °F)] with closed windows at 0.5 air change/h

$K_{cond} = 145$ W/K due to conduction alone

Heat gains $\dot{Q}_{gain} = 1200$ W (4094 Btu/h)

$T_i = 24.0°C$ (75.2°F)

Average air change rate with open windows = 10 air changes/h (a reasonable value in view of Figure 7.2)

Find: Q_c

Assumptions: K_{tot}, K_{max}, $\dot{Q}_{gain}$, and T_i are constant.

SOLUTION

The balance-point temperature for cooling is, from Eq. (8.2),

$$T_{bal} = T_i - \frac{\dot{Q}_{gain}}{K_{tot}} = 24.0°C - \frac{1200 \text{ W}}{205 \text{ W/K}} = 18.1°C$$

The corresponding number of cooling degree-days already found in Example 8.2 is

$$D_c(T_{bal}) = 734 \text{ K} \cdot \text{days}$$

If windows were always *closed*, the annual cooling load would be

$$Q_c = K_{tot}D_c(T_{bal}) = 205 \text{ W/K} \times 734 \text{ K} \cdot \text{days} \times 24 \text{ h/day} \times 3600 \text{ s/h} = 13.0 \text{ GJ}$$

If windows are *opened for* ventilative cooling, we first have to find K_{max}. For this we note that at 0.5 air change/h, the contribution of the air change is

$$K_{tot} - K_{cond} = 205 - 145 \text{ W/K} = 60.0 \text{ W/K}$$

Hence at 10 air changes/h this term must be multiplied by 10/0.5, giving a value

$$K_{max} = K_{cond} + 60.0 \text{ W/K} \times \frac{10}{0.5} = 145 + 1200 \text{ W/K} = 1345 \text{ W/K}$$

Next we find T_{max} from Eq. (8.10):

$$T_{max} = T_i - \frac{\dot{Q}_{gain}}{K_{max}} = 24.0°\text{C} - \frac{1200 \text{ W}}{1345 \text{ W/K}} = 23.1°\text{C} \ (73.6°\text{F})$$

When T_o is between $T_{bal} = 18.1°\text{C}$ and $T_{max} = 23.1°\text{C}$, the building can be maintained at comfortable conditions by opening the windows.

The required number of cooling degree-days was calculated in Example 8.2 as

$$D_c(23.1°\text{C}) = 208 \text{ K} \cdot \text{days}$$

To find the number of days N_{max} when T_o is above T_{max}, we take the data for the number of hours in each temperature bin, as shown by the labels in Fig. 8.3a. Adding these numbers, we find that there is 7756 h when T_o is below 75.5°F(24.2°C). Since the total number of hours in a year is 8760 h, this makes 8760−7756 h = 1004 h *above* 75.5°F. Around this temperature range, the distribution of hours per bin is quite uniform at 926 h/5°F, a fact which allows us to interpolate the number of hours above 73.6°F as 1004 h + 926 h × [(75.5 − 73.6)°F]/5°F = 1355 h. Hence $N_{max} = 56.5$ days.

Finally, we can insert these values into Equation 8.11 to obtain the cooling load as

$$Q_c = 205 \text{ W/K} \times [208 \text{ K} \cdot \text{days} + (23.1 - 18.1)°\text{C} \times 56.5 \text{ days}] \times 24 \times 3600$$
$$= 8.7 \text{ GJ}$$

COMMENTS

The difference between the loads with and without a window opening is quite large, on the order of 50 percent in this case (very dependent on K_{tot}, K_{max}, $\dot{Q}_{max}$, T_i, and the climate). This is one of several reasons why residential cooling loads are highly sensitive to occupant behavior.

A basic assumption of the degree-day method, the constancy of T_{bal}, is, of course, not well satisfied in practice. Solar gains are zero at night, and internal gains tend to be highest during the evening. The pattern for a typical house is shown in Fig. 8.5. As long as T_o stays always below T_{bal}, the variations average out without changing the consumption. But for the situation in Fig. 8.5, T_o rises above T_{bal} from shortly after 10:00 a.m. to 10:00 p.m.; the consequences for the energy consumption depend on the thermal inertia and on the control of the HVAC system. If this building had low inertia and temperature control were critical, heating would be needed at night and cooling during the day. In practice, this effect is reduced by thermal inertia and by the deadband of the thermostat which allows T_i to float somewhat.

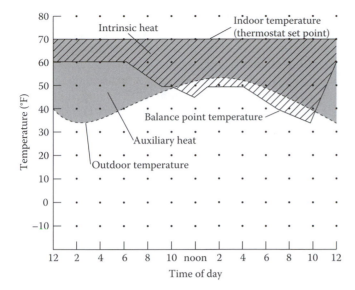

FIGURE 8.5
Variation of balance-point temperature and internal gains for a typical house. (Courtesy of Nisson, J.D.N. and Dutt, G., *The Superinsulated Home Book*, Wiley, New York, 1985.)

The closer T_o is to T_{bal}, the greater the uncertainty. If the occupants keep the windows closed during mild weather, T_i will rise above the set point. If they open the windows, the potential benefit of the heat gain is reduced. In either case, the true values of T_{bal} become quite uncertain. Therefore, the degree-day method is unreliable for estimating the consumption during mild weather, and so is any steady-state method, for that matter. In fact, the consumption becomes very sensitive to occupant behavior and cannot be predicted with certainty.

Despite these problems, the degree-day method can give remarkably accurate results for the annual heating energy of single-zone buildings. There are several reasons. Typical buildings have time constants that are roughly on the order of 1 day, and their thermal inertia averages in effect over the diurnal variations, especially if T_i is allowed to float. Furthermore, the uncertainty of the energy consumption in mild weather is small in absolute terms; hence a relatively large error here does not carry much weight in the total for the season.

An important point to note about annual energy calculations, by any of the steady-state methods described in this section, is that the heat gains must be the average values of the period in question, *not* the peak values. In particular, the solar radiation should be based on the averages (see Sec. 6.3.4), *not* the peak (Sec. 6.3.3 or the SHGF tables on the CD-ROM).

8.1.2 A Model for Degree-Days with Variable Base

The calculation of Q_h from the degree-days $D_h(T_{bal})$ in Example 8.1 was easy since T_{bal} happened to have the conventional value. That should not be counted on. The average value of T_{bal} varies widely from one building to another, because of widely differing personal preferences for the settings of thermostat and of thermostat setback and because of different building characteristics. In response to the fuel crises of the 1970s, heat transmission coefficients have been reduced, and thermostat setback has become a common practice. At the same time, the energy use by appliances has increased. These trends reduce T_{bal}, and evidence for this has indeed been found by Fels and Goldberg (1986). Hence degree-days with a base of 65°F or 18°C must be employed with caution.

Several authors have proposed formulas for estimating the degree-days relative to an arbitrary base, when detailed data are not available for evaluating the integral in Eq. (8.4). The basic idea is to assume a typical probability distribution of temperature data, characterized by its average $\bar{T}_o$ and by its standard deviation σ. Erbs et al. (1983) have developed a model that needs as input only the averages $\bar{T}_o$ for each month of the year. The standard deviations σ_m for each month are then estimated from the correlation

$$\sigma_m = 1.45 - 0.0290\bar{T}_o + 0.0664\sigma_{yr} \quad \text{(dimensional equation, } T \text{ and } \sigma \text{ in } °C) \quad \text{(8.12SI)}$$

where σ_{yr} is the standard deviation of the monthly mean temperatures

$$\sigma_{yr} = \sqrt{\frac{1}{12}\sum_{n=1}^{12}(\bar{T}_o - \bar{T}_{o,yr})^2} \quad \text{(8.13)}$$

about the annual average $\bar{T}_{o,yr}$. To obtain a simple expression for the degree-days, a normalized temperature variable θ is defined as

$$\theta = \frac{T_{bal} - \bar{T}_o}{\sigma_m\sqrt{N}} \quad \text{(8.14)}$$

with N = number of days in the month (N and θ are dimensionless). While temperature distributions can be quite different from month to month and location to location, most of this variability can be accounted for by the average and the standard deviation of T_o. Being centered around $\bar{T}_o$ and scaled by σ_m, the quantity θ eliminates these effects. In terms of θ, the monthly heating degree-days for any location are very well approximated by the following distribution:

$$D_h(T_{bal}) = \sigma_m N^{3/2}\left\{\frac{\theta}{2} + \frac{\ln[\exp(-a\theta) + \exp a\theta]}{2a}\right\} \quad \text{with } a = 1.698 \quad \text{(8.15)}$$

Erbs et al. (1983) have verified, for nine locations spanning most climatic zones of the United States, that the annual heating degree-days can be estimated with a maximum error of 175 K·days if one uses this equation for each month. For cooling degree-days, they found that the largest error is 150 K · days. Such errors are quite acceptable, representing less than 5 percent of the total.

Example 8.4

Find the heating degree-days for New York, using the model of Erbs et al. (1983).

   Given: Bin data for monthly averages of T_o from the HCB software as reproduced below in the second column.

   Find: $D_h(T_{bal})$ for $T_{bal} = 18.3°C$ (65°F)

SOLUTION

The results have been calculated with a spreadsheet and are printed below. The second column lists the values of monthly average outdoor temperature that are given (from HCB), and N is the number of days in the month. Intermediate quantities are shown in the fourth and fifth columns. And $\bar{T}_{o,yr}$ and σ_{yr} are shown at the bottom. The sixth column shows the results, and for comparison the seventh column shows the values calculated by HCB [direct evaluation of Eq. (8.5), using bin data].

Month	$\bar{T}_o$	N	$\sigma_{m\prime}$ Eq. (8.12)	$\theta,$ Eq. (8.14)	$D_h(T_{bal})$ Eq. (8.15)	$D_h(T_{bal})$ HCB
Jan	0.10	31	2.031	1.610	565	565
Feb	0.80	28	2.010	1.645	490	492
Mar	5.10	31	1.886	1.257	411	412
Apr	11.20	30	1.709	0.759	219	215
May	16.80	31	1.546	0.174	81	76
Jun	22.00	30	1.396	−0.484	12	0
Jul	24.80	31	1.314	−0.888	3	0
Aug	23.80	31	1.343	−0.735	5	0
Sep	20.20	30	1.448	−0.240	26	16
Oct	14.80	31	1.604	0.392	128	116
Nov	8.60	30	1.784	0.993	294	293
Dec	1.90	31	1.978	1.489	509	508
$\bar{T}_{o,yr}$	12.51			**Sum:**	**2742**	**2693**
σ_{yr}	8.789					

COMMENTS

The agreement is excellent, except for the summer months where the true values are zero.

For some applications, one needs the temperature distribution during the day, e.g., to account for the effect of occupancy schedules on the consumption of commercial buildings. Erbs et al. (1983) offer a formula for estimating the average temperature $\bar{T}_{o,t}$ for any hour t of the month:

$$\frac{\bar{T}_{o,t} - \bar{T}_o}{A} = 0.4632 \cos(t^* - 3.805) + 0.0984 \cos(2t^* - 0.360)$$

$$+ 0.0168 \cos(3t^* - 0.822) + 0.0138 \cos(4t^* - 3.513) \tag{8.16}$$

where

$$t^* = \frac{2\pi(t - 1\ \text{h})}{24\ \text{h}}$$

($t = 1$ h corresponding to 1:00 a.m.), and A is the diurnal temperature swing (peak to peak), correlated via

$$A = 25.8\bar{K}_T - 5.21 \quad \text{(dimensional equation, in °C)} \tag{8.17SI}$$

$$A = 1.8(25.8\bar{K}_T - 5.21) \quad \text{(dimensional equation, in °F)} \tag{8.17US}$$

with the monthly average solar clearness index $\bar{K}_T$ [defined in Equation 6.27 as the ratio of terrestrial and extraterrestrial solar radiation, with data on the CD Rom].

8.1.3 Bin Method

There are many applications where the degree-day method should not be used, even with variable base: The heat loss coefficient K_{tot}, the efficiency η of the HVAC system, or the balance-point temperature may not be sufficiently constant. The efficiency of a heat pump, e.g., varies strongly with outdoor temperature (discussed in Chapter 9). Or the efficiency of the HVAC equipment may be affected indirectly by T_o when the efficiency varies with the load, a common situation for boilers and chillers. Furthermore, in most commercial buildings, the occupancy has a very pronounced pattern, which affects the heat gains, the indoor temperature, and the ventilation rate.

In such cases, a steady-state calculation can yield good results for the annual energy consumption, if different temperature intervals and time periods are evaluated separately. This approach is called the *bin method*, because the consumption is calculated for several values of the outdoor temperature T_o and multiplied by the number of hours N_{bin} in the temperature interval ($=$bin) centered on that temperature:

$$Q_{bin} = N_{bin} \frac{K_{tot}}{\eta} (T_{bal} - T_o)_+ \tag{8.18}$$

The plus subscript on the parentheses indicates that only positive values are to be counted; no heating is needed when T_o is above T_{bal}. This equation is evaluated for each bin, and the total consumption is the sum of the Q_{bin} values over all bins.

In the United States, the necessary data, called *bin data*, are widely available in temperature intervals of 5°F (2.8°C). For enhanced flexibility, the data may be given for each month and for several periods of the day. For the data included with the HCB software for this book, the day is divided into six intervals of 4 h each; e.g., during the time from midnight to 4:00 a.m. the average February in New York has 24 h when the temperature is between 29°F and 34.5°F.

Example 8.5

Find the heating energy consumption of a simple commercial building in New York in February, for the following conditions:

Given: Occupied 7 days per week, 8:00 to 18:00; unoccupied otherwise

$$K_{tot} = 2.0 \text{ kW/K } [3.8 \text{ kBtu/(h} \cdot \text{°F)]} \quad \text{occupied}$$
$$= 1.0 \text{ kW/K } [1.9 \text{ kBtu/(h} \cdot \text{°F)} \quad \text{unoccupied (reduced because ventilation systerm shut off)}$$

$$\dot{Q}_{gain} = 20.0 \text{ kW (68.2 kBtu/h)} \quad \text{occupied}$$
$$= 1.0 \text{ kW (3.4 kBtu/h)} \quad \text{unoccupied}$$
$$T_i = 20.0 \text{°C} = \text{constant}$$

Efficiency of heating system $\eta = 1.0$. Bin data of HCB software, reproduced below.

Find: Q_h

Assumptions: Steady state

SOLUTION

The results have been calculated with a spreadsheet and are printed below.

T	Bin	1 to 4 Number of Hours Q_h	5 to 8 Number of Hours Q_h	9 to 12 Number of Hours Q_h	13 to 16 Number of Hours Q_h	17 to 20 Number of Hours Q_h	21 to 24 Number of Hours Q_h	Sum (kWh)
11.1°C		0	0	0	5	1	0	
(52°F)		0.0	0.0	0.0	0.0	3.9	0.0	3.9
8.3°C		0	1	7	16	12	1	
(47°F)		0.0	10.7	23.3	53.3	84.0	10.7	182.0
5.6°C		10	13	25	30	21	15	
(42°F)		134.4	174.8	222.2	266.7	234.5	201.7	1234.3
2.8°C		33	32	27	22	31	33	
(37°F)		535.3	519.1	390.0	317.8	475.3	535.3	2772.9
0.0°C		24	16	20	25	30	34	
(32°F)		456.0	304.0	400.0	500.0	585.0	646.0	2891.0
−2.8°C		13	18	12	3	9	12	
(27°F)		283.1	392.0	306.7	76.7	213.0	261.3	1532.8
−5.6°C		17	11	16	9	0	9	
(22°F)		417.4	270.1	497.8	280.0	0.0	221.0	1686.3
−8.3°C		11	17	2	2	8	4	
(17°F)		300.7	464.7	73.3	73.3	256.0	109.3	1277.3
−11.1°C		3	0	3	0	0	4	
(12°F)		90.3	0.0	126.7	0.0	0.0	120.4	337.4
−13.9°C		1	4	0	0	0	0	
(7°F)		32.9	131.6	0.0	0.0	0.0	0.0	164.4
Sum (kWh)		2250.2	2266.9	2040.0	1567.8	11851.8	2105.8	12,082.4

First, the balance-point temperatures are calculated:

$$T_{bal} = 20°C - \frac{20 \text{ kW}}{2.0 \text{ kW/K}} = 10°C \quad \text{occupied}$$

$$T_{bal} = 20°C - \frac{1 \text{ kW}}{1.0 \text{ kW/K}} = 19°C \quad \text{unoccupied}$$

Then for each bin (temperature and time of day), the heating load Q_h is calculated according to Eq. (8.18). For instance, the bin of 5.6°C from 1:00 to 4:00 contains 10 h, and so we have $Q_h = 1.0$ kW/K × (19°C − 5.6°C) × 10 h = 134 kWh, taking the K_{tot} and T_{bal} values for unoccupied periods weighted according to the number of hours of each type.

The sums of the rows and columns are shown at the right and the bottom; the grand total comes to

$$12{,}082.4 \text{ kWh} = 12{,}082.4 \text{ kWh} \times \frac{3.6}{1000} \text{GJ/kWh} = 43.5 \text{ GJ} (41.2 \text{ MBtu})$$

If the indoor temperature is set back during unoccupied periods, the bin method becomes less reliable because the exact temperature profile during setback is not known. One can obtain an approximate answer for the heating energy by assuming that the temperature will instantaneously drop to the setback value at the start and jump back to the normal occupancy value at the end of the unoccupied period; with that assumption the bin calculation of Example 8.5 can be used (with the appropriate values of T_i during occupied and unoccupied periods). The error of this approximation can be significant for mild climates or deep thermostat setbacks.

An analogous calculation can be carried out for the cooling season. Of course, the heat input, in particular the solar radiation, should correspond to average—not peak—conditions. Thermostat setup is more problematic for the calculation of cooling energy than for heating energy, because the temperature profile during the setup period is poorly known (most of the time, the building does not even reach the upper temperature limit, if any, at which the cooling equipment turns on again).

There are various refinements that can be included in a bin calculation. Of special interest is the seasonal variation of solar gains. If, as in Example 8.5, a separate calculation is done for each month, $\dot{Q}_{gain}$ could be based on the average solar heat gain of the month. The diurnal variation of solar gains can be accounted for by calculating the average solar gain for each of the hourly time periods of the bin method.

If such a detailed calculation of solar gains is considered too tedious or if great accuracy in the treatment of solar gains is not required, one can take a shortcut and assume a linear correlation of monthly average solar heat gains with monthly average outdoor temperature T_o. Here, too, one can work at several levels of detail.

We begin by illustrating this procedure in Fig. 8.6a for the simplest case: a building with constant operating conditions every hour of the day and every day of the year (that is, T_i, K_{tot}, and nonsolar heat gains $\dot{Q}_{nonsol}$ are all constant). In this case, it suffices to calculate the daily average heating load $\dot{Q}_1$ for the peak winter day and the daily cooling load $-\dot{Q}_2$ for the peak summer day:

$$\dot{Q}_1 = K_{tot}(T_i - T_{o,1}) - \dot{Q}_{nonsol} - \dot{Q}_{sol,1} \tag{8.19}$$

and

$$\dot{Q}_2 = K_{tot}(T_i - T_{o,2}) - \dot{Q}_{nonsol} - \dot{Q}_{sol,2} \tag{8.20}$$

where $T_{o,1}$ and $T_{o,2}$ are the corresponding values of the outdoor temperature T_o, and $\dot{Q}_{sol,1}$ and $\dot{Q}_{sol,2}$ are the corresponding daily average solar gains. Then these values are plotted versus the corresponding values $T_{o,1}$ and $T_{o,2}$ of the outdoor temperature T_o. Connecting these two points by a straight line, one obtains a simple linear approximation for the daily

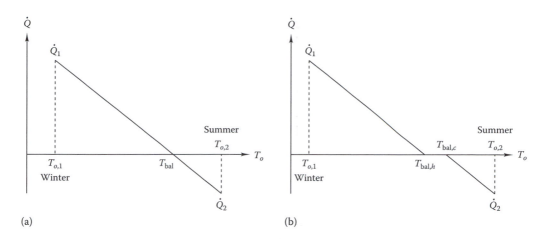

FIGURE 8.6
Load as function of outdoor temperature T_o (ventilative cooling is not shown). (a) Simplest case, with T_i constant year-round. (b) Higher set point in summer than in winter.

load as a function of T_o. The balance-point temperature of Eq. (8.2) is the value of T_o where this line crosses the T_o axis.[2]

After dividing by the equipment efficiency, we find the annual heating energy by summing over all temperature bins below the balance point, each bin weighted by the corresponding number of days. The same can be done for cooling above the balance-point temperature, although ventilative cooling should be taken into account according to Fig. 8.4.

Next we show in Fig. 8.6b how this procedure is modified to take into account the common practice of setting the thermostat a few degrees higher in summer than in winter. Evaluating $\dot{Q}_2$ for the summer value of T_i, one finds that the cooling part of the line is shifted to the right.

As a further refinement, one can carry out a separate calculation of the solar gain for the transition season, when T_o is close to the values of T_{bal} found above. Then the corrected values of the points $T_{bal,h}$ and $T_{bal,c}$ where the lines intersect the T_o axis are calculated with the midseason solar gain. Now the slopes for heating and for cooling can be different.

Finally, the correlations shown in Fig. 8.6b can be developed separately for different time periods of the day and for different occupancy conditions, say, 12:00 to 16:00 Saturdays. By using the bin data for the corresponding periods, the calculation can thus take the operating schedules of commercial buildings into account. We will take up the bin method again in Chaps. 9 through 11 to evaluate the annual performance of HVAC equipment with variable efficiency.

8.1.4 Dynamic Effects

So far we have discussed only steady-state methods for energy consumption. They are correct when the indoor temperature T_i is constant. But how about cases where T_i varies? Savings from thermostat setback in winter and setup in summer can be appreciable.

---

[2] As an alternative method for calculating solar gains in the bin method, we mention recent correlations developed by Vadon et al. (1991) that yield directly the solar gains coincident with each bin of T_o.

For example, ASHRAE (1989) cites typical savings in the range of 7 to 13 percent of the annual heating energy if the thermostat is set back by 10°F (5.6°C) for 8 h each winter night; of course, the precise numbers in a given case are highly dependent on climate and on building construction.

Besides variable thermostat settings, there is another effect that can make T_i vary: During the milder parts of the heating season, the instantaneous balance-point temperature can rise above T_o for part of the day, as already described in Fig. 8.5. Assuming the windows are kept closed, much of the excessive heat gain may be stored in the mass of the building, thus reducing the heating load when T_o finally drops below T_{bal}. Such storage implies an increase of T_i above the thermostat set point, even if the latter is constant. This effect is particularly important in buildings with large solar gains and high inertia. Obviously an uncorrected steady-state method should not be used for passive solar buildings.

If T_i at each hour were known, one could calculate the corresponding values of T_{bal} and proceed with the bin method as before. Unfortunately the evolution of T_i is not known in general; it depends on the thermal inertia of the building and on the temperature difference $T_i - T_o$. By examining some typical temperature profiles, one can derive certain correlations for correction terms. Several shorthand methods have been developed that include correction terms for variable T_i in steady-state methods. As an example, we mention the *solar load ratio (SLR) method* (see Sec. 14.5.3) for passive solar buildings; it modifies the degree-day result by a correction term that has been derived from a large number of dynamic computer simulations.

Of course, the introduction of correction terms tends to spoil the simplicity of the steady-state approach, and with the evolution of computer technology one may reach a point where a dynamic calculation will be easier than the old shorthand methods with correction terms.

The principles of a dynamic calculation of energy consumption are the same as for a dynamic calculation of peak loads. One simply repeats the calculation, time step by time step, for the entire year, including the efficiency of the HVAC equipment as appropriate. An hourly time step is commonly used. We have already illustrated this procedure with the examples in Sec. 7.7. For an annual energy calculation, one uses hourly weather data for each day of the year instead of a sequence of identical days. This procedure is realized in the DOE2.1 computer simulation program (Birdsall et al., 1990) which is essentially based on the transfer function method that we described in Sec. 7.7.

The weather data must be representative of long-term averages. One can save some calculational steps by using less than 365 days for a year, scaling up the results at the end; of course, the days for the calculation must be chosen carefully to be representative. For example, for a number of European sites, a set of so-called *short reference year* data is available that replaces the year by 8 weeks.

8.2 Thermal Networks

In Chap. 2, we presented the basic physical phenomena of heat transfer by conduction, convection, and radiation. Applying them to load calculations, we have, until now, remained essentially in the steady-state domain (our presentation of the transfer function method did not show the link with those laws). In the following sections, we outline a general approach for the dynamic modeling of a building, based on thermal networks.

They are intuitive, and they allow a systematic formulation and solution of general and complicated problems.

In this section, we describe how to draw diagrams of thermal networks and how to obtain the corresponding equations. Then we examine the simplest case, the *RC* network; it lumps the entire heat capacity of a building into a single massive node. This model is instructive for explaining the important concept of the time constant of a building. Even though the *RC* network is quite crude, it gives the designer a simple tool for estimating the warm-up and cooldown times associated with thermostat setback and setback recovery. In fact, most controllers for optimizing the start-up time after setback are based on this network.

The generalization to more accurate models is discussed in Sec. 8.3. There we also explain the connection with transfer functions and with the transient heat conduction equation.

8.2.1 Network Diagrams and Equations

In this approach, one approximates a building as being composed of a finite number of parts *N*, called *nodes*, each of which is assumed to be isothermal. To model heat exchange, the nodes are connected by resistances, thus forming a thermal network. Neighboring nodes are nodes that are directly coupled by conduction, convection, or radiation. The heat flow between neighbors is given by

$$\dot{Q}_{n'-n} = \frac{T_{n'} - T_n}{R_{n'n}} \tag{8.21}$$

where $R_{n'n}$ is the resistance between n' and n. In addition, there may be direct heat input $\dot{Q}_n$ at node n, from heat sources such as solar radiation, lights, or electric resistance heating. Let us designate the heat capacity of node n by C_n and its temperature by T_n. Assuming constant C_n, the rate of change of the heat stored in node n is $C_n \dot{T}_n$, and by the first law of thermodynamics it must be equal to the total rate of heat input. Thus the heat balance of node n is a first-order differential equation in T_n:

$$C_n \dot{T}_n = \sum_{n'=1}^{N} \frac{T_{n'} - T_n}{R_{n'n}} + \dot{Q}_n \tag{8.22}$$

As for signs, we note that if $T_{n'} - T_n$ is positive, heat flows from n' to n, making a positive contribution to $\dot{T}_n$. In most cases, a given node can interact directly with only a relatively small number of nodes, and so the number of nonzero terms in this sum is much smaller than *N*. For example, a homogeneous wall can be modeled as a one-dimensional network where each node has only two neighbors.

Equation 8.22, one for each node, form a system of *N* first-order differential equations with *N* unknowns, the node temperatures T_n. By analogy with electric circuits it is convenient to represent thermal networks by diagrams, where[3]

---

[3] Compared to electric circuits, one can note two simplifications: There are no inductances, and the heat stored in a capacitance depends only on a single temperature derivative: Thus the loose end of each capacitance can be taken as "ground" (zero voltage).

—|⊢ represents a capacitance C.

—/\/\/— represents a resistance R.

Temperatures T are analogous to voltages.

Heat flows $\dot{Q}$ are analogous to currents.

There is a one-to-one correspondence between the diagram and the set of equations of a thermal network. The diagram has the advantage of being much easier to grasp, but the equations are needed for finding the solutions. Once the diagram has been drawn, one can easily write down the equations. There is one first-order differential equation for each node.

For a simple example, consider a single-pane window between indoor air at T_i and outdoor air at T_o, with surface heat transfer coefficients h_i and h_o (combining radiation and convection). The glass can be approximated as a single isothermal node, with capacitance

$$C = \rho c_p A \Delta x \tag{8.23}$$

where
$\rho = $ density
$c_p = $ specific heat
$A = $ area
$\Delta x = $ thickness

The corresponding network is shown in Fig. 8.7a.

If there is no absorption of radiation in the glass, $\dot{Q}$ is zero and Eq. (8.22) for the glass temperature T becomes

$$C\dot{T} = \frac{T_i - T}{R_i} + \frac{T_o - T}{R_o} \tag{8.24}$$

where the resistances are related to the surface heat transfer coefficients by

$$\frac{1}{R_i} = Ah_i \quad \text{and} \quad \frac{1}{R_o} = Ah_o$$

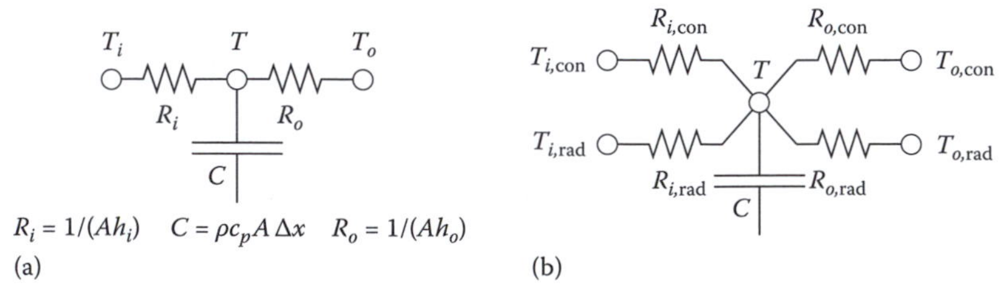

$R_i = 1/(Ah_i) \quad C = \rho c_p A \Delta x \quad R_o = 1/(Ah_o)$

(a)   (b)

FIGURE 8.7

Thermal network for heat flow through a pane of glass, treating the glass as a single node with heat capacity C and neglecting the resistance of the glass relative to the surface resistances R_i and R_o. (a) With combined heat transfer coefficients (convection plus radiation). (b) With separate heat transfer coefficients for convection and for radiation, coupled to different temperatures.

Using combined heat transfer coefficients for convection plus radiation is a simplification that is not appropriate when radiation and convection couple to nodes at different temperatures, e.g., radiation to the sky and convection to outdoor air. Figure 8.7b shows how the network has to be modified if a pane of glass faces different radiative and convective temperatures on both sides, e.g., sky and outdoor air on one side and indoor air and indoor wall surfaces on the other sides. The corresponding heat balance equation is

$$C\dot{T} = \frac{T_{i,\text{rad}} - T}{R_{i,\text{rad}}} + \frac{T_{i,\text{con}} - T}{R_{i,\text{con}}} + \frac{T_{o,\text{rad}} - T}{R_{o,\text{rad}}} + \frac{T_{o,\text{con}} - T}{R_{o,\text{con}}} \tag{8.25}$$

As a more complicated example, Fig. 8.8 shows the thermal network for a homogeneous wall that is approximated by N layers of equal thickness $\Delta x = w/N$, where $w =$ thickness of wall. With the same notation as above, the capacitance of the nodes is again given by Eq. (8.23). The resistance between nodes is

$$R = \frac{\Delta x}{kA} \tag{8.26}$$

where $k =$ conductivity {W/(m · K) [Btu/(h · ft · °F)]}. For the heat balance of the internal nodes, we have

$$C\dot{T}_n = \frac{T_{n-1} - T_n}{R} + \frac{T_{n+1} - T_n}{R} \quad \text{for } 2 \leq n \leq N - 1 \tag{8.27}$$

The equations for the external nodes are

$$C\dot{T}_1 = \frac{T_i - T_1}{R_i} + \frac{T_2 - T_1}{R} \tag{8.28}$$

and

$$C\dot{T}_N = \frac{T_{N-1} - T_N}{R} + \frac{T_o - T_N}{R_o} \tag{8.29}$$

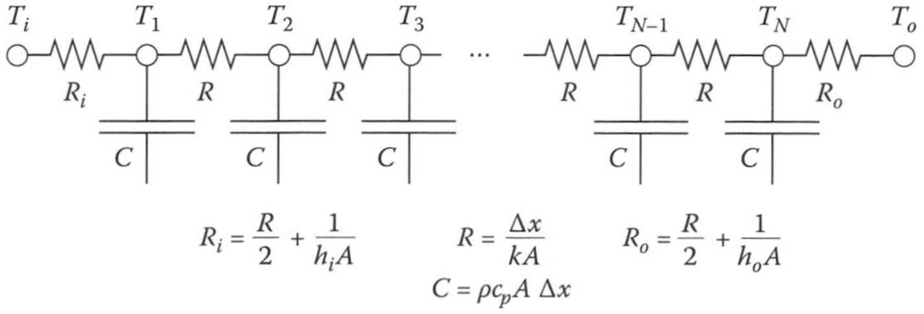

$$R_i = \frac{R}{2} + \frac{1}{h_i A} \qquad R = \frac{\Delta x}{kA} \qquad R_o = \frac{R}{2} + \frac{1}{h_o A}$$
$$C = \rho c_p A \, \Delta x$$

FIGURE 8.8
Thermal network for one-dimensional heat flow through homogeneous slab. Surfaces are in contact with media at temperatures T_i and T_o, the surface heat transfer coefficients being h_i and h_o. Slab has area A and thickness w, and is divided into N layers of thickness $\Delta x = w/N$ each.

with

$$R_i = \frac{R}{2} + \frac{1}{Ah_i} \quad \text{and} \quad R_o = \frac{R}{2} + \frac{1}{Ah_o}$$

where T_i and T_o are the air temperatures on the two sides of the wall and h_i and h_o are the surface conductances.

This represents a total of N first-order differential equations for the N unknowns T_n, $n = 1$ to N. Sources of heat, e.g., solar radiation absorbed at the surfaces, can easily be included: If $\dot{Q}_1$ is absorbed at node 1 and $\dot{Q}_N$ at node N, one adds $\dot{Q}_1$ and $\dot{Q}_N$ to the right-hand side of Eqs. (8.28) and (8.29), respectively.

A moderately detailed model for a one-zone building is shown in Fig. 8.9. It treats the roof, opaque wall, glazing, and floor in the manner of the above equations. Heat $\dot{Q}$ from the HVAC system is added to the air; the air is assumed to be perfectly mixed. The heat capacity of the air in the building is C_{air}. There is no additional heat capacity associated with the air exchange, because in the energy balance the act of replacing a certain quantity

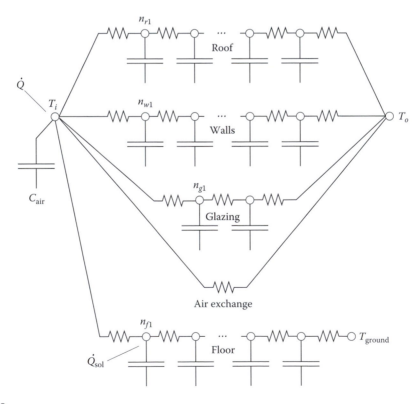

FIGURE 8.9

Thermal network for simple one-zone building whose envelope contains elements with distinct compositions: roof, wall, double glazing, plus floor. Heat input $\dot{Q}$ is added to interior air with heat capacity C_{air}. Solar gains $\dot{Q}_{sol}$ are absorbed by the floor.

of indoor air by outdoor air at T_o has the same effect as changing the temperature of that quantity of air from T_o to T_i. Solar radiation $\dot{Q}_{sol}$ is absorbed by the floor; the bottom of the floor is assumed to have the temperature of the ground.

Further features could be added. For instance, in winter the inside surface of a ceiling is likely to be warmer than the surface of a window, even if the air temperature in the room is uniform. The resulting radiative exchange between different interior surfaces can be treated by adding radiative conductances between nodes n_{w1}, n_{r1}, n_{f1}, and n_{g1} in Fig. 8.9. More complex buildings can be modeled similarly, including as necessary several different wall and window types or several zones at different temperatures. The number of nodes depends on the desired accuracy. Even three-dimensional effects can be taken into account, e.g., for a more accurate treatment of the conduction through the ground.

This approach is suitable for any linear thermal system, i.e., one where resistances and capacitances are independent of temperature. That is the case for the envelopes of buildings because the range of temperatures is usually small enough to permit approximation by linearized equations. Some of the parameters may vary with time, to account for variable ventilation or variable heat transfer coefficients.

In this manner, any building can be modeled as a thermal network or, equivalently, as a system of coupled first-order differential equations in time. At this point, the formulation is hybrid in the sense that space is treated as a discrete variable, while time is continuous. For numerical solution, one frequently resorts to discretization of time as well, as we explain in Sec. 8.3.2.

8.2.2 *RC* Network and Time Constant

To model thermal inertia, a network must contain at least one capacitance. The simplest possibility is the network shown in Fig. 8.10, containing one resistance R and one capacitance C. The node C is at the indoor temperature T_i, and the resistance R is the inverse of the total heat transmission coefficient K_{tot} of the building

$$R = \frac{1}{K_{tot}} \tag{8.30}$$

Writing down the heat balance equation Eq. (8.22) for this network, we find

$$C\dot{T}_i = \frac{T_o - T_i}{R} + \dot{Q} \tag{8.31}$$

This is a first-order differential equation in the variable T_i. Let us define a quantity τ, called the *time constant*,

$$\tau = RC \tag{8.32}$$

To explain the reason for the name, we rewrite the equation in the form

$$\tau\dot{T}_i + T_i = T_o + R\dot{Q} \tag{8.33}$$

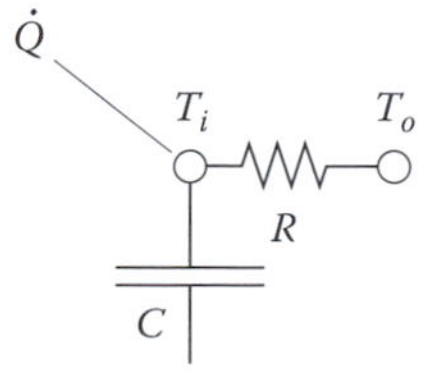

FIGURE 8.10
RC network.

and solve it for the special case when the driving terms T_o and $\dot{Q}$ are constant. Since in that case the derivative of the variable

$$T(t) = T_i(t) - T_o - R\dot{Q} \tag{8.34}$$

is equal to $\dot{T}_i$, the equation becomes

$$\tau\dot{T} + T = 0 \tag{8.35}$$

Its solution is a simple exponential

$$T(t) = T(0)\exp\left(\frac{-t}{\tau}\right) \tag{8.36}$$

where $T(0)$ is the initial value. The time constant sets the time scale for the cooldown; after one time constant, $T(t)$ has decayed to $1/e \approx 0.368$ of its initial value $T(0)$. The longer the time constant of a building, the longer it takes to cool down or warm up. Now we have a model, crude but better than nothing, for estimating the effects of thermostat setback and recovery.

Example 8.6

Suppose a building can be described by the RC network of Fig. 8.10. And T_i, T_o, and the heat input $\dot{Q}$ have been constant until $t=0$, when the thermostat is set back. How long does it take for T_i to drop from 20°C (68°F) to 15°C (59°F)?

Given: Network of Fig. 8.10, with

$R = 5.0$ K/kW, implying $K_{tot} = 200$ W/K [379 Btu/(h · °F)]

$C = 10$ MJ/K (5.27 kBtu/°F)

$T_o = 0$°C (32°F)

$\dot{Q}$ as necessary to maintain $T_i = 20$°C (68°F) until $t=0$ ($\dot{Q} = 0$ for $t > 0$)

Find: Time t at which $T_i(t) = 15$°C (59°F)

SOLUTION

Let $T(t) = T_i(t) - T_o$. For $t > 0$, it satisfies Eq. (8.35), and the solution according to Eq. (8.36) is

$$T(t) = T(0)\exp\left(\frac{-t}{\tau}\right)$$

with $T(0) = 20$°C $- 0$°C $= 20$ K and time constant

$$\tau = RC = 5.0 \text{ K/kW} \times 10 \text{ MJ/K} = 5.0 \times 10^4 \text{ s} = 13.889 \text{ h}$$

We are looking for the value of t at which $T(t) = 15$°C $- 0$°C $= 15$ K; hence we have to find the value of t that solves

$$15 \text{ K} = 20 \text{ K}\exp\left(\frac{-t}{\tau}\right)$$

The solution is

$$t = -\tau \ln\left(\frac{15}{20}\right) = 13.889 \text{ h} \times 0.28768 = 4.00 \text{ h}$$

Example 8.7

Suppose the heating system of the building in Example 8.6 has been built for design temperatures $T_{i,des} = 20°C$ (68°F) and $T_{o,des} = -20°C$ (−4°F). How long is the setback recovery time for the conditions of Example 8.6?

Given: Conditions of Example 8.6, with peak heat input $\dot{Q}$ designed for $T_i = 20°C$ and $T_o = -20°C$. Heating system has simple on/off control: It runs at full capacity until the desired temperature is reached (this type of control is typical of residential heating systems).

Find: Time t for bringing $T_i(t)$ from 15°C back to 20°C.

SOLUTION

First find the heat input $\dot{Q}$,

$$\dot{Q} = \frac{T_{i,des} - T_{o,des}}{R} = \frac{20°C - (-20°C)}{5.0 \text{ K/kW}} = 8.0 \text{ kW (27.3 kBtu/h)}$$

Take the start of the warm-up as $t = 0$. During warm-up, the conditions of constant driving terms are again satisfied, and so we can again use the exponential solution (8.36)

$$T(t) = T(0) \exp\left(\frac{-t}{\tau}\right)$$

if we take

$$T(t) = T_i(t) - T_o - R\dot{Q} \quad \text{with } R\dot{Q} = 5.0 \text{ K/kW} \times 8.0 \text{ kW} = 40 \text{ K}$$

The initial value is

$$T(0) = T_i(0) - T_o - R\dot{Q} = 15°C - 0°C - 40 \text{ K} = -25 \text{ K}$$

We want to know at what time t we reach $T_i(t) = 20°C$, or

$$T(t) = 20°C - 0°C - 40 \text{ K} = -20 \text{ K}$$

hence we have to solve

$$-20 \text{ K} = -25 \text{ K} \exp\left(\frac{-t}{\tau}\right) \quad \text{for } t$$

The solution is

$$t = -\tau \ln\left(\frac{-20}{-25}\right) = (-13.889 \text{ h})(-0.2231) = 3.10 \text{ h}$$

COMMENTS

The warm-up time depends on T_o and on $\dot{Q}$. The lower T_o and the smaller $\dot{Q}$, the slower the warm-up. At $T_o = -10°C$, we would have found

$$T(0) = T_i(0) - T_o - R\dot{Q} = 15°C - (-10°C) - 40\text{ K} = -15\text{ K}$$

and

$$T(t) = 20°C - (-10°C) - 40\text{ K} = -10\text{ K}$$

and a warm-up time of

$$t = -\tau \ln\left(\frac{-10}{-15}\right) = (-13.889\text{ h})(-0.40546) = 5.63\text{ h}$$

How long would the warm-up be when T_o equals the design value of $-20°C$ ($-4°F$)? In that case we would have to solve the equation with

$$T(0) = T_i(0) - T_o - R\dot{Q} = 15°C - (-20°C) - 40\text{ K} = -5\text{ K}$$

and

$$T(t) = 20°C - (-20°C) - 40\text{ K} = 0\text{ K}$$

Now the solution for t diverges—the warm-up would take infinitely long. This illustrates our remarks about pickup loads in Sec. 7.5.

Equation (8.31) can also be solved if the right side is not constant. In that case, the solution is the sum of two terms, one that depends on the initial condition $T_i(t_0)$ and one that depends on the driving term

$$T_i(t) = \exp\left(\frac{t_0 - t}{\tau}\right) T_i(t_0) + \int_{t_0}^{t} dt' \exp\left(\frac{t' - t}{\tau}\right) u(t') \tag{8.37}$$

with the driving term $u(t) = [T_o(t) + R\dot{Q}(t)]/\tau$. That this is indeed a solution can be verified by inserting it into the differential equation.

Unfortunately the RC network is not sufficient for all purposes. It treats the building as if all the thermal mass were in the interior, always at the same temperature as the indoor air. In real buildings the arrangement of masses is much more complicated, as suggested by the network of Fig. 8.9. For applications to thermostat control, the RC network can give acceptable results if the value of the heat capacity C is chosen appropriately. With a judicious choice of C, one can make the RC network imitate the response of a building to typical changes in heat input (see the discussion of system identification in Sec. 8.4). This value of C is not the static heat capacity, and its determination is not entirely obvious. It can be approximated by the effective diurnal heat capacity, as discussed in Sec. 7.3 [which can be estimated by the method of Jones (1983)]. Next we discuss the inclusion of further nodes in a network.

8.3 Other Models

8.3.1 A Network with Two Nodes

To obtain a more realistic model, let us add another resistance and another capacitance, as shown in Fig. 8.11. Compared to the simple RC network, an extra node with capacitance C_e has been added between T_i and T_o; it can represent the heat capacity of the envelope while C_i represents the heat capacity of the interior of the building (interior walls and floors, furniture, air). The resistances R_i and R_o on each side of C_e can be different. The total resistance is, of course, the inverse of the heat transmission coefficient

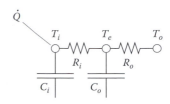

FIGURE 8.11
Thermal network with two resistances and two heat capacities, for Example 8.8.

$$R_i + R_o = \frac{1}{K_{tot}} \tag{8.38}$$

It is instructive to solve this case. Now we have the network equations

$$C_i \dot{T}_i = \frac{T_e - T_i}{R_i} + \dot{Q} \tag{8.39}$$

and

$$C_e \dot{T}_e = \frac{T_i - T_e}{R_i} + \frac{T_o - T_e}{R_o} \tag{8.40}$$

They are not very convenient in this form because they contain the variable T_e that is neither observed nor needed in most applications. It would be preferable to have a model that involves only T_i and the driving terms T_o and $\dot{Q}$. To eliminate T_e, we take the derivative of Eq. (8.39) and multiply by R_i:

$$R_i C_i \ddot{T}_i = \dot{T}_e - \dot{T}_i + R_i \ddot{Q}_i \tag{8.41}$$

and on its right-hand side we insert Eq. (8.40) for $\dot{T}_e$. This yields an equation that contains T_e but no derivatives of T_e. Now we can eliminate T_e by means of Eq. (8.39). After rearranging terms, we obtain the desired result

$$T_i + (C_i R_i + C_i R_o + C_e R_o)\dot{T}_i + C_i R_i C_e R_o \ddot{T}_i$$
$$= T_o - (R_i + R_o)\dot{Q} + R_i R_o C_e \ddot{Q}_i \tag{8.42}$$

The price for the elimination of variable T_e has been the introduction of higher derivatives in the other variables T_i and $\dot{Q}$ We now have a second-order differential equation in T_i.

To find out the time constant(s), let us consider once more the special case when the driving terms T_o and $\dot{Q}$ are constant. Then the variable

$$T(t) = T_i(t) - T_o - (R_i + R_o)\dot{Q} \tag{8.43}$$

satisfies the equation

$$T + (C_iR_i + C_iR_o + C_eR_o)\dot{T} + C_iR_iC_eR_o\ddot{T} = 0 \qquad (8.44)$$

Trying an exponential solution

$$T(t) = T_i(0)\exp\left(\frac{-t}{\tau}\right) \qquad (8.45)$$

we see that the time constant τ must satisfy

$$\tau^2 - (C_iR_i + C_iR_o + C_eR_o)\tau + C_iR_iC_eR_o = 0 \qquad (8.46)$$

This is a quadratic equation, and it has two solutions for τ:

$$\tau_{\pm} = \frac{1}{2}(C_iR_i + C_iR_o + C_eR_o)\left[1 \pm \sqrt{1 - \frac{4C_iR_iC_eR_o}{(C_iR_i + C_iR_o + C_eR_o)^2}}\right] \qquad (8.47)$$

Example 8.8

Find the time constants for the network of Fig. 8.11 with the following parameter values. The values of R_o and R_i are chosen to give the same K_{tot} as in Example 8.6, with equal resistance on the inside and on the outside of the capacity C_e of the envelope.

   Given: $R_o = 2.5$ K/kW

   $R_i = 2.5$ K/kW

   $C_e = 5$ kWh/K = 18.0 MJ/K

   $C_i = 1$ kWh/K = 3.6 MJ/K

   Find: τ_+ and τ_-

SOLUTION

Inserting the numbers into Eq. (8.47), we obtain

$$\tau_+ = 15.481 \text{ h} \approx 15.5 \text{ h} \qquad \tau_- = 2.019 \text{ h} \approx 2.0 \text{ h}$$

COMMENTS

We have tried to choose the numbers so that the models of Example 8.6 and Example 8.8 could represent approximately the same building. However, one cannot compare the heat capacities and time constants directly with the single-node model of Example 8.6, because the masses are distributed differently. In the two-node model, the capacity C_e of the envelope is almost twice as large as C_i of the single-node model, but since it is not directly coupled to T_i, its effect on the time constant is reduced. Hence the value of the larger time constant is comparable to the value 13.889 h found in Example 8.6.

The fact that there are two time constants implies that the general solution of Eq. (8.44) is a superposition of two exponentials

$$T_i(t) = A_1 \exp\left(\frac{-t}{\tau_1}\right) + A_2 \exp\left(\frac{-t}{\tau_2}\right) \tag{8.48}$$

The coefficients A_1 and A_2 are determined by the initial conditions, for instance, $T_i(0)$ and $\dot{T}_i(0)$.

This passage from the *RC* network to a network with two nodes illustrates some general phenomena that are encountered in the development of more complicated models. The model with two nodes has two time constants. As a general theorem, one finds that a network with N nodes has N time constants (usually different from each other). In buildings there is usually one time constant that is much larger than all the others; this large time constant determines the response of the building to slow changes. When talking about "the time constant" of a building, one refers to this value.

The solution of the *RC* network involves a single exponential and one initial condition [Eq. (8.36) or (8.37)]. The solution of the two-node network involves two exponentials and two initial conditions [Eq. (8.48)]. This, too, has a natural generalization in the theory of linear differential equations: A network with N nodes involves N exponentials and N initial conditions.

In a complicated network, one will rarely (if ever) be interested in the values of most of the node temperatures. Rather one would like to know the output (or response) as a function of the input (or driving terms), e.g., the indoor air temperature T_i as a function of the outdoor temperature T_o and the heat input $\dot{Q}$. When the uninteresting variables are eliminated, higher derivatives appear in the other variables. If one eliminates all but a single output variable T_i from a network with N nodes, one obtains a differential equation of order N in T_i.

The procedure we have used in this section for eliminating a variable and for finding time constants and solutions becomes unwieldy when the number of nodes is large. As the number of nodes is increased, matrix algebra provides a simpler and more systematic approach. That is beyond the scope of this book. Instead we use the two-node network to illustrate how transfer functions can be derived.

8.3.2 Transfer Function

Until now we have treated time t as a continuous variable. For numerical solution it is often convenient to employ discrete variables instead, replacing t by $k\,\Delta t$:

$$x(t) \rightarrow x_k \quad \text{with } k = \frac{t}{\Delta t} \tag{8.49}$$

The time step Δt is usually taken as 1 h (although shorter intervals may become necessary for modeling systems with fast response times, e.g., the control of air distribution within a room). Such a discretization is an approximation; information on the detailed behavior within the time step is lost. Time derivatives are replaced by differences. There are several options; e.g., the first derivative can be taken as

$$\dot{x}(t) \rightarrow \frac{x_{k+1} - x_k}{\Delta t} \quad \text{forward differencing} \tag{8.50}$$

or

$$\dot{x}(t) \rightarrow \frac{x_k - x_{k-1}}{\Delta t} \quad \text{backward differencing} \tag{8.51}$$

or

$$\dot{x}(t) \rightarrow \frac{x_{k+1} - x_{k-1}}{2\Delta t} \quad \text{symmetric differencing} \tag{8.52}$$

One could also use a symmetric differencing scheme:

$$x(t) \rightarrow \frac{x_{k+1} + x_k}{2}$$
$$\dot{x}(t) \rightarrow \frac{x_{k+1} - x_k}{\Delta t} \tag{8.53}$$

In any case, the nth derivative requires at least $n+1$ time steps.

As an example, let us take the two-node network of the preceding section. It is described by the differential equation in Eq. (8.42), which we rewrite here in terms of the time constants as

$$T_i + (\tau_1 + \tau_2)\dot{T}_i + \tau_1\tau_2\ddot{T}_i = T_o - (R_i + R_o)\dot{Q} + R_iR_oC_e\ddot{Q}_i \tag{8.54}$$

since the coefficients of $\dot{T}_i$ and $\ddot{T}_i$ are related to the time constants by

$$\tau_1 + \tau_2 = C_iR_i + C_iR_o + C_eR_o \tag{8.55}$$

and

$$\tau_1\tau_2 = C_iR_iC_eR_o$$

Using the differencing scheme

$$x(t) \rightarrow x_{k-1}$$
$$\dot{x}(t) \rightarrow \frac{x_k - x_{k-2}}{2\Delta t} \tag{8.56}$$
$$\ddot{x}(t) \rightarrow \frac{x_k - 2x_{k-1} + x_{k-2}}{(\Delta t)^2}$$

we obtain

$$a_{i,0}T_{i,k} + a_{i,1}T_{i,k-1} + a_{i,2}T_{i,k-2} = a_{o,1}T_{o,k} + a_{Q,0}\dot{Q}_k + a_{Q,1}\dot{Q}_{k-1} + a_{Q,2}\dot{Q}_{k-2} \tag{8.57}$$

with

$$a_{i,0} = \frac{\tau_1 + \tau_2}{2\Delta t} + \frac{\tau_1\tau_2}{(\Delta t)^2}$$
$$a_{i,1} = 1 - \frac{2\tau_1\tau_2}{(\Delta t)^2}$$

$$a_{i,2} = -\frac{\tau_1 + \tau_2}{2\Delta t} + \frac{\tau_1 \tau_2}{(\Delta t)^2}$$

$$a_{o,1} = 1 \quad (a_{o,0} \text{ and } a_{o,2} \text{ are both } 0)$$

$$a_{Q,0} = \frac{\tau_1 \tau_2}{2C_i \Delta t} = -a_{Q,2}$$

$$a_{Q,1} = -(R_i + R_o)$$

Such a relation between function values at discrete time steps is called a *time series*, and its coefficients are called *transfer function coefficients*. This approach is convenient for numerical calculations on a computer.

Example 8.9

Use the transfer functions for Example 8.8 with the numbers of Example 8.8 to calculate the time evolution of T_i.

Given: Two-node network with

$R_i = 2.5$ K/kW $= R_o$

$C_i = 1.0$ kWh/K, $C_e = 5.0$ kWh/K

$\tau_1 = 15.481$ h $\tau_2 = 2.019$ h (from end of Example 8.8)

$T_{o,k} = 0 \ \dot{Q}_k = 0$ for $k \geq 0$

Initial conditions $T_{i,k} = 1$ K for $k \leq 0$

Find: $T_{i,k}$ for $k > 0$

SOLUTION

The coefficients of the transfer function in Eq. (8.57) are

$$
\begin{array}{lll}
a_{i,0} = 40.00 & a_{o,0} = 0.00 & a_{Q,0} = 15.63 \\
a_{i,1} = -61.50 & a_{o,1} = 1.00 & a_{Q,1} = -5.00 \\
a_{i,2} = 22.50 & a_{o,2} = 0.00 & a_{Q,2} = -15.63
\end{array}
$$

The a_i and a_o are dimensionless, while the a_Q have dimension of kelvins per kilowatt. Here we need only the coefficients a_i, and T_i can be calculated by repeated use of

$$T_{i,k} = \frac{-(a_{i,1} T_{i,k-1} + a_{i,2} T_{i,k-2})}{a_{i,0}}$$

For example,

$$
\begin{aligned}
T_{i,1} &= \frac{-(a_{i,1} T_{i,0} + a_{i,2} T_{i,-1})}{a_{i,0}} \\
&= \frac{-(-61.50 \times 1 \text{ K} + 22.50 \times 1 \text{ K})}{40.00} = 0.98
\end{aligned}
$$

The results listed below show good agreement with the exact solution [Eq. (8.48) for $T_i(0) = 0$ and $\dot{T}_i(0) = 0$] in the first column.

k	$T_{i,k-2}$	$T_{i,k-1}$	$T_{i,k}$	$T_{i,\text{exact}}$
0	1.00	1.00	1.00	1.00
1	1.00	1.00	0.98	0.99
2	1.00	0.98	0.94	0.95
3	0.98	0.94	0.89	0.91
4	0.94	0.89	0.84	0.87
5	0.89	0.84	0.80	0.82
10	0.66	0.62	0.58	0.60
20	0.35	0.33	0.31	0.32
30	0.18	0.17	0.16	0.17

We already presented time series in Sec. 7.7, but without indicating how the transfer function coefficients can be calculated for a given configuration. Here we have shown how the coefficients can be calculated for a simple thermal network. For networks with a large number of nodes, this method would yield a large number of coefficients, but one can approximate the resulting transfer function by a simpler one with fewer coefficients, such as the ones in Sec. 7.7.

8.3.3 Periodic Heat Flow

The driving terms of a building are dominated by the diurnal frequency: Outdoor temperature, solar radiation, and occupancy all follow a basic 24-h cycle. According to the principles of Fourier analysis, any function that repeats itself after 24 h can be represented by a sum of sine and cosine terms with period 2×24 h, 3×24 h, 4×24 h, etc. Therefore it is appropriate to consider the response of a linear thermal system to sinusoidal driving terms (i.e., sines or cosines of a single frequency). This is helpful for an intuitive understanding of the behavior of real buildings.

Quite generally, if the input of a linear system is sinusoidal, the response is sinusoidal with the same frequency; the amplitude and phase of the response depend on the system parameters. This fact can be seen from the solution of the RC network, Eq. (8.37), by inserting sinusoidal driving functions on the right-hand side: The contribution of the initial state, whatever it was, vanishes exponentially with time, and a purely sinusoidal response remains, as the reader can verify with some algebra and a table of integrals.

Applying this idea to the conduction of heat through the envelope of a building, we suppose the outdoor sol-air temperature T_{os} is periodic while the indoor temperature is constant at T_i. Then the conductive heat gain (or loss) at time t through an element of the envelope with area A and a U value of U is

$$\dot{Q}_t = (UA)(\text{TETD}_t) \tag{8.58}$$

where TETD is the *total equivalent temperature differential*

$$\text{TETD}_t = T_{os,\text{av}} - T_i + \lambda(T_{os,t-\delta} - T_{os,\text{av}})$$

The first term represents the daily average; the second, the periodically varying term. As far as the heat gain is concerned, the variation of T_{os} appears reduced by the *decrement factor* λ

and delayed by the *time lag* δ. Data for λ and δ can be found in Table 7.7. For example, the wall in the third line of Table 7.7b (4-in heavyweight concrete with 1-in insulation) has $U = 0.191$ Btu/(h $\cdot$ ft^2 $\cdot$ °F), $\lambda = 0.78$, and $\delta = 3.33$ h. If $T_{os,av} = 76$°F with a peak of 96°F at 16:00, while $T_i = 76$°F is constant, then the peak conductive gain occurs at $16:00 + 3.33$ h $= 19:20$ and its value per unit area is

$$\frac{\dot{Q}}{A} = [0.191 \text{ Btu/(h} \cdot \text{ft}^2 \cdot °\text{F)}](0.78)(96°\text{F} - 76°\text{F}) = 2.98 \text{ Btu/(h} \cdot \text{ft}^2)$$

A methodology of building energy analysis based on this approach is described by Shurcliff (1984).

8.3.4 Transient Heat Conduction Equation

Finally we show the connection between thermal networks and the transient heat conduction equation. Let us consider Eq. (8.27) in the limit where the slab thickness Δx goes to zero (and number of layers N to infinity, keeping $N\Delta x$ fixed). To simplify, note that R and C enter only in the combination

$$RC = \frac{\rho c_p}{k} \Delta x^2 \tag{8.59}$$

Defining the thermal diffusivity κ as

$$\kappa = \frac{k}{\rho c_p} \tag{8.60}$$

one can write Eq. (8.27) in the form

$$\dot{T}_n = \frac{\kappa}{\Delta x} \left(\frac{T_{n-1} - T_n}{\Delta x} + \frac{T_{n+1} - T_n}{\Delta x} \right) \tag{8.61}$$

where T_n is the temperature at $x = (n - \frac{1}{2})\Delta x$. As $\Delta x \to 0$, the terms on the right-hand side approach the spatial derivatives:

$$\frac{T_{n+1} - T_n}{\Delta x} \to \frac{\partial T}{\partial x} \quad \text{at } x = \left(n + \frac{1}{2}\right)\Delta x$$

$$\frac{T_{n-1} - T_n}{\Delta x} \to -\frac{\partial T}{\partial x} \quad \text{at } x = \left(n - \frac{1}{2}\right)\Delta x \tag{8.62}$$

The notation with partial derivatives reflects the fact that in this limit T becomes a continuous function of t and of x. The difference of the derivatives at $x = (n + \frac{1}{2})\Delta x$ and at $x = (n - \frac{1}{2})\Delta x$, divided by Δx, becomes the second derivative. Therefore, Eq. (8.27) turns into

$$\frac{\partial T}{\partial t} = \kappa \frac{\partial^2 T}{\partial x^2} \tag{8.63}$$

the well-known equation for transient heat conduction in one dimension. Conversely, Eq. (8.27) is a finite-difference approximation of the heat conduction equation.

8.4 System Identification

At the design stage, one needs to calculate the performance of a building, based on its detailed description (blueprint). This is an instance of what is sometimes called the *forward problem*, by contrast to the *inverse problem* where one has performance data and needs to deduce a building description. The latter is actually an example of a very general subject, arising in fields as diverse as engineering and economics, and it is also known as *system identification*. In existing buildings, this problem appears when one wants to answer the following questions:

1. How do the data compare with design predictions (and in case of discrepancies, are they due to anomalous weather, unintended thermostat settings, malfunctioning equipment, or other causes)?
2. How will the energy consumption change if thermostat settings or ventilation rates are changed (in other words, what is the cost of comfort)?
3. How much could be saved by retrofits?
4. If retrofits are implemented, how can one verify the savings?
5. How can one optimize control and operation of the HVAC equipment?

This last question is also of interest at the design stage.

One could try to go back to the blueprints (if they can still be found), but the process of coding the input of a simulation program requires much labor. Also, materials as actually installed are often different from the bulk properties reported for new materials, and the builder may have deviated from the blueprints without leaving any documentation. In such cases, it is better to base a model on measured data.

A model for the inverse problem can be quite different from the conventional forward models. It should contain only a small number of adjustable parameters, because the information content of the data is very limited, being collected under fairly repetitive conditions and subject to errors. Thus the model itself can be quite simple, although a fairly sophisticated procedure is needed to ensure reliable determination of the adjustable parameters. While pure forward and pure inverse models are extremes, a hybrid approach promises the greatest accuracy: Use as much a priori information as is available to develop an initial model, and calibrate it with measured performance data (Subbarao, 1988).

8.4.1 A Steady-State Inverse Model

Models known as the *energy signature* are the simplest example of inverse modeling (Hammarsten, 1987). The name refers to the behavior of the energy consumption as a function of outdoor conditions, as shown in a plot like Fig. 8.6. There are several variants, of which we would like to mention PRISM (Princeton scorekeeping method), developed by Fels and coworkers and available as software for mainframe or personal computers. A review, including the results of its applications, has been published as a special issue of *Energy and Buildings*, edited by Fels (1986). Its principal application is weather normalization of energy consumption data, and it can be used if one has data for at least six periods spanning most of the year. In the heating-only version, it contains three adjustable parameters:

$$C_{base} = \text{base level (energy other than heating or cooling, i.e., hot water,}$$
$$\text{cooking, lights), GJ/yr (Btu/yr)}$$

$$\beta = \frac{K_{tot}}{\eta} = \text{ratio of total heat loss coefficient and efficiency of}$$
$$\text{HVAC system, W/K [Btu/(h} \cdot {}^\circ\text{F)]}$$

$$T_{bal} = \text{balance-point temperature [outdoor temperature above which}$$
$$\text{no heating is needed, see Eq. (8.2)], } {}^\circ\text{C (}{}^\circ\text{F)}$$

These parameters are assumed constant throughout the year.

The parameters are determined by the classic statistical procedure of minimizing the sum (over all periods i) of the squared deviations between data and the model

$$C_i = \frac{C_{base} n_i}{365} + \beta D_{h,i}(T_{bal}) \tag{8.64}$$

where n_i = number of days in period and $D_{h,i}(T_{bal})$ = heating degree-days in period, based on T_{bal}. Once the parameters have been found, one can calculate the weather-corrected ("normalized") annual consumption based on the long-term average annual degree-days $D_{h,yr}(T_{bal})$ of the site

$$C_{yr} = C_{base} + \beta D_{h,yr}(T_{bal}) \tag{8.65}$$

Weather correction is crucial if one wants to compare measured and predicted consumption data.

When the data are reasonably good and the basic assumptions of the model are close to the truth, this approach has been found to work very well (in the sense of yielding good correlation coefficients, R^2 above 0.8, and small standard errors, a few percent, for C_{yr}). That is likely to be the case in residential buildings heated by one of the commercial energy sources (i.e., no significant use of firewood). In that case, the normalized consumption can be determined with a standard error of 1 to 3 percent. But the standard error of the individual parameters tends to be much larger (10 percent or more), because of compensating errors. That is a general phenomenon in system identification: For the quantity chosen as the dependent variable for the regression, the error is small, while the parameters of the model can remain quite uncertain.

The extension of PRISM to air conditioning requires at least two additional parameters. Of course, adding parameters does not guarantee an improvement of the results, even if the standard error or R^2 appears better. Residential cooling is often too erratic for a meaningful determination of the corresponding PRISM parameters. PRISM can also be used for commercial buildings, but attention should be paid to the role of weekends, variable ventilation rates (economizer), and simultaneous heating and cooling; in particular, the interpretation of the parameters may not be straightforward (Rabl et al., 1986).

8.4.2 Application of Static Inverse Models[4]

The principal applications of inverse models are in the following areas:

- Evaluation of energy conservation programs
- Prescreening indices for energy auditing
- Building energy management
- Optimal control
- In situ characterization of HVAC
- Components

In each of these applications both steady-state and dynamic inverse models have been applied. In general, steady-state inverse models are used with monthly and daily data containing one or more independent variables. Dynamic inverse models are usually used with hourly or subhourly data in cases where the thermal mass of a building is significant enough to delay the heat gains or losses.

Evaluation of Energy Conservation Programs   Aside from simply regressing energy use against temperature (e.g., often a two-parameter model with a slope and y-axis intercept) other widely used steady-state inverse methods for the evaluation of energy conservation retrofits include three-, four-, and five-parameter change-point models previously described (Fels, 1986; Kissock et al., 1993). Such models have been shown to be useful for statistically determining average weather-dependent and weather-independent energy use for buildings. Three-parameter change-point models can yield baseline energy use, the temperature at which weather-dependent energy use begins to increase energy use above the baseline (i.e., the change point), and the linear slope of the temperature dependency above (cooling model) or below (heating model) the temperature change point.

The existence of a change point in heating or cooling data that are plotted against ambient temperature can be physically justified since most HVAC systems use a thermostat that turns systems on or off above or below a set point temperature. Change-point regressions work best with heating data from buildings with systems that have little or no part-load nonlinearities (i.e., systems that become less efficient as they begin to cycle on and off with part loads). In general, change-point regressions for cooling loads exhibit less of a good fit because of changes in outdoor humidity which influence latent coil loads. Other factors that decrease the goodness of fit of change-point models include solar effects, thermal lags, and on/off HVAC schedules. In buildings with continuous, year-round cooling or heating, four-parameter models exhibit a better statistical fit than three-parameter models (i.e., grocery stores and office buildings with high internal loads). However, results of every modeling effort should be inspected for reasonableness (i.e., make sure that the regression is not falsely indicating an unreasonable relationship).

One of the main advantages of using a steady-state inverse model to evaluate the effectiveness of energy conservation retrofits lies in its ability to factor out year-to-year weather variations. This can be accomplished by using a *normalized annual consumption* (NAC) (Fels, 1986). Basically, once the regression parameters have been calculated for both pre-retrofit and post-retrofit periods, the annual energy conservation savings can be calculated by comparing the difference one obtains by multiplying the pre-retrofit and

---

[4] From Kreider (2001), *Handbook of HVAC*, with permission.

post-retrofit parameters by the weather conditions for the average year. Typically, 10 to 20 years of average daily weather data from a nearby National Weather Service site are used to calculate 365 days of average weather conditions, which are then used to calculate the average pre-retrofit and post-retrofit conditions.

Energy Management Steady state and dynamic inverse models can be used by energy management and control systems to predict energy use (Kreider and Haberl, 1994). Hourly or daily comparisons of measured energy use against predicted energy use can be used to determine if systems are being left on unnecessarily or are in need of maintenance. Combinations of predicted energy use and a knowledge-based system have been shown to be capable of indicating above-normal energy use and diagnosing the possible cause of the malfunction if sufficient historical information about malfunction signatures has been previously gathered (Haberl and Claridge, 1987). Hourly systems that utilize artificial neural networks have also been constructed (Kreider and Wang, 1992).

Hybrid Modeling Forward plus inverse models or hybrid models encompass everything that does not neatly fit into the exact definition of forward or inverse models. For example, when a traditional fixed-schematic simulation program such as DOE2 or BLAST (or even a component-based model) is used to simulate the energy use of an existing building, then one has a *forward* analysis method that is being used in an *inverse* application; i.e., the forward simulation model is being calibrated or fit to the actual energy consumption data from a building in much the same way that one fits a linear regression of energy use to temperature. Such an application is a *hybrid model*.

Although at first thought this might appear to be a simple process, there are several practical difficulties in achieving a "calibrated simulation," including the measurement and adaptation of weather data for use by the simulation programs (i.e., converting global horizontal solar into beam and diffuse solar radiation), the choice of methods used to calibrate the model, and the choice of methods used to measure the required input parameters for the simulation (i.e., the weight of the building, infiltration coefficients, and shading coefficients). In the scientific sense, truly "calibrated" models have only been achieved in a very few applications since they require a very large number of input parameters, a high degree of expertise, and enormous amounts of computing time, patience, and financial resources—much more than most practical applications would allow.

Classification of Methods In Table 8.1 different methods of analyzing building energy use are classified using an expanded version of Rabl's definitions (Rabl, 1988). Simple linear regression and multiple linear regression are the most widely used forms of inverse analysis. In the proper application, multiple linear regression must adequately address intercorrelations among the independent parameters as discussed above.

How to Select an Approach Table 8.2 presents a decision diagram for selecting an inverse model where usage of the model (diagnostics D, energy savings calculations ES, design DE, and control C), degree of difficulty in understanding and applying the model, time scale for the data used by the model (hourly H, daily D, monthly M, and subhourly S), calculation time, and input variables used by the models (temperature T, humidity H, solar S, wind W, time t, thermal mass tm) are the criteria used to determine the choice of a particular model.

Space does not permit an examination of each modeling approach; but based on the guidance in the two tables, the advanced student can make a selection and apply it to measured building data sets. The interested reader is referred to ASHRAE (2001) for more details of the methods.

TABLE 8.1

Classification of Methods for Thermal Analysis of Buildings

Method	Forward	Inverse	Hybrid	Comments
Steady-state methods				
Simple linear regression		X		One dependent parameter, one independent parameter. May have slope and y intercept
Multiple linear regression		X	X	One dependent parameter, multiple independent parameters
Modified degree-day method	X			Based on fixed reference temperature of 65°F
Variable-base degree-day method	X			Variable reference temperatures
ASHRAE bin method and inverse bin method	X	X	X	Hours in temperature bin time load for that bin
Change-point models: three-parameter (PRISM CO, HO), four-parameter, five-parameter (PRISM HC)		X	X	Uses daily or monthly utility billing data and average period temperatures
ASHRAE TC 4.7 modified bin method	X		X	Modified bin method with cooling load factors
Dynamic methods				
Thermal network (Sonderegger, 1977)	X	X	X	Uses equivalent thermal parameters (inverse mode)
Response factors (Stephenson and Mitalas, 1967)	X			Tabulated or as used in simulation programs
Fourier analysis (Shurcliff, 1984; Dhar, 1995)	X	X	X	Frequency domain analysis convertible to time domain
ARMA model (Subbarao, 1986)		X		Autoregressive moving-average model
ARMA model (Reddy, 1989)		X		Multiple-input autoregressive moving-average model
BEVA, PSTAR (Subbarao, 1986)	X	X	X	Combination of ARMA and Fourier series, includes loads in time domain
Modal analysis (Bacot et al., 1984)	X	X	X	Building described by diagonalized differential equation using nodes
Differential equation (Rabl, 1988)		X		Analytical linear differential equation
Computer simulation (DOE2, BLAST)	X		X	Hourly simulation programs with system models
Computer emulation (HVACSIM+, TRNSYS)	X		X	Subhourly simulation programs
Artificial neural networks (Kreider and Wang, 1992; Kreider, 1992; Kreider and Haberl, 1994)		X	X	Connectionist models

8.4.3 Dynamic Inverse Models

Dynamic building models (for a review see Rabl, 1988) are needed if one wants to account for phenomena with shorter than diurnal time scale. An important example is the optimal control of the thermostat with the goal of minimizing the sum of energy and demand charges. Developing an inverse model involves the following steps:

TABLE 8.2

Decision Diagram for Selection of Inverse Models

Method	Usage*	Difficulty	Time Scale[†]	Calc. Time	Variables[‡]	Accuracy
Simple linear regression	ES	Simple	D, M	Very fast	T	Low
Multiple linear regression	D, ES	Moderate	D, M	Fast	T, H, S, W, t	Medium
ASHRAE bin method and inverse bin method	ES	Moderate	H	Fast	T	Medium
Change-point models	D, ES	Moderate	H, D, M	Fast	T	Medium
ASHRAE TC 4.7 modified bin method	ES, DE	Moderate	H	Medium	T, S, tm	Medium
Thermal network	D, ES, C	Complex	S, H	Fast	T, S, tm	High
Fourier series analysis	D, ES, C	Complex	S, H	Medium	T, H, S, W, t, tm	High
ARMA model	D, ES, C	Complex	S, H	Medium	T, H, S, W, t, tm	High
Modal analysis	D, ES, C	Complex	S, H	Medium	T, H, S, W, t, tm	High
Differential equation	D, ES, C	Very complex	S, H	Fast	T, H, S, W, t, tm	High
Computer simulation (component-based)	D, ES, C, DE	Very complex	S, H	Slow	T, H, S, W, t, tm	Medium
Computer simulation (fixed schematic)	D, ES, DE	Very complex	H	Slow	T, H, S, W, t, tm	Medium
Computer emulation	D, C	Very complex	S, H	Very slow	T, H, S, W, t, tm	High
Artificial neural networks	D, ES, C	Complex	S, H	Fast	T, H, S, W, t, tm	High

* Usage shown includes diagnostics D, energy savings calculations ES, design DE, and control C.
[†] Time scales shown are hourly (H), daily (D), monthly (M), and subhourly (S).
[‡] Variables include temperature T, humidity H, solar S, wind W, time t, thermal mass tm.

1. Choose a model, making sure that it captures the crucial features of the situation.
2. Take one variable (usually T_i) as the dependent variable y_{model}.
3. Determine the parameters of the model by minimizing the squared differences between y_{model} and y_{data}, summed over all data points (usually this will be done by linear least squares, using some commercially available regression software).
4. Test the model against another set of data.
5. As much as possible, express the results in terms of quantities with direct physical interpretation, such as heat transmission coefficient, admittances, and time constants, to see if they make sense.

There is no unique procedure that will guarantee the best results. The main difficulty lies in the choice of variables and derivatives to be included. Interactive regression software is advisable (i.e., programs that allow the user to change the number of variables without having to reread the data). There are several guidelines for the selection of the most pertinent variables:

- The correlation coefficient R^2 (should be close to unity)
- Contribution of the variable to R^2 (should be significant)
- The standard error of the regression (should be small)
- The standard error of the parameters (should be relatively small)
- The F statistic or the t statistic [for interpretation and use, see any text on regression analysis, e.g., Draper and Smith (1981)].

For control applications it is often desirable to update the parameter estimates continually as new data are becoming available. This can be accomplished with recursive algorithms; the theory is discussed by Gelb (1974), and it has been applied to buildings by Norlen et al. (1987), and Pryor and Winn (1982).

Another important application of inverse modeling is the development of shortcuts to speed up large computer programs. Much time can be saved if, instead of recalculating a certain process in full detail at each step, one can approximate it by simpler expressions. An example is the use of transfer functions instead of the original transient heat conduction equations. In a similar spirit, the behavior of a large network can often be approximated by an equivalent and much simpler network. For instance, simulating a house with a network model containing some one hundred nodes, Neveu et al. (1986) found that the results for the heating loads could be reproduced within a fraction of a percent by a two-node approximation.

8.4.4 Foundation Heat Gain/Loss

A simplified design tool for calculating heat gain/loss for building foundations was developed by Chuangchid and Krarti (2000).[5] This tool is based on the analytical results of the interzone temperature profile estimation (ITPE) solutions and is suitable for estimating the design heat gain from building foundations as a function of a wide range of variables.

This model assumes that there is a periodic variation of the annual ground-coupled heat gain $Q(t)$

$$Q(t) = Q_m + Q_a \sin(\omega t + \phi) \tag{8.66}$$

where Q_m is the annual mean heat gain/loss and Q_a and ϕ describe the annual heat gain/loss amplitude and phase shift. If the indoor (space) temperature remains constant, then the amplitude and phase shift are driven by the outdoor temperature and are characterized by the building dimensions, soil properties, and foundation insulation. The outdoor temperature can, in turn, be described as a periodic function with an annual mean temperature T_m and amplitude T_a. In this method, the period of the heat gain/loss (ω in units of radians per second) for the slab is the same as for the ambient temperature.

It can be seen from Eq. (8.66) that the maximum (design) foundation heat gain/loss occurs at the crest of the sine wave

$$Q_{des} = Q_m + Q_a \tag{8.67}$$

The simplified method presented here offers a way of estimating Q_m and Q_a. At this point, it is convenient to define a normalized slab U value, U_o, as

$$U_o = \frac{k_s P}{A} \tag{8.68}$$

---

[5] We thank Drs. Chuangchid and Krarti for contributing to this section.

where k_s is the soil thermal conductivity [in $W/(m \cdot °C)$], P is the perimeter length of the slab (in meters), and A is the slab area (in square meters). An equivalent R value for the slab foundation floor R_{eq} is calculated as follows:

- For uniform insulation configurations (placed horizontally beneath the slab floor)

TABLE 8.3

Coefficients m and a for Foundation Heat Gain/Loss Calculations

Insulation Placement	m	a
Uniform—horizontal	0.40	0.25
Partial—horizontal	0.34	0.20
Partial—vertical	0.28	0.13

$$R_{eq} = R_f + R_i \tag{8.69}$$

- For partial insulation configurations (both horizontal and vertical)

$$R_{eq} = R_f \times \frac{1}{1 - [c/(AP)][R_i/(R_i + R_f)]} \tag{8.70}$$

where R_f and R_i are the thermal resistances, respectively, of the floor and insulation in units of $m^2 \cdot K/W$ and c is the partial insulation width or height (depending on orientation). The normalized multiplier is found from

$$D = \ln\left[(1 + H)\left(1 + \frac{1}{H}\right)^H\right] \quad \text{where } H = \frac{A}{Pk_sR_{eq}} \tag{8.71}$$

The annual mean heat gain/loss from the foundation can then be found from

$$Q_m = mU_oDA(T_r - T_m) \tag{8.72}$$

and the annual amplitude of heat gain/loss is estimated from

$$Q_a = aU_oD^{0.16}\left(k_sR_{eq}\sqrt{\frac{\omega}{\alpha_s}}\right)^{-0.6}(AT_a) \tag{8.73}$$

where α_s is the thermal diffusivity of the soil. Per Eq. (8.67), summing these two values provides the design heat gain/loss.

The coefficients a and m depend on the insulation placement configurations and are provided in Table 8.3. This model can estimate slab heat gain/loss within 10 percent when the fraction A/P is larger than 0.5 m.

8.5 Comparison of Calculations with Each Other and with Measured Data

The ultimate test of any theory is, of course, the confrontation with observation. How do load calculations fare in this regard? Much work has been done to ensure confidence in the validity of building models, and here we present only a few remarks.

To begin, there is no uncertainty about the principles; the relevant laws of physics are known beyond a shadow of a doubt. How to apply them is another matter. It is not obvious which phenomena are important in real buildings and in how much detail they should be modeled. For instance, what are the roles played by air infiltration, moisture absorption, or zone-to-zone heat transfer? Are the necessary data available? Is hourly time resolution sufficient? Some of the processes are extremely complicated and difficult to calculate; one needs to know which approximations are acceptable in practice. Finally, most buildings are occupied by people, and their behavior can have great bearing on energy use, yet it eludes prediction.

Improving the accuracy of models is a long and involved process, with continuing iterations. Advances are possible in the formulation of the models and in their numerical resolution, aided by progress in computer technology.

For the validation there are basically two approaches: comparison of models with each other and comparison with measured data. Such comparisons can be carried out at different levels, from tests of components (e.g., a wall segment or a fan) to entire buildings. The difficulties grow with the size of the system. While the validation of component models by laboratory tests is relatively straightforward, the instrumentation and evaluation of a large occupied building are a formidable undertaking. In recent years, the collection of data has become easier thanks to computerized energy management systems, but the analysis still requires a great deal of effort.

Test cells and small laboratory buildings are an intermediate solution, and they have been used for testing the simulation of dynamic behavior. The main challenge is the modeling of rapid changes, e.g., during setback recovery. The tests have shown that models such as DOE2.1 (Birdsall et al., 1990) and BLAST (1986) are acceptable for modeling phenomena with hourly time resolution.

One reason why validation by real buildings is so difficult lies in the near impossibility of performing observations under strict control of all variables. Therefore one usually resorts to the alternative of validating different simulation programs for buildings against each other. An interesting exercise of that type has been reported by Kusuda (1981). In an attempt to evaluate the accuracy of a steady-state method (the TC4.7 method) for calculating annual energy consumption, the same building was analyzed by seven different analysts; each analyst used a different dynamic simulation program and compared his results with the TC4.7 method (Knebel, 1983). As it turned out, the discrepancies between the results of the seven analysts were generally greater than the differences that the analysts found between their own simulation and their own TC4.7 result. The explanation can be found in the complexity of the input needed to give a reasonably realistic description of a typical building—and in the fact that different simulations use different approaches and require different information. So it is not surprising that different analysts would come up with different interpretations, even though the ground rules are the same.

For comparisons with measured data, we cite a study by Piette et al. (1986), who have compared the observed and predicted energy consumption of 33 new commercial buildings. Figure 8.12 shows the results in the classic format, with observed energy intensity on the x axis, predicted energy intensity on the y axis, and the diagonal corresponding to perfect agreement. There is much scatter, and the average observed energy intensity, 57 kBtu/(ft^2 · yr), is significantly higher than the prediction, 46 kBtu/(ft^2 · yr). For many of the buildings, there is significant overconsumption. By and large, such differences are not due to inaccuracies in the models; rather they arise from factors that were not known at the time of the prediction. Actual operating conditions can turn out quite different from the assumptions made during the design, in particular with regard to thermostat settings, base

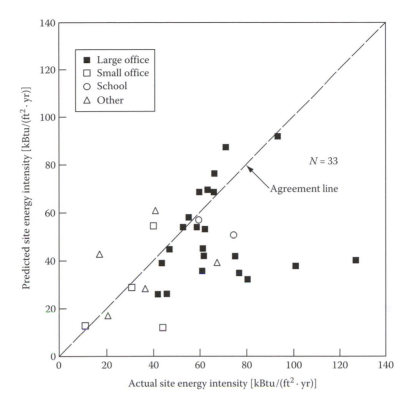

FIGURE 8.12

Comparison of predicted and observed energy intensity for 33 new commercial buildings. Dashed line corresponds to perfect agreement. 1 kBtu/(ft$^2 \cdot$ yr) = 11.36 MJ/(m$^2 \cdot$ yr). (Courtesy of Piette, M.A. et al., *ASHRAE J.,* January, 72, 1986.)

load, or occupancy. Often consumption is higher during the start-up year(s) as the operating personnel learn how to run the building more efficiently. Warm-up of soil and drying of building materials could also play a role, especially in houses.

For another comparison, we show results for a group of 82 new single-family houses in France (Marchio and Rabl, 1991). On average, the agreement is quite good, but again there is enormous scatter. Many detailed investigations have found that behavioral differences play a large role in residential energy consumption. Figure 8.13 contains an interesting illustration of this effect: The six houses at the predicted consumption $C_{the} = 10,300$ kWh have the same construction, yet consumption varies by almost a factor of 2. This appears to be quite typical. In apartment buildings, the scatter can be even larger because of "heat theft" between neighboring units with different thermostat settings.

8.6 Summary of Methods

Since the number of methods and models is quite large, we present a brief survey in Table 8.4. There is no single "best" method; rather the choice depends on the application.

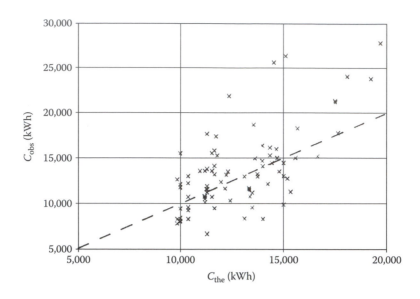

FIGURE 8.13

Comparison of predicted and observed energy consumption for 82 new single-family houses in France. C_{the} = predicted and C_{obs} = observed consumption. The dashed line corresponds to perfect agreement (note suppressed zero and different scales). 1 kWh = 3.6 MJ = 3.412 kBtu. (Courtesy of Marchio, D. and Rabl, A., *Energy Build.*, 17, 131, 1991.)

TABLE 8.4

Survey of Methods for Thermal Analysis of Buildings

Method	Suitability		Comments	Section in Book
	Peak Loads	Annual Energy Consumption		
Dynamic methods			Short time resolution (typically 1 h)	
Coupled differential equations	*	*	A large set of coupled partial differential equations for transient heat conduction through all components and materials of building Accuracy limited only by detail of description and computation time	8.3.4
			In practice, several approximations chosen, singly or in combination	
Fourier method	*		Analytical solution for periodic boundary conditions: convenient for peak loads if no thermostat setback	8.3.3
Thermal networks	*	*	System of ordinary first-order differential equations (in time), obtained by discretization of spatial variables of partial differential equations	8.2, 8.3
Differential equation linking output directly to input	*	*	The nth-order differential equation for response (output) T_i, obtained by eliminating the nonobserved variables of an n-node network	8.3
			Convenient for analytical manipulations	

TABLE 8.4 (continued)

Survey of Methods for Thermal Analysis of Buildings

Method	Suitability		Comments	Section in Book
	Peak Loads	Annual Energy Consumption		
Transfer function (also known as time series or ARMA model)	*	*	Time series, linking present value of output to past values of input and output	8.3.2 (theory)
			Obtained by also discretizing the time variable of the differential equations	
			Convenient for numerical work	7.7
			Basis for CLF/CLTD method and for computer simulation DOE2.1	(application)
Inverse models (system identification)			For example, networks with equivalent thermal parameters	8.4
			Useful for analysis of consumption data and for optimal control	
Quasi-static methods			Account for some dynamic effects, yet simple enough for hand calculation	
CLF/CLDT method	*		Peak cooling loads only	7.6
			For standard construction	
			Assumes constant T_i	
Static methods			Exact in limit where all variables are constant Simple input (design conditions and K_{tot})	
	*		Good for peak heating loads if constant T_i (but underestimate if thermostat is set back during design conditions)	
			Significant underestimate for peak cooling loads (and their time lags), even if T_i constant	
		*	For annual loads, static calculation yields excellent results if T_i constant; otherwise errors could be 10% or larger	
			There are the following variations:	
Fixed-base degree-day method		*	Eq. (8.6) with $T_{bal} = 18°C$ or 65°F. Temperature data simple and widely available, but if T_{bal} not close to 18°C, method is unreliable (commercial, superinsulated, or solar buildings).	8.1.1
Variable-base degree-day method		*	Eq. (8.6) with correct T_{bal} Good if building operation and equipment efficiency independent of T_{ext}	8.1.2
Bin method		*	The integral in Eq. (8.4) is approximated by temperature bins of 2.8 K (5°F).	8.1.3
			Can account for variation of equipment efficiency with T_{ext}	
Energy signature models		*	For analysis of consumption data	8.4.1

Problems

The problems in this book are arranged by topic. The approximate degree of difficulty is indicated by a parenthetic italic number from 1 to 10 at the end of the problem. Problems are stated most often in USCS units; when similar problems are presented in SI units, it is done with approximately equivalent values in parentheses. The USCS and SI versions of a problem are not exactly equivalent numerically. Solutions should be organized in the same order as the examples in the text: given, figure or sketch, assumptions, find, lookup values, solution. For some problems, the Heating and Cooling of Buildings (HCB) software included in this book may be helpful. In some cases it is advisable to set up the solution as a spreadsheet, so that design variations are easy to evaluate.

8.1 For a house with total heat loss coefficient $K_{tot} = 200$ W/K [379 Btu/(h · °F)], calculate the magnitude of the internal heat gains for the following conditions:

- There are four occupants.
- Average heat gain from lights and equipment is 600 W (2047 Btu/h).
- The average (24-h) solar flux through the windows is 25 W/m² [7.93 Btu/(h · ft²)].

Find the balance-point temperature if the indoor temperature is 21°C (69.8°F).   (5)

8.2 Find the annual energy consumption of a house with total heat loss coefficient $K_{tot} = 700$ Btu/(h · °F) (369.6 W/K) in a climate with 7000°F · days (3889 K · days). Assume that the balance-point temperature is 65°F and the heating system efficiency is 80 percent.   (3)

8.3 Consider a house with total heat loss coefficient $K_{tot} = 200$ W/K [379 Btu/(h · °F)], indoor temperature 21°C (69.8°F), average heat gains 1200 W (4094 Btu/h), and heating system efficiency of 80 percent. Calculate the annual energy consumption for Washington, D.C., using (a) the variable-base degree-day method with the data of Fig. 8.3 and (b) the bin data on the CD-ROM.   (6)

8.4 Solve Prob. 8.3, using the HCB software (if this case is not treated directly, you can adapt the part-load boiler performance calculation or the heat pump performance calculation).   (5)

8.5 A commercial building in Boston has the following specifications:

- T_i constant 24 h/day at 70°F (21.1°C) in winter
- Size 30 ft × 30 ft × 8 ft (9.14 m × 9.14 m × 2.44 m), orientation due south
- Open floor plan
- Equal windows on all facades, with window/wall ratio = 0.25
- Double-pane glass with $U = 0.5$ Btu/(h · ft²) [2.84 W/(m² · K)] and SC = 0.82
- 0.5 air change/h from 8:00 a.m. to 6:00 p.m.; 0.2 air change/h the rest of the time
- Lights and office equipment 3.0 W/ft² (32.28 W/m²) from 8:00 a.m. to 6:00 p.m.; 0.5 W/ft² (5.38 W/m²) the rest of the time
- 0.01 occupant per ft² (0.108 per m²) from 8:00 a.m. to 6:00 p.m.
- Heating system efficiency 80 percent

Neglect coupling to the ground. Calculate (a) the peak heating load, (b) the balance-point temperatures corresponding to daytime and nighttime conditions, and (c) the heating energy for January (use bin data of the HCB software). (10)

8.6 Consider a house in New York with $K_{tot} = 200$ W/K [379 Btu/(h · °F)], heat gains of 1000 W (3412 Btu/h), and heating system efficiency of 80 percent. Use the following approach to estimate the annual savings if the thermostat setting is reduced from 21 to 16°C (69.8 to 60.8°F) for 12 h of the day. Assume that the heat capacity of a house is negligible, and use the variable-base degree-day method or the bin method. In which direction would the result change if the heat capacity were taken into account? (5)

8.7 Use the following steps to estimate the annual heating energy of a building you know (e.g., the building where you live; it is preferable to choose a building that does not use air conditioning during the heating season).

(a) Estimate the areas of roof, walls, and windows.

(b) Estimate the U values by assuming typical construction.

(c) Estimate the air change rate.

(d) Estimate the balance-point temperature.

(e) Use the variable-base degree-day method to obtain the annual heating energy.

(f) Compare with utility bills, if available. Which assumptions or parameters would you change to get agreement? (*Note:* Agreement of overall results does not guarantee correctness of a model—there could be compensating errors.) (7)

8.8 Use Fig. 8.3 and the bin data on the CD-ROM to find the cooling degree-days for a balance-point temperature of 72°F (22.2°C) in Washington, D.C. (4)

8.9 What is the time constant of a building with total heat loss coefficient $K_{tot} = 30$ kW/K [56.85 kBtu/(h · °F)] and effective heat capacity $C_{eff} = 2.4$ GJ/K (1.26 MBtu/°F)? Estimate how long it takes the indoor temperature to drop from 20°C to 15°C after the heating system is shut off when $T_o = 0$°C. (5)

8.10 To speed up an hourly simulation program, someone proposes replacing the weather tape (with 365 days) by 2×3 typical days, arguing that three seasons (winter, spring/fall, and summer), with one clear and one cloudy day each, suffice to represent all the essential features. Discuss by considering the following points:

(a) What features of the weather are needed for peak loads, and what features for annual energy?

(b) If a single day can represent the summer (winter) design conditions, is it acceptable to calculate the summer (winter) peak load by running the simulation for just 24 h, or does the procedure have to be modified?

(c) How about buildings that are unoccupied during weekends? (5)

8.11 In energy signature models, why is it better to correlate the consumption with degree-days than with T_o? *Hint:* Consider a period where one-half of the days have T_o above and one-half below the balance-point temperature T_{bal}. (5)

8.12 Toward the end of the Roman Empire, the technology of glassmaking had advanced to the point where some villas could be built with a sunspace (called a *solarium*). Make reasonable assumptions about the construction, and use a simple thermal network to estimate the temperature in a solarium during a sunny day in Rome when

$$T_o = \left[1 + \cos\left(\pi \frac{t - 16\,\text{h}}{12\,\text{h}}\right)\right] \times 4°\text{C}$$

and

$$I(t) = \max\left[0,\ -0.2 + \cos\left(\pi \frac{t - 12\,\text{h}}{12\,\text{h}}\right)\right] \times 800\ \text{W/m}^2$$

as a function of time of day t. Plot T_i and T_o.   (10)

8.13 Suppose the house of Example 7.5 (Fig. 7.7) has a forced-air heating system with some air ducts placed directly along the outside walls. The total duct surface is 20 m², and the portion in contact with the outside wall is 4 m²; the surface heat transfer coefficient inside the duct is 20 W/m². The hot air is distributed at 60°C when the furnace is running; when the furnace is off, assume for simplicity that the duct is at $T_i = 20$°C. Thus on average the temperature rise above T_i of the duct is proportional to the load.

(a) How much do these duct losses increase the peak heating load, in absolute and in relative terms?

(b) How much does the annual energy consumption for heating increase, in absolute and in relative terms?

(c) How would these answers change if the duct were insulated with 2.5 cm of glass wool?

(d) What is the payback time of adding this glass wool insulation if the house is heated with natural gas at $5/GJ, the furnace efficiency is 90 percent, and the glass wool costs $25?

8.14 Suppose the house of Example 7.5 (Fig. 7.7) has a hydronic heating system with radiators placed along the outside walls. The total wall surface facing the radiators is 10 m². The surface heat transfer coefficient from radiator to wall is 10 W/m². The difference between radiator temperature and T_i is proportional to the load, the peak radiator temperature being 80°C.

(a) Calculate how much the temperature of the wall next to the radiators increases.

(b) How much does this effect increase the peak heating load, in absolute and in relative terms?

(c) How much does the annual energy consumption for heating increase, in absolute and in relative terms?

(d) How do these numbers change if 5 cm of fiberglass insulation is placed on these portions of the wall?

(e) What is the payback time of adding this fiberglass insulation if the house is heated with natural gas at $5/GJ, the furnace efficiency is 90 percent, and the glass wool costs $25?

8.15 Consider a house with $K_{\text{tot}} = 500$ Btu/(h · °F) and $T_i = 70$°F, assumed constant during the entire heating season.

(a) What average internal heat gain due to solar, lights, people, and equipment would be required to give a balance temperature of 65°F?

(b) Using the degree-day data of Fig. 8.3, what is the annual heating energy consumption if the house is heated with a furnace having a constant efficiency of 82 percent?

(c) What would be the annual heating energy consumption if the house were moved to a location with the bin data in the following table?

Temperature, °F	N_{bin}, h
52.5	32
57.5	180
62.5	300
67.5	620

8.16 Consider the building from Example 8.1 and Fig. 7.7. Assume that the internal temperature remains constant through the year and that the house is slab-on-grade construction with uniform horizontal insulation under the slab. The insulation is 2.5 cm of polyisocyanurate [conductivity of 0.02 W/(m · K)]. What are the annual mean and peak heat gain/loss from this slab?

8.17 Consider the residential building of Example 8.1. How does the annual heating bill change if you take into account the ground-coupled heat loss from Prob. 8.16?

References

Alamdari, F., and G. P. Hammond (1982). Time-dependent Convective Heat Transfer in Warm-air Heated Rooms. Energy Conservation in the Built Environment: Proceedings of the CIB W67 Third International Symposium, Dublin, Ireland, pp. 209–220.

Alamdari, F., and G. P. Hammond (1983). Improved Data Correlations for Buoyancy-Driven Convection in Rooms. *Building Services Engineering Research and Technology*, vol. 4, no. 3, pp. 106–112.

Altmayer, E. F., A. J. Gadgil, F. S. Bauman, and R. C. Kammerud (1983). Correlations for Convective Heat Transfer from Room Surfaces. *ASHRAE Transactions*, vol. 89, pt. 2A, pp. 61–77.

Anstett, M., and J. F. Kreider (1993). "Application of Artificial Neural Networks to Commercial Building Energy Use Prediction." *ASHRAE Transactions*, vol. 99, pt. 1, pp. 505–517.

ASHRAE (1984). *Bin Weather Data* (RP 385, L. Degelman)

ASHRAE (1989, 2001). *Handbook of Fundamentals*. American Society of Heating, Refrigerating and Air-Conditioning Engineers, Atlanta.

ASHRAE (1993). *HVAC2 Toolkit: Algorithms and Subroutines for Secondary HVAC System Energy Calculations*. American Society of Heating, Refrigerating, and Air Conditioning Engineers, Inc, Atlanta.

Ayres, M. J., and E. Stamper (1995). "Historical Development of Building Energy Calculations," *ASHRAE Transactions*, vol. 101, pt. 1.

Bacot, P., A. Neveu, and J. Sicard (1984). "Analyse Modale Des Phenomenes Thermiques en Regime Variable Dans le Batiment," *Revue Generale de Thermique*, no. 267, p. 189.

Balcomb, J. D., R. W. Jones, R. D. McFarland, and W. O. Wray (1982). Expanding the SLR method, *Passive Solar Journal* 1(2).

Bauman, F., A. Gadgil, R. Kammerud, E. Altmayer, and M. Nansteel (1983). Convective Heat Transfer in Buildings: Recent Research Results. *ASHRAE Trans.* vol. 89, pt. 1a, pp. 215–232.

Bohn, M. S., A. T. Kirkpatrick, and D. A. Olson (1984). Experimental Study of Three-Dimensional Natural Convection High-Rayleigh Number. *Journal of Heat Transfer*, vol. 106, pp. 339–345.

Bou Saada, T., and J. Haberl (1995a). "A weather-daytyping Procedure for Disaggregating Hourly End-use Loads in an Electrically Heated and Cooled Building from Whole-building Hourly Data," *Proc. 30th IECEC*, July 31–Aug 4, 1995, Orlando FL, pp. 349–356.

Birdsall, B., W. F. Buhl, K. L. Ellington, A. E. Erdem, and F. C. Winkelmann (1990). "Overview of the DOE2.1 Building Energy Analysis Program." Report LBL-19735, rev. 1. Lawrence Berkeley Laboratory, Berkeley, Calif. 94720.

BLAST (1986). "The Building Load Analysis and System Thermodynamics Program." Version 3.0. User's Manual. BLAST Support Office, University of Illinois, Urbana-Champaign.

Carroll, J. A. (1980). AN "MRT Method" of Computing Radiant Energy Exchange in Rooms. Systems Simulation and Economic Analysis, San Diego, California. pp. 343–348.

Chandra, S., and A. A. Kerestecioglu (1984). Heat Transfer in Naturally Ventilated Rooms: Data from Full-scale Measurements. *ASHRAE Transactions*, vol. 90, pt. 1b, pp. 211–224.

Chuangchid, P., and M. Krarti. "Parametric Analysis and Development of a Design Tool for Foundation Heat Gain for Coolers," *ASHRAE Trans.*, vol. 106, pt. 2, pp. 240–250.

Clark, D. R. (1985). *HVACSIM + Building Systems and Equipment Simulation Program: Reference Manual*. NBSIR 84-2996, U.S. Department of Commerce, Washington, D.C.

Clarke, J. A. (1985). *Energy Simulation in Building Design*. Boston: Adam Hilger Ltd.

Claridge, D. E., M. Krarti, and M. Bida (1987). A validation study of variable-base degree-day cooling calculations, *ASHRAE Transactions*, 93(2), 90–104.

Claridge, D. E., J. S. Haberl, R. Sparks, R. Lopez, and K. Kissock (1992). "Monitored Commercial Building Energy Data: Reporting the Results." 1992 *ASHRAE Transactions Symposium Paper*, vol. 98, pt. 1, pp. 636–652.

Cole, R. J. (1976). The Longwave Radiation Incident Upon the External Surface of Buildings. *The Building Services Engineer*, vol. 44, pp. 195–206.

Cooper, K. W., and D. R. Tree (1973). A Re-evaluation of the Average Convection Coefficient for Flow Past a Wall. *ASHRAE Transactions*, vol. 79, pp. 48–51.

Corson, G. C. (1992). "Input-output Sensitivity of Building Energy Simulations." *ASHRAE Transactions*, vol. 98, pt. 1, p. 618.

Davies, M. G. (1988). Design Models To Handle Radiative and Convective Exchange in a Room. *ASHRAE Transactions*, vol. 94, pt. 2, pp. 173–195.

Dhar, A. (1995). "Development of Fourier Series and Artificial Neural Network Approaches to Model Hourly Energy Use in Commercial Buildings," PhD dissertation, ME Dept. Texas A&M University, May.

Draper, N., and H. Smith (1981). *Applied Regression Analysis*, 2d ed. Wiley, New York.

Draper, N. R., and H. Smith (1981). *Applied Regression Analysis*, 2d ed. Wiley, New York.

Elmahdy, A. H., and R. C. Biggs (1979). Finned tube heat exchanger: Correlations of dry surface heat transfer data, *ASHRAE Transactions* 85(2).

Erbs, D. G., S. A. Klein, and W. A. Beckman (1983). "Estimation of Degree-Days and Ambient Temperature Bin Data from Monthly-Average Temperatures." *ASHRAE J.* (June), pp. 60–65.

Fels, M. F., ed. (1986). "Measuring Energy Savings: The Scorekeeping Approach." Special double issue of *Energy and Buildings*, vol. 9, nos. 1 and 2.

Fels, M. F., and M. Goldberg (1986). "Refraction of PRISM Results into Components of Saved Energy." *Energy and Buildings*, vol. 9, pp. 169–180.

Gelb, A., ed. (1974). *Applied Optimal Estimation*. M.I.T. Press, Cambridge, Mass.

Haberl, J. S., and D. E. Claridge (1987). "An Expert System for Building Energy Consumption Analysis: Prototype Results." *ASHRAE Trans.* vol. 93, pt. 1, pp. 979–998.

Hammarsten, S. (1987). "A Critical Appraisal of Energy Signature Models." *Applied Energy*, vol. 26, pp. 97–110.

Jones, R. W., ed. (1983). *Passive Solar Design Handbook*, vol. 3. American Solar Energy Society, Boulder, Colo.

Kissock, J. K., T. A. Reddy, J. S. Haberl, and D. E. Claridge (1993). "E-model: A New Tool for Analyzing Building Energy Use Data." *Proc. Intl. Indust. Energy Tech. Conf.*, March, Texas A&M University.

Knebel, D. E. (1983). *Simplified Energy Analysis Using the Modified Bin Method*. American Society of Heating, Refrigerating and Air-Conditioning Engineers, Atlanta. See also D. Knebel and S. Silver, "Upgraded Documentation of the TC4.7 Simplified Energy Analysis Procedure," *ASHRAE Trans.*, vol. 91, pt. 2A, 1985.

Knebel, D. E. (1983). Simplified Energy Analysis Using the Modified Bin Method, ASHRAE Special Publication, ISBN 0910110395.

Knebel, D. E. (1995). ASHRAE technical committee 4.7 development of simplified energy analysis procedures, *AHSRAE Transactions*, 101(1).

Kreider, J. F., and J. Haberl (1994). "Predicting Hourly Building Energy Usage: The Great Predictor Shootout—Overview and Discussion of Results." *ASHRAE Trans.* vol. 100, pt. 2, pp. 1104–1118.

Kreider, J. F., and X. A. Wang (1992). "Improved Artificial Neural Networks for Commercial Building Energy Use Prediction." *J. Solar Energy Eng.* 92, ASME, New York, pp. 361–366.

Kreider, J. F., and X. A. Wang (1992). "Improved Artificial Neural Networks for Commercial Building Energy Use Prediction," *Solar Energy Eng.*

Kreider, J. F., and J. Haberl (1994). "Predicting Hourly Building Energy Usage: The Great Predictor Shootout–Overview and Discussion of Results," ASHRAE Trans.

Kreider, J. F., and A. Rabl (1994). *Heating and Cooling of Buildings*, McGraw-Hill Book Company, New York, 890+xv11 pp.

Kreider, J. F., and X. A. Wang (1991). "Artificial Neural Networks Demonstration for Automated Generation of Energy Use Predictors for Commercial Buildings." *ASHRAE Transactions*, vol. 97, pt. 1.

Kreider, J. (2001). *Handbook of HVAC*, CRC Press, Boca Raton, FL.

Kusuda, T. (1969). Thermal Response Factors for Multi-Layer Structures of Various Heat Conduction Systems. *ASHRAE Transactions*, vol. 75, pt. 1, 246–271.

Kusuda, T. (1981). "A Comparison of Energy Calculation Procedures." *ASHRAE J.* (August), p. 21.

Lawrence Berkeley Laboratory (LBL) (1982). *DOE2 Engineers Manual*, Report number LBL-11353 (LA-8520-M, DE83004575), National Technical Information Services, Springfield, VA.

Lewis, P. T., and D. K. Alexander (1990). HTB2: A Flexible Model for Dynamic Building Simulation. *Building and Environment*, vol. pp 7–16.

Liu, M., and D. Claridge (1995). "Application of Calibrated HVAC System Models to Identify Component Malfunctions and to Optimize the Operation and Control Strategies," Proceedings of the 1995 ASME Solar Engineering Conference, vol. 1., pp. 209–218.

MacDonald, J. M., and D. M. Wasserman (1989). "Investigation of Metered Data Analysis Methods for Commercial and Related Buildings," ORNL Rept. ORNL/CON-279, May.

McAdams, W. H. (1954). *Heat Transmission*, 3rd ed. McGraw-Hill Book Co., Inc., New York.

McClelland, J. L., and D. E. Rumelhart (1988). EXPLORATION IN PARALLEL DISTRIBUTED PROCESSING. MIT Press, Cambridge.

Marchio, D., and A. Rabl (1991). "Energy-Efficient Gas-Heated Housing in France: Predicted and Observed Performance." *Energy and Buildings*, vol. 17, pp. 131–139.

Melo, C., and G. P Hammond (1991). Modeling And Assessing The Sensitivity Of External Convection From Building Facades in Heat And Mass transfer In Building Materials And Structures. New York: Hemisphere. pp. 683–695.

Miller, R., and J. Seem (1991). "Comparison of Artificial Neural Networks with Traditional Methods of Predicting Return from Night Setback." ASHRAE Trans. vol. 97, pt. 2, pp. 500–508.

Mitalas, G.P. (1968). Calculations of Transient Heat Flow Through Walls and Roofs. ASHRAE Transactions, vol. 74, pt. 2, pp. 182–188.

Mitalas, G. P., and D. G. Stephenson (1967). Room Thermal Response Factors. ASHRAE Transactions, vol. 3, pt. 1, pp. 2. 1–2.10.

Neter, J., W. Wasseran, and M. Kutner (1989). *Applied Linear Regression Models*, 2nd Edition, Richard C. Irwin, Inc., Homewood, IL.

Neveu, A., P. Bacot, and R. Regas (1986). "Modèles d'évolution thermique des bâtiments: Conditions pratiques d'identification." *Revue Générale de Thermique*, no. 296, p. 413.

Nisson, J. D. N., and G. Dutt (1985). *The Superinsulated Home Book*. Wiley, New York.

Norlen, U., D. von Hattem, S. Hammarsten, and F. Conti (1987). "Evaluating the Performance of Passive Solar Components Using Kalman Filtering." *Proc. Eur. Conf. on Architecture*, Munich, Germany, April 6–10.

Park, C., D. R. Clark, and G. E. Kelly (1985). An Overview of HVACSIM+, a Dynamic Building/HVAC Control Systems Simulation Program. Proceedings of the First Building Energy Simulation Conference, Seattle, Washington.

Piette, M. A., L. W. Wall, and B. L. Gardiner (1986). "Measured Performance." *ASHRAE J.* (January), pp. 72–78.

Pryor, D. V., and C. B. Winn (1982). "A Sequential Filter Used for Parameter Estimation in a Passive Solar System." *Solar Energy*, vol. 28, p. 65.

Rabl, A. (1988). "Parameter Estimation in Buildings: Methods for Dynamic Analysis of Measured Energy Use." *ASME J. Solar Energy Eng.*, vol. 110, pp. 52–66.

Rabl, A., L. K. Norford, and G. V. Spadaro (1986). "Steady State Models for Analysis of Commercial Building Energy Data." *American Council for Energy Efficient Economy Summer Study*, August, Santa Cruz, Calif.

Rabl, A. (1988). "Parameter Estimation in Buildings: Methods for Dynamic Analysis of Measured Energy Use," *Journal of Solar Energy Engineering*, vol. 110, pp. 52–66.

Rabl, A., and A. Riahle (1992). "Energy Signature Model for Commercial Buildings: Test With Measured Data and Interpretation," *Energy and Buildings*, vol. 19, pp 143–154.

Reddy, T. (1989). "Application of Dynamic Building Inverse Models to Three Occupied Residences Monitored Non-intrusively," Proceedings of the Thermal Performance of Exterior Envelopes of Buildings IV, ASHRAE/DOE/BTECC/CIBSE.

Reddy, T., and D. Claridge (1994). Using Synthetic Data to Evaluate Multiple Regression and Principle Component Analyses for Statistical Modeling of Daily Building Energy Consumption, *Energy and Building*. 24. 35–44.

Rich, D. G. (1966). The efficiency and thermal resistance of annular and rectangular fins. Proceedings of the Third International Heat Transfer Conference, AICHE 111:281–89.

Ruch, D., L. Chen, J. Haberl, and D. Claridge (1993). A Change-Point Principal Component Analysis (CP/PCA) Method for Predicting Energy Usage in Commercial Buildings: The PCA Model, The *Journal of Solar Energy Engineering*, vol. 115, no. 2, May.

Ruch, D., and D. Claridge (1991). "A Four Parameter Change-point model for Predicting Energy Consumption in Commercial Buildings," *Proceedings of the ASME-JSES-JSME International Solar Energy Conference*, Reno, NV, March 17–22, pp. 433–440.

SERI (1981). "Solar Radiation Energy Resource Atlas of the United States." Solar Energy Research Institute, Report SERI/SP-642-1037, Golden, Colo.

Shavit, G. (1995). Short-time-step analysis and simulation of homes and buildings during the last 100 years, *ASHRAE Transactions*, 101(1).

Shurcliff, W. A. (1984). *Frequency Method of Analyzing a Building's Dynamic Thermal Performance*. W. A. Shurcliff, 19 Appleton Street, Cambridge, Mass. 02138.

Sonderegger, R. C. (1977). "Dynamic Models of House Heating Based on Equivalent Thermal parameters," Ph.D. Thesis, Center for Energy and Environmental Studies Report No. 57, Princeton University, Princeton, N. J.

Sowell, E. F., and G. N. Walton (1980). Efficient Computation of Zone Loads. *ASHRAE Transactions*, vol. 86, pt. 1, pp. 49–72.

Sowell, E. F. (1988). Classification of 200,640 parametric zones for cooling load calculations, *ASHRAE Transactions*, 94(2).

Sowell, E. F. (1990). Lights: A Numerical Lighting/HVAC Test Cell. *ASHRAE Transactions*, vol. 96, pt. 2, pp. 780–786.

Sowell, E. F., and M. A. Moshier (1995). HVAC Component Model Libraries for Equation-based Solvers, *ASHRAE Transactions*, 101(1).

Sowell, E. F., and D. C. Hittle (1995). Evolution of Building Energy Simulation Methodology, *ASHRAE Transactions*, 101(1).

Spitler, J. D. (1996). *Annotated Guide to Load Calculation Models and Algorithms*, (Atlanta, Georgia: ASHRAE).

Spitler, J. D., C. O. Pedersen, and D. E. Fisher (1991). Interior Convective Heat Transfer in Buildings with Large Ventilative Flow Rates. *ASHRAE Transactions*, vol. 97, pt. 1, pp. 505–515.

Steinman, M., L. N. Kalisperis, and L. H. Summers (1989). The MRT-Correction Method: A New Method of Radiant Heat Exchange. *ASHRAE Transactions*, vol. 95, pt. 1, pp. 1015–1027.

Stephenson, D. G., and G. P. Mitalas (1967). "Cooling Load Calculations by Thermal Response Factor Method," *ASHRAE Trans.*, vol. 73.

Subbarao, K., J. Burch, and C. E. Hancock (1990). "How to accurately measure the load coefficient of a residential building," *Journal of Solar Energy Engineering*, in preparation.

Subbarao, K. (1986). "Thermal Parameters for Single and Multi-zone buildings and their determination from performance data," SERI Report SERI/TR-253–2617, Golden, CO.

Subbarao, K. (1988). "PSTAR—Primary and Secondary Terms Analysis and Renormalization: A Unified Approach to Building Energy Simulations and Short-Term Monitoring." Solar Energy Research Institute, Report SERI/TR-254–3175, Golden, Colo.

Vadon, M., J. F. Kreider, and L. K. Norford (1991). "Improvement of the Solar Calculations in the Modified Bin Method." *ASHRAE Trans.*, vol. 97, pt. 2.

9

Heat Generation and Transfer Equipment

9.1 Introduction

This chapter discusses equipment used for producing heat from fossil fuels, electricity, or solar power. The emphasis is on design-oriented information including system characteristics, operating efficiency, the significance of part-load characteristics, and criteria for selecting from the vast array of heat-producing equipment available.

The heating plants discussed in this chapter are often called the *primary systems*. Systems intended to distribute heat produced by the primary systems are called *secondary systems* and include duct and pipes, fans and pumps, terminal devices, and auxiliary components. Such secondary systems for heating and cooling are described in Chap. 11 and are best understood after the primary heating and cooling systems described in this and the next chapter are well comprehended. The terms *primary* and *secondary* are equivalent to the terms *plant* and *system* used by some building analysts and HVAC system modelers.

The goals of this chapter are to have the reader understand the operation of various heat generation or transfer systems and their performance:

- Furnaces
- Boilers
- Heat pumps
- Heat exchangers
- Part-load performance and energy calculations for each

The primary sources of heat for building heating systems are fossil fuels, natural gas, and various grades of fuel oil and coal. Electricity is used under certain circumstances for heat in commercial buildings, although the economic penalties for so doing are significant. Solar power can be converted to heat for applications in commercial buildings including perimeter zone heating and service water heating. (The direct use of sunlight for lighting is described in Chap. 13.)

9.2 Natural Gas and Fuel Oil–Fired Equipment

This section describes fossil fuel–fired *furnaces and boilers*—devices that convert the chemical energy in fuels to heat. Furnaces heat airstreams that are used in turn for heating the interior of buildings. Forced-air heating systems supplied with heat by furnaces are the most common type of residential heating system in the United States. Boilers are pressure vessels used to transfer heat, produced by burning a fuel, to a fluid. The most common fluid used for this purpose in buildings is liquid water or water vapor. The key distinction between furnaces and boilers is that air is heated in the former and water is heated in the latter.

The fuels used for producing heat in boilers and furnaces include natural gas (i.e., methane), propane, fuel oil (at various grades numbered through 6), wood, coal, and other fuels including refuse-derived fuels. It is beyond the scope of this book to describe in detail the design of boilers and furnaces or how they convert chemical energy to heat. Rather we provide the information needed by HVAC designers for these two classes of equipment. Since boilers and furnaces operate at elevated temperatures (and pressures for boilers), they are hazardous devices. As a result, a body of standards has been developed to ensure the safe operation of this equipment. We briefly describe this aspect of this equipment as well.

9.2.1 Furnaces

Modern furnaces use forced convection to remove heat produced within the firebox from its outer surface. There are very many designs to achieve this; four residential classifications based on airflow type are shown in Fig. 9.1. The *upflow* furnace shown in Fig. 9.1a has a blower located below the firebox heat exchanger with heated air exiting the unit at the top. Return air from the heated space enters this furnace type at the bottom.

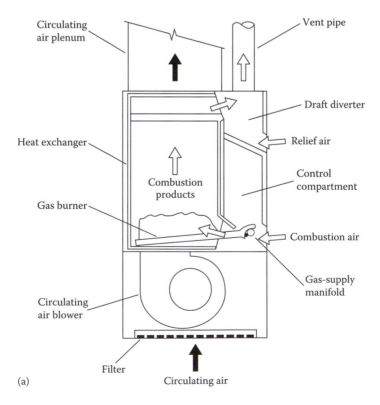

FIGURE 9.1
(a) Examples of furnaces for residential space heating: vertical.

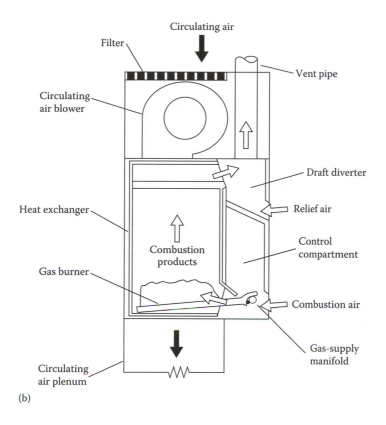

(b)

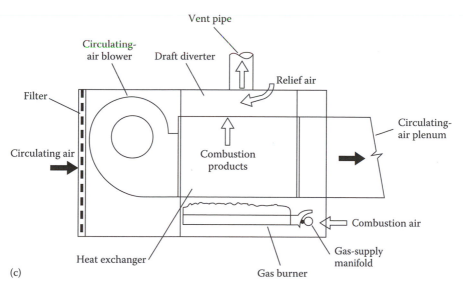

(c)

FIGURE 9.1 (continued)

(b) Examples of furnaces for residential space heating: downflow. (c) Examples of furnaces for residential space heating: horizontal.

(*continued*)

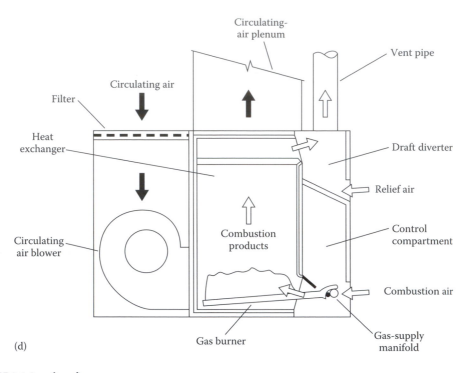

(d)

FIGURE 9.1 (continued)

(d) Examples of furnaces for residential space heating: low boy.

The upflow design is used in full-size mechanical rooms where sufficient floor-to-ceiling space exists for the connecting ductwork. This is the most common form of residential furnace.

Downflow furnaces (Fig. 9.1b) work in reverse: Air flows downward as it is heated by passing over the heat exchanger. This design is used in residences without basements or in upstairs mechanical spaces in two-story buildings. *Horizontal* furnaces of the type shown in Fig. 9.1c use a horizontal airflow path with the air mover located beside the heat exchanger. This design is especially useful in applications where vertical space is limited, e.g., in attics or crawl spaces of residences.

A combination of upflow and horizontal furnaces is available and is named the *basement* (low-boy) furnace (Fig. 9.1d). With the blower located beside the firebox, air enters the top of the furnace, is heated, and exits from the top. This design is useful in applications where headroom is restricted.

The combustion side of the heat exchanger in gas furnaces can be at either atmospheric pressure (the most common design for small furnaces) or superatmospheric pressures produced by combustion air blowers. The latter are of two kinds: forced-draft (blower upstream of combustion chamber) or induced-draft (blower downstream of combustion chamber) furnaces that have better control of parasitic heat losses through the stack. As a result, efficiencies are higher for such *power combustion furnaces*.

In addition to natural gas, liquefied propane gas (LPG) and fuel oil can be used as energy sources for furnaces. LPG furnaces are very similar to natural gas furnaces. The only differences between the two arise due to the difference in energy content (1000 Btu/ft^3 for natural gas and 2500 Btu/ft^3 for propane) and supply pressure to the burner.

Gas furnaces can be adapted for LPG use, and vice versa in many cases. Fuel oil burner systems differ from gas burner systems owing to the need to atomize oil before combustion. The remainder of the furnace is not much different from a gas furnace except that heavier construction is often used.

Other furnaces for special applications are also available. These include (1) un-ducted space heaters located within the space to be heated and relying on natural convection for heat transfer to the space; (2) wall furnaces attached to walls and requiring very little space; and (3) direct-fired unit heaters used for direct space heating in commercial and industrial applications. Unit heaters are available in sizes from 25,000 to 320,000 Btu/h (7 to 94 kW).

On commercial buildings one often finds furnaces incorporated into *package units* (or "rooftop units") consisting of air conditioners and gas furnaces (or electric resistance coils). Typical sizes of these units range from 5 to 50 tons of cooling (18 to 175 kW) with a matched capacity to 50 percent oversized furnace. Smaller units are designed to be used for a single zone in either the heating-only or the cooling-only mode. Larger units above 15 tons (53 kW) can operate simultaneously in heating and cooling modes to condition several zones. In the heating mode, these commercial-size package units operate with an air temperature rise of about 85°F (47 K).

Figure 9.2 shows a furnace intended for residential heating applications.

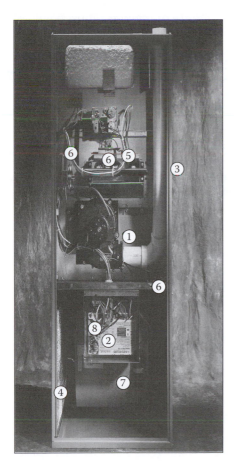

FIGURE 9.2

Photograph of a gas-fired furnace. (1) Blower; (2) controls; (3) housing; (4) air filters; (5) igniter; (6) flame sensor, door interlock safety switch, pressure switch; (7) circulating air blower; (8) viewing port. (Courtesy of Goodman Manufacturing Company, L.P., Houstan, TX. 2009. With permission.)

9.2.1.1 Furnace Design and Selection for HVAC Applications

Selection of a furnace is straightforward once the fuel source and heat load (see Chap. 7) are known. The following factors must be accounted for in furnace sizing and type selection:

- *Design heat loss* of area to be heated, in Btu per hour or kilowatts (see Chap. 7)
- Morning *recovery capacity from night setback*
- Constant *internal gains or waste heat recovery* that reduces the needed heat rating of a furnace
- *Humidification load* (see Chaps. 4 and 7)
- *Fan and housing size* sufficient to accommodate air conditioning system, if any
- *Duct heat losses* if heat so lost is external to the heated space
- *Available space* for furnace location

Residential furnaces are available in sizes ranging from 35,000 to 175,000 Btu/h (10 to 51 kW). Commercial sizes range up to 1,000,000 Btu/h (300 kW).

Economic criteria including initial cost and life cycle operating cost must be considered.[1] Although high efficiency may cost more initially, it is very often worthwhile to make the investment when the overall economic picture is considered. However, in many cases first cost is the primary determinant of selection. In these cases the HVAC engineer must point out to the building owner or architect that the building lifetime penalties of using inexpensive but inefficient heating equipment are considerable, many times the initial cost difference.

The designer is advised to avoid the customary tendency to oversize furnaces. An over-sized furnace operates at a lower efficiency than a properly sized one owing to the penalties of part-load operation. If a correct heat load calculation is done (with proper attention to the recognized uncertainty in infiltration losses), only a small safety factor should be needed, say, 10 percent. The safety factor is applied to account for heat load calculation uncertainties and possible future, modest changes in building load due to usage changes. Oversizing of furnaces also incurs other penalties including excessive duct size and cost along with poorer control of comfort due to larger temperature swings in the heated space.

9.2.1.2 Furnace Efficiency and Energy Calculations

The steady-state efficiency η_{furn} is defined as the fuel supplied less flue losses, all divided by the fuel supplied:

$$\eta_{\text{furn}} = \frac{\dot{m}_{\text{fuel}}h_{\text{fuel}} - \dot{m}_{\text{flue}}h_{\text{flue}}}{\dot{m}_{\text{fuel}}h_{\text{fuel}}} \tag{9.1}$$

in which the subscripts identify the fuel input and flue gas exhaust mass flow rates $\dot{m}$ and enthalpies h. Gas flows are usually expressed in cubic feet per hour (liters per second). To find the mass flow rate, one must know the density, which in turn depends on the gas main pressure. The ideal gas law can be used for such calculations. Efficiency values are specified by the manufacturer at a single value of fuel input rate. Of course, one could use either greater or lesser flow rates, but the design flow rate is that at which the manufacturer's efficiency value applies. At higher fuel input rates, the furnace could overheat with hazardous results. Industry standards dictate the temperature limits allowable in furnaces and thereby limit the fuel input rates.

---

[1] The techniques used to make the final selection appear in Chap. 15 on economics.

This instantaneous efficiency is of limited value in selecting furnaces owing to the fact that furnaces often operate in a cyclic, part-load mode where instantaneous efficiency may be lower than that at peak operating conditions. Part-load efficiency is low since cycling causes inefficient combustion, cyclic heating and cooling of the furnace heat exchanger mass, and thermal cycling of the distribution ductwork. A more useful performance index is the *annual fuel utilization efficiency* (AFUE), which also accounts for other loss mechanisms over a season. These include stack losses (sensible and latent), cycling losses, infiltration, and pilot losses (ASHRAE, 2000). An ASHRAE standard (103-1982R) is used for finding the AFUE for residential furnaces.

Table 9.1 shows typical values of AFUE for residential furnaces. The table shows that efficiency improvements can be achieved by eliminating standing pilots, by using a forced-draft design, or by condensing the products of combustion to recover latent heat normally lost to the flue gases. Efficiency can also be improved by using a vent damper to reduce stack losses during furnace-off periods. Although this table is prepared from residential furnace data, it can be used for commercial-size furnaces as well. Few data have been published for commercial systems since this has not been mandated by law, as it has been for residential furnaces. The AFUE has the shortcoming that a specific usage pattern and equipment characteristics are assumed. In the next section, we discuss a more accurate method for finding annual performance of heat-producing primary systems.

The AFUE can be used to find the annual energy consumption directly from its definition below. The fuel consumption during an average year $Q_{fuel,yr}$ is given by

$$Q_{fuel,yr} = \frac{Q_{yr}}{AFUE} \text{ MBtu/yr (GJ/yr)} \tag{9.2}$$

where Q_{yr} is the annual heat load. By using this approach it is a simple matter to find the savings one might expect, on average, by investing in a more efficient furnace.

TABLE 9.1

Typical Values of AFUE for Furnaces

Type of Gas Furnace	AFUE, Percent
1. Atmospheric with standing pilot	64.5
2. Atmospheric with intermittent ignition	69.0
3. Atmospheric with intermittent ignition and automatic vent damper	78.0
4. Same basic furnace as type 2, except with power vent	78.0
5. Same as type 4 but with improved heat transfer	81.5
6. Direct vent with standing pilot, preheat	66.0
7. Direct vent, power vent, and intermittent ignition	78.0
8. Power burner (forced-draft)	75.0
9. Condensing	92.5
Type of Oil Furnace	**AFUE, Percent**
1. Standard	71.0
2. Same as type 1 with improved heat transfer	76.0
3. Same as type 2 with automatic vent damper	83.0
4. Condensing	91.0

Source: Courtesy of ASHRAE, *Handbook of Systems and Equipment*, American Society of Heating, Refrigerating and Air-Conditioning Engineers, Atlanta, GA, 2000. With permission.

Example 9.1: Energy Savings from Using a Condensing Furnace

A small commercial building is heated by an old atmospheric-type gas furnace. The owner proposes to install a new pulse-type (condensing) furnace. If the annual heat load Q_{yr} on the warehouse is 200 GJ, what energy savings will the new furnace produce?

    Given: Q_{yr}, furnace types

    Assumptions: AFUE is an adequate measure of seasonal performance, and furnace efficiency does not degrade with time.

    Find: $\Delta Q_{fuel} = Q_{fuel,old} - Q_{fuel,new}$

    Lookup values: AFUEs from Table 9.1

$$AFUE_{old} = 0.645 \quad AFUE_{new} = 0.925$$

SOLUTION

Equation (9.2) is used to find the solution. The energy saving is given by

$$\Delta Q_{fuel} = Q_{yr}\left(\frac{1}{AFUE_{old}} - \frac{1}{AFUE_{new}}\right)$$

Substituting the tabulated values for AFUE, we have

$$\Delta Q_{fuel} = 200\left(\frac{1}{0.645} - \frac{1}{0.925}\right) = 93.9 \text{ GJ/yr}$$

COMMENTS

The saving of energy from using the modern furnace is substantial, almost equivalent to 50 percent of the annual heating load. For a more precise calculation for a given furnace, one could use the so-called bin method, described in Chap. 7. This approach requires more detailed part-load data from a manufacturer but will give a more accurate result. We present such an example in Sec. 9.2.2 on boilers. You can calculate the economic payback using the results of this example by methods given in Chaps. 1 and 15.

In addition to energy consumption, the designer must be concerned with myriad other factors in furnace selection. These include

- Air-side temperature rise; duct design and airflow rate affected
- Airflow rate; duct design affected
- Control operation (for example, will night or unoccupied day-night setback be used? Is fan control by thermal switch or time-delay relay?)
- Safety issues (combustion gas control, fire hazards, high-temperature limit switch)

Chapter 11 will deal with other aspects of furnace-based heating system design including duct sizing and layout, fan sizing, and cooling coil design in combination commercial rooftop units.

9.2.2 Boilers

A boiler is a device made from copper, steel, or cast iron to transfer heat from a combustion chamber (or electric resistance coil) to water in the liquid phase, vapor phase, or both.

Boilers are classified both by the fuel used and by the operating pressure. Fuels include gas, fuel oils, wood, coal, refuse-derived fuels, and electricity. In this section we focus on fossil fuel-fired boilers.

Boilers produce either hot water or steam at various pressures. Although water does not literally boil in hot-water "boilers," they are called boilers, nevertheless. Steam is an exceptionally effective heat transport fluid due to its very large heat of vaporization, as noted in Chap. 3.

Boilers for buildings are classified as

- *Low-pressure*: Steam boilers with operating pressures below 15 psig (100 kPa). Hot-water boilers with pressures below 150 psig (1000 kPa); temperatures are limited to 250°F (120°C).

- *High-pressure*: Steam boilers with operating pressures above 15 psig (100 kPa). Hot-water boilers with pressures above 150 psig (1000 kPa); temperatures are above 250°F (120°C).

Heat rates for steam boilers are often expressed in pound-mass of steam produced per hour (or kilowatts). The heating value of steam for these purposes is rounded to 1000 Btu/lb$_m$. Steam boilers are available at heat rates of 50 to 50,000 lb$_m$/h of steam (15 to 15,000 kW). This overlaps the upper range of furnace sizes noted in Sec. 9.2.1. Steam produced by boilers is used in buildings for space heating, water heating, and absorption cooling.

Water boilers are available in the same range of sizes as steam boilers: 50 to 50,000 kBtu/h (15 to 15,000 kW). Hot water is used in buildings for space and water heating.

Since the energy contained in steam and hot water within and flowing through boilers is very large, an extensive codification of regulations has evolved to ensure safe operation. In the United States, the ASME Boiler and Pressure Vessel Code governs construction of boilers. For example, the code sets the limits of temperature and pressure on low-pressure water and steam boilers listed above.

Large boilers are constructed from steel or cast iron. Cast-iron boilers are modular and consist of several identical heat transfer sections bolted and gasketed together to meet the required output rating. Steel boilers are not modular but are constructed by welding various components together into one assembly. Heat transfer occurs across tubes containing either the fire or the water to be heated. The former are called *fire-tube boilers*, and the latter are *water-tube boilers*. Either material of construction can result in equally efficient designs. Small, light boilers of moderate capacity are sometimes needed for use in buildings. For these applications, the designer should consider the use of copper boilers.

Figure 9.3 shows a cross section of a steam boiler of the type used in buildings, and Fig. 9.4 is a photograph of a steam boiler.

9.2.2.1 Boiler Design and Selection for Buildings

The HVAC engineer must specify boilers based on a few key criteria. In this section we list these but do not discuss the internal design of boilers and their construction. Boiler selection is based on the following criteria:

- *Boiler fuel*: type, energy content, heating value including altitude effects if gas-fired (no effect for coal or fuel oil boilers)
- *Required heat output*: net output rating, kBtu/h (kW)

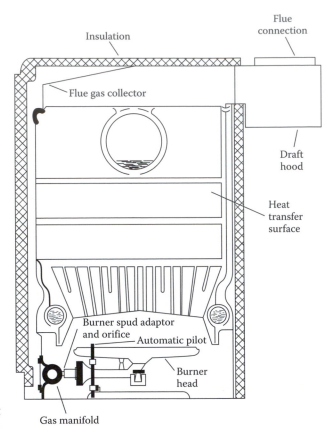

FIGURE 9.3
Cross-sectional drawing of boiler showing burner, heat exchanger, and flue connection.

- *Operating pressure and working fluid*
- *Efficiency and part-load characteristics*
- *Other*: space needs, control system, combustion air requirements, safety requirements, ASME code applicability

The boiler heat output required for a building is determined by summing the *maximum heating requirement* of all zones or loads serviced by the boiler during peak demand for steam or hot water and adding to that (1) parasitic losses including piping losses and (2) initial loop fluid warm-up. As described in Chap. 11, simply adding all the *peak heating unit capacities* of all the zones in a building can result in an oversized boiler, since not all zones require peak heating simultaneously. The ratio of the total of all zone loads under peak conditions to the total heating capacity installed in a building is called the *diversity*. This matter is also described in greater detail in Chap. 11.

Additional boiler capacity may be needed to recover from night setback in massive buildings. This transient load is called the *pickup* load and must be accounted for in both boiler and terminal heating unit sizing. Chapters 7 and 8 contain a detailed description of pickup loads along with an example of estimating the needed extra capacity.

Boilers are often sized by their *sea-level input* fuel ratings. Of course, this rating must be multiplied by the applicable efficiency to determine the gross output of the boiler. In addition, if a gas boiler is not to be located at sea level, the effect of altitude must be

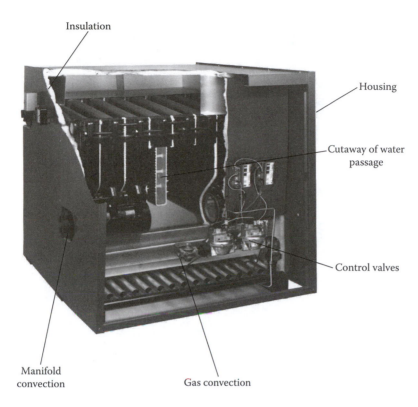

Insulation

Housing

Cutaway of water passage

Control valves

Manifold convection

Gas convection

FIGURE 9.4
Cutaway photograph of boiler. (Courtesy of Weil-McLain, Inc., Michigan City, IN. With permission.)

accounted for in the rating. Some boiler designs use a forced-draft burner to force additional combustion air into the firebox, to offset part of the effect of altitude. Also enriched or pressurized gas may be provided at high altitude so that the heating value per unit volume is the same as that at sea level. If no accommodation to altitude is made, the output of a gas boiler drops by approximately 4 percent per 1000 ft (13 percent per 1 km) of altitude above sea level. Therefore, a gas boiler located in Denver, Colorado (5000 ft, 1500 m), will have a capacity of only 80 percent of its sea-level rating.

Table 9.2 shows the type of data provided by manufacturers for the selection of boilers for a specific project. Reading across the table, the fuel input needs are first tabulated for the 13 boiler models listed. Column 5 is the sea-level boiler output at the maximum design heat rate. Columns 6 to 9 convert the heat rate to steam and hot-water production rates. Column 10 expresses the heat rate in still a different way, by using units of boiler horsepower (= 33,475 Btu/h or 9.8 kW). Columns 11 to 14 provide information needed for designing the combustion air supply system and the chimney.

This is a rule of thumb to check boiler selection in heating climates in the United States: The input rating (e.g., columns 2 and 3 of Table 9.2) in Btu per hour expressed on the basis of *per heated square foot* of building is usually in the range of one-third to one-fifth of the design temperature difference (difference between indoor and outdoor winter design temperatures). For example, if the design temperature difference for a 100,000 ft² building is 80°F, the boiler input is expected to range between 1600 kBtu/h [(80/5) × 100,000 ft²] and 2700 kBtu/h [(80/3) × 100,000 ft²]. The difference between the two depends on the energy

TABLE 9.2

Example of Manufacturer's Boiler Capacity Table

Boiler Unit Number, Steam, or Water (1)	IBR Burner Capacity				Net IBR Ratings			Net Heat Transfer Area, ft² H₂O (9)	Boiler hp (10)	Net Firebox Volume, ft³ (11)	Stack Gas Volume, ft³/min (12)	Positive Pressure in Firebox, inWG (13)	IBR Chimney Size Vent Dia., in (14)
	Light Oil, gal/h (2)	Gas, kBtu/h (3)	Min. Gas Press. Req'd., inWG (4)	Gross IBR Output, Btu/h (5)	Steam, ft²/h (6)	Steam, Btu/h (7)	Water, Btu/h (8)						
486*F	6.30	882	5.5	720,000	2,250	540,100	626,100	4,175	21.5	11.02	395	0.34	10
586*F	8.25	1,155	7.0	940,000	2,940	705,200	817,400	5,450	28.1	14.45	517	0.35	10
686*F	10.20	1,428	5.5	1,160,000	3,625	870,200	1,008,700	6,725	34.6	18.08	640	0.35	10
786*F	12.15	1,701	6.0	1,380,000	4,355	1,044,700	1,200,000	8,000	41.2	21.61	762	0.36	12
886*F	14.10	1,974	5.0	1,600,000	5,115	1,227,900	1,391,300	9,275	49.6	25.14	884	0.37	12
986*F	16.05	2,247	6.0	1,820,000	5,875	1,409,800	1,582,600	10,550	54.3	28.67	1,006	0.38	14
1086*F	18.00	2,520	6.5	2,040,000	6,600	1,583,900	1,773,900	11,825	60.9	32.20	1,128	0.39	14
1186*F	19.95	2,793	7.0	2,260,000	7,310	1,754,700	1,965,200	13,100	67.5	35.73	1,251	0.40	14
1286*F	21.95	3,073	7.0	2,480,000	8,025	1,925,500	2,156,500	14,375	74.1	39.26	1,376	0.41	14
1386*F	23.90	3,346	6.5	2,700,000	8,735	2,096,300	2,347,800	15,650	80.6	42.79	1,498	0.42	14
1486*F	25.90	3,626	7.5	2,920,000	9,445	2,267,100	2,539,100	16,925	87.2	46.32	1,623	0.43	16
1586*F	27.85	3,899	7.5	3,140,000	10,160	2,437,900	2,730,400	18,200	93.8	49.85	1,746	0.44	16
1686*F	29.75	4,165	8.5	3,350,000	10,835	2,600,900	2,913,000	19,420	100.1	53.38	1,865	0.45	16

Source: Courtesy of Weil-McLain, Inc., Michigan City, IN. With permission.

Note: 1 bhp = 33,475 Btu/h – 9.8 kW.

efficiency of the building envelope and its infiltration controls. Boiler efficiency also has an effect on this design check.

Proper control of boilers in response to varying outdoor conditions can improve efficiency and occupant comfort. A standard feature of boiler controls is the *boiler reset* system. Since full boiler capacity is needed only at peak heating conditions, better comfort control results if capacity is reduced with increasing outdoor temperature. Capacity reduction of zone hot-water heating is easy to accomplish by simply reducing the water temperature supplied by the boiler. For example, the reset schedule might specify boiler water at 210°F at an outdoor temperature of −20°F and at 50°F outdoors a water temperature of 140°F. This schedule is called a one-to-one schedule because the boiler output drops by 1.0°F for every degree rise in outdoor temperature.

9.2.2.2 Auxiliary Steam Equipment

Steam systems have additional components needed to provide safety or adequate control in building thermal systems. In this section we provide an overview of the most important of these components, including steam traps and relief valves.

Steam traps are used to separate both steam condensate and noncondensable gases from live steam in steam piping systems and at steam equipment. Steam traps "trap" or confine steam in heating coils, e.g., while releasing condensate to be revaporized in the boiler. The challenge in trap selection is to ensure that the condensate and gases are removed promptly and with little to no loss of live steam. For example, if condensate is not removed from a heating coil, it will become waterlogged and will have a much reduced heating capacity. We briefly describe the most common types of traps and which ones should be used in HVAC applications.

Thermodynamic traps are simple and inexpensive. The most common type, called the *disk trap*, is shown in Fig. 9.5a. This type of trap operates on kinetic energy changes as condensate flows through and flashes into steam within the trap. Steam flashed

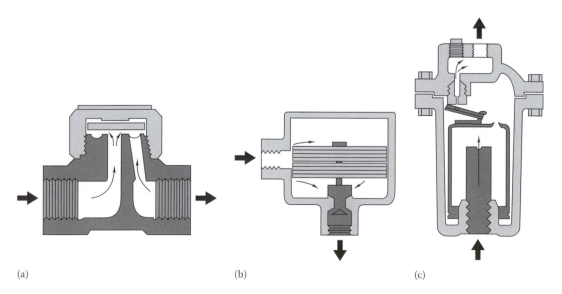

(a)   (b)   (c)

FIGURE 9.5
Steam traps: (a) thermodynamic disk trap; (b) thermostatic trap; (c) mechanical, inverted-bucket trap.

(i.e., converted from hot liquid to vapor) from hot condensate above the disk holds the trap closed until the disk is cooled by cooler condensate. Steam line pressure then pushes the disk open. It remains open until all cool condensate has been expelled and hot condensate is again present and flashes again to close the valve. The disk action is made more rapid by the flow of condensate beneath the disk; the high velocities produce a low-pressure area there in accordance with Bernoulli's equation, and the disk slams shut. These traps are rugged and make a characteristic clicking sound, facilitating operational checking. They can stick open if a particle lodges in the seat. This design has relatively high operating cost due to its live-steam loss.

Thermostatic traps use the temperature difference between steam and condensate to control condensate flow. One type of thermostatic trap is shown in Fig. 9.5b. The bimetal unit within the housing opens the valve as condensate cools, thereby allowing condensate to exit from the trap. Significant subcooling of the condensate is needed to open the valve, and operation can be slow. Other, more complex designs have more rapid response and reduced need for subcooling. The bimetal element can be replaced with a bellows filled with an alcohol-water mixture, permitting closer tracking of release setting as the steam temperature changes. A trap that has a temperature/pressure characteristic greater than the temperature/pressure of saturated steam (see saturation curve in Fig. 3.2 or saturated-steam data in Tables A3.1 and A3.2) will lose live steam, whereas a trap with a T/p characteristic lying below the steam curve will build up condensate. The ideal trap has an opening T/p characteristic identical with the T/p curve of saturated steam.

Mechanical traps operate on the density difference between condensate and live steam to displace a float. Figure 9.5c shows one type of mechanical trap, called the *inverted-bucket trap*, that uses an open, upside-down bucket with a small orifice. Steam flowing with the condensate (that fills the housing outside of the bucket) fills the inverted bucket and causes it to float, since the confined steam is less dense than the liquid water surrounding the bucket. Steam bleeds through the small hole in the bucket and condenses within the trap housing. As the bucket fills with condensate, it becomes heavier, eventually sinks, and opens the valve. Steam pressure forces condensate from the trap. The design of this trap continuously vents noncondensable gases, although the capacity for noncondensable gas flow (mostly air) rejection is limited by the size of the small hole in the top of the bucket. This hole is limited in size by the need to control parasitic steam loss through the same hole. Dirt can block the hole, causing the trap to malfunction. The trap must be mounted vertically. An inverted bucket trap will have significantly smaller parasitic live-steam losses than the thermodynamic disk trap.

Steam traps are used to drain condensate from steam headers and from equipment where condensing steam releases its heat to another fluid. Steam piping is sloped so that condensate flows to a collecting point, where it is relieved by the trap. At equipment condensate collection points, the trap is placed below the equipment where the condensate drains by gravity. Figure 9.6 shows a typical trap application for both purposes. The left trap drains the header, and the right trap drains the condensate produced in the heating coil. The design of steam piping systems is discussed in greater detail in Chaps. 5 and 11.

Selection of traps requires knowledge of the condensate rejection rate (lb_m/h, kg/s) and the suitability of various trap designs to the application. Table 9.3 summarizes the applications of the three types of traps discussed above (Haas, 1990).

The operating penalties for malfunctioning steam traps (clogged, dirty, or corroded) can dwarf the cost of a trap because expensive heat energy is lost if live steam is lost from malfunctioning traps. One of the first things to inspect in an energy audit of a new or existing steam system is the condition of the traps. For example, if steam is produced in

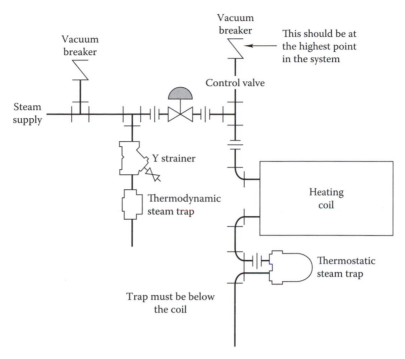

FIGURE 9.6
Piping arrangement for heating coil steam trap application.

TABLE 9.3

Operating Characteristics of Steam Traps

System Needs	Thermodynamic	Float Thermostatic	Inverted Bucket
Maximum pressure, psig	1740	465	2755
Maximum capacity, lb/h	5250	100,000	20,500
Discharge temperature, °F	Hot	Hot	Hot
		(Close to saturated-steam temperature)	
Discharge	On/off	Continuous	On/off
Air venting	Good	Excellent	Fair
Dirt handling	Fair	Good	Good
Freeze resistance	Good	Poor	Poor
Superheat	Excellent	Poor	Fair
Waterhammer	Excellent	Fair	Excellent
Varying load	Good	Excellent	Good
Change in psi	Good	Excellent	Fair
Backpressure	Maximum 80%	Good	Good
Usual failure	Open	Closed/air vent open	Open

Source: Courtesy of Haas, J.H., *Chem. Eng.*, 97, 151, January 1990.

a gas-fired boiler of typical efficiency, a 0.25-in (0.64-cm) leak in a steam trap will lose about $2000 worth of steam in a year. [The cost of gas in this example is $3.00 per 1000 ft^3 ($0.11 per 1 m^3), the usual units used by utilities; this converts to approximately $3.00 per 1 MBtu or $2.84 per 1 GJ.]

A pressure *relief valve* is needed to control possible overpressure in boilers for safety reasons. Valves are specified by their ability to pass a given amount of steam or hot water at the boiler outlet condition. This dump rate can be specified in units of either mass per time or energy flow per time. Pressure relief valves must be used wherever heat can be added to a confined volume of water. Water could become confined in the piping of an HVAC system, e.g., if automatic control valves failed while closed or if isolation valves were improperly closed by a system operator. Not only boilers must be protected, but also heat exchangers and water pipe lengths that are heated externally by steam tracing or solar heat. The volume expansion characteristics of water can produce tremendous pressures if heat is added to confined water. For example, water warmed by only 30°F (17°C) will increase in pressure by 1100 psi (7600 kPa)! The method for sizing boiler relief valves is outlined in Wong (1989). The discharge from boiler relief valves must be piped to a drain or other location where injury from live steam will be impossible. Expansion of fluid in piping is also accommodated by *expansion tanks;* they are described in Chap. 11.

9.2.2.3 Combustion Calculations—Flue Gas Analysis

The combustion of fuel in a boiler is a chemical reaction and as such is governed by the laws of stoichiometry. In this section we discuss the combustion of natural gas (for our purposes it is assumed to be 100 percent methane) in boilers as an example of fuel burning for heat production. We also outline how the flue gas from a boiler can be analyzed to ascertain the efficiency of the combustion process. Continuous monitoring of flue gases by a building's energy management system can result in early identification of boiler combustion problems. In a new building, one should test a boiler to determine its efficiency as installed and to compare output to that specified by the designer.

Combustion analysis involves using the basic chemical reaction equation and the known composition of air to determine the composition of flue gases. The inverse problem—finding the precombustion composition—is also important in analyzing flue gases. The chemical reaction for stoichiometric combustion of methane is

$$CH_4 + 2O_2 \rightarrow CO_2 + 2H_2O \tag{9.3}$$

Recalling that the molecular weights are

- Hydrogen (H_2): 2
- Methane (CH_4): 16
- Oxygen (O_2): 32
- Carbon dioxide (CO_2): 44
- Water (H_2O): 18

we can easily determine that 4.0 lb of oxygen per pound of methane is required for complete combustion. Since air is 23 percent oxygen by weight, 17.4 lb_m (or kg) of air per lb_m (or kg) of fuel is required theoretically. It is easy to show that on a volumetric basis (recall Avogadro's law, which states that 1 mol of any gas at the same temperature and pressure occupies the same volume) the equivalent requirements are 2.0 ft^3 of oxygen per cubic foot of methane for complete combustion. This oxygen requirement is equivalent to 8.7 ft^3 of air per cubic foot of methane. Here is a rule of thumb to check the preceding calculation: 0.9 ft^3 of air is required for 100 Btu of fuel heating value (about 0.25 m^3 of air per 1 MJ of heating value). For example, the heating value of natural gas is about 1000 Btu/ft^3, requiring 9 ft^3 of air according to the above rule. This compares well with the value of 8.7 ft^3 previously calculated.

Combustion air is often provided in excess of this theoretical amount to guarantee complete combustion. Incomplete combustion yields toxic carbon monoxide (CO) in the flue gas. This is to be avoided not only as energy waste but also as air pollution. The amount of excess air involved in combustion is usually expressed as the *excess air fraction* $f_{exc\,air}$:

$$f_{exc\,air} = \frac{\text{Air supplied} - \text{stoichiometric air}}{\text{stoichiometric air}} \tag{9.4}$$

In combustion calculations for gaseous fuels, the air amounts in Eq. (9.4) are usually expressed on a volumetric basis, whereas for all other fuels a mass basis is used.

The amount of excess air provided is critical to the efficiency of a combustion process. Excessive air both reduces combustion temperature (reducing the heat transfer rate to the working fluid) and results in excessive heat loss through the flue gases. Insufficient excess air results in incomplete combustion and loss of chemical energy in the flue gases. The amount of excess air provided varies with the fuel and with the design of the boiler (or furnace). Recommendations of the manufacturer should be followed. The optimum excess air fraction is usually between 10 and 50 percent.

Flue gas analysis is a method of determining the amount of excess air in a combustion process. This information can be used to find an approximate value of boiler efficiency. Periodic, regular analysis can provide a trend of boiler efficiency with time, indicating possible problems with the burner or combustion equipment in a boiler or furnace. Flue gas analysis is often expressed as the volumetric fraction of flue gases—oxygen, nitrogen, and carbon monoxide. If these three values are known, the excess air (%) can be found from (ASHRAE, 1989)

$$f_{exc\,air} = \frac{O_2 - 0.5CO}{0.264N_2 - (O_2 - 0.5CO)} \tag{9.5}$$

in which the chemical symbols represent the volume fractions (units of percent) in the flue gas analysis. Example 9.2 indicates how this expression is used.

Example 9.2: Flue Gas Analysis

The volumetric analysis of flue gas from combustion of methane in a gas boiler is measured to be 10.5 percent carbon dioxide, 3.2 percent oxygen, 86.3 percent nitrogen, and 0 percent carbon monoxide. Find the amount of excess air. Is it within the recommended range suggested above?

Given: Flue gas composition tabulated above

Find: $f_{exc\,air}$

SOLUTION

Equation 9.5 is used as follows.

$$f_{exc\,air} = \frac{3.2\% - (0.5 \times 0\%)}{0.264(86.3\%) - [3.2\% - (0.5 \times 0\%)]} = 0.163$$

The excess air is 16.3 percent, within the 10 to 50 percent range above.

According to the Boiler Efficiency Institute (1991), the efficiency of a steam boiler η_{boil} can be found from field measurements by

$$\eta_{boil} = \frac{\dot{Q}_{steam}}{\dot{m}_{fuel}(HHV)}$$ (9.6)

where
$\dot{Q}_{steam}$ = steam output rate, Btu/h (kW)
$\dot{m}_{fuel}$ = fuel supply rate, lb_m/h (kg/s)
HHV = higher heating value of fuel, Btu/lb (kJ/kg)

An additional efficiency is defined for boilers, the combustion efficiency. It can be determined experimentally to ascertain the condition of the fuel combustion equipment, including burners, heat transfer surfaces, and combustion air supply equipment.

The previous discussion has described the combustion of methane and at what rate air is to be supplied for proper combustion. Of course, many other fuels are used to fire boilers. Table 9.4 contains data that can be used to quickly estimate the excess air from a flue gas analysis for other fuels.

Coal and fuel oil contain carbon and hydrogen along with sulfur, the combustion of all of which produces heat. However, sulfur oxide formed during combustion is a corrosive acid if dissolved in liquid water. To avoid corrosion of boilers and stacks, liquid water must be avoided anywhere in a boiler by maintaining sufficiently high stack temperatures to prevent condensation. (Stainless-steel stacks and fireboxes provide an alternative solution, since they are not subject to corrosion; but they are very costly.) In addition, sulfur oxides are one of the sources of acid rain. Therefore, these emissions must be carefully controlled.

TABLE 9.4

Stoichiometric and Excess Air Values of CO_2 for Combustion of Common Fossil Fuels

Type of Fuel	Theoretical or Maximum CO_2, Percent	CO_2 at Given Excess Air Values		
		20 Percent	40 Percent	60 Percent
Gaseous fuels				
Natural gas	12.1	9.9	8.4	7.3
Propane gas (commercial)	13.9	11.4	9.6	8.4
Butane gas (commercial)	14.1	11.6	9.8	8.5
Mixed gas (natural and carbureted water gas)	11.2	12.5	10.5	9.1
Carbureted water gas	17.2	14.2	12.1	10.6
Coke oven gas	11.2	9.2	7.8	6.8
Liquid fuels				
No. 1 and No. 2 fuel oil	15.0	12.3	10.5	9.1
No. 6 fuel oil	16.5	13.6	11.6	10.1
Solid fuels				
Bituminous coal	18.2	15.1	12.9	11.3
Anthracite	20.2	16.8	14.4	12.6
Coke	21.0	17.5	15.0	13.0

Source: Courtesy of ASHRAE, *Handbook of Systems and Equipment*, American Society of Heating, Refrigerating and Air-Conditioning Engineers, Atlanta, GA, 2000. With permission.

9.2.2.4 Boiler Efficiency and Energy Calculations

A simpler, overall efficiency equation can be used for boiler energy estimates if the steam rate required for using Eq. (9.6) is not measurable:

$$\eta_{\text{boil}} = \frac{\text{HHV} - \text{losses}}{\text{HHV}} \qquad (9.7)$$

The loss term includes five parts:

1. Sensible heat loss in flue gases
2. Latent heat loss in flue gases due to combustion of hydrogen
3. Heat loss in water in combustion air
4. Heat loss due to incomplete combustion of carbon
5. Heat loss from unburned carbon in ash (coal and fuel oil)

These losses can be estimated by using information in Boiler Efficiency Institute (Dyer and Maples 1991). Alternatively, boilers can be tested in laboratories and rated in accordance with standards issued by the Hydronics Institute (formerly IBR, the Institute of Boiler and Radiator Manufacturers, and the SBI, the Steel Boiler Institute), the American Gas Association (AGA), and other industry groups. In addition to cast-iron boiler ratings, IBR ratings are industry standards for baseboard heaters and finned-tube radiation. The ratings of the Hydronics Institute apply to steel boilers. (IBR and SBI are trademarks of the Hydronics Institute.) SBI and IBR ratings apply to oil- and coal-fired boilers while gas boilers are rated by the AGA.

As noted earlier, efficiency under specific test conditions has very limited usefulness in calculating the annual energy consumption of a boiler because of significant dropoff of efficiency under part-load conditions. For small boilers [up to 300 kBtu/h (90 kW)], the Department of Energy has set forth a method for finding the AFUE (defined in Sec. 9.2.1.2 on furnaces). The annual energy consumption must be known in order to perform economic analyses for optimal boiler selection.

For larger boilers, data specific to a manufacturer and an application must be used to determine annual consumption. Efficiencies of fossil fuel boilers vary with heat rate depending on the internal design. If the boiler has only one or two firing rates, the continuous range of heat inputs needed to meet a varying heating load is achieved by cycling the boiler on and off. However, as the load decreases, efficiency decreases since the boiler spends progressively more time in transient warm-up and cooldown modes, during which relatively less heat is delivered to the load. At maximum load, the boiler cycles very little, and efficiency can be expected to be near the rated efficiency of the boiler. Part-load effects can reduce average efficiency to less than one-half of the peak efficiency. Of course, for an oversized boiler, the average efficiency is well below the peak efficiency since it operates at part load for the entire heating season. This operating-cost penalty persists for the life of a building long after the designer who oversized the system has forgotten the error.

To quantify part-load effects, we define the *part-load ratio* (PLR), a quantity between 0 and 1, as

$$\text{PLR} \equiv \frac{\dot{Q}_o}{\dot{Q}_{o,\text{full}}} \qquad (9.8)$$

where $\dot{Q}_o$ is the boiler heat output at part load, Btu/h (kW) and $\dot{Q}_{o,full}$ is the rated heat output at full load, Btu/h (kW).

It is not practical to calculate from basic principles how boiler *input* depends on the value of PLR since the processes to be modeled are very complex and nonlinear. The approach used for boilers (and other heat-producing equipment in this chapter) involves using test data to calculate the boiler input needed to produce an output $\dot{Q}_o$. If efficiency were constant and if there were no standby losses, the function relating input to output would merely be a constant, the efficiency. For real equipment, the relationship is more complex. A common function used to relate input to output (i.e., to PLR) is a simple polynomial (at least for a boiler) such as[2]

$$\frac{\dot{Q}_i}{\dot{Q}_{i,full}} = A + B(\text{PLR}) + C(\text{PLR})^2 + \cdots \qquad (9.9)$$

where $\dot{Q}_i$ is the fuel input (or input energy) required to meet the part-load level corresponding to PLR and $\dot{Q}_{i,full}$ is the fuel input at rated full load on the boiler.

The first term of Eq. (9.9) represents standby losses, e.g., those resulting from a standing pilot light in a gas boiler. Since the part-load characteristic is not far from linear for most boilers, a quadratic or cubic expression is sufficient for annual energy calculations. Part-load data are not as readily available as standard peak ratings. If available, the data may often be in tabular form. The designer will need to make a quick regression of the data to find A, B, and C, using commonly available spreadsheet or statistical software in order to be able to use the tabular data for annual energy calculations, as described in the following paragraphs.

In the remainder of this section, we examine a particularly simple application of a boiler—building space heating—to see how important part-load effects can be. The *annual* energy input $Q_{i,yr}$ of a space heating boiler can be calculated from the basic equation

$$Q_{i,yr} = \int_{yr} \frac{\dot{Q}_o(t)}{\eta_{boil}(t)} dt \qquad (9.10)$$

where η_{boil} is the boiler efficiency—a function of time since the load on the boiler varies with time—and $\dot{Q}_o(t)$ is the *boiler heat output*, which varies with time as well. The argument of the integral is just the instantaneous, time-varying energy input to the boiler. However, since the needed output—not the input—is usually known as a result of building load calculations, the form in Eq. (9.10) is that practically used by designers. The time dependence in this expression is determined, in turn, by the temporal variation of load on the boiler as imposed by the HVAC system in response to climatic, occupant, and other time-varying loads.

A simple case is a boiler used solely for space heating. As described in Chap. 7, the heating load is determined to first order by the difference between indoor and outdoor temperatures, all characteristics of the building's load and use remaining fixed. Therefore, the heat rate in Eq. (9.10) is determined by the outdoor temperature if the interior

---

[2] This simple polynomial form is used for illustration purposes in this book, recognizing that more complex forms such as those used in ASHRAE Standard 90 may be more accurate.

temperature remains constant. In this very simple case, one could replace the integral in Eq. (9.10) with a sum, utilizing the bin approach, as follows:

$$Q_{i,yr} = \sum_{j=0}^{N} \frac{\dot{Q}_o(T_j) n_j(T_j)}{\eta_{boil}(T_j)} \tag{9.11}$$

where

$\eta_{boil}(T_j)$ = efficiency of boiler in a given ambient temperature bin j; the efficiency depends strongly but indirectly on ambient temperature T_j since the load, which determines PLR, depends on temperature

$\dot{Q}_o(T_j)$ = boiler load (i.e., building heat load) that depends on ambient temperature as described above

$n_j(T_j)$ = number of hours in temperature bin j for which the values of efficiency and heat input apply

This expression assumes that the sequence of hours during the heating season is of no consequence. Example 9.3 illustrates how the bin weather data described in Chap. 8 can be used to take proper account of part-load efficiency of a boiler used for space heating.

Example 9.3: Annual Energy Consumption of a Gas Boiler

A gas boiler is used to supply space heat to a building. The load varies linearly with ambient temperature, as shown in Table 9.5. If the efficiency of the boiler is 80 percent at peak-rated conditions, find the seasonal average efficiency, annual energy input, and annual energy output,

TABLE 9.5

Summary of Solution for Example 9.3, Boiler Energy Analysis

				Calculating Annual Boiler Energy Use			
Bin Range, °F	Bin Size, h	Heating Load, kBtu/h	PLR	$\dot{Q}_i$, kBtu/h	Boiler Effic.	Fuel Used, MBtu	Net Output, MBtu
55 to 60	762	0	0.00	875	0.000	667	0
50 to 55	783	500	0.07	1844	0.271	1444	391
45 to 50	716	1000	0.14	2750	0.364	1969	716
40 to 45	665	1500	0.21	3594	0.417	2390	997
35 to 40	758	2000	0.29	4375	0.457	3316	1516
30 to 35	713	2500	0.36	5094	0.491	3632	1782
25 to 30	565	3000	0.43	5750	0.522	3249	1695
20 to 25	399	3500	0.50	6344	0.552	2531	1396
15 to 20	164	4000	0.57	6875	0.582	1127	656
10 to 15	106	4500	0.64	7344	0.613	778	477
5 to 10	65	5000	0.71	7750	0.645	504	325
0 to 5	80	5500	0.79	8094	0.680	647	440
−5 to 0	22	6000	0.86	8375	0.716	184	132

using the data in the table. The boiler input at rated conditions is 8750 kBtu/h corresponding to −12.5°F, temperature bin at which the load is 7000 kBtu/h.

This boiler is turned off in temperature bins higher than 57.5°F, roughly corresponding to the limit of the heating season; therefore, the standby losses above this temperature are zero.

The values of the coefficients in the part-load characteristic, Eq. (9.9), are

$$A = 0.1 \quad B = 1.6 \quad C = -0.7$$

Given:

- Bin data in the first two columns of Table 9.5 (HCB software contains bin data files)
- Load data in the third column of Table 9.5 (note the linearity of load with ambient temperature)
- Part-load characteristic equation

$$\frac{\dot{Q}_i}{\dot{Q}_{i,\text{full}}} = 0.1 + 1.6(\text{PLR}) - 0.7(\text{PLR})^2 \tag{9.12}$$

Assumptions: The bin approach is sufficiently accurate for this problem.

Find: $\bar{\eta}_{\text{boil}}$, $Q_{i,\text{yr}}$, $Q_{o,\text{yr}}$

Lookup values: All load data are given.

SOLUTION

The key equation for the solution is Eq. (9.11). Since we are given the part-load energy input equation instead of the efficiency at part load, this expression takes a somewhat simpler form

$$Q_{i,\text{yr}} = \sum_{j=0}^{j=N} \dot{Q}_i(T_j) n_j(T_j)$$

The solution is outlined in the right half of Table 9.5. We will work through in detail the calculations for the 37.5°F bin to see how the table is completed. The load at this bin is 2000 kBtu/h.[3] The value of PLR is the ratio of this load to the peak load 7000 kBtu/h

$$\text{PLR} = \frac{2000\,\text{kBtu/h}}{7000\,\text{kBtu/h}} = 0.286$$

Equation (9.12) is now used to find the energy input. This expression is the dimensionless form of the argument of the summation in Eq. (9.11):

$$\frac{\dot{Q}_i}{\dot{Q}_{i,\text{full}}} = 0.1 + 1.6(0.29) - 0.7(0.29)^2 = 0.50$$

Hence the heat input is

$$\dot{Q}_i = 0.50\dot{Q}_{i,\text{full}} = 0.50(8750)\,\text{kBtu/h} = 4375\,\text{kBtu/h}$$

---

[3] kBtu = 1000 Btu; MBtu = 1000 kBtu.

This is the value entered in the fifth column of the table. The sixth column is the boiler efficiency in each bin, defined as the ratio of the load (2000 kBtu/h) to the heat input just calculated (4375 kBtu/h)

$$\eta_{boil,37.5} = \frac{2000 \, \text{kBtu/h}}{4375 \, \text{kBtu/h}} = 0.457$$

The part-load effect is immediately obvious—the efficiency is much less than the rating at full load of 0.80.

Finally the fuel used is the product of the fuel input rate (4375 kBtu/h) and the number of hours in the 37.5°F bin (758 h):

$$Q_{i,37.5} = (4375 \, \text{kBtu/h})(758 \, \text{h}) = 3316 \, \text{MBtu}$$

The heat produced by the boiler in 758 h in the 37.5°F bin is

$$Q_{o,37.5} = (2000 \, \text{kBtu/h})(758 \, \text{h}) = 1516 \, \text{MBtu}$$

These last two numbers are the entries in the rightmost two columns in Table 9.5.

Finally, to find the total energy used by the boiler, one sums the "fuel used" column of the table to find that 22,439 MBtu is used to meet the annual load of 10,525 MBtu. The ratio of these two numbers is the overall annual boiler efficiency: 47 percent. This value is 41 percent less than the peak efficiency of 80 percent. Clearly one must take part-load effects into account in annual energy calculations.

COMMENTS

Calculations of this type are well suited to spreadsheet solutions. This problem was particularly simple since the load depended in a simple way on the ambient temperature. If the boiler had been used to supply heat for an absorption chiller whose load depended on solar flux and dry- and wet-bulb temperatures, the same approach could have been used but would have been more complex.

One method of avoiding the poor efficiency of this system is to use two (or more) smaller boilers, the combined capacity[4] of which totals the needed 7000 kBtu/h. Properly chosen, the smaller boilers will operate more nearly at full load for more of the time, resulting in higher seasonal efficiency. Multiple-boiler systems also offer standby security; if one boiler should fail, the other could carry at least part of the load. A single-boiler system would entirely fail to meet the load. However, smaller boilers cost more than one large boiler with the same total capacity.

The final decision must be made based on economics, giving proper account to the increased reliability of a system composed of several smaller boilers. Constraints are imposed on such decisions by initial budget, fuel type, owner, and architect decisions and available space.

9.2.3 Service Hot Water

Heated water is used in buildings for various purposes including basins, sinks for custodial service, showers, and specialty services including kitchens in restaurants and the like. In this section, we give an overview of service (or *domestic*) water heating methods for buildings. For details refer to ASHRAE (1999).

---

[4] The careful reader will note that the boiler design point of −12.5°F does not occur in the bin data since bin data are based on a typical year, whereas the design temperature extreme will not occur in an average year. The boiler "oversizing" due to this effect is also part of the cause for reduced annual boiler efficiency.

Water is heated either by equipment that is part of the space heating system (i.e., the boiler) or by a stand-alone water heater. The stand-alone equipment is similar to a small boiler except that water chemistry must be accounted for by use of anodic protection for the tank and by water softening in geographic areas where hardness can cause scale (lime) deposits in the water heater tank.

Two types of systems are used for water heating: *instantaneous* and *storage*. The former heats water on demand as it passes through the heater, which uses either steam or hot water. Output temperatures can vary with this system unless a control valve is used on the heated water (not the heat supply) side of the water heater (usually a heat exchanger). Instantaneous water heaters are best suited to relatively uniform loads. They avoid the cost and heat losses of the storage tank but require larger and more expensive heating elements.

Storage-type systems are used to accommodate varying loads or loads where large peak demands make it impractical to use instantaneous systems. Water in the storage tank is heated by an immersion steam coil, by direct firing, or by an external heat exchanger. In sizing this system, the designer must account for standby losses from the tank jacket and connected hot-water piping. For any steam-based system, cold supply water can be preheated by using the steam condensate.

To size the equipment, two items must be known:

1. Hourly peak demand for the year, gal/h (L/h)
2. Daily consumption, gal/day (L/day)

Of course, the volumetric usage rates must be converted to energy terms by multiplying by the specific heat and water temperature rise:

$$\dot{Q}_{water} = \dot{m}_{water} c_{water} (T_{set} - T_{source}) \tag{9.13}$$

where

$\dot{Q}_{water}$ = water heat rate, on daily or hourly basis, Btu/day or Btu/h (kWh/day or W)

$\dot{m}_{water}$ = water mass flow rate, on daily or hourly basis, calculated from volumetric flow listed above

T_{set}, T_{source} = required hot-water supply temperature and water source temperatures, respectively

c_{water} = specific heat of water

Table 9.6 summarizes water demands for various types of buildings, and Table 9.7 lists nominal setpoints of water heaters for several end uses. When using the lower settings in the table, the designer must be aware of the potential for *Legionella pneumophila* (Legionnaires' disease). This microbe has been traced to infestations of showerheads; it is able to grow in water maintained at 115°F (46°C). This problem can be limited by using domestic water temperatures near 140°F (60°C).

Hot water can be supplied from a storage-type system at the maximum rate

$$\dot{V}_{water} = \dot{V}_r + \frac{f_{useful} V_{tank}}{\Delta t} \tag{9.14}$$

TABLE 9.6

Hot-water Demands and Use for Various Types of Buildings

Type of Building*	Maximum Hour	Maximum Day	Average Day
Men's dormitories	3.8 gal (14.4 L)/ student	22.0 gal (83.4 L)/ student	13.1 gal (49.7 L)/ student
Women's dormitories	5.0 gal (19 L)/ student	26.5 gal (100.4 L)/ student	12.3 gal (46.6 L)/ student
Motels: No. of units[†]			
20 or less	6.0 gal (22.7 L)/unit	35.0 gal (132.6 L)/unit	20.0 gal (75.8 L)/unit
60	5.0 gal (19.7 L)/unit	25.0 gal (94.8 L)/unit	14.0 gal (53.1 L)/unit
100 or more	4.0 gal (15.2 L)/unit	15.0 gal (56.8 L)/unit	10.0 gal (37.9 L)/unit
Nursing homes	4.5 gal (17.1 L)/bed	30.0 (113.7 L)/bed	18.4 gal (69.7 L)/bed
Office buildings	0.4 gal (1.5 L)/person	2.0 gal (7.6 L)/person	1.0 gal (3.8 L)/person
Food service establishments:			
Type A: full-meal restaurants and cafeterias	1.5 gal (5.7 L)/max meals/h	11.0 gal (41.7 L)/max meals/h	2.4 gal (9.1 L)/average meals/day[‡]
Type B: drive-ins, grills, luncheonettes, sandwich and snack shops	0.7 gal (2.6 L)/max meals/h	6.0 gal (22.7 L)/max meals/h	0.7 gal (2.6 L) average meals/day[‡]
Apartment houses: No. of apartments			
20 or less	12.0 gal (45.5 L)/apt.	80.0 gal (303.2 L)/apt.	42.0 gal (159.2 L)/apt.
50	10.0 gal (37.9 L)/apt.	73.0 gal (276.7 L)/apt.	40.0 gal (151.6 L)/apt.
75	8.5 gal (32.2 L)/apt.	66.0 gal (250 L)/apt.	38.0 gal (144 L)/apt.
100	7.0 gal (26.5 L)/apt.	60.0 gal (227.4 L)/apt.	37.0 gal (140.2 L)/apt.
200 or more	5.0 gal (19 L)/apt.	50.0 gal (195 L)/apt.	35.0 gal (132.7 L)/apt.
Elementary schools	0.6 gal (2.3 L)/student	1.5 gal (5.7 L)/student	0.6 gal (2.3 L)/student.[†]
Junior and senior high schools	1.0 gal (3.8 L)/student	3.6 gal (13.6 L)/student	1.8 gal (6.8 L)/student[†]

Source: Courtesy of ASHRAE, *Handbook of Systems and Equipment*, American Society of Heating, Refrigerating and Air-Conditioning Engineers, Atlanta, GA, 2000. With permission.

* The average usage of a U.S. residence is 60 gal/day (227 L/h) with a peak usage of 6 gal/h (22.7 L/h) (ASHRAE, 1999).

[†] Interpolate for intermediate values.

[‡] Per day of operation. Temperature basis: 140°F.

where

$\dot{V}_{water}$ = volumetric hot-water supply rate, gal/h (L/s)

$\dot{V}_r$ = water heater recovery rate, gal/h (L/s)

f_{useful} = useful fraction of hot water in tank before dilution lowers temperature excessively, 0.60–0.80

$\dot{V}_{tank}$ = tank volume, gal (L)

Δt = duration of peak demand, h (s)

Jacket losses are assumed to be small.

TABLE 9.7

Representative Hot-Water Use Temperatures

	Temperature	
Use	°F	°C
Lavatory		
Handwashing	105	40
Shaving	115	45
Showers and tubs	110	43
Therapeutic baths	95	35
Commercial and institutional laundry	180	82
Residential dishwashing and laundry	140	60
Surgical scrubbing	110	43
Commercial spray-type dishwashing		
Single or multiple tank hood(s) or rack(s)		
Wash	150 min	65 min
Final rinse	180–195	82–90
Single tank conveyor		
Wash	160 min	71 min
Final rinse	180–195	82–90

Source: Courtesy of ASHRAE, *Handbook of Systems and Equipment*, American Society of Heating, Refrigerating and Air-Conditioning Engineers, Atlanta, GA, 2000. With permission.

Note: Table values are water use temperatures, not necessarily water heater setpoints.

9.3 Electric Resistance Heating

Electricity can be used as the heat source in both furnaces and boilers. Electric units are available in the full range of sizes from small residential furnaces (5 to 15 kW) to large boilers for commercial buildings (200 kW to 20 MW). Electric units have four attractive features:

- Relatively lower initial cost
- Efficiency near 100 percent
- Near-zero part-load penalty
- Flue gas vents not necessary

The high cost of electricity (both energy and demand charges) diminishes the apparent advantage of electric boilers and furnaces, however. Nevertheless, they continue to be installed where first cost is a prime concern. The prudent designer should consider the overwhelming life cycle costs of electric systems, however. Electric boiler and furnace sizing follows the methods outlined above for fuel-fired systems. In many cases, the thermodynamic and economic penalties of pure resistance heating can be reduced by using electric heat pumps, the subject of Sec. 9.4.

Environmental concerns must also be taken into account in considering electric heating. Low conversion and transmission efficiencies (relative to direct combustion of fuels for water heating) result in relatively higher CO_2 emissions. Also SO_2 emissions from coal power plants are an environmental concern.

9.4 Electric Heat Pumps

Heat pumps are based on refrigeration systems, described in Sec. 3.4.3. Refrigeration and cooling systems are used to produce a cooling effect as their output. However, the heat rejected to the environment by cooling systems can be of value when heat is needed. A *heat pump* extracts heat from environmental or other medium-temperature sources (such as the ground, groundwater, or building heat recovery systems), raises its temperature sufficiently to be of value in meeting space heating or other loads, and delivers it to the load. In this chapter, we emphasize heat pumps used for space heating with outdoor air as the heat source.

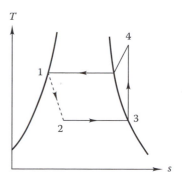

FIGURE 9.7
Heat pump *T-s* diagram showing four steps of the simple heat pump process.

   Figure 9.7 shows a heat pump cycle on the *T-s* diagram; Fig. 9.8 shows it on the more frequently used *p-h* diagram. It is exactly the refrigeration cycle discussed in Chap. 3. Vapor is compressed in step 3-4, and heat is extracted from the condenser in step 4-1. This heat is used for space heating in the systems discussed in this section. In step 1-2, isenthalpic throttling takes place to the low-side pressure. Finally, heat extracted from the environment, or other low-temperature heat source, is used to boil the refrigerant in the evaporator in step 2-3.

   An ideal Carnot heat pump would appear as a rectangle in the *T-s* diagram (recall Fig. 3.10). The coefficient of performance (COP, i.e., output divided by input) of a Carnot heat pump is given in Eq. (3.30) and was shown there to be inversely proportional to the difference between the high- and low-temperature reservoirs. The same result applies generally to heat pumps using real fluids. Although the high-side

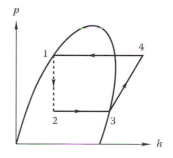

FIGURE 9.8
Heat pump *p-h* diagram showing four steps of the simple heat pump process.

temperature (T_4 in Fig. 9.7) remains essentially fixed (ignoring for now the effect of night thermostat setback), the low-side temperature closely tracks the widely varying outdoor temperature. As a result, the *capacity and COP of air source heat pumps are strong functions of outdoor temperature*. This feature of heat pumps must be accounted for by the designer, since heat pump capacity diminishes as the space heating load on it increases. Heat pumps can be supplemented by fuel heat or electric resistance heating, depending on the cost of each. Figure 9.9 shows a water source heat pump system that is not subject to outdoor temperature variations if groundwater or a heat recovery loop is used as the heat source.

   The attraction of heat pumps is that they can deliver more thermal power than they consume electrically during an appreciable part of the heating season. In moderate climates requiring both heating and cooling, the heat pump can also be operated as an air conditioner, thereby avoiding the additional cost of a separate air conditioning system. Figure 9.10 shows one way to use a heat pump system for both heating and cooling by reversing flow through the system.

9.4.1 Typical Equipment Configurations

Heat pumps are available in sizes ranging from small residential units (10 kW) to large central systems (up to 15 MW) for commercial buildings. Large systems produce heated water at temperatures up to 220°F (105°C). Central systems can use both environmental

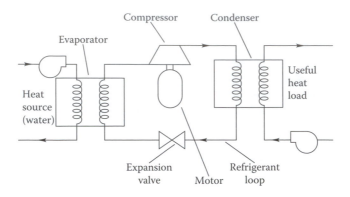

FIGURE 9.9
Liquid source heat pump mechanical equipment schematic diagram, showing motor-driven centrifugal compressor, condenser, and evaporator.

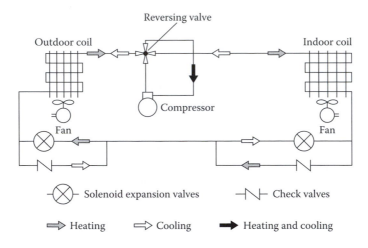

FIGURE 9.10
Air-to-air heat pump diagram. A reciprocating compressor is used. This design allows operation as a heat pump or an air conditioner by reversing the refrigerant flow.

and internal building heat sources. In many practical circumstances, the heat gains in the core zones of a commercial building could satisfy the perimeter heat losses in winter. A heat pump can be used to efficiently condition both types of zones simultaneously.

Heat pumps are made of components that we discuss in greater detail in Chap. 10. In short, heat pumps require a *compressor and two heat exchangers*. In the energy bookkeeping that one does for heat pumps, the power input to the compressor is added to the heat removed from the low-temperature heat source, to find the heat delivered to the space to be heated (see Chap. 3 and Sec. 14.3). Increased heating capacity at low air source temperatures can be achieved by oversizing the compressor. To avoid part-load penalties in moderate weather, a variable-speed compressor drive can be used.

The *outdoor and indoor heat exchangers* use forced convection on the air side to produce adequate heat transfer coefficients. In the outdoor exchanger, the temperature difference between the boiling refrigerant and the air is between 10 and 25°F (6°C and 14°C). If the heat source is internal building heat, water is used to transport heat to the heat pump evaporator, and smaller temperature differences can be used.

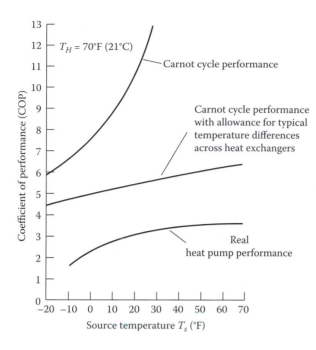

FIGURE 9.11
COP of ideal Carnot, Carnot (with heat exchanger penalty), and real heat pumps.

A persistent problem with air source heat pumps is the accumulation of frost on the outdoor coil at coil surface temperatures just above the freezing point. The problem is most severe in humid climates; little defrosting is needed for temperatures below 20°F (−7°C) where humidities are below 60 percent. *Reverse-cycle defrosting* can be accomplished by briefly operating the heat pump as an air conditioner (by reversing the flow of refrigerant) and turning off the outdoor fan. Hot refrigerant flowing through the outside melts the accumulated frost. This energy penalty must be accounted for in calculating the COP of heat pumps. Defrost control can be initiated by time clock or, better, by a sensor measuring either the refrigerant condition (temperature or pressure) or ideally the air pressure drop across the coil.

The realities of heat pump performance as discussed above reduce the capacity of real systems from the Carnot ideal. Figure 9.11 shows ideal Carnot COP values as a function of source temperature for a high-side temperature of 70°F (21°C). The intermediate curve shows performance for a Carnot heat pump with real (i.e., finite temperature difference) heat exchangers. Finally, the performance of a real heat pump is shown in the lower curve. Included in the lower curve are the effects of heat exchanger losses, use of real fluids, compressor inefficiencies, and pressure drops. The COP of real machines is much lower (about 50 percent) than that for an ideal Carnot cycle with heat exchanger penalties. Chapter 14 discusses this matter in greater detail.

9.4.2 Heat Pump Selection

The strong dependence of heat pump output on ambient temperature must be accounted for in selecting central plant equipment. If outdoor air is used as the heat source, peak heating requirements will invariably exceed the capacity of any economically feasible unit. Therefore, auxiliary heating is needed for such systems. Supplemental heat should always be added downstream of the heat pump condenser. This ensures that the condenser will operate at as low a temperature as possible, thereby improving the COP.

The amount of auxiliary heat needed and the type (electricity, natural gas, oil, or other) must be determined by an economic analysis and fuel availability (heat pumps are often used when fossil fuels are unavailable). The key feature of such analysis is the combined effect of part-load performance and ambient source temperature on system output and efficiency. In Sec. 9.4.3 we show how the temperature bin approach can be used for such an assessment. Figure 9.12 shows the conflicting characteristics of heat pumps and buildings in the heating season. As the ambient temperature drops, loads increase but heating capacity drops. The point at which the two curves intersect is called the *heat pump balance point*. To the left, auxiliary heat is needed; to the right, the heat pump must be modulated since excess capacity exists.

Recovery from night thermostat setback must be carefully thought out by the designer if an air source heat pump is used. A step change up in the thermostat setpoint on a cold winter morning will inevitably cause the auxiliary heat source to come on. If this heat source is electricity, high electric demand charges may result, and the possible economic advantage of the heat pump will be reduced. One approach to avoid activation of the electric resistance heat elements uses a longer warm-up period with gradually increasing thermostat setpoint. A smart controller could control the setup time based on known heat pump performance characteristics and the outdoor temperature. Alternatively, fuel could be used as the auxiliary heat source. During building warm-up, all outside air dampers remain closed, as is common practice for any commercial building heating system.

Heat pump efficiency is greater if lower delivery temperatures can be used. To produce adequate space heat in such conditions, a larger coil may be needed in the airstream. However, if the coil is sized for the cooling load, it will nearly always have adequate capacity for heating. In such a case, adequate space heat can be provided at relatively low air temperatures of 95 to 110°F (35 to 43°C). Table 9.8 summarizes advantages and disadvantages of air and water source heat pumps.

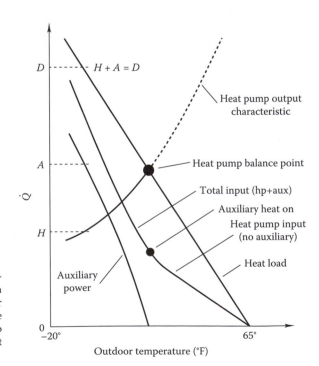

FIGURE 9.12
Graph of building heat load, heat pump capacity, and auxiliary heat quantity as a function of outdoor air temperature for a typical air source residential heat pump. Note that the heat pump balance point has no relation to the heating and cooling load balance-point temperature.

TABLE 9.8

Advantages and Disadvantages of Air and Water Source Heat Pumps

Type	Advantages	Disadvantages
Air source	Indoor distribution permits air conditioning and humidity control	Defrost required
	Outdoor air source readily available	Low capacity at cold outdoor temperature
	Simple installation	Lower efficiency because of large evaporator $\Delta T \approx 30°F$
	Least expensive	Indoor air distribution temperature must be high for comfort reasons
	Established commercial technology	Reliability at low temperature is only fair, due to frosting effects
		Must keep evaporator clear of leaves, dirt, etc.
Water source	Multiple family and commercial installations as central system	Needs water source at useful temperature
	In commercial installations, good coupling to cooling towers	Efficiency penalty due to space heat exchanger ΔT
	No refrigerant reversal needed; reverse water flow instead	

Controls for heat pumps are more complex than those for fuel-fired systems since outdoor conditions, coil frosting, and heat load must all be considered. In addition, to avoid excessive demand charges, the controller must avoid *coincident* operation of resistance heat and the compressor at full capacity (attempting to meet a large load on a cold day).

9.4.3 Part-Load Performance

As discussed in detail above, air source heat pumps are particularly sensitive to the environment. In Sec. 9.2.2.4 we examined how the performance of boilers changed when the heating load changed with outdoor temperature. The COP of an air source heat pump has an even greater dependence on environmental conditions. In this section we work an example to illustrate the magnitude of the effect. Since air source heat pumps are often used on residences, we work a residential-scale example.

Example 9.4: Seasonal Heat Pump Performance Calculated by the Bin Method

A residence in a heating climate has a total heat transmission coefficient $K_{tot} = 650$ Btu/(h · °F) (343 W/K). An air source heat pump with a capacity of 39,900 Btu/h (11.7 kW) at 47°F (8.3°C) (standard rating point in the United States) is to be evaluated. Find the heating season electric energy usage, seasonal COP [often called the *seasonal performance* factor (SPF)], and energy savings relative to electric resistance heating. Use the bin data and heat pump performance data given in Table 9.9. The house heating base temperature is 65°F (18.3°C), accounting for internal gains.

  Given: $K_{tot} = 650$ Btu/(h · °F), bin data in Table 9.9

  Figure: See Fig. 9.13.

  Find: SPF, $Q_{yr,elect}$ (with and without heat pump)

TABLE 9.9

Heat Pump and Building Load Data

Bin Temp., °F	Heating Load, Btu/h	Heat Pump COP	Heat Pump Output, Btu/h	Heat Pump Input, Btu/h	Auxiliary Power, Btu/h	Heating System COP
62	1,950	2.64	1,950	739	0	2.64
57	5,200	2.68	5,200	1,940	0	2.68
52	8,450	2.64	8,450	3,201	0	2.64
47	11,700	2.63	11,700	4,449	0	2.63
42	14,950	2.50	14,950	5,980	0	2.50
37	18,200	2.39	18,200	7,615	0	2.39
32	21,450	2.23	21,450	9,619	0	2.23
27*	24,700	2.07	24,700	11,932	0	2.07
22	27,950	1.97	25,100	12,741	2,850	1.79
17	31,200	1.80	22,400	12,444	8,800	1.47
12	34,450	1.70	19,900	11,706	14,550	1.31
7	37,700	1.54	17,600	11,429	20,100	1.20
2	40,950	1.39	15,400	11,079	25,550	1.12
−3	44,200	1.30	13,500	10,385	30,700	1.08
−8	47,450	1.17	11,700	10,000	35,750	1.04

* Heat pump balance point.

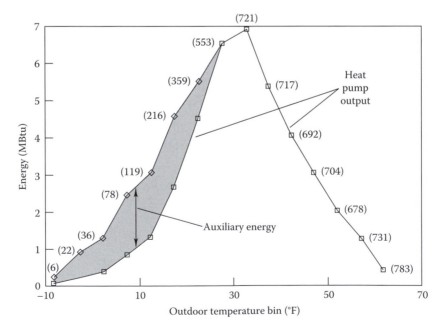

FIGURE 9.13

Heat pump energy use by the bin method, Example 9.4. The numbers at each bin temperature indicate the number of hours of occurrence in each bin.

SOLUTION

The solution will use weather and performance data from Table 9.9 to solve this problem. The seasonal results are presented in Table 9.10.

The contents of Table 9.9 by column are

1. Center point of temperature bin T_{bin}
2. Heating demand $\dot{Q} = K_{tot}(65°F - T_{bin})$
3. COP from manufacturer's data, a function of temperature, including defrost
4. Heat pump output: above the balance point, Q; below the heat pump balance point, manufacturer's data
5. Heat pump input, the heat pump output divided by COP
6. Auxiliary power; the positive difference, if any, between Q and heat pump output
7. Heating *system* COP given by $\dot{Q}$ divided by the sum of auxiliary power and heat pump input

The energy calculations are summarized in Table 9.10. We will work through the calculations for the 22°F bin in detail to clarify the process. The first two columns of this table are the bin weather data. The third column is the heating energy by bin. For the subject bin we have

$$Q_{22} = \frac{(27,950\ \text{Btu/h})(359\ \text{h})}{1,000,000} = 10.03\ \text{MBtu}$$

Since 22°F is below the balance point, the heat pump capacity is less than the load, as shown in Table 9.9. The heat pump output is

$$Q_{o,22} = \frac{(25,100\ \text{Btu/h})(359\ \text{h})}{1,000,000} = 9.01\ \text{MBtu}$$

TABLE 9.10

Heat Pump Energy Calculations

Bin Temp., °F	Bin Time, h	Heating Energy, MBtu	Heat Pump Output, MBtu	Heat Pump Input, MBtu	Aux. Heat Input, MBtu	Total Input, MBtu
62	783	1.53	1.53	0.58	0.00	0.58
57	731	3.80	3.80	1.42	0.00	1.42
52	678	5.73	5.73	2.17	0.00	2.17
47	704	8.24	8.24	3.13	0.00	3.13
42	692	10.35	10.35	4.14	0.00	4.14
37	717	13.05	13.05	5.46	0.00	5.46
32	721	15.47	15.47	6.94	0.00	6.94
27*	553	13.66	13.66	6.60	0.00	6.60
22	359	10.03	9.01	4.57	1.02	5.60
17	216	6.74	4.84	2.69	1.90	4.59
12	119	4.10	2.37	1.39	1.73	3.12
7	78	2.94	1.37	0.89	1.57	2.46
2	36	1.47	0.55	0.40	0.92	1.32
−3	22	0.97	0.30	0.23	0.68	0.90
−8	6	0.28	0.07	0.06	0.21	0.27
Total		98.36	90.33	40.66	8.03	48.70

* Heat pump balance point.

From Table 9.9 the COP at this temperature is 1.97. Therefore, the electricity input in this bin is

$$Q_{i,22} = \frac{Q_{o,22}}{COP} = \frac{9.01\,MBtu}{1.97} = 4.57\,Mbtu$$

Because the heat pump cannot meet the load, some auxiliary heat is needed

$$Q_{aux,22} = \frac{(2850\,Btu/h)(359\,h)}{1,000,000} = 1.02\,MBtu$$

Finally, the total electric input is the sum of the heat pump and supplemental electricity requirements

$$Q_{tot,22} = 4.57 + 1.02 = 5.6\,MBtu$$

The rest of Table 9.10 is completed in this manner. The bottom line in the table contains energy totals. With the heat pump, the total electricity requirement is 48.7 MBtu/yr (51.4 GJ/yr). If pure resistance heating were used, the total electricity requirement would be 98.36 MBtu/yr (103.8 GJ/yr).

The SPF for the heat pump is the seasonal output divided by the seasonal input to the heat pump:

$$SPF_{hp} = \frac{Q_{o,yr}}{Q_{i,yr}} = \frac{90.33\,MBtu}{40.66\,MBtu} = 2.22$$

The SPF for the heating system is the seasonal heat load divided by the seasonal input to the heat pump and the auxiliary heater:

$$SPF_{sys} = \frac{Q_{o,yr}}{Q_{i,yr} + Q_{i,aux,yr}} = \frac{98.36\,MBtu}{(40.66 + 8.03)\,MBtu} = 2.02$$

COMMENTS

This example has shown how the part-load characteristics of air source heat pumps are influenced by ambient temperature, and that this characteristic of the equipment must be considered in annual energy calculations.

The advantage of a constant-temperature heat source is apparent from this example. If groundwater or building exhaust air (both essentially at constant temperature) were used as the heat source rather than outdoor air, the dropoff in capacity that occurs in the air source device just when heat is most needed would not occur.

The HCB software can be used to solve heat pump problems of this type. You are encouraged to try it out for this problem and to study the effect of heat pump size on the seasonal COP. Oversized heat pumps will have lower SPF values. The software heat pump routine includes a defrost penalty on capacity above the balance point, calculated by way of

$$\text{Actual capacity} = 0.75\,(\text{nominal capacity}) + 0.25\,(\text{bin load})$$

(ASHRAE, 1989, Chap. 28). Below the balance point there is no adjustment needed.

This section has discussed practical aspects of heat pumps with an illustrative application of space heating. Heat pumps have other applications in buildings:

- Operation in buildings with simultaneous heating and cooling needs. A system of this type is discussed in Sec. 14.3.5.
- Heat pump water heater. This system heats water by operating a heat pump using indoor air as the source and the water to be heated as the sink. These systems are small and primarily for residential applications.

- Ventilation air heat recovery. Since all buildings with mechanical ventilation exhaust some warmed air in winter, a heat pump can be operated between this exhaust and the fresh air supply. The COP for this application is constant.

Although the COP of a heat pump is the natural, dimensionless performance index that arises in the thermodynamic analysis of heat pumps and air conditioners, the industry sometimes uses a dimensional performance measure. The *energy efficiency ratio* (EER) is the ratio of heating capacity (Btu per hour) to the electric input rate (watts). EER thus has the units of Btu per watt per hour. The dimensionless COP is found from the EER by dividing it by the conversion factor 3.413 Btu/(W · h).

9.5 Heat Exchanger Design and Selection

9.5.1 Introduction

Various types of heat exchangers are used in buildings for

- Heating or cooling of liquids
- Sensible heating or cooling of air
- Cooling and dehumidification of airstreams

A very common configuration is the shell-and-tube exchanger shown in Fig. 9.14 for liquid heat transfer. Figure 9.15 shows a cross-flow air-liquid heat exchanger. The shell-and-tube device could be used for service water heating in a building (see Example 2.7), whereas the coil could also be used in the main air handler of a building to heat or cool outdoor air.

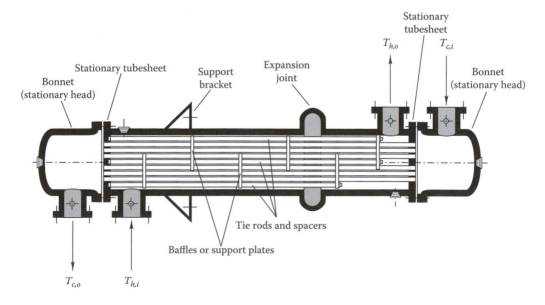

FIGURE 9.14
Typical shell-and-tube heat exchanger showing piping connections, baffles, tube sheets, and supports.

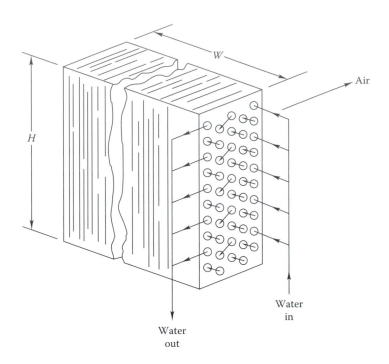

FIGURE 9.15
Cross-flow air heating coil showing 10 tube passes per row. Heat exchangers with at least this number of tube passes can be well modeled as counterflow heat exchangers.

In Chap. 2 we discussed the analysis of common heat exchangers that transfer heat without phase change (i.e., without condensation or evaporation of water vapor). In this section we discuss the application of these analytical techniques to the three types of heat exchanges described above. The ultimate task of the designer is to select the optimal heat exchanger for a given building design. This involves a measure of economic analysis that is described in Chap. 5. In this section we limit ourselves to the heat and mass transfer questions involved in heat exchanger design.

The designer of HVAC systems during *preliminary design* will often first calculate the required heat rate of an exchanger based, e.g., on heating or cooling load calculations, known heat rejection rates from chillers, or domestic hot-water loads. Next, typical fluid temperature drops are specified based on experience; a typical water-side temperature drop in an air heating coil in a duct is 20°F. Finally, required flows are calculated from the known heat rate and the specified temperature drop. These data are entered into the heat exchanger specification section of the building's contract documents.

During *final design*, other details must be considered including heat exchanger pressure drop, effects of fouling on performance, avoidance of tube vibration in shell-and-tube exchangers, water quality and potential scaling problems, physical location and mounting, service access, piping arrangements, insulation needs (if any), materials, and flow control. Many of these items can be handled with the assistance of manufacturers, who often have computer programs to assist the designer with final selection; manufacturers also provide all physical dimensions needed for piping layout. A list of potential heat exchanger problems that should be avoided at final design is given in Yokell (1983).

9.5.2 Methods for Selection

The heat exchanger *industry* (setting aside, for the moment, direct-contact heat exchangers) specifies the heat rate by using this basic equation:

$$\dot{Q} = UA(\text{LMTD}) \tag{9.15}$$

where

$\dot{Q}$ = heat rate, Btu/h (W)

UA = product of overall U value and heat transfer area A of active heat exchanger surface (see Fig. 2.12), Btu/(h · °F) (W/K)

LMTD = log mean temperature difference, °F (K)

The LMTD is the logarithmic average temperature difference between the two fluid streams. In simple heat exchangers without the change of phase, it is given by

$$\text{LMTD} = \frac{\Delta T_1 - \Delta T_2}{\ln(\Delta T_1/\Delta T_2)} \tag{9.16}$$

where ΔT_1 and ΔT_2 are the temperature differences between the two streams at the inlet and the outlet of the heat exchanger. For example, in the counterflow, shell-and-tube heat exchanger shown in Fig. 9.14,

$$\Delta T_1 = T_{h,i} - T_{c,o} \tag{9.17}$$

The LMTD approach has a number of disadvantages, including more complex calculations, compared to the ϵ-NTU approach developed in Chap. 2. However, the designer must be familiar with the LMTD method since it is the approach used by manufacturers.

Equation 9.15 must be modified by a correction factor if the flow geometry is not one of the few simple types shown in Table 2.10. This factor F is used as follows:

$$\dot{Q} = F(UA)(\text{LMTD}) \tag{9.18}$$

This factor can be read from charts in a number of sources including Turton et al. (1986) and Bowman and Turton (1990). Of course, F is embodied in the software used by manufacturers to assist designers with heat exchanger selection.

Manufacturers also provide performance data in extensive charts. Table 9.11 shows an example of such data. The body of the table shows heat transfer rates (at sea level) for a range of airflow rates and water flow rates. The heat rates are for a given difference between the entering air temperature (EAT) and the entering water temperature (EWT), as shown in Chap. 2 in the discussion of heat exchanger effectiveness.

The small table at the bottom of Table 9.11 includes an adjustment factor for other temperature differences. The numbers in this table are just the ratio of the temperature difference shown to the rating-point temperature difference of 145°F. The assumption is made that the heat exchanger effectiveness is independent of the operating temperature—a good assumption for air and water. (Note that Table 9.11 is valid only at sea level; for high altitudes, the heat rates shown must be multiplied by the ratio of local air density to standard air density.) Table 9.11 also shows the water-side pressure drop at various flow rates. If other liquids are used, the pressure drop will differ—glycols are often used for

TABLE 9.11

Example of Manufacturer's Heating Coil Capacity Table

Gal/min	H$_2$O Pressure Drop, ft	Capacity, kBtu/h @ ft³/min Fan, ft³/min									
		200	250	300	350	400	450	500	550	600	650
0.3	<0.1	12.6	13.4	13.9	14.4	14.7	15	15.2	15.4	15.5	15.7
0.5	0.1	17.1	18.8	20.2	21.3	22.3	23	23.7	24.3	24.8	25.3
0.8	0.2	19.3	21.6	23.5	25.1	26.4	27.6	28.6	29.5	30.4	31.1
1	0.3	20.6	23.3	25.5	27.4	29	30.5	31.8	33	34	35
1.5	0.5	22	25.1	27.8	30.1	32.2	34	35.6	37.1	38.5	40
2	0.9	22.8	26.2	29.1	31.7	34	36	37.9	40	41.2	42.6
2.5	1.3	23.3	26.8	30	32.7	35.1	37.3	39.4	41.2	43	44.5
3	2.6	23.6	27.3	30.5	33.4	35.9	38.3	40.4	42.4	44.2	45.9
(EWT − EAT)	160°	150°	140°	130°	120°	110°	100°	90°	80°	70°	60°	50°	40°
Facto	1.10	1.03	0.97	0.90	0.83	0.76	0.69	0.62	0.55	0.48	0.41	0.34	0.28

Source: Courtesy of Anemostat, Inc., Scranton, PA. With permission.

Above data for two row coils are based on an EWT − EAT value of 145°F. For other EWT − EAT values, use this table.

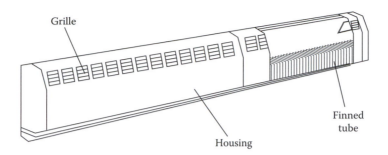

FIGURE 9.16
Baseboard convector. (Courtesy of Sterling Radiator, Westfield, MA. With permission.)

freeze protection in coils; pressure drops with glycol are larger than those for water. Hot-water coils are usually designed in the following operating ranges:

- Air-side velocity: 3 to 25 ft/s (1 to 8 m/s)
- Hot-water temperatures: 120 to 250°F (50°C to 120°C)
- Water velocities: 0.5 to 8 ft/s (0.2 to 2.5 m/s)

Cooling coils are operated in similar ranges (ASHRAE, 2000) except that the chilled water supply to coils is in the range of 40 to 60°F. When steam is used in heating coils, it is usually low-pressure steam at 2 to 10 psig (14 to 70 kPa). Water-side temperature drops are in the range of 10 to 20°F (5 to 10°C) for both heating and cooling coils.

Coils and shell-and-tube heat exchangers rely on forced convection for effective heat transfer. Free convection in air can also be effective for direct space heating. Figure 9.16 shows an example of a baseboard unit operated with hot water. High water temperatures (215°F is a typical rating point for convectors) induce free convection in the air surrounding the pipes containing the heating water. To enhance heat transfer, the pipes are usually fitted with air-side fins. Still better performance can be achieved by using a small fan to force air over the fin tubes in baseboard heaters.

The principles of coil selection and sizing used for heating and cooling coils and for shell-and-tube heat exchangers are also used for other HVAC equipment, such as air-cooled condensers, water tank immersion heaters using steam, chiller condensers and evaporators, and air-to-air heat exchangers for heat recovery in ventilation air. Example 9.5 shows how a coil is selected for space heating.

Example 9.5: Heating Coil Selection

The heat load calculation for a space to be heated by a forced-air heating coil is 20,000 Btu/h (5860 W) in a building at an altitude where the air density is 80 percent that at sea level. The space is maintained at 70°F (21°C), and hot water is supplied from the heating plant at 180°F (82°C). Select a coil from Table 9.11 for this application. For proper air distribution in the space, the airflow rate must be at least 500 ft³/min (236 L/s).

Given: $\dot{Q} = 20{,}000$ Btu/h, $T_{i,\text{air}} = 70°F$ (21°C), $T_{w,i} = 180°F$ (82°C),
$\dot{V}_{\text{air}} = 500$ ft³/min (236 L/s)

Find: Coil water flow

Lookup values: See Table 9.11.

SOLUTION

The heat rate must be adjusted by two factors before the table can be used. First, since the temperature difference is 110°F, not the standard 145°F used for the tabular values, the heat rates in the table must be multiplied by the factor 0.76, read from the bottom of Table 9.11. Second, the heat rate is 20 percent less than that at sea level due to the high altitude of the site. Hence the value of heat rate to be read from the table is

$$\dot{Q}_{\text{table}} = \frac{20,000\,\text{Btu/h}}{0.76(0.80)} = 32,900\,\text{Btu/h}$$

In the table under the 500 ft³/min column, one finds a heat rate of 35.6 kBtu/h corresponding to a water flow rate of 1.5 gal/min. A higher airflow rate, say 550 ft³/min, could be used. For this airflow, one finds a heat rate of 33 MBtu/h at a lower water flow rate of 1 gal/min. Which operating point is selected from the table ultimately depends on the airflow or water flow available at the location of this terminal heating unit and the operating costs.

COMMENTS

When using tables for coil selection, the designer should always pick a value equal to or larger than the heat load. This provides a small capacity margin if the coil specified should fall below specifications. The water-side pressure drop shown in the second column of Table 9.11 is needed for piping system design. Recall that the larger the product of pressure drop and flow, the larger the pump operating energy needed. The same rule applies for fan energy use; if the 550 ft³/min coil is used, the fan energy will increase by the cube of the flow rate (see the discussion of fan laws in Chap. 5).

9.5.3 Heat Transfer with Phase Change

Heat transfer processes in Sec. 9.5.2 involved only sensible heat exchanges. For heating and cooling coils, this is always the case if the dew point of entering air is less than the dew point of the surface of the coil. However, the surface of cooling coils is often below the dew point of the air passing over them during the cooling season in moderately humid and humid climates. In this section, we discuss briefly how coil calculations are done if the surface is wet. A third case also exists: a partially wet and partially dry coil. This intermediate case is handled by treating the coil as two heat exchangers—one wet and the other dry—with a common boundary (ASHRAE, 2000). Although this chapter is primarily concerned with heating, we cover cooling coils under Sec. 9.5.1 above that concludes this chapter.

The driving force for combined heat and mass transfer at the surface of a wet cooling coil is the air enthalpy rather than the air temperature. For example, the local heat transfer from a wet surface is given by (Kuehn, et al., 1998)

$$\dot{Q}_{\text{wet}} = (\tilde{U}A)_{\text{wet}}(h_a - h_{w,\text{sat}}) \qquad (9.19)$$

where

$\dot{Q}_{\text{wet}}$ = heat transfer rate from cooling coil liquid, via wetted surface, to airstream

$(\tilde{U}A)_{\text{wet}}$ = *enthalpy-based* overall heat transfer coefficient, $\text{lb}_{\text{m}}/\text{h}$ (kg/s). Note that the units differ from other UA quantities used elsewhere in this book

h_a = enthalpy of air flowing over coil, Btu/lb_{m} (kJ/kg)

$h_{w,\text{sat}}$ = saturated air enthalpy at *coil water temperature* inside coil tubes, Btu/lb_{m} (kJ/kg)

The overall (between coil fluid and airstream) enthalpy-based heat transfer coefficient is calculated from

$$(\tilde{U}A)_{\text{wet}} = \frac{A_{\text{wet}}}{b(R_i + R_m) + c_{p,\text{da}}(R_{\text{fin}} + R_t)} \tag{9.20}$$

where

A_{wet} = external wet surface area of coil
R_i = internal water-side thermal resistance, $(\text{h} \cdot \text{ft}^2 \cdot °\text{F})/\text{Btu}$ $[(\text{m}^2 \cdot \text{K})/\text{W}]$
R_m = metal tube conductive resistance
b = slope of approximate, locally linear equation $h_s = a + bT_{\text{sat}}$ relating saturated air enthalpy to saturation temperature T_{sat}, $\text{Btu}/(\text{lb}_{\text{m}} \cdot °\text{F})$ $[\text{kJ}/(\text{kg} \cdot \text{K})]$
$c_{p,\text{da}}$ = specific heat of dry air
$R_{\text{fin}} = \frac{1-\eta_o}{\eta_o h_o}$ = thermal resistance of fins in cooling coil
$\eta_o = 1 - \frac{A_{\text{fin}}}{A_{\text{fin}}+A_{\text{pipe}}}(1 - \eta_f)$, in which η_f is standard fin efficiency
$R_t = \frac{1}{h_o}$ = thermal resistance at outer surface of tube

This $\bar{U}A$ value can be used in a *log mean enthalpy* expression analogous to Eq. (9.16) to find the total heat rate. The enthalpy differences in this expression are the difference between air enthalpy and *saturated* air enthalpy (evaluated at the *cooling water* temperatures) at the inlet and outlet of the coil, as shown in Fig. 9.17.

Alternatively, the ϵ-NTU method developed in Chap. 2 can be used to find the heat transfer rate. An expression similar to Eq. (2.26) is used except that the UA value in NTU is replaced by UA above. The overall driving force (i.e., temperature difference) is replaced

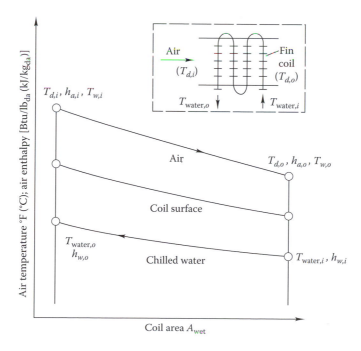

FIGURE 9.17
Diagram showing driving potentials for heat and mass transfer in a wet coil (T_w denotes wet-bulb temperature).

by the enthalpy difference between coil inlet air and that of saturated air at the coil water inlet temperature. Finally, the capacitance ratio (assuming that the airstream is the lower-capacitance stream) is replaced by $\dot{m}_{air}/(\dot{m}_w c_{p,w}/b)$. Example 9.6 indicates how this method can be applied to a wet coil.

Example 9.6: Performance of a Wet Cooling Coil

A testing and balancing contractor tests a coil to determine if its performance is as specified. The coil specified by the designer had the following characteristics (refer to Fig. 9.17 for nomenclature) at the design point:

1. Entering air: $T_{d,i} = 95°F$ (35°C), $T_{w,i} = 75°F$ (23.9°C)
2. Leaving air: $T_{d,o} = 66°F$ (18.9°C), $T_{w,o} = 61°F$ (16.1°C)
3. Entering coil water temperature: $T_{water,i} = 55°F$ (12.7°C)
4. Airflow rate: 20,000 ft³/min (9440 L/s)

During the coil test, the contractor measures the following data. Since test conditions rarely are the same as design conditions, the operating condition is different. Use these measured data to determine if the coil has the specified cooling capacity.

- Entering air: $T_{d,i} = 90°F$ (32.2°C), $T_{w,i} = 73°F$ (22.8°C)
- Leaving air: $T_{d,o} = 63°F$ (17.2°C), $T_{w,o} = 59°F$ (15.0°C)
- Entering coil water temperature: $T_{water,i} = 52°F$ (11.1°C)
- Airflow rate: 20,000 ft³/min (9440 L/s)

The air side is the minimum-capacitance side of the coil, and the airflow during the test is the same as that at design conditions.

Given: Design and measured data as above

Figure: See Fig. 9.17.

Assumptions: The coil is tested at sea level. Its surface is wet—confirmed by plotting the process on a psychrometric chart.

Find: Cooling capacity of coil

Lookup values: Enthalpies are read from the psychrometric chart as follows:

$$h_{a,i,des} = 38.7 \text{ Btu/lb}$$

$$h_{a,o,des} = 27.2 \text{ Btu/lb}$$

$$h_{a,i,test} = 36.8 \text{ Btu/lb}$$

$$h_{a,o,test} = 25.8 \text{ Btu/lb}$$

$$h_{w,i,des} = 23.2 \text{ Btu/lb}_m \quad \text{(saturated at 55°F)}$$

$$h_{w,i,test} = 21.5 \text{ Btu/lb}_m \quad \text{(saturated at 52°F)}$$

SOLUTION

The way to compare the performance of a coil to the design specifications is to calculate the effectiveness for both cases. For the *design* conditions, the enthalpy form of Eq. (2.32) is

$$\epsilon_{des} = \frac{\dot{Q}_{des}}{\dot{m}_a(h_{a,i} - h_{w,i})} \qquad (9.21)$$

$$\dot{m}_a = (20{,}000\,\text{ft}^3/\text{min})(0.075\,\text{lb}_\text{m}/\text{ft}^3)(60\,\text{min}/\text{h}) = 90{,}000\,\text{lb}_\text{m}/\text{h}$$

$$\dot{Q}_\text{des} = \dot{m}_a\,\Delta h_a = (90{,}000\,\text{lb}_\text{m}/\text{h})(38.7\,\text{Btu}/\text{lb}_\text{m} - 27.2\,\text{Btu}/\text{lb}_\text{m}) = 1035\,\text{kBtu}/\text{h}$$

The design effectiveness is

$$\epsilon_\text{des} = \frac{1{,}035{,}000\,\text{Btu}/\text{h}}{(90{,}000\,\text{lb}_\text{m}/\text{h})[(38.7 - 23.2)\,\text{Btu}/\text{lb}_\text{m}]} = 0.742$$

Under *test* conditions, the same calculations give these results:

$$\dot{Q}_\text{test} = \dot{m}_a\,\Delta h_a = (90{,}000\,\text{lb}_\text{m}/\text{h})[(36.8 - 25.8)\,\text{Btu}/\text{lb}_\text{m}] = 990\,\text{kBtu}/\text{h}$$

The calculated effectiveness is

$$\epsilon_\text{test} = \frac{990{,}000\,\text{Btu}/\text{h}}{(90{,}000\,\text{lb}_\text{m}/\text{h})[(36.8 - 21.5)\,\text{Btu}/\text{lb}_\text{m}]} = 0.719$$

The measured effectiveness is 97 percent of that specified by the designer, a small difference indeed.

COMMENTS

This type of calculation is necessary whenever test and balance contractor test results are used. Testing will almost never be done at design conditions for either heating or cooling coils. The concept of heat exchanger effectiveness provides a convenient method for using the results at test conditions to find coil performance at the design condition.

9.6 Low-Temperature Radiant Heating

Heating systems in many parts of the world use warmed floors and/or ceilings for space heating in buildings. Although this system is rare in the United States, the good comfort and quiet operation provided by this approach make it worth considering for some applications. In Europe it is far more common. Radiant systems are well suited to operation with heat pump, solar, and other low-temperature systems. In this section, we discuss the principles of low-temperature space heating. This is distinct from high-temperature radiant heating using either electricity or natural gas to provide a high-temperature source from which radiation can be directed for localized heating.

Figure 9.18 shows how a radiant floor might be configured in a residence. The same concept can also be used in the ceiling in both residential and commercial buildings. The term *radiant* is a misnomer since between 30 percent (ceilings) and 50 percent (floors) of the heat transferred from "radiant" panels is actually by convection. However, we use the industry's nomenclature for this heating system.

The radiation heat output of radiant panels is given by the Stefan-Boltzmann equation, discussed in Chap. 2,

$$\dot{Q} = \epsilon_\text{eff} F_{h,u}\sigma\left(T_h^4 - T_u^4\right) \tag{9.22}$$

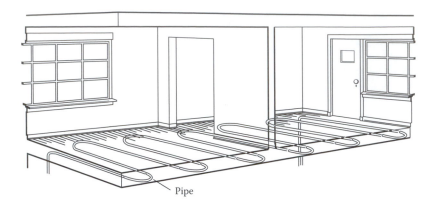

FIGURE 9.18
Residential radiant floor heating system.

where

$\epsilon_{\text{eff}} = 1/(1/\epsilon_h + 1/\epsilon_u - 1) =$ effective emittance of space; subscripts h and u refer to heated and unheated (by radiant panels) surfaces of space; ϵ_{eff} is approximately 0.8

$F_{h,u} =$ view factor between heating surface and unheated surfaces; its value is 1.0 in present case

$T_h =$ heating surface temperature

$T_u =$ mean of unheated surface temperatures

$\sigma =$ Stefan-Boltzmann constant (see Chap. 2)

The convection from the heating surface can be found by using the standard free-convection expressions in Chap. 2.

The designer's job is to determine the panel area needed, its operating temperature, the heating liquid flow rate, and construction details. The panel size is determined based on standard heat load calculations (Chap. 7). Proper account should be made of any losses from the back of the radiant panels to unheated spaces. Panel temperatures should not exceed 85°F (29.5°C) for floors and 115°F (46°C) for ceilings.

Water temperatures are typically 120°F (49°C) for floors and up to 155°F (69°C) for ceilings. Panels can be piped in a series configuration if pipe runs are not excessively long (the final panels in a long series run will not perform up to specifications due to low fluid temperatures). Long series loops also have excessively high pressure drops. If large areas are to be heated, a combination of series and parallel connections can be used. Manufacturer's advice should be sought regarding the number of panels that can be connected in series without performance penalties.

If radiant floors are to be built during building construction rather than used as prefabricated panels in ceilings, the following guidelines can be used: Tubing spacing for a system of the type shown in Fig. 9.18 should be between 6 and 12 in (15 and 30 cm). The tubing diameter ranges between 0.5 and 1.0 in (1.2 and 2.5 cm). Flow rates are determined by the rate of heat loss from the panel, which in turn depends on the surface temperature and hence the fluid temperature. This step in the design is iterative. Panel design follows this process:

1. Determine the room heat load.

2. Decide on the location of panels (roof or floor).

3. Find panel heat flux, including both radiative and convective contributions at 80°F (27°C) for floor panels and 110°F (43°C) for ceiling panels.

4. Divide heat load by heat flux to find needed panel area.

5. If panel area exceeds available floor or ceiling area, raise panel temperature (not exceeding temperatures noted earlier) and repeat steps 3 and 4.

6. If the panel area is still insufficient, consider both floor and ceiling panels to improve thermal quality of room insulation.

Control of radiant heating systems has proved to be a challenge in the past due to the large time constant of these systems. Both under- and overheating are problems. If the outdoor temperature drops rapidly, this system will have difficulty responding quickly. On the other hand, after a morning warm-up followed by high solar gains on a sunny winter day, the radiant system may overshoot. The current generation of "smart" controls should help improve the comfort control of these systems.

9.7 Solar heating

Solar energy is a source of low-temperature heat that has selected applications to buildings. In this section we consider two. A third—passive solar space heating—often is used in residences and is described in Chap. 14. Solar water heating is a particularly effective method of using this renewable resource since low- to moderate-temperature water [up to 140°F (60°C)] can be produced by readily available flat-plate collectors (Goswami, Kreith and Kreider, 2000).

Figure 9.19 shows one system for heating service water for residential or commercial needs by using solar collectors. The system consist of three loops; it is instructive to describe the system's operation based on these three. First, the collector loop (filled with a nonfreezing solution, if needed) operates whenever the DHW (domestic hot-water) controller determines that the collector is warmer, by a few degrees, than the storage tank. Heat is transferred from the solar-heated fluid by a counterflow or plate heat exchanger to the storage tank in the second loop of the system. Storage is needed since the availability of solar heat rarely matches the instantaneous water-heating load. The check valve in the collector loop is needed to prevent reverse flow at night in systems where the collectors (which are cold at night) are mounted above the storage tank.

The third fluid loop is the hot-water delivery loop. Hot water drawn off to the load is replaced by cold water supplied to the solar preheat tank, where it is heated as much as possible by solar heat. If solar energy is insufficient to heat the water to its setpoint, conventional fuels can finish the heating in the water heater tank, shown on the right of the figure. The tempering valve in the distribution loop is used to limit the temperature of water dispatched to the building if the solar tank should be above the water heater setpoint in summer.

The energy delivery of DHW systems can be found by using the *f*-chart method described in Goswami et al. (2000). As a rough rule of thumb, 1 ft^2 of collector can provide 1 gal of hot water per day (45 L/m^2) on average in sunny climates. Design pump flows are to be 0.02 gal/min per square foot of collector [0.01 L/(s · m^2)], and a heat exchanger effectivenesses of at least 0.75 can be justified economically. Tank should be insulated so that no more than 2 percent of the stored heat is lost overnigh (Goswami et al., 2000).

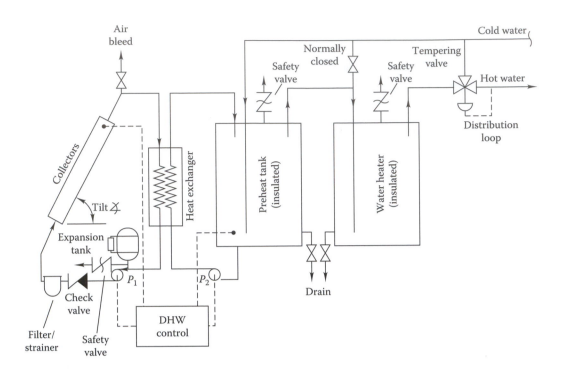

FIGURE 9.19
Solar water-heating system including collectors, pumps, heat exchanger, and storage tanks along with piping and ancillary fittings. Collectors are tilted up from the horizontal at a fixed angle roughly equal to the local latitude.

The second practical solar system for HVAC applications—ventilation ai preheating—is shown in Fig. 9.20. In this application, ventilation air is passed through a filter and through an air-type solar collector prior to being introduced to the HVAC system. In this way, solar heat can provide some preheating of all outdoor air in winter, regardless of the solar flux available. With a typical flat-plate collector, about 50 percent of the incident sunlight can be converted to useful heat. Flow rates fo these systems range between 1 and 2 ft^3/(min $\cdot$ ft^2) [5 and 10 L/(s $\cdot$ m^2)].

No special controls are needed; whenever ventilation air is needed, it is drawn through the collectors, and whatever heat is then available is used to heat the air. The storage shown in the figure is optional. If solar availability exceeds the load, heat can be stored for use at night. If insufficient solar heat is available, steam is used to heat the ventilation air, as shown. Air preheat is also accomplished by recovering heat from exhaust air, as described in Chap. 11.

Solar heating should be assessed on an economic basis. If the cost of delivered solar heat, including the amortized cost of the delivery system and its operation, is less than that of competing energy sources, then an incentive exists for using the solar resource. The collector area needed on commercial buildings can be large; if possible, otherwise unused roof space can be used to hold the collector arrays. Of course, large high-rise buildings in urban locations are not likely candidates for solar heating, since very little roof or nearby ground area is likely to be available for collector mounting.

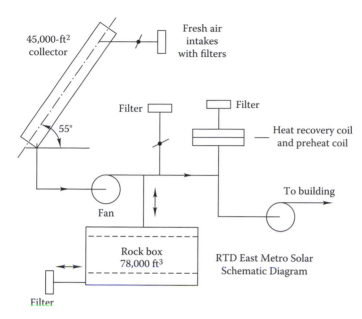

45,000-ft² collector

55°

Fresh air intakes with filters

Filter

Filter

Heat recovery coil and preheat coil

To building

Fan

Rock box 78,000 ft³

RTD East Metro Solar Schematic Diagram

Filter

FIGURE 9.20
Solar air preheating system for ventilation air with optional storage. The collector and storage sizes shown in the figure represent those used on the first system of this type built in Denver, Colorado. (Courtesy of Jan F. Kreider and Associates, Inc., Boulder, CO. With permission.)

9.8 Cogeneration Definition and Overview[5]

Cogeneration or combined heat and power (CHP) systems capture the heat energy from electric generation for a wide variety of thermal needs, including hot water, steam, and process heating or cooling. Figure 9.21 gives an example of the efficiency difference between separate and combined heat and power. A typical U.S. CHP system converts 80 out of 100 units of input fuel to useful energy—30 to electricity and 50 to heat. By contrast, traditional separated heat and power components require 163 units of energy to provide the same amount of heat and power. Thus, with today's technologies, CHP systems can cut fuel use nearly 40 percent (Roop and Kaarsberg, 1999).

9.8.1 Available Technologies

Commercially available CHP technologies for distributed generation include diesel engines, natural gas engines, steam turbines, gas turbines, microturbines, and phosphoric acid fuel cells. Table 9.12 summarizes the characteristics of commercial CHP prime movers. The table shows the wide range in CHP capacity—from 1-kW Stirling engine CHP systems to 250-MW gas turbines.

9.8.2 Typical CHP Applications

CHP systems can provide cost savings as well as substantial emissions reductions for industrial, institutional, and commercial users. This section reviews some of the primary issues faced by the design engineer in selecting and designing an optimized CHP system. Selecting the right CHP technology for a specific application depends on many factors,

---

[5] Contributed by Borbely and Kreider from *Distributed Generation*, CRC Press, with permission.

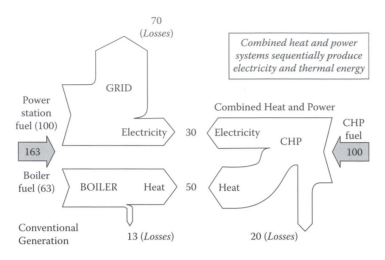

FIGURE 9.21
Cogenerated vs. separate power production.

TABLE 9.12

Comparison of CHP Technologies

	Diesel Engine	Natural Gas Engine	Gas Turbine	Microturbine	Fuel Cells	Stirling Engine
Electric efficiency (LHV)	30–50%	25–45%	25–40 (simple) 40–60% (combined)	20–30%	40–70%	25–40%
Part load	Best	OK	Poor	Poor	N/A	OK
Size, MW	0.05–5	0.05–5	3–200	0.025–0.25	0.2–2	0.001–0.1
CHP installed cost, $/kW	800–1500	800–1500	700–900	500–1300	>3000	>1000
Start-up time	10 s	10 s	10 min–1 h	60 s	3–48 h	60 s
Fuel pressure, psi	<5	1–45	120–500	40–100	0.5–45	N/A
Fuels	Diesel, residual oil	Natural gas, biogas, propane	Natural gas, biogas, propane, distillate oil	Natural gas, biogas, propane, distillate oil	H_2, natural gas, propane	All
Uses for heat recovery	Hot water, LP steam, district heating	Hot water, LP steam, district heating	Heat, hot water, LP-HP steam hotwater, district heating	Heat, hot water, LP steam	Hot water, LP steam	Direct heat, hot water, LP steam
CHP output, Btu/kWh	3400	1000–5000	3400–12,000	4000–15,000	500–3700	3000–6000
Usable temp. for CHP, °F	180–900	300–500	500–1100	400–650	140–700	500–1000

Source: ONSITE SYCOM Energy Corporation, "Market Assessment of CHP in the State of California," draft report to the California Energy Commission, September 1999 (except for Stirling data).

including the amount of power needed, the duty cycle, space constraints, thermal needs, emission regulations, fuel availability, utility prices, and interconnection issues.

Designing a technically and economically feasible CHP system for a specific application requires detailed engineering and site data. Engineering information should include electric and thermal load profiles, capacity factor, fuel type, and performance characteristics of the prime mover. Site-specific criteria such as maximum noise levels and footprint constraints must be taken into account. See Borbely and Kreider (2001) for details.

9.8.3 Benefits of CHP for HVAC

CHP Efficiency   Power generation systems create large amounts of heat in the process of converting fuel to electricity. More than two-thirds of the energy content of the input fuel is converted to heat and wasted in many older central generating plants. As an alternative, an end user with significant thermal and power needs can generate both thermal and electric energy in a single combined heat and power system located at or near its facility. Figure 9.21 showed how a well-balanced CHP system outperforms a traditional remote electricity supply and on-site boiler combination. The chart illustrates that out of 100 units of input fuel, CHP converts 80 to useful work—30 to electricity and 50 to steam or some other useful thermal output for HVAC purposes such as space or water heating or absorption cooling; these values are typical but vary from one plant to another depending on loads.

While future central station plants will be able to generate electricity more efficiently than the 30 percent average rate used in developing the chart, CHP installations with proper thermal/electric balance have design efficiencies of 80 to 90 percent and will still result in significant overall energy savings. On-site use of CHP systems also reduces transmission and distribution system line losses to zero from typical central unit line losses of 4 to 7 percent.

9.9 Summary

This chapter has discussed the design-oriented details of boilers, furnaces, and heat pumps. Practical details of heat exchangers have also been discussed. Particular emphasis has been placed on the importance of part-load performance of heat-generating and transport equipment. CHP systems were shown to have significant benefits, as well.

Problems

The problems in this book are arranged by topic. The approximate degree of difficulty is indicated by a parenthetic italic number from 1 to 10 at the end of the problem. Problems are stated most often in USCS units; when similar problems are presented in SI units, it is done with approximately equivalent values in parentheses. The USCS and SI versions of a problem are not exactly equivalent numerically. Solutions should be organized in the same order as the examples in the text: given, figure or sketch, assumptions, find, lookup values, solution. For some problems, the Heating and Cooling of Buildings (HCB) software may be helpful. In some cases it is advisable to set up the solution as a spreadsheet, so that design variations are easy to evaluate.

9.1 A residence in Chicago uses a type 3 (Table 9.1) gas furnace for space heating with a thermostat setting of 72°F (22°C). During the immediately past winter with 6500°F · days (3600°C · days), 100 MBtu (105 GJ) of gas energy was used. If the thermostat had been set at 68°F (20°C), what would the gas consumption have been?   (4)

9.2 The owner of a residence in Boston decides to replace her failed, old, standard oil-fired furnace with a new condensing type. How many gallons of oil could be saved if the annual heat load on this residence is 120 MBtu (127 GJ) on average? The heating value of 1 gal of oil is 140,000 Btu (39,000 kJ/L). If the cost of oil is $1.05/gal (27¢/L), what is the annual cost saving due to the furnace upgrade, and what is the payback period if the more efficient furnace costs $800 more than would replacement with the original design?   (5)

9.3 Work Prob. 9.2 for Denver, Colorado, where the annual load on the building is 125 MBtu (130 GJ).   (4)

9.4 Flue gas analysis of a boiler gives the following results: CO_2, 9 percent; O_2, 2 percent; N_2, 87 percent; CO, 2 percent. What is the excess air fraction? Would you suggest any adjustments to the boiler? If so, what is needed?   (5)

9.5 Flue gas analysis of a boiler gives the following results: CO_2, 11 percent; O_2, 5.0 percent; N_2, 83.5 percent; CO, 0.5 percent. What is the excess air fraction? Would you suggest any adjustments to the boiler? If so, what is needed?   (5)

9.6 Equation (9.3) is the chemical equation for the complete combustion of methane, the principal constituent of natural gas. How many pounds (kilograms) of (a) oxygen and (b) standard air are needed to completely burn 10 lb (4.5 kg) of methane?   (4)

9.7 Derive the result stated following Eq. (9.3) that 8.7 ft^3 of standard air is required to combust 1 ft^3 of methane at sea level.   (5)

9.8 If the heating value of natural gas is 1000 Btu/ft^3 (37 MJ/m^3) at standard conditions, what is the heating value at 4000 ft (1200 m)?   (3)

9.9 Using a rule of thumb, estimate *for preliminary purposes only* what the *input* heat rating for a boiler would be in Chicago for a well-insulated, 200,000-ft^2 (20,000-m^2) warehouse. Check the calculation, assuming that the warehouse is square in plan, 10 ft (3 m) high, flat-roofed, and insulated to R 20 (R 3.5) and has an air exchange rate of 0.7 per hour. Ignore floor slab losses because the perimeter is small relative to the building surface area, and use a reasonable boiler efficiency. What combination of boilers (specify the model numbers) would you select from Table 9.2 for this building if the owner requested that at least two boilers be installed, to ensure some building heat even if one boiler were to fail?   (8)

9.10 Work Prob. 9.9 for Denver, Colorado, accounting for elevation effects on both boiler capacity and heat load.   (8)

9.11 A large boiler is supplied with 90,000 ft^3/h (2550 m^3/h) of natural gas with a higher heating value of 1000 Btu/ft^3 (37 MJ/m^3) and 40 percent excess air. If combustion is complete and the stack gases leave the boiler at 300°F (105°C), how much superheated steam at 20 psia and 300°F (140 kPa and 150°C) is produced? What is the boiler efficiency? Water is supplied to the boiler at 200°F (94°C), and boiler thermal losses (other than stack gas losses) are 5 percent of the heat rate. (8)

9.12 Describe in words the function of a steam trap. Why is the proper operation of steam traps so important in an energy-conserving building? (5)

9.13 A boiler has a peak efficiency of 90 percent at the peak load of 5 MBtu/h (1465 kW). Figure P9.13 shows the part-load characteristic. If the actual load is 45 percent of the peak load, what is the efficiency? If the load is 80 percent of the peak? (4)

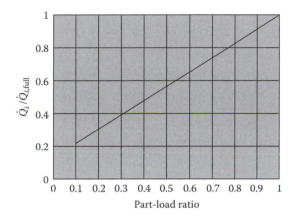

Part-load ratio

FIGURE P9.13

9.14 The part-load characteristics of a boiler are given in the table below. Find the coefficients in the part-load equation, Eq. (9.9), assuming a linear characteristic (that is, C and higher-order coefficients are zero). Plot the efficiency of the boiler as a function of PLR. (5)

PLR	0.2	0.4	0.6	0.8	1.0
$\dot{Q}_{i,full}$	0.31	0.44	0.63	0.81	1.00

9.15 A hospital in Denver, Colorado, requires heating to 72°F (22°C) during the entire heating season. The heat load at −2.5°F (−19°C) is 10.0 MBtu/h (2950 kW), and the balance-point temperature is estimated to be 62°F (17°C). The boiler capacity (corrected for altitude) is 11.0 MBtu/h (3240 kW) with a full-load efficiency of 82 percent. Either write a computer program or use spreadsheet software to solve this problem using the bin data for Denver. The part-load equation for the boiler is

$$\frac{\dot{Q}_i}{\dot{Q}_{i,full}} = 0.10 + 0.90(\text{PLR})$$

What are the annual energy consumption, the annually averaged efficiency, and the annual PLR of the boiler? (8)

9.16 Solve Prob. 9.15 using the HCB software. (4)

9.17 A college dormitory in Pittsburgh requires heating to 69°F (20°C) during the entire heating season. The heat load at 1°F (−17°C) is 4.0 MBtu/h (1180 kW), and the balance-point temperature is estimated to be 62°F (17°C). The boiler capacity is 4.6 MBtu/h (1400 kW) with a full-load efficiency of 84 percent. Either write a computer program or use a spreadsheet with the bin data from the appendix CD-ROM to solve this problem. The part-load equation for the boiler is

$$\frac{\dot{Q}_i}{\dot{Q}_{i,full}} = 0.12 + 0.88(\text{PLR})$$

What are the annual energy consumption, the annually averaged efficiency, and the annual PLR of the boiler?   (8)

9.18 Solve Prob. 9.17, using the HCB software.   (4)

9.19 The building in Example 9.3 can be heated either by the boiler specified in the example or by three boilers each of one-third the capacity but with the same part-load characteristics. Since the boiler in the example has more capacity than needed most of the time, less energy may be used with smaller boilers operating more closely to their full capacity. Add the columns needed to the spreadsheet in Table 9.5 to study this option. For simplicity, assume that the lead boiler (the first one to be operated) operates up to full capacity before the second is activated; the same strategy is used to operate the second boiler relative to the third.   (10)

9.20 The building in Example 9.3 can be heated either by the boiler specified in the example or by two boilers, one with one-third the example boiler capacity and one with two-thirds the capacity. Both have the same part-load characteristics as the boiler in Example 9.3. Since the boiler in the example has more capacity than needed most of the time, less energy may be used with smaller boilers operating more closely to their full capacity. Add the columns needed to the spreadsheet in Table 9.5 to study this option. For simplicity, assume that the smaller boiler operates up to full capacity before the second is activated.   (10)

9.21 What are the peak (Btu/h, kW) and daily (Btu/day, MJ/day) water-heating loads for a full-service restaurant that serves 40 meals per hour (during the 3-h lunch peak) and 280 meals per day (daily total)? The water source temperature is 48°F (9°C). How large should the storage tank be to meet the peak load if the recovery rate is 40 gal/h (150 L/h)? State any assumptions needed.   (6)

9.22 What are the peak (Btu/h, kW) and daily (Btu/day, MJ/day) water-heating loads for a 150-unit apartment in which the 2-h peak occurs in the morning? The water source temperature is 50°F (10°C). How large should the storage tank be to meet the peak load if the recovery rate is 900 gal/h (3500 L/h)? State any assumptions needed.   (6)

9.23 A Carnot heat pump is operated between a constant indoor temperature (heat sink) at 70°F (21°C) and a varying outdoor temperature (heat source). Ignoring the temperature difference required for heat transfer from the source and sink, use a spreadsheet to determine the seasonal COP of this ideal device in Nashville, Tennessee, on a building with a heat loss coefficient of 500 Btu/(h · °F) (265 W/K) if the building balance-point temperature is 64°F (18°C). The capacity of this ideal heat pump is 16,000 Btu/h (4.7 kW) at 32°F (0°C). Whenever it is on, the heat pump power input is at its 32°F level.   (7)

9.24 Work Prob. 9.23 with an ideal Carnot heat pump, but assume that a 10°F (5.5°C) temperature difference is required for heat transfer at both the heat source and the sink.   (9)

9.25 Figure P9.25 shows the capacity of several residential heat pumps. If the heat loss coefficient of a duplex residence is 1700 Btu/(h · °F) (900 W/K) with a heating balance-point temperature of 66°F, what are the heat pump balance points for each of the four heat pumps shown?   (5)

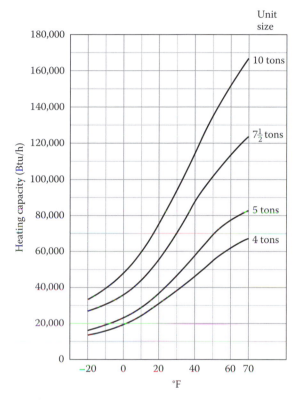

FIGURE P9.25

9.26 Figure P9.25 shows the capacity of several residential heat pumps. Each can meet the heating requirements of a building with a heat pump balance point in the range between 20 and 40°F (−7 and 5°C). Describe how you would select one unit from the four shown to minimize the homeowner's cost, including both the first cost for the heat pump (increases with size) and the ongoing operating costs for electric power and auxiliary resistance heat (this last cost decreases with increasing heat pump size).   (8)

9.27 (a) Figure P9.27 is a table of capacities (including fans and defrost effects) for several sizes of heat pump. If a residence has a heat loss coefficient of 700 Btu/(h · °F) (365 W/K) and a balance-point temperature of 68°F (20°C), what will be the SPF of the model C heat pump in Boise, Idaho? What is the overall heating system SPF, including resistance heat?   (6)

(b) Repeat (a) with the next-larger- and next-smaller-capacity (at 32°F) heat pumps given in the table. Comment on the results and how this influences the selection of a heat pump.   (4)

Outdoor Dry-Bulb Temperature

Model	72° kBtu/h	67°	62°	57°	52°	47°	42°	37°	32°	27°	22°	17°	12°	7°	2°	-3°	-8°	-13°	-18°
	kW																		
A	27.6	27.6	27.1	26.4	25.5	24.3	22.4	20.4	18.3	16.4	14.6	13.0	11.7	10.6	9.5	8.6	7.7	7.1	6.6
	3.7	3.6	3.6	3.5	3.4	3.3	3.2	3.1	3.0	2.9	2.8	2.7	2.6	2.5	2.4	2.3	2.2	2.1	2.0
B	35.6	35.4	35.0	34.1	32.9	31.4	29.2	27.0	24.6	22.5	20.6	18.5	16.6	14.8	13.2	11.7	10.3	9.2	8.2
	4.3	4.3	4.2	4.1	3.9	3.8	3.7	3.6	3.5	3.4	3.3	3.2	3.1	3.0	2.9	2.8	2.7	2.5	2.4
C	34.4	34.2	33.8	33.0	31.8	30.3	28.2	25.9	23.7	21.5	19.6	17.6	15.8	14.0	12.6	11.1	9.8	8.9	7.9
	4.2	4.1	4.0	3.9	3.8	3.7	3.6	3.5	3.4	3.3	3.2	3.1	3.0	2.9	2.8	2.7	2.5	2.4	2.3
D	45.2	45.0	44.5	43.4	41.8	39.9	37.0	33.8	30.7	27.8	25.1	22.4	19.9	17.6	15.4	13.5	11.7	10.3	9.0
	5.1	5.0	4.9	4.7	4.6	4.4	4.3	4.1	4.0	3.9	3.7	3.6	3.4	3.3	3.2	3.0	2.9	2.7	2.6
E	44.1	43.9	43.3	42.3	40.7	38.8	36.0	32.8	29.7	26.8	24.1	21.5	19.1	16.9	15.0	13.1	11.5	10.2	9.0
	5.1	5.0	4.9	4.7	4.6	4.4	4.3	4.1	4.0	3.9	3.7	3.6	3.4	3.3	3.2	3.0	2.9	2.7	2.6
F	50.4	50.2	49.6	48.4	46.6	44.4	41.3	37.9	34.5	31.4	28.5	25.6	22.9	20.5	18.3	16.2	14.4	13.0	11.7
	6.2	6.0	5.9	5.7	5.6	5.4	5.2	5.1	4.9	4.7	4.6	4.4	4.3	4.1	4.0	3.8	3.6	3.5	3.3

FIGURE P9.27

9.28 Work Prob. 9.27 for Washington, D.C.   (6)

9.29 Work Probs. 9.27 and 9.28, using the HCB software. You will need to enter data from Fig. P9.27 and adjust the envelope losses and internal gains in the software to match the problem values of both K_{tot} and the balance-point temperature.   (7)

9.30 Figure P9.30 shows how the *indoor* temperature of a building affects the performance of a heat pump. For the building described in Prob. 9.27, find the annual electric energy use (including resistance heat) at two indoor temperatures—70 and 75°F (21 and 24°C)—in Denver, Colo. Use the HCB software if you prefer. You will need to work several key assumptions. State them all.   (7)

Outdoor Temp.	Heating Capacity, kBtu/h, at Indicated Indoor Dry-Bulb Temp.				Compressor Power, kW, at Indicated Indoor Dry-Bulb Temp.			
	60	70	75	80	60	70	75	80
−20	10.4	9.3	8.8	8.2	1.8	1.9	1.9	2.0
−10	11.9	10.5	9.8	9.1	2.1	2.2	2.2	2.3
0	15.0	13.4	12.7	11.9	2.3	2.5	2.5	2.6
10	19.4	17.7	16.9	16.0	2.6	2.7	2.8	2.9
20	24.8	23.0	22.1	21.2	2.9	3.0	3.1	3.2
30	30.7	28.8	27.9	27.0	3.1	3.3	3.3	3.4
40	36.9	35.0	34.0	33.0	3.3	3.5	3.6	3.7
145	39.6	38.0	37.2	36.4	3.4	3.6	3.7	3.8
510	41.6	40.0	39.2	38.3	3.5	3.7	3.8	3.9
60	44.6	43.0	42.1	41.3	3.7	3.9	4.0	4.1
70	45.9	44.1	43.3	42.4	3.8	4.1	4.2	4.3
80	44.8	43.0	42.1	41.3	4.0	4.2	4.3	4.4

Correction factors for other airflows
(value at 1200 ft³/min times corr factor = value at new airflow)

Airflow	1050	1200	1350
Heating capacity	0.980	1.000	1.020
Compressor, kW	1.025	1.000	0.975

FIGURE P9.30

9.31 Define a building in the HCB software such that it has a heat loss coefficient of 800 Btu/(h · °F) (420 W/K) with a balance temperature of approximately 64.5°F (18.1°C) if the thermostat setting is 68°F (20°C). Consider heat pump models D and F in Fig. P9.27. What is the balance point for each heat pump applied to this building?   (3)

9.32 Define a building in the HCB software such that it has a heat loss coefficient of $K_{tot} = 800$ Btu/(h · °F) (422 W/K); with a specified thermostat setting of 68°F (20°C) it is to have a balance temperature of approximately 64.5°F (18°C). Consider heat pump models D and F in Fig. P9.27. If this building is located in Boston, find the total electric consumption and the overall heating system SPF, using the HCB software. On the basis of these calculations, which heat pump would you recommend based on energy consumption? (Of course, the final selection must include the effect of the costs of the two heat pumps.)   (6)

9.33 Work Prob. 9.32 for Kansas City, Mo.   (6)

9.34 Work Prob. 9.32 for Salt Lake City, Utah.   (6)

9.35 If the building conductance in Prob. 9.32 were reduced 18 percent by improving the thermal integrity of the envelope, what would the cost savings in electricity be if the model D heat pump were used? The cost of electricity is 10¢/kWh including demand charges.   (6)

9.36 Repeat Prob. 9.27a for a *ground source heat pump* for which the heat source temperature is constant year-round at 48°F (9°C). Note that this problem is considerably simpler than Prob. 9.27 because the COP is constant.   (4)

9.37 A double-pipe counterflow heat exchanger transfers heat between two water streams. What is the overall heat transfer conductance U_oA_o if 300 gal/min (19 L/s) is heated in the tubes from 50 to 100°F (10 to 38°C) with 500 gal/min (31 L/s) of inlet water on the shell side at 115°F (46°C)? What are the effectiveness and LMTD?   (7)

9.38 Repeat Prob. 9.37, except use saturated steam at 220°F (105°C) as the heat source. The outlet of the steam side of the heat exchanger is saturated water at 220°F (105°C).   (7)

9.39 Repeat Prob. 9.37, but assume that the fluid being heated is 50 percent propylene glycol. Account for both the differences in specific heat and density and the difference in heat transfer coefficient for the glycol. Assume that 90 percent of the resistance to heat transfer in the heat exchanger of Prob. 2.20 is attributed to convection, equally split between the two water-to-tube surfaces.   (9)

9.40 A counterflow heat exchanger has an entering hot-water stream at 12 kg/s and 60°C and a cold stream at 15 kg/s and 8°C. What is the effectiveness if the cold-water stream leaves at 42°C? What is the overall heat transfer conductance U_oA_o?   (6)

9.41 If the heat transfer surface in Prob. 9.40 becomes fouled by a substance that decreases the overall heat transfer conductance by 18 percent (a common event if water treatment is not continuously applied), what are the effectiveness and heat transfer rate under the same entering-water conditions? By how much must the hot water temperature be increased to produce the same heat transfer rate as with the clean heat exchanger?   (7)

9.42 Suppose that the heat exchanger in Prob. 9.40 is connected by mistake as a parallel-flow heat exchanger instead of a counterflow exchanger. What will the effectiveness be if the value of U_oA_o is unaffected by the change?   (7)

9.43 Figure P9.43 is a schematic diagram of a heat recovery system, consisting of a cross-flow/counterflow heat exchanger and filters, on a commercial laundry that operates 24 h/day, 7 days/week. Hot, moist air leaves the large clothes dryer at 10,000 *standard* ft³/min (4700 L/s) and 220°F (105°C). If the effectiveness of the heat recovery unit is 0.75, what is the annual energy savings produced by the heat recovery unit? Assume that the air mass flows on both sides of the unit are equal. Use bin data for Boise, Idaho. Comment on the possibilities of freezing the heat exchanger at the lowest temperature in the bin data set.   (9)

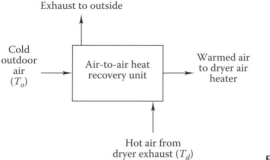

FIGURE P9.43

9.44 Work Prob. 9.43 for Nashville, Ky., and Omaha, Neb. (9)

9.45 Calculate the part-load heat transfer curve for a tube-in-tube, water-based counterflow heat exchanger in which flow is varied from 10 to 100 percent of design on the hot-water side. Refer to Fig. P9.45. Flow is constant on the cold side. Include the effect of water velocity on the heat transfer coefficient on the hot side, using equations from Chap. 2. For the purposes of this problem, assume a simple heat exchanger consisting of a 30-ft (10-m) length of nominal $\frac{3}{4}$-in (19-mm) copper pipe inserted in the center of a larger pipe, so that the flow areas of the hot and cold streams are the same. At full flow, the fluids on both sides of the heat exchanger flow at 6 ft/s (2 m/s), and the heat transfer coefficients on both sides are the same at full flow. The hot-water inlet temperature is 120°F, and the LMTD is 15°F (8°C) at full flow. Ignore the heat transfer resistance of the common pipe wall and the effect of fluid temperature on water properties. Express your results as a graph of the percent of full-flow heat transfer rate on the ordinate versus percent of hot fluid flow on the abscissa. If the results for this problem are typical of counterflow heat exchanges in general, what are the key findings regarding ϵ (PLR)? (10)

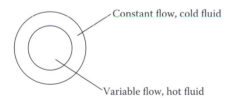

Constant flow, cold fluid

Variable flow, hot fluid   **FIGURE P9.45**

References

ASHRAE (1999). *Handbook of HVAC Applications*. American Society of Heating, Refrigerating and Air-Conditioning Engineers, Atlanta.

ASHRAE (2000). *Handbook of Systems and Equipment*. American Society of Heating, Refrigerating and Air-Conditioning Engineers, Atlanta.

ASHRAE (1989, 2001). *Handbook of Fundamentals*. American Society of Heating, Refrigerating and Air-Conditioning Engineers, Atlanta.

Borbely, A. M., and J. F. Kreider (2001). *Distributed Generation: The Power Paradigm for the New Millennium*. CRC Press, Boca Raton, FL.

Bowman, J., and R. Turton (1990). "Quick Design and Evaluation: Heat Exchangers." *Chem. Eng.*, vol. 97 (July), pp. 92–99.

Brandemuehl, M. J., S. Gabel, and I. Andresen (1992). *A Toolkit for Secondary System Energy Calculations.* American Society of Heating, Refrigerating and Air-Conditioning Engineers, Atlanta.

Dyer, David F., and G. Maples (1991). *Boiler Efficiency Improvement*, 5th ed. Boiler Efficiency Institute, Auburn University, Auburn, Ala.

Goswami, J., F. Kreith, and J. F. Kreider (2000). *Principles of Solar Engineering*, 2d ed. Taylor and Francis, New York.

Haas, J. H. (1990). "Steam Traps—Key to Process Heating." *Chem. Eng., vol.* 97 (January), pp. 151–156.

Kuehn, T. H., J. L. Threlkeld, and J. W. Ramsey (1998). *Thermal Environmental Engineering*, 3d ed., Upper Saddle River: Prentice-Hall, Inc.

Roop, J. M., and T. M. Kaarsberg (1999). "Combined Heat and Power: A Closer Look." *Proceedings of the 21st National Industrial Energy Technology Conference*, Houston, Tex., May 12.

Thamilseran, S., and J. Haberl (1995). "A Bin Method for Calculating Energy Conservation Retrofit Savings in Commercial Buildings," *Proc. 1995 ASME/JSME/JSES Intl. solar Energy Conf.*, Lahaina HI, pp. 111–124, March.

Turton R., D. Ferguson, and O. Levenspiel (1986). "Charts for the Performance and Design of Heat Exchangers." *Chem. Eng.*, vol. 93 (August), pp. 81–88.

U.S. Air Force (1978). Engineering Weather Data, AF Manual AFM 88–29, U.S. Government Printing Office, Washington, D.C.

U.S. Army (1979). *BLAST, The Building Loads Analysis and System Thermodynamics Program-Users Manual*. U.S. Army Construction Engineering Research Laboratory Report E-153.

U.S. Department of Energy. (1981). *DOE Reference Manual Version 2.1A*. Los Alamos Scientific Laboratory Report LA-7689-M, Version 2.1A. Lawrence Berkeley Laboratory, Report LBL-8706 Rev. 2.

Walton, G. (1983). *Thermal Analysis Research Program Reference Manual*. National Bureau of Standards. NBSIR 83–2655.

Walton, G. N. (1980). "A New Algorithm for Radiant Interchange in Room Loads Calculations." *ASHRAE Transactions*, vol. 86, pt. 2, pp. 190–208.

Walton, G. N. (1993). *Computer Programs for Simulation of Lighting/HVAC Interactions*. National Institute of Standards and Technology. NISTIR 5322.

Wang, X. A., and J. F. Kreider (1992). "Improved Artificial Neural Networks for Commercial Building Energy Use Prediction." *Solar Engineering—1992*, ASME, New York.

Wasserman, P. D. (1989). *Neural Computing, Theory and Practice*. Van Nostrand Reinhold, New York.

Wong, W. Y. (1989). "Safer Relief Valve Sizing." *Chem. Eng.*, vol. 96 (May), pp. 137–140.

Yokell, S. (1983). "Troubleshooting Shell and Tube Heat Exchangers." *Chem. Eng.*, vol. 90 (July), pp. 57–75.

Yuill, G. K. (1990). *An Annotated Guide to Models and Algorithms for Energy Calculations Relating to HVAC Equipment*. American Society of Heating, Refrigerating, and Air Conditioning Engineers, Inc., Atlanta.

10

Cooling Equipment

10.1 Introduction

This chapter describes the equipment used to produce a cooling effect in residential and commercial buildings. Two approaches are widely used. The first uses a closed thermodynamic cycle with the input of shaft power. The second uses heat to cause a different type of thermodynamic cycle to operate. The former is a *vapor compression* system whereas the latter is an *absorption* system.

Another method for producing a cooling effect uses the ability of low-humidity air to evaporate water in an adiabatic process, with the result that the dry-bulb temperature of the air is lowered. This process is called the *evaporative cooling process*. It is restricted to drier areas of the world using outdoor air or to systems using desiccants to dry moist outdoor air. Many buildings in desert climates can have all their sensible cooling loads provided by evaporative equipment that uses significantly less power than either the vapor compression method or the absorption method, if a water supply is available.

This chapter first reviews the details of the vapor compression and absorption cycles. Following these analytical sections, the design of equipment used to produce mechanical cooling is described. Components over which the designer has little control (e.g., the internal design of a chiller or expansion device) are not discussed in detail in this book. Finally, the evaporative cooling cycle and its equipment are covered.

10.2 Rankine Refrigeration Cycle

In Sec. 3.4 we briefly discussed the ideal vapor compression refrigeration cycle. There we described it as a Rankine heat engine cycle operated in reverse. In this section we discuss the real *vapor compression* (VC) refrigeration cycle in detail.

Figure 10.1 shows the ideal cycle schematically on a *p-h* diagram. Recall from Chap. 3 that these thermodynamic coordinates are particularly useful since three of the four processes appear on it as straight lines. We call this process *ideal* because pressure and temperature drops are ignored, superheating in the evaporator and subcooling in the condenser that are present in real equipment are ignored, and compressor inefficiencies are omitted. These factors are all included in the treatment in Sec. 10.3.

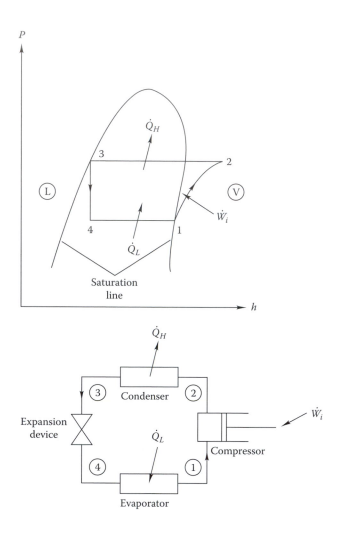

FIGURE 10.1
Ideal Rankine refrigeration cycle plotted on *p-h* diagram and corresponding schematic arrangement of mechanical equipment.

The four steps in the ideal VC cycle are

1. Isentropic compression (1-2)
2. Constant-pressure refrigerant condensation (2-3)
3. Throttling in expansion valve (or other device) (3-4)
4. Constant-pressure heat input in the evaporator (4-1)

Point 1 in the process represents the inlet to the compressor. Compressor work $\dot{W}_i$ raises the pressure and temperature of the working fluid to a level—point 2—at which heat $\dot{Q}_H$ can be readily rejected to the environment. The heat rejection step results in a phase change of the working fluid from superheated vapor to saturated liquid. The liquid is throttled from the high-side pressure to the low-side pressure in an isenthalpic process. At the end of this irreversible process, the refrigerant is a saturated mixture of liquid and vapor at a temperature selected to enable effective transfer of heat from the space to be cooled. This heat input $\dot{Q}_L$ causes the liquid phase to evaporate completely, and the fluid is in a state of saturated vapor at the compressor inlet, by the definition of the ideal cycle. Heat transferred in the condensation and evaporation steps is proportional to the lengths of the horizontal lines in Fig. 10.1.

The most common measure of refrigeration cycle performance is the coefficient of performance (COP). For the ideal VC cycle, it is given by

$$\text{COP} = \frac{\dot{Q}_L}{\dot{W}_i} = \frac{h_1 - h_4}{h_2 - h_1} \tag{10.1}$$

The enthalpies in Eq. (10.1) are identified with the subscripts in Fig. 10.1. Example 10.1 illustrates how cycle calculations are done for the ideal cycle.

Example 10.1: Ideal Vapor Compression Cycle

A refrigeration cycle uses refrigerant 12 (also denoted by R12) and operates between a low-side pressure of 0.15 MPa (21.8 psia) and a high-side pressure of 1.0 MPa (145 psia). The refrigerant mass flow is 0.050 kg/s (396 lb/h). Find the cooling effect, work input, and COP of this machine.

Given: $p_2 = p_3 = 1.0$ MPa

$p_4 = p_1 = 0.15$ MPa

$\dot{m} = 0.050$ kg/s

Figure: See Fig. 10.1.

Assumptions: All processes are steady. The cycle is ideal as defined above.

Find: $\dot{W}_i$, $\dot{Q}_L$, COP

Lookup values: From the R12 chart in the HCB software thermodynamic properties are read, as shown in Table 10.1. Italicized values are given: the remainder are read from the thermodynamic chart. Values in parentheses are not needed for the solution but are included as reference data.

SOLUTION

Part of the solution is embodied in Table 10.1. One begins at point 1 where the enthalpy and entropy are first found. The next step uses the entropy from point 1 and the known high-side pressure to find the properties at 2. Since point 3 represents saturated liquid at p_2, the properties at 3 can be read from the saturated-liquid line in the property chart. Finally the conditions at 4 can be found since the pressure and enthalpy are both known. (The entropy values at points 3 and 4 are

TABLE 10.1

R12 Properties for Example 10.1

Point	p, MPa	T, °C	h, kJ/kg	s, kJ/(kg · K)
1	0.15	(−20)	343	1.57
2	1.0	(52)	377	1.57
3	1.0	(42)	242	(1.14)
4	0.15	(−20)	242	(1.17)

not needed for the cycle calculation and need not be determined unless one is concerned about the irreversibility inherent in the throttling process 3-4.)

The heat removal is

$$\dot{Q}_L = \dot{m}(h_1 - h_4) = (0.050\,\text{kg/s})[(343 - 242)\,\text{kJ/kg}] = 5.1\,\text{kW}$$

The power input is

$$\dot{W}_i = (0.050\,\text{kg/s})[(377 - 343)\,\text{kJ/kg}] = 1.7\,\text{kW}$$

Finally, the COP is the ratio of $\dot{Q}_L / \dot{W}_i$:

$$\text{COP} = \frac{5.1}{1.7} = 3$$

COMMENTS

The careful reader will observe that the throttling process 3-4 represents an irreversible process in the ideal VC cycle as indicated by the entropy increase. This irreversibility is forced on us by the difficulties of trying to build a wet-mixture turbine that could survive in a refrigeration machine. However, if such a turbine could be made, some of the power input to the compression process could be avoided, since it would be provided by this hypothetical turbine.

Thermodynamic properties can be determined more accurately and conveniently by using the HCB software accompanying this text. You are encouraged to try the software for the problems in this chapter. The software makes it particularly easy to compare the operating characteristics of various refrigerants. An interesting software exercise is to compare the evaporator and condenser pressures, and the refrigerant flow rates for fixed evaporator and condenser temperatures. Even for the ideal cycle the differences are striking. For five common refrigerants used between the same two operating temperatures, $-15°C$ and $30°C$, the software shows these quite different characteristics:

Refrigerant	Evap. Pressure, kPa	Cond. Pressure, kPa	Flow, kg/(s · kW)
R12	183	745	0.033
R22	296	1192	0.024
R114	47	250	0.034
R134a	164	771	0.025
R502	349	1319	0.033

These results are significant. Flow rates, and therefore compressor sizes, vary by as much as 40 percent. Pressures show a range significant enough to affect the thickness of condenser and evaporator tubing walls. Although not shown in the table, the COP values for these five refrigerants vary by only 5 percent from the average of 3.75 because the primary effect is the specified operating temperatures, which are all the same. Note that the R114 evaporator is subatmospheric.

10.2.1 Real Cycle Analysis

Figure 10.2 is a schematic diagram of a real air conditioning system showing the actual equipment used—compressor, condenser, expansion valve, and evaporator—and typical operating conditions for refrigerant 22 (see below). We discuss each of these components

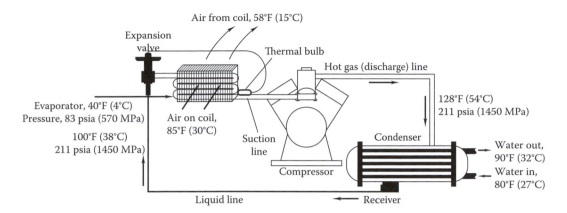

FIGURE 10.2
VC cycle equipment with typical R22 operating temperatures and pressures. A direct expansion (DX) evaporator is used. The state points are based on a compressor efficiency of 85 percent and an evaporator outlet superheat of 9°F (5°C). Pressure drops are not noted.

separately later in the chapter. In this section, we will learn what effect the use of *real components* has on the efficiency of the VC cycle.

Each mechanical component in Fig. 10.2 has thermodynamic irreversibilities associated with it. These include

1. *Compressor*—friction losses, heat losses, or heat gains
2. *Condenser*—friction losses, subcooling to ensure pure liquid at throttling valve inlet
3. *Evaporator*—friction losses, superheating to ensure pure vapor at compressor inlet (the thermal bulb shown in Fig. 10.2 controls the expansion valve flow and hence the superheat)
4. *System losses*—pressure drops, heat gains, and heat losses in refrigerant lines; compressor shaft friction

These effects distort the ideal cycle shown in Fig. 10.1 to the real cycle shown in Fig. 10.3. Each of the above irreversibilities except one causes inefficiencies. The one exception is the possible *cooling of refrigerant in the compressor*. If cooled, the refrigerant specific volume will be reduced, and the work, too, will be less. For this reason, one may find cooling fins on compressors in small VC units.

Analysis of a real cycle requires considerably more data than for the ideal cycle. To complete the analysis, the following are required:

- Compressor efficiency
- Liquid and vapor line heat loss coefficients
- Liquid and vapor line pressure drop values
- Compressor heat loss or gain rate
- Condenser and evaporator pressure drops
- Amount of refrigerant superheating at evaporator outlet
- Amount of refrigerant subcooling at condenser outlet

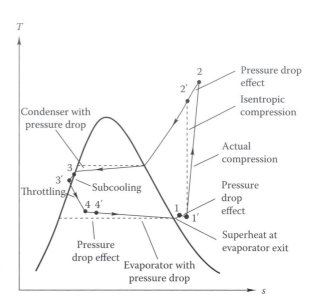

FIGURE 10.3
The *T-s* diagram of actual VC cycle showing friction losses, compressor inefficiencies, enthalpy changes, superheating, and subcooling in a real cycle. Line segments such as 1-1′ or 2-2′ account for piping losses, and sloped evaporator and condenser paths include pressure drop effects.

The most significant factors of those listed are the compressor efficiency, compressor suction line heat gains and pressure losses, and condenser and evaporator pressure drops. It is beyond the scope of this book to examine the real VC process in such detail, but the same approach used in Example 10.1 can be used if all the data listed immediately above are known.

10.2.2 Refrigerants

The ASHRAE *Handbook of Fundamentals* (ASHRAE, 1989, 2001) lists dozens of refrigerants and their thermodynamic properties. These include compounds ranging from ammonia, air, and water vapor to propane, sulfur dioxide, and dichlorodifluoromethane. Only a few of these are practical for building systems due to toxicity restrictions, high cost, flammability, inappropriate thermodynamic or transport properties, or environmental concerns.

The key requirement of a refrigerant is that it have properties that match the needs of building cooling systems. As a compressed vapor, a refrigerant must not have excessive pressures at temperatures needed for heat rejection to the environment on a hot summer day. It is desirable to have the evaporator pressure above atmospheric pressure to avoid refrigerant contamination by atmospheric moisture due to leaks. The evaporator temperature at this pressure should be near 40°F (5°C) to produce cooled air in direct-expansion equipment at temperatures in the range of 50 to 55°F (10 to 13°C); chillers and ice makers will operate at lower temperatures. With these thermal restrictions plus the concerns listed earlier, it is not surprising that only a few inexpensive refrigerants exist that meet all criteria.

During the early 1990s, 90 percent of all refrigerants used consisted of R11, R12, R22, and R502 (an azeotropic mixture of R22 and R115; an azeotrope is a mixture that cannot be separated by distillation and that has properties different from either constituent). These chlorofluorocarbons (CFCs)[1] are chemically stable for terrestrial applications, inexpensive,

---

[1] Refrigerant 22 is an HCFC (hydrogenated CFC) in which the carbon atoms are not fully populated with halogens. It is less harmful in the upper atmosphere than true CFCs.

and nontoxic compounds. However, they have been shown to destroy the stratospheric ozone layer which protects us from ultraviolet (UV) solar radiation. Hence, damaging UV flux levels are predicted to increase at the earth's surface in the future. To avoid these serious developments, the use of these common CFCs is being phased out worldwide.

Refrigerants 11 and 12 are the most destructive to the ozone layer and have been the first to be phased out. They will be replaced by R22, R123, and R134a. Some of these replacements, in turn, are likely to be replaced by more benign compounds in the future. Some of the compounds proposed for the first round of replacements have problems, including toxicity, higher cost, reduced equipment capacity, and incompatibility with existing compressor seal designs and lubricating oils. Thermodynamic properties for several common refrigerants are given in charts and tables in Chap. 3 and in the CD-ROM appendix. A complete set of properties is also included in the HCB software.

The method of identifying CFC refrigerants with the numbering system used above is based on the number of fluorine atoms (right digit of number), number of hydrogen atoms plus 1 (center digit), and number of carbon atoms minus 1 (left digit, unless zero, which is suppressed). For example, the chemical formula for dichlorodi-fluoromethane is CCl_2F_2. Therefore, the first digit is zero (suppressed); the center digit is 1 since there are no hydrogen atoms. The right digit is 2 since there are two fluorine atoms. The symbol is R12. Inorganic compounds are numbered by adding 700 to their molecular weight. Therefore, ammonia (MW = 17) is numbered R717. Mixtures of refrigerants use a serial numbering system beginning with 500 are not related to the chemical composition.

10.3 Absorption Cycle

Absorption cooling equipment replaces the mechanical power input to the compressor of VC machines with a small liquid pump and heat addition. The absorption and desorption of the refrigerant from the pumped liquid are achieved by heat transfer alone. Therefore, the use of electric power is minimal with absorption equipment. This approach has considerable merit when a source of heat between 100°C and 200°C is available at competitive cost. In this section we discuss the ideal single-stage *absorption cooling* (AC) cycle. Nearly all modern absorption machines use two- or three-stage cycles. In multiple-stage equipment, generators at different temperatures, heated by different sources, are used to improve the COP.

Figure 10.4 is a schematic diagram of the absorption cycle. To the left of the dashed line, this system is identical to the VC cycle in Fig. 10.1. In the AC cycle, refrigerant pressurization is achieved by dissolving the refrigerant in a liquid absorbent, pressurizing the solution by using a liquid pump that requires very little work input, and finally driving the low-boiling refrigerant from the high-pressure, refrigerant-rich absorbent in the generator by the addition of heat. The two most common refrigerant-absorbent pairs are water–lithium bromide (LiBr) and ammonia-water. In the former, the refrigerant is low-pressure water vapor; in the latter, it is ammonia. The LiBr system is simpler since it does not require a rectifying column to guarantee complete separation of the refrigerant and absorbent. But LiBr has two drawbacks, given the use of water as the refrigerant: The evaporator temperature cannot be used much below 40°F (5°C), and care must be taken with this system to operate the generator at temperatures sufficiently high to avoid crystallization of LiBr salts.

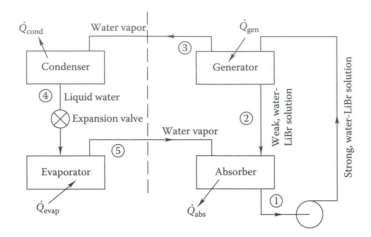

FIGURE 10.4

Schematic diagram of simple LiBr absorption cycle. To the left of the vertical dashed line, the system is the same as the VC cycle.

After the refrigerant is released at high pressure, the weak absorbent returns to the absorber where refrigerant is again reabsorbed into the absorbent. The absorption of refrigerant into absorbent is an exothermic reaction. Hence heat is removed from the AC equipment in two places—the condenser (as with VC equipment) and the absorber.

Performance of the AC system is measured by its COP. However, the principal input to this process is not shaft power, but heat. Hence, the COP is given by

$$\mathrm{COP_{AC}} = \frac{\dot{Q}_{\mathrm{evap}}}{\dot{Q}_{\mathrm{gen}}} \tag{10.2}$$

where

$\dot{Q}_{\mathrm{evap}}$ = heat removal rate at evaporator, Btu/h (kW)

$\dot{Q}_{\mathrm{gen}}$ = heat supply rate to generator, Btu/h (kW)

The COP for LiBr systems ranges from 0.5 to slightly above 1.2 for the most sophisticated, steam-fired systems.

Ammonia absorption systems are older but less widely used than LiBr systems. They have the advantage that the evaporator operates at pressures above atmospheric and that the evaporator can operate at temperatures below those possible with the LiBr system (below 0°C). However, extra components are needed, as noted above, and due to the higher operating pressures, heavier construction is necessary. The toxicity of ammonia is also a concern. The COPs of both of these common absorption systems are about the same.

10.3.1 Analysis of Lithium Bromide Cycle

The analysis of the absorption cycle is somewhat different from that for the VC cycle, which uses a pure substance. Instead, a homogeneous mixture of absorbent and refrigerant is the working fluid. The concentration of the refrigerant is expressed as the *mass fraction X* [lb LiBr/lb mixture (kg LiBr/kg mixture)]. This is one of the thermodynamic state variables used to conduct cycle analyses for this equipment.

Figure 10.5 is an example of the charts one can use to find the enthalpy of the mixture as a function of concentration and pressure (or temperature). Plotted along the abscissa is the mass fraction X or concentration of LiBr in solution. The ordinate is the enthalpy h of the solution at the value of either temperature or pressure read from the curves in the figure. The chart applies for saturation and subcooled conditions.

To solve LiBr cycle problems, one often knows a pressure and temperature condition but not the mass fraction or enthalpy. In those cases, find the intersection of the known pressure and temperature lines in the figure, and then read the corresponding values of enthalpy (needed for energy balance calculations) and concentration (needed for mass balance calculations) from the axes of Fig. 10.5. Example 10.2 illustrates the method.

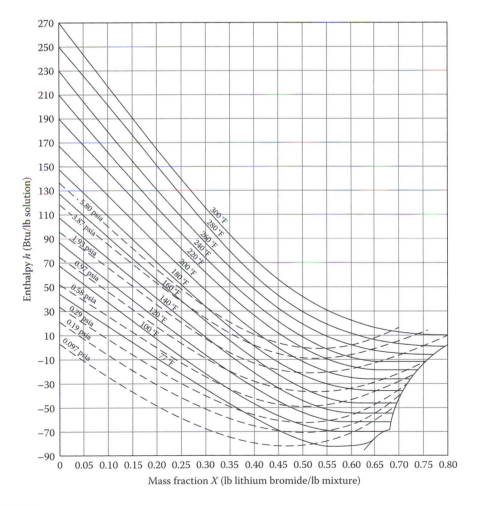

FIGURE 10.5

The LiBr enthalpy-concentration diagram. 1 psia = 6.89 kPa; 1 Btu/lb = 2.32 kJ/kg. (Courtesy of Institute of Gas Technology, Chicago, IL. With permission.)

Example 10.2: Lithium Bromide Absorption Cycle

A lithium bromide chiller operates between a condensing temperature of 104°F (40°C) and an evaporator temperature of 50°F (10°C). Heat is added to the generator at 212°F (100°C) and removed from the absorber at 86°F (30°C). If the pump flow rate is 4800 lb_m/h (0.605 kg/s), what are the COP and the heat rates at each component of the cycle—generator, absorber, condenser, and evaporator?

Given: $T_{gen} = 212°F$, $T_{con} = 104°F$, $T_{evap} = 50°F$, $T_{abs} = 86°F$ (italicized in Table 10.2) $\dot{m}_{pump} = 4800$ lb/h (0.61 kg/s)

Figure: See Fig. 10.4.

Assumptions: All components operate at equilibrium. State points 1 to 5 are at saturation. Ignore all pressure and heat losses in piping.

Find: $\dot{Q}_{cond}$, $\dot{Q}_{evap}$, $\dot{Q}_{abs}$, $\dot{Q}_{gen}$, COP

Solution

The solution consists of two parts. First, one finds thermodynamic properties at all state points 1 to 5. Second, one uses mass and energy balance equations to find the heat and mass flow terms.

The table is completed as follows. First, the known temperatures in the second column are entered. The given temperatures within each component are assumed to be the temperatures of fluid streams *exiting* that component.

Second, pressures are entered in the table. The condenser pressure p_4 is set by the known condenser saturation temperature and is interpolated from the CD-ROM LiBr, property for LiBr tables. This pressure is then the pressure for all high-side points 2, 3, and 4. Likewise, the evaporator pressure p_5 is set by the known evaporator saturation temperature. This pressure read from the CD-ROM LiBr, property for LiBr tables is the low-side pressure for points 1 and 5. Note that pressures in the LiBr cycle are subatmospheric and small.

Third, find the LiBr mass fractions X. Of course, for states 3, 4, and 5 consisting of pure water vapor, the mass fraction is zero. For state 1, the intersection of the known pressure and temperature lines is used to find X_1 by using Fig. 10.5. At the same time, the enthalpy can be read from the figure and entered into the table. Finally, the mass fraction and enthalpy at point 2 are found in the same way. The enthalpies at points 3, 4, and 5 are read from the the CD-ROM LiBr, property for LiBr tables.

At this point, a mass balance is written on the generator, and then each heat flow term can be found. The total mass balance and the LiBr mass balance written on the generator are

$$\dot{m}_2 + \dot{m}_3 = \dot{m}_1 = 4800\,lb_m/h \tag{10.3}$$

$$\dot{m}_1 X_1 = \dot{m}_2 X_2 \tag{10.4}$$

TABLE 10.2

Thermodynamic Properties for Example 10.2

State	Temp., °F	p, psia	X	h, Btu/lb
1	*86*	0.178	0.50	−71*
2	212	1.070	0.67	−24*
3 (superheated vapor)	212	1.070	0.0	1150.5*
4 (saturated liquid)	*104*	1.070	0.0	72*
5 (saturated vapor)	*50*	0.178	0.0	1083.3*

* Read from Fig. 10.5.

From these expressions, the two unknown mass flow rates can be found:

$$\dot{m}_2 = \dot{m}_1 \left(\frac{X_1}{X_2}\right) = 4800 \left(\frac{0.50}{0.67}\right) = 3582 \, lb_m/h \tag{10.5}$$

$$\dot{m}_3 = \dot{m}_1 - \dot{m}_2 = 4800 - 3582 = 1218 \, lb/h = \dot{m}_4 = \dot{m}_5 \tag{10.6}$$

The four heat rate terms can now be found from energy balances on the four components of the AC cycle. The arrows in Fig. 10.4 show the direction of the heat flow:

$$\dot{Q}_{gen} = \dot{m}_3 h_3 + \dot{m}_2 h_2 - \dot{m}_1 h_1$$
$$= 1218(1150.5) + 3582(-24) - 4800(-71) = 1656 \, kBtu/h \tag{10.7}$$

$$\dot{Q}_{cond} = \dot{m}_3 (h_3 - h_4)$$
$$= 1218(1150.5 - 72) = 1314 \, kBtu/h \tag{10.8}$$

$$\dot{Q}_{abs} = \dot{m}_2 h_2 + \dot{m}_5 h_5 - \dot{m}_1 h_1$$
$$= 3582(-24) + 1218(1083.3) - 4800(-71) = 1574 \, kBtu/h \tag{10.9}$$

$$\dot{Q}_{evap} = \dot{m}_5 (h_5 - h_4)$$
$$= 1218(1083.3 - 72) = 1232 \, kBtu/h$$

As a check, by the first law of thermodynamics, since pumping power is negligible,

$$\dot{Q}_{evap} = -\dot{Q}_{gen} + \dot{Q}_{abs} + \dot{Q}_{cond} = -1656 + 1574 + 1314 = 1232 \, kBtu/h$$

Finally, the COP, defined as the ratio of cooling produced in the evaporator to the heat added in the generator, is

$$COP = \frac{\dot{Q}_{evap}}{\dot{Q}_{gen}} = \frac{1232 \, kBtu/h}{1656 \, kBtu/h} = 0.74$$

COMMENTS

The solution to AC cycle problems involves the use of two sets of thermodynamic tables or charts—those for pure water and those for LiBr. Although the reference states for various LiBr charts may vary from those in Fig. 10.5, the energy flows will be the same. Figure 10.5 has the advantage that all thermodynamic variables are included in a single LiBr chart rather than two charts, as in other references.

Although the COP of an absorption cycle is much lower than that for a VC cycle, remember that the heat input to a power plant that produces the electricity to power VC machines is about 3 times the power plant's electric output. Therefore, the ratio of cooling effect to *thermal* energy input (to the power source) is not much different for VC and AC cycles.

This section has described the basic AC system. A number of refinements are often used to improve performance. For example, a heat exchanger is often used across the LiBr streams entering and leaving the generator. By extracting heat from the generator exit stream and adding it to the inlet stream, the amount of external heat required for the generator can be reduced. Higher generator temperatures can improve the COP within limits. Two-stage systems with interstage heat exchange can have better efficiency than single-stage systems. These refinements are discussed in ASHRAE (2001, 2000, 1989).

10.4 Mechanical Cooling Equipment—Chillers

A *chiller* is an assembly of equipment used to produce chilled water for cooling of spaces within buildings. The equipment comprising a chiller depends on whether the AC or VC cycle is used. At a minimum, an evaporator, condenser, expansion device, and compressor (for VC cycle) or pump-generator-absorber subsystem (for the AC cycle) is involved. Heat exchangers between the heat rejection subsystem and cooling distribution subsystem are also included in the chiller assembly. Figure 10.6 is a simple diagram of a liquid chiller based on the VC cycle. In this section we describe the *equipment* that makes up a chiller. This includes compressors of various types and heat rejection equipment. Unitary systems in which all components in Fig. 10.6 are included in one assembly are also discussed. Controls are left for Chap. 12.

In USCS units, the rate of refrigeration is often expressed as *tons of cooling*. This unit of cooling effect is equivalent to 12,000 Btu/h (3.5 kW). It was originally defined on the basis of melting of 1 ton (2000 lb$_m$) of ice (heat of fusion 144 Btu/lb) per day (24 h), giving a cooling rate equal to 12,000 Btu/h (3515 W).

10.4.1 Compressors

Compressors of several types are used to compress gases (not liquids) in VC cycles in building equipment. In this section, we discuss the three most common types of compressors—centrifugal, reciprocating, and screw. Prior to that, we will describe the generic compressor from a thermodynamic point of view. Rotary and scroll devices are also described.

The work input for the open-cycle polytropic compression cycle ($pv^n = \text{const}$) shown in Fig. 10.7 is given by

$$
w = \int_i^0 v\,dp = \frac{np_iv_i}{n-1}\left[\left(\frac{p_o}{p_i}\right)^{(n-1)/n} - 1\right]
\tag{10.10}
$$

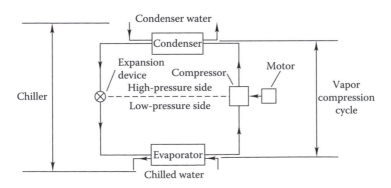

FIGURE 10.6
Schematic diagram showing essential components of basic liquid chiller and relation to vapor compression cycle.

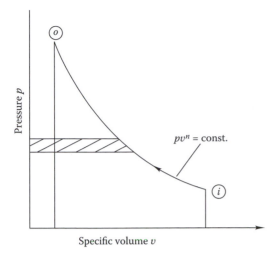

Pressure p

$pv^n = $ const.

Specific volume v

FIGURE 10.7
Polytropic compression process p-v diagram used for chiller compressor work calculation.

where

$w = $ work input, ft $\cdot$ lb$_f$/lb$_m$ (kJ/kg)

$n = $ polytropic exponent

$p_i, p_o = $ compressor inlet and outlet pressures, lb$_f$/ft^2 (Pa)

$v_i = $ fluid inlet specific volume, ft^3/lb$_m$ (m^3/kg)

This form of the polytropic work expression is the most useful of the several available since the inlet and outlet pressures are most often known in compressor calculations. Refrigerants cannot always be well represented by the ideal gas law. Therefore, property tables must be used. The polytropic exponent, in general, will not be the ratio of specific heats, as discussed in Chap. 3. Example 10.3, indicates how one can find the efficiency of a compressor defined as the work for the ideal compressor divided by the work input for the actual compressor.

Example 10.3: Vapor Compressor Analysis

A compressor using R12 operates between an inlet condition of 45 psia (310 kPa) and a discharge pressure of 200 psia (1379 kPa). Refrigerant enters the compressor at 45°F (7.2°C). The polytropic exponent for the real process is 1.35. Find the compressor efficiency compared to an isentropic compressor.

Given: $p_i = 45$ psia $= 6480$ lb/ft^2 (0.31 MPa)

$p_o = 200$ psia (1379 kPa)

$T_i = 45°F$ (7.2°C)

$n = 1.35$

Figure: See Fig. 10.7.

Assumptions: The polytropic model is sufficiently accurate.

Find: η_{comp}

Lookup values: From the HCB software interpolate at the inlet condition (at 45°F and 45 psia) to find the specific volume $v_i = 0.9$ ft^3/lb$_m$.

SOLUTION

The actual work input is found from Eq. (10.10)

$$W_{actual} = 1.35\left(6480\;\frac{lb_f}{ft^2}\right)\left(\frac{0.90\;ft^3/lb_m}{1.35-1}\right)\left[\left(\frac{200}{45}\right)^{(1.35-1)/1.35}-1\right]$$
$$= 10,620\;(ft \cdot lb_f)/lb_m\;(31.7\;kJ/kg)$$

The ideal work is based on an isentropic compression. The enthalpy (and entropy) at the start and end of this process can be read from the HCB software:

$$h_i = 82.6\;Btu/lb_m\;[s_i = 0.170\;Btu/(lb_m \cdot °R)]$$

From the entropy and the outlet pressure, one can find the ideal outlet enthalpy

$$h_o = 94.3\;Btu/lb_m$$

The ideal (isentropic) compressor work is

$$W_{ideal} = h_0 - h_i = [(94.3 - 82.6)\;Btu/lb_m][778\;(ft \cdot lb_f)/Btu]$$
$$= 9103\;(ft \cdot lb_f)/lb_m\;(27.2\;kJ/kg)$$

Finally, the compressor efficiency is the ratio of ideal to actual work:

$$\eta_{comp} = \frac{9,103}{10,620} = 0.86$$

COMMENTS

The *isentropic* efficiency calculated above indicates that fundamental improvements can be made in the compression *process* itself. The efficiency of the compressor *machinery* is also affected by other losses such as flow losses in the centrifugal compressor blading or in the reciprocating compressor valves, driveshaft losses, and piping pressure losses.

10.4.1.1 Reciprocating Compressors

The most common refrigerant compressor is the reciprocating compressor. It is used in cooling systems ranging from small residential units on the order of a few hundreds of Btu per hour (tenths of a kilowatt) output to large units used in commercial buildings rated at hundreds of tons (hundreds of kilowatts) of cooling. Fig. 10.8a shows the schematic arrangement used in each cylinder. The inlet refrigerant stream enters through a mechanical valve during the suction stroke. At the end of the stroke, the valve closes and the compression begins. Near top dead center, the outlet valve opens, allowing the compressed refrigerant to enter the outlet manifold. Small pressure drops occur across each valve, as indicated by the dashed horizontal lines in the figure. Note that the abscissa of Fig. 10.8b is *extensive* volume. The volume changes shown are due to the piston's movement. The specific volume varies with mass flow into and from the cylinder volume.

Figure 10.8b is idealized to the extent that the intake (1-2) and exhaust (3-4) processes are shown to be constant-pressure. Process 4-1 is an expansion of the compressed

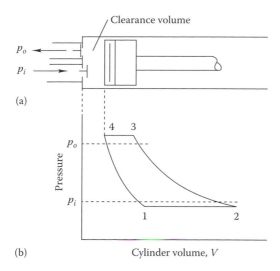

p_o

p_i

(a)

4   3

p_o

Pressure

p_i

1   2

(b)   Cylinder volume, V

Clearance volume

FIGURE 10.8
Reciprocating chiller: (a) piston-and-valve arrangement;
(b) corresponding p-V diagram.

refrigerant remaining in the *clearance volume* $V_{cl} = V_4$ after the exhaust valve closes. This refrigerant must be expanded to a sufficiently low pressure that refrigerant will flow through the inlet valve. The expanded refrigerant is then mixed with inlet refrigerant at a different thermodynamic condition. Process 2-3 is the poly-tropic compression described in Sec. 10.3. Hot, compressed gases may undergo heat transfer during the exhaust process 3-4.

A common measure of the effectiveness of a compressor is the *volumetric efficiency*, defined as the ratio of actual mass of vapor compressed to that which would be compressed if the cylinder were filled with refrigerant at the inlet condition:

$$\eta_{vol} = \dot{m}_{ref} \frac{v_i}{\dot{V}_{swept}} \qquad (10.11)$$

where

η_{vol} = volumetric efficiency
$\dot{m}_{ref}$ = refrigerant mass flow, lb_m/h (kg/s)
v_i = inlet refrigerant specific volume, ft^3/lb_m (m^3/kg)
$\dot{V}_{swept}$ = rate at which cylinder volume is swept by piston, ft^3/h (m^3/s)

This theoretical volumetric efficiency depends on the clearance volume, cylinder displacement, and inlet and exhaust conditions. The refrigerant used also has an effect on the efficiency value. For a well-designed compressor matched with its refrigerant, theoretical volumetric efficiencies will exceed 90 percent. Actual volumetric efficiencies including piston ring and valve leakage, and thermal effects can be 10 percent to 15 percent lower, depending on the pressure ratio and mechanical details. The volumetric efficiency is a measure of the effectiveness of a compressor as a refrigerant "mover," while the isentropic compressor efficiency is a measure of the thermodynamic efficacy of the compression process itself. The part-load performance of reciprocating chillers at various speeds is better than that for some other designs, such as centrifugal compressors, since the effect of rotational speed on the various sources of inefficiency is relatively smaller.

The various efficiencies just described can be combined to find the COP of an actual VC cycle by using the following expression (terms have been described previously):

$$COP = \frac{\dot{Q}_L}{[(\Delta h)_{s=const}/\eta_{comp}](\eta_{vol}\dot{V}_{swept}/v_i)} \tag{10.12}$$

Actual compressor capacity data can be presented as shown in Fig. 10.9 (called a compressor map). The lower family of curves represents the power input for R22 machines at various evaporator and condensing temperatures. The upper curves represent the cooling capacity produced at the same set of conditions applying to the lower family of curves. The ratio of the two ordinates is the COP (taking proper account of units for the USCS calculation). Capacity increases with increasing evaporator temperature and with decreasing condenser temperature. Power input increases with increasing evaporator temperature or condensing temperature.

Example 10.4: Actual Reciprocating Compressor Performance

Find the capacity and COP of a four-cylinder, R22 compressor operating at 1725 r/min (180.6 rad/s). The bore of the compressor is 60.3 mm (2.38 in), and the stroke is 44.4 mm (1.75 in). The evaporator temperature is 5°C (41°F), and the condensing temperature is 47.5°C (117.5°F), a case included in Fig. 10.9.

   Given: Compressor physical data
   $T_{cond} = 47.5°C$ (117.5°F)
   $T_{evap} = 5°C$ (41°F)
   Figure: See Fig. 10.9.
   Find: COP, $\dot{Q}_{cool}$

SOLUTION

Fig. 10.9 can be used directly since it applies to a compressor of the exact type specified in the problem statement. Along the right ordinate at 47.5°C condensing and 5°C evaporating temperature (abscissa at top of figure), the input power is read as

$$\dot{W}_{in} = 13\,kW$$

From the upper curves the output is read:

$$\dot{Q}_{cool} = 42\,kW\ (143,350\,Btu/h)$$

Finally, the COP is the ratio of these two values:

$$COP = \frac{42\,kW}{13\,kW} = 3.2$$

COMMENTS

To solve problems of this type for other chillers, a figure similar to Fig. 10.9 is needed. Alternatively, a table of values is often supplied by the compressor (or chiller) manufacturer.

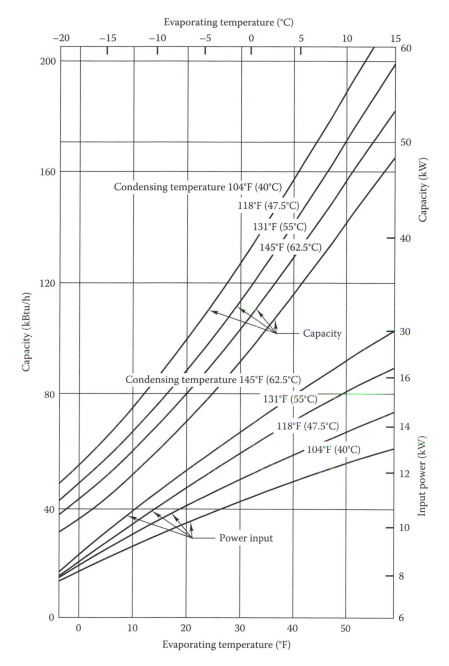

FIGURE 10.9
Example of performance "map" for reciprocating chiller; based on R22, 10°F subcooling, 20°F superheat, and 1725 r/min compressor speed. (Courtesy of ASHRAE, *Handbook of Systems and Equipment*, American Society of Heating, Refrigerating and Air-Conditioning Engineers, Atlanta, GA, 2000. With permission.)

Two configurations of reciprocating compressors are the most common. *Hermetic* compressors consist of a motor and compressor sealed in a single unit using a common driveshaft. The motor is exposed to the refrigerant; therefore, heat produced by motor inefficiencies is added to the circulating refrigerant, increasing heat rejection requirements.

However, the motors themselves in hermetic compressors may be more efficient due to better cooling. Chillers of this type using CFCs may be difficult to retrofit with new refrigerants having reduced ozone-layer impact. *Open* compressors use a separate external motor to drive the compressor. The drive can be by direct coupling, belts, or gears. These are easier to repair in the field than are hermetic compressors. Open compressors are able to incorporate more cylinders and higher outputs than hermetic units.

Control of the capacity of reciprocating compressors can be achieved by reducing the motor drive speed or by unloading the compressor by holding the intake valves open on part of the cylinders of a multicylinder device. The most common method uses mechanical or electrical unloading. Cylinders are unloaded, depending on the compressor design, in such a way that rotational imbalance is not induced.

10.4.1.2 Centrifugal Compressors

Centrifugal compressors are used when larger capacities than those offered by reciprocating compressors are needed. These compressors can produce between 200 kW and 10 MW of cooling output. The design concept is at least 80 years old and is similar to that of a centrifugal pump. Fluid to be compressed enters at the center of the rotor and flows to the periphery of the rotor under the centrifugal force produced by backward-curved blading in the rotor. Several stages of compression are common in one machine. Overall compression efficiencies of 70 to 80 percent are normal. Figure 10.10a is a cutaway drawing of a large centrifugal chiller.

Figure 10.10b shows one stage of a centrifugal compressor that rotates counterclockwise. The shaft work input is the product of the torque and rotational speed:

$$\dot{W}_i = T\omega = \dot{m}_{\text{ref}}(u_{to}r_o)\omega \tag{10.13}$$

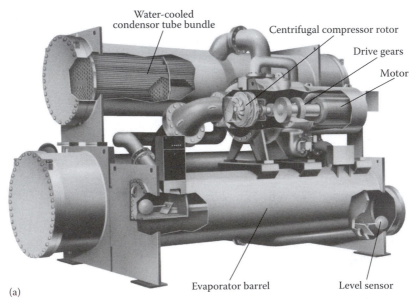

(a)

FIGURE 10.10
(a) Centrifugal chiller cutaway drawing. (Courtesy of United Technologies/Carrier, Farmington, CT. With permission.)

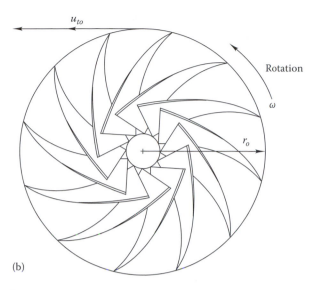

(b)

FIGURE 10.10 (continued)
(b) Centrifugal chiller rotor velocity diagram.

where
$\dot{W}_i$ = shaft power input, $(\text{ft} \cdot \text{lb}_f)/s$ (W)
T = shaft torque, $\text{lb}_f \cdot \text{ft}$ (N · m)
ω = angular velocity, s^{-1}
$\dot{m}_{ref}$ = refrigerant flow rate, lb_m/s (kg/s)
u_{to} = fluid exit tangential velocity, ft/s (m/s)
r_o = rotor outer radius, ft (m)

[For USCS units, the conversion constant g_c must be used in Eq. (10.13) as a divisor of the mass flow rate.]

With proper aerodynamic design, the refrigerant exit tangential velocity from the blading is little different from the rotor tip velocity. Therefore,

$$u_{to} = \omega r_o \tag{10.14}$$

Using this result in Eq. (10.13), we have

$$\dot{W}_i = \dot{m}_{ref} u_{to}^2 \tag{10.15}$$

By the first law of thermodynamics, one can also find the required compressor input power. All terms drop out of Eq. (3.19) except the enthalpy terms, with the result that

$$\dot{W}_i = \dot{m}_{ref}(h_o - h_i) \tag{10.16}$$

Equating the two expressions provides the following result for velocity needed to produce a given enthalpy increase. The required enthalpy increase is known from a cycle analysis such as that performed in Sec. 10.2.1. This result is approximate due to idealizations made above:

$$u_{to} = \sqrt{h_o - h_i} \tag{10.17}$$

Example 10.5: Rotational Speed for Centrifugal Chiller

What is the diameter of a centrifugal compressor rotor needed to isentropically compress R12 from a saturated suction temperature of 5°C (41°F) to a condensing temperature of 50°C (122°F) and a pressure of 1.0 MPa (145 psia)? The compressor rotates at 1800 r/min (188 rad/s) at design conditions.

Given: $\omega = (1800 \text{ r/min})(2\pi \text{ rad/r})(1 \text{ min/60 s})$

$\qquad = 188 \text{ s}^{-1}$

$T_i = 5°C \ T_o = 50°C \ p_o = 1.0 \text{ Mpa}$

Assumptions: The idealizations of the previous section apply.

Find: r_o

Lookup values: From the HCB software in the appendix CD-ROM:

$$h_i = 355 \text{ kJ/kg (saturated)}$$

$$h_o = 376 \text{ kJ/kg (superheated)}$$

SOLUTION

Equation 10.17 is used to find the tangential velocity. From the known angular velocity, the rotor diameter can be found. The tangential velocity is

$$u_o = \sqrt{[(376 - 355) \text{ kJ/kg}](1000 \text{ J/kJ})}$$

$$= 144 \text{ m/s}$$

The rotor outer radius is

$$r_o = \frac{U_o}{\omega} = \frac{144 \text{ m/s}}{188 \text{ s}^{-1}} = 0.76 \text{ m (2.5 ft)}$$

COMMENTS

This result is refrigerant-specific, but for the common CFCs there is not much difference. However, for ammonia operating between the same two conditions, the enthalpy change would have been 52 kJ/kg and the rotor radius some 1.20 m, 60 percent larger. This is a relatively large rotor size to be economically manufactured. Of course, the rotor size can be reduced if the rotor is spun faster. Centrifugal forces increase with the square of angular velocity and linearly with the rotor radius.

   The mass flow rate of refrigerant through centrifugal compressors is set by the *depth* of the impeller, i.e., the dimension into the plane of Fig. 10.10b.

A number of practical considerations affect centrifugal chiller performance. Part-load *control* can be accomplished either by prerotation vanes at the inlet or by variable-speed drive. The former is relatively effective at moderate reductions of capacity but is a source of inefficiency at low capacity, where the vanes act as little more than flow obstructions. Variable-speed drives are more effective at maintaining good efficiency at reduced capacity.

Surging is the term applied to unstable flow in centrifugal compressors. It results from flow separation at reduced flow rate, i.e., at reduced capacity. Operation in this range is to be avoided to escape equipment damage. Manufacturer's automatic chiller controls usually make operation in this area impossible.

10.4.1.3 Other Compressor Designs

Screw compressors are intermediate in size between small reciprocating machinery and large centrifugal units. Typical sizes are on the order of a few hundreds to a few thousands of cooling tons (kilowatts) of output. In this range, they are more compact than reciprocating equipment of the same capacity.

Figure 10.11a shows the double-screw design with its two mating counterrotating rotors and housing. The compression effect is accomplished by differing rotation rates of the two rotors. In the compressor shown, the male rotor has four lobes while the female rotor has six. Therefore, the female rotor rotates at two-thirds the rate of the male rotor. Refrigerant enters one end of the compressor at the top and leaves the other end at the bottom.

Compression occurs as follows. Two lobes facing the inlet manifold separate, forming a volume of increasing size into which refrigerant flows. As the rotor continues to turn, two volumes of gas are trapped between the rotors and the casing. When two lobes remesh, the volume of gas shrinks and the refrigerant is compressed. Finally, the trapped compressed volume moves to the discharge port, where it is squeezed out axially and radially. A seal is maintained between the two rotors by an oil film.

The part-load efficiency of the screw compressor is relatively flat with load if variable-speed drive is used. If capacity is controlled by a slide valve that controls the discharge port size and location, performance penalties are not negligible at small part-load ratios even though this chiller design can operate down to 10 percent of rated capacity.

Scroll compressors are positive displacement devices that are quite widely used for small air conditioners including residential coolers and heat pumps as well as some automotive applications. Sizes range from 1.5 to 10 tons (5 to 35 kW). Their improved efficiency versus earlier designs is the prime motivation for expected wider usage in the future in small systems. Fig. 10.11b shows a cross-sectional view of a scroll compressor at several positions.

The scroll itself is an involute spiral confined on its sides by flat planes. There are a pair of scrolls, one fixed and the other moving. The moving scroll is driven in eccentric translation by the crankshaft. The moving scroll does not rotate. The seal between the moving and fixed scrolls is achieved by precision machining and seals at the involute tips.

Compression is accomplished as follows. Gas enters the scroll on the outer circumference. The meshed involutes form crescent-shaped pockets that reduce in volume as

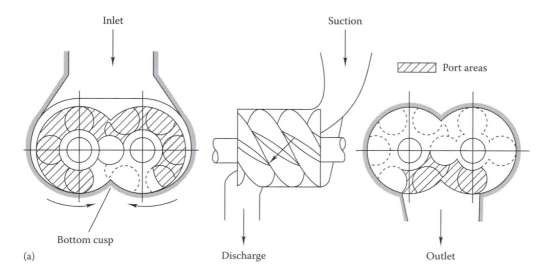

FIGURE 10.11

(a) Dual-screw compressor. (Courtesy of ASHRAE, *Handbook of Systems and Equipment*, American Society of Heating, Refrigerating and Air-Conditioning Engineers, Atlanta, GA, 2000. With permission.)

(*continued*)

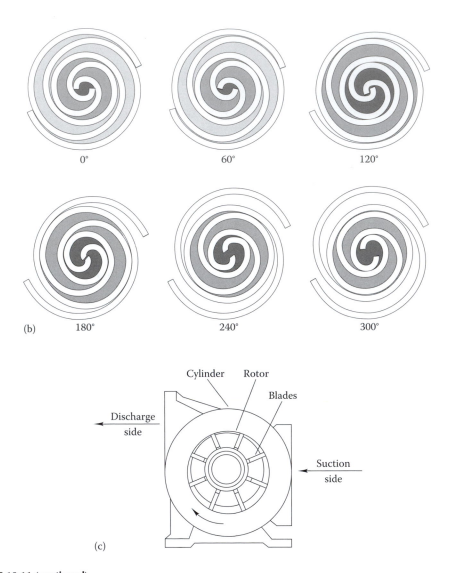

FIGURE 10.11 (continued)
(b) Scroll compressor. (Courtesy of ASHRAE, *Handbook of Systems and Equipment*, American Society of Heating, Refrigerating and Air-Conditioning Engineers, Atlanta, GA, 2000. With permission.) (c) Rotary compressor. (Courtesy of ASHRAE, *Handbook of Systems and Equipment*, American Society of Heating, Refrigerating and Air-Conditioning Engineers, Atlanta, GA, 2000. With permission.)

the moving scroll translates. The trapped gas is compressed as the crescents reduce in size from the outer periphery inward, as shown in subsequent positions in the figure. At the limit of inward movement, the discharge port is uncovered, and the high-pressure gas is discharged. Off-peak performance is good if deviation from the design pressure ratio is not too large. At standard rating conditions, the COP of air conditioners using scroll compressors is about 2.9 to 3.2 (ASHRAE, 1989).

Rotary compressors are used in applications from home refrigerators, to automotive air conditioning, to low-temperature refrigeration such as freezing plants. Fig. 10.11c is a cross-sectional drawing of a rotary compressor. A series of blades slide in slots as the rotor turns. Near the inlet port the gap between the rotor and housing is large. As the rotor turns, refrigerant trapped between each pair of blades, the rotor, and the housing is compressed because the rotor-to-housing gap decreases as the discharge port is approached. The blades basically move in translation relative to the rotor while the rotor itself is driven by the compressor drive system.

Small rotary machines have the advantage that they can be started with inexpensive motors that ordinarily do not have high starting torque abilities. These machines are inherently balanced so that rotational balance is not a concern. Rotary compressors are also relatively light and wear well. Compression ratios are limited to about 7:1 due to the excessive leakage and high blade and driveshaft stresses that occur at higher pressures. Oil lubricates the blades and bearings as well as cools the compressor via an external heat exchanger on larger units. Part-load performance is good down to 20 percent part-load ratio.

10.4.2 Heat Rejection Equipment

All cooling equipment must reject to a heat sink the total of the heat removed from the cooled building spaces and the compressor work (or the absorber heat input for an AC system). There are two standard methods for rejecting heat. A *cooling tower* passes condenser water in direct contact with outdoor air to cool the water by evaporation. An *air-cooled condenser* cools refrigerant directly by flowing outdoor air over the outer surface of tubes in which the condensing refrigerant flows. In this section, we describe both forms of heat rejection to outdoor air from central plant cooling systems. Although other heat sinks are used, outdoor air is by far the most common.

10.4.2.1 Cooling Towers

Moderate to large chillers [larger than 150 to 200 tons (525 to 700 kW)] use water as a condenser cooling medium. The lower condensing temperatures achieved with water cause the chiller to operate more efficiently than if an air-cooled condenser were used. The most common method of producing cooling water for a condenser is by use of a cooling tower. A cooling tower in its most basic and simplest form is shown in Fig. 10.12.

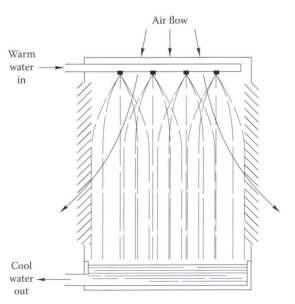

Air flow

Warm water in

Cool water out

FIGURE 10.12
Basic atmospheric cooling tower.

Water exits the spray manifold, falls under the influence of gravity, and entrains atmospheric air with it. While in direct contact with the air, some of the liquid water is evaporated. The heat of evaporation is extracted from the water stream, thereby cooling it. Cooled water collects at the bottom of the tower, where it is removed and pumped to the chiller for condenser cooling. The tower shown in Fig. 10.12 is called an *atmospheric tower* since it uses entrained atmospheric-pressure air for liquid cooling. This simple design is relatively ineffective for large heat rejection rates and requires considerable energy input to create the water spray at the inlet. It is rarely used.

Mechanical draft towers use fans or blowers to cause air to counterflow upward against the down-flowing water droplets. Fig. 10.13a shows a *forced-draft* cooling tower. In this

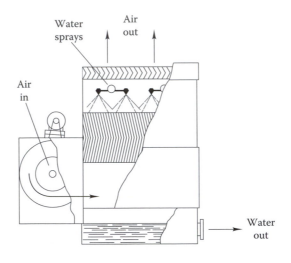

(a)

(b)

FIGURE 10.13
(a) Forced-draft counterflow cooling tower; (b) photograph of induced draft cooling tower installation. (Courtesy of SPX Cooling Technologies, Inc., Overland Park, KS. With permission.).

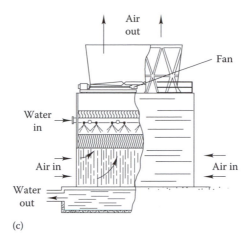

(c)

FIGURE 10.13 (continued)
(c) Induced-draft counterflow cooling tower.

equipment, air is forced to flow against the water droplet stream by use of an external centrifugal fan. One problem with this device is the relatively close proximity of the humid air exiting the tower and the high-velocity fan inlet. This can cause short-circuiting, whereby humid air is drawn back through the cooling tower, and reduced performance will result. As we will learn later, cooling towers can be used for "tower cooling" of buildings in winter to reduce energy consumption (compared to the former practice of operating chillers in winter). However, in subfreezing conditions, the fan near the base of a forced-draft tower can freeze or collect ice on the fan blading, thereby unbalancing it. This tower design is being used less in northern locations for this reason. Fig. 10.13b shows a mechanical draft cooling tower installation.

Induced-draft towers use a fan to draw air through the tower fill, as shown in Fig. 10.13c. This approach solves both the short-circuiting and the freezing problems of the forced-draft tower. The former is avoided since exit velocities are several times higher for this design than for forced-draft towers. The exit air plume will, therefore, be less likely to mix with the air at the fan inlet. Fan freezing is avoided since the fan is located in the warm exit airstream.

A steady energy balance on a cooling tower involves two principal terms and a small third term:

1. Enthalpy increase in airstream due to heat transfer to the warmer water
2. Enthalpy decrease in water stream due to heat transfer to cooler air
3. Enthalpy decrease in water stream due to evaporation of water (small term, the water loss to evaporation is 1 to 4 percent of the water throughflow)

In equation form, the energy balance must be expressed in differential form since the mass flow rate of *liquid water* changes with the location in the tower. The local energy balance at any point in the tower, as shown in Fig. 10.14a, is (ignoring heat transfer through the tower casing)

$$\dot{m}_{da}dh_{da} = \dot{m}_w c_w dT_w + \dot{m}_{da} h_w dW \qquad (10.18)$$

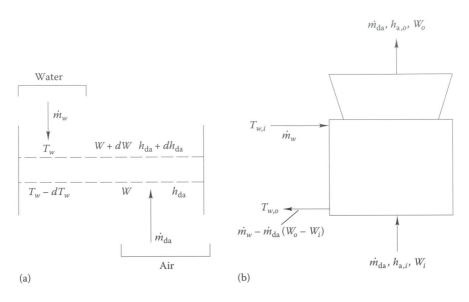

FIGURE 10.14
(a) Cooling tower differential volume energy and mass balance terms; (b) cooling tower overall mass and energy balance terms.

where all terms are defined relative to the differential volume in Fig. 10.14. This expression is solved by numerical integration as shown in ASHRAE (2000) after ignoring the second term on the right-hand side of Eq. (10.18). This cumbersome, iterative approach can be avoided by using the method described by Lowe and Christie (1961) and Braun et al. (1989a). We will briefly describe their approach in this section. It is particularly convenient since it uses the heat exchanger effectiveness idea first described in Chap. 2 and avoids the complication of a differential equation.

An overall energy balance on the tower shown in Fig. 10.14b is

$$\dot{m}_{da}(h_{a,o} - h_{a,i}) = \dot{m}_w c_w (T_{w,i} - T_{w,o}) + \dot{m}_{da}(W_0 - W_i)c_w T_{w,o} \qquad (10.19)$$

where
$\dot{m}_{da}$ = airflow through tower on dry-air basis, lb_m/h (kg/s)
$h_{a,o}, h_{a,i}$ = exit and inlet moist air enthalpies, Btu/lb_m (kJ/kg)
$\dot{m}_w$ = net tower water supply rate to make up for evaporation, lb_m/h (kg/s)
c_w = specific heat of liquid water, $Btu/(lb_m \cdot °F)$ [kJ/(kg $\cdot$ °C)]
$T_{w,i}, T_{w,o}$ = tower water inlet and exit temperatures, °F (°C)
W_o, W_i = tower outlet and inlet airstream absolute humidity ratios, $lb_m/lb_{m,da}$ (kg/kg$_{da}$)

It is assumed that water is supplied to the tower sump at $T_{w,o}$.

According to Chap. 2, the effectiveness of a heat exchanger is the ratio of the actual heat transfer rate to the maximum rate permitted by the second law of thermodynamics. In the context of a cooling tower, the maximum heat transfer rate is limited by the enthalpy difference between inlet air and outlet air *if it were saturated at the temperature of the warm inlet water* (this is the upper bound of enthalpy theoretically achievable by the airstream in a tower). The tower heat transfer effectiveness ϵ_{tower} is then

$$\epsilon_{\text{tower}} = \frac{\dot{Q}}{\dot{m}_{\text{da}}(h_{a,\text{sat},i} - h_{a,i})} \tag{10.20}$$

in which $h_{a,\text{sat},i}$ is the enthalpy of saturated air at the cooling tower inlet water temperature. The tower inlet water is at essentially the same temperature as the chiller condenser outlet water stream.

By analogy with a counterflow heat exchanger, the tower effectiveness is also given by

$$\epsilon_{\text{tower}} = \frac{1 - \exp\left[-\text{NTU}(1 - R)\right]}{1 - R\exp\left[-\text{NTU}(1 - R)\right]} \tag{10.21}$$

The *capacity ratio R* is defined as

$$R \equiv \frac{\dot{m}_{\text{da}}c_{a,\text{sat}}}{\dot{m}_w c_w} \tag{10.22}$$

where c_w is the specific heat of liquid water and $c_{a,\text{sat}}$ is the effective specific heat of *saturated air*, given by

$$c_{a,\text{sat}} = \frac{h_{a,\text{sat},i} - h_{a,\text{sat},o}}{T_{w,i} - T_{w,o}} \tag{10.23}$$

where $h_{a,\text{sat},i}$ is the enthalpy of saturated air at the tower water inlet (warm) temperature and $h_{a,\text{sat},o}$ is the enthalpy of saturated air at the tower water outlet (cool) temperature.

The specific heat of saturated air depends on temperature. Hence it must be evaluated at the known tower operating conditions at the chiller design point. The number of transfer units (NTU) must be calculated from tower manufacturer data. Braun et al. (1989a) found that a simple power law was adequate for this purpose. They suggest

$$\text{NTU} = a\left(\frac{\dot{m}_w}{\dot{m}_{\text{da}}}\right)^m \tag{10.24}$$

in which a and m are dimensionless constants found from curve-fitting catalog data. The value of a is between 1.0 and 3.0 for towers, and m ranges between 0.2 and 0.6.

When the effectiveness has been determined, the tower air exit conditions and heat rate can be found. For example, the exit air enthalpy is

$$h_{a,o} = h_{a,i} + \epsilon_{\text{tower}}(h_{a,\text{sat},i} - h_{a,i}) \tag{10.25}$$

If water loss is ignored, the tower water outlet temperature is found from

$$T_{w,o} = T_{w,i} - \frac{\dot{m}_{\text{da}}(h_{a,o} - h_{a,i})}{\dot{m}_w c_w} \tag{10.26}$$

The value of $c_{a,\text{sat}}$ is a weak function of tower outlet temperature, so one or two iterations of Eqs. (10.21) to (10.26) may be needed for precise calculations. The basis for the first estimate of $c_{a,\text{sat}}$ could be the entering air wet-bulb temperature.

The attractiveness of the preceding method is that it is closed-form and can be used for a wide range of operating conditions with good accuracy. Such calculations are needed when annual energy consumption for building cooling systems must be found. All that is required for a tower heat rejection rate determination is the air inlet condition (e.g., wet- and dry-bulb temperatures) and the tower water inlet temperature, which is known from the chiller condenser outlet. An alternate closed-form solution to the cooling tower problem is given by Whillier (1967) and is used in an example in Braun et al. (1987).

Example 10.6: Cooling Tower Design and Part-Load Operation

Use the effectiveness method to evaluate a cooling tower at design conditions and to check its operation at off-peak conditions. At peak load, an efficient[2] 525-ton (1845-kW) chiller consumes 0.67 kW/ton (a mixture of units used in the industry in the United States; this power input is equivalent to a COP of 5.25).[3] Using the NTU relation given below, find the required airflow rate at full load and several additional points of lower heat rejection rate. Additional details of the load, climate, and heat rejection system design are given below.

Given: At design conditions:

$T_{w,i} = 105°F$

$T_{w,o} = 85°F$

$T_{wet,i} = 78°F$

$T_{d,i} = 91°F$

$NTU = 2(\dot{m}_w/\dot{m}_{da})^{0.3}$ (from manufacturer's data for a specific tower, this is not a generally applicable eq.; each tower is different)

Figure: See Fig. 10.14.

Assumptions: In accordance with good cooling tower design, the water flow rate remains constant to ensure proper water flow distribution. Only airflow is varied to change the tower capacity. The tower is at sea level.

Find: $\dot{m}_{da}$ at full and part load

Lookup values: $h_{a,i} = 41.2$ Btu/lb$_m$ (Fig. 4.6)

$h_{a,sat,i}$ @ 105°F $= 81.4$ Btu/lb$_m$ (CD-ROM)

$h_{a,sat,o}$ @ 85°F $= 49.5$ Btu/lb$_m$ (CD-ROM)

$c_w = 1.0$ Btu/(lb · °F)

SOLUTION

First, the tower heat rejection is found. It is the sum of the heat removed from the space and the compressor power:

$$\dot{Q}_{tower} = (525 \text{ tons})\{[12{,}000 + 0.67(3413)] \text{ Btu/(h} \cdot \text{ton)}\} = 7500 \text{ kBtu/h}$$

The tower water flow rate can be found since the tower water range is known ($105 - 85 = 20°F$):

$$\dot{m}_w = \frac{7{,}500{,}000 \text{ Btu/h}}{[1 \text{ Btu/(lb} \cdot °F)](20°F)} = 375{,}000 \text{ lb/h}$$

---

[2] See Austin (1991).

[3] This power input is to the compressor only; it does not include pumps, controls, or the cooling tower fans.

Equations (10.20) to (10.24) are now used to find the peak-load performance of the tower. Since the airflow rate appears implicitly in these equations, an iterative solution is needed. We set up the equations, substituting all known numerical values, and then we present the results of the iterative solution. The iteration is started with an estimate of the tower effectiveness, which is refined at each iteration until convergence is reached.

The effective specific heat of *saturated air* $c_{a,sat}$ is given by Eq. (10.23):

$$c_{a,sat} = \frac{81.4 - 49.5}{105 - 85} = 1.59\,\text{Btu/(lb} \cdot {}^\circ\text{F)}$$

The coupled equations governing tower performance are solved in the following order. First, Eq. (10.20) is used to find the first estimate of the tower airflow rate. If we use an initial estimate of effectiveness halfway between the thermodynamic limits of 0.0 and 1.0

$$\epsilon_{tower} = 0.500 \tag{10.27}$$

then the airflow rate by rearranging Eq. (10.20) is

$$\dot{m}_{da} = \frac{7,500,000}{0.500(81.4 - 41.2)} = 373,000\,\text{lb/h}$$

Second, the capacitance ratio R can be found from Eq. (10.22):

$$R = \frac{373,000 \times 1.59}{375,000 \times 1} = 1.58$$

Third, the NTU value is found from Eq. (10.24) as follows:

$$\text{NTU} = 2\left(\frac{375,000}{373,000}\right)^{0.3} = 2.00$$

Finally the *new value* of the tower effectiveness can be found by inserting these values of R and NTU into Eq. (10.21). If it is sufficiently close to that given in Eq. (10.27), the iteration is complete. If the difference between the two values is less than the acceptable tolerance, say, ± 0.1 percent, the iteration is at an end. If not, the iteration is repeated, beginning from Eq. (10.27) and using the new value of the effectiveness found below:

$$\epsilon_{tower} = \frac{1 - \exp\left[-2.00(1 - 1.58)\right]}{1 - 1.58\,\exp\left[-2.00(1 - 1.58)\right]} = 0.542$$

This value of ϵ_{tower} is too far from the estimate to be considered accurate, so the iteration is repeated. The final results for this problem after several iterations[4] are

NTU $= 2.131$

$R = 1.288$

$\dot{m}_{da} = 303,700$ lb/h (38.35 kg/s)

$\epsilon_{tower} = 0.614$

---

[4] This "simple iteration" method converges, but slowly. Better methods that converge more quickly include the Newton-Raphson method. Computer software packages, such as Mathematica, exist specifically for this purpose.

TABLE 10.3

Cooling Tower Part Load Performance for Example 10.6

Airflow, lb/h	R	NTU	ϵ_{tower}	$\dot{Q}_{tower}$, Btu/h	$h_{a,o}$, Btu/lb	$T_{w,o}$, °F	$h_{a,sat,o}$, Btu/lb	$c_{a,sat}$, Btu/(lb · °F)
350,000	1.48	2.04	0.56	7,944,742	63.90	83.81	47.91	1.59
303,700*	1.29	2.13	0.61	7,500,000	65.90	85.00	49.30	1.61
300,000	1.28	2.14	0.62	7,426,315	65.95	85.20	49.53	1.61
250,000	1.07	2.26	0.68	6,790,050	68.36	86.89	51.63	1.64
200,000	0.88	2.42	0.74	5,931,386	70.86	89.18	54.66	1.69
150,000	0.68	2.63	0.81	4,860,225	73.60	92.04	58.74	1.75
100,000	0.47	2.97	0.88	3,534,868	76.55	95.57	64.26	1.82
50,000	0.24	3.66	0.95	1,913,422	79.47	99.90	71.73	1.90

* Design point, i.e., heat rejection is 7,500,000 Btu/h.

The fan airflow rate is the key result of the calculation. The designer will specify this along with the cooling tower heat rate, range, and approach.

At part load, the tower performance can be found as above by using a new value of the airflow rate and finding the corresponding heat rejection rate. At reduced heat rejection rates at part load, the tower exit water temperature is higher than the design point of 85°F; this has a small effect on $c_{a,sat}$. Table 10.3 summarizes the performance of this tower at flow rates ranging from slightly above the design flow rate to 50,000 lb/h, representing about 15 percent of full load.

COMMENTS

The effectiveness increases as the airflow decreases at part load because the residence time of air within the tower is longer. However, the heat rate decreases with airflow since Equation 10.22 shows that the heat rate decreases linearly with decreasing flow, whereas the effectiveness increases with decreasing flow but at a less than linear rate. Problem 10.24 requires the reader to calculate the entries in Table 10.3.

Tower design requires specification of several quantities related to the maximum capacity of the tower:

1. *Heat rejection rate*—set by peak condenser load of chiller at design cooling load conditions. Tower size increases linearly with load.
2. *Water flow rate*—determined by chiller condenser specifications.

   Manufacturers (Marley, 1983) suggest that water flow not be modulated at part load because proper flow distribution is difficult to maintain and freezing could occur in stagnant flow areas in winter. The fan power in a cooling tower is much larger than the pumping power; therefore, adjustment of the water flow rate does not reduce power consumption much anyway. Typical flow rates are 1.5 to 3 gal/min per ton (0.025 to 0.05 L/s per kW of cooling).

3. *Water temperature range* (temperature drop across the tower)—determined from items 1 and 2. Tower size decreases with increasing range subject to constraints on condenser water temperature. The range is usually selected between 10 and 20°F (6 and 11°C).
4. *Tower water outlet temperature*—determined by chiller manufacturer's recommendation.
5. *Design entering air condition*—The 1 percent wet-bulb design temperature (see Sec. 7.2.1) is recommended for tower design (Grimm and Rosaler, 1990).

The cooling tower capacity is controlled by adjusting the fan speed or less efficiently by bypassing either air or water around the tower at part load to control the water outlet temperature. Internal tower water flow itself should not be modulated to vary capacity since distribution within the cells and the fill of the tower depend on proper water flow rates. In multiple-cell cooling towers with variable-fan-speed capability, the following control strategy should be used: Operate as many cells as needed to meet heat rejection requirements with all fans on the lowest speed setting. When capacity is added, increment the lowest-speed [with off (i.e., the natural-draft mode) being the "lowest" of the low-speed settings] fan first (Braun et al., 1989b). All fans should be on at low speed before any fan is operated at a higher speed as the load is increased. Equal water flow distribution among tower cells leads to lowest energy consumption.

Cooling towers can be used in the absence of any mechanical cooling operation to cool buildings in swing seasons and in winter (Murphy, 1991). Fig. 10.15 shows a simple method of implementing *tower cooling*. When weather conditions are appropriate, the cooling tower operates with the chiller off and provides cold water to the chilled water distribution system by bypassing the chiller, as shown by the solid lines in the figure (dashed lines show the usual mechanical cooling configuration). A serious problem that must be addressed with this approach is the mixing of cooling tower water with treated chilled water loop fluid. Filtration of the former can help alleviate contamination and heat exchanger surface fouling in the cooling coils. Alternatively and much more commonly, a heat exchanger can be placed between the tower loop and the chiller water loop.[5] A very

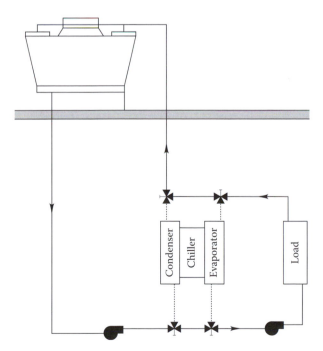

FIGURE 10.15
Schematic diagram of tower cooling system piped in parallel with mechanical chiller.

---

[5] Since the quality of water passing through the tower side of the heat exchanger is not tightly controlled, heat exchanger fouling may occur. The HVAC designer must leave sufficient space to disassemble and clean this heat exchanger periodically.

effective heat exchanger such as a plate heat exchanger can transfer the cooling effect from the tower loop to the chilled water loop with little temperature penalty. This system is the *indirect* tower cooling design.

Episodes of *Legionella pneumophila* (Legionnaires' disease) have been associated with cooling towers on buildings. Prudent preventive maintenance to avoid this disease includes keeping cooling tower sumps clean, using chemical inhibitors, circulating water daily (even if tower is not required for cooling purposes), and avoiding contamination of cooling water from other streams in a building.

Cooling tower maintenance is very important. Formerly, various organic and inorganic compounds using heavy metals were used, and they performed very well. It is not possible to use these materials any longer, but ozone produced on site and new organic materials can help control pH, scaling, corrosion, and biological growth.

10.4.2.2 Air-Cooled Condensers

Small chillers [less than 200 tons (700 kW)] often use air-cooled condensers rather than water-cooled condensers and cooling towers, as discussed above. Air-cooled condensers are simply cross-flow heat exchangers with refrigerant flowing and condensing inside of finned tubes. Fans force air over the exterior surface of the tubes and the fins. These condensers have modest maintenance requirements but may result in higher energy consumption and shorter compressor life, because condensing pressures are relatively high on summer days given typical air-to-refrigerant temperature differences, as discussed below.

Although much of the heat transfer involves isothermal phase change, some initial desuperheating and final liquid subcooling are usual. Fig. 10.16a shows an example of how the three heat exchange processes occur, and Fig. 10.16b is a photograph of an air-cooled condenser. The ϵ-NTU method described in Chap. 2 can be used to calculate the performance of an air-cooled condenser of this type by dividing it into three heat exchange sections—two sensible and one latent.

The design of air-cooled condensers requires the specification of

1. *Heat rejection rate*—determined by chiller or air conditioner peak heat discharge rate.
2. *Airflow rate*—a balance between excessive pressure drop for high flow rates and high initial cost for large heat transfer surface. Usual values are 600 to

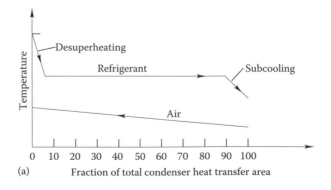

(a)

FIGURE 10.16
(a) Refrigerant and air temperature profiles within an air-cooled condenser.

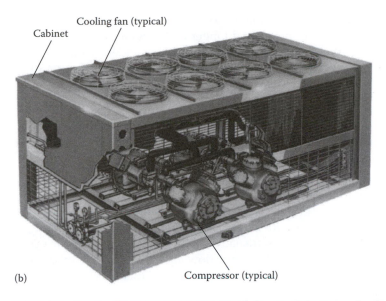

(b)

FIGURE 10.16 (continued)
(b) Photograph of air-cooled condenser bank located above compressor. (Courtesy of United Technologies/ Carrier, Farmington, CT. With permission.)

1200 ft³/min per ton [80 to 160 L/s per kW]. Fan power consumption is typically 0.1 to 0.2 hp/ton (20 to 40 W/kW).

3. *Temperature difference* (refrigerant to entering air)—affected by decision in item 2 and heat rate from item 1. Typical values are 15 to 40°F (8 to 22°C).

4. *Noise*—large air-cooled condensers can be noisy. They should be located so that noise produced is not a nuisance to building occupants.

5. *Unobstructed airflow*—a supply of outdoor air is needed for proper operation. Short-circuiting of warm air leaving the condenser back to the condenser inlet must be avoided. The area near the condenser air inlet must be kept clean so that the coils do not become blocked off or fouled.

In addition to problems associated with high-temperature operation, air-cooled condensers require special controls at low outdoor temperatures to avoid cooling condensed refrigerant to such an extent that it will not expand completely through the expansion valve. The fan speed can be controlled to avoid excessive subcooling, or the condenser fan can be cycled on and off. Dampers to control condenser airflow have also been used, but not entirely successfully.

High altitude causes a reduction in air-cooled condenser capacity since the mass flow rate is reduced (because of lower air density), even though the fan produces the same volumetric flow. At 5000-ft (1500-m) elevation, air-cooled condenser capacity is reduced by 10 percent.

The performance of air-cooled condensers can be improved if the air-side surface is kept wet with purified water. Evaporation from the condenser will enhance performance markedly because the driving potential for a coil cooled by evaporation is the wet-bulb, not the dry-bulb, temperature. Since the wet-bulb temperature is 15 to 25°F (8 to 14°C) below the dry-bulb temperature, *evaporative condensers* will operate at temperatures substantially lower than those of air-cooled equipment and somewhat lower than those of water-cooled condensers employing a cooling tower. Low condenser temperatures result in lower compressor power needs and longer compressor life.[6]

The key consideration in the design of evaporative condensers is the water composition. If minerals are not controlled, they will accumulate on the condenser surface and foul it, reducing heat transferability in relatively short order. Biological growth can also foul the surface. The cost of water must also be considered in arid climates.

10.4.3 Evaporators

Evaporators are heat exchangers within the tubes of which boiling refrigerant flows to cool either water (for use in cooling coils) or building supply air directly. The method of analyzing the performance of these components was discussed in Sec. 9.5. In centrifugal chiller evaporators, the refrigerant boils on the shell side of the evaporator.

Refrigerant also boils to cool air in *direct-expansion* (DX) evaporators. The thermal performance of DX evaporators (usually shell-and-finned-tube heat exchangers) can be enhanced by using fins. The refrigerant flow rate is controlled by a thermal expansion valve at the exit of the evaporator, as shown in Fig. 10.2. In systems for buildings, a few degrees of refrigerant superheat are desirable at the outlet, and the expansion valve ensures this condition. Generally DX machines have lower COP values than do water-based chillers, because compressors are often less efficient and air-cooled condensers are used.

In general, the evaporator heat rate decreases with decreasing chilled water or airflow rate at a given inlet temperature. The capacity increases with reduced evaporation temperatures or with higher incoming air or water temperature. These trends are apparent from the referenced, prior discussion of heat exchangers in Chap. 9. Details of evaporator design are not discussed in this book.

10.4.4 Expansion Devices

The final component in the basic cooling systems shown in Figs. 10.1, 10.2, 10.4, and 10.6 is the expansion device which

- Causes the pressure drop between high and low sides of the system
- Controls refrigerant flow rate

In this section we will briefly describe how two common devices—capillary tubes and thermostatic expansion valves (TXVs)—achieve these goals.

Capillary tubes are small-bore, long tubes used in small cooling systems up to a few tons (few kilowatts) in size. Tubes 1 to 2 mm in diameter and up to a few meters long are used to produce the high- to low-side pressure drop. As liquid refrigerant passes through the tube, the pressure drops due to friction; the reduced pressure causes evolution of refrigerant gas. As liquid is converted to vapor, the velocity increases; this acceleration causes additional pressure drop. The tube is essentially a passive device and cannot accommodate a large range of load and system pressures.

---

[6] A small cooling effect (up to 20 to 25 percent of full load) can be achieved by opening all valves in the refrigerant loop when the compressor is off. The free migration of refrigerant between the indoor evaporator (warm) and the outdoor condensor (cool) can produce some cooling without operating the compressor for a small fraction of the year.

From basic fluid mechanics it is clear that a capillary tube will flow more refrigerant under higher pressure differences (unless the flow becomes choked). High head pressures can be caused by high condenser temperatures (as a result of hot weather or a large liquid inventory there with the effect of reducing the useful condensing heat transfer area). However, if the load is unchanged as high-side pressure increases, some accommodation between the constant refrigerant enthalpy rate is needed for constant load and the increased flow produced by the increasing system pressure difference. The remaining degree of freedom is the evaporator temperature and corresponding suction pressure. The suction pressure will adjust itself within limits by rising to reduce system flow.

There is also the situation of reduced load under cooler condensing conditions. In this case the liquid refrigerant inventory in the evaporator will increase to the limit permitted by the size of the refrigerant charge. Of course, the charge must be selected so that all refrigerant can be held in the evaporator without any liquid entering the compressor inlet, a destructive event that must be avoided. Capillaries are sized, by a measure of trial and error, for one condition, and deviation therefrom will reduce system efficiency. Stoecker and Jones (1982) describe the details of capillary selection.

Capillaries have the advantage of being inexpensive and passive. They have the disadvantages of a relatively narrow operating range, susceptibility to clogging by small particles, and the requirement of proper charging within rather narrow limits. Instead of a capillary system, a fixed orifice can be used to produce the needed pressure drop. Its behavior is similar to that of a capillary, but it is less likely to clog because of its larger bore.

Figure 10.2 shows a TXV used as the expansion device. It is the most popular method of controlling flow and producing the pressure drop in medium-capacity systems. The sensor bulb is filled with a small amount of the refrigerant in the system to be controlled. Since the bulb is in close thermal contact with the suction line, the thermodynamic state of the controller fluid represents the state of superheat refrigerant in the evaporator outlet.

Figure 10.17 helps make clear how the pressure in the sensor bulb activates the valve. Recall that the pressure at the outlet of the TXV is the evaporator pressure; thus, the evaporator force plus the spring force must just equal the pressure exerted by the controller fluid at the point where the valve just begins to open. For the valve to actually open, the pressure in the bulb must exceed the evaporator pressure, the difference representing several degrees of superheat at the evaporator outlet. The spring selected determines the amount of superheat needed to open the valve. If the evaporator has a large pressure drop, the TXV design must be modified, but the basic operation is the same.

10.4.5 Unitary Equipment

Unitary equipment for cooling consists of all components needed for air conditioning, factory-assembled into one or two pieces of equipment. These assemblies include an evaporator, condenser, and compressor with fans needed for cool air distribution and condenser cooling. A heating subsystem can also be included. Standard units up to 50 tons (175 kW) are available from many manufacturers. The attraction of unitary systems is that they require little design or field-assembly work. The basic function of unitary systems is not much different from that of the systems described previously. Unitary commercial and residential cooling systems have standard rating methods in the United States that prescribe testing and calculation procedures. The most common, single rating number is the *seasonal energy efficiency ratio* (SEER). It is essentially a seasonally averaged COP. However, the ratio is not dimensionless, being expressed as the ratio of cooling effect (Btu/h) to the electric power input (kW). The SEER is discussed in Chap. 3.

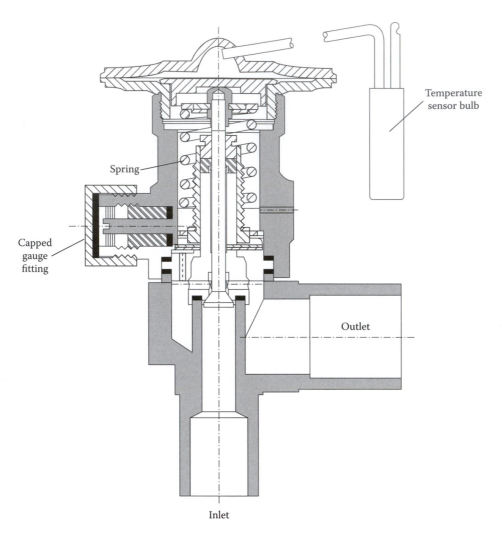

Temperature sensor bulb

Spring

Capped gauge fitting

Outlet

Inlet

FIGURE 10.17
Thermostatic expansion valve.

10.5 Part-Load Performance of Chillers

Chillers rarely operate at their peak design capacity. At loads below peak, reduced capacity can be accomplished in various ways, as described previously. The usual procedure reduces the capacity of the compressor in response to a reduced load on the chiller evaporator. Automatic controls on the condenser or cooling tower then reduce the heat rejection rate in response to the reduced heat produced by the compressor work and the evaporator heat rate. Previous sections have described the method for reducing capacity of the three principal types of compressors, of cooling towers, and of condensers. In this section we consider the *overall effect* of all capacity reduction actions in a given chiller as they manifest themselves in power input needed to supply a given part-load condition on VC chillers. A similar approach would be used with AC chillers.

The reader is referred to Sec. 9.2.2 regarding the part-load performance of boilers. A similar approach can be used for chillers. First, we define the *part-load ratio* (PLR) as

$$\text{PLR} = \frac{\dot{Q}_{\text{cool}}}{\dot{Q}_{\text{full}}} \tag{10.28}$$

where $\dot{Q}_{\text{cool}}$ is the cooling demand on chiller, i.e., cooling load, Btu/h (kW), and $\dot{Q}_{\text{full}}$ is the rated full-load capacity of chiller, Btu/h (kW). The chiller power input is expressed in terms of PLR as follows:

$$\dot{W}_i = \frac{\dot{Q}_{\text{full}}}{\text{COP}_{\text{full}}} \left[A + B(\text{PLR}) + C(\text{PLR})^2 \right] \tag{10.29}$$

where
$\dot{W}_i = $ shaft power input
$\text{COP}_{\text{full}} = $ COP at rated full-capacity point
$A, B, C = $ chiller-specific part-load coefficients

The part-load coefficients A, B, and C must be determined from manufacturers' data. An example of typical values is given in Table 10.4. Although these averaged values have been used for some time by building designers, it is preferable to use actual data from chiller manufacturers. This is done in several problems. The simplified PLR approach is valid only for the case when the cooling tower return temperature and chilled water temperatures remain constant.[7] If they do not, a more complex treatment beyond the scope of this book must be used (LBL, 1982). The appendix in the CD-ROM contains PLR coefficients for six specific chillers from three different manufacturers.

If part-load equation coefficients are known, the seasonally averaged COP can be found if the cooling-season load profile is known (the method used is the same as that used in Example 9.3). In addition, the COP at part load can be found. This latter use of part-load data is illustrated in Example 10.7.

TABLE 10.4

Part-Load Coefficients for Sample VC Chillers

Chiller Type	*A*	*B*	*C*
Hermetic compressor	0.160	0.316	0.519
Reciprocating compressor	0.023	1.429	−0.471
Centrifugal compressor	0.049	0.545	0.389

Source: Courtesy of LBL, *DOE-2 Engineers Manual,* Lawrence Berkeley Laboratory Report, 1982.

---

[7] Two kinds of part-load data are available. The first and the less useful is the power input for various capacities at constant condenser water entering temperature; these data are not representative of actual operation because condenser temperature drops with load as both are associated with reduced outdoor temperature. A better set of part-load data to use is that found from the ARI (1988) procedure, in which condenser temperature is adjusted downward as load is reduced. On the other hand, the ARI data should not be used for evaluating performance at a specific operating point.

Example 10.7: Part-Load Performance of VC Chiller

Find the part-load COP of a hermetic chiller, using the data in Table 10.4, if the full-load $COP_{full} = 4.0$. Plot the results in the range of PLR = [0.1, 1.0].

    Given: $A = 0.160$

    $B = 0.316$

    $C = 0.519$

    Find: COP(PLR)

SOLUTION

The chiller COP at part load COP(PLR) is given by

$$COP(PLR) = \frac{\dot{Q}_{cool}}{\dot{W}_i} \tag{10.30}$$

The denominator of Eq. (10.30) can be found from Eq. (10.29) at given levels of cooling load $\dot{Q}_{cool}$. When Eqs. (10.29) and (10.30) are combined, an expression for COP(PLR) is found:

$$COP(PLR) = \frac{(COP_{full})PLR}{A + B(PLR) + C(PLR)^2} \tag{10.31}$$

Table 10.5 summarizes the solution, and Fig. 10.18 shows the COP as a function of PLR.

TABLE 10.5

Chiller Part-Load COP

PLR	0.00	0.10	0.20	0.30	0.40	0.50	0.60	0.70	0.80	0.90	1.00
COP	0.00	2.03	3.28	3.98	4.33	4.47	4.47	4.41	4.30	4.16	4.00

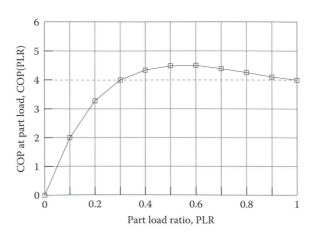

FIGURE 10.18
Chiller COP at part load (Example 10.7).

COMMENTS

The performance of this chiller is relatively uniform in the PLR range between 0.3 and 1.0 in contrast to boilers and furnaces described in Chap. 9 where efficiency drops strongly for PLR < 1. Note that the COP at part load appears to be larger than at full load. This is any artifact of using PLR coefficients based on constant evaporator and condenser temperatures. The reader should be aware that the part-load data used in this example are not universally applicable to this generic chiller type. Manufacturer's data must be used for reliable assessments of part-load performance. See Probs. 10.28 to 10.36 for further part-load calculations.

   When several chillers of different sizes are available in a building, part-load information can be used to determine which chillers should be operating. Proper sequencing of chillers to minimize energy use is the job of the control system, discussed in Chap. 12.

   There is a consistency check that the PLR coefficients must meet since at PLR = 1 the input power must match the peak rating. This requires that

$$A + B + C + \cdots = 1 \tag{10.32}$$

Notice that this requirement is met nearly exactly for the coefficients in this example:

$$0.160 + 0.316 + 0.519 = 0.995 \approx 1.0$$

Finally, we observe that the PLR equation may not be accurate for PLR < 0.1. Chillers do not often operate at such low capacities, so this does not present a major problem, when viewed in the context of year-long energy consumption calculations. At such low PLR values the chiller controller may shut off the unit because of lubrication or other criteria.

Absorption cycle chiller part-load performance can be calculated in a manner similar to that used above for VC chillers. For example, the steam rate $\dot{Q}_{in}$ at part load is given by (LBL, 1982)

$$\frac{\dot{Q}_i}{\dot{Q}_{i,\text{full}}} = \frac{A + B(\text{PLR}) + C(\text{PLR}^2)}{\text{PLR}} \quad 0 < \text{PLR} \leq 1.0 \tag{10.33}$$

10.6 Rules for Chiller System Operation and Control in Commercial Buildings

Taking into account the part-load characteristics of chillers, cooling towers, fans, and pumps, Braun et al. (1989b) have prepared a set of rules for minimizing the energy consumption of chiller plants in commercial buildings. In a case study, the authors showed that chiller energy consumption could be reduced from 26 to 43 percent compared to conventional fixed-speed equipment, if these rules are followed. Some rules have been mentioned earlier, but the complete set is given here. The designer will use this information in preparing the chiller control specification.

- If the cooling tower is constructed of several cells with variable-speed drives, operate all cells at the same speed.
- If the cooling tower is constructed of several cells with multiple-speed fans, increment the lowest-speed fans first when additional tower capacity is needed.
- Variable-speed condenser water pumps should be controlled with their associated chillers to give peak pump efficiency. If chillers have more than one fixed-speed

pump, all pumps should be controlled so that they all operate at the same speed (assuming the common practice that the several pumps are all the same model). As described earlier, *great care must be used if cooling tower water flow is to be modulated*, since flow imbalance and freezing may occur.

- Multiple-chiller plants should be operated so that they all have the same chilled water set points and so that condenser and evaporator flows are proportional to each chiller's cooling capacity. There is not much difference in energy consumption for multiple-chiller plants operated in series or parallel.
- Parallel air handlers should all operate with the same air set points.
- Optimal sequencing of multiple chillers depends on the specific characteristics of the chillers, as described in Sec. 10.5. No general rules exist.

This relatively compact and simple set of rules was found because all the key characteristics of an *optimally operating chiller plant*—chilled water temperature, supply air temperature, cooling tower airflow rate, and cooling tower water flow rate (if allowed to vary)—are related nearly linearly to cooling load and wet-bulb temperature. Since control of building relative humidity (rarely done except for critical applications) imposes an additional constraint on the chiller plant, humidity control will always consume more energy than not controlling it will under typical commercial building conditions.

10.7 Evaporative Cooling Equipment

The evaporation processes described in Chap. 4 can be applied to cooling of airstreams if a source of sufficiently dry air is available. In dry parts of the world, significant air cooling can be produced with this equipment. Alternatively, dry air can be produced with air-drying (desiccant) systems. This very dry air can be humidified by direct contact with water to produce sensible cooling. In this section, we give an overview of the operation of commercial evaporative cooling equipment. These systems have become increasingly attractive because

- CFCs are not involved.
- Energy consumption is significantly less in large areas of the United States.
- Ventilation required for human health is easily provided (during the cooling season).

Another evaporative process for reducing mechanical cooling needs in buildings (particularly buildings with large roof areas in dry climates) is the roof spray system, in which a thin, evaporating layer is maintained on flat-roof buildings. For example, evaporation from the roof can reduce that cooling load by 3 to 4 Btu/(h · ft^2) (9 to 13 W/m^2), amounting to a 40 to 50 percent reduction in peak roof cooling loads in a 95°F (dry-bulb)/60°F (wet-bulb) climate. Water flow to the roof sprays is controlled by roof temperature, so standing water is avoided.

10.7.1 Direct Evaporative Coolers

Figure 10.19 shows the essential components of a small, side-draft *direct* evaporative cooler. The evaporative medium (one panel of which is shown behind the fan in Fig. 10.19) is constructed of specially fabricated paper, fiberglass, or aspen wood fibers. The pads are kept saturated by the small pump that raises water from the sump, through a distribution system, to the top of the pads on a continuous basis. The water then falls through the pads back to the sump. Evaporation occurs adiabatically to air drawn through the pads by the centrifugal blower. Systems of this basic design are available in flow ranges between 2000 and 20,000 ft^3/min (1 and 10 m^3/s). Water treatment (softening, filtration) is important

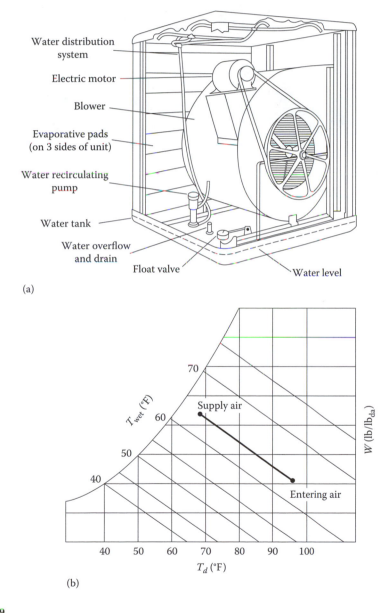

(a)

(b)

FIGURE 10.19
(a) Direct evaporative cooler showing key components, including wetted pads, blower, motor, and water supply system. (b) Typical direct evaporative cooling process plotted on the psychrometric chart.

(continued)

(c)

FIGURE 10.19 (continued)
(c) Sample of evaporative cooling media. (Courtesy of Munters, Inc., Kista, Sweden. With permission.)

if pad lifetime is to be maximized. The face velocity should be limited to 5 ft/s (1.5 m/s) to avoid entraining excessive water in the airstream, from which it could be carried into the building being cooled. In a draw-through system as shown, the cooled air also cools the motor, improving its efficiency and lengthening its life.

Larger direct evaporative coolers use different media with greater thicknesses. However, the same fundamental approach is used. Pads on larger units can be made from rigid cellulose or fiberglass (Fig. 10.19c) and may be up to 1.5 ft (46 cm) thick. Coolers using rigid media can produce cooled airflows up to 200,000 ft^3/min (94 m^3/s).

The temperature of evaporatively cooled air cannot be below the wet-bulb temperature of the entering airstream. In humid climates, this places a limit on the amount of sensible cooling that can be realistically expected from evaporation processes. The *effectiveness* of evaporative coolers (sometimes called the *saturation efficiency*) is defined as the dry-bulb temperature depression divided by the difference between the entering dry- and wet-bulb temperatures:

$$\epsilon_{\text{evap}} = \frac{T_{d,i} - T_{d,o}}{T_{d,i} - T_{w,i}} \qquad (10.34)$$

The effectiveness of practical direct systems ranges from 80 to 90 percent. Of course, evaporative coolers provide no latent cooling since the relative humidity of the direct cooler outlet air is higher than that of the inlet air,

Heat and mass transfer coefficients for evaporative cooler media have been measured and correlated by Dowdy et al. (1986) and by Liesen and Pederson (1991). The heat and mass transfer coefficients were found to depend on the pad thickness and on the Reynolds and Prandtl numbers. Figure 10.20 shows example effectiveness data for various thicknesses

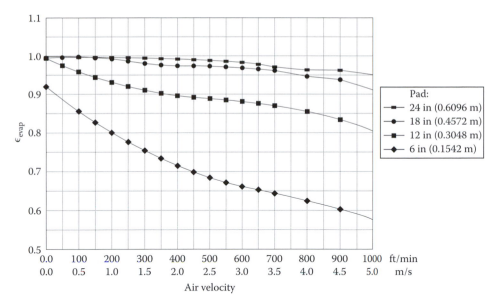

FIGURE 10.20
Example of evaporative cooler effectiveness for various pad thicknesses and airflow rates. (Courtesy of ASHRAE, *Handbook of Systems and Equipment*, American Society of Heating, Refrigerating and Air-Conditioning Engineers, Atlanta, GA, 2000; ASHRAE, *Handbook of Fundamentals*, American Society of Heating, Refrigerating and Air-Conditioning Engineers, Atlanta, GA, 2001, 1989. With permission.)

of pad and various airflow rates (at sea level).[8] For thick pads, effectivenesses approaching 100 percent can be achieved, but the pressure drop and associated fan power are quite high. Inlet air temperature does not have a strong effect on effectiveness.

Example 10.8: Direct Evaporative Cooler Exit Air Condition

What is the relative humidity of air leaving a direct evaporative cooler in the Arizona desert where outdoor conditions are 105°F (40.6°C) dry-bulb and 65°F (18.3°C) wet-bulb temperatures? The cooler effectiveness is 80 percent.

Given: $T_{d,i} = 105°F$

$T_{w,i} = 65°F$

Find: Relative humidity ϕ

SOLUTION

Equation 10.34 can be solved for the exit dry-bulb temperature. The exit wet-bulb temperature is known since evaporative cooling is a constant wet-bulb process, as described in Chaps. 3 and 4. The wet- and dry-bulb temperatures determine the relative humidity; its value can be read from the psychrometric chart.

From Eq. (10.34) the outlet dry-bulb temperature is

$$T_{d,o} = T_{d,i} - \epsilon_{evap}(T_{d,i} - T_{w,i})$$

---

[8] Problem 10.59 gives an empirical equation for a common evaporative cooler medium. It can be used instead of reading data from Fig. 10.20.

Inserting numerical values, we get

$$T_{d,o} = 105 - 0.80(105 - 65) = 73°F$$

From the psychrometric chart for sea level (Fig. 4.6), we read

$$\phi = 66 \text{ percent}$$

COMMENTS

Although the humidity is not excessive, the dry-bulb temperature is not sufficiently low to accomplish much sensible cooling to comfort conditions. The space could be held at 82 to 85°F (28 to 29°C) with this source of supply air. Long-term evaporative effectiveness of 80 percent relies on proper cooler maintenance. Although 80 percent is easily achievable with well-designed, new units, it may be difficult to achieve this over 3 years or more unless pads are replaced or well maintained.

Figure 10.21 shows summer wet-bulb temperature data at the 5 percent frequency level for the United States (wet-bulb temperatures shown will not be exceeded more than 5 percent of the cooling season—June through September). The rule of thumb is that evaporative cooling should be considered if the 5 percent wet-bulb temperature is less than 75°F (24°C). According to this rule, the map shows that all the western United States, the north central states, and the northeastern states are good candidates for evaporative cooling.

In addition to space cooling, direct evaporative precoolers can be used to reduce the dry-bulb temperature of outdoor air used to cool air-cooled condensers. Lower condenser

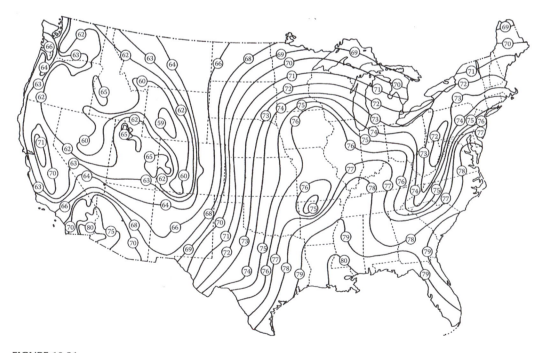

FIGURE 10.21
Map of lines of constant summer wet-bulb temperature in the United States at the 5% frequency level. (Courtesy of SPX Cooling Technologies, Inc., Overland Park, KS. With permission.)

temperatures improve COP values. The modest energy input to an evaporative cooler upstream of an air-cooled condenser essentially consists of extra fan power needed to overcome the additional pressure drop offered by the evaporative medium. This usually amounts to 0.2 inWG (0.5 cmWG). In nearly all cases where the 5 percent wet-bulb temperature is less than 75°F (24°C), evaporative precooling of condenser air should be considered. The improvement in compressor efficiency will nearly always outweigh the small extra electric power needs of the direct cooler stage. Costs of water and its treatment must also be considered before a final decision is made.

10.7.2 Indirect Evaporative Coolers

Indirect evaporative coolers sensibly cool an outdoor airstream without humidifying it. Fig. 10.22 shows the indirect system schematically. The secondary airstream is evaporatively cooled by using a direct evaporative cooler. The sensible cooling effect so produced is transferred to the primary stream by means of a heat exchanger; this cooled air is used to meet cooling loads. The primary stream may consist of some return air from the building plus the outside ventilation air. Unlike the adiabatic process, in a direct cooler, the indirect cooling of the primary airstream is not an adiabatic process, since no moisture is added. A true cooling effect is produced without the tradeoff of added humidity in the building. Indirect coolers can be used by themselves or with mechanical cooling (or with a direct evaporative cooler, as described shortly) depending on the climate.

The exit dry-bulb temperature from an indirect cooler cannot be less than the entering wet-bulb temperature of the entering secondary airstream. Therefore, the performance of an indirect cooler is rated by a factor similar to the effectiveness, called the *performance factor* (PF). It is the ratio of the primary air dry-bulb temperature drop to the difference

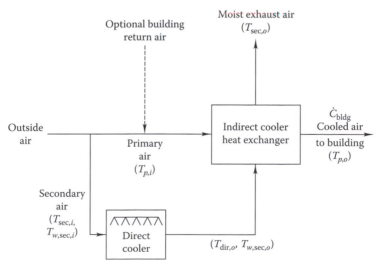

Note: $T_{p,i} = T_{sec,i}$; $T_{w,sec,i} = T_{w,sec,o}$

FIGURE 10.22
Indirect evaporative cooler system diagram. Return air from building is optional.

between the entering air dry-bulb temperature and the entering secondary air *wet*-bulb temperature. Referring to Fig. 10.22, we see that

$$PF = \frac{T_{p,i} - T_{p,o}}{T_{p,i} - T_{w,\text{sec},i}}$$ (10.35)

Typical values of PF range from 60 to 80 percent.

The PF can be expressed in a working equation rather than the less convenient definition above by using the more familiar direct cooler effectiveness ϵ_{evap} and primary-to-secondary heat exchanger effectiveness ϵ_{hx} as follows. The indirect cooler heat exchanger effectiveness is (for the case of no building return air)

$$\epsilon_{hx} = \frac{\dot{C}_{\text{bldg}}(T_{p,i} - T_{p,o})}{\dot{C}_{\text{min}}(T_{p,i} - T_{\text{dir},o})}$$ (10.36)

where $\dot{C}_{\text{bldg}}$ is the capacitance rate for building supply air, Btu/(h · °F) (W/K), and $\dot{C}_{\text{min}}$ is the minimum capacitance rate in the heat exchanger. Combining Eqs. (10.34) to (10.36) gives

$$PF = \epsilon_{hx}\epsilon_{\text{evap}}\frac{\dot{C}_{\text{min}}}{\dot{C}_{\text{bldg}}}$$ (10.37)

Finally, the cooling rate of an indirect cooler is

$$\dot{Q}_{\text{cool}} = (PF)(\dot{C}_{\text{bldg}})(T_{p,i} - T_{w,\text{sec},i})$$ (10.38)

Indirect coolers can be used as air precoolers for chiller coils if the local outdoor air is not of sufficiently low enthalpy to carry the entire cooling load by direct evaporative cooling. This approach is beneficial since chiller power input can be reduced during the entire cooling season. In addition, a smaller chiller can be purchased, initially reducing capital investment. Indirect systems can provide more of the annual cooling load of buildings than one might think, even in humid climates. For example, Supple (1982) reports that 30 percent of the annual cooling load in Chicago on a typical building can be provided by this means. An evaporative cooler, even in a humid climate, could be used in the return airstream to remove heat produced by lighting and return fans, as described in Chap. 11.

Indirect cooling systems can use exhaust building air (relatively dry and cool) instead of outdoor air in humid climates. Such systems can be as effective as indirect systems in dry climates. There are limitations due to the air quantities involved, but this system should be considered instead of the usual "heat" recovery system, which retrieves only (a fraction of) the sensible cooling effect in building exhaust air. The latent cooling effect is lost when only sensible heat exchange is used between intake and exhaust air.

Direct and indirect evaporative coolers can be used together to produce additional air cooling. Primary air first passes through the indirect system within which the dry-bulb and wet-bulb temperatures are reduced (at constant humidity ratio), as shown on the psychro-metric chart in Fig. 10.23. At the exit of the indirect stage, the cooled air passes through a direct evaporative cooler in which the dry-bulb temperature drops further as the humidity

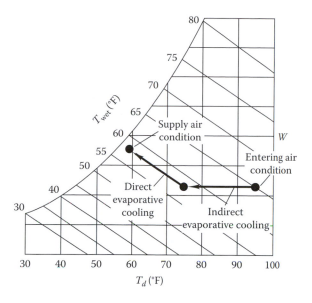

FIGURE 10.23

Indirect and direct evaporative cooling processes plotted on a psychrometric chart. Indirect process is at constant humidity ratio whereas direct process is at constant enthalpy, i.e., nearly constant wet-bulb temperature.

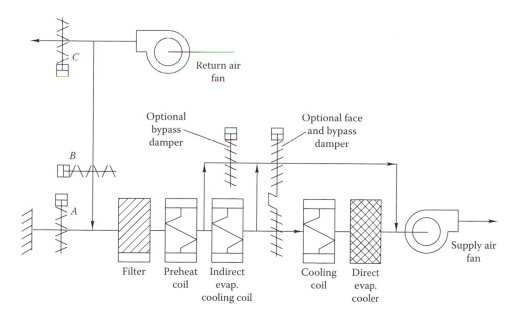

FIGURE 10.24

Evaporatively assisted mechanical cooling system with bypass arrangements to permit 100 percent evaporative operation or economizer cycle operation.

ratio increases in a constant-wet-bulb temperature process. In dry climates, COP values up to 15 can be achieved with *indirect-direct* systems. The COP is defined as the air cooling produced divided by the electric input, primarily to the air handler. Fig. 10.24 shows an indirect-direct system used to assist mechanical cooling. When outdoor conditions are appropriate, total building cooling can be accomplished without mechanical cooling; the cooling coil is bypassed in these cases. Both direct and indirect stages are bypassed when the economizer cycle can carry the cooling load in winter.

Other enhancements to evaporative coolers can further improve performance. For example, the water used to wet the pads can be cooled by a small chiller. Colder water will produce both lower air temperatures and lower humidities in the cooler outlet stream. This simple step can reduce cooling costs by one-third in favorable climates.

Control of indirect-direct systems follows this sequence: On the first call for cooling, outdoor air is introduced directly into the building if the air temperature is low enough to offset cooling loads (this is the economizer cycle described in Chaps. 3 and 11). With increasing cooling load the *indirect* stage is activated and operates until still more cooling is needed. Finally, the *direct* stage is operated to reduce the dry-bulb temperature still more. This is the limit of evaporative cooling that can be provided. At this point, mechanical cooling as shown in Fig. 10.24 is activated. During the heating season, the direct stage can be operated if needed to humidify building air; Proper attention must be paid to freeze protection in control system design for this winter use of evaporative cooling equipment.[9]

10.7.3 Design Considerations

The sizing of evaporative coolers can be accomplished by various methods ranging from rules of thumb regarding volumetric flow per unit floor area (for residences) to perform-ance specifications at the design cooling temperature and the coincident peak amount of latent and sensible heating needed. A typical rule for residences (Watt, 1986) states that between 15 and 30 air changes per hour should be used depending on the local difference between the dry-bulb and wet-bulb temperatures. This method assumes that evaporatively cooled air should be supplied at a temperature 8°F (4.5°C) below the room-temperature set point. The required flow rate is then the sensible load divided by the product of the density, specific heat, and temperature rise:

$$\dot{V} = \frac{\dot{Q}_{cool}}{\rho c_p \Delta T} \tag{10.39}$$

where

$$\Delta T = 8°F\ (4.4°C)$$

More careful sizing is needed for commercial buildings. Often the evaporative cooler sizing is based not on meeting the peak load (as is the case for mechanical cooling) but on reducing operating costs by an optimal amount. In such a design exercise, the cooling effect produced by several sizes of coolers is computed and compared to the load. When capacity is sufficient to meet the load, the mechanical chiller is not operated, thereby saving electricity charges. It is rarely economical (except in arid locations) to invest in an evaporative cooler sufficiently large to meet the design cooling load. A full-capacity mechanical chiller is assumed to be installed, and a moderately sized evaporative cooler acts as an energy saver when load and climatic conditions so allow. Example 10.9 indicates how a designer would go about finding the seasonal cooling effect produced by an evaporative cooler.

---

[9] A heat exchange loop from building exhaust air to the outside air supply duct can reduce the pad-freezing potential.

Example 10.9: Use of Bin Method for Evaporative Cooler Performance Calculations

Find the total sensible cooling produced by an indirect evaporative cooler with performance factor $PF = 70$ percent during the cooling season at the site for which the weather data are given below. The data are for June, July, and August in Denver, Colo.[10] The outdoor air supply rate is 50,000 ft^3/min (23,600 L/s) as might be appropriate for a small commercial building of 25,000 ft^2 (2320 m^2).

Given: $\dot{V}_{da} = 50{,}000$ ft^3/min

$PF = 0.70$

Bin dry-bulb (and mean coincident wet-bulb) temperature data in Table 10.6 (data aggregated into large bins in the interest of economy of calculations in this example)

Figure: See Fig. 10.22.

Find: Cooling effect produced for the months of June, July, and August

SOLUTION

It is the custom in evaporative cooler calculations to define the indices of performance on a temperature basis, not on an enthalpy basis, as is done for the evaporative process in cooling towers. This has been reflected in the derivation of Eqs. (10.35) to (10.38). We use this approach in this solution. Since the value of PF is given, only Eq. (10.38) need be used to solve this problem. The solution is summarized in Table 10.6.

For the 90 to 99°F bin, the calculations are given below. The cooling rate is found from Eq. (10.38) (the building airstream is the minimum capacitance rate stream in this example):

$$\dot{Q}_{cool} = 0.7 \left(50{,}000 \ \frac{ft^3}{min} \right) \left(0.060 \ \frac{lb}{ft^3} \right) \left(0.24 \ \frac{Btu}{lb \cdot °F} \right)$$
$$\times \left(60 \ \frac{min}{h} \right) \left[\frac{(95 - 60)°F}{1000} \right] = 1058 \ kBtu/h$$

The total cooling effect is the product of the cooling rate and the number of bin hours:

$$Q_{bin} = n_{bin} \dot{Q}_{cool} = \frac{(78 \ h)(1058 \ kBtu/h)}{1000} = 82.6 \ MBtu$$

The total cooling effect produced in the 3-month period is 879.2 MBtu (927.6 GJ) or 73,200 ton · h.

TABLE 10.6

Indirect Evaporative Cooler Performance in Denver

Bin, °F	Hours per Bin	$T_{d,av}$ $(T_{p,i})$, °F	T_{wet} $(T_{w,sec,i})$, °F	$\dot{Q}_{cool}$, kBtu/h	Q_{bin}, Mbtu	$T_{p,o}$, °F
90–99	78	95	60	1058	82.6	70
80–89	371	85	59	786	291.7	67
70–79	541	75	58	514	278.1	63
60–69	750	65	55	302	226.8	58
				Total	879.2	

---

[10] The HCB software also contains dry- and wet-bulb temperature bin data (based on the TMY database). The data are somewhat different for some sites from those used in this example, which are taken from *Engineering Weather Data* (AFM 88–29, 1978) published by the Government Printing Office.

By using a typical rule of thumb for evaporative coolers of 0.2 kW/ton of electric input at peak conditions (the 90 to 99°F bin), the COP of this unit is

$$COP_{peak} = \frac{12{,}000 \text{ Btu/h}}{(0.2 \text{ kW})(3413 \text{ Btu/kWh})} = 17.6$$

At off-peak conditions, less cooling is produced since the difference between the wet- and dry-bulb temperatures is smaller. However, the fan power is the same. Therefore, the COP is lower in the low-temperature bins.

Another performance characteristic also needs to be checked. The temperature supplied by the indirect cooler must be found to determine whether additional mechanical cooling is needed for space conditioning. Of course, the most critical period is the warmest part of the cooling season. The direct cooler outlet temperature can be found from Eq. (10.38) and from the first law:

$$\dot{Q}_{cool} = (PF)(\dot{C}_{bldg})(T_{p,i} - T_{w,\sec,i}) = \dot{C}_{bldg}(T_{p,i} - T_{p,o})$$

Solving the right-hand expression for the outlet temperature gives

$$T_{p,o} = (1 - PF)(T_{p,i}) + (PF)(T_{w,\sec,i})$$

The rightmost column in Table 10.6 shows the result for the four temperature bins of this example. The outlet temperature in the 85 and 95°F bins is too warm to be used directly in a space. However, the cooling produced can offset some of the mechanical cooling needed if the direct cooler is used in a system of the type shown in Fig. 10.24. In the cooler bins (75 and 65°F), the evaporatively cooled air could be used directly to cool a space.

COMMENTS

This example is the first step in selecting an evaporative cooler to reduce mechanical cooler operating time and to save on operating costs. The next step would compare evaporative cooler capacity to the load to determine when chiller operation can be avoided. Finally, chiller power savings are calculated by using the known chiller part-load characteristics. Included in this comparative assessment must be the costs of water and its demineralization, moving relatively larger volumes of air, and larger duct requirements for evaporative systems. In 1991 dollars, evaporative cooling systems cost about $0.35 per cubic foot per minute of airflow.

There is an additional, little-known economic benefit to using evaporative coolers. Since the water is evaporated and not returned to the sewer system, many municipal water districts permit submetering of evaporative cooler water and do not make the usual sewer tax assessment on the submetered usage! The sewer cost is about 50 percent of the water bill in the United States for water returned to the sewer by whatever means.

Note that the bin data were not given for temperatures below 60°F. For temperatures below this, evaporative cooling is not used—outdoor air is used directly in an economizer cycle if cooling is required. Of course, most of the hours below 60°F in June through August occur at night, when cooling may not be needed at all, but precooling could be accomplished during nighttime.

If an evaporative cooler is to be used as the *only cooling source* for a building, one must select the proper *design wet-bulb* temperature. Cooling loads are often determined by using the 1.0 percent design condition described in Chap. 7. For example, the HCB software or ASHRAE (1989) gives the 1.0 percent values of dry-bulb and mean coincident wet-bulb temperatures as 90 and 59°F (32 and 15°C) for Denver, Colo., a location well suited to evaporative cooling. However, the *1 percent wet-bulb condition* is quite different at 63°F (17°C). This difference represents roughly a 4°F (2.2°C) increase in evaporative cooler outlet, a major reduction in performance compared to what would be expected if the 59°F (15°C) value had been used. Experienced designers design evaporative coolers with at least the 1 percent wet-bulb temperature value and the 1 percent dry-bulb design temperature. More conservative design would use the 0.4 percent not 1.0 percent conditions. This sizing should be checked against two other points: (1) maximum wet-bulb and mean coincident dry-bulb temperatures and (2) maximum dry-bulb and mean coincident wet-bulb temperatures.

According to Watt (1986), the tendency for *Legionella pneumophila* (Legionnaires' disease) to multiply in basins of evaporative coolers is nil. The primary reason is that the temperature of this water is always very cool (even when the cooler is turned off in direct sun) and rarely stagnant during the cooling season. This bacterium breeds at 120°F to 140°F (49°C to 60°C) in stagnant water, according to Watt. However, Barbaree (1991) noted that growth can occur at as low as 77°F (25°C). Control can be achieved by using biocides, pH adjustment, and periodic cooler cleaning. The same preventive maintenance is recommended for cooling towers. Alternatively, the amount of standing water in the sump can be minimized by using a sump moisture sensor to control water supplied to the evaporative media.

Another application of evaporative cooling is the evaporative water chiller. Here a conventional chiller system equipped with an air-cooled condenser has the condenser cooled by evaporatively cooled, not atmospheric, air. The lowered compressor head pressure reduces power needs, and the lower outlet temperature of the chillers evaporator increases capacity. Both of these results are achieved with a modest increase in power in suitable climates. Details are given in the section on air-cooled condensers.

10.8 Summary

This chapter has considered the quantitative aspects of the design of mechanical and evaporative cooling central plants. Vapor compression and absorption cycles are both competitive and should be considered by the designer. Evaporative processes using water can reduce energy consumption in many ways, including reducing cooling loads, meeting part of the sensible load, or improving the performance of air-cooled condensers on chiller plants.

Part-load performance of mechanical or evaporative cooling equipment must be considered to find the annual energy consumption needed for system optimization. The approach presented has used bin weather data and part-load curves found from manufacturer's data to perform energy calculations for chillers, cooling towers, and evaporative coolers.

Problems

The problems in this book are arranged by topic. The approximate degree of difficulty is indicated by a parenthetic italic number from 1 to 10 at the end of the problem. Problems are stated most often in USCS units; when similar problems are presented in SI units, it is done with approximately equivalent values in parentheses. The USCS and SI versions of a problem are not exactly equivalent numerically. Solutions should be organized in the same order as the examples in the text: given, figure or sketch, assumptions, find, lookup values, solution. For some problems, the Heating and Cooling Buildings (HCB) software may be helpful. In some cases it is advisable to set up the solution as a spreadsheet, so that design variations are easy to evaluate.

10.1 What are the ASHRAE numbers for refrigerants having chemical formulas CCl_2FCClF_2 and $CClF_3$?   (3)

10.2 What are the ASHRAE numbers for refrigerants having chemical formulas CCl_2F_2, $CHClF_2$, and SO_2?   (3)

10.3 A refrigerator uses refrigerant 134a and operates between a low-side pressure of 43 psia (0.30 MPa) and a high-side pressure of 180 psia (1.2 MPa). The refrigerant mass flow is 1000 lb/h (0.126 kg/s). Find the cooling produced, power input, and COP of this machine.   (5)

10.4 Repeat Prob. 10.3 for a real cycle in which the compressor efficiency is 85 percent and 9°F (5°C) superheat is used at the compressor inlet. Find the cooling produced, power input, and COP of this machine.   (6)

10.5 The throttling step of the Rankine refrigeration cycle represents a loss of potential power generation because a turbine could be inserted in place of the throttling valve to produce additional shaft power that could increase the cycle refrigeration effect. Setting aside the practical problems with a turbine for a wet vapor, how much work could be produced by an isentropic turbine for the cycle in Prob. 10.3? What would the COP be if the power so produced were used for additional power input for vapor compression?   (8)

10.6 An ideal chiller uses refrigerant 22 and operates between a low-side pressure of 76 psia (0.52 MPa) and a high-side pressure of 300 psia (2.1 MPa). The chiller capacity is 100 tons (350 kW). Find the refrigerant flow rate, power input, and COP of this machine as well as the evaporator and condensing temperatures.   (5)

10.7 Work Prob. 10.6 with a real compressor efficiency of 85 percent and 9°F (5°C) of superheat at the compressor inlet. You can check your solution with the HCB software.   (6)

10.8 An ideal chiller uses refrigerant 22 and operates between a low-side pressure of 65 psia (0.45 MPa) and a high-side pressure of 250 psia (1.7 MPa). The chiller capacity is 200 tons (700 kW). Find the refrigerant flow rate, power input, and COP of this machine as well as the evaporator and condensing temperatures.   (5)

10.9 Work Prob. 10.8 for the same chiller capacity with a real compressor efficiency of 85 percent and 9°F (5°C) of superheat at the compressor inlet. You can check your solution with the HCB software.   (6)

10.10 An ideal chiller uses refrigerant 134a and operates between a low-side pressure of 36 psia (0.25 MPa) and a high-side pressure of 200 psia (1.4 MPa). The chiller capacity is 500 tons (1750 kW). Find the refrigerant flow rate, power input, and COP of this machine as well as the evaporator and condensing temperatures.   (5)

10.11 Work Prob. 10.10 with a real compressor efficiency of 85 percent and 9°F (5°C) of superheat at the compressor inlet. You can check your solution with the HCB software.   (6)

10.12 An ideal R134a chiller operates with a condensing temperature of 125°F (52°C) and an evaporator temperature of 35°F (2°C). If the cooling rate is 20 tons (70 kW), what are the refrigerant flow rate, condenser pressure, evaporator pressure, and COP? Compare the COP with that of a Carnot refrigeration cycle operating between the same condensing and evaporating temperatures. Discuss the reason(s) for the difference.[11]   (8)

---

[11] See Prob. 10.29 for an extension of this problem.

10.13 What is the power input [kW/ton (kW/kW)] required for an ideal R22 chiller operating between a condensing temperature of 130°F (54°C) and an evaporating pressure of 55 psia (380 kPa)?   (5)

10.14 What is the power input [kW/ton (kW/kW)] required for an R22 chiller operating between a condensing temperature of 130°F (54°C) and an evaporating pressure of 55 psia (380 kPa) if the compressor efficiency is 85 percent and if 9°F (5°C) of superheat exists at the compressor inlet? You can use the HCB software to check your solution.   (6)

10.15 The outlet ("head") pressure of Rankine cycle chillers is very sensitive to the condensing temperature, which in turn is determined by the outdoor weather conditions where heat is rejected. Excessive head pressure causes excessive compressor wear and may even cause physical damage. Discuss methods of controlling head pressure on very warm days when the load is high.   (4)

10.16 Derive an equation for the maximum COP of an absorption refrigeration cycle with the same components used for the Carnot refrigeration cycle, namely, isothermal reservoirs (four) and reversible, isentropic processes. Ignore pump work.   (6)

10.17 Using the approach of Example 10.2, evaluate the performance of a LiBr absorption cycle with the following temperatures: condensing temperature, 100°F (38°C); evaporating temperature, 44°F (7°C); generator temperature, 210°F (99°C); and absorber temperature, 93°F (34°C). Find the COP and cooling rate if the pump flow rate is 100 lb/min (0.76 kg/s).   (10)

10.18 The R22 compressor in the chiller of Prob. 10.8 has a polytropic exponent of 1.35. What is the compressor efficiency?   (3)

10.19 The R134a compressor in the chiller of Prob. 10.10 has a polytropic exponent of 1.32. What is the compressor efficiency?   (3)

10.20 What is the diameter of the centrifugal chiller (isentropic) compressor rotor for the R22 chiller in Prob. 10.8? Assume a rotor speed of 1800 r/min.   (4)

10.21 What is the diameter of the centrifugal chiller (isentropic) compressor rotor for the R134a chiller in Prob. 10.10? Assume a rotor speed of 3600 r/min.   (4)

10.22 A 300-ton (1050-kW) chiller is operated with 4 gal/min of condensing water per ton (0.072 L/s per kW) of refrigeration. Under design conditions of 95°F dry-bulb and 75°F wet-bulb (35°C and 24°C) temperatures, the cooling tower is designed to cool condensing water from 100 to 85°F (38 to 29°C). The air under these conditions leaves the tower at 92°F (33°C) and 90 percent relative humidity. What is the required air mass flow rate? How much water is consumed? What are the tower range and the inlet air to water temperature difference (sometimes called the *approach*)?   (6)

10.23 A 1000-ton (3400-kW) chiller is operated with 3.5 gal/min of condensing water per ton (0.063 L/s per kW) of refrigeration. Under design conditions of 93°F dry-bulb and 71°F wet-bulb (34 and 21°C) temperatures, the cooling tower is designed to cool condensing water from 100 to 83°F (38 to 28°C). The air under these conditions leaves the tower at 95°F (35°C) and 85 percent relative humidity (RH). What is the required air mass flow rate? How much water is consumed?   (6)

10.24 Construct the part-load curve for the cooling tower in Example 10.6 for a range of airflow rates between 50,000 and 350,000 lb$_m$/h (6 and 45 kg/s). Plot the heat rejection rate and effectiveness versus air mass flow, assuming that the water flow rate remains constant. Check your solution with Table 10.3.[12] (8)

10.25 Rework Example 10.6 for a tower range of 16°F (9°C) instead of the value used in the example. The tower heat load and water flow rate remain the same. (7)

10.26 Suppose that a more effective cooling tower is used instead of the tower specified in Example 10.6. Specifically, the tower constant is 2.7 instead of 2.0. How does this affect the needed airflow for the same heat rejection rate? (6)

10.27 What is the effectiveness of a cooling tower that rejects 600,000 Btu/h (180 kW) under design condition of 105°F (41°C) entering water temperature with 92°F (33°C) and 40 percent RH entering air if the airflow rate is 300 lb/min (2.2 kg/s)? (4)

10.28 What is the effectiveness of a cooling tower that rejects 150 kW under design conditions of 39°C entering water temperature with 32°C and 28 percent RH entering air if the airflow rate is 2.2 kg/s? (4)

10.29 The condensing temperature of the chiller in Prob. 10.12 is reduced by 12°F (6.6°C) due to the use of an evaporative condenser. What is the COP of the chiller with the addition of the evaporative feature? Is the change significant? The refrigerant mass flow rate is the same as for Prob. 10.12. (6)

10.30 An HVAC engineer selects the design condition of return air from building zones during the cooling season to be 75°F (24°C) and 50 percent relative humidity. The coil load calculation reveals that the sensible heat ratio is 0.6. Comment on this design in the context of achievable apparatus dew point. (5)

10.31 An HVAC engineer selects the design condition of return air from building zones during the cooling season to be 75°F (24°C) and 50 percent relative humidity. The coil load calculation reveals that the sensible heat ratio is 0.7. Comment on this design in the context of achievable apparatus dew point. (5)

10.32 In Chap. 4 we observed that certain latent plus sensible air cooling processes that can be drawn on a psychrometric chart cannot actually be achieved because no *apparatus dew point* (ADP) exists. A coil entering air condition that illustrates this is 70°F (21°C) and 50 percent RH with SHR = 0.5 (relatively high latent load). If you were faced with such a return air condition, describe how you could make system modifications that would allow an ADP to be achieved and the desired coil outlet condition to be reached. (6)

10.33 Plot the COPs of two chillers whose part-load characteristics are given in the CD-ROM chiller part-load ratio table: the 925-ton (0.625 kW/ton) unit by manufacturer A and the 900-ton (0.624 kW/ton) unit by manufacturer B. Plot the COP as a function of PLR in the interval (0.10, 1.0). Although the full-load COP of each is almost the same, the part-load performance is not. Comment on the part-load characteristics and which chiller is likely to use less energy during a cooling season. (5)

10.34 Plot the COP of an absorption chiller whose part-load characteristic is given by Eq. (10.33) with $A = 0.11$, $B = 0.36$, and $C = 0.53$ with a full-load COP = 1.2. Plot the

---

[12] Problems 10.19 to 10.21 require the enthalpy of saturated air as a function of temperature. In USCS units, an equation relating the two is $h_{a,\text{sat}} = 98.92 - 2.362T + 0.02092T^2$, where the temperature is in degrees Fahrenheit and enthalpy is in Btu per pound. Use this equation only within the temperature range of these three problems.

COP as a function of PLR in the interval (0.10, 1.0). Comment on your plot relative to operation at part load.   (4)

10.35 What is the power input to the chiller in Example 10.7 for a part-load ratio of 0.64 if the peak chiller capacity is 50 tons (175 kW)?   (3)

10.36 A commercial building has a peak cooling load of 875 tons (3080 kW) at 95°F (35°C) and zero cooling load at 55°F (13°C). Assume that the cooling load is a linear function of the outdoor dry-bulb temperature. A single, large centrifugal chiller with a capacity of 925 tons (3250 kW) (5 percent safety factor) is to be evaluated. It is the unit made by manufacturer A (0.692 kW/ton) in the CD-ROM chiller PLR table. What are the annual energy consumption and the annually averaged COP for a building in Chicago equipped with this chiller?   (7)

10.37 Work Prob. 10.36 for Kansas City, Mo.   (7)

10.38 The building in Prob. 10.36 can be cooled either by the specified chiller or by three chillers, each capable of meeting one-third of the load (use three 300-ton chillers by manufacturer A). Since the large chiller has more capacity than needed most of the time, it may use more energy than the small chillers operating more closely to their full capacity. Rework Prob. 10.36, using the three-chiller strategy. For simplicity, assume that the lead chiller (the first one to be operated) operates up to full capacity before the second is activated; the same strategy is used to operate the third chiller relative to the second.   (10)

10.39 A commercial building has a peak cooling load of 740 tons (3080 kW) at 100°F (38°C) and zero cooling load at 50°F (10°C). Assume that the cooling load is a linear function of the outdoor dry-bulb temperature. A single, large chiller with a capacity of 925 tons (3250 kW) (25 percent safety factor) is to be evaluated. It is the unit made by manufacturer A (0.692 kW/ton) in the CD-ROM chiller PLR table. What are the annual energy consumption and the annually averaged COP in a building in Albuquerque, N. Mex., equipped with this chiller? Study the effects of the 25 percent oversizing of this chiller on annual energy consumption by using a chiller exactly matching the load [740 tons (3080 kW)] but having the same part-load characteristics as the 925-ton (3250-kW) chiller.   (8)

10.40 Work Prob. 10.36 with a 925-ton (3250-kW) absorption chiller described in Prob. 10.34 instead of with a vapor compression chiller.   (7)

10.41 The building in Prob. 10.40 can be cooled either by the single specified absorption chiller or by two absorption chillers—one capable of meeting one-third of the load and the other capable of meeting two-thirds. Rework Prob. 10.40, using the two-chiller strategy. For simplicity, assume that the lead chiller (the first one to be operated) operates up to full capacity before the second is activated. Depending on the load in a particular bin, either chiller may be the lead chiller. (In actual practice, one would use a more sophisticated control approach to further minimize energy consumption.)   (10)

10.42 What is the exit air condition (temperature and humidity) from a direct evaporative cooler at sea level with inlet air at 100°F (38°C) dry-bulb temperature and wet-bulb temperature of 60°F (16°C) if the effectiveness is 85 percent?   (3)

10.43 What is the exit air condition (temperature and humidity) from a direct evaporative cooler at 5000-ft (1500-m) altitude with inlet air at 100°F (38°C) dry-bulb temperature and wet-bulb temperature of 60°F (16°C) if the effectiveness is 85 percent?   (3)

10.44 What is the sensible cooling produced by a direct evaporative cooler with 9000 standard ft^3/min (4230 L/s) of inlet air at 95°F (35°C) dry-bulb temperature and wet-bulb temperature of 59°F (15°C) if the effectiveness is 84 percent? (4)

10.45 What is the water consumption in the evaporative cooler in Prob. 10.44? (5)

10.46 What is the exit air temperature from an indirect evaporative cooler at sea level if the inlet air is 93°F (34°C) and 60°F (16°C) if the PF is 65 percent? (3)

10.47 A classroom is to be sensibly cooled with a direct evaporative cooler (87 percent effectiveness) when the outside air condition is 95°F dry-bulb and 67°F wet-bulb (38 and 19°C) temperatures. The room temperature is 78°F (25°C), and the sensible cooling load is 62,000 Btu/h (18 kW). What airflow is required to cool the space? (6)

10.48 List the 10 best states in the United States for considering evaporative cooling based on historical summer wet-bulb temperatures. (4)

10.49 Describe how one could calculate the reduction in flat-roof cooling load during a clear summer day that could be achieved by the use of roof sprays in a dry climate. Include the use of sol-air temperature, the psychrometric chart, and local weather data along with the roof properties. (5)

10.50 Repeat Example 10.9 for a direct evaporative cooler that has an effectiveness of 87 percent. Compare the results with the indirect cooler considered in Example 10.9. (5)

10.51 The table below summarizes the summer bin data for a location in the western United States (assume sea-level elevation). Find the sensible cooling energy provided by an 85 percent effective, direct evaporative cooler with an airflow rate of 3000 ft^3/min. The peak building sensible cooling load is 36,000 Btu/h at 95°F and decreases linearly with dry-bulb temperature to zero at 70°F. Note that the capacity of the cooler increases with decreasing outdoor temperature while the load decreases; therefore, the cooler will cycle on and off at temperatures below 95°F. Summarize your solution by completing the last three lines of the table. What is the seasonal COP if the fan motor is rated at 0.25 hp? (8)

	Temperature Bin		
	1	2	3
Dry-bulb range, °F	91–100	81–90	71–80
Coincident wet-bulb temperature, °F	62	57	54
Bin hours	120	420	490
Cooling load, Btu/h			
Available sensible cooling rate, Btu/h			
Bin total sensible cooling, Btu			

10.52 The table below summarizes the summer bin data for a location in the western United States (assume sea-level elevation). Find the sensible cooling provided by an 85 percent effective, direct evaporative cooler with an airflow rate of 1300 L/s. The peak building load is 10 kW at 38°C and decreases linearly with the dry-bulb temperature to zero at 21°C. Note that the capacity of the cooler increases with decreasing outdoor temperature while the load decreases; therefore, the cooler will cycle on and off at temperatures below 38°C. Summarize your solution by completing the last three lines of the table. What is the seasonal COP if the fan motor is rated at 180 W? (8)

	Temperature Bin		
	1	2	3
Dry-bulb range, °C	33–38	27–32	22–26
Coincident wet-bulb temperature, °C	17	14	12
Bin hours	120	420	490
Cooling load, kW			
Available sensible cooling rate, kW			
Bin total sensible cooling, kWh			

10.53 Repeat Example 10.9 for Chicago, a more humid site for which the relevant weather data are summarized below. (6)

Bin, °F	Hours per Bin	T_{wet}, °F
90–99	51	76
80–89	380	71
70–79	794	66
60–69	716	61

10.54 Repeat Example 10.9, using the HCB software bin data for Denver, Colo. Are the differences significant for the TMY database versus the Air Force data used in Example 10.9? (6)

10.55 The effectiveness of one of the most common evaporative media used in large evaporative coolers for commercial buildings can be calculated from

$$\epsilon_{evap} = 1 - e^{-4.76t/v^{0.33}}$$

in which v is the air velocity in feet per second and t is the media pad thickness in feet. Plot the effectiveness for a 6-in-thick pad for a range of air velocities between 2 and 20 ft/s. Make a similar plot for a fixed velocity of 8 ft/s for pad thicknesses varying between 2 and 12 in. In view of these results (assuming that the pressure drop through the pad increases linearly with thickness), discuss the tradeoffs involved in selecting the optimal pad thickness, including the effects of fan power and cooling produced. (8)

10.56 Consider the chiller in Prob. 10.8 to be coupled to an evaporative cooler as shown in Fig. 10.24. If outdoor air at 95°F (38°C) dry-bulb and 65°F (18°C) wet-bulb temperatures passes through an 85 percent effective, direct evaporative cooler prior to entering the air-cooled condenser, what improvements in chiller performance will result? Assume, for every degree of condenser inlet air temperature drop, that the condenser temperature also drops one degree. Find the change in capacity and COP resulting from the use of the cooler. (10)

References

ARI (1988). *Standard 550: Centrifugal or Rotary Screw Water Chilling Packages*. American Refrigeration Institute, Arlington, Va.

ASHRAE (2000). *Handbook of Systems and Equipment*. American Society of Heating, Refrigerating and Air-Conditioning Engineers, Atlanta.

ASHRAE (2001, 1989). *Handbook of Fundamentals*. American Society of Heating, Refrigerating and Air-Conditioning Engineers, Atlanta.

Austin, S. B. (1991). "Optimum Chiller Loading." *ASHRAE J.*, vol. 33, pp. 40–43.

Barbaree, J. (1991). "Controlling *Legionella* in Cooling Towers." *ASHRAE J.*, vol. 33, pp. 38–42.

Braun, J. E., J. W. Mitchell, S. A. Klein, and W. A. Beckman (1987). "Performance and Control Characteristics of a Large Cooling System." *ASHRAE Trans.*, vol. 93, pt. 1, pp. 1830–1852.

Braun, J. E., S. A. Klein, and J. W. Mitchell (1989a). "Effectiveness Models for Cooling Towers and Cooling Coils." *ASHRAE Trans.*, vol. 95, pt. 2, pp. 164–174.

Braun, J. E., J. W. Mitchell, S. A. Klein, and W. A. Beckman (1989b). "Applications of Optimal Control to Chilled Water Systems without Storage." *ASHRAE Trans.*, vol. 95, pt. 1, pp. 663–675.

Dowdy, J. A., R. L. Reid, and E. T. Handy (1986). "Experimental Determination of Heat- and Mass-Transfer Coefficients in Aspen Pads." *ASHRAE Trans.*, vol. 92, pt. 2, pp. 60–70.

Grimm, N. R., and R. C. Rosaler (1990). *Handbook of HVAC Design*. McGraw-Hill, New York.

LBL (1982). "*DOE-2 Engineers Manual*." Lawrence Berkeley Laboratory Report

LBL-11353 (LA-8520-M, DE83004575). National Technical Information Service, Springfield, Va.

Liesen, R. J., and C. O. Pederson (1991). "Development and Demonstration of an Evaporative Cooler Simulation Model for the BLAST Energy Analysis Computer Program." *ASHRAE Trans.*, vol. 97, pt. 2.

Lowe, H. J., and D. G. Christie (1961). "Heat Transfer and Pressure Drop Data on Cooling Tower Packings and Model Studies of the Resistance of Natural Draft Towers to Airflow." *ASME Heat Transfer Proc.*, Paper 113, pp. 933–950.

Marley (1983). *Cooling Tower Fundamentals*. Marley Cooling Tower Co., Mission, Kan.

Murphy, D. (1991). "Cooling Towers Used for Free Cooling." *ASHRAE J.*, vol. 33, no. 6, pp. 16–22.

Stoecker, W.F., and J.W. Jones (1982). *Refrigeration and Air Conditioning*. McGraw-Hill, New York.

Supple, R. (1982). "Evaporative Cooling for Comfort." *ASHRAE J.*, vol. 24, no. 8, p. 42.

Watt, J. R. (1986). *Evaporative Air Conditioning Handbook*. Chapman & Hall, New York.

Whillier, A. (1967). "A Fresh Look at the Calculation of Performance of Cooling Towers." *ASHRAE Trans.*, vol. 82, pt. 1, pp. 269–282.

11

Secondary Systems for Heating and Cooling

In this chapter we will learn how to synthesize much of the material developed previously in this book into the design of efficient heating, ventilating, and air conditioning systems for buildings. Since we have discussed the primary systems for heating and cooling in Chaps. 9 and 10, we focus on the secondary HVAC systems for buildings in this chapter. *Secondary systems* are those that meet the HVAC needs of specific zones. They consist of air and liquid handling equipment, duct and pipe systems, and heating and cooling terminal devices including coils, mixing boxes, and baseboard heating units. The control of primary and secondary systems is an important topic—all of Chap. 12 is devoted to it.

The energetic or economic measure of the effectiveness of the HVAC system depends almost solely on its efficiency of operation at part-load conditions since the system spends so much of its time at less than peak-load conditions. In this chapter, we will see that peak-load conditions are used to initially size equipment but that the much larger, long-term operating cost of HVAC systems depends on their operation at off-design conditions. For that reason we will spend considerable time on this topic as well.

The proper design of HVAC systems is subject to several criteria:

1. Standards of comfort—temperature, humidity, air motion, noise, cleanliness
2. Standards of economic and energetic efficiency—low initial and operating costs and energy consumption subject to the criteria in item 1
3. Standards of safety—embodied in codes and laws applicable to buildings
4. Desires of the owner as specified initially in the architectural program and on an ongoing basis during the design process at design meetings
5. Effective communication between all design professionals and ongoing documentation of the design process

We will be concerned in this chapter primarily with the first two items.

This chapter is divided into four sections. Sections 11.1 and 11.2 cover the basics of *subsystems* for handling heating or cooling air and liquid flows. Section 11.3 discusses the most common, *complete HVAC systems* widely used today. They include both air- and water-based heating and/or cooling systems. Section 11.4 discusses design, sizing, and energy consumption calculations of secondary HVAC systems.

11.1 Air Distribution Systems

Air is the medium used in the majority of HVAC systems in the United States to condition spaces of many types. Either air can be supplied from a central plant, as described in Chaps. 9 and 10, or a heated or cooled liquid can heat or cool air within a space. In any event, it is air that is the final heat transfer medium used to condition a building.[1] The earliest methods for cooling a building used natural or mechanical ventilation with outdoor air. We begin with the basics of simple ventilation before we consider mechanically cooled or heated air and liquids.

In Chap. 5, we discussed the basic principles of fluid mechanics and the machinery used to move air and liquids. In Secs. 11.1 and 11.2 we will apply these principles to fluid transport subsystems in buildings. The first section deals with air movement in duct systems and within building zones; the second, with liquid systems.

11.1.1 Basics of Mechanical Ventilation

Ventilation is defined as the supply or removal of air from a space by mechanical or natural means. Meeting comfort conditions described in Chap. 4 will often require conditioning of this air by heating or cooling it or by humidifying or dehumidifying it or by both. Ventilation serves two purposes:

- Addition or removal of heat and/or humidity from occupied spaces
- Supply of fresh air to meet health requirements

Systems of the type shown in Fig. 11.1 are commonly used for supplying air for both purposes in buildings. We will learn later in this chapter that there are many enhancements to this basic system. For the purposes of discussion, at this point the simple system will serve adequately. When the system in Fig. 11.1 is used for cooling, the air leaves the cooling coil (connected to the primary system chiller) at about 55°F (13°C) in conventional systems. Ordinarily some dehumidification is also achieved in the cooling coil. The heating coil is not active. On the other hand, when heating is required, the air delivery temperature for heating is approximately 90 to 105°F (32 to 40°C) or above in conventional systems. During heating, the cooling coil is inactive in this simple system.

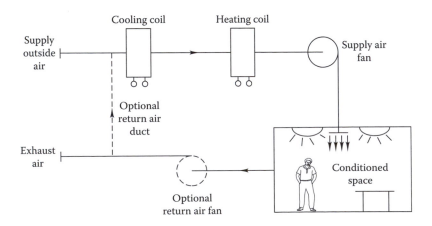

FIGURE 11.1
Simplified secondary HVAC system for a single zone.

---

[1] The only exceptions are radiant systems used for heating or cooling.

The supply air fan draws air through the coils and causes it to flow into the space to be conditioned. According to the conservation of mass, the amount of air supplied to the space must also be removed from it, except for differences which may be caused by infiltration or exfiltration. The return air duct and fan serve to remove air from the zone. Under conditions of significant heating or cooling, considerable energy and cost have been expended to condition room air. Instead of discarding this air after a single pass through the space, a portion can be reused. This is the purpose of the return air duct. The amount of room air that must be discarded depends on the fresh air supply requirements, which we discuss next.

During intermediate seasons when neither extreme heating nor extreme cooling is needed, it is possible to use more outdoor air than the minimum required for health conditions. We will discuss this shortly.

11.1.1.1 Fresh Air Ventilation Rates

The requirement for fresh air in buildings is set by the need to dilute indoor contaminants so that their concentration never exceeds established threshold levels. Building occupants and materials are the main source of pollutants that require control. (The amount of oxygen consumed by human metabolism is very small in typical occupancy densities in buildings and is not the main determinant for fresh air supply.) Contaminants in buildings include radon, formaldehyde, various other organic compounds, particulates, carbon dioxide, tobacco smoke, odors, and nitrogen oxides. Of course, each of these evolves at different rates in different buildings. Since the control of each contaminant in a building is impossible from a practical point of view, carbon dioxide concentration can be taken as an approximate surrogate for all other contaminants, at least those related to human activity. To maintain a minimum CO_2 concentration indoors, C_i in the presence of its evolution by humans requires a ventilation rate $\dot{V}$ given by

$$C_i = C_o + \frac{\dot{V}_g}{\dot{V}} \tag{11.1}$$

where
C_o = outdoor volumetric concentration of CO_2; dimensionless, e.g., percent or parts per million commonly used
$\dot{V}_g$ = generation rate of CO_2 by occupants, ft^3/min (L/s)
$\dot{V}$ = needed outdoor ventilation rate, ft^3/min (L/s)

Given the known rate of evolution and the threshold level of CO_2 to be permitted (about 0.1 percent) and a measure of conservatism, ASHRAE Standard 62-1999 specifies that each person in a building should be supplied with at least 15 ft^3/min (7.5 L/s) of fresh air.[2] Table 11.1 shows the amount of fresh air to be supplied based on the type of space and level of activity. Details are given in Table 4.6.

"Fresh" air can be either outdoor air or treated indoor air with minimal contaminant levels. In some cities, outdoor air is quite contaminated and unsuitable for use in buildings. In such cases, other techniques such as particulate control by mechanical filtration, electronic filtration, air washing to remove gaseous contaminants, or use of activated charcoal are needed to provide fresh air.

---

[2] Rather than supply this amount of air, the standard allows an alternate method of control whereby the concentration of certain contaminants can be controlled by other means such as filtration and dilution.

TABLE 11.1

Fresh Air Ventilation Requirements.

Application	ft³/min per Person	L/s per Person
Dining rooms	20	10
Hotel room	30 (per room)	15 (per room)
Offices, conference rooms	20	10
Public smoking lounge	60	30
Retail stores	0.20–0.30 ft³/(min · ft²)	1.0–1.5 L/(s · m²)
Auditorium	15	8
School classroom	15	8
Hospital patient room	25	13
Residential living areas	0.35 air change/h	0.35 air change/h
	but >15 ft³/min per occupant	but >7.5 L/s per occupant

Source: Courtesy of ASHRAE, *Standard 62–1999: Ventilation for Acceptable Indoor Air Quality*, American Society of Heating, Refrigerating and Air-Conditioning Engineers, Atlanta, GA, 1999b.

Note: See Table 9.6 for a complete listing of ventilation air supply rates.

11.1.1.2 Economizer

The previous concerns regarding indoor air quality are most critical during "deep" heating and cooling seasons, when energy consumed by conditioning outdoor air at extreme conditions must be minimized. However, in the spring and fall, extreme outdoor air conditions do not often occur, but cooling may still be needed, e.g., due to high internal gain levels in commercial office buildings. Cooling that uses outdoor air without operating a cooling coil is called the *economizer cycle*. The economizer is not a cycle in the thermo-dynamic sense, but the name is in common use; it is discussed in Chap. 3.

Whenever the outdoor dry-bulb temperature is below the nominal coil discharge tem-perature (typically 55°F), then outdoor air can be used in the amount required for cooling in the dry-bulb *temperature economizer* cycle. However, the economizer approach can be used under a broader set of conditions than just those meeting the dry-bulb temperature criterion. In fact, some cooling effect (sensible plus latent) can theoretically be achieved any time that the enthalpy of outdoor air is less than that of the required indoor air condition. This leads to the more general case of the *enthalpy economizer*. To achieve proper interior wet-bulb temperature conditions, supplementary mechanical cooling may be needed. However, the amount of mechanical cooling will be less than that needed if the economizer approach was not used at all.

Figure 11.2 shows the portions of the psychrometric chart where the two economizer cycles could be used if a building is to be maintained at 75°F and 50 percent relative humidity with a 55°F cooling coil outlet temperature. To the left of the 75°F dry-bulb temperature line, the dry-bulb temperature economizer will use outdoor air for cooling. Note that area D above the room condition enthalpy line has higher enthalpy than room air and will result in a cooling penalty if a dry-bulb temperature economizer is used. To the left of 75°F line *and* below the room enthalpy line (regions A and C combined), the enthalpy economizer will introduce outdoor air. Region C is ignored by the dry-bulb temperature economizer. Although the enthalpy here is lower than the room enthalpy, cooling it to the cooling coil dry-bulb setpoint temperature of 55°F results in an energy penalty; therefore, area C is of marginal cooling value. Of course, in the remainder of the weather domain

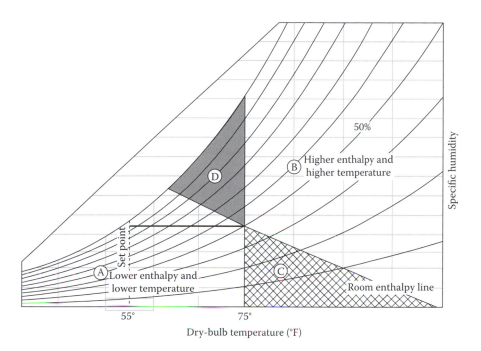

FIGURE 11.2
Enthalpy and temperature economizer operating ranges. The constant-enthalpy line represents the room enthalpy criterion.

covering the balance of the chart to the right (region B), mechanical cooling is needed and minimal outdoor air is used just to satisfy ventilation.

There is a practical problem with enthalpy-based economizers that limits, their use. It is the unavailability of a reliable, durable, and inexpensive enthalpy sensor. Such a sensor does not exist, but from a value of relative humidity and dry-bulb temperature one can find the enthalpy, as discussed in Chap. 4. Unfortunately a humidity sensor meeting the listed criteria does not exist either. Hence, most designers settle for dry-bulb temperature economizers. Spitler et al. (1987) have shown by means of computer simulation that most of the benefits of an enthalpy economizer can be achieved by a temperature economizer. One reason for this is that cooling coils are controlled based on the dry-bulb outlet temperature as noted above (see earlier discussion of area C), whereas the enthalpy economizer is controlled on enthalpy. Economizers of either type are more effective in moderate, dry climates than in hot, humid ones.

Economizers can be applied to many HVAC systems (but not all) that we will discuss in this chapter. This introductory section applies to all such systems.

11.1.2 The Prime Movers—Fans

The movement of air through HVAC systems in buildings is caused by pressure forces produced by fans. In the past, the energy consumption of fans has often been overlooked, yet it can be very significant. According to Coad (1989), it represents about 30 percent of the electricity used in commercial buildings. Fan electricity consumption can easily be larger than the consumption of the chiller because fans operate for many more hours per year, albeit at lower power rates, than chillers do. It is, therefore, very important that every feasible means be used to control fan power and to design duct systems that are energy-efficient. In this and following sections, you will be given the tools to do this. The designer must also specify the most efficient motors that are economically feasible for fans (and pumps) used in secondary HVAC systems (Greenberg et al., 1988). In Chap. 5 we developed the fan laws which govern the performance of fans at various conditions if the efficiency remains constant. For design purposes, the fan laws are assumed to apply.

It is worthwhile to reemphasize that fans are *volumetric devices* causing the flow of a given volume of air under specified conditions. However, at high altitudes, the mass flow produced by fans is lower because the density is lower, even though the volumetric flow remains the same at the same fan speed. One can imagine two identical air distribution systems operating with identical fan speeds at two different altitudes. The system at lower altitude will have a flow rate depending on the fan capacity, duct system, and air density. The system at high altitude will have the same volumetric rate since the fan speed is the same, but the pressure drop will be proportionally less by the same amount as the density is less (since the system pressure drop depends nearly linearly on the density). Since the horsepower is the product of the pressure rise and flow rate, it, too, is reduced by the density ratio. Figure 11.3 shows the performance of a fan and duct system at sea level and at an

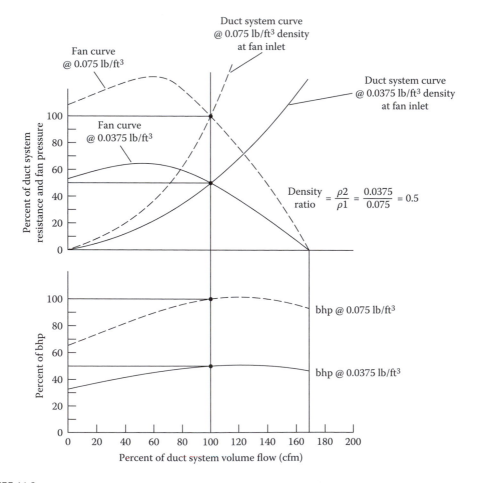

FIGURE 11.3
Effect of altitude on duct resistance and fan power. The low-density case is an extreme corresponding to the density of process air at 10,000-ft altitude and 250°F. (The bhp is the brake horsepower input to the fan.)

altitude where the density is one-half that at sea level. The reduced *mass* flow at high altitude results in reduced coil performance, as we will discuss later in the chapter. The appendix in the CD-ROM contains tabulated data for air density at various altitudes.

11.1.2.1 Fixed-Speed Operation

Manufacturers present "fan curves" under fixed-speed conditions, as discussed in Chap. 5. Figure 11.3 shows an example of a centrifugal fan curve in the upper plot. The shape of the fan curve depends upon the type of blading arrangement, as depicted in Fig. 11.4.

Forward-curved blading results in relatively lower efficiency. As a result, this design is used for low-speed, low-pressure-rise cases. Because of the dip in the pressure curve, this fan must always be operated to the right of the maximum pressure point for stability. It is also important to note that the power curve continues to rise with the volume. If this fan is operated with inadequate load (during system construction or later for service with ducts disconnected), the motor can be overloaded since the input power increases continuously with the flow. The low operating speed is an advantage, however, since stresses in the rotor will be smaller and bearing life longer.

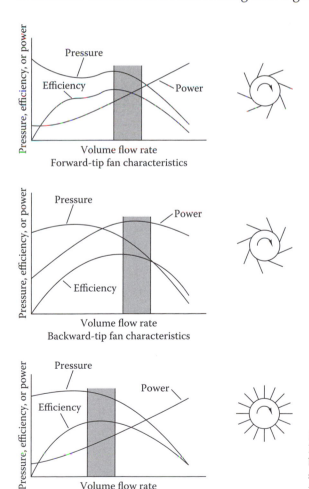

FIGURE 11.4
Forward-curved, backward-curved, and radial fan blading arrangements and performance curves. The shaded region is the suggested operating range near peak efficiency, but to the right of the peak pressure point to ensure stability.

Backward-curved blading results in the best efficiency of any centrifugal HVAC fan design. The noise level is relatively low, and the power curve is well behaved and stable over a broad flow range. This is the most common design in HVAC systems, even though the fan speed must be higher for a given flow rate than for other designs.

Radial blading designs lie between the two previous approaches. The power curve is not self-limiting, however. The efficiency is about the same as for forward-curved designs while the speed is lower than that for backward-curved fans. The noise level is moderate.

Many hundreds of fans are available for a given flow rate and pressure rise application. If several fans of similar cost are available for a project, the one having the highest efficiency at the operating point will be selected. One always wishes to operate near the maximum efficiency point of the fan curve and to have the duct system curve intersect the fan curve (thus defining the operating point, as shown in Chap. 5) near the maximum efficiency point.

The *static pressure* plotted in fan curves such as those in Figs. 11.3 and 11.4 has a special meaning. Fans are tested under specific conditions including a static pressure tap at the fan outlet and a smooth transitional fitting (not connected to a duct system) at the fan inlet. The fan static pressure rise, which is plotted on fan curves, is defined as the outlet static pressure minus the inlet *total* pressure, since the fan outlet static pressure is referenced to the *test room pressure* in which the fan is housed, not to an inlet duct. In equation form,

$$\Delta P_{static} = P_{static,out} - P_{total,in} \qquad (11.2)$$

You should recall this special definition when using this pressure rise to match calculated system static pressure drop. Also, if as-tested performance is to be realized in the field, the inlet condition must be the same as the test—uniform flow with no prerotation.

Finally, for proper fan operation, the fan outlet duct should be a straight section of sufficient length that the nonuniform flow at the fan outlet becomes approximately uniform. As a rule of thumb, the designer should allow at least 2.5 duct diameters of downstream straight duct to achieve uniform flow for air velocities less than 2500 ft/min (13 m/s), as shown in Fig. 11.5. For each additional 1000 ft/min above 2500, an additional duct diameter in the outlet duct length must be added. This duct length required to achieve fully developed flow is called the *effective duct length*. If space restrictions in the mechanical

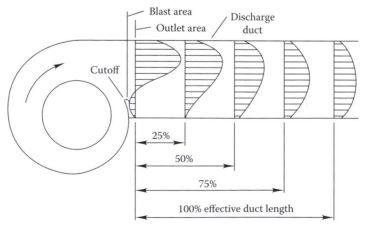

FIGURE 11.5
Adjustment of fan outlet velocity profile with distance from fan outlet connection.

room limit duct sizes to less than this value, the *additional pressure drop* must be accounted for, as described in AMCA (1973), where several hundred outlet (and inlet) conditions are documented and their extra pressure drops are listed.

The designer must be aware that parasitic pressure losses due to inadequate straight ducts at the fan itself can result in pressure drops of the same order of magnitude as total system friction losses, and that those losses will cost the owner extra operating energy for the life of the system. One method of quantifying the effect of less-than-ideal duct connections to fans uses the *system effect factor* (SEF). The system effect is based on an expression similar to that for duct pressure drops developed in Chap. 5:

$$\Delta p_{SEF} = K_{SEF} \left(\frac{1}{2} \rho v^2 \right)$$ (11.3)

The value of K_{SEF} depends on the fan inlet and outlet duct arrangement in a complicated way. The appendix in the CD-ROM contains the data needed to find K_{SEF} for some common fan connection geometries by using the idea of effective duct length described above. For those not included there, the reader is referred to ASHRAE (2000).

Example 11.1: Fan Operating Point

A curve fit of data from a fan manufacturer shows that the equation for the *fan curve* of a backward-curved fan at 1100 r/min (115 rad/s) is given by the dimensional equation (pressure is in inches water gauge and flow is in cubic feet per minute)

$$\Delta P = 4 - 0.2 \left[\left(\frac{\dot{V}}{1000} \right)^2 - \frac{\dot{V}}{1000} \right]$$ (11.4a)

This expression properly accounts for the fact that the fan curve is plotted by using the pressure rise from Eq. (11.2).

Find the operating point if the *system curve*, including the system effect factor, is given by the following dimensional equation:

$$\Delta P = 0.8 \left(\frac{\dot{V}}{1000} \right)^2$$ (11.4b)

If the fan efficiency is 61 percent, what is the required motor size?

  Given: Fan curve, system curve, and $\eta_{fan} = 0.61$
  Figure: See Fig. 11.6.
  Find: Operating point (P and $\dot{V}$), $\dot{W}$

SOLUTION

The operating point for this fan system occurs where the system pressure drop and flow exactly match the available fan pressure rise at the same flow. To find the solution, one can either plot the two curves, as shown in Fig. 11.6, or equate the fan curve and the system curve analytically. If the latter approach is used, we have the equation

$$0.8 \left(\frac{\dot{V}}{1000} \right)^2 = 4 - 0.2 \left[\left(\frac{\dot{V}}{1000} \right)^2 - \frac{\dot{V}}{1000} \right]$$ (11.5)

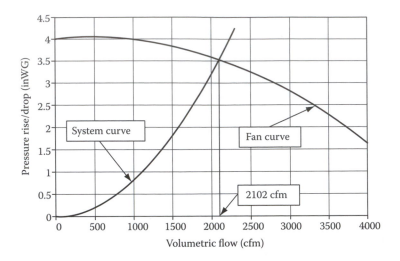

FIGURE 11.6
Pump and system curves for Example 11.1.

The quadratic formula is used to find the flow rate at the operating point:

$$\dot{V} = 2102 \, \text{ft}^3/\text{min} \ (992 \, \text{L/s})$$

The system pressure at this flow rate can be found from Eq. (11.4b):

$$\Delta P = 0.8 \left(\frac{2102}{1000} \right)^2 = 3.54 \, \text{inWG}$$

The ideal power from the basic equation from Chap. 3 is:

$$\dot{W}_{\text{ideal}} = \dot{V} \Delta P$$

$$= (2102 \, \text{ft}^3/\text{min}) \frac{(3.54 \, \text{in})[5.2 \, \text{lb}_f/(\text{ft}^2 \cdot \text{in})]}{33{,}000 \ (\text{ft} \cdot \text{lb}_f)/(\text{min} \cdot \text{hp})}$$

$$= 1.17 \, \text{hp} \, (0.87 \, \text{kW})$$

The actual power is the ideal power divided by the efficiency:

$$\dot{W}_{\text{act}} = \frac{\dot{W}_{\text{ideal}}}{\eta_{\text{fan}}} = \frac{1.17}{0.61} = 1.92 \, \text{hp}$$

COMMENTS

The method for finding the system curve has been discussed in Chap. 5 (see Fig. 5.12 and discussion). Essentially all pressure drop terms were expressed as a function of duct velocity (or, equivalently, volumetric flow rate). The analytical method of expressing the system curve is particularly convenient for finding the operating point under various fan speeds. We will see this in the next section.

The analytical fan curve is readily found from tables of fan data provided by manufacturers using common, preprogrammed calculator or regression methods for personal computers.

The system curve and fan curve must intersect to the right of the maximum pressure point to ensure stable operation, as shown in Fig. 11.4. The physical reason for instability is that there is always a tendency for air from the high-pressure side of the fan (outlet) to flow toward the low-pressure side (inlet). To the right of the maximum pressure point, this presents no problem since the momentum in the air ensures throughflow. However, at flow conditions to the left of the pressure maximum, backflow can occur in a cyclic fashion. This can cause large intermittent forces on the fan blading with subsequent physical damage.

Given the constraints on maximum efficiency, flow stability, and power, the operating range of a given fan is relatively narrow compared to the entire span of the usual plotted fan curve. This restriction is further complicated by the fact that a nominally constant-volume system does not actually operate at constant volume, since filters become dirty or cooling coils have varying amounts of condensed moisture. The careful designer will evaluate the expected range of system curves to ensure that all possible operating points lie to the right of the pressure maximum and not far from the efficiency optimum. Fortunately we have a wide selection of fans from which to choose, so that meeting the seemingly large number of criteria is readily accomplished in practice.

In a "blow-through" system, heat is added in winter to the building air system by a coil downstream of the fan. Since the density of the heated air is less than that of air flowing through the fan, the duct velocity downstream of the heating coil must be higher than if unheated air were flowing, as required by the conservation of mass. As a result, the system pressure drop will be higher by the ratio of density at the fan outlet to density at the heating coil outlet. To select a fan under these conditions, the designer must specify the volumetric flow rate of unheated air but the system pressure drop for the heated air. The effects of density on fan performance are discussed in Chap. 5 in detail.

11.1.2.2 Variable-Volume Operation

One of the most common air systems installed in commercial buildings uses varying amounts of air (within limits) to meet varying loads. The variation in flow is of interest since the power used to move air varies as the cube of the flow. For a factor-of-2 reduction in flow, the power needed and energy consumption theoretically decrease by a factor of 8 *if* efficiency remains unchanged. This variable-air-volume HVAC system approach will be discussed in detail below. In this section we compare three methods of providing variable airflow.

The first method of flow control introduces restrictions (e.g., dampers) in the airflow circuit. By thus increasing the resistance in a flow loop, the system curve becomes steeper and the flow lower, as shown in the left system curve in Fig. 11.7a. In this figure, three operating points are shown (1 to 3) for two fan speeds (see below) and two system resistance curves with different damper settings. For either fan speed, it is seen that flow is reduced as resistance increases with partially closed dampers. The amount of reliable control achievable by this method is limited, given the constraints on operating range discussed earlier. The dampers most often used with this approach are *fan outlet dampers* to throttle the flow. However, the pressure drop across the dampers represents wasted fan power. This method is inexpensive from an equipment viewpoint but costly operationally.

Adjustable *inlet vanes* can be used to more efficiently control flow. These vanes impart a spin to the air prior to its entry into the fan blading. The result is a modified fan curve,

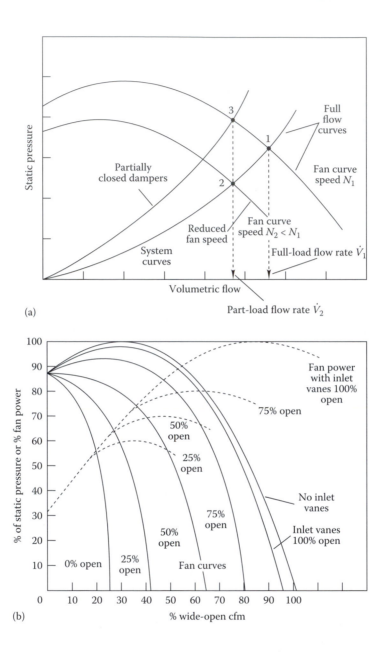

FIGURE 11.7
(a) Variable-volume fan and system curves for variable-speed fan and outlet damper flow control methods.
(b) Variable-volume fan and system curves for inlet vane control method.

as shown by solid lines in Fig. 11.7b. This approach is more efficient than the outlet damper approach since fan power is not wasted across dampers, but rather is partly used to produce angular momentum at the fan inlet. The dashed curves show how the fan power decreases with flow, with the result that electricity savings can be achieved due to the use of inlet vanes.

The most efficient method for fan flow control uses *variable-speed*[3] fan motors, but it has the highest cost. The result of speed changes is to change the fan curve, which now intersects the system curve at progressively lower flow rates as the fan speed slows. Figure 11.7a shows how flow is reduced by reducing the fan rotational speed. On the undampered (right) system curve, reducing fan speed reduces flow from point 1 to point 2. In practice, one would not use both dampers and a variable-speed drive to control flow.

One measure of the energy savings that can be expected from the three flow control methods is the part-load power consumption, expressed as a function of the part-load ratio (in this case defined as the fraction of design fan flow). A third-order polynomial is appropriate according to Knebel (1983).

For variable-speed drives, he suggested the expression[4]

$$\frac{\dot{W}}{\dot{W}_{rate}} = 0.00153 + 0.0052(\text{PLR}) + 1.1086(\text{PLR})^2 - 0.1164(\text{PLR})^3 \tag{11.6}$$

where

$\dot{W} =$ power input at part load, hp (kW)
$\dot{W}_{rate} =$ power input at rated design flow, hp (kW)
$\text{PLR} = \frac{\dot{V}}{\dot{V}_{rate}} =$ ratio of flow at part load to design rated flow

A similar expression applies to *outlet dampers*:

$$\frac{\dot{W}}{\dot{W}_{rate}} = 0.371 + 0.973(\text{PLR}) - 0.342(\text{PLR})^2 \tag{11.7}$$

For *inlet vane control* we have

$$\frac{\dot{W}}{\dot{W}_{rate}} = 0.351 + 0.308(\text{PLR}) - 0.541(\text{PLR})^2 + 0.872(\text{PLR})^3 \tag{11.8}$$

Figure 11.8 compares the power inputs of the three methods of fan flow control as a function of PLR. It is clear that the variable-speed approach has the largest savings over a wider operating range. The equations given above are not universal, only typical. If part-load data from manufacturers become available for flow control methods, they should be used instead of the preceding expressions. The data that do exist are consistent; in order of increasing efficiency, the three flow control systems are dampers, inlet vanes, and variable-speed drives.

11.1.3 Duct Design

The design of duct systems for air distribution uses fundamental material presented in Chap. 5 to size ducts and terminal devices for heating and cooling. Several design methods are described in this section, and the most common is discussed in detail with examples.

---

[3] The term *variable frequency* is sometimes used interchangeably with *variable speed* to describe this method of flow control because speed reduction of the most common electric motors is accomplished by reducing the frequency of the motor power supply.

[4] The fan laws in Chap. 5 would suggest a part-load curve involving only a cubic term. However, not all assumptions involved in deriving the fan laws, e.g., no flow separation or constant efficiency at any speed, necessarily apply to actual fans in actual systems.

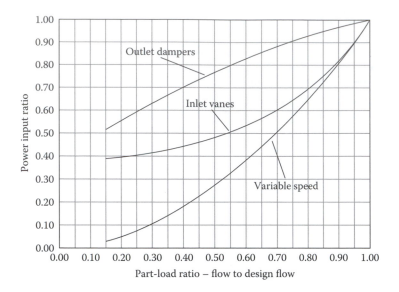

FIGURE 11.8
Part-load fan characteristics for outlet damper, inlet vane, and variable-speed control methods. (Courtesy of DOE2.1, Several manuals for the DOE2.1 software—*2.1D Building Description Language Summary* (DE-890-17726) and *2.1A Engineers' Manual* (DE-830-04575), The National Technical Information Service, Springfield, VA, 1981.)

The objective of duct design is to deliver the amount of air (at the proper condition) needed to meet the loads in each zone of a building. Duct design is constrained by many factors. One of the most important—available space—is often beyond the responsibility of the HVAC engineer. Other constraints include the need to (1) meet loads in a variety of zones, (2) meet economic criteria, (3) minimize operating energy subject to item 2, and (4) control noise levels.

The sequence of events in duct design is as follows. A preliminary system is laid out on a set of preliminary drawings including all structural members. Once a layout has been made, ducts are sized based on needed air quantities in each zone and for each terminal device. Pressure drop calculations are made at this point, and a fan is selected. The next iteration of the design will need to account for potential flow imbalances in the original design, duct runs of excessive pressure drop, and noise problems. After one or more iterations to accommodate these criteria, a set of final design drawings is prepared. At least one cost estimate is prepared as part of the design process, when the design is sufficiently complete. Computer-aided design (CAD) of duct systems is commonplace in large and medium-size design offices; it replaces the formerly tedious, manual iterations needed for duct design. In addition, various design options are easy to compare by using CAD approaches.

Duct designs are divided into two generic classes. Low-velocity systems have velocities below about 2500 ft/min (13 m/s), whereas high-velocity systems have velocities up to 4500 ft/min (23 m/s). Figure 11.9 shows the recommended velocity and friction rates for the two regimes. High-velocity systems can use smaller ducts if space is a problem, but fan power levels are higher. Low-velocity systems are used if fan operating costs are to be lower and if adequate building space for larger ducts exists. High-velocity systems require special duct-work (round or oval spiral ducts) to control leaks and properly confine the high-pressure air.

There are four methods of duct design. However, only one is widely used for low-pressure systems in most buildings, and we will emphasize it in this section. This method is

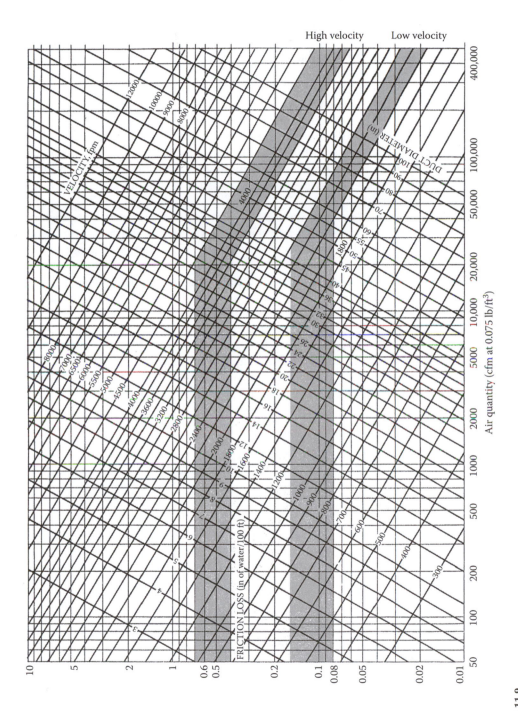

FIGURE 11.9
Recommended operating ranges for low- and high-velocity air systems; pressure drop versus flow rate.

the *equal-friction* method that attempts to maintain the same pressure gradient throughout the system. A well-balanced design can be produced with this approach if runs are of similar capacity. *Well-balanced* means that required flows and pressure drops are achieved without resort to brute-force methods of flow control such as excessive dampering.

Very large high-velocity systems are usually sized by the *static regain* method. This method is based on the requirement that the system static pressure remain about the same throughout a system. Specifically, ducts are sized so that the increase in static pressure in one section of duct exactly balances the pressure loss in the next section of duct. This is accomplished by progressively reducing the velocity in the duct system; the conversion of velocity pressure to static pressure balances the total pressure loss due to friction in the duct section. A nearly uniform static pressure throughout a duct system is desirable, since system balancing is simplified and the same air distribution equipment can be used anywhere in a system so designed. This method is best implemented by using commercially available computer software, since the iterative calculations are tedious and time-consuming to perform by hand.

The *velocity reduction* method progressively reduces air velocity in ducts with distance from the fan. Table 11.2 lists recommended velocities for various parts of ventilation systems. It is seen that the highest velocity occurs at the fan outlet, and progressively lower speeds are used downstream. This duct design method requires considerable experience and judgment as to what the appropriate velocities must be. A system designed with this method will not be self-balancing in general. A similar method called the *balanced-pressure method* progressively changes velocity so that the pressure drop in each branch is the same, thereby guaranteeing system flow balance. If the system has various duct lengths, this method will give a different solution from the equal-friction method, since the pressure gradient will differ in accordance with duct length differences. This method is suggested for use in systems such as the variable-air-volume system (Sec. 11.3.2) where precise flow balance is not necessary, since other means of flow control are used.

TABLE 11.2

Recommended and Maximum Duct Velocities

	Recommended Velocity, ft/min			Maximum Velocity, ft/min		
Designation	Residences	Schools, Theaters, Public Buildings	Industrial Buildings	Residences	Schools, Theaters, Public Buildings	Industrial Buildings
Outside air intakes*	500	500	500	800	900	1200
Filters*	250	300	350	300	350	350
Heating coils*	450	500	600	500	600	700
Air washers	500	500	500	500	500	500
Suction connections	700	800	1000	1000	1400	1400
Fan outlets	1000–1600	1300–2000	1600–2400	1500–2000	1700–2800	1700–2800
Main ducts	700–900	1000–1300	1200–1800	800–1200	1100–1600	1300–2200
Branch ducts	600	600–900	800–1000	700–1000	800–1300	1000–1800
Branch risers	500	600–700	800	650–800	800–1200	1000–1600

Source: Courtesy of Reynolds Metal Company, Richmond VA. With permission.

* The velocities are for the total face area, not the net free area. Other velocities are for the net free area. Divide ft/min by 197 to convert to m/s.

The fourth method for duct design, the *constant-velocity* method, sizes ducts so that the velocity is constant everywhere. Of course, flow imbalance can result, and special measures may be needed to achieve balance. Constant velocities are often needed in industrial exhaust systems where particulates must remain entrained until the air is exhausted from a facility.

In summary, use the equal-friction method for intermediate-size systems and for low-pressure branch ducts in large systems. Use the static regain method in large buildings with high-pressure systems or for the main distribution ducts in medium-size systems. For small buildings with simple systems, the velocity reduction method is adequate. Constant-velocity designs are usually restricted to industrial systems.

The duct designer will specify other features besides its size. Duct leakage is a significant problem (unsealed metal ducts can lose up to 15 percent of the airflow at 1-inWG duct pressure whereas sealed ducts will lose less than 1 percent); therefore, sealed ductwork must be included in the specifications. The thickness gauge of duct metalwork must be determined; it depends on air pressure and duct dimensions. Duct supports, turning vanes, layout to avoid building structural members, and other fabrication details are the responsibility of the design engineer. During and after construction, the design engineer should also inspect the system to ensure conformance with the design documents.

Example 11.2: Equal-Friction Duct Design

Use a *design pressure drop* of 0.1 inWG/100 ft to design the simple duct system shown in Fig. 11.10. The fitting pressure drops are taken from tables in the Appendix in the CD-ROM. Use round ductwork in standard 1-in increments.[5] The pressure loss at each branch outlet grille is equivalent to 20 ft of duct according to the grille manufacturer. The elbows at G and F are full-radius elbows $(r/D=1.0)$. The tees are of conical design, and the location is at sea level.

Although they are inconsistent, use the units of the problem statement in the solution since they are those used by duct designers in the United States.

Given: $\dot{V}$ (per figure)

$$\frac{\Delta p}{L} = \frac{0.1 \text{ inWG}}{100 \text{ ft}}$$

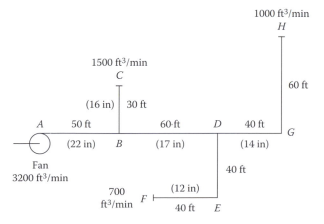

FIGURE 11.10
Duct layout for Example 11.2.

---

[5] If round ducts cannot be fitted above ceilings in a building, their equivalent in rectangular ducts can be found from Eq. (5.34). Different C values will apply and can be found in the Appendix CD-ROM.

TABLE 11.3

Results of Example 11.2—Duct Design

Section	Length, ft	V, ft³/min	$\Delta p/L$, inWG/100 ft	Duct Loss, inWG	Dia., in	Velocity, ft/min	p_v, inWG	C	Fitting Loss, inWG	Total, inWG
A-B	50	3200	0.10	0.05	22	1210	0.091	—	—	0.050
B-C	30	1500	0.10	0.03	16	1070	0.071	0.48	0.034	0.064
C	20	1500	0.10	0.02	—	1070	0.071	—	—	0.020
B-D	60	1700	0.10	0.06	17	1080	0.073	0.011	0.001	0.061
D-G	40	1000	0.10	0.04	14	935	0.055	0.013	0.001	0.041
G-H	60	1000	0.10	0.06	14	935	0.055	0.22	0.012	0.072
H	20	1000	0.10	0.02	—	935	0.055	—	—	0.020
D-E	40	700	0.10	0.04	12	890	0.049	0.51	0.025	0.065
E-F	40	700	0.10	0.04	12	890	0.049	0.22	0.011	0.051
F	20	700	0.10	0.02	—	890	0.049	—	—	0.020

Figure: See Fig. 11.10.

Assumptions: Ignore pressure losses due to duct size transitions. (The reader is asked to add these to the present solution in Prob. 11.7.) The location is at sea level.

Find: Branch pressure drops Δp

Lookup values: From the appendix CD-ROM we will need to look up several pressure drop coefficients C for tees and elbows. For example, $C = 0.22$ for an $r/D = 1.0$ elbow. Tabulating of the tee coefficients will need to await duct velocities calculated below. The values of C for all fittings are shown in the ninth column of Table 11.3.

SOLUTION

The results of the solution are tabulated in Table 11.3. The table is completed from left to right as follows. First the duct length and volumetric flow values are entered from information in Fig. 11.10. The next column (pressure drop per 100 ft) is specified in the problem statement. It is constant everywhere since we are using the equal-friction approach. Multiplying the duct length by this factor gives the values in the fifth column.

From Fig. 11.9 (or Fig. 5.9a) the duct size is read from the known volumetric flow rate and design pressure drop. These values are entered in the sixth column of the table. For example, a 3200 ft³/min flow at 0.1 inWG/100 ft pressure drop occurs if a 22-in-diameter duct is used. From the volumetric flow and the diameter, it is easy to find the air velocity in each branch. From the velocity, the sea-level velocity pressure for standard air is found from

$$p_v = \frac{1}{2}\rho V^2 = \left(\frac{V}{4005}\right)^2 \tag{11.9}$$

(The rightmost portion of this expression is dimensional, giving a value of velocity pressure p_v in inches water gauge for values of velocity in feet per minute.)

From the appendix fitting loss coefficients tables, C can now be found, since the velocities are known. For the straight-through loss in the tees at B and D, one needs the velocity ratios. At B the ratio is $1080/1210 = 0.89$, and at D the ratio is $935/1080 = 0.89$. From these ratios the respective C coefficients are found to be 0.011 and 0.013, from the table labeled "main" in the "wye, diverging" table. These C values are entered into the ninth column of Table 11.3.

The branch line loss coefficients are found in a similar manner. At B the velocity ratio is $1070/1210 = 0.88$, and at D it is $890/1080 = 0.82$. From the appendix table "tee, diverging, round, conical branch" entries, the C coefficients are interpolated as 0.48 and 0.51, respectively. At this point the fitting pressure losses can be calculated from the basic equation

$$\Delta p_{\text{fit}} = C\left(\frac{1}{2}\rho V^2\right) = Cp_V \tag{11.10}$$

The final column of the table is the sum of straight duct and fitting pressure loss.
   For the three branches the pressure drops are

$$\Delta p_{ABDGH} = 0.050 + 0.061 + 0.041 + 0.072 + 0.020 = 0.244 \text{ in}$$
$$\Delta p_{ABC} = 0.050 + 0.064 + 0.020 = 0.134 \text{ in}$$
$$\Delta p_{ABDEF} = 0.050 + 0.061 + 0.065 + 0.051 + 0.020 = 0.247 \text{ in}$$

The maximum loss is 0.247 inWG (61 Pa). This is the value that would be used for fan selection (along with other pressure drops including coils, the system effect factor, and filters; the velocity pressure p_V, dissipated in the conditioned room, must also be added to the total for fan sizing).

COMMENTS

The two longer branches in this example have essentially the same pressure drop and are therefore self-balancing. The shorter duct with only one-half the pressure drop of the other branches will require a *balancing damper* to provide approximately another 0.113-inWG pressure drop. This will result in a system in which the pressure drop in each branch is balanced, i.e., essentially the same. In practice, all ducts in an air distribution system will have balancing dampers. Even when fully open, these dampers will have an associated pressure drop. This additional flow resistance of the balancing dampers is treated in Prob. 11.5.
   Return air systems are designed by using the same technique as in this example. The only difference lies in the pressure coefficient for tees because the flow converges in return ducts while it diverges in supply ducts. The appendix on the CD-ROM contains data for both cases. As shown in Fig. 11.1, a separate return fan is often used in the return air system.
   Recall that we have ignored the duct size transitions that occur at points B and D in both downstream branches. These four resistances are small but will be included in Prob. 11.7. Abrupt transitions are to be avoided due to excessive pressure loss.

Although the majority of this section has had to do with airflow design of ducts, their thermal design is also important. Since conditioned air from the central plant is often at a temperature quite different from that of the space in which ducts are located, heat losses or gains must be controlled by the use of duct insulation. The methods of heat flow estimation given in Chap. 2 apply. Warm air ducts in unheated spaces (e.g., crawl spaces in residences) and cool air ducts in uncooled spaces (residential attics, commercial building zone plenums) are prime candidates for added insulation. The heat loss or gain from ducts must be accounted for in secondary system design. Insulation either can be applied to the duct exterior or can be used as the duct liner, where it also serves to absorb sound.

Typically, 1- to 2-in (2.5- to 5-cm) thicknesses of rigid duct insulation provide the needed insulation. ASHRAE Standard 90 recommends that the R value (USCS units) of duct insulation be chosen according to $R = \Delta T_{\text{duct}}/15$, where ΔT_{duct} is the air temperature difference from inside to outside of the duct; this R value is for the required insulation only and does not include inner and outer surface convection resistances. Ducts conducting cold air may require external vapor barriers (with sealed seams) to avoid condensation and insulation degradation. Most building codes also require that duct insulation meet standard fire hazard requirements.

11.1.4 Air Distribution in Rooms

After having been delivered to a room by the duct system described above, conditioned air must be effectively distributed in the space to ensure the comfort of its occupants. The design of air diffusers and grilles relies in large part on manufacturer's data, since the design of these components varies so much. In this section we discuss the basic principles of room air distribution.

Figure 11.11 shows how a jet of air issuing from a duct outlet behaves upon entering a room. The upper diagram shows that an isothermal jet slowly diffuses into the surrounding still air. If the jet is round, the velocity v at any point in the jet at standard conditions in air is given by Schlichting (1979) as

$$v = \frac{7.41 v_0 \sqrt{A_0}}{x(1 + 57.5 r^2 / x^2)^2} \tag{11.11}$$

where
v_0 = velocity at jet source
A_0 = jet flow area at source
x = distance from jet source
r = distance from jet centerline

This equation shows that jet centerline velocity decreases essentially inversely with distance from the source. One can define a jet radius in several ways; e.g., it could be the distance from the centerline at which the velocity is 1 percent of the centerline velocity. The jet radius r increases with distance from the source.

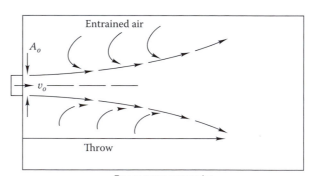

Room-temperature jet

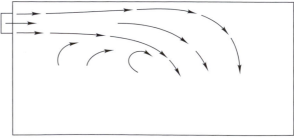

Cool jet

FIGURE 11.11
Isothermal and cool jet diffusion patterns.

Example 11.3

Air at standard conditions issues from a 15-cm-diameter (6-in-diameter) circular opening with a uniform velocity of 3 m/s (9.8 ft/s). What is the centerline velocity 3 m (9.8 ft) from the jet source?

Given: $D_o = 0.15$ m, $v_o = 3$ m/s

Figure: See Fig. 11.11.

Find: $v(x = 3, r = 0)$

SOLUTION

Equation (11.11) can be used to solve the problem by setting $r = 0$ m. The jet source area is

$$A_o = \frac{\pi}{4}D^2 = 0.7854(0.15)^2 = 0.0177\,\text{m}^2$$

The jet centerline velocity is

$$V = \frac{7.41(3.0)\sqrt{0.0177}}{3.0} = 0.985\,\text{m/s}\,(3.2\,\text{ft/s})$$

COMMENTS

Since conditioned air jets are usually either warmer or cooler than room air, they will not have a horizontal centerline. They will either rise if warm or drop if cool (see Fig. 11.11). This example shows that jets at typical duct air velocities will penetrate too far into a room for proper air circulation. Cool air should not have a velocity of more than 45 to 50 ft/min (0.25 m/s) in occupied rooms, e.g., for good comfort. Therefore, special fittings called diffusers are fitted to the ends of ducts to better distribute airflow in rooms.

Figure 11.12 shows a *diffuser* which is used to distribute air through a room without the jet effect described in Example 11.3. Diffusers are normally mounted in the ceiling plane and are connected to the distribution duct system above the ceiling. They can be either square or round. The performance of diffusers depends on design details. As a result, their performance must be determined from manufacturer's data.

Figure 11.13 shows an example of manufacturer's information for a round diffuser. The throw of the diffuser must be related to zone size to ensure that no dead airspaces exist.

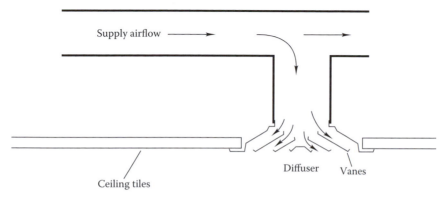

FIGURE 11.12
Cross-sectional sketch of ceiling diffuser.

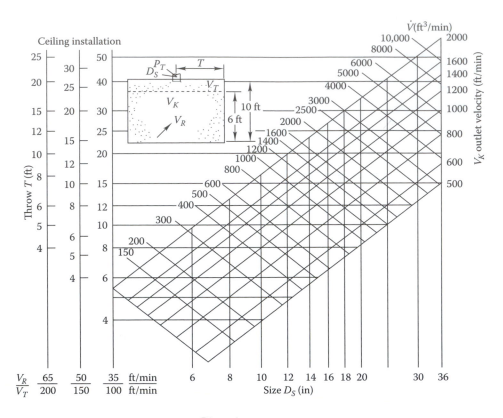

Dimensions

Listed size D_S(in)	6	8	10	12	14	16	18	20	24	30	36
A_K outlet area	0.16	0.28	0.44	0.66	0.91	1.2	1.5	1.9	2.8	4.3	6.2
A_N neck area	0.20	0.35	0.55	0.79	1.1	1.4	1.8	2.2	3.1	4.9	7.1

Pressure drop table

V_K outlet velocity	500	600	700	800	900	1000	1200	1400	1600	1800	2000
P_T w/#4 damper	0.02	0.03	0.04	0.05	0.07	0.09	0.12	0.17	0.22	0.28	0.34
P_S w/#4 damper	0.01	0.01	0.02	0.02	0.03	0.04	0.06	0.08	0.11	0.14	0.17
P_T w/o #4 damper	0.02	0.02	0.03	0.04	0.05	0.06	0.09	0.12	0.16	0.20	0.25
P_S w/o #4 damper	<0.01	<001	0.01	0.01	0.02	0.02	0.03	0.04	0.05	0.07	0.08

Pressure accuracy is ±0.01 in or 10%, whichever is greater.
When diffusers are used on exposed duct, multiply the throw (T) by 0.07.

Symbols: V_T – Terminal velocity (ft/min) T – Throw (ft) P_T – Total pressure (inWG)
V_R – Room velocity (ft/min) A_K – Outlet area (ft²) P_S – Static pressure (inWG)
V_K – Outlet velocity (ft/min) A_N – Neck area (ft²)

FIGURE 11.13
Example of air diffuser performance data table for an NC level of 35. Symbols are defined in the figure. (Courtesy of Allied Thermal Corporation. With permission.)

The terminal velocity V_T shown in Fig. 11.13 is the average air velocity at the throw distance T (i.e., the half-width of the room, as shown). The figure is used by entering it with the design value of air volumetric flow (known from load calculations) and throw (based on room size). At the intersection of the throw and flow lines, one drops down vertically to find the size (i.e., throat diameter D_s). From the size the designer reads the diffuser pressure drop, neck size, and diffuser outlet velocity. Most persons involved in sedentary activities will not be uncomfortable at room air velocities of 50 ft/min (0.25 m/s) (ASHRAE, 2001, 1989). The three room velocities V_R shown in the left center of the figure are typical for offices.

In the absence of manufacturer's data, during preliminary design the throw can be estimated from the dimensional equation

$$T = K \frac{\dot{V}}{\sqrt{A_{\text{eff}}}} \tag{11.12}$$

with the throw T in feet, the net flow area A_{eff} in square feet, and the flow rate $\dot{V}$ in cubic feet per minute. The dimensional constant K varies approximately linearly between 0.012 and 0.0075 between terminal velocities of 100 and 200 ft/min, respectively.

For rooms with ceilings higher than the nominal 10 ft (3 m), special diffuser designs are needed. This is particularly critical when one is attempting to heat tall spaces such as auditoriums and gymnasiums from duct systems located near the ceiling.

ASHRAE (2001, 1989) discusses the allowable air noise levels for various spaces in terms of the *noise criteria* (NC) value. Briefly, the NC value of a diffuser is the emitted sound pressure weighted by the octave of the sound emitted. For a quiet room the NC value is low (25 to 30), but for a factory it is high (above 50). These values are also supplied in manufacturer's data such as those shown in Fig. 11.13, which is based on an NC rating of 35.

11.2 Piping Design

This section discusses the application of liquid flow fundamentals from Chap. 5 to heating and cooling systems for buildings. Since many of the basic equations are similar to those used for fans, this section will make reference to the discussion on fans in Sec. 11.1. For example, the fan laws are replaced by *affinity laws* relating pump flow, speed, and pressure rise at various operating conditions. The equations given in Chap. 5 are exactly the same as the fan laws. System and pump curves have the same interpretation as for air systems. Their point of intersection determines the operating point for pumped systems.

Variable-speed drives are often used for pumps to reduce operating costs. Fig. 11.14 shows how the pump curve is affected by reducing the speed. The power drops with the cube of the pump speed (if efficiency remains constant), as is the case with fans. Variable flow can also be achieved by throttling flow with a valve analogous to damper control of airflow with fan systems.

Piping is manufactured from various materials including steel, copper, polyvinyl chloride (PVC), and stainless steel. For hot water and steam, black steel piping is used; for chilled water (and refrigerants such as R22), either steel or hard copper is used. Cold water and condensate drainage use either metal or PVC piping. Chemically reactive liquids require stainless-steel or glass-lined steel piping.

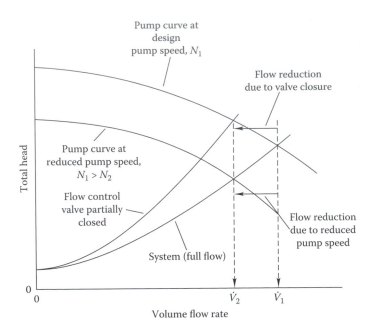

FIGURE 11.14
Pump curves for variable-speed drives also show effect of partial closure of flow control valve.

This section considers only the flow of water or of aqueous solutions (including anti-freeze mixtures) in the context of building secondary systems for heating and cooling. Steam and refrigerant flow are described in Chap. 5.

11.2.1 Design Considerations of Water Systems

Since hot water system design is more involved than cold water design, we present it first. Much of the information presented on this topic also applies to chilled water system design. The objective of pipe design is to deliver the amount of water[6] at the proper temperatures needed to meet the loads in each zone of a building. As with duct design, there are a number of constraints to piping design. They include available space, the need to meet loads in a wide variety of zones under all load conditions, economic criteria, and minimal operating costs. The piping designer's task is to determine needed water flow in a building including the primary heating and cooling systems as well as the water-based secondary system, if any. All-air HVAC systems may not need a secondary liquid system.

After the first flow estimation step based on zone loads, a preliminary primary and secondary piping system is laid out on a set of preliminary drawings. Once a layout has been made, pipes are sized based on needed heating or cooling rates at each central plant component and in each zone for each terminal device. The appendix CD-ROM contains a table that can be used to assist with quick sizing of water, steam, condensate, and natural gas pipe sizing. Pressure drop calculations can be made at this point, and a pump is selected. A cost estimate is made. The next design refinement accounts for potential flow imbalances in the original design. Ultimately a set of final design drawings is prepared.

---

[6] We will use the term *water* generically to include antifreeze solutions as well.

Piping systems consist of pumps, filters, valves, piping runs, pipe fittings, tanks, heating and cooling generation and transfer devices, and specialty items depending on the application. The velocity through piping systems is limited either by noise or by erosion criteria. In piping runs in or near occupied spaces, velocities should be limited to 4 ft/s (1.2 m/s) to control noise. Even if noise is not a problem, velocities in copper piping should be limited to 7 ft/s (2 m/s) to avoid erosion. For preliminary design, a pressure drop of 4 ftWG/100 ft (0.4 kPa/m) of pipe is usually used. The final design may deviate from this value depending on an economic analysis that considers pipe cost, pumping power, and insulation prices.

Figures 5.4a and b shows the pressure drop to be expected in commercial steel pipe at various volumetric flows. This pipe pressure loss can also be calculated from Eqs. (5.13) and (5.15). Added to the loss for straight pipe runs is the loss for fittings, coils, and other components, which is calculated based on the number of velocity heads lost:

$$\Delta p_{\text{fit}} = K \left(\frac{1}{2} \rho v^2 \right) \tag{11.13}$$

in which v is the velocity and K is the resistance coefficient, tabulated in Chap. 5. Addition of the fitting and straight-pipe resistances results in the *system characteristic* for a piping circuit

$$\Delta P = \left[f \left(\frac{L}{D} \right) + \sum K \right] \left(\frac{1}{2} \rho v^2 \right) \tag{11.14}$$

in which f is the pipe friction factor.

Some designers use the equivalent-length method instead of K. However, equivalent lengths are valid for only one (usually unspecified or unknown) velocity. Therefore, the results may be less accurate than if one uses the K approach of Chap. 5.

Since piping systems operate within a relatively narrow flow range, the values of f and K are approximately constant, with the result that the system characteristic for a pipe loop can be expressed as

$$\Delta P = K_{\text{sys}} \dot{V}^2 \tag{11.15}$$

This form is particularly useful since pump curves are also expressed in terms of Δp and $\dot{V}$. A typical pump curve is shown in Fig. 11.15. In addition to the flow versus pressure characteristic with which we are now familiar, other data are given. The dashed lines are lines of constant horsepower whereas the solid lines labeled with percentages are constant-efficiency lines. Of course, one selects a pump so that the required flow/pressure rise condition occurs near the maximum efficiency point. For example, suppose the pump of Fig. 11.15 is needed to provide 750 gal/min (47 L/s) at 85 ft (26 m) of head. From the figure it is seen that a 10-in- (25-cm-) diameter impeller can meet this condition with a power input of 20 hp (15 kW). The efficiency is shown to be 81 percent. A dimensional equation for pump power is sometimes useful for designers using the inconsistent USCS units of gallons per minute and feet of head. The motor horsepower can be found from

$$\dot{W}_{\text{shaft}}(\text{hp}) = \frac{\dot{V}(\text{gal/min})H(\text{ft})}{3960 \eta_{\text{pump}}} \tag{11.16US}$$

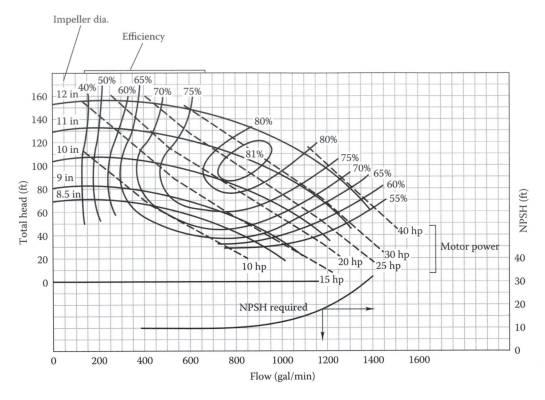

FIGURE 11.15
Example pump curve showing constant-efficiency lines (labeled with %), NPSHR, and input power (labeled in hp units) for five impeller sizes (labeled in inches).

$$\dot{W}_{shaft}(\text{kW}) = \frac{\dot{V}(\text{m}^3/\text{h})H(\text{m})}{369\eta_{pump}} = \frac{\dot{V}(\text{L/s})H(\text{m})}{102\eta_{pump}} \qquad (11.16\text{SI})$$

To illustrate the use of this equation, consider the operating point noted above and the efficiency of 81 percent read from Fig. 11.15:

$$\dot{W} = \frac{750(85\,\text{ft})}{3960(0.81)} = 19.9\,\text{hp} \qquad (11.17)$$

The value read from the curve matches the calculated power very closely.

Below the pump capacity, power, and efficiency curves, a curve is shown to find the *net positive suction head* (NPSH). This design parameter is the value of static pressure (at the pump inlet flange) that the manufacturer specifies to avoid cavitation—the damaging flash of hot liquid to vapor (or the release of soluble gases from the pumped liquid) and its recondensation in pump blading. Cavitation is likely to occur at the pump inlet due to the local acceleration of fluid that occurs at the pump impeller entry. Cavitation must be avoided in any proper piping system design. The designer must guarantee that the *available NPSH*, called the *NPSHA*, is greater than the *required NPSH*, called the *NPSHR*, read from the pump curve. The NPSHR increases approximately as the square of the pump flow rate.

The design pump NPSH establishes the reference pressure above which the entire piping system operates.

An expansion tank is often connected to the pump suction piping in closed systems not far from the pump, to establish a constant-pressure reference point in the system. The expansion tank pressure is set to equal at least the NPSHR plus all friction losses between the expansion tank fitting and the pump inlet, as shown in Fig. 11.16. Adequate NPSHA is not often a problem with closed systems but must be carefully considered with pumps in open systems, such as cooling tower sump pumps where the amount of static head at the pump inlet is limited. To meet the NPSHR criterion for a cooling tower, the pump's inlet must be located a distance below the tower sump equal to the NPSHR plus any pipe friction losses between the sump outlet and the pump inlet.

Figure 11.16 shows an expansion tank and air separator connected near the inlet to a pump. We have explained the function of the expansion tank but not its sizing. The purpose of an expansion tank is to allow water to expand into an airspace to avoid high stresses that would otherwise occur in a piping system if no space for water expansion existed. Unconfined water expands almost linearly with temperature in the range of temperatures encountered in HVAC systems. For example, water expands about 2 percent between 40 and 150°F (5 and 65°C); it expands a little more than 3 percent between 40 and 200°F (5 and 94°C). Expansion tanks can be sized by using the expression

$$V_{\text{exp}} = V_{\text{sys}} \frac{v_2/v_1 - 1 - 3\alpha\Delta T}{1 - p_1/p_2} \tag{11.18}$$

where

V_{exp} = needed working volume of tank
V_{sys} = volume of water in entire system
v_1, v_2 = specific volumes of water at minimum and maximum design systemtemperatures, respectively. See discussion of p1 and p2 below

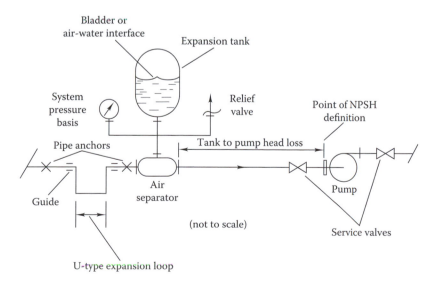

FIGURE 11.16
Expansion tank connection upstream of pump.

$\alpha =$ linear coefficient of thermal expansion for piping material; 0.000065/°F (0.000117/K) for steel and 0.000093/°F (0.000167/K) for copper

$\Delta T =$ difference between maximum and minimum piping system operating temperatures

$p_1, p_2 =$ tank gas pressures, respectively, at (1) system initial fill temperature (for hot water systems) or operating temperature (for chilled water systems) and (2) maximum acceptable system pressure corresponding to condition for v_2

Equation (11.18) assumes that the air (or nitrogen) charge is added to the tank after the system is filled at its operating pressure. To avoid dissolving the gas charge in the system water, a flexible, rubber diaphragm often separates the charge and the system water. Further, it is assumed that the tank is isothermal and does not track system operating temperature; this is accomplished by using an uninsulated tank thermally isolated from the active pipe loop by a small-diameter pipe.

Another method of pressurizing an expansion tank without a bladder is to simply fill the system. Air trapped in the tank will serve as an air cushion for the system. Expansion tanks of this type are larger than the bladder tank for the same service since the denominator in Eq. (11.18) is replaced, for these tanks, by the quantity $p_a/p_1 - p_a/p_2$, in which p_a is the local atmospheric pressure. Even though this type of tank will be larger, it is preferred for hot water systems since air that comes out of solution at the air separator will collect at the top of the tank. Air is continuously introduced into hot water systems along with boiler makeup water. For chilled water systems, a bladder-type design is used since cold water retains air in solution.

The air separator shown in Fig. 11.16[7] is a special fitting where low velocities and a vortex are produced to separate entrained air from system water. The air then rises to the expansion tank, where it collects and can be periodically released if necessary. The amount of air in a water loop is limited; if there are no water leaks requiring continuous makeup water, the amount of air in the tank will stabilize.

Chilled water piping design is less complex since the conditions under which chilled water is used are less critical. For example, systems are often closed and operated at conditions where the NPSH needs of pumps are easy to meet. The temperature swings of chilled water piping are often less than those for hot water piping since the pipes are insulated (with moistureproof insulation) and not exposed to the sun. Hence the expansion tanks can be smaller than those in hot water systems. It is important to recognize, however, that chilled water systems can become relatively warm when the chiller is not operating. Fluid expansion under these conditions must be considered in the expansion tank sizing [see discussion of Eq. (11.18)].

In addition to fluid expansion, the designer must account for expansion (and contraction) of piping itself due to temperature changes in either hot or chilled water systems. The forces produced by pipe expansion can become extremely large. Figure 11.16 shows one method of accommodating pipe expansion by using an elastically deformable expansion loop. The size of expansion loops depends on the distance between expansion loops, pipe

---

[7] Good HVAC piping practice requires that each component be able to be isolated by service valves for replacement or service, hence the valves shown at the inlet and outlet of the pump in Fig. 11.16. In this chapter we do not always show the isolation valves in system diagrams, but the designer must include them. Likewise equipment such as chillers and boilers must be equipped with drain valves for service purposes.

diameter, expected temperature change, and loop design—U bend (as shown in the figure), L bend, or Z bend. The Appendix in the CD-ROM contains nomographs for sizing expansion loops in steel and copper piping networks.

11.2.2 Equipment Arrangement in Liquid Loops

In this section we discuss the assembly of components described above and in Chap. 5 into systems for heating or cooling of buildings. Various arrangements of equipment can be used. This section discusses the key parameters for a successful liquid loop design. An example illustrates the key points.

Systems for large buildings are more complex than the simple systems shown to this point in this section. They often contain several pumps, control valves, and various pieces of terminal heating or cooling equipment. The flow through multiple loops is relatively more complex as a result. For example, large systems will often require standby capacity if a pump should fail. One method of accomplishing this is to use two identical, *parallel-connected* pumps, each with the capacity to provide one-half the design flow. Figure 11.17 shows the arrangement to be used and the resulting pump curves. The curve of parallel-connected identical pumps is just the curve of a single pump except that the flow rate is

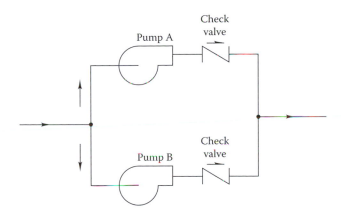

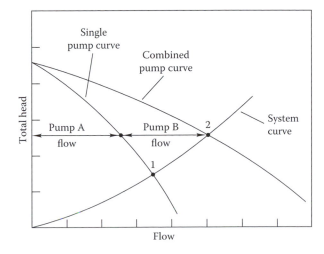

FIGURE 11.17
Parallel-pump arrangement with combined pump curve.

doubled. The system curve intersects the dual-pump curve at operating point 2 (denoting two pumps) with corresponding flow $\dot{V}_2$.

If one pump should fail, the operating point is determined by the system curve intersection with the single-pump curve at point 1 (the check valve ensures that reverse flow through the inactive pump does not occur). It is seen that the flow rate is not one-half of that which exists for two active pumps. In fact, the flow for a single pump in this situation is about two-thirds of the flow of both pumps taken together. For most HVAC systems, this flow rate is adequate temporarily for most conditions, since the full capacity is needed only at peak-load conditions.

Series pumping arrangements are sometimes used when high-pressure drops must be overcome without the use of special positive-displacement pumps. If identical pumps are used in series, the combined, effective pump curve is that of a single pump except that the developed head at any flow is double that of a single pump. The series arrangement is less common than the parallel one in HVAC systems.

Downstream of the multiple-pump station, multiple terminal units are also common in large buildings. A single loop may heat or cool many zones on several floors of a skyscraper. The layout of a piping loop to minimize installation and operating costs involves several tradeoffs. Figure 11.18 shows the two generic designs of systems used for connecting multiple cooling coils to a central plant (this discussion applies to heating systems as well). The system in Fig. 11.18a employing the *direct-return* approach uses less piping than the other, but has flow imbalance problems since the flow resistance through branch 1 is much less than that through branch 4 with its greater length of piping. It is poor design to rely on the control valves that respond to load to also balance flow. Therefore, the additional balance valves shown will be needed to balance flow with the control valves serving their purpose of flow control to meet the cooling load.

The *reverse-return* approach, shown schematically in Fig. 11.18b, has approximately equal pipe lengths for each cooling coil. As a result, the flow is balanced automatically. In fact, if the pressure drop through the header is one-tenth that through the branch lines, flow will balance to within 3 percent in each branch. Balancing valves can often be

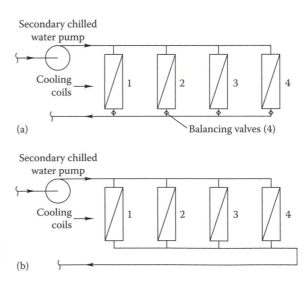

FIGURE 11.18
(a) Direct-return and (b) reverse-return piping arrangements (for example of multiple cooling coils).

eliminated, but an additional length of pipe is needed on the longer-return pipe run. Control of each cooling coil will be proper if one uses this approach.

As discussed earlier, the expansion tank is placed at the pump inlet to provide the pump NPSHR. The tank pressure sets the entire system pressure since the pump inlet is the lowest pressure point of the system. Expansion tanks are needed in all *closed* systems but not in systems *open* to the atmosphere. All liquid loops in HVAC systems are closed systems except for the cooling tower or evaporative cooling circuit, if either is present. In open loops, the design is not much different from that for closed systems, except that system pressurization must rely on gravity. The major practical effect is that pumps must be positioned so that they do not lose their prime or have inadequate inlet NPSH.

In complex systems, there may be conflicting flow requirements in various parts of a circuit. For example, a chiller may be designed to operate at a 10°F temperature change in the evaporator, but the designer may specify a 20°F temperature drop across the cooling coils to save pumping power. Or a variable-flow distribution system may be used to save pumping costs with a fixed-chiller-flow requirement. One method of satisfying both criteria in these two cases is to use a *primary and secondary* loop design, as shown in Fig. 11.19.

The basic concept of primary-secondary system design is to have a relatively short common section of pipe between the two systems. The primary circuit flow is determined by *central plant* flow specifications, and the secondary circuit flow is determined by requirements of the *secondary* distribution system. If the common-pipe section is short and of appropriate diameter, its pressure drop will be very small and the flow in one loop will be independent of flow in the other. The flow in the common pipe can be in either direction depending on the relative magnitudes of the primary and secondary loop flows. The direction can change during a day depending on loads or on operating system on and off times.

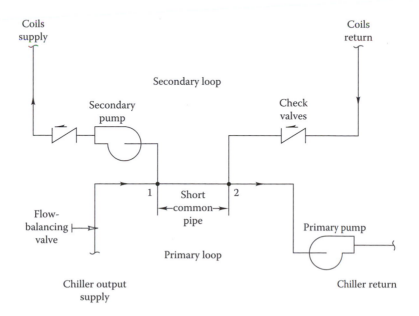

FIGURE 11.19
Primary and secondary loop concept (shown for chilled water loop). Common pipe is designed to be short and large so that pressure drop is very small relative to system pressure drop.

Example 11.4: Primary and Secondary Loop Flows

The flow through a chiller is 500 gal/min (32 L/s). At a given time of day, the load on the secondary system is 625,000 Btu/h (183 kW) with a temperature rise of 10°F (5.5°C). Find the secondary loop flow and the temperature drop across the chiller-evaporator under these operating conditions.

Given: $\dot{V}_{chill} = 500 \, \text{gal/min}$

$\dot{Q}_{chill} = 625{,}000 \, \text{Btu/h}$

$\Delta T_{coils} = 10°F$

Figure: See Fig. 11.19.

Assumptions: Steady flow is assumed, and heat transfer from the surface of primary and secondary piping is ignored.

Find: $\dot{V}_{coils}, \Delta T_{chill}$

Lookup values: $\rho_{water} = 8.33 \, \text{lb}_m/\text{gal}$

$c = 1 \, \text{Btu}/(\text{lb}_m \cdot °F)$

Solution

The temperature drop across the chiller-evaporator can be found from the first law of thermodynamics

$$\dot{Q} = \rho \dot{V} c \Delta T_{chill} \tag{11.19}$$

from which the chiller water temperature drop is

$$\Delta T_{chill} = \frac{625{,}000 \, \text{Btu/h}}{(500 \, \text{gal/min})(8.33 \, \text{lb}_m/\text{gal})(60 \, \text{min/h})[1 \, \text{Btu}/(\text{lb}_m \cdot °F)]}$$
$$= 2.5°F$$

By the conservation of energy we can write

$$\dot{Q}_{chill} = \dot{Q}_{coils}$$

from which we can find the coil flow rate (the density of water in primary and secondary loops is the same)

$$\dot{V}_{coils} = \dot{V}_{chill} \frac{\Delta T_{chill}}{\Delta T_{coils}}$$
$$= (500 \, \text{gal/min}) \left(\frac{2.5°F}{10°F}\right) = 125 \, \text{gal/min} \tag{11.20}$$

The flow in the common pipe can be found from this value. By the conservation of mass at either tee joining the two loops, we can see that

$$\dot{V}_{common} = 500 - 125 = 375 \, \text{gal/min} \, (23.7 \, \text{L/s})$$

The direction of flow is from tee 1 to tee 2.

11.3 Complete HVAC Systems for Commercial Buildings

This section discusses complete *secondary* systems for heating and cooling of buildings. There are two generic classes of secondary systems: those using air for heating and cooling and those using water and air. The former include fixed- and variable-air-volume systems while the latter include combined systems using air for ventilation along with coils at each zone for heating and cooling. There are many combinations of these systems, but an understanding of a few basic systems will permit the proper design of hybrids of the basic systems. We discuss three basically different systems in this section from a functional point of view. In Sec. 11.4.1 we will provide a detailed design and sizing procedure for one of the most common of these.

The *anatomy* of a building's HVAC system consists of four main parts:

1. Central plant[8]
2. Distribution system
3. Terminal devices
4. Controls

In Chaps. 9 and 10, we discuss central plants for heating and cooling, and in Chap. 12 we discuss controls in detail. In this section, we will learn how components discussed in Secs. 11.1 and 11.2 can be integrated into various secondary systems consisting of items 2 and 3 of this list.

11.3.1 All-Air Fixed-Volume Systems

Building loads vary greatly during even a single day, as shown in Chaps. 7 and 8. It is a considerable challenge for the HVAC designer to conceive of and execute a system design that will provide comfort during not only a single day but also an entire year under widely varying climatic and occupancy conditions that typify modern building use. In this section we discuss the design of fixed-volumetric-flow-rate air systems for heating and cooling a single zone and then multiple zones.

Figure 11.20a shows an air-based system for heating or cooling a single zone. Air flows through the filter, cooling coil, heating coil, humidifier, and fan to the conditioned space. This secondary system equipment along with a return/outside air mixing chamber and flow control dampers is customarily housed in one custom-built assembly, called the *air handling unit* (AHU). The air handler is usually located near the central plant (i.e., chiller

---

[8] Some systems such as those used in schools and hotels are self-contained, unitary systems without a central plant per se.

and boiler) to offer convenient piping connections to the chilled and hot water coils. Insulated ducts carry conditioned air to the single zone in this system. The zone load depends on both internal loads and loads driven by external conditions such as solar gain and ambient temperature that affect both the transmission load and the ventilation load. The coil load $\dot{Q}_{coil}$ for the space shown in the figure is given on the *air side* of the coil by

$$\dot{Q}_{coil} = \dot{m}_{air}(h_{coil,o} - h_{coil,i}) \tag{11.21}$$

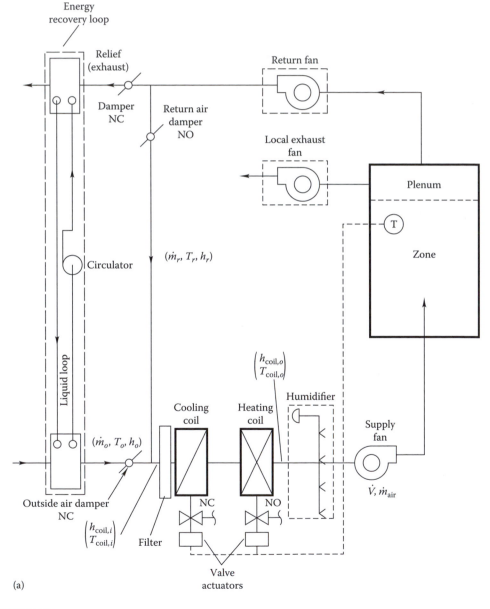

FIGURE 11.20
(a) Single-zone fixed-airflow-volume HVAC system. Optional components are shown in dashed boxes.

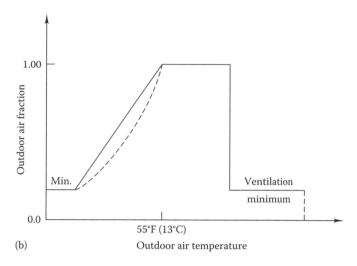

(b)

FIGURE 11.20 (continued)
(b) Outside airflow characteristic for fixed-volume system with economizer.

in which h denotes the airstream enthalpies. The coil inlet enthalpy is determined by the outside air fraction at the air mixing station upstream of the coils and by both the return and outdoor air enthalpies:

$$h_{\text{coil},i} = \frac{\dot{m}_r h_r + \dot{m}_o h_o}{\dot{m}_r + \dot{m}_o} \tag{11.22}$$

The cooling load on the coil can also be equated to the temperature rise on the *liquid side* (connected to the chilled water loop) by

$$\dot{Q}_{\text{coil}} = \dot{m}_{\text{liq}} c_{\text{liq}} (T_o - T_i) \tag{11.23}$$

Equations (11.21) and (11.23) show that as coil loads $\dot{Q}_{\text{coil}}$ vary, the coil output may be made to vary in two ways:

- By changing the airflow rate while fixing the liquid inlet temperature
- By varying the coil inlet liquid temperature (or flow) while holding volumetric airflow constant

The latter approach is called the *fixed-volume* (or variable-temperature) method. Control of the fixed-volume system must ensure that heating and cooling coils not operate simultaneously to avoid energy waste; when both coils operate, they are said to "fight" or "buck" each other. If only the heating or cooling coil is permitted to operate by the control system, the fixed-volume single-zone system can be rather energy-efficient. However, very few commercial buildings can be considered single-zone buildings, as described in Chap. 7. Comfort control is poor in this system with its single pair of coils when used in a building with several thermally distinct zones.

The supply fan shown downstream of the coils in Fig. 11.20a (draw-through arrangement) could be placed upstream of the coils in the blow-through configuration. In draw-through units, cool air from the cooling coil is heated by flow work produced by the fan (and further heated by the fan motor if it is the airstream), thereby requiring that the cooling coil subcool the air to provide expected room supply air. Blow-through systems avoid this problem by having the cooling coils remove fan and motor heat directly, so that subcooling is not needed. However, this design is more expensive, requires more floor area, and can have problems with cooling coil condensate entrainment in the airstream. Draw-through units are the most common design.

The fixed airflow in the single-zone system is determined by the heating and cooling load. The cooling load often requires the greater flow. By default, since only one fan with a fixed flow rate is available, the heating season airflow is, therefore, the same as the cooling season airflow. For sensible *cooling*, the required airflow rate is

$$\dot{V}_{cool} = \frac{\dot{Q}_{cool}}{\rho_{air} c_{p,air} (T_{coil,i} - T_{coil,o})}$$ (11.24)

where
$\dot{V}_{cool}$ = cooling airflow rate, ft^3/h (m^3/s)
$\dot{Q}_{cool}$ = sensible cooling load, Btu/h (W)
ρ_{air} = density of zone air, lb_m/ft^3 (kg/m^3)
$c_{p,air}$ = specific heat of air, $Btu/(lb_m \cdot °F)$ $[kJ/(kg \cdot K)]$
$T_{coil,i}$ = room temperature, °F (°C)
$T_{coil,o}$ = cooling coil outlet temperature, °F (°C)

If both latent and sensible loads are to be considered, Eq. (11.24) must be based on enthalpy, not temperature:[9]

$$\dot{V}_{cool} = \frac{\dot{Q}_{cool}}{\rho_{air} (h_{coil,i} - h_{coil,o})}$$ (11.25)

where
$\dot{Q}_{cool}$ = *total* (sensible plus latent) cooling load, Btu/h (W)
$h_{coil,i}$ = cooling coil inlet air enthalpy, Btu/lb_m (kJ/kg)
$h_{coil,o}$ = cooling coil outlet enthalpy, Btu/lb_m (kJ/kg)

The enthalpy of moist air is found from equations developed in Chap. 4 or from Appendix CD-ROM tables.

The airflow required for single-zone *heating* is

$$\dot{V}_{heat} = \frac{\dot{Q}_{heat}}{\rho_{air} c_{p,air} (T_{coil,o} - T_{coil,i})}$$ (11.26)

where $\dot{V}_{heat}$ is the heating airflow rate, ft^3/h (m^3/s); $\dot{Q}_{heat}$ is the heating load, Btu/h (W); and the coil temperatures refer to the heating coil.

---

[9] Of course, this equation will give an identical result to Eq. (11.24) if only sensible cooling is required.

The fan for a single-zone fixed-volume system is sized by selecting the larger of the two flow rates calculated above:

$$\dot{V} = \max\left(\dot{V}_{\text{cool}}, \dot{V}_{\text{heat}}\right) \tag{11.27}$$

In the single-zone system in Fig. 11.20a, several optional components are shown within dashed lines. The first, a humidifier, is needed to avoid excessive air dryness in the heating season but does not operate in the cooling season. The humidifier is located downstream of all coils and filters to avoid liquid moisture collection on them if water droplets from the humidifier do not completely evaporate within the air handling unit. The humidification load is added to the zone sensible heat load to size the heating coil since the heating coil must provide the energy to evaporate moisture for humidification. Otherwise the evaporating water would cool the heated air below design conditions. Either steam or evaporative-media humidifiers are the most common sources for humidity in commercial buildings.

The second optional component is an exhaust fan that may be needed for local exhaust from a fume hood, stove, or toilet. In many cases, the third option, the return fan, can be avoided in single-zone systems if air is relieved from the space in a low-pressure-drop return system. If large outdoor air fractions are used, however, a return fan will be needed to avoid overpressuring the space in the single zone.

An energy recovery system is the fourth optional subsystem shown in Fig. 11.20a. The leaving conditioned exhaust air in either heating or cooling systems can be used to precondition the entering outside air. In winter, the entering outdoor air will be warmed; in summer, cooled. One method for accomplishing this is via the fluid loop, with two air-to-liquid heat exchangers, shown in Fig. 11.20a. This system, called a "run-around" system, is most often used when the distance between exhaust and supply ducts is large. If the distance is not great, heat pipes, plate heat exchangers, or "heat wheels" can be used for energy recovery purposes.

As described earlier in this chapter, for energy conservation and cost reasons, a portion of the zone return air is reused by mixing it with the required outdoor air upstream of the coils. The use of return air should be maximized, subject to constraints of ventilation needs, to minimize coil loads and energy consumption. During the cooling season, however, one may wish to use increased outdoor air if (1) in dry climates, internal latent loads are high or (2) the *economizer cycle* can provide cooling. Figure 11.20b shows the amount of outside air to be used as a function of outdoor temperature.[10] The relative flows are controlled by the joint action of the three dampers, shown in Fig. 11.20a. The exhaust and intake dampers operate in the same sense whereas the damper in the return air duct is reverse-acting.

The characteristic in Fig. 11.20b is constructed as follows. During hot weather, ventilation minimum outside air is used. As the outdoor temperature drops to the point where the economizer can begin to save cooling energy, maximum outdoor air is used since it is less costly to condition the outdoor air than to cool and dehumidify warm, humid indoor air at the same dry-bulb temperature.[11] At still cooler outdoor temperatures below the nominal

---

[10] The sloped line is straight if dry-bulb temperature control is used. If enthalpy control were used instead, the line would be concave to the left.

[11] The outdoor condition where increased outdoor air should be introduced instead of using return air varies from one location to the next. The designer must make this evaluation on a case-by-case basis. Although the common criterion for the switchover is dry-bulb temperature, to be strictly correct, the comparison of indoor and outdoor enthalpy should be used instead.

building cold air supply temperature (approximately 55°F or 13°C), the amount of outdoor air needed for cooling is gradually reduced to the minimum. During cold weather, only the ventilation minimum outdoor air is used. Although this usage schedule of outdoor air has been discussed in the context of a single zone, it applies to large multiple-zone buildings as well.

Example 11.5: Outdoor Temperature for Minimum Flow

If the return temperature from a zone is 25°C (77°F) and if at least 25 percent outdoor air is required at all times to meet ventilation requirements, below which outdoor temperature will the outdoor inlet damper be at its minimum setting? The design cooling system air supply temperature is to be 13°C (55°F).

  Given: $T_i = 25$°C

$T_{sup} = 13$°C

  Figure: See Fig. 11.20b.

  Assumptions: Ignore temperature rise across the fan.

  Find: T_{min}

SOLUTION

The problem is solved by writing the first law of thermodynamics at the junction of the return and outside air inlet ducts. The only nonzero terms are enthalpy terms. Since the specific heat of air in each duct is the same, the supply air temperature is just the flow-weighted average of the outdoor and return temperatures, shown as follows. The first law for this problem is

$$T_{sup} = \frac{C_{p,\,air}(\dot{m}_r T_r + \dot{m}_o T_o)}{C_{p,\,air}(\dot{m}_r + \dot{m}_o)}$$

The minimum outdoor temperature found by substitution is (after canceling the specific heat of air)

$$0.75(25°C) + 0.25\,T_{o,\,min} = 13°C$$

from which

$$T_{o,\,min} = -23°C\,(-9.4°F)$$

COMMENTS

This outdoor temperature is well below the freezing point of water; hence the designer must guarantee that this very cold air is properly mixed with warm return air. Cold air and warm air do not mix easily. A special air handler component called an *air blender* is needed, or else antifreeze must be used in the coils. Antifreeze is expensive, needs periodic replacement, and has poorer heat transfer characteristics than water. Whatever approach is used, it is the designer's responsibility to avoid coil freezing by specification of controls, outside air dampers, coil fluids, and use of redundant safety measures.

  In the problem statement, the basis of the 25 percent outside requirement was not stated—whether on a mass or a volume basis. In the United States, it is customary to specify airflow on a volumetric basis, whereas the first law is properly used only on a mass basis. The difference between a volumetric basis and a mass basis would be the presence of density terms in the above equations for the mass basis. Since air density depends on absolute temperature and the temperatures involved in HVAC systems are not widely different on an absolute basis, the densities of all airstreams are not much different. This argument is the basis for the customary use of volumetric flow percentages by HVAC designers.

An *energy-inefficient* version of the constant-volume system is the constant-volume reheat system. It can be used for multiple- or single-zone space conditioning. Its use is curtailed by energy codes except for special circumstances where accurate zone temperature or humidity control is needed. This system operates with a fixed-volume flow rate and fixed supply fan outlet temperature, typically 55°F (13°C) selected to meet the peak sensible and latent cooling load. Variation of load within a zone or among zones below the peak is accounted for by adding heat to the air at each zone with a reheat coil (electrical, steam, or hot water). The source of energy waste is obvious—cooled air is reheated prior to release to the zone(s). To minimize energy waste, the cold air temperature should be reset to the highest possible temperature that will just meet the cooling load.

The single-zone single-duct system is the simplest secondary system that one can imagine. However, multiple-zone systems are the norm in commercial buildings because of the diversity of loads and relatively stringent comfort requirements. One could imagine adding zones to the single-zone system in Fig. 11.20a, but control would be poor unless the energy-inefficient reheat system described in the previous paragraph were used.

The *dual-duct* constant-volume system shown in Fig. 11.21 has a single supply and return fan but two sets of ducts—one for cold air and the other for hot air. The heating and cooling coils are energized by the central plant as in the single-zone system. The two sets of ducts terminate at a mixing box at each zone. The relative amount of hot and cold air passed to each zone is controlled by reverse-acting dampers operated by the thermostat in each zone. The airflow is maintained constant to each zone by the action of these dampers. The dual-duct system has several advantages:

1. All space conditioning needs are taken care of at the central plant—no HVAC primary heating or cooling equipment is located in the zones.
2. Since warm air is available year-round, temperature control of lightly loaded zones is good during the cooling season. (If heat were not available, leaks in cold duct dampers could cause lightly loaded zones to be too cold.) Control response to load changes is rapid.
3. Duct sizing calculations are not as critical since the dampers at each zone absorb any pressure imbalance caused by design inaccuracy or part-load operation.
4. If humidity control is needed, this system is a good candidate.

However, there are a number of *compelling disadvantages* to the dual-duct system. Therefore, few dual-duct systems are built today. The problems include these:

1. Ducts leak.
2. Zone damper leakage requires oversizing both the heating plant (to account for cold air damper leakage during peak heating) and the cooling plant (to account for warm air damper leakage during peak cooling).
3. The space required for two full-size ducts and the requirement that each zone have two duct connections make duct design more difficult. Duct velocities and pressures are higher because two ducts must fit into the space of one.

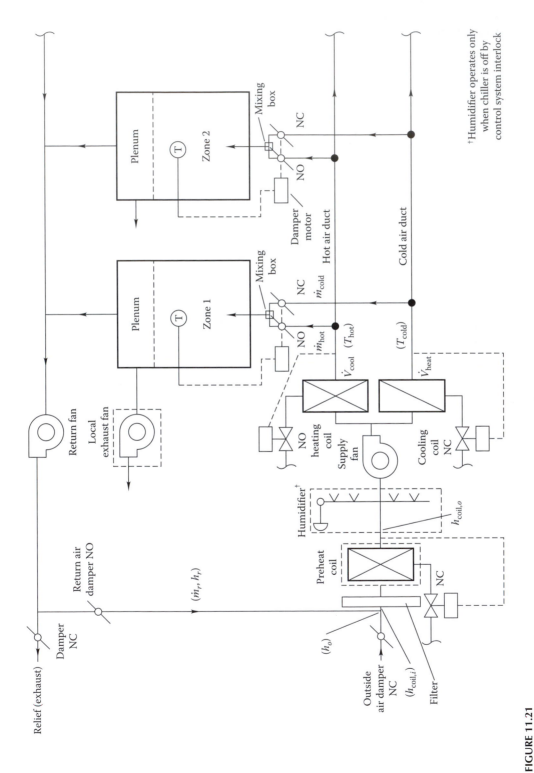

FIGURE 11.21
Dual-duct fixed-volume HVAC system in which mixing occurs at zone mixing box.

4. Overall system control stability requires that the zone mixing boxes be capable of constant volume control. A separate volume control damper is sometimes needed at each zone's mixing box.

5. The system is *not energy-efficient* since reheating of cold air by warmed air and the reverse occur.

6. Initial cost is higher than that for single-duct systems.

7. An economizer should not be applied to the basic dual-duct system because the heating coil inlet air will be at the cold deck setpoint, not at the mixed-air temperature. Heating energy requirements will be much higher and can exceed the cooling energy savings. Hence the benefits of an economizer are not available to dual-duct systems.

The *system* airflow in a dual-duct system is most often determined by the full cooling airflow requirement (see Example 11.6). Therefore, an equation similar to Eq. (11.25) applies except on a systemwide basis. The total system airflow required by a cooling system under peak cooling conditions is

$$\dot{V} = \frac{\dot{Q}_{cool}}{\rho_{air}(h_{coil,i} - h_{coil,o})} \tag{11.28}$$

in which the peak coil cooling load $\dot{Q}_{cool}$ is taken to be the total sensible and latent load for the entire collection of zones served by one air handler.

Under part-load conditions, the dual-duct system air handler flow is the same as under full-load conditions but is split between the hot and cold "decks." The conservation of mass requires that[12]

$$\dot{V} = \dot{V}_{cool} + \dot{V}_{heat} \tag{11.29}$$

To find the cold and hot deck flows individually, the zone loads must be known. Example 11.6 describes the approach.

Example 11.6: Dual-Duct System Part-Load Operation

A dual-duct conditioned zone has a peak heating load of 7.50 kW (25,600 Btu/h) and a peak sensible cooling load of 6.50 kW (22,200 Btu/h). The zone temperature is to be held at 25°C (77°F) (also assumed for simplicity to be the return air temperature).

The two duct supply temperatures at the zone are 40°C (104°F) and 14°C (57°F). If the cooling load is 3.50 kW (11,950 Btu/h) at a part-load point, what are the heating and cooling rates at the zone?

Given: Hot and cold deck and room temperatures:

$T_{hot} = 40°C$ $T_{cold} = 14°C$ $T_{zone} = 25°C$

Design peak loads: $\dot{Q}_{max,cool} = 6.5$ kW $\dot{Q}_{max,heat} = 7.5$ kW

Part load: $\dot{Q}_{part,cool} = 3.5$ kW

Figure: See Fig. 11.21 (consider only one of the zones).

---

[12] Actually mass is conserved, not volume. See Comments in Example 11.5 for an explanation.

Assumptions: Ignore the air temperature rise across the fan and the heat losses and gains in the ductwork. Consider only sensible loads for simplicity. For better accuracy, use a mass basis, rather than a volume basis, for the calculations.

Find: $\dot{Q}_{cool}$, $\dot{Q}_{heat}$

SOLUTION

The solution involves two steps. First, the airflow at peak load is determined. Since the dual-duct system is a constant-volume system, this same flow will apply to part-load conditions. Second, solve for the heating and cooling rates at part load.

The airflow rate needed for peak heating is found from Eq. (11.24):

$$\dot{m}_{hot} = \frac{7.5 \, kW}{[(40 - 25)°C][1.0 \, kJ/(kg \cdot K)]} = 0.50 \, kg/s$$

and for peak cooling the airflow rate is

$$\dot{m}_{cold} = \frac{6.5 \, kW}{[(25 - 14)°C][1.0 \, kJ/(kg \cdot K)]} = 0.59 \, kg/s$$

The higher flow rate (0.59 kg/s) will be that used for both heating and cooling since the constant-volume dual-duct system uses a single constant-volume fan.

The supply air temperature entering the zone at part load (3.5-kW cooling) is found from

$$\dot{Q}_{part, cool} = 3.5 \, kW = (0.59 \, kg/s)[1.0 \, kJ/(kg \cdot K)](25 - T_{supply})$$

from which the supply air temperature is found to be

$$T_{supply} = 19.1°C$$

At the part-load condition, the mixing box energy balance is

$$\dot{m}_{cold}(14°C) + \dot{m}_{hot}(40°C) = (0.59 \, kg/s)(19.1°C)$$

The mass balance requires that the total of hot and cold deck flows be

$$\dot{m}_{hot} + \dot{m}_{cold} = 0.59 \, kg/s$$

Solving the two equations, we have the following cold and hot deck flow mass flow rates:

$$\dot{m}_{cold} = 0.475 \, kg/s \quad \dot{m}_{hot} = 0.115 \, kg/s$$

Finally, the heating and cooling rates are

$$\dot{Q}_{cool} = (0.475 \, kg/s)[1.0 \, kJ/(kg \cdot K)](25 - 14)$$
$$= 5.23 \, kW \, (17,850 \, Btu/h)$$

$$\dot{Q}_{heat} = (0.115 \, kg/s)[1.0 \, kJ/(kg \cdot K)](40 - 25)$$
$$= 1.73 \, kW \, (5900 \, Btu/h)$$

The difference between these two coil loads is the 3.5-kW cooling rate, as required by the problem statement.

COMMENTS

This example illustrates the key problem with dual-duct systems from an energy viewpoint. The heating energy in this example is not needed at all. However, it is present due to the system's basic configuration requiring constant airflow. To compound the problem, not 3.50 kW but 5.23 kW of cooling is needed, since the unwanted 1.7 kW of heat must be extracted by the cooling system in addition to the zone's load. The energy waste is obvious.

The fundamental reason that this waste occurs is that heating is needed to warm a larger-than-necessary volume of air to maintain zone temperatures at part load. The cold deck temperature cannot be adjusted upward since 14°C is needed for proper dehumidification. However, if the air volume could be reduced at part load, the energy waste could be avoided. This is the motivation for variable-volume systems described in Sec. 11.3.2.

One method of simplifying the dual-duct system and reducing cost is to mix hot and cold airstreams at the central air handler, instead of at each zone, and to use only one duct for distribution to the zones. This *multizone* system is shown in Fig. 11.22a. Zone thermostat signals control dampers at the main air handler (see Fig. 11.22b) rather than at the zone mixing box, as in the dual-duct system. Zone control precision is moderately good with this system.

Damper leakage can be a problem since leakage affects the entire hot or cold deck airflow. Leakages of 10 percent are common, although 3 to 5 percent dampers are available. Leakage forces the oversizing of cooling and heating plants by the leakage amount. In hot, humid climates, multizone systems should not be used, since hot deck damper leakage bypasses humid air and can increase the humidity of zone air to unacceptable levels. Other practical problems include the need for running many small ducts—one to each zone—and problems with properly mixing hot and cold airstreams among various zone ducts at part load. The energy penalties of dual-duct systems also apply to multizone systems.

The multizone system is less flexible than the dual-duct system since at most 12 to 14 zones can be served by packaged multizone units. A small zone connected to a multizone unit with a number of large zones with changing loads can have unpredictable airflow and comfort control. Finally, it is more difficult to add a new zone to a multizone unit than to a dual-duct system. As with dual-duct systems, few multizone systems are being built.

Another modification of the dual-duct system is the variable-flow-volume version. In Sec. 11.3.2 we discuss variable-volume systems. Still another refinement to the dual-duct system is the dual-fan dual-duct system. This more complex system consumes less energy since only the amount of cold or warm air needed to meet zone loads is supplied. Cold air and warm air are not mixed, as in the basic dual-duct or multizone approach. Both of the separate hot and cold systems are variable-volume in this design.

11.3.2 All-Air Variable Air Volume

The various fixed- or constant-volume systems described in the preceding paragraphs have significant inefficiencies and energy waste at part load. The air handlers are also expensive to operate since full, peak flow rates are used at part-load conditions. One method of reducing these penalties associated with fixed-volume systems is to reduce the airflow at part-load conditions. In this section, we discuss *variable-air-volume* (VAV) systems for space heating and cooling. In the United States, they are the most frequently installed systems in large commercial buildings. Although this design can solve some problems associated with fixed-volume systems, it can have problems of its own that can test the ingenuity of the HVAC design engineer.

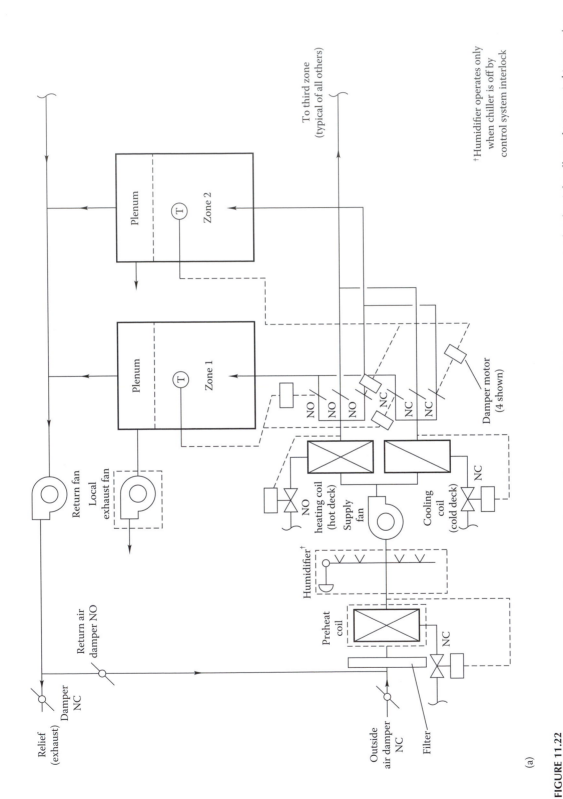

FIGURE 11.22

(a) Multizone HVAC system. Three (of several) zone connections are shown for illustration. Cold air and hot air are mixed at the air handler and transported to zones in a single duct.

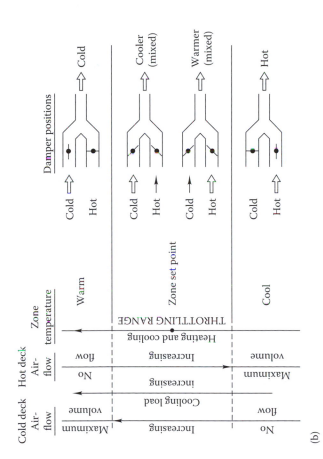

FIGURE 11.22 (continued)

(b) Multizone system damper operation under various heating and cooling load conditions.

The basic concept of a VAV system is to reduce system airflow from full-load levels whenever loads are less than peak loads. Since flow is reduced, energy transfer at the air handler coils is reduced, and fan power is also reduced markedly. The reduction in energy transfer is essentially proportional to the airflow reduction. One would also expect the fan power reduction to be essentially proportional to the cube of the airflow reduction; but fan power is not reduced this much because of the need to maintain duct pressure; as described later. The VAV system is the only commonly used design that reduces energy consumption significantly as the load is reduced. The vast majority of these systems are designed for cooling. (The conditions that might lead one to adopt a variable-volume heating-only system are rarely encountered in commercial buildings.) Some automotive air conditioning systems are variable-air-volume designs with the flow being chosen by the blower speed selector.

Figure 11.23 shows a typical VAV system with several zones. The fundamental VAV system is a cooling-only system that modulates *system* airflow in response to cooling loads as sensed by a dry-bulb temperature thermostat. The basic system is a cooling-only system since the air supply is cool and at a fixed temperature. Therefore, a separate system is needed for zones with heating loads.

Under peak cooling conditions, the VAV system operates identically to a fixed-volume system with the air handler operating at maximum flow and maximum cooling coil capacity. However, at reduced cooling load, the system airflow is reduced by the combined action of the closing of zonal VAV box dampers and the fan speed controller, as described later, using one of the three fan volume control methods discussed in Sec. 11.1. The air supplied to each zone is at a constant temperature, set by the main cooling coil controller. This VAV flow control method is distinct from the constant-volume dual-duct approach, where airflow remains constant but the air temperature supplied to the zone varies. If heating is needed in a perimeter zone, it is supplied by a separate subsystem (such as baseboard radiation or reheat coils) in the simple VAV system; in the heating mode the system then provides only ventilation air, which may itself require some heating. We will discuss other methods of providing heat to zones as an integral part of VAV systems shortly. Since the VAV system is primarily a cooling system, it should be used where cooling is required during the majority of the year. Buildings with significant internal gains and those located in warm climates are good candidates.

Because distribution of air within a zone depends on the velocity of air entering a zone (see Sec. 11.1.4), the reduced airflow needed to meet off-peak cooling loads can drastically affect zone airflow patterns, and hence comfort. As a result, the simple VAV system is often modified to provide constant airflow in a *zone* with varying conditioned air input to each zone. The conditioned air supplied to a zone from the HVAC system air handler is called *primary air*. Constant room airflow with varying primary airflow is accomplished by mixing room air (sometimes called *secondary air*) with primary air within a VAV box. The constant total airflow (conditioned outdoor air plus air recycled within the zone by the VAV box) is typically 4 to 6 air changes/h [or 0.6 to 1.0 $ft^3/(min \cdot ft^2)$] (Holness, 1990) in commercial buildings, irrespective of the primary air supply volume.

The first method for combining primary and secondary air is illustrated in Fig. 11.24a. In this *induction* method, primary air entrains secondary air in a specially designed VAV terminal box located in the ceiling plenum of a zone. Return air from zones normally passes through grilles located in ceiling tiles into the plenum, from which it flows to a return fan. An adequate supply of return air is therefore available in the plenum for mixing with primary air. Since the room air volume flow rate remains essentially constant in induction systems due to the action of the dampers shown, satisfactory room air circulation is

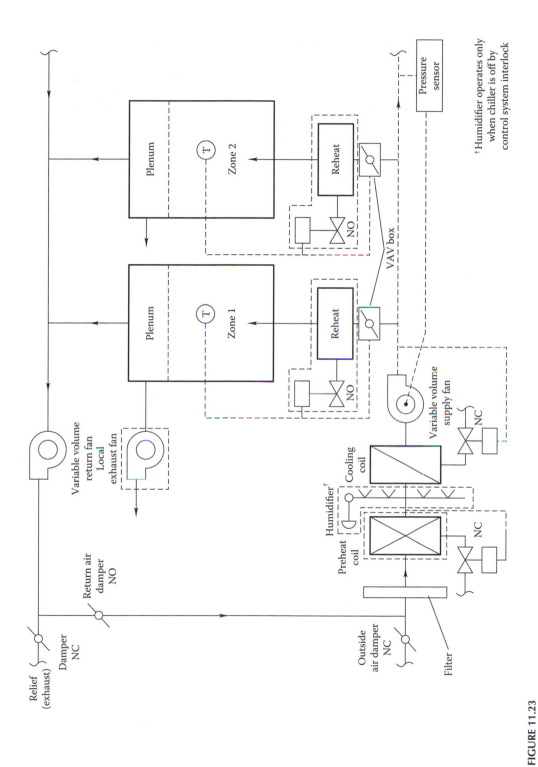

FIGURE 11.23
Variable-air-volume system with optional reheat.

maintained as cooling loads vary. Effective induction systems require increased duct pressure to accomplish entrainment compared to the basic VAV system without induction. Therefore, operating costs are higher.

Another method to guarantee good room air distribution uses *fan-powered* VAV mixing boxes. In this method, a small fan is used to mix primary air (the amount controlled by the primary air damper which is controlled in turn by the room thermostat) and secondary air and to ensure proper distribution throughout the area served by a VAV box. The mixed air thus produced has a varying temperature but a constant airflow as the load varies. Fan-powered units are either of *parallel* or *series* design. Series systems are constant-volume flow systems that use a continuously operating fan located in the mixed airstream to

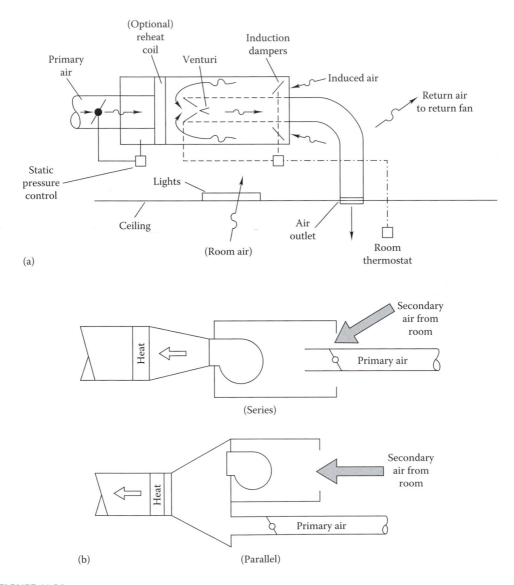

FIGURE 11.24
(a) Schematic diagram of induction VAV box. (b) Schematic diagrams of series and parallel fan-powered VAV box.

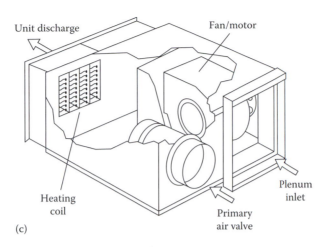

Unit discharge

Fan/motor

Plenum inlet

Heating coil

Primary air valve

(c)

FIGURE 11.24 (continued)
(c) Cutaway drawing of parallel VAV box. (Courtesy of Titus, Inc., Plymouth, IN. With permission.)

provide essentially constant zone airflow. Constant flow is needed in some large zones not only during part-load cooling conditions but also during heating conditions to avoid stratification. Figure 11.24b shows fan-powered series and parallel VAV boxes. Figure 11.24c is a cutaway of a parallel box.

The parallel approach uses an intermittently operating fan. Parallel boxes rely on primary cooling airflow to adequately distribute zone air during most of the cooling season. However, when cooling needs are low or during the heating season, the fan operates in parallel boxes to provide air distribution. Figure 11.25 shows primary and room air characteristics of both types of fan-powered VAV boxes. The key distinction is that series boxes produce constant room airflow for either heating or cooling, whereas the parallel design does not. The operation of the parallel box fan is controlled by the room thermostat.

Another important flow characteristic of VAV boxes is the dependence of flow on supply duct pressure. As described below, duct pressure at one point is controlled by the fan speed controller. However, pressure can vary at other points along the duct as more or less air is required by various zones. It is desirable to have the primary flow to all VAV boxes be independent of local supply duct pressure, to ensure stable system operation. Mixing boxes designed to operate in this way are called *pressure-independent* VAV boxes. Pressure independence is provided by local microprocessors or by mechanical devices responding to duct pressure and zone air velocity. The designer must verify that claims of pressure independence are in fact borne out by independent test data. VAV boxes that are sensitive to supply duct pressure are called *pressure-dependent*; the air valve position in these boxes is controlled by the zone thermostat.

The primary airflow in any VAV system is determined by two considerations. First, the load must be sustained. The airflow rate needed to do so is given by

$$\dot{V}_{cool} = \frac{\dot{Q}_{cool}}{\rho_{air}(h_{room} - h_{coil,o})} \tag{11.30}$$

In this equation the denominator is essentially constant. As a result, the flow rate is seen to be linearly related to the load $\dot{Q}$. However, during some combinations of weather and internal loads, essentially no heating or cooling is needed to maintain comfort.

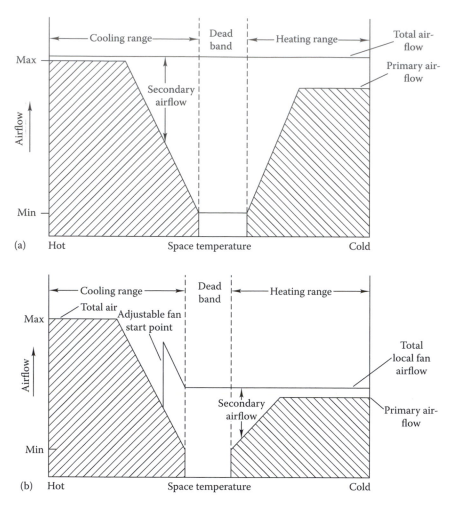

FIGURE 11.25
(a) Flow characteristics for series fan-powered VAV boxes. The filled areas represent primary airflow. The total airflow is constant. (b) Flow characteristics for parallel fan-powered VAV boxes. Filled areas represent primary airflow. Total box airflow is higher during cooling than heating. Fan operates only at zone temperatures below the adjustable setpoint.

Nevertheless air must still be provided to zones for fresh air ventilation purposes. This air amount is shown as the minimum in Fig. 11.25 between the heating and cooling season flow levels. The minimum primary air setting on VAV boxes is usually 15 to 20 percent of the maximum flow.

The need to supply ventilation air and to meet perimeter (or top-floor ceiling) heating loads complicates the cooling-only VAV system discussed to this point. Heat can be added by a coil in the VAV box, as shown in Fig. 11.24b, or by perimeter baseboard or fan coil units. Some heating can be done from heat sources within a zone, such as lights and computers, but this cannot be relied on to warm up exposed zones after periods of no occupancy in winter. Hence, local heat for warm-up, at least, is needed. There may also be a need for heat in internal rooms which are not continuously occupied since the dampers on VAV boxes cannot close entirely, particularly if there is a physical minimum damper closure to ensure adequate ventilation airflow. Without internal gains and without heat, such intermittently occupied spaces can drop in temperature. Heating is also needed if humidity control is a requirement in a zone.

The control of VAV systems is discussed in detail in Chap. 12. However, it is important to understand the general features of the VAV system control in order to design the secondary system properly. Control of these systems is complex and has been one of the sources of difficulty in some installed systems. First, one must ensure adequate flow at the zone most remote from the air handler. This is traditionally accomplished by controlling the supply fan speed with a pressure signal measured near the end of the duct, as shown in Fig. 11.23. The actual airflow through the zones is controlled by each thermostat's control of the damper position. Because of the need to maintain static pressure at the end of a duct run, the fan power savings that one might estimate by using the fan laws are not achievable (Englander and Norford, 1988).[13] However, if the VAV system supply fan is controlled instead in such a way that one box is fully open at all times, then greater fan power savings can be realized because the extra fan power used to pressurize the end of a duct system is no longer needed.

Because of exhaust and exfiltration from zones, the return flow will be less than the supply flow. Return fans are variable-speed and can be controlled on a flow basis if the supply and exhaust flows are known. For cost reasons, these flows are not always measured. An alternative method of controlling the return fan (and exhaust dampers) is by building pressure control.[14] This approach can be difficult since measuring a representative building pressure is sometimes problematical. A variable-volume relief fan is sometimes used in VAV systems. It is located in the exhaust duct (shown in the upper left-hand corner of Fig. 11.23). It operates only when the amount of outside air exceeds the minimum (i.e., during economizer cycle operation) and is controlled by the building static pressure sensor or by airflow measurements in the outside and exhaust air ducts.

The engineering of VAV systems requires *very careful attention to design*. First, as for all systems, the zone loads must be determined accurately to minimize part-load penalties. If calculated loads are higher than actual loads, the VAV system will always operate at part load, and potential savings will not be realized due to part-load inefficiencies; this penalty applies even more strongly to constant-volume systems. Contrariwise, if zone loads are undercalculated, the VAV box will remain open more than expected, thereby starving other zones in the system. A particular problem exists with zones having high latent loads controlled by dry-bulb temperature thermostats. The dry-bulb temperature thermostat may close off the VAV box flow without meeting the latent load.

Flow balancing is also essential if adequate air is to be supplied to all zones. Avery (1986) and Guntermann (1986) describe a number of other VAV system problems that can be avoided by careful design.

---

[13] One also needs to consider safety. For example, if a variable-speed fan drive were to fail to full speed in a system with normally closed (NC) dampers, duct seams could be ruptured. A duct maximum pressure sensor, interlocked with the fan drive, could protect the system. Also, NC boxes open more slowly than the fan reaches full speed; this also requires design attention such as start-up fan speed controller.

[14] Building pressure control is important for infiltration management. A negative building (with exhaust exceeding outdoor air supply) will be less expensive to construct since the air handling equipment and ducts are smaller. However, uncontrolled infiltration may cause discomfort and local freezing of pipes if they are near winter air leaks. A neutral to slightly positive building pressure with exhaust about equal to or less than supply will much reduce infiltration problems.

Example 11.7: VAV System Part-Load Performance

Rework Example 11.6 for a VAV system rather than a dual-duct system. Calculate the primary airflow at the zone and the cooling energy requirement. Assume that the same zone and cooling coil outlet temperatures apply. (The same nomenclature is used here as in Example 11.6.)

Given: $T_{cold} = 14°C$

$T_{zone} = 25°C$

Figure: See Fig. 11.23.

Assumptions: Ignore air temperature rise across fan.

$\dot{Q}_{part,cool} = 3.5$ kW

Find: $\dot{V}_{zone}$

SOLUTION

The solution is simpler than that for Example 11.6. All that is required is to find the volumetric flow from an energy balance based on the 14°C cold air supply temperature:

$$\dot{m}_{cold} = \frac{3.5\,\text{kW}}{[(25 - 14)°C][1.0\,\text{kJ}/(\text{kg}\cdot\text{K})]} = 0.32\,\text{kg/s}$$

The volumetric flow rate is given by

$$\dot{V}_{cold} = \frac{\dot{m}}{\rho_{air}}$$

The density can be found from the ideal gas law:

$$\rho_{air} = \frac{P}{RT} = \frac{101,325\,\text{Pa}}{[287\,\text{kJ}/(\text{kg}\cdot\text{K})][(273 + 14)\,\text{K}]}$$
$$= 1.23\,\text{kg/m}^3$$

Finally, the needed volumetric flow rate is

$$\dot{V}_{cold} = \frac{0.32\,\text{kg/s}}{1.23\,\text{kg/m}^3} = 0.26\,\text{m}^3/\text{s}$$

COMMENTS

This flow rate is only 39 percent of that required by the dual-duct system of Example 11.6. According to the fan laws, the fan power is proportional to the cube of the volumetric flow rate if the flow reduction is achieved by slowing the fan. Hence, the fan power required at this volumetric flow rate is theoretically only 6 percent of that for the fixed-volume dual-duct system! In addition, no heating is required. Finally, the amount of cooling needed is only 3.5 kW, not the 5.2 kW needed in Example 11.6 for the constant-volume dual-duct system. The energy efficiency advantages of the VAV system are obvious compared to those of the dual-duct (or multizone) system.

The *advantages* of properly designed VAV systems over other all-air systems include these:

1. Fan power and central plant energy savings can be significant when VAV systems are applied correctly to appropriate buildings. Cumulative savings of operating costs are usually much greater than the initial cost of installing VAV boxes at the zones.
2. System flexibility is large. Future load changes are easy to accommodate.

3. Balancing of VAV systems is readily achieved if adequate fan capacity exists. If loads exceed design estimates, poor flow balance will probably occur. (The velocity reduction method described earlier in this chapter is adequate for sizing most VAV systems, since flows vary so much from peak to minimum load.)

4. Noise levels are lower most of the time than for fixed-volume systems since maximum flow occurs only infrequently under maximum load conditions.

ASHRAE (1999a) discusses additional advantages and design precautions for VAV systems.

11.3.3 Air–Water Systems

Air is ultimately the principal medium for heating or cooling spaces in all buildings except those that are conditioned by radiant systems. In Sec. 11.3.2, systems that provide warmed or cooled air from the central plant were discussed. In this section, systems that use heated or cooled water from a central primary system to heat or cool zones are discussed. The energy transfer from liquid to air in *air-water systems* occurs at terminal devices in each zone. Temperature control is achieved by controlling water or air temperatures (or both). Humidity control is often a by-product of cooling, but humidification during the heating season is not employed.

The reason for interest in using water for a heat transfer medium is that it has a significantly higher volumetric heat capacity than air—a factor-of-3,500 difference, in fact. As a result, less water need be moved, conduits can be much smaller, and pumping power is much reduced. If a central air system is used, only minimum ventilation air must be provided, thereby reducing duct sizes significantly. This *primary* air is provided by a constant-volume central system. In some systems, outdoor air is not provided by the central system at all—it is introduced into the zone at the zone itself through outside openings. If wind and stack effects are present, this approach can be unreliable for ventilation air supply and zone pressurization. To achieve proper interior air distribution, zone air is recirculated along with the supplied fresh air. Return air systems can be eliminated in many cases since the modest amount of outdoor air supplied to a zone offsets exfiltration, provides needed zone pressurization, or is balanced with local exhaust flows.

Figure 11.26 shows the simplest air-water system. This *two-pipe system* uses an air-to-water heat exchanger for both heating and cooling the zone. The coil is housed in a *fan coil* (or unit ventilator) system located in a zone. The two-pipe system is provided with either hot or cold water for heating or cooling depending on the load and season of the year. The change from heating to cooling is made at the central plant for the entire building. Control is achieved by a zone thermostat that controls coil water flow.

The two-pipe system has inherent practical difficulties in some climates. For example, on a sunny fall or spring day, a zone with southern exposure may need cooling while a zone on the north side may need heating. Simultaneous heating and cooling are not possible. If a single pump is used, as shown in Fig. 11.26, only one flow rate is available. Coil or central plant characteristics may dictate different heating and cooling flows for proper performance. During very cold weather, outdoor air supply dampers at each unit ventilator will be closed (if a separate ventilation air system is not used), limiting the ventilation air supply. Coil freezing must be prevented under low-temperature conditions. Two-pipe systems are not considered practical or energy-efficient in climates where frequent heating-to-cooling changeovers are needed.

One method of solving some of the problems of the two-pipe system is to use the somewhat more complex *four-pipe system*, as shown in Fig. 11.27. The configuration of

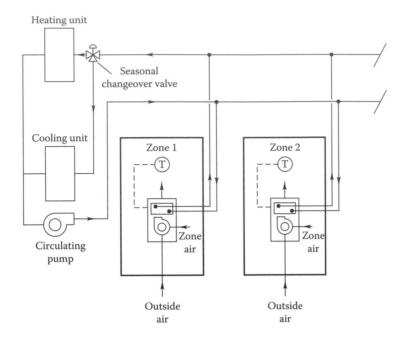

FIGURE 11.26
Two-pipe air-water system with local outside air supply and fan coil terminal units. A single circulation pump is used. Outdoor air can be supplied locally (by unit ventilators) or centrally via a constant-volume system.

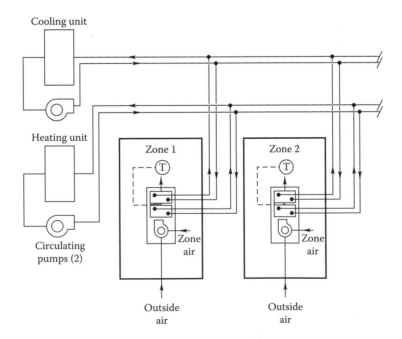

FIGURE 11.27
Four-pipe air-water system with local outside air supply and fan coil terminal units. Separate chilled and hot water pumps are used.

the unit ventilator for a four-pipe system is shown in Fig. 11.28. With the four-pipe system, heating and cooling are both available for different zones that may need each at the same time. In addition, since two pumps are used, the proper flow for each loop (and its primary system) can be used. Of course, the cost is higher, but control is much better.

All water systems have several inherent disadvantages that cannot be avoided with either design. First, maintenance (e.g., changing filters, repairing piping) must be done in the zone space housing the fan coil unit instead of in the equipment room, as in air-based systems. Second, if significant latent cooling loads are present, a means is needed for disposing of condensate from zone latent loads at each unit ventilator. These condensate disposal systems must be cleaned periodically. In addition, filtration at local fan coil units is problematic since filters are generally small (and often inefficient), requiring frequent replacement. As noted above, ventilation can be questionable unless provided by a separate central ventilation system. Humidity control is unreliable in many cases without special measures.

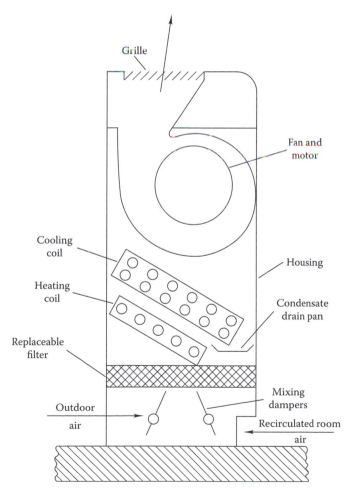

FIGURE 11.28
Cross-sectional diagram of four-pipe fan coil unit showing outside air supply and two separate coils.

In both two-pipe and four-pipe systems, central plant controls may adjust water supply temperatures based on outdoor weather. For example, on a mild day, chilled water can be warmer than on a hot day and still meet the load. Energy savings can be significant if water temperatures are *reset* to match the load conditions in both air-water and all-air systems.

11.4 HVAC System Design Sizing and Energy Calculations

Section 11.3 describes the operation of secondary HVAC systems commonly used in buildings. The *sizing* of the *secondary* HVAC system and the annual energy consumption calculations are summarized in this section. Since this topic is complex and involved, it is not possible to treat it in detail. Therefore, we illustrate the method used to size secondary equipment including coils, fans, and ducts only for the VAV system. Sizing for dual-duct and other systems follows the same procedure and requires the same knowledge of loads as for the VAV system. Annual and hourly energy consumption calculations complete this section.

11.4.1 VAV System Sizing

Selection of secondary equipment is driven by loads imposed by the building on the primary and secondary systems. The calculation of loads is described in Chaps. 7 and 8. Once the *peak* loads are known, the secondary system can be sized as follows. The steps listed will be used in Example 11.8.

Steps in sizing the secondary system components in a VAV system (refer to Fig. 11.23) are as follows:

1. Determine the peak heating and cooling (sensible and latent) loads for all zones, including pickup loads and safety factors (Chap. 7). Establish minimum ventilation airflow rate (Table 11.1) and cooling coil leaving humidity ratio (related to cooling coil effectiveness and hence to coil cost) during the cooling season (Chap. 9). The outdoor air loads are included in steps 8 and 10.

Cooling design day

2. Compute each zone's peak supply airflow rate based on the zonal supply air temperature difference—the difference between the zone air temperature and the zone cooling air supply temperature. This temperature difference is the designer's decision, usually 20 to 22°F (11 to 12°C). Total all zone flow rates to find the total system airflow. This is the *fan size. Duct sizing* can be now completed by using the techniques described in Sec. 11.1.

3. Determine the cooling air supply temperature at the zones by subtracting the supply air temperature difference from the zone temperature.

4. Calculate the required cooling coil leaving dry-bulb temperature to meet the criterion of step 3, accounting for duct heat gains and the air temperature rise through the fan.

5. Find each zone's return air temperature, accounting for return duct heat gains.

6. Calculate the flow-weighted average return air humidity ratio and temperature, including return fan temperature rise (if any).

7. Compute the mixed-air (return and outside airstreams) temperature and humidity ratio. This is the cooling coil inlet air condition.

8. Find the cooling coil sensible and latent loads. This is the *cooling coil rating* (accounting for piping heat gains, this is also the chiller size when safety factors and pickup loads are added).

Heating design day

9. Compute the preheat coil load. The preheat coil heats outdoor air from the design temperature to the coil leaving temperature, determined in step 4. This step sizes the *preheat coil*.

10. Compute the zone heating requirement. It is the sum of the zone heat load from step 1 and the heat required to warm zone supply air from the supply air temperature (see step 3) to the zone dry-bulb temperature. Account for any supply duct heat losses. This step sizes the zone *reheat coil and/or zone baseboard* heating units.

The heat load from steps 9 and 10 along with piping losses, pickup loads, and safety factors determines the central system *boiler size*. Often a diversity factor is applied to the same zone loads, reflecting the fact that the peak loads on all zones are not coincident.

Example 11.8 shows how a secondary system is sized for a small, two-zone commercial building for which the loads are known.

Example 11.8: VAV Secondary System Sizing

A two-zone building is to be equipped with a VAV system with preheat and reheat under the load conditions noted below. Size the supply fan, cooling coil, preheat coil, and reheat coil (or baseboard heating) for this building.

Given: The following data are known from the load calculation and by the designer's specification (these loads exclude the contribution from outdoor air):

	Zone 1 (Exterior)	Zone 2 (Interior)
Sensible peak cooling load	$\dot{Q}_{s_1} = 100{,}000$ Btu/h	$\dot{Q}_{s_2} = 200{,}000$ Btu/h
Latent peak cooling load	$\dot{Q}_{l_1} = 30{,}000$ Btu/h	$\dot{Q}_{l_2} = 45{,}000$ Btu/h
Heating peak load	$\dot{Q}_{h_1} = 200{,}000$ Btu/h	$\dot{Q}_{h_2} = 50{,}000$ Btu/h
Zone temperature T_{zone}	75°F	75°F

Other:

- Summer design dry- and wet-bulb temperatures: 90 and 74°F (resulting in 0.0144 lb/lb$_{da}$ humidity ratio at peak conditions)
- Winter design temperature: 10°F
- Supply system pressure drop at full airflow: 3.0 inWG
- Fan efficiency: 60 percent
- Supply fan air temperature rise: 1°F

- Return air fan air temperature[15] rise: 0.5°F
- Ventilation fresh air flow rate: 2400 ft³/min
- Cooling coil outlet humidity ratio (at typical 90 percent relative humidity): 0.0077 lb/lb$_{da}$
- Zone primary supply air temperature difference ΔT_{sa}: 20°F

Figure: See Fig. 11.23.

Assumptions: Ignore factors not included in the list given above, such as duct heat losses and gains. The location is assumed to be at sea level (the air density is 0.075 lb$_m$/ft³). Peak loads are coincident; no diversity adjustment is used.

Find: Coil loads (three) and fan size

SOLUTION

The solution will follow the numbering system used in the preceding discussion.

   1. *Loads.* The loads and other required design data are given and are listed above.

Cooling design day

   2. *Airflow rates* [Eq. (11.24)]

$$\dot{V}_1 = \frac{100,000 \text{ Btu/h}}{(0.075 \text{ lb/ft}^3)[0.24 \text{ Btu/(lb} \cdot {}^\circ\text{F)}](60 \text{ min /h})(20 \,{}^\circ\text{F})}$$
$$= 4630 \text{ ft}^3/\text{min}$$

$$\dot{V}_2 = \frac{200,000 \text{ Btu/h}}{(0.075 \text{ lb/ft}^3)[0.24 \text{ Btu/(lb} \cdot {}^\circ\text{F)}](60 \text{ min /h})(20 \,{}^\circ\text{F})}$$
$$= 9260 \text{ ft}^3/\text{min}$$

The total airflow is 13,890 ft³/min. The system pressure drop is 3.0 inWG. Therefore, the fan power is [see Eq. (5.50US)]

$$\dot{W} = \frac{13,890 \times 3.00}{0.6 \times 6356} = 10.9 \text{ hp}$$

   3. *Cooling air supply temperature*

$$T_{sa} = T_{zone} - \Delta T_{sa} = 75 - 20 = 55°F$$

   4. *Cooling coil leaving temperature*

$$T_{coil,o} = T_{sa} - \Delta T_{fan} - \Delta T_{duct} = 55 - 1 - 0 = 54°F$$

   5. *Zone return air condition.* According to the problem statement, there are no return duct heat gains; therefore, the return dry-bulb temperature for each zone is 75°F.

   6. *Average return air humidity and temperature.* The return air average moisture content under steady state is the zone supply air humidity ratio (0.0077, given) increased by the total latent load (see Chapter 7):

---

[15] The return fan heat is often less than the supply fan heat because airflow is less (local exhausts remove air locally) and return ducting via ceiling plenums has smaller pressure drop than found in supply ducts.

$$W_r = \frac{(30{,}000 + 45{,}000)\,\text{Btu/h}}{(60\,\text{min/h})(1075\,\text{Btu/lb})(0.075\,\text{lb/ft}^3)(13{,}890\,\text{ft}^3/\text{min})} + 0.0077$$

$$= 0.00883\,\text{lb/lb}_{da}$$

The average return air temperature is 75°F plus the return fan temperature rise of 0.5°F:

$$T_{ra} = 75 + 0.5 = 75.5°\text{F}$$

7. *Mixed-air condition.* To find the mixed-air temperature entering the cooling coil, an energy balance is performed at the point of connection of the return and outside air ducts. Although mass flow rates are theoretically required for this, engineers conventionally use volumetric flow rates instead, since these values are already known from previous steps in the sizing process. Because all three airstreams involved are at nearly the same *absolute* temperature, the density of all airstreams is about the same and the inaccuracy introduced is small. Since the ventilation airflow rate is 2400 ft³/min, the mixed-air temperature under design conditions is

$$T_{ma} = 75.5°\text{F}\left(1 - \frac{2400}{13{,}890}\right) + 90°\text{F}\left(\frac{2400}{13{,}890}\right) = 78.0°\text{F}$$

Likewise, the mixed-air humidity ratio is found by a water vapor mass balance:

$$W_{ma} = (0.00883\,\text{lb/lb}_{da})\left(1 - \frac{2400}{13{,}890}\right) + (0.0144\,\text{lb/lb}_{da})\left(\frac{2400}{13{,}890}\right)$$

$$= 0.00979\,\text{lb/lb}_{da}$$

8. *Coil loads.* The coil sensible load is found from $\dot{Q} = \rho \dot{V} c \Delta T_{coil}$:

$$\dot{Q}_{c,s} = 13{,}890\,\frac{\text{ft}^3}{\text{min}} \times 60\,\frac{\text{min}}{\text{h}} \times 0.24\,\frac{\text{Btu}}{\text{lb}\cdot°\text{F}} \times 0.075\,\frac{\text{lb}}{\text{ft}^3} \times (78 - 54)°\text{F}$$

$$= 360{,}000\,\text{Btu/h}$$

The latent load is found from $\dot{Q} = \rho \dot{V} h_{fg}\, \Delta W_{coil}$:

$$\dot{Q}_{c,l} = 13{,}890\,\frac{\text{ft}^3}{\text{min}} \times 60\,\frac{\text{min}}{\text{h}} \times 0.075\,\frac{\text{lb}}{\text{ft}^3} \times 1075\,\frac{\text{Btu}}{\text{lb}} \times (0.00979 - 0.0077)\,\frac{\text{lb}}{\text{lb}_{da}}$$

$$= 140{,}400\,\text{Btu/h}$$

The total coil load and coil cooling rate is

$$\dot{Q}_{tot} = 360{,}000 + 140{,}400 = 500{,}400\,\text{Btu/h}$$

Heating design day

9. *Preheat coil load.* The peak load on the preheat coil occurs on the winter design day when the minimum outside airflow must be heated to the nominal coil outlet temperature (54°F in this example). Therefore, the preheat (subscript ''ph'') coil size is given by

$$\dot{Q}_{ph} = 2400\,\frac{\text{ft}^3}{\text{min}} \times 60\,\frac{\text{min}}{\text{h}} \times 0.24\,\frac{\text{Btu}}{\text{lb}\cdot°\text{F}} \times 0.075\,\frac{\text{lb}}{\text{ft}^3} \times (54 - 10)°\text{F}$$

$$= 114{,}000\,\text{Btu/h}$$

10. *Zone heating.* The zone heating system must carry the total zone heat load (given) and the heat needed to warm 2400 ft³/min of 55°F supply air to the zone temperature of 75°F:

$$\dot{Q}_h = 200,000 + 50,000 + 2400 \frac{ft^3}{min} \times 60 \frac{min}{h} \times 0.24 \frac{Btu}{lb \cdot {}^{\circ}F} \times 0.075 \frac{lb}{ft^3} \times (75 - 55){}^{\circ}F$$

$$= 301,800 \, Btu/h$$

The total peak heating requirement is

$$\dot{Q}_{tot} = 301,800 + 114,000 = 415,800 \, Btu/h$$

This heat rate must be increased by pickup loads from night setback, safety factors, and piping losses is used to finally size the coils and the boiler.

COMMENTS

Since this is a simple two-zone building, no diversity was assumed in the cooling or heating loads. Although diversity is rarely taken into account in heating loads unless internal or solar gains coincide with peak heating conditions, some diversity can be assumed in cooling loads when the building consists of a larger number of zones. Given two zones, no diversity should be assumed.

This example shows that the secondary systems have a considerable effect on the sizing of the primary heating and cooling plants. In any building more complex than a single-family residence, the relation between zone loads and the secondary system is a first-order effect and must be considered in sizing the central plant. In this example, the *zone cooling load* is 375,000 Btu/h while the *coil cooling load* is 500,400 Btu/h. The difference is due to the outdoor air that is conditioned at the central air handler coil.

In this example we have used a given value of fan temperature rise. The actual rise can be found from the first law of thermodynamics [see Eq. (5.52)]. Although the return fan has not been sized in this example, it can be done exactly as it was done with the supply fan. The duct pressure drop and return fan efficiency must be known to complete the calculation.

11.4.2 Secondary System Energy Calculations—Bin Method

Although the peak condition sizing procedure described in Sec. 11.4.1 determines the size of primary and secondary system equipment once the system concept has been decided upon (it was a VAV system in Sec. 11.4.1), it does not help the designer select the system concept itself, nor does it assist with assessing the relative economic and energy efficiency of various manufacturers' equipment offerings once a system concept has been selected. In this section we describe how the year-long performance of an HVAC system can be calculated if the loads are known.

First, we will show by example the use of the "bin" weather data approach (see Chaps. 8, 9, and 10 for examples and for bin method discussions) for finding the annual energy consumption of a VAV system. This approach is a powerful one due to its computational efficiency. However, the methodology does not permit the accurate calculation of transient loads and time-of-day performance indices such as peak electric demand. The bin approach is also not able to assess the performance of any HVAC system including either hot or cold storage subsystems. However, in many cases, the bin approach can give an indication of the *comparative* performance of various systems on a given building. At least six commercial versions of bin-based building analysis computer programs were available at this writing.

If the shortcomings of the bin approach preclude its use in a given problem, an hourly simulation can be used instead. The computational run time needed is considerably greater, but the flexibility is commensurately greater. Although the subject of simulation is beyond the scope of this book, we will present an example of a partial secondary system simulation of a VAV system and a reheat system. The results of even one day of such a simulation point out clearly the differences in efficiency between the two systems. The Energy Plus, DOE2.1, and BLAST computer simulation programs are standard, publicly available software packages for studying the performance of many HVAC systems in a wide variety of buildings.

In the balance of this section, we will solve two examples to demonstrate the bin and hourly simulation methods of assessing secondary system performance. With either method, the number of calculations is large. Therefore, the examples have been simplified to illustrate the methodology without imposing an excessive computational burden.

Example 11.9: VAV System Calculations Based on the Bin Approach

A two-zone building similar to that studied in Example 11.8 is to have its annual cooling and heating energy determined. To simplify the calculations, ignore latent loads (they are treated separately in Prob. 11.39). Use the bin method to make this calculation.

Given: The zone envelope loads along with the corresponding binned dry-bulb temperature data (for Andrews Air Force Base, Maryland) are given in Table 11.4. Chiller and boiler part-load coefficients are included in part-load equations below.

The system parameters that are needed for the solution include the following:

- Zone temperatures: 77°F (25°C)
- Coil outlet temperature: 55°F (12.8°C)
- Supply fan air temperature rise: 1°F (0.55°C)
- Fan size: 20,000 ft^3/min (9440 L/s); equipped with outlet dampers for flow control
- Fan motor power: 17.5 hp (13 kW)
- Chiller output capacity: 600,000 Btu/h (50 tons, 176 kW)
- Chiller full-load COP: 3.9
- Boiler size: 440,000 Btu/h (129 kW)
- Boiler full-load efficiency: 0.80
- Minimum airflow: zone 1: 1100 ft^3/min (519 L/s); zone 2: 1,300 ft^3/min (613 L/s)

Figure: See Fig. 11.23.

Assumptions: The following simplifying assumptions are made:

1. For the purposes of chiller performance estimation, latent loads can be taken to be 27 percent of the sensible load.
2. Ducts are assumed to be adiabatic.
3. Economizer cycle will be used.
4. The system is located at sea level.
5. The building is conditioned 24 h/day.
6. Cooling tower and pump energy calculations will not be made.

Find: Chiller, fan, and boiler annual energy consumption

SOLUTION

Problems of this type lend themselves to spreadsheet solution. The solution to this problem is shown in a spreadsheet in Table 11.4. We will work through the entries in the nonpeak part-load

TABLE 11.4

VAV System Bin Calculation Summary

	1 (Perim.)	2 (Core)
T_{zone}	77°F	77°F
Min. zone flow	1100 ft³/min	1300 ft³/min
$T_{c,o}$	55°F	55°F
ΔT_{fan}	1°F	1°F
Outside air airflow	2400 total ft³/min	
Supply air temp.	56°F	56°F
Return air temp.	77°F	77°F

			Part-Load Quadratic Constants		
			A	B	C
Fan size	20,000 ft³/min	Fan	0.371	0.973	−0.342
Fan power	17.5 hp				
Chiller size	600,000 Btu/h	Chiller	0.179	0.739	0.082
Chiller full-load COP	3.9				
Boiler size	440,000 Btu/h	Boiler	0	1.368	−0.368
Boiler, full-load eff.	0.8				

Basic data

$\rho_{air} = 0.075$ lb/ft³

$c_{p,air} = 0.24$ Btu/(lb·°F)

		Zone 2			Zone 1			Total System					
Bin, °F	Hours	Sens. Load, Btu/h	Zone Flow, ft³/min	$Q_{reh,2}$ Btu/h	Sens. Load, Btu/h	Zone Flow, ft³/min	$Q_{reh,1}$ Btu/h	Tot. Flow, ft³/min	T_{mix} °F	$T_{c,i}$ °F	Q_{ph} Btu/h	Q_c Btu/h	$Q_{h,tot}$ Btu/h
97	6	103,308	4,555	0	224,844	9,914	0	14,469	80.32	80.32	0	395,618	0
92	72	100,296	4,422	0	204,408	9,013	0	13,435	79.68	79.68	0	358,094	0
87	243	97,284	4,289	0	183,972	8,112	0	12,401	78.94	78.94	0	320,569	0
82	428	94,272	4,157	0	163,536	7,211	0	11,367	78.06	78.06	0	283,045	0
77	631	91,260	4,024	0	143,100	6,310	0	10,333	77.00	77.00	0	245,520	0
72	925	88,248	3,891	0	122,664	5,408	0	9,299	75.71	75.71	0	207,995	0
67	858	85,236	3,758	0	102,228	4,507	0	8,266	74.10	74.10	0	170,471	0
62	755	82,224	3,625	0	81,792	3,606	0	7,232	72.02	72.02	0	132,946	0
57	688	79,212	3,493	0	61,356	2,705	0	6,198	69.26	69.26	0	95,422	0
52	686	76,200	3,360	0	40,920	1,804	0	5,164	65.38	65.38	0	57,897	0
47	665	73,188	3,227	0	20,484	1,100	4,464	4,327	60.36	60.36	0	25,049	4,464
42	734	70,176	3,094	0	48	1,100	24,900	4,194	56.97	56.97	0	8,934	24,900
37	708	67,164	2,961	0	−20,388	1,100	45,336	4,061	53.36	55.00	7,182	0	52,518
32	621	64,152	2,829	0	−40,824	1,100	65,772	3,929	49.51	55.00	23,297	0	89,069
27	362	61,140	2,696	0	−61,260	1,100	86,208	3,796	45.39	55.00	39,413	0	125,621

22	212	58,128	2,563	0	−81,696	1,100	106,644	3,663	40.96	55.00	55,528	0	162,172
17	101	55,116	2,430	0	−102,132	1,100	127,080	3,530	36.21	55.00	71,643	0	198,723
12	51	52,104	2,297	0	−122,568	1,100	147,516	3,397	31.08	55.00	87,759	0	235,275
7	13	49,092	2,165	0	−143,004	1,100	167,952	3,265	25.54	55.00	103,874	0	271,826
2	1	46,080	2,032	0	−163,440	1,100	188,388	3,132	19.52	55.00	119,990	0	308,378
Totals	8,760												

Part-Load Calculations

Bin, °F	Hours	Fan PLR	Fan Power, kW	Fan en., kWh	Approx. clg. Load sens. + 27%	Chiller PLR	Chiller Power, kW	Chiller en., kWh	Boiler PLR	Boiler Power, Btu/h	Boiler en., kBtu
97	6	0.723	11.70	70	502,435	0.837	38.56	231	0.000	0	0
92	72	0.672	11.36	818	454,779	0.758	35.44	2,552	0.000	0	0
87	243	0.620	11.00	2,674	407,123	0.679	32.37	7,867	0.000	0	0
82	428	0.568	10.62	4,546	359,467	0.599	29.35	12,563	0.000	0	0
77	631	0.517	10.21	6,445	311,810	0.520	26.38	16,645	0.000	0	0
72	925	0.465	9.78	9,051	264,154	0.440	23.45	21,692	0.000	0	0
67	858	0.413	9.33	8,006	216,498	0.361	20.57	17,649	0.000	0	0
62	755	0.362	8.85	6,684	168,842	0.281	17.74	13,390	0.000	0	0
57	688	0.310	8.35	5,746	121,186	0.202	14.95	10,284	0.000	0	0
52	686	0.258	7.83	5,368	73,529	0.123	12.21	8,374	0.000	0	0
47	665	0.216	7.38	4,909	31,812	0.053	9.85	6,547	0.010	7,613	5,062
42	734	0.210	7.31	5,366	11,346	0.019	8.70	6,386	0.057	41,931	30,777
37	708	0.203	7.24	5,125	0	0.000	0.00	0	0.119	86,922	61,541
32	621	0.196	7.17	4,450	0	0.000	0.00	0	0.202	144,014	89,433
27	362	0.190	7.09	2,568	0	0.000	0.00	0	0.286	198,313	71,789
22	212	0.183	7.02	1,488	0	0.000	0.00	0	0.369	249,819	52,962
17	101	0.177	6.95	702	0	0.000	0.00	0	0.452	298,531	30,152
12	51	0.170	6.87	350	0	0.000	0.00	0	0.535	344,450	17,567
7	13	0.163	6.80	88	0	0.000	0.00	0	0.618	387,575	5,038
2	1	0.157	6.72	7	0	0.000	0.00	0	0.701	427,906	428
Totals	8,760			74,461				124,179			364,749

(Note that assumptions such as 24-h occupancy, no economizer, and a rule-of-thumb method for finding latent loads are *not to be used in actual building design*. They are used here only in the interest of conciseness of the example.)

32°F bin in detail (since both heating and cooling are needed in this bin) and generally describe the balance of the table entries. Note that two parts of the primary plant are also included in the table to the right. The solution has three parts—fan calculations, cooling system calculations, and heating system calculations.

Fan energy calculations
From the table we see that the interior zone has a cooling load of 64,152 Btu/h, and the perimeter zone has a heating load of 40,824 Btu/h in the 32°F bin. The airflow rate for the internal zone is

$$\dot{V}_2 = \frac{64,152}{0.075(0.24)(60)(77-56)} = 2829 \, \text{ft}^3/\text{min}$$

The airflow for the perimeter zone is at the minimum, since this zone is in a heating mode:

$$\dot{V}_1 = 1100 \, \text{ft}^3/\text{min}$$

The total airflow is

$$\dot{V}_{\text{tot}} = 2829 + 1100 = 3929 \, \text{ft}^3/\text{min}$$

Since the fan capacity is 20,000 ft³/min, the fan part-load ratio is

$$\text{PLR}_{\text{fan}} = \frac{3929}{20,000} = 0.196$$

The fan power $\dot{W}_{\text{fan}}$ is given by Eq. (11.7) for outlet damper control:

$$\dot{W}_{\text{fan}} = (17.5 \, \text{hp})[0.371 + 0.973(0.196) - 0.342(0.196)^2]$$
$$= (9.60 \, \text{hp})(0.746 \, \text{kW/hp}) = 7.17 \, \text{kW}$$

In the 32°F bin there is 621 h, therefore the fan energy W_{fan} is

$$W_{\text{fan}} = (7.17 \, \text{kW})(621 \, \text{h}) = 4450 \, \text{kWh}$$

Cooling system calculations
Because of the low ambient temperature of the bin under consideration, no mechanical cooling is needed. In higher-temperature bins where mechanical cooling is required, the chiller power is found from Eq. (10.29) with part-load coefficients from the manufacturer for the selected chiller:

$$\dot{W}_{\text{chill}} = \frac{600,000}{3.9}[0.179 + 0.739(\text{PLR}) + 0.082(\text{PLR})^2]$$

As noted earlier, this example has been abbreviated by using a rule of thumb for the latent load on the chiller. In Prob. 11.39, we will do the latent load calculation completely.[16]

---

[16] Even though there is a zone latent load, it need not be addressed in the 32°F bin since economizer cooling is used.

Heating system calculations

In the 32°F bin, the perimeter zone has a heating load. The heat supplied[17] to zone 1 is the zone heat load itself plus the heat required to raise the supply air temperature from 56°F to the zone temperature of 77°F:

$$\dot{Q}_{reh} = 40{,}824 + 1100(0.075)(0.24)(60)(77 - 56) = 65{,}772\ \text{Btu/h}$$

In addition to the zone heating requirement, preheat is needed to raise the mixed-air temperature to the design system supply temperature of 55°F. The mixed-air temperature for the 32°F bin is

$$T_{mix} = 77\left(1 - \frac{2400}{3929}\right) + 32\left(\frac{2400}{3929}\right) = 49.51°\text{F}$$

And the preheat coil heat rate is

$$\dot{Q}_{ph} = 3929(0.075)(0.24)(60)(55 - 49.51) = 23{,}297\ \text{Btu/h}$$

The total boiler heat rate is the sum of the preheat and reheat requirements:

$$\dot{Q}_{h,tot} = 65{,}772 + 23{,}297 = 89{,}069\ \text{Btu/h}$$

Since the boiler capacity is 440,000 Btu/h, the boiler part-load ratio is

$$PLR_{boil} = \frac{89{,}069}{440{,}000} = 0.202$$

The boiler input heat rate is given by

$$\dot{Q}_{boil} = \frac{440{,}000}{0.80}[0.0 + 1.368(0.202) - 0.368(0.202)^2] = 144{,}014\ \text{Btu/h}$$

The boiler energy use in the 32°F bin is the heat rate multiplied by 621 h in this bin:

$$\dot{Q}_{boil} = 144{,}014(621\ \text{h}) = 89{,}433\ \text{kBtu}$$

The total fan, chiller, and boiler energy requirements for this example are shown along the bottom line of Table 11.4. Also note that the total number of bin hours must be 8760.

COMMENTS

This example forms the basis of a number of homework problems that will complete the energy consumption picture. These will include cooling tower and pump bin calculations. Once you have set up the spreadsheet, it is a simple matter to consider more efficient means of airflow control such as variable-speed drives and inlet vane control. Various chiller and boiler types with different COPs are easy to study by changing the appropriate value in the input data list.

---

[17] For best winter comfort control with VAV systems, heat is often added in two ways. The outside ventilation air heating load is met by the reheat coil while the zone transmission load (if any) is met by baseboard heat. When significant zone heating is needed, it is not possible to accomplish it with only the preheat coil because outlet air temperatures would be excessive.

11.4.3 Secondary System Energy Calculations—Simulation Method

This section shows how one might conduct an hourly computer simulation of a three-zone VAV system during one day of the cooling season. This example is provided to suggest to the reader how secondary system simulations are constructed. In practice, the simulation of secondary systems, as well as primary systems and loads, is considerably more complex, requiring computer codes on the order of tens of thousands of lines in length.

Example 11.10: Simulation for 24-h Period for a VAV System

Table 11.5 contains a tabulation of outdoor weather data and zone sensible cooling loads for a three-zone building. Find the hourly values of the following:

- Zone airflow rate $\dot{V}$
- Zone average return temperature $\bar{T}_r$
- Coil inlet temperature $T_{c,i}$
- Coil sensible load $\dot{Q}_{c,s}$

The system characteristics needed to complete the simulation include

- Zone temperature: 25°C (77°F)
- Cooling coil outlet temperature: 12.5°C (54.5°F)
- Fan air temperature rise: 0.5°C (0.9°F)
- Total outside ventilation airflow: 57 m³/min (2010 ft³/min)
- Supply and return duct heat gain: 0.0 W
- Return fan air temperature rise: 0.0°C
- Minimum zone flow rate: none

Given: Sensible loads and system characteristics listed in Table 11.5 and above. (Latent loads are calculated in Table 11.5.)

Figure: See Fig. 11.23.

Assumption: The location is at sea level.

Find: $\dot{V}$, $\bar{T}_r$, $T_{c,i}$, $\dot{Q}_{c,s}$

SOLUTION

The solution of this example follows the same sequence for each hour. First find the volumetric flow rate needed to meet the load for each zone. The average return air temperature is used to find the flow-weighted mixed-air temperature. This, along with the specified coil outlet temperature, is used to find the coil sensible load each hour. At this point, one could also do the primary system and fan energy calculations from known part-load relationships. However, this procedure was demonstrated in Example 11.9 and will not be repeated here.

The simulation procedure will be completed in detail for the first hour of loads shown in Table 11.5.

The supply air (13°C) density is given by

$$\rho_{air} = \frac{101,325\,\text{Pa}}{[287\,\text{kJ}/(\text{kg}\cdot\text{K})][(273+13)\,\text{K}]} = 1.23\,\text{kg/m}^3$$

The airflow rate needed to meet the load for zone 1 is

$$\dot{V}_1 = \frac{7.3\,\text{kW}}{[1\,\text{kJ}/(\text{kg}\cdot\text{K})][(1.23\,\text{kg/m}^3)][(25-13)\,\text{K}]} = 0.49\,\text{m}^3/\text{s}$$

TABLE 11.5

Hourly VAV System Simulation Results

	1	2	3
T_{zone}	25°C	25°C	25°C
Coil outlet $T_{c,o}$	12.5°C	12.5°C	12.5°C
Coil outlet $W_{c,o}$		0.0077 kg/kg$_{da}$	
ΔT_{fan}	0.5°C	0.5°C	0.5°C
Total Outside Air (OSA) airflow		57 m³/min	
h_{fg}		2500 kJ/kg	

				Zone 1				Zone 2				Zone 3					Coil calculations			
Hour	T_{amb} °C	$T_{s,}$ °C	$\rho_{s,}$ kg/m³	Sens. Load, kW	$\dot{V}_{1,}$ m³/s	$T_{sa,1}$ °C	$Q_{r,1r}$ kW	Sens. Load, kW	$\dot{V}_{2,}$ m³/s	$T_{sa,2}$ °C	$Q_{r,2}$ kW	Sens. Load, kW	$\dot{V}_{3,}$ m³/s	$T_{sa,3}$ °C	$Q_{r,3r}$ kW	$\bar{T}_{rr}$ °C	ρ_{or} kg/m³	$\dot{m}_{or}$ kg/s	$T_{c,ir}$ °C	$Q_{ch,sr}$ kWh
1	23.3	13.0	1.23	7.3	0.49	13.00	0.0	5.9	0.40	13.00	0.0	4.4	0.30	13.00	0.00	25	1.19	1.13	23.69	16
2	22.8	13.0	1.23	6.6	0.45	13.00	0.0	5.9	0.40	13.00	0.0	4.4	0.30	13.00	0.00	25	1.19	1.13	23.23	15
3	22.2	13.0	1.23	5.9	0.40	13.00	0.0	5.9	0.40	13.00	0.0	4.4	0.30	13.00	0.00	25	1.20	1.14	22.64	14
4	21.1	13.0	1.23	4.7	0.32	13.00	0.0	5.9	0.40	13.00	0.0	4.4	0.30	13.00	0.00	25	1.20	1.14	21.44	11
5	20.0	13.0	1.23	3.7	0.25	13.00	0.0	5.9	0.40	13.00	0.0	4.4	0.30	13.00	0.00	25	1.20	1.14	20.09	9
6	19.4	13.0	1.23	2.9	0.20	13.00	0.0	5.9	0.40	13.00	0.0	4.4	0.30	13.00	0.00	25	1.21	1.15	19.16	7
7	20.0	13.0	1.23	11.0	0.74	13.00	0.0	8.8	0.59	13.00	0.0	7.3	0.49	13.00	0.00	25	1.20	1.14	22.47	23
8	22.2	13.0	1.23	22.0	1.49	13.00	0.0	29.3	1.98	13.00	0.0	26.4	1.78	13.00	0.00	25	1.20	1.14	24.51	78
9	23.9	13.0	1.23	26.4	1.78	13.00	0.0	32.2	2.18	13.00	0.0	29.3	1.98	13.00	0.00	25	1.19	1.13	24.83	90
10	26.1	13.0	1.23	33.7	2.28	13.00	0.0	32.2	2.18	13.00	0.0	29.3	1.98	13.00	0.00	25	1.18	1.12	25.16	100
11	28.3	13.0	1.23	41.0	2.77	13.00	0.0	32.2	2.18	13.00	0.0	29.3	1.98	13.00	0.00	25	1.17	1.11	25.43	110
12	30.6	13.0	1.23	49.8	3.36	13.00	0.0	26.4	1.78	13.00	0.0	24.9	1.68	13.00	0.00	25	1.16	1.10	25.73	111
13	32.8	13.0	1.23	56.4	3.81	13.00	0.0	32.2	2.18	13.00	0.0	29.3	1.98	13.00	0.00	25	1.15	1.10	25.87	131
14	34.4	13.0	1.23	64.4	4.35	13.00	0.0	32.2	2.18	13.00	0.0	29.3	1.98	13.00	0.00	25	1.15	1.09	25.98	141
15	36.1	13.0	1.23	70.3	4.75	13.00	0.0	32.2	2.18	13.00	0.0	29.3	1.98	13.00	0.00	25	1.14	1.09	26.10	149
16	36.1	13.0	1.23	70.3	4.75	13.00	0.0	32.2	2.18	13.00	0.0	29.3	1.98	13.00	0.00	25	1.14	1.09	26.10	149
17	36.1	13.0	1.23	70.3	4.75	13.00	0.0	29.3	1.98	13.00	0.0	24.9	1.68	13.00	0.00	25	1.14	1.09	26.16	142
18	35.0	13.0	1.23	58.6	3.96	13.00	0.0	20.5	1.38	13.00	0.0	17.6	1.19	13.00	0.00	25	1.15	1.09	26.35	112
19	32.8	13.0	1.23	44.0	2.97	13.00	0.0	14.7	0.99	13.00	0.0	10.3	0.70	13.00	0.00	25	1.15	1.10	26.49	80
20	30.6	13.0	1.23	26.4	1.78	13.00	0.0	11.7	0.79	13.00	0.0	7.3	0.49	13.00	0.00	25	1.16	1.10	26.64	53
21	28.3	13.0	1.23	24.9	1.68	13.00	0.0	8.8	0.59	13.00	0.0	5.9	0.40	13.00	0.00	25	1.17	1.11	26.11	45
22	26.1	13.0	1.23	11.7	0.79	13.00	0.0	7.3	0.49	13.00	0.0	4.4	0.30	13.00	0.00	25	1.18	1.12	25.63	26
23	25.0	13.0	1.23	8.8	0.59	13.00	0.0	5.9	0.40	13.00	0.0	4.4	0.30	13.00	0.00	25	1.18	1.13	25.00	20
24	23.9	13.0	1.23	7.3	0.49	13.00	0.0	5.9	0.40	13.00	0.0	4.4	0.30	13.00	0.00	25	1.19	1.13	24.15	17
							0				0				0					1652

The same procedure is used for zones 2 and 3 to find

$$\dot{V}_2 = 0.40\,\text{m}^3/\text{s}$$

$$\dot{V}_3 = 0.30\,\text{m}^3/\text{s}$$

Contrary to the solution of Example 11.9, the mixed-air temperature is found by using the first law on a mass basis, instead of on the approximate volumetric basis widely used by designers. The total return air mass flow rate is

$$\dot{m}_r = [(0.49 + 0.40 + 0.30)\,\text{m}^3/\text{s}](1.23\,\text{kg/m}^3) = 1.46\,\text{kg/s}$$

Next, the mixed-air temperature is found by an energy balance at the return air/outside air junction. The outside air density is required to convert the given fresh air volumetric flow to a mass basis:

$$\rho_{\text{air}} = \frac{101{,}325\,\text{Pa}}{[287\,\text{kJ}(\text{kg} \cdot \text{K})][(273 + 23.3)\text{K}]} = 1.19\,\text{kg/m}^3$$

The outside air mass flow rate is then

$$\dot{m}_o = (57\,\text{m}^3/\text{min})(1.19\,\text{kg/m}^3)(1\ \text{min}/60\,\text{s}) = 1.13\,\text{kg/s}$$

The mixed-air temperature (and cooling coil inlet temperature) is found from

$$T_{\text{mix}} = T_{c,i} = 25°\text{C}\left(1 - \frac{1.13}{1.46}\right) + 23.3°\text{C}\left(\frac{1.13}{1.46}\right) = 23.69°\text{C}$$

Finally, the coil *sensible* load is given by

$$\dot{Q}_{c,s} = (1.46\,\text{kg/s})[1.0\,\text{kJ}/(\text{kg} \cdot \text{K})](23.69 - 12.5)\,\text{K]}$$
$$= 16.3\,\text{kW}$$

The process outlined for this hour is repeated for each hour of the day to produce the results shown in Table 11.5. The total sensible coil load is 1652 kWh.

The supply air temperature (T_s at cooling coil outlet, T_{sa} at zone boxes) and reheat (Q_r) entries in a number of columns in Table 11.5 do not change during the day. They are included in the table for comparison with a 24-h reheat system calculation described shortly.

COMMENTS

To find the *annual* energy consumption results similar to those found in Example 11.9, the preceding series of calculations must be repeated for each of the 8760 h of a typical year. If the system to be simulated is not sensitive to the sequence of hours in a year (negligible building or other thermal storage is involved, no time-of-day information on energy use is needed, no thermostat setback), the bin method is clearly superior from a computational point of view.

Table 11.6 contains the result of a simulation of a reheat system (thermodynamically the same as a dual-duct or multizone system) for the same three-zone case. With this constant-volume system, air at a constant-temperature (13°C in the example) is supplied to each zone. To avoid over-cooling, the air is reheated at the zone to the temperature required to meet the zonal loads. The table shows that the sensible cooling coil load is 4255 kWh or *257 percent greater* than for the VAV case. In addition, 2499 kWh of reheat is required, even though this is a hot summer day with outdoor temperatures between 19.4 and 36.1°C (67 and 97°F)! No reheat is needed in the VAV case. Reheat systems are very inefficient and are forbidden by some energy codes.

TABLE 11.6

Hourly Reheat System Simulation Results

	1	2	3
T_{zone}	25°C	25°C	25°C
Airflow	340 m³/min	198 m³/min	142 m³/min
$T_{c,o}$	12.5°C	12.5°C	12.5°C
$W_{c,o}$	0.0077 kg/kg$_{da}$		
ΔT_{fan}	0.5°C	0.5°C	0.5°C
OSA airflow	57 m³/min		
h_{fg}	2500 kJ/kg		

	Zone 1							Zone 2				Zone 3				Coil Calculations				
Hour	T_{amb} °C	T_s °C	ρ_s kg/m³	Sens. Load, kW	$\dot{m}_1$, kg/s	$T_{sa,1}$ °C	$Q_{r,1}$ kW	Sens. Load, kW	$\dot{m}_2$ kg/s	$T_{sa,2}$ °C	$Q_{r,2}$ kW	Sens. Load, kW	$\dot{m}_3$ kg/s	$T_{sa,3}$ °C	$Q_{r,3}$ kW	$\bar{T}_{rr}$ °C	ρ_{or} kg/m³	$\dot{m}_{or}$ kg/s	$T_{c,ir}$ °C	$Q_{ch,sr}$ kW
1	23.3	13.0	1.23	7.3	7.0	23.96	76.6	5.9	4.1	23.55	42.9	4.4	2.9	23.49	30.6	25	1.19	1.13	24.86	173
2	22.8	13.0	1.23	6.6	7.0	24.06	77.3	5.9	4.1	23.55	42.9	4.4	2.9	23.49	30.6	25	1.19	1.13	24.82	172
3	22.2	13.0	1.23	5.9	7.0	24.16	78.0	5.9	4.1	23.55	42.9	4.4	2.9	23.49	30.6	25	1.20	1.14	24.77	172
4	21.1	13.0	1.23	4.7	7.0	24.33	79.2	5.9	4.1	23.55	42.9	4.4	2.9	23.49	30.6	25	1.20	1.14	24.68	170
5	20.0	13.0	1.23	3.7	7.0	24.47	80.2	5.9	4.1	23.55	42.9	4.4	2.9	23.49	30.6	25	1.20	1.14	24.59	169
6	19.4	13.0	1.23	2.9	7.0	24.59	81.0	5.9	4.1	23.55	42.9	4.4	2.9	23.49	30.6	25	1.21	1.15	24.54	168
7	20.0	13.0	1.23	11.0	7.0	23.43	72.9	8.8	4.1	22.84	40.0	7.3	2.9	22.50	27.7	25	1.20	1.14	24.59	169
8	22.2	13.0	1.23	22.0	7.0	21.85	61.9	29.3	4.1	17.80	19.5	26.4	2.9	15.96	8.6	25	1.20	1.14	24.77	172
9	23.9	13.0	1.23	26.4	7.0	21.22	57.5	32.2	4.1	17.09	16.6	29.3	2.9	14.96	5.7	25	1.19	1.13	24.91	174
10	26.1	13.0	1.23	33.7	7.0	20.18	50.2	32.2	4.1	17.09	16.6	29.3	2.9	14.96	5.7	25	1.18	1.12	25.09	176
11	28.3	13.0	1.23	41.0	7.0	19.13	42.9	32.2	4.1	17.09	16.6	29.3	2.9	14.96	5.7	25	1.17	1.11	25.26	178
12	30.6	13.0	1.23	49.8	7.0	17.88	34.1	26.4	4.1	18.51	22.4	24.9	2.9	16.47	10.1	25	1.16	1.10	25.44	181
13	32.8	13.0	1.23	56.4	7.0	16.93	27.5	32.2	4.1	17.09	16.6	29.3	2.9	14.96	5.7	25	1.15	1.10	25.61	183
14	34.4	13.0	1.23	64.4	7.0	15.79	19.5	32.2	4.1	17.09	16.6	29.3	2.9	14.96	5.7	25	1.15	1.09	25.73	185
15	36.1	13.0	1.23	70.3	7.0	14.94	13.6	32.2	4.1	17.09	16.6	29.3	2.9	14.96	5.7	25	1.14	1.09	25.86	187
16	36.1	13.0	1.23	70.3	7.0	14.94	13.6	32.2	4.1	17.09	16.6	29.3	2.9	14.96	5.7	25	1.14	1.09	25.86	187
17	36.1	13.0	1.23	70.3	7.0	14.94	13.6	29.3	4.1	17.80	19.5	24.9	2.9	16.47	10.1	25	1.14	1.09	25.86	187
18	35.0	13.0	1.23	58.6	7.0	16.62	25.3	20.5	4.1	19.96	28.3	17.6	2.9	18.97	17.4	25	1.15	1.09	25.78	186
19	32.8	13.0	1.23	44.0	7.0	18.71	39.9	14.7	4.1	21.39	34.1	10.3	2.9	21.47	24.7	25	1.15	1.10	25.61	183
20	30.6	13.0	1.23	26.4	7.0	21.22	57.5	11.7	4.1	22.13	37.1	7.3	2.9	22.50	27.7	25	1.16	1.10	25.44	181
21	28.3	13.0	1.23	24.9	7.0	21.44	59.0	8.8	4.1	22.84	40.0	5.9	2.9	22.98	29.1	25	1.17	1.11	25.26	178
22	26.1	13.0	1.23	11.7	7.0	23.33	72.2	7.3	4.1	23.21	41.5	4.4	2.9	23.49	30.6	25	1.18	1.12	25.09	176
23	25.0	13.0	1.23	8.8	7.0	23.74	75.1	5.9	4.1	23.55	42.9	4.4	2.9	23.49	30.6	25	1.18	1.13	25.00	175
24	23.9	13.0	1.23	7.3	7.0	23.96	76.6	5.9	4.1	23.55	42.9	4.4	2.9	23.49	30.6	25	1.19	1.13	24.91	174
							1285.0				743.0				472.0					4255
															2499.0					

Total reheat: 2499.0

11.5 Summary

In this chapter we have discussed secondary systems that provide heating and cooling to individual zones. Secondary systems couple the primary systems—chillers and boilers—to the load points, the zones. Both air- and liquid-based systems are used. Hence, the designer must use tools for both liquid and air system design. Included among these are fan, pump, and system flow–pressure drop curves and coil and zone energy balance equations along with part-load characteristics for secondary system components.

Problems

The problems in this book are arranged by topic. The approximate degree of difficulty is indicated by a parenthetic italic number from 1 to 10 at the end of the problem. Problems are stated most often in USCS units; when similar problems are presented in SI units, it is done with approximately equivalent values in parentheses. The USCS and SI versions of a problem are not exactly equivalent numerically. Solutions should be organized in the same order as the examples in the text: given, figure or sketch, assumptions, find, lookup values, solution. For some problems, the Heating and Cooling of Buildings (HCB) software may be helpful. In some cases it is advisable to set up the solution as a spreadsheet, so that design variations are easy to evaluate.

11.1 Compare the pressure loss coefficient C for flow into the two input connections of a converging, equal-main-area, round tee for which the branch flows are equal. What is the effect of the branch flow area ratio (between 0.3 and 1.0) on this result?   (3)

11.2 Which of three designs of equal-area 90° elbows (consider round or square) has the smallest pressure drop; which, the largest? Consider three types: smooth-radius, round; three-, four-, and five-piece round; and mitered rectangular elbows. Discuss only the case for an elbow radius/duct diameter ratio of 1.0.   (3)

11.3 Identical round duct transitions with an area ratio of 2:1 are used in the supply (transition is to a smaller duct size as air is diverted to branches) and return ducting (transition is to a larger duct size) in a building air supply and return loop. In which application is the pressure drop through the transition greater if it is fabricated with an angle of 15°?   (3)

11.4 Identical tees are used in identical converging and diverging flows. In which application is the *main* branch pressure drop larger?   (3)

11.5 One branch of a duct system has one less three-piece, $r/D = 1$, round 90° elbow than the other. To balance the flow, a butterfly damper (whose diameter is one-half the duct's diameter) is used in the former duct. At what angle should the butterfly damper be positioned to balance the flow?   (4)

11.6 Rework Example 11.2 at 4000-ft (1,220-m) altitude with the same $\Delta p/L$ value. Note that the only entries that need to be changed in Table 11.3 are those that are affected by the velocity pressure. What is the replacement for the velocity pressure, Eq. (11.9), at 4000 ft (1220 m)?   (5)

11.7 Rework Example 11.2 for the greatest pressure drop branch, but only include 15°, round duct transitions at each point of duct diameter change. How significant is the addition of the transitions to the fan rating? (4)

11.8 Figure P11.8 is the schematic diagram for an air distribution system at sea level that is to be analyzed by the equal-friction method [$\Delta p/L = 0.25$ inWG/100 ft (2.0 Pa/m)]. The ductwork is round; and smooth-radius, round, $r/D = 1$ elbows are used. What pressure rise must the fan develop to deliver air in this system to branch *ABCD* if the duct outlets are of negligible pressure drop? The diverging tees at *B* and *C* are round, conical branch designs; all size transitions are 15° converging designs. (7)

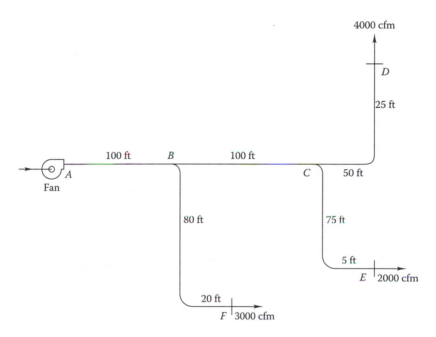

FIGURE P11.8

11.9 The pressure drop in duct *CE* in Figure P11.8 is to be made the same as in branch duct *CD* (refer to Prob. 11.8 for the design static drop to be used in *CD*; point *C* is upstream of the tee) to balance the system. The key difference between the two branches, other than the flow rate, is that the *CD* flow is straight through the tee and the *CE* flow is into the branch line. The duct lengths are about the same. What should the diameter of branch *CE* be to produce the same pressure drop as in branch *CD*? (6)

11.10 Figure P11.10 is the schematic diagram for an air distribution system at sea level that is to be analyzed by the equal-friction method [$\Delta p/L = 0.15$ inWG/100 ft (1.2 Pa/m)]. The ductwork is round; and smooth-radius, round, $r/D = 1$ elbows are used. What pressure rise must the fan develop to deliver air in this system to air outlet 5 if the duct outlets have a pressure drop equivalent to 25 ft (8 m) of straight duct? The diverging tees at *B*, *C*, *D*, and *E* are round, conical branch designs; all size transitions are 15° converging designs, and the balancing damper downstream of tee *E* is a wide-open butterfly type. (7)

11.11 Work Prob. 11.10 at an altitude of 3000 ft (915 m).   (7)

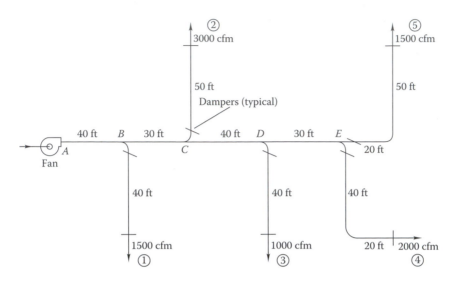

FIGURE P11.10

11.12 Select the five round duct sizes for the residential heating system shown in Fig. P11.12. The main supply duct velocity should not exceed 1000 ft/min (5 m/s), and the branch duct air velocity should be no more than 600 ft/min (3 ft/s).   (3)

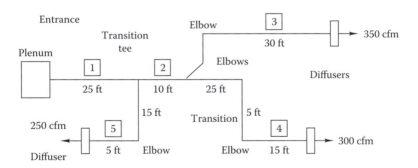

FIGURE P11.12

11.13 Rework Prob. 11.12, assuming that rectangular ducts no deeper than 8 in (20 cm) can be accommodated between the floor joists of this residence.   (4)

11.14 Write a spreadsheet template or computer program for determining the pressure drop in the round ductwork system of Prob. 11.10, using the equal-friction method. It should be general enough to accommodate various values of design $\Delta p/L$, airflow rates, air density, number of tees (main pressure drop), duct lengths, number of elbows, and outlet diffuser pressure drops.   (10)

11.15 Select a diffuser from Fig. 11.13 to distribute 2000 ft^3/min (950 L/s) of air at a low velocity of 35 ft/min (0.2 m/s) in a 40-ft (7-m) square room. What are the recommended diffuser throat diameter and its static pressure drop if the diffuser is not equipped with a damper?   (4)

11.16 A 15-kW fan will be equipped with either a VSD or inlet vanes to control flow. If the application requires that the fan operate the following number of hours at the indicated part-load levels, what is the kilowatthour savings due to use of the VSD? If the VSD controller costs $700 more than adjustable inlet vanes, what is the payback period for the VSD approach, assuming that electric power costs 8¢/kWh? (5)

PLR	0.20	0.40	0.60	0.80	1.00
Hours	300	700	900	250	50

11.17 Work Prob. 11.16, except compare the VSD approach with outlet dampers for flow control. (5)

11.18 The flow through a pump is 500 gal/min (30 L/s) against a head of 20 ftWG (60 kPa). What is the electric power required if the motor efficiency is 91 percent and the pump efficiency is 72 percent?[18] (2)

11.19 Figure P11.19 is a pump curve for one manufacturer's series of 6-in (15.2-cm) to 9.75-in (24.8-cm) pumps. The 100 percent system curve intersects the 7.5-in (19.3-cm) rotor diameter curve at 100 gal/min (6.3 L/s). At this full-load operating point, what are the values of the pump power, efficiency, and NPSH? What are the same parameters for the system curve in which flow is reduced by 50 percent by using a flow control valve? (4)

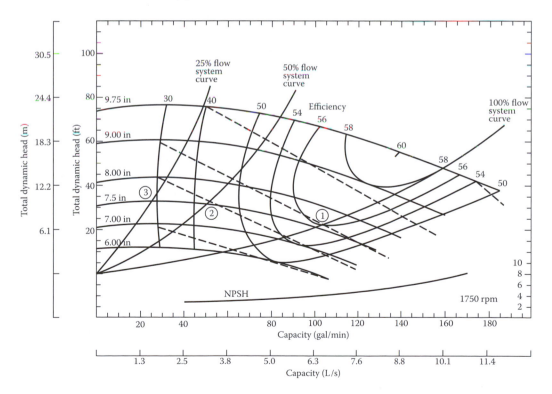

FIGURE P11.19

---

[18] Additional problems on the basics of pipe design are Probs. 5.11 to 5.20.

11.20 Flow is reduced to 50 percent of the full flow of 90 gal/min (5.7 L/s) using the 7.5-in (19.3-cm) pump in Fig. P11.19 by partially closing a flow control valve. Rather than use this energy-inefficient approach, an engineer proposes to reduce the pump rotational speed from the nominal value shown in the figure. What will be the new pump speed as well as the new values of pump power and efficiency? How will the NPSH value be affected? (6)

11.21 What is the available net positive suction head (NPSHA) for the system shown in Fig. P11.21? Particular care is needed in the design of open, cooling tower pump systems. (3)

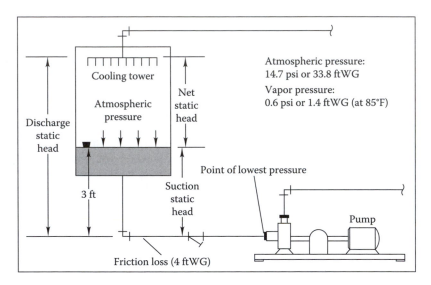

FIGURE P11.21
Water sump pump for cooling tower.

11.22 One hundred gal (380 L) of water is confined in a length of nominal 4-in (10-cm) steel pipe. If the pressure in the pipe is maintained essentially constant by an expansion tank while the water warms from 100 to 200°F (38 to 93°C), how much water will move into the expansion tank? (5)

11.23 Five hundred gal (1900 L) of water is confined in a completely filled steel tank, the pressure in which is maintained essentially constant by an expansion tank. If the water is heated from 50 to 250°F (10 to 121°C), how much water will move into the expansion tank? (5)

11.24 If the pipe in Prob. 11.22 were to expand freely, what would the length change be? Design a U-type expansion loop for the pipe in Prob. 11.22 so that this expansion can be accommodated without any length change between the fixed anchors attached to the ends of the pipe. (5)

11.25 Work Prob. 11.24 for copper pipe with anchors placed 50 ft (15 m) apart. (5)

11.26 Size an expansion tank for a residential heating system at sea level containing 150 gal (550 L) of water if the water temperature increases from 50 to 230°F (10 to 110°C). The system maximum pressure is 60 psig (420 kPa), and the initial expansion tank nitrogen charge pressure is 15 psig (105 kPa). (4)

11.27 Repeat Prob. 11.26, but assume that the expansion tank was filled with atmospheric air prior to charging. No special gas charge was added to the tank. (4)

11.28 Make a dimensional plot of the economizer outside airflow characteristic similar to Figure 11.20b for a 20000 ft³/min (9500 L/s) fixed-volume system with a minimum flow setpoint of 18 percent, assuming dry-bulb temperature control. The supply air condition is required to be 55°F (13°C), and the room air return condition is 77°F (25°C) and 50 percent relative humidity. For simplicity, ignore fan heat. At 10°F (−12°C) the outside air setting is the 18 percent minimum. How is the high-temperature return to minimum flow determined? (5)

11.29 Repeat Prob. 11.28 with enthalpy control. To identify the point at which one will switch to and from the minimum airflow, you will need to make an assumption regarding outside air conditions of temperature and humidity. (7)

11.30 In a VAV system, find the primary airflow requirements range for a 75°F (24°C) and 40 percent relative humidity zone whose load (entirely sensible) varies from 2000 to 5000 Btu/h (600 to 1500 W). Air is supplied to the zone at 56°F (13°C). (3)

11.31 In a VAV system, find the primary airflow requirement range for a 75°F (24°C) zone whose load of 2000 to 5000 Btu/h (600 to 1500 W) has an SHR that varies from 0.7 to 0.9. Air is supplied to the zone at 56°F (14°C) and 85 percent relative humidity, and the control is by dry-bulb temperature. (3)

11.32 Determine the air handler fan flow rate for cooling a small commercial building consisting of four zones, the peak load of each of which is 30000 Btu/h (9 kW) including ventilation air. Coil conditions are 84°F (29°C) and 38 percent relative humidity inlet and 53°F (12°C) and 90 percent relative humidity outlet. The diversity among the four zones is 80 percent. What is the size of the main duct and each of the four branch ducts, if all are square in cross section, that you would suggest for this building? (6)

11.33 Size the preheat coil for a VAV system with 100000 actual ft³/min (47000 actual L/s) system flow if the outside air fraction is 19 percent in Billings, Mont. (4)

11.34 Complete a VAV system design, following the process used in Example 11.8, for the following two-zone building:

	Zone 1 (Exterior)	Zone 2 (Interior)
Sensible peak cooling load	$\dot{Q}_{s_1} = 150{,}000$ Btu/h	$\dot{Q}_{s_2} = 250{,}000$ Btu/h
Latent peak cooling load	$\dot{Q}_{l_1} = 40{,}000$ Btu/h	$\dot{Q}_{l_2} = 60{,}000$ Btu/h
Heating peak load	$\dot{Q}_{h_1} = 300{,}000$ Btu/h	$\dot{Q}_{h_2} = 70{,}000$ Btu/h
Zone temperature T_{zone}	75°F (24°C)	75°F (24°C)

Other: Design conditions are for Denver [altitude 5280 ft (1610 m)]. The supply system pressure drop at full airflow is 3.5 inWG (28 Pa), and the fan efficiency is 63 percent. Take the supply fan air temperature rise to be 1.5°F (0.8°C) at peak conditions; and the return fan air temperature rise is 1.0°F (0.6°C). The required fresh air ventilation flow rate is 3200 ft³/min (1500 L/s), and the cooling coil outlet humidity is 90 percent. Use a zone supply air temperature difference ΔT_{sa} of 20°F (11°C). (9)

11.35 Work Prob. 11.34 for St. Louis, Mo. (9)

11.36 Work Prob. 11.34 for Albuquerque, N. Mex. (9)

11.37 A constant-volume reheat system is needed to accurately control conditions in a building designed to the specifications of Prob. 11.34. Size the cooling, reheat, and preheat coils and other equipment for this system, to be located in New York City. (9)

11.38 Freeze protection of a reheat coil in either constant- or variable-volume systems can be accomplished by the use of antifreeze, but at the expense of reduced heat transfer, higher pumping power, and increased maintenance costs. However, another approach is to locate it in the return duct downstream of the building exhaust, where freezing air does not exist. The shortcomings of antifreeze are thereby avoided. Discuss this approach. Can you identify any difficulties? (5)

11.39 Latent loads can be included in the bin method (needed in the cooling season only) by using the *mean coincident wet-bulb* (MCWB) temperature corresponding to each dry-bulb temperature bin. Replace the crude estimation of cooling loads in Example 11.9 with a correct calculation, using the following weather data for Andrews Air Force Base. Internal latent gains are from 120 persons at 150 Btu/h per person. (10)

Bin, °F	97	92	87	82	77	72	67	62	57	52	47	42
MCWB, °F	76	74	72	70	68	66	61	56	51	46	42	37

11.40 Work Example 11.9 with the bin dry-bulb temperature data listed below for Denver, Colo. (10)

Bin, °F	97	92	87	82	77	72	67	62	57	52	47	42
Hours	7	71	174	291	384	494	618	794	776	739	729	752

37	32	27	22	17	12	7	2	−3	−8	−13
724	704	555	394	243	137	84	54	22	13	9

11.41 Use the calculated airflow rates in Example 11.9 to assess the performance of a VSD instead of the outlet dampers used in the example. What is the savings in fan power that would be achieved if this more efficient method of volume control were used? (5)

11.42 If the boiler in Example 11.9 were oversized by 20 percent to 528000 Btu/h (155 kW), what would the annual energy penalty be compared to the original design? (4)

11.43 If the chiller in Example 11.9 were oversized by 20 percent to 60 tons (210 kW), what would the annual energy penalty be compared to the original design? (4)

11.44 If the air condition to be maintained in a building is 72°F and 50 percent relative humidity and the outdoor humidity ratio remains constant at 0.009 $lb_m/lb_{m,da}$, evaluate the performance of two methods of economizer control: dry-bulb temperature and enthalpy. The outdoor dry-bulb temperature for the day in question is given by $T_a = 58 - 12\cos(15t)$, in which t is the time measured in military time and T_{da} is in degrees Fahrenheit. The design air delivery temperature is 55°F in the dry-bulb temperature control approach; above 55°F the outside air damper is fully open, and below 55°F it is modulated as needed down to its 20 percent flow minimum.

The enthalpy controller operates the same as a dry-bulb temperature controller below 55°F, whereas above 55°F the outside air damper remains fully open whenever the outside air enthalpy is below the room enthalpy; otherwise it is at its minimum position. What is the percentage of savings in cooling energy during this day? You may wish to specify a numerical value of total flow rate to solve the problem, although it is not necessary. Comment on the feasibility of enthalpy control for this situation. (10)

11.45 Occasionally it is necessary to control both dry-bulb temperature and humidity in a zone. This can be accomplished by adjusting the apparatus dew point (ADP) and supply airflow rate so that both latent and sensible loads are met. Find the ADP and coil airflow rate to control a zone to 75°F (24°C) and 40 percent relative humidity if the sensible load is 300000 Btu/h (88 kW), the latent load is 30000 Btu/h (8.8 kW), and the design outdoor conditions are 95°F (35°C) dry-bulb temperature and 40 percent relative humidity. Outside ventilation airflow is 3000 ft^3/min (1400 L/s). What are the recommended coil outlet air temperature and the airflow needed to match the zone requirements? (10)

References

AMCA (1973). *Fans and Systems*. Publication 201–73, Air Movement and Control Association, Inc., Arlington Heights, Ill.

ASHRAE (1999a). *HVAC Applications*. American Society of Heating, Refrigerating and Air-Conditioning Engineers, Atlanta.

ASHRAE (1999b). *Standard 62–1999: Ventilation for Acceptable Indoor Air Quality*, American Society of Heating, Refrigerating and Air-Conditioning Engineers, Atlanta.

ASHRAE (2000). *Systems and Equipment Handbook*. American Society of Heating, Refrigerating and Air-Conditioning Engineers, Atlanta.

ASHRAE (2001, 1989). *Handbook of Fundamentals*, American Society of Heating, Refrigerating and Air-Conditioning Engineers, Atlanta.

Avery, G. (1986). "VAV—Designing and Controlling an Outside Air Economizer." *ASHRAE J.* (December), pp. 26–29. See also G. Avery (1989). "The Myth of Pressure-Independent VAV Terminals," *ASHRAE J.*, pp. 28–30.

Coad, W. J. (1989). "The Air System in Perspective." *Heating, Piping and Air Conditioning*, vol. 61, pp. 124, 128.

DOE2.1 (various dates). Several manuals for the DOE2.1 software are available from the National Technical Information Service, Springfield, Va.; they include the 2.1D Building Description Language Summary (DE-890–17726) and the 2.1A Engineers' Manual (DE-830–04575).

Englander, S. K., and L. K. Norford (1988). "Fan Energy Savings: Analysis of a Variable-Speed Drive Retrofit." *Proceedings of the 1988 ACEEE Summer Study on Energy Efficiency in Buildings*, American Council for an Energy Efficient Economy, Washington, D.C.

Greenberg, S., J. P. Harris, H. Akbari, and A. DeAlmeida (1988). *Technology Assessment: Adjustable-Speed Motors and Motor Drives*. Lawrence Berkeley Laboratory Report LBL-25080 (March).

Guntermann, A. E. (1986). "VAV System Enhancements." *Heating, Piping and Air Conditioning* (August), pp. 67–78.

Holness, G. V. R. (1990). "Human Comfort and IAQ." *Heating, Piping and Air Conditioning*, vol. 62 (February), pp. 43–52.

Knebel, D. (1983). *Simplified Energy Analysis Using the Modified Bin Method*. American Society of Heating, Refrigerating and Air-Conditioning Engineers, Atlanta.

Schlichting, H. (1979). *Boundary Layer Theory*, 7th ed. McGraw-Hill, New York.

Spitler, J. D., D. C. Hittle, D. L. Johnson, and C. O. Pederson (1987). "A Comparative Study of the Performance of Temperature-Based and Enthalpy-Based Economy Cycles." *ASHRAE Trans.*, vol. 93, pt. 2, pp. 13–22.

Waller, B. (1990). "Piping—From the Beginning." *Heating, Piping and Air Conditioning*, vol. 62 (October), pp. 51–70.

12

Heating, Ventilating, and Air-Conditioning Control Systems

12.1 Introduction—The Need for Control

In earlier chapters, we assumed that the indoor environmental conditions are maintained at the desired setpoints by the HVAC system sized for that purpose. We discussed how to size systems and made calculations of energy consumption at peak and off-peak conditions. However, we have not discussed exactly how equipment is operated to meet uniform indoor comfort requirements (ventilation and temperature) under varying load and weather conditions. HVAC system controls are the information link between varying energy demands on a building's primary and secondary systems and the (usually) approximately uniform demands for indoor environmental conditions. Without an adequately designed and properly functioning control system, the most expensive, most thoroughly designed HVAC system will not operate as expected. It simply will not control indoor conditions to provide comfort.

The HVAC designer must design a control system that

- Sustains a comfortable building interior environment
- Maintains an acceptable indoor air quality
- Is as simple and inexpensive as possible and yet meets HVAC system operation criteria reliably for the system lifetime
- Results in efficient HVAC system operation under all conditions

According to Coad (1990), the "majority" of HVAC control systems installed in the past 50 years in buildings (in the United States) are, if not maintained, not "performing as intended." A considerable challenge is presented by this state of affairs to the HVAC system designer. One reason for inadequate control operation is inadequate design or unclear assignment of responsibility for control system design. In this chapter, devoted entirely to controls, we will learn the rudiments of control design from the point of view of the HVAC system designer. The reader is encouraged to do additional study in the following references on the subject: ASHRAE (1987), Haines (1987), Honeywell (1988), and Letherman (1981).

The HVAC engineer identifies the control sequences at an advanced stage of the design process. At that point the HVAC system (both primary and secondary) concept is known, HVAC sizing has been done at least in a preliminary way, and zoning has been completed. The HVAC engineer's or control designer's key task is to crystallize every detail of the control system's logic under all operating conditions. The products of the control design exercise are

1. Control specification
2. Control drawings
3. Control system equipment lists (schedules)—controllers, sensors, actuators, control equipment (valves, dampers, etc.), software

To achieve proper control based on the control system design, the HVAC system itself must be constructed and calibrated according to the mechanical system drawings. These must include properly sized primary and secondary systems. In addition, air stratification must be avoided, proper provision for control sensors is required, freeze protection is necessary in cold climates, and proper attention must be paid to minimizing energy consumption subject to reliable operation and occupant comfort. Proper maintenance is essential as well.

The principal, ultimate controlled variable in buildings is the zone temperature (and, to a lesser extent, air quality in some buildings). Therefore, in this chapter we focus on temperature control. Of course, the control of zone temperature involves many other types of control within the primary and secondary HVAC systems, including boiler and chiller control, pump and fan control, liquid and airflow control, humidity control, and auxiliary system control (e.g., thermal storage control). This chapter discusses only *automatic control* of these subsystems. Honeywell (1988) defines an *automatic control system* as "a system that reacts to a change or imbalance in the variable it controls by adjusting other variables to restore the system to the desired balance."

Figure 12.1 shows a familiar control problem. It is necessary to maintain the water level in the tank under varying inlet flow rates. The float operates a valve controlling flow from the tank. This simple system includes all the elements of a control system:

Sensor—float; reads the water level

Controlled system characteristic—water flow; often termed the *controlled variable*

Controller—linkage connecting float to valve stem

Actuator (controlled device)—internal valve stem mechanism; sets valve (the final control element) position

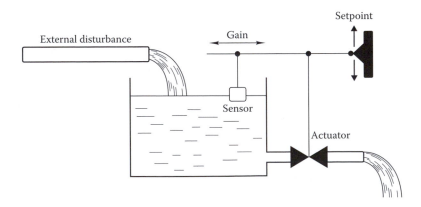

FIGURE 12.1
Simple water-level controller. The setpoint is the full water level; the error is the difference between the full level and the actual level.

This system is called a *closed-loop* or *feedback* system because the sensor (float) is directly affected by the action of the controlled device (valve). In an *open-loop* system, there is no sensor to measure the controlled variable. An example would be a method of controlling the valve based on an external parameter such as the time of day that may have an indirect relation to water consumption from the tank.

There are four common methods of control, of which Fig. 12.1 shows but one. In the next section, we describe each with relation to an HVAC system example.

12.1.1 Modes of Feedback Control

The four common modes of relating the error (difference between desired setpoint and sensed value of controlled variable—see Fig. 12.1) to the corrective action to be taken by the controller are

- Two-position (on/off)
- Proportional
- Integral
- Derivative

The last three are usually used in a variety of combinations with one another. Figure 12.2a shows a steam coil used to heat air in a duct. The simple control system shown includes an air temperature sensor, a controller that compares the sensed temperature to the setpoint, a steam valve controlled by the controller, and the coil itself. We will use this example system as a point of reference when discussing the various control system types below. Figure 12.2b is the *control diagram* corresponding to the physical system shown in Fig. 12.2a.

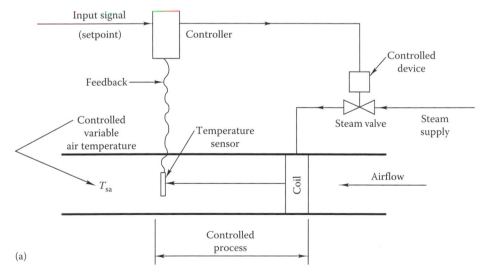

FIGURE 12.2
(a) Simple heating coil control system showing the process (coil and short duct length), controller, controlled device (valve and its actuator), and sensor. The setpoint entered externally is the desired coil outlet temperature.

(continued)

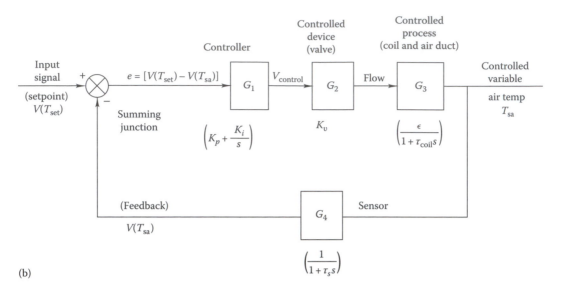

(b)

FIGURE 12.2 (continued)
(b) Equivalent control diagram for heating coil. The G's represent functions relating the input to the output of each module (the expressions in parentheses are examples of the G's discussed in the last section of this chapter on Laplace transforms). Voltages V represent both temperatures (setpoint and coil outlet) and the controller output to the valve in electronic control systems.

Two-position control applies to an actuator that is either fully open or fully closed. In Fig. 12.2a, the valve is a two-position valve if two-position control is used. The position of the valve is determined by the value of the coil outlet temperature. Figure 12.3 depicts two-position control of the valve as follows. If the air temperature drops below 95°F (35°C), the valve opens and remains open until the air temperature reaches 100°F (38°C). The differential is usually adjustable, as is the temperature setting itself. Two-position control is the least expensive method of automatic control and is suitable for control of HVAC systems with large time constants. Examples include residential space and water heating systems. Systems that are fast-reacting should not be controlled by using this approach since inaccurate control may result.

Proportional control corrects the controlled variable in proportion to the difference between the controlled variable and the setpoint. For example, a proportional controller would make a 10 percent increase in the coil heat output rate in Fig. 12.2 if a 10 percent decrease in the coil outlet air temperature were sensed. The proportionality constant between the error and the controller output is called the *gain* K_p. Equation (12.1) shows the characteristic of a proportional controller

$$V = V_o + K_p e \tag{12.1}$$

where
    V = controller output; symbol V is used since in electronic controls the controller output is often a voltage
    V_o = constant value of controller output when no error exists at control range midpoint
    e = error

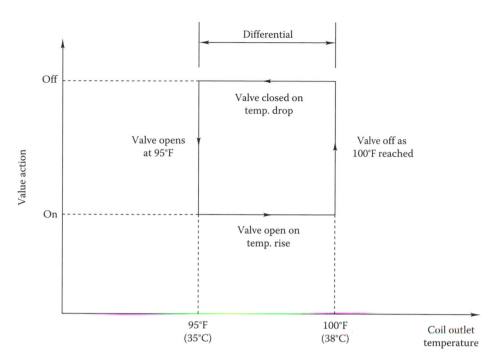

FIGURE 12.3
Two-position (on/off) control characteristic for steam valve in Fig. 12.2a.

In the case of the steam coil, the *error e* is the difference between the sensed air temperature T_{sensed} and the needed air temperature, the setpoint T_{set}:

$$e = T_{set} - T_{sensed} \qquad (12.2)$$

A progressively lower coil air outlet temperature results in a greater error e and hence a greater control action—a greater steam flow rate.

The *throttling range* ΔV_{max} is the total change in the controlled variable that is required to cause the actuator or controlled device to move between its limits. For example, if the nominal temperature of a zone is 72°F (22°C) and the heating controller throttling range is 6°F (3°C), then the heating control undergoes its full travel between a zone temperature of 69 and 75°F (21 and 24°C). This control, whose characteristic is shown in Fig. 12.4, is *reverse-acting*; i.e., as the temperature (controlled variable) increases, the heat output of the coil decreases (the heating valve position decreases).

Another way of expressing the control characteristic is by its gain. The gain is inversely proportional to the throttling range, as shown in Fig. 12.4. The narrower the throttling range (i.e., the steeper the characteristic), the greater the numerical value of the slope of the line and the greater the gain. Beyond the throttling range, the system is out of control. In actual hardware, one can set the setpoint and either the gain or the throttling range (most common), but not both of the latter two since one determines the other. Proportional control by itself is not capable of reducing the error in heat output (except at the midpoint) to zero, since an error is needed to produce the capacity required for meeting a load, as we will see in Example 12.1. This unavoidable value of the error in proportional systems is

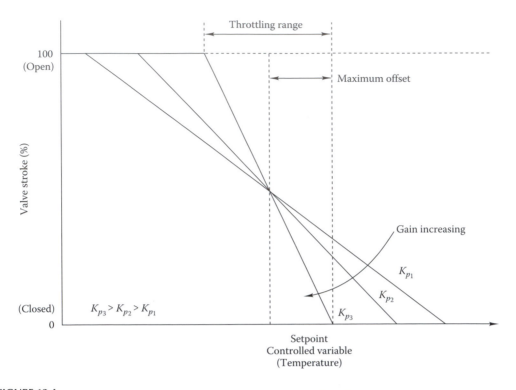

FIGURE 12.4
Proportional control characteristic showing various throttling ranges and the corresponding proportional gains K_p. This characteristic is typical of a heating coil temperature controller.

called the *offset*. From Fig. 12.4 it is easy to see that the offset is larger for systems with smaller gains. As we will see later, there is a limit to which one can increase the gain to reduce offset, because high gains can produce control instability, i.e., unpredictable and unsteady control.

Example 12.1: Proportional-Gain Calculation

If the steam heating coil in Fig. 12.2a has a heat output that varies from 0 to 20 kW as the outlet air temperature varies from 35 to 45°C in an industrial process (with fixed input temperature), what is the coil gain and what is the throttling range? Find an equation relating the heat rate at any sensed air temperature to the maximum rate in terms of the gain and setpoint.

Given: $\dot{Q}_{max} = 20$ kW

$\dot{Q}_{min} = 0$ kW

$T_{max} = 45°C$

$T_{min} = 35°C$

$T_{set} = 40°C$ (coil output at actuator midtravel)

Figure: See Fig. 12.2a.

Assumption: Steady-state operation

Find: $K_p, \Delta V_{max}$

Solution

The throttling range is the range of the controlled variable (air temperature) over which the controlled system (heating coil) exhibits its full-capacity range. The temperature varies from 35 to 45°C; therefore the throttling range is

$$\Delta V_{max} = 45 - 35 = 10°C \tag{12.3}$$

The proportional gain is the ratio of the controlled system (coil) output to the throttling range. For this example, the gain is

$$K_p = \frac{\dot{Q}_{max} - \dot{Q}_{min}}{\Delta V_{max}} = \frac{(20 - 0) \text{ kW}}{10 \text{ K}} = 2.0 \text{ kW/K} \tag{12.4}$$

The controller characteristic can be found by inspecting Fig. 12.4. It is assumed that the average air temperature T_{set} (40°C) occurs at the average heat rate (10 kW). The equation of the sloped lines shown in Fig. 12.4 is

$$\dot{Q} = K_p(T_{set} - T_{sensed}) + \frac{\dot{Q}_{max}}{2} = K_p e + \frac{\dot{Q}_{max}}{2} \tag{12.5}$$

The quantity $T_{set} - T_{sensed}$ is the *error e* and is the signal to the control system that a valve-opening change is needed to meet the desired setpoint.

Inserting the numerical values, we have

$$\dot{Q} = (2.0 \text{ kW/K})(40 - T_{sensed}) + 10 \text{ kW}$$
$$= (2 \text{ kW/K})(T_{set} - T_{sensed}) + 10 \text{ kW} \tag{12.6}$$

Comments

In an actual steam coil control system, it is the steam valve that is controlled directly to indirectly control the heat rate of the coil. This is typical of many HVAC system controls in that the desired control action is achieved indirectly by controlling another variable which in turn accomplishes the desired result. That is why the controller and controlled device are often shown separately, as in Fig. 12.2b.

This example illustrates in a simple system why proportional control always requires an error signal with the result that there is always an *offset* (i.e., the desired setpoint and the actual temperature are always different except at one point when $\dot{Q} = \dot{Q}_{max}/2$) with this control mode. Equation (12.5) shows that for any coil heat output other than $\dot{Q}_{max}/2$, an error must exist. This error is the offset and is smaller for higher coil gains.

Proportional control is used with stable, slow systems that permit the use of a narrow throttling range and resulting small offset. Fast-acting systems need wide throttling ranges to avoid instability and large offset results. In a later section, we will examine the quantitative criteria for determining the stability of proportional control systems.

Integral control is often added to proportional control to eliminate the offset inherent in proportional-only control. The result, *proportional plus integral control*, is identified by the acronym *PI*. Initially the corrective action produced by a PI controller is the same as for a proportional-only controller. After the initial period, a further adjustment due to the integral term reduces the offset to zero. The rate at which this occurs depends on the time scale of the integration. In equation form, the PI controller is modeled by

$$V = V_o + K_p e + K_i \int e \, dt \tag{12.7}$$

An alternate form is sometimes used:

$$V = V_o + K_p \left(e + K_i' \int e \, dt \right) \quad \text{where } K_i' = \frac{K_i}{K_p}$$

in which K_i is the integral gain constant. It has units of reciprocal time and is the number of times that the integral term is calculated per unit time. This is also known as the *reset rate*; *reset control* is an older term used by some to identify integral control.

The integral term in Eq. (12.7) has the effect of adding a correction to the output signal V as long as the error term exists. The continuous offset produced by the proportional-only controller can thereby be reduced to zero because of the integral term. The time scale $K_i' = K_p/K_i$ of the integral term is on the order of 1 to 60 min. PI control is used for almost all applications in which the sensor is near the process and for fast-acting systems for which accurate control is needed. Examples include mixed-air controls, duct static pressure controls, and coil controls. Because the offset is eventually eliminated with PI control, the throttling range can be set rather wide, to improve stability under a wider range of conditions than good control would permit with only proportional control. We will learn in Sec. 12.5 that integral control can cause stability problems under certain conditions; therefore it has mixed benefits.

Derivative control is used to speed up the action of PI control. When derivative control is added to PI control, the result is called *PID control*. The derivative term added to Eq. (12.7) generates a correction signal proportional to the time rate of the change of error. This term has little effect on a steady proportional system with uniform offset (time derivative is zero) but initially, after a system disturbance, produces a larger correction more rapidly. Equation (12.8) includes the derivative term in the mathematical model of the PID controller:

$$V = V_o + K_p e + K_i \int e \, dt + K_d \frac{de}{dt} \tag{12.8}$$

or alternatively,

$$V = V_o + K_p \left(e + K_i' \int e \, dt + K_d' \frac{de}{dt} \right) \quad \text{where } K_d' = \frac{K_d}{K_p}$$

in which K_d is the derivative gain constant. The time scale K_d/K_p of the derivative term is typically in the range of 0.2 to 15 min. Since HVAC systems do not often require rapid control response, the use of PID or PD control is less common than the use of PI control. However, it is used, for example, whenever there is a long delay between the process (e.g., a heating coil) and the sensor (e.g., a zone's thermostat). Since a derivative is involved, any noise in the error (i.e., sensor) signal must be avoided to maintain good control. One application in buildings where PID control has been effective is in duct static pressure control, a fast-acting subsystem that could be unstable otherwise.

Figure 12.5 compares the reaction of the three systems discussed above to a step change in load on a coil. The proportional system never achieves the setpoint. However, since the integral term generates a correction proportional to the area under the error curve, it slowly

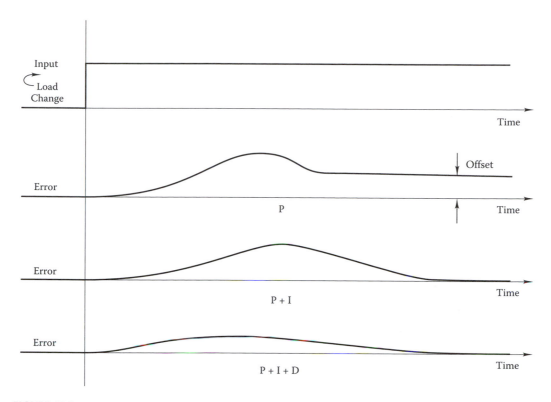

FIGURE 12.5
Performance comparison of P, PI, and PID controllers when subjected to a uniform input step change.

reduces the error to zero, but the maximum error is greater than for the PID approach which takes action faster to reduce the error than does the PI system alone. The PD system (not shown) never achieves zero error since the zero derivative of the constant offset produces no control action.

12.2 Basic Control Hardware

In this section, we describe the various physical components needed to achieve the actions required by the control strategies of the previous section. Since there are two fundamentally different control approaches—pneumatic and electronic—the following material is so divided. Sensors, controllers, and actuators for principal HVAC applications are described. In Sec. 12.3 the design of these components is discussed, while Sec. 12.4 describes the design of several complete control systems.

12.2.1 Pneumatic Systems

The first widely adopted automatic control systems used compressed air as the operating medium. Although a transition to electronic controls is occurring, about 50 percent of the sales of one large controls company in the early 1990s was pneumatic equipment.

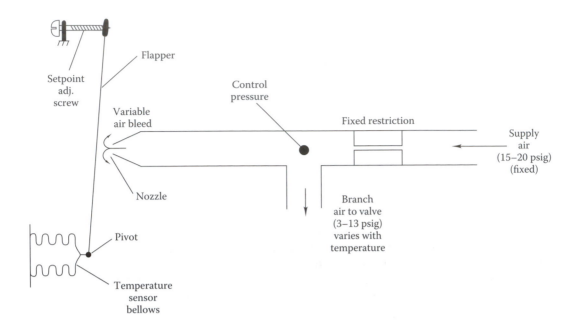

FIGURE 12.6
Drawing of pneumatic thermostat showing adjustment screw used to change temperature setting.

Pneumatic controls use compressed air [approximately 18 psig (125 kPa gauge) in the United States] for operation of sensors and actuators. In this section we give an overview of how these devices operate.

Since temperature is the most common parameter controlled, the most common pneumatic *sensor* is the temperature sensor. Figure 12.6 shows one method of sensing temperature and producing a control signal from it. Main supply air is supplied by a compressor to the zone thermostat. An amount of this air is bled from the nozzle depending on the position of the flapper and the size of the restrictor [diameters on the order of thousandths of an inch (hundredths of millimeters) are typical]. The pressure in the branch line that controls the valve ranges between 3 and 13 psig (20 and 90 kPa) typically. In simple systems, this pressure from a thermostat could operate an actuator such as a control valve for a room heating unit. In this case, the thermostat is both the sensor and the controller—a rather common configuration. The consumption of air in this sensor is quite small.

Many other temperature sensor approaches can be used. For example, the bellows shown in Fig. 12.6 can be eliminated, and the flapper can be made of a bimetal strip. As the temperature changes, the bimetal strip changes curvature, opening or closing the flapper-nozzle gap. Another approach uses a remote bulb filled with either liquid or vapor that pushes a rod (or a bellows) against the flapper to control the pressure signal. This device is useful if the sensing element must be located where direct measurement of temperature by a metal strip or bellows is not possible, such as in a water stream or high-velocity ductwork. The bulb and connecting capillary size may vary considerably by application. Bulbs can be up to several feet long in freeze control sensors, and capillaries up to 30 ft long are available.

Pressure sensors may use either bellows or diaphragms to control the branch line pressure. For example, the motion of a diaphragm may replace that of the flapper in Fig. 12.6 to control the bleed rate. A bellows similar to that shown in the same figure may be internally pressurized to produce a displacement that can control the air bleed rate. A bellows produces significantly greater displacements than a single diaphragm.

Humidity sensors in pneumatic systems are made from materials that change size with the moisture content. Nylon or other synthetic hygroscopic fibers that change size significantly (that is, 1 to 2 percent) with humidity are commonly used. Since the dimensional change is relatively small, mechanical amplification of the displacement is used. The materials that exhibit the desired property include nylon, hair, and cotton fibers. Since the properties of hair vary with age, the more stable material—nylon—is preferred and most widely used (Letherman, 1981). Humidity sensors for electronic systems are quite different and are discussed in Sec. 12.2.2.

An *actuator* converts pneumatic energy to motion—either linear or rotary. It creates a change in the controlled variable by operating control devices such as dampers or valves. Figure 12.7 shows a pneumatically operated control valve. The valve opening is controlled by the pressure in the diaphragm acting against the spring. The spring is essentially a linear device. Therefore, the motion of the valve stem is essentially linear with air pressure. However, this does not necessarily produce a linear effect on flow, as discussed in Sec. 12.3. Figure 12.8 shows a pneumatic damper actuator. Linear actuator motion is converted to rotary damper motion by the simple mechanism shown. The details of control valve and damper design are included in Sec. 12.3.

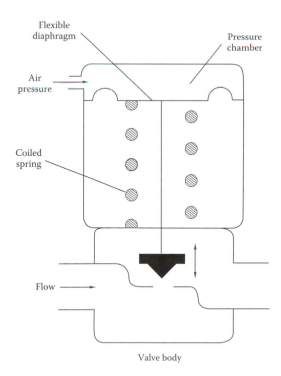

FIGURE 12.7
Pneumatic control valve showing counterforce spring and valve body. Increasing pressure closes the valve.

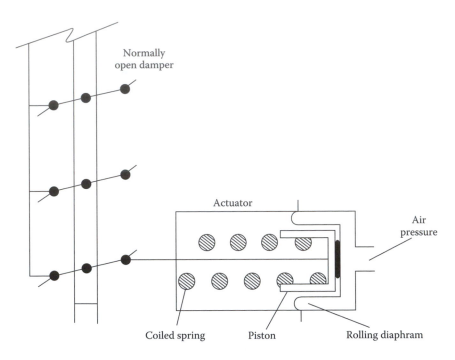

FIGURE 12.8
Pneumatic damper actuator. Increasing pressure closes the parallel-blade damper.

Pneumatic *controllers* produce a branch line (see Fig. 12.6) pressure that is appropriate to produce the needed control action for reaching the setpoint. They are manufactured by a number of control firms for specific purposes. Classifications of controllers include the sign of the output (direct- or reverse-acting) produced by an error, by the control action (proportional, PI, or two-position), or by the number of inputs or outputs. Figure 12.9 shows the essential elements of a dual-input single-output controller. The two inputs could be the heating system supply temperature and outdoor temperature used to control the output water temperature setting of a boiler in a building heating system. This is essentially a boiler *temperature reset* system that reduces heating water temperature with increasing ambient temperature for better system control and reduced energy use. (Note that *reset* has a different meaning than when used with an integral controller.)

The air supply for pneumatic systems must produce very clean, oil-free, dry air. A compressor producing 80 to 100 psig is typical. Compressed air is stored in a tank for use as needed, avoiding continuous operation of the compressor. The air system should be oversized by 50 to 100 percent of estimated, nominal consumption. The air is then dried to avoid moisture freezing in cold control lines in air handling units and elsewhere. Dried air should have a dew point of −30°F (−34°C) or less in severe heating climates. In deep-cooling climates, the lowest temperature to which the compressed air lines are exposed may be the building cold air supply. Next, the air is filtered to remove water droplets, oil (from the compressor), and any dirt. Finally, the air pressure is reduced in a pressure regulator to the control system operating pressure of approximately 18 psig (124 kPa gauge). Control air piping uses either copper or nylon (in accessible locations).

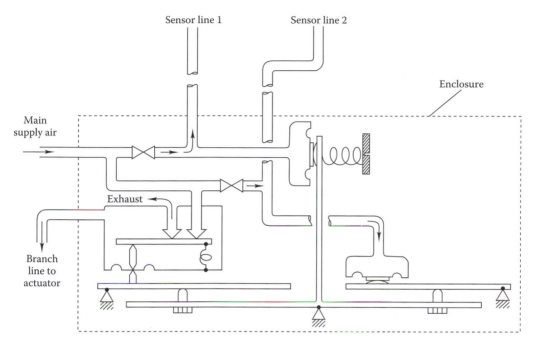

FIGURE 12.9

Example of pneumatic controller with two inputs (sensors 1 and 2) and one control signal output (branch line to actuator).

12.2.2 Electronic Control Systems

Electronic controls are used increasingly more widely in HVAC systems in commercial buildings. Their precise control, flexibility, compatibility with microcomputers, and reliability are advantages, but they may have increased cost per control loop. With the continuous decrease in microprocessor cost and associated increase in capability, the cost penalty, if any, is expected to virtually disappear, especially when it is calculated on a per-function basis. In this section, we survey the sensors, actuators, and controllers used in modern electronic control systems for buildings.

Figure 12.10 shows a simple *analog* electronic control system used to control the temperature of a hot-wire (or hot-film) anemometer for airflow measurement. Since the heat loss from the heated wire (a few ten-thousandths of an inch in diameter) is a function of the velocity, the velocity can be found by measuring the I^2R loss in the hot wire if the wire temperature remains constant. The resistance, and hence temperature, will remain constant if its value is the same as that of R_3, a precision isothermal reference resistor. The controller produces a current output proportional to the difference in voltage e between the two branches of the Wheatstone bridge. An increase in current results from an increase in airflow over and heat transfer from the heated wire. After calibration, the voltage drop across the hot wire (or across the vertical diagonal of the bridge) is a direct and accurate indication of the airflow over the wire.

The hot-wire system embodies all features of an electronic control system. The sensor component is the differential voltage measurement e across the bridge. The proportional controller and actuator is the voltage-to-current device shown in the figure. The controlled

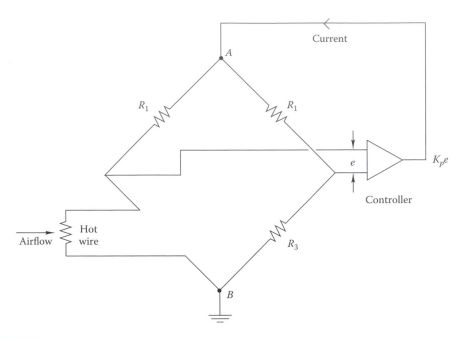

FIGURE 12.10
Hot-wire anemometer analog controller. The voltage between points *A* and *B* is related to the airflow value over the hot wire.

variable is the wire resistance which indirectly accomplishes the desired control goal—a constant-temperature hot-wire anemometer.

Direct digital control (DDC) enhances the previous analog-only electronic system with digital computer features. The term *digital* refers to the use of digital computers in these systems. Modern DDC systems use analog sensors (converted to digital signals within a computer) along with digital computer programs to control HVAC systems. The output of this microprocessor-based system can be used to control electronic, electric, or pneumatic actuators or a combination of them. DDC systems have the advantage of reliability and flexibility that others do not. For example, it is easier to accurately set control constants in computer software than to make adjustments at a controller panel with a screwdriver. DDC systems offer the option of operating *energy management systems* (EMSs) and HVAC diagnostic knowledge-based systems since the sensor data used for control are very similar to those used in EMSs. Pneumatic systems do not offer this ability. Figure 12.11 shows a schematic diagram of a direct digital controller. The entire control system must include sensors and actuators not shown in this controller-only drawing.

Temperature measurements for DDC applications are made by three principal methods:

- Thermocouples
- Resistance temperature detectors (RTDs)
- Thermistors

Each has its advantages for particular applications. Thermocouples consist of two dissimilar metals chosen to produce a measurable voltage at the temperature of interest. The voltage output is low (millivolts) but is a well-established function of the junction temperature. Except for flame temperature measurements, thermocouples produce

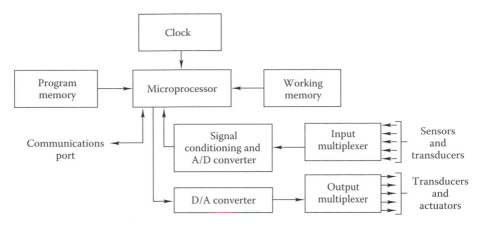

FIGURE 12.11
Block diagram of direct digital controller.

voltages too small to be useful in most HVAC applications (e.g., a type-J thermocouple produces only 5.3 mV at 100°C).

RTDs use small, responsive sensing sections constructed from metals whose resistance-temperature characteristic is well established and reproducible. To first order,

$$R = R_o(1 + kT) \tag{12.9}$$

where
    R = resistance, Ω
    R_o = resistance at reference temperature (0°C), Ω
    K = temperature coefficient of resistance, °C^{-1}
    T = RTD temperature, °C

This equation is easy to invert to find the temperature as a function of resistance. Although complex higher-order expressions exist, their use is not needed for HVAC applications.

Two common materials for RTDs are platinum and Balco (a 40 percent nickel, 60 percent iron alloy). Nominal values of k, respectively, are 3.85×10^{-3} and 4.1×10^{-3} K^{-1}.

Resistance is measured indirectly by measurements of voltage and current. Therefore, the controller must supply current to the RTD. The current can cause self-heating and consequent errors. These are avoidable by using higher-resistance RTDs. In addition, lead wire resistance can cause lack of accuracy for the class of platinum RTDs whose nominal resistance is only 100 Ω because the lead resistance of 1 to 2 Ω is not negligible by comparison with that of the sensor itself.

Thermistors are semiconductors with the property that resistance is a strong but nonlinear function of temperature, given approximately by

$$R = Ae^{B/T} \tag{12.10}$$

where A is related to the nominal value of resistance at the reference temperature and is on the order of 0.06 to 0.07 for a nominal 10 K sensor (at 20°C). The exponential coefficient B (a weak function of temperature) is on the order of 5400 to 7200°R (3000 to 4000 K). The nonlinearity inherent in thermistors can be reduced by connecting a properly selected fixed resistor in parallel with it. The resulting linearity is desirable from a control system design viewpoint. Thermistors can have a problem with long-term drift and aging; the designer and control manufacturer should consult on the most switchable thermistor design for HVAC applications.

Humidity measurements are needed for control of enthalpy economizers, as discussed in Chap. 11. Humidity may also need to be controlled in special environments such as clean rooms, hospitals, and spaces that house computers. The relative humidity, dew point, and humidity ratio are all indicators of the moisture content of air. The psychrometer described in Chap. 4 or the sensors described in the previous section (for pneumatic system applications) can be used with DDC systems but are impractical. Instead, an electrical, capacitance-based approach is preferred. The high dielectric constant of water absorbed into a polymer causes a significant change in capacitance. This is used along with a local electronic circuit to produce a linear voltage signal with humidity. If not saturated by excessive exposure to very high humidities, these devices produce reproducible signals without excessive hysteresis (Huang, 1991). Response times are on the order of seconds if the local air velocity over the sensor is above a few feet per second. Another solid-state device for humidity measurement uses the variation in resistance of a thin film of lithium chloride. The resistance change is significant but also

FIGURE 12.12
Photograph of humidity and temperature sensor. (Courtesy of Phys-Chem Scientific Corp., New York. With permission.)

depends on the temperature. The most common sensor type used presently employs the capacitance approach. Figure 12.12 is a photograph of a commercial humidity sensor.

Pressure measurements are made by electronic devices that depend on a change of resistance or capacitance with imposed pressure. Figure 12.13 shows a cross-sectional drawing of each. In the resistance type, stretching of the membrane lengthens the resistive element, (a strain gauge), thereby increasing resistance. This resistor is an element in a Wheatstone bridge; the resulting bridge voltage imbalance is linearly related to the imposed pressure. The less common capacitative type unit has a capacitance between a fixed and a flexible metal plate that decreases with pressure. The capacitance change is amplified by a local amplifier that produces an output signal proportional to the pressure.

Flow measurement or indication is also needed in DDC systems. Pitot tubes (or arrays of tubes), hot films, thermistors, and other flow measurement devices described in Chap. 5 can be used to measure either airflow or liquid flow in secondary HVAC systems. Airflow information is important for proper control of building pressure, VAV system control, and outside air supply. Water flow rates are needed for chiller and boiler control and for secondary-system liquid-loop control. In some cases, the quantitative value of flow is not needed, only a knowledge that flow exists. Sensors for this are called *flow switches*. They are electromechanical switches that change from open to closed (or vice versa) upon the existence of flow. Similar switches are used in some control system designs to sense the damper position (open or closed).

Temperature, humidity, and pressure *transmitters* are often used in HVAC systems. They amplify signals produced by the basic devices described in the preceding paragraphs, and they produce a standardized electric signal over a standard range, thereby permitting standardization of this aspect of DDC systems. The standard ranges are

- Current: 4 to 20 mA (dc)
- Voltage: 0 to 10 V (dc)

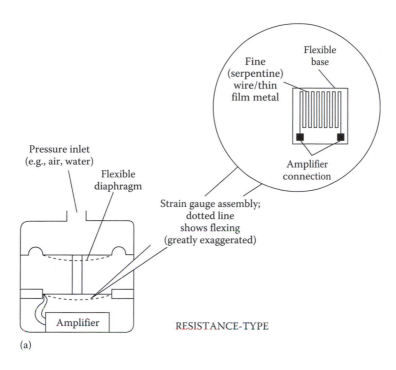

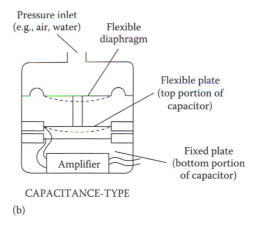

FIGURE 12.13
(a) Resistance- and (b) capacitance-type pressure sensors.

Although the majority of transmitters produce such signals, the noted values are not universally used. Stand-alone transmitters cost more than those integrated with the controller.

Figure 12.11 shows the elements of a direct digital controller. The heart of the controller is the microprocessor that can be programmed in either a standard or a system-specific language. Control algorithms (linear or not), sensor calibrations, output signal shaping, and historical data archiving can all be programmed as the user requires. A number of firms have constructed controllers on standard personal-computer platforms. It is beyond the scope of this book to describe the details of programming HVAC controllers, since each manufacturer uses a different approach. The essence of any DDC system, however, is the same as shown in the figure. Honeywell (1988) discusses DDC systems and their programming in greater detail.

Actuators and positioners for electronic control systems include

- *Motors*—operate valves, dampers
- *Variable-speed controls*—pump, fan, chiller drives
- *Relays and motor starters*—operate other mechanical or electric equipment (pumps, fans, chillers, compressors), electric heating equipment
- *Transducers*—convert, e.g., electric signal to pneumatic (EP transducer)
- *Visual displays*—not actuators in the usual sense but used to inform system operator of control and HVAC system function

We discuss several of these in the context of design in Sec. 12.3. Figure 12.14 shows an example of a DDC system actuator—an electric motor used for damper control. The motor is controlled by a signal from the direct digital controller. As an option, it can be equipped with a rotary potentiometer to feed back the damper position to the controller microprocessor. Pneumatic actuators with EP (electric to pneumatic) transducers are often used since they are more powerful and less costly than electrical actuators.

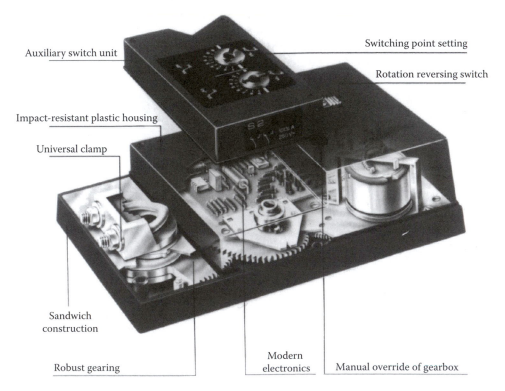

FIGURE 12.14
Electric damper motor. Cutaway showing motor, gear drive, shaft attachment, and circuit board. (Courtesy of Belimo Company, New York. With permission.)

The power supply to a DDC system and critical actuators must be maintained in case of a power outage. This is accomplished by use of an *uninterruptible power supply* (UPS) that is able to operate the DDC central processor and important peripherals for a period until power is restored. The length of the UPS capacity duration must be decided by the designer in view of local power quality and the time needed to conduct an orderly shutdown of the HVAC system.

Pneumatic and DDC systems have both advantages and disadvantages. Pneumatic systems have the advantage of lower cost, inherently modulating actuators and sensors, more powerful actuators, explosion-proof components, and diagnostic simplicity. Disadvantages include the need for a compressor producing clean and dry air, the cost of air piping, and the need for regular component calibration. DDC systems can be very precise (limited by sensor and actuator accuracy), can accommodate complex control algorithms and scheduling, easily accept changes to control constants, produce data usable by EMSs, and allow central control of a group of buildings. Present disadvantages include cost (this penalty is decreasing, as discussed earlier, but wires and power supplies cost more than pneumatic piping) and training needs of maintenance personnel who may be more familiar with pneumatic controls.

12.3 Basic Control System Design Considerations

This section discusses selected topics in control system design, including control system zoning, valve and damper selection, and control logic diagrams. Section 12.4 shows several HVAC system control design concepts.

The ultimate purpose of an HVAC control system is to provide adequate ventilation while controlling the zone temperature (and secondarily air motion and humidity) to conditions that ensure maximum comfort and productivity of the occupants. In other chapters of this book, we have discussed topics bearing on this issue including heating and cooling loads, zone air supply, and zone air distribution. From a controls viewpoint, the HVAC system is assumed to be able to provide comfort conditions if controlled properly.

Thermal criteria for zone selection are discussed in Chap. 7. Basically, a zone is a portion of a building whose loads differ in magnitude and timing sufficiently from other areas so that separate portions of the secondary HVAC system and control system are needed to maintain comfort. Typical zoning is done based on *exposure*—four cardinal directions and a core zone with little outside exposure—and *schedule*. Corner rooms with double exposure and top-floor spaces require special attention. In very tall buildings, upper floors may have sufficiently different exposure (wind and solar) that different zoning may be needed. Internal gain distributions within the core zone may dictate separate zones within the core. Rooms with large electrical loads (printing, computing, manufacturing machinery) will be controlled separately.

Having specified the zones, the designer must select the location of the thermostat (and other sensors, if used). According to Madsen (1989), the best placement for a thermostat is in a central location within a zone rather than at the typical wall locations used. It was found that zone temperature control was better (smaller temperature swings) and that energy consumption was reduced for a central location rather than for a wall location. Both field and laboratory tests supported this conclusion. Electronic sensors can be easily located anywhere; pneumatic thermostats are less easy to locate anywhere but on a wall.

If a central location is not possible, the usual wall location must be used. The thermostat should be located so that it senses the average temperature. Therefore, cold outside walls, areas exposed to direct sunlight, and parts of the zone directly in the airstream of either heating or cooling equipment should be avoided. A nominal mounting height that places the thermostat within the occupied level of a zone should also be used; height should be about eye level.

Thermostat signals either are passed to the central controller or are used locally to control the amount and temperature of conditioned air or coil water introduced into a zone. The air is conditioned either locally (e.g., by unit ventilator or baseboard heater) or centrally (e.g., by the heating and cooling coils in the central air handler). In either case, a flow control actuator is controlled by the thermostat signal. In addition, airflow itself may be controlled in response to zone information in variable-air-volume (VAV) systems. Except for variable-speed drives used in variable-volume air or liquid systems, flow is controlled by valves or dampers. The design selection of valves and dampers is discussed next.

12.3.1 Steam and Liquid Flow Control

The flow-through valve such as that shown in Fig. 12.7 is controlled by the valve stem position that determines the flow area. The variable flow resistance offered by valves depends on their design. The flow characteristic may be linear with position or not. Figure 12.15 shows common flow characteristics. Note that the plotted characteristics

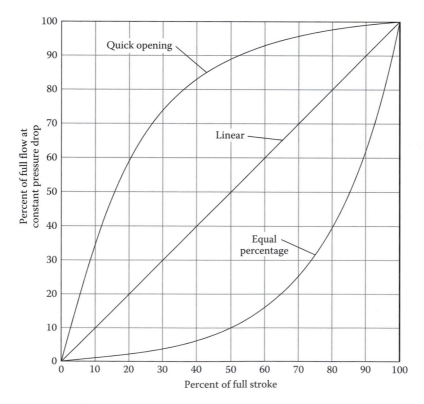

FIGURE 12.15
Quick-opening, linear, and equal-percentage valve characteristics.

apply only for *constant valve pressure drop*. The characteristics shown are idealizations of actual valves. Commercially available valves will resemble but not necessarily exactly match the curves shown.

The *linear* valve has a proportional relation between the volumetric flow $\dot{V}$ and the valve stem position z:

$$\dot{V} = kz \tag{12.11}$$

where k is the proportionality constant. The flow in *equal-percentage valves* increases by the same fractional amount for each increment of opening. In other words, if the valve is opened from 20 to 30 percent of full travel, the flow will increase by the same percentage as if the travel had increased from 80 to 90 percent of its full travel. However, the absolute volumetric flow increase for the latter case is much greater than for the former. The equal-percentage valve flow characteristic is given by

$$\dot{V} = Ke^{kz} \tag{12.12}$$

where K is a valve size constant. Quick-opening valves do not provide good flow control but are used when rapid action is required with little stem movement for on/off control.

Example 12.2: Equal-Percentage Valve

A valve at 30 percent of travel has a flow of 4 gal/min. If the valve opens another 10 percent and the flow increases by 50 percent to 6 gal/min, what are the constants in Eq. (12.12)? What will the flow be at 50 percent of full travel?

Figure: See Fig. 12.15.

Assumptions: Pressure drop across the valve remains constant.

Find: k, K, $\dot{V}_{50}$

SOLUTION

Equation (12.12) can be evaluated at the two flow conditions. If the results are divided by each other, we have

$$\frac{\dot{V}_2}{\dot{V}_1} = \frac{6}{4} = e^{k(z_2 - z_1)} = e^{k(0.4 - 0.3)} \tag{12.13}$$

In this expression, the travel z is expressed as the dimensionless *fraction* of the total travel and is dimensionless. Solving this equation for k gives

$$k = 4.05 \quad \text{(no units)}$$

From the known flow at 30 percent travel, we can find the second constant K:

$$K = \frac{4\,\text{gal/min}}{e^{4.05 \times 0.3}} = 1.19\,\text{gal/min} \tag{12.14}$$

Finally, the flow is given by

$$\dot{V} = 1.19e^{4.05z} \tag{12.15}$$

At 50 percent travel, the flow can be found from this expression:

$$\dot{V}_{50} = 1.19e^{4.05 \times 0.5} = 9.0 \, \text{gal/min} \qquad (12.16)$$

COMMENTS

This result can be checked since the valve is an equal-percentage valve. At 50 percent travel the valve has moved 10 percent beyond its 40 percent setting at which the flow was 6 gal/min. Another 10 percent stem movement will result in another 50 percent flow increase from 6 to 9 gal/min, confirming the solution.

The plotted characteristics of all three valve types assume constant pressure drop across the valve. However, in an actual system, the pressure drop across a valve will not remain constant. But, for the valve to maintain its control characteristics, the pressure drop across it must be the majority of the entire loop pressure drop. If the valve is designed to have a full-open pressure drop equal to that of the balance of the loop, good flow control will exist. This introduces the concept of *valve authority*, defined as the valve pressure drop as a fraction of total system pressure drop:

$$A \equiv \frac{\Delta p_{v, \text{open}}}{\Delta p_{v, \text{open}} + \Delta p_{\text{sys}}} \qquad (12.17)$$

For proper control, the full-open valve authority should be at least 0.50. If the authority is 0.5 or more, control valves will have installed characteristics not much different from those shown in Fig. 12.15. If not, the valve characteristic will be distorted at low flow since the majority of the system pressure drop will be dissipated across the valve.

Valves are further classified by the number of connections or ports. Figure 12.16 shows sections of typical *two-way* and *three-way* valves. Two-port valves control flow through coils or other HVAC equipment by varying the valve flow resistance as a result of flow area changes. As shown, the flow must oppose the closing of the valve. If not, near closure the valve would slam shut or oscillate, both of which cause excessive wear and noise. The three-way valve shown in the figure is configured in the *diverting* mode. That is, one stream is split into two depending on the valve opening. The three-way valve shown is double-seated (single-seated three-way valves are also available); therefore, it is easier to close than a single-seated valve, but tight shutoff is not possible.

Three-way valves can also be used as *mixing* valves. In this application, two streams enter the valve and one leaves. Mixing and diverting valves *cannot be used interchangeably* since their internal design is different to ensure that they can each seat properly. Particular attention must be paid by the installer to be sure that connections are made properly; arrows cast in the valve body show the proper flow direction. Figure 12.17 shows an example of a three-way valve for mixing and one for diverting applications.

The *valve capacity* is denoted in the industry by the dimensional flow coefficient C_v, defined by

$$\dot{V} \, (\text{gal/min}) = C_v \sqrt{\Delta p \, (\text{psi})} \qquad (12.18\text{US})$$

where C_v is specified as the flow rate of 60°F water that will pass through the fully open valve if a pressure difference of 1.0 psi is imposed across the valve. If SI units (liters per second and megapascals) are used, the numerical value of C_v is 24 percent larger than that

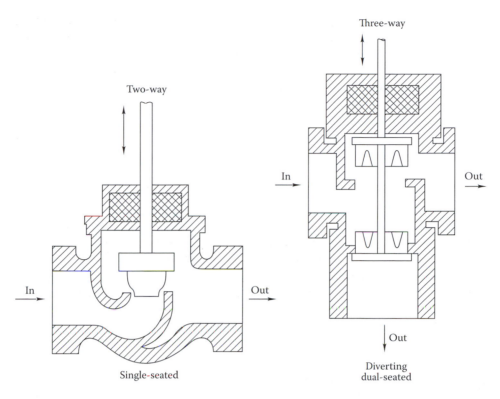

FIGURE 12.16
Cross-sectional drawings of direct-acting, single-seated, two-way valve and dual-seated, three-way, diverting valve.

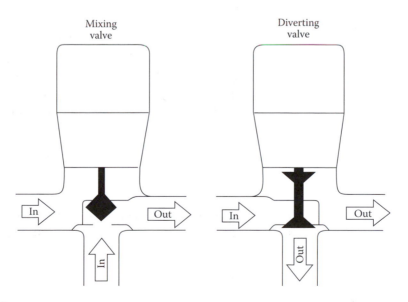

FIGURE 12.17
Three-way mixing and diverting valves. Note the significant difference in internal construction. Mixing valves are more commonly used.

in USCS units. Once the designer has determined a value of C_v, manufacturers' tables can be consulted to select a valve for the known pipe size. If a fluid other than water is to be controlled, the C_v value found from Eq. (12.18) should be multiplied by the square root of the fluid's specific gravity.

Steam valves are sized by using a similar dimensional expression

$$\dot{m} \text{ (lb/h)} = 63.5 C_v \sqrt{\frac{\Delta p \text{ (psi)}}{v \text{ (ft}^3/\text{lb)}}} \qquad (12.19\text{US})$$

in which v is the steam specific volume. If the steam is highly superheated, multiply C_v found from Eq. (12.19) by 1.07 for every 100°F of superheat. For wet steam, multiply C_v by the square root of the steam quality. Honeywell (1988) recommends that the pressure drop across the valve for use in the equation be 80 percent of the difference between the steam supply and return pressures (subject to the sonic flow limitation discussed below). Tables in the appendices on the CD-ROM can be used for preliminary selection of control valves for either steam or water.

The type of valve (linear or not) for a specific application must be selected so that the controlled system is as nearly linear as possible. Control valves are very commonly used to control the heat transfer rate in coils. For a linear system, the *combined characteristic of the actuator, valve, and coil should be linear*. This will require quite different valves for hot water and steam control, e.g., as we shall see. This difference is the reason for using linear and equal-percentage values.

Figure 12.18 shows the part-load performance of a hot water coil used for air heating; at 10 percent of full flow, the heat rate is 50 percent of its peak value. In Chap. 2 we learned that the heat rate in a cross-flow heat exchanger increases roughly in exponential fashion with the flow rate, a highly nonlinear characteristic. This heating coil nonlinearity follows from the longer water residence time in a coil at reduced flow and the relatively large temperature difference between air being heated and the water heating it.

However, if we were to control the flow through this heating coil by an equal-percentage valve (positive exponential increase of flow with valve position), the combined valve-plus-coil characteristic would be roughly linear. Referring to Fig. 12.18, we see that 50 percent of stem travel corresponds to 10 percent flow. The third graph in the figure is the combined characteristic. This near-linear subsystem is much easier to control than if a linear valve were used with the highly nonlinear coil. Hence the rule: Use equal-percentage valves for heating coil control.

Linear two-port valves are to be used for steam flow control to coils, since the transfer of heat by steam condensation is a linear constant-temperature process—the more steam supplied, the greater the heat rate, in exact proportion. Note that this is a completely different coil flow characteristic from that for hot water coils. However, steam is a compressible fluid, and the sonic velocity sets the flow limit for a given valve opening when the pressure drop across the valve is more than 60 percent of the steam supply line absolute pressure. As a result, the pressure drop to be used in Eq. (12.19) is the *smaller* of (1) 50 percent of the absolute steam line pressure upstream of the valve or (2) 80 percent of the difference between the steam supply and return line pressures. The 80 percent rule gives good valve modulation in the subsonic flow regime (Honeywell, 1988).

Chilled water control valves should also be linear since the performance of chilled water coils (smaller air-water temperature difference than in hot water coils) is more similar to that of steam coils than to that of hot water coils.

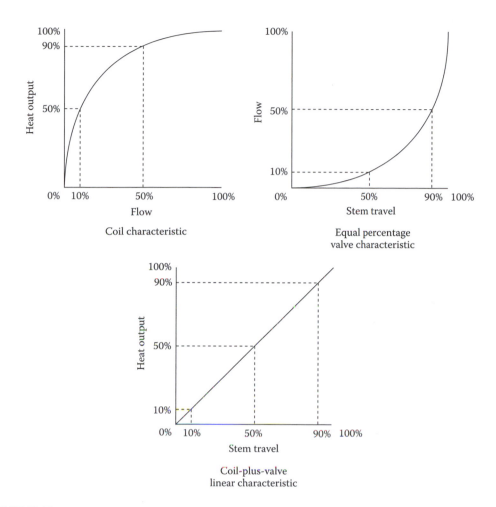

FIGURE 12.18
Heating coil, equal-percentage valve, and combined coil-plus-valve linear characteristic.

Either two- or three-way valves can be used to control flow at part load through heating and cooling coils, as shown in Fig. 12.19. The control valve can be controlled from the coil outlet water or air temperature. Two- or three-way valves achieve the same local result at the coil when used for part-load control. However, the designer must consider effects on the balance of the secondary system (pump size and power, flow balance) when selecting the valve type.

In essence, the two-way valve flow control method results in variable flow (tracking variable loads) with constant coil water temperature change, whereas the three-way valve approach results in roughly constant secondary loop flow rate but smaller coil water temperature change (beyond the local coil loop itself). Since some chillers and boilers require that the flow remain within a rather narrow range, the energy and cost savings that could accrue due to the two-way valve, variable-volume system are difficult to achieve in small systems unless the two-pump, primary/secondary loop approach described in Chap. 11 is employed. If this dual-loop approach is not used, the three-way valve method is required to maintain constant boiler or chiller flow. In large systems, a primary/secondary design with two-way valves is preferred.

The location of the three-way valve at a coil must also be considered by the designer. Figure 12.19b shows the valve used downstream of the coil in a mixing bypass mode. If a balancing valve is installed in the bypass line and is set to have the same pressure drop as the coil, the local coil loop will have the same pressure drop for both full and zero coil flow. However, at the valve midflow position, the overall flow resistance is less, since two parallel paths are involved, and the total loop flow increases to 25 percent more than that at either extreme.

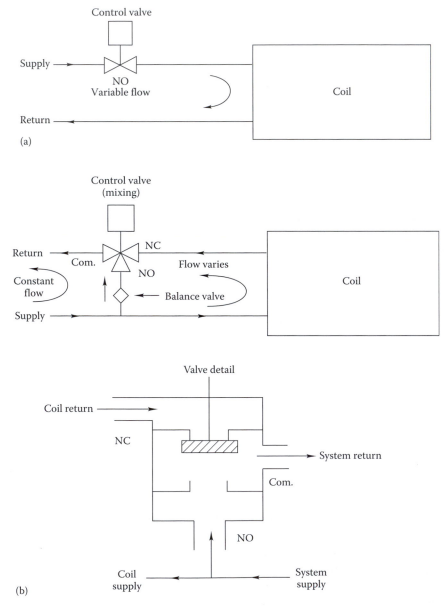

FIGURE 12.19
(a) Various control valve piping arrangements: two-way valve. (b) Various control valve piping arrangements: three-way mixing valve.

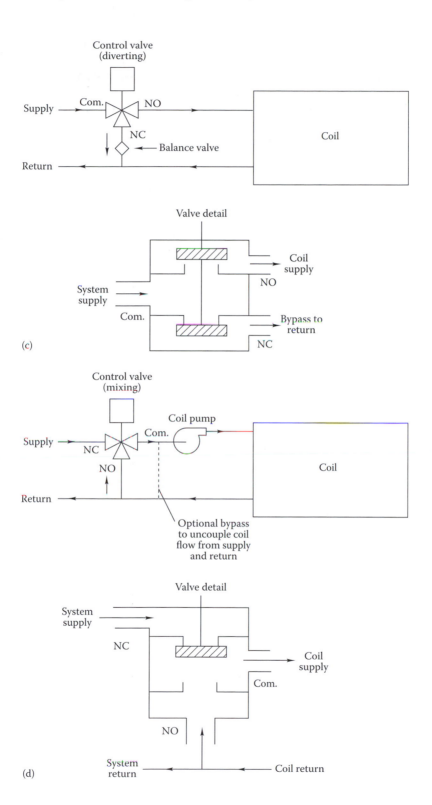

FIGURE 12.19 (continued)

(c) Various control valve piping arrangements: three-way diverting valve. (d) Various control valve piping arrangements: pumped coil with three-way mixing valve.

Alternatively, the three-way valve can also be used in a diverting mode, as shown in Fig. 12.19c. In this arrangement, essentially the same considerations apply as for the mixing arrangement discussed above.[1] However, if a circulator (small pump) is inserted as shown in Fig. 12.19d, the direction of flow in the branch line changes and a mixing valve is used. Pumped coils are used because control is improved. With constant coil flow, the highly nonlinear coil characteristic shown in Fig. 12.18 is reduced, since the residence time of hot water in the coil is constant, independent of the load. However, this arrangement appears to the external secondary loop the same as a two-way valve. As load is decreased, flow into the local coil loop also decreases. Therefore, the uniform secondary loop flow normally associated with three-way valves is not present unless the optional bypass is used.

For HVAC systems requiring precise control, high-quality control valves are required. The best controllers and valves are of "industrial quality." The additional cost for these valves compared to conventional building hardware results in more accurate control and longer lifetime.

12.3.2 Airflow Control

Dampers are used to control airflow in secondary HVAC air systems in buildings. As discussed in Chap. 11, variable airflow rates are also produced by variable fan speeds. In this section, we discuss the characteristics of dampers used for flow control in systems where constant-speed fans are involved. Figure 12.20 shows cross sections of the two common types of dampers used in commercial buildings. Parallel-blade dampers use blades that all rotate in the same direction. They are most often applied to two-position locations—either open or closed. Their use for flow control is not recommended. The blade rotation changes the airflow direction, a characteristic that can be useful when airstreams at different temperatures are to be effectively blended.

Opposed-blade dampers have adjacent counterrotating blades. The airflow direction is not changed with this design, but pressure drops are higher than for parallel blading. Opposed-blade dampers are preferred for flow control. Figure 12.21 shows the flow characteristics of these dampers to be closer to the desired linear behavior. The parameter α on the curves is the ratio of system pressure drop to fully open damper pressure drop.

Damper leakage is always a concern of the designer in critical locations such as at outdoor air intakes. Extra expense should be expected for low-leakage dampers in such applications. Damper leakage is expressed in units of leakage volumetric flow per unit damper area (e.g., cubic feet per minute per square foot) at a specified pressure difference. The manufacturer's literature should be consulted. A typical damper may leak 40 $ft^3/(min \cdot ft^2)$ at 1.0 inWG while a low-leakage damper may leak only 10 $ft^3/(min \cdot ft^2)$ at the same pressure difference. The designer will specify acceptable leakage rates depending on the application. The actuator torque, air velocity, materials of construction, and maximum pressure difference (damper fully closed) must also be considered by the HVAC engineer.

---

[1] A little known disadvantage of three-way valve control has to do with the *conduction* of heat from a closed valve to a coil. For example, the constant flow of hot water through two ports of a *closed* three-way heating coil control valve keeps the valve body hot. Conduction from the closed hot valve mounted close to a coil can cause sufficient air heating to actually decrease the expected cooling rate of a downstream cooling coil during the cooling season. Three-way valves have a second practical problem: Installers often connect three-way valves incorrectly, given the choice of three pipe connections and three pipes to be connected. Both these problems can be avoided by using two-way valves.

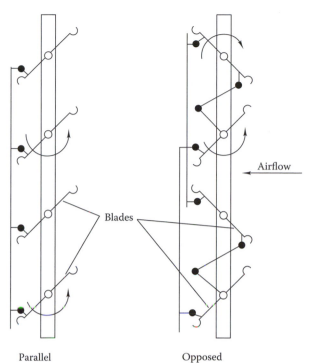

Parallel   Opposed

FIGURE 12.20
Diagram of parallel- and opposed-blade dampers.

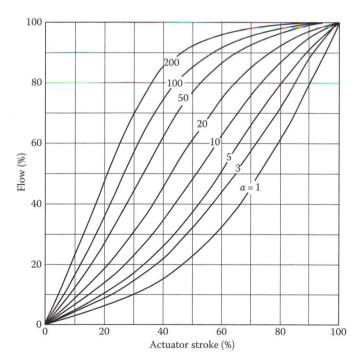

FIGURE 12.21
Flow characteristics of opposed-blade dampers. The parameter α is the ratio of system resistance (not including the damper) to damper resistance. The ideal linear damper characteristic is achieved if this ratio is about 10 for opposed-blade dampers.

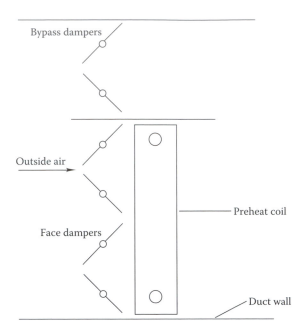

FIGURE 12.22
Face and bypass dampers used for preheat coil control.

   A common application of dampers controlling the flow of outside air uses two sets in a
face and bypass configuration, as shown in Fig. 12.22. The preheat coil is used for preheating
cold outdoor air upstream of the system heating coil, as described in Chap. 11. For
full heating, all air is passed through the coil, and the bypass dampers are closed. If no
heating is needed in mild weather, the coil is bypassed (for minimum flow resistance and
fan power cost, flow through fully open face and bypass dampers can be used if the
preheat coil water flow is shut off). Between these extremes, flow is split between the
two paths. The face and bypass dampers are sized so that the pressure drop in full-bypass
mode (damper pressure drop only) and full-heating mode (coil plus damper pressure
drop) is the same.

12.4 Examples of HVAC System Control Systems

Several widely used control configurations for specific tasks are described in this section.
These have been selected from the hundreds of control system configurations that have
been used for buildings. The goal of this section is to illustrate how control components
described previously are assembled into systems and what design considerations are
involved. For a complete overview of HVAC control system configurations, the reader is
referred to ASHRAE (1987), Grimm and Rosaler (1990), and Honeywell (1988). The illustrative systems in this section are drawn in part from the last reference.
   In this section, we will discuss seven control systems in common use. Each system will
be described by using a schematic diagram, and its operation and key features will be
discussed in the accompanying text.

12.4.1 Outside Air Control

Figure 12.23 shows a system for controlling outside and exhaust air from a central air handling unit equipped for economizer cooling when available. In this and the following diagrams, the following symbols are used:

C	Cooling coil
DA	Discharge air (supply air from fan)
DX	Direct-expansion coil
EA	Exhaust air
H	Heating coil
LT	Low-temperature limit sensor or switch; must sense lowest temperature in air volume being controlled
M	Motor or actuator (for damper or valve), variable-speed drive
MA	Mixed air
NC	Normally closed
NO	Normally open
OA	Outside air
PI	Proportional plus integral controller

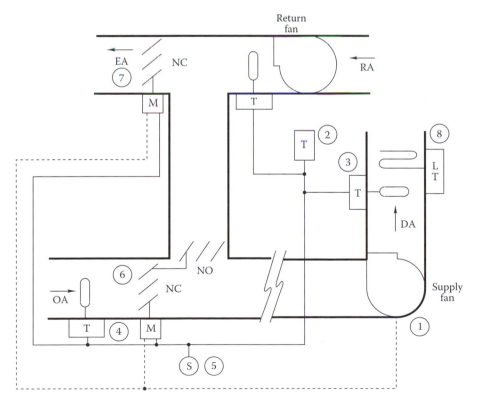

FIGURE 12.23
Outside air control system with economizer capability.

R   Relay

RA   Return air

S   Switch

SP   Static pressure sensor used in VAV systems

T   Temperature sensor; must be located to read average temperature representative of volume being controlled

This system is able to provide the minimum outside air during occupied periods, to use outdoor air for cooling when appropriate by means of a temperature-based economizer cycle, and to operate fans and dampers under all conditions. The numbering system used in the figure indicates the sequence of events as the air handling system begins operation after an off period.

1. The fan control system turns on when the fan is turned on. This may be initiated by a clock signal or a low-temperature space condition. The OA dampers are at the minimum position before the fan starts.

2. The space temperature signal determines if the space is above or below the setpoint. If above, the economizer feature will be activated and will control the outdoor and mixed-air dampers. If below, the outside air damper is set to its minimum position.

3. The mixed-air PI controller controls both sets of dampers (OA/RA and EA) to provide the desired mixed-air temperature.

4. When the outdoor temperature rises above the cutoff point for economizer operation, the outdoor air damper is returned to its minimum setting.

5. Switch S is used to set the minimum setting on outside and exhaust air dampers manually. This is ordinarily done only once during building commissioning and flow testing.

6. When the supply fan is off, the outdoor air damper returns to its NC position and the return air damper returns to its NO position.

7. When the supply fan is off, the exhaust damper also returns to its NC position.

8. Low temperature sensed in the duct will initiate a freeze protect cycle. This may be as simple as turning on the supply fan to circulate warmer room air. Of course, the OA and EA dampers remain tightly closed during this operation.

12.4.2 Heating Control

If the minimum air setting is large in the preceding system, the amount of outdoor air admitted in cold climates may require preheating, as discussed above and in Chap. 11. Figure 12.24 shows a preheat system using face and bypass dampers. [A similar arrangement is used for direct-expansion (DX) cooling coils.] The equipment shown is installed upstream of the fan in Fig. 12.23. This system operates as follows:

1. The preheat subsystem control is activated when the supply fan is turned on.

2. The preheat PI controller senses temperature leaving the preheat section. It operates the face and bypass dampers to control the exit air temperature between 45 and 50°F.

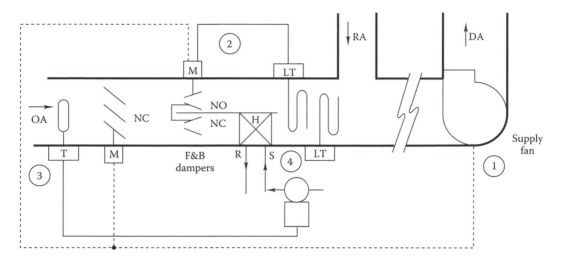

FIGURE 12.24
Preheat control system. Counterflow of air and hot water in the preheat coil results in the highest heat transfer rate.

3. The outdoor air sensor and associated controller control the water valve at the preheat coil. The valve may be either a modulating valve (better control) or an on/off valve (less costly).

4. The low-temperature (LT) sensors activate coil freeze protection measures, including closing dampers and turning off the supply fan.

Note that the preheat coil (and all coils in this section) is connected so that the hot water (or steam) flows counter to the direction of airflow. In Chap. 2, we learned that counterflow provides a higher heating rate for a given coil than parallel flow does. Mixing of heated and cold bypass air must occur upstream of the control sensors. Stratification can be reduced by using sheet-metal *air blenders* or by propeller fans in the ducting. The preheat coil should be located in the bottom of the duct. Steam preheat coils must have adequately sized traps and vacuum breakers to avoid condensate buildup that could lead to coil freezing at light loads.

The face and bypass damper approach enables air to be heated to the required system supply temperature without endangering the heating coil. (If a coil were as large as the duct—no bypass area—it could freeze when the hot water control valve cycles opened and closed to maintain discharge temperature.) The designer should consider pumping the preheat coil as shown in Fig. 12.19d to maintain water velocity above the 3 ft/s needed to avoid freezing. If glycol is used in the system, the pump is not necessary, but heat transfer will be reduced.

During winter, in cold climates, heat must be added to the mixed airstream to heat the outside-air portion of mixed air to an acceptable discharge temperature. Figure 12.25 shows a common heating subsystem controller used with central air handlers. (It is assumed that the mixed-air temperature is above freezing by action of the preheat coil, if needed.) This system has the added feature that the coil discharge temperature is adjusted for ambient temperature since the amount of heat needed decreases with increasing

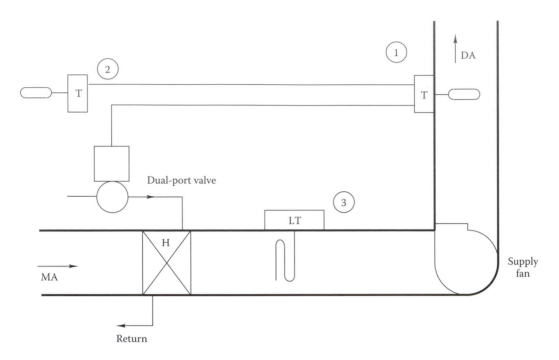

FIGURE 12.25
Heating coil control subsystem using two-way valve and optional reset.

outside temperature. This feature, called the coil discharge *reset*, provides better control and can reduce energy consumption. The system operates as follows:

1. During operation, the discharge air sensor and PI controller control the hot water valve.
2. The outside-air sensor and controller resets the *setpoint* of the discharge air PI controller up as the ambient temperature drops.
3. Under sensed low-temperature conditions, freeze protection measures are initiated as discussed above.

Reheat at zones in VAV or other systems uses a system similar to that just discussed. However, the boiler water temperature is reset, and no freeze protection is normally included. The air temperature sensor is the zone thermostat for VAV reheat, not a duct temperature sensor.

12.4.3 Cooling Control

Figure 12.26 shows the components in a cooling coil control system for a single-zone system. Control is similar to that for the heating coil discussed above, except that the zone thermostat (not a duct temperature sensor) controls the coil. If the system were a central system serving several zones, a duct sensor would be used. Chilled water supplied to the coil partially bypasses and partially flows through the coil, depending on the coil load. The use of three- and two-way valves for coil control was discussed in detail earlier.

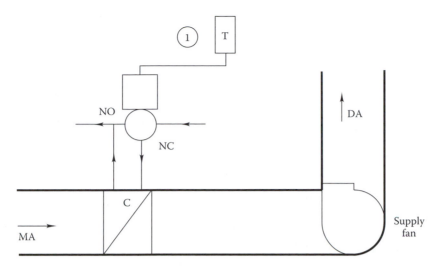

FIGURE 12.26
Cooling coil control subsystem using three-way diverting valve.

The valve-NC connection is used as shown so that valve failure (control power outage or loss of air supply) will not block secondary loop flow.

Figure 12.27 shows another common cooling coil control system. In this case, the coil is a DX coil, and the controlled medium is refrigerant flow. DX coils are used when precise

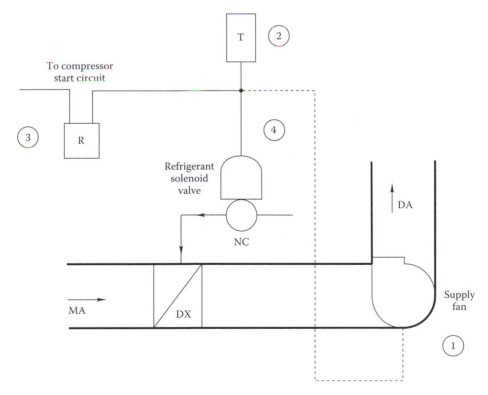

FIGURE 12.27
DX cooling coil control subsystem (on/off control).

temperature control is not required, since the coil outlet temperature drop is large when-ever refrigerant is released into the coil because refrigerant flow is not modulated; it is most commonly either on or off. The control system sequences as follows:

1. The coil control system is energized when the supply fan is turned on.
2. The zone thermostat opens the two-position refrigerant valve for temperatures above the setpoint and closes it in the opposite condition.
3. At the same time, the compressor is energized or deenergized. The compressor has its own internal controls for oil control and pump-down.
4. When the supply fan is off, the refrigerant solenoid valve returns to its NC position and the compressor relay to its NO position.

At light loads, bypass rates are high, and ice may build up on coils. Therefore, control is poor at light loads with this system.

12.4.4 Complete Systems

The preceding five example systems are actually control subsystems that must be inte-grated into a single control system for the HVAC system's primary and secondary systems. In the remainder of this section, we briefly describe two complete HVAC control systems widely used in commercial buildings. The first is a fixed-volume system, and the second is a VAV system.

Figure 12.28 shows a constant-volume, central system air-handling system equipped with supply and return fans, heating and cooling coils, and an economizer for a single-zone

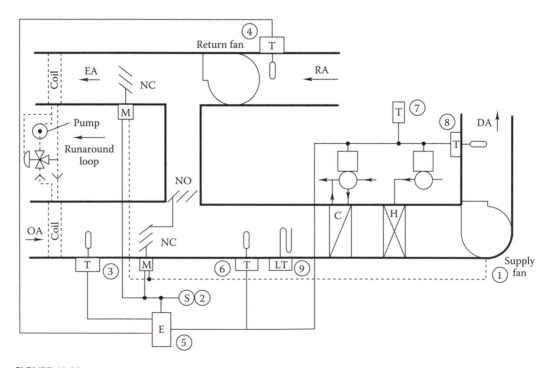

FIGURE 12.28
Control for a complete fixed-volume HVAC system. Optional runaround heat recovery system is shown to left in dashed lines.

application. If the system were to be used for multiple zones, the zone thermostat shown would be replaced by a discharge air temperature sensor. This fixed-volume system operates thus:

1. When the fan is energized, the control system is activated.
2. The minimum air setting is set (usually only once during commissioning, as described earlier).
3. The OA temperature sensor supplies a signal to the damper controller.
4. The RA temperature sensor supplies a signal to the damper controller.
5. The damper controller positions the dampers to use outdoor or return air, depending on which is cooler.
6. The mixed-air low-temperature controller controls the OA dampers to keep excessively low-temperature air from entering the coils. If a preheat system were included, this sensor would control it.
7. Optionally the space sensor could reset the coil discharge air PI controller.
8. The discharge air controller controls the
   a. Heating coil valve
   b. Outdoor air damper
   c. Exhaust air damper
   d. Return air damper
   e. Cooling coil valve after economizer cycle upper limit is reached
9. The low-temperature sensor initiates freeze protection measures, as described previously.

A method for reclaiming either heating or cooling energy is shown by dashed lines in Fig. 12.28. This "runaround" system extracts energy from exhaust air and uses it to precondition outside air. For example, the heating season exhaust air may be at 75°F while outdoor air is at 10°F. The upper coil in the figure extracts heat from the 75°F exhaust and transfers it through the lower coil to the 10°F intake air. To avoid icing of the air intake coil, the three-way valve controls this coil's liquid inlet temperature to a temperature above freezing. In heating climates, the liquid loop should also be freeze-protected with a glycol solution. Heat reclaim systems of this type can also be effective in the cooling season, when outdoor temperatures are well above the indoor temperature.

A VAV system has additional control features including a motor speed (or inlet vane) control and a duct static pressure control. Figure 12.29 shows a VAV system serving both perimeter and interior zones. It is assumed that the core zones always require cooling during the occupied period. The system shown has a number of options and does not include every feature present in all VAV systems. However, it is representative of VAV design practice. The sequence of operation is as follows:

1. When the fan is energized, the control system is activated. Prior to activation, during unoccupied periods the perimeter zone baseboard heating is under the control of room thermostats.
2. Return and supply fan interlocks are used to prevent pressure imbalances in the supply air ductwork.

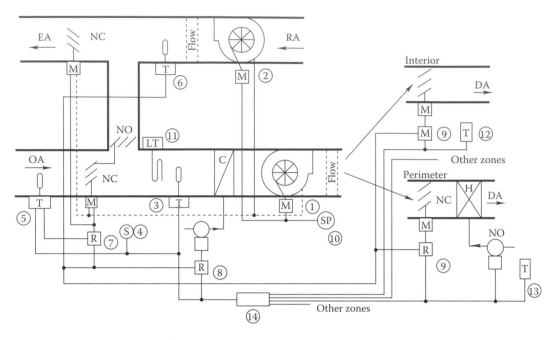

FIGURE 12.29
Control for complete VAV system. Optional supply and return flow stations are shown with dashed lines.

3. The mixed-air sensor controls the outdoor air dampers (and/or preheat coil, not shown) to provide the proper coil air inlet temperature. The dampers will be at their minimum position at about 40°F typically.

4. The damper's minimum position controls the minimum outdoor airflow.

5. As the upper limit for economizer operation is reached, the OA dampers are returned to their minimum position.

6. The return air temperature is used to control the morning warm-up cycle after night setback. (Option is present only if night setback is used.)

7. The OA damper is not permitted to open during morning warm-up by action of the relay shown.

8. Likewise, the cooling coil valve is deenergized (NC) during morning warm-up.

9. All VAV box dampers are moved full open during morning warm-up by action of the relay override. This minimizes warm-up time because the boxes will almost always be in the closed position. Perimeter zone coils and baseboard units are under the control of the local thermostat.

10. During operating periods, the PI static pressure controller controls both supply and return fan speeds (or inlet vane positions) to maintain approximately 1.0 inWG of static pressure at the pressure sensor location (or optionally to maintain building pressure). An additional pressure sensor (not shown) at the supply fan outlet will shut down the fan if fire dampers or other dampers should close completely and block airflow. This sensor overrides the duct static pressure sensor shown. (Alternatively, the supply fan can be controlled by duct pressure and the return fan by building pressure.)

11. The low-temperature sensor initiates freeze protection measures.

12. At each zone, room thermostats control VAV boxes (and fans, if present); as the zone temperature rises, the boxes open more.

13. At each perimeter zone, room thermostats close VAV dampers to their minimum settings and activate zone heat (coil and/or perimeter baseboard) as the zone temperature falls.

14. The controller, using temperature information for all zones (or at least for enough zones to represent the characteristics of all zones), modulates OA dampers (during economizer operation) and the cooling control valve (above economizer cycle cutoff) to provide air sufficiently cooled to meet the load of the warmest zone.

The duct static pressure controller is critical to proper operation of VAV systems. As discussed in Chap. 11, if proper duct pressurization at the end of long runs is not present, flow starvation can occur at VAV terminals located far from the air handler. The static pressure controller must be of PI design since a proportional-only controller would permit the duct pressure to drift upward as cooling loads dropped due to the unavoidable offset in P-type controllers. In addition, the control system should position inlet vanes closed during fan shutdown to avoid overloading on restart.

Return fan control is best achieved in VAV systems by an actual flow measurement in supply and return ducts, as shown by dashed lines in the figure. The return airflow rate is the supply rate less local exhausts (fume hoods, toilets, etc.) and exfiltration needed to pressurize the building. Then the duct pressure sensor is used only for controlling maximum duct pressure.

VAV boxes are controlled locally, assuming that adequate duct static pressure exists in the supply duct and that supply air is at an adequate temperature to meet the load (this is the function of the controller described in item 14 above). Figure 12.30 shows a *local* control system used with a series-type fan-powered VAV box. This particular system delivers a constant flow rate to the zone, to ensure proper zone air distribution, by action of the airflow controller. Primary air varies with the cooling load, as shown in the lower part of the figure and as discussed in Chap. 11. Optional reheat is provided by the coil shown.

12.4.5 Other Systems

This section has not covered the control of central plant equipment such as chillers and boilers. Most primary system equipment controls are furnished with the equipment and as such do not offer much flexibility to the designer. However, Braun et al. (1989) verified what has long been used in practice by showing that considerable energy savings can be accrued by properly sequencing cooling tower stages on chiller plants and by properly sequencing chillers themselves in multiple-chiller plants.

Fire and smoke control is important for human safety in large buildings. The design of smoke control systems is controlled by national codes. The principal concept is to eliminate smoke from the zones where it is present while keeping adjacent zones pressurized to avoid smoke infiltration. Some components of space conditioning systems, e.g., fans, can be used for smoke control, but HVAC systems are generally not smoke control systems by design.

Electric systems are primarily the responsibility of the electrical engineer on a design team. However, HVAC engineers must ensure that the electrical design accommodates the HVAC control system. Interfaces between the two occur where the HVAC controls activate motors on fans or chiller compressors, pumps, electric boilers, or other electric equipment.

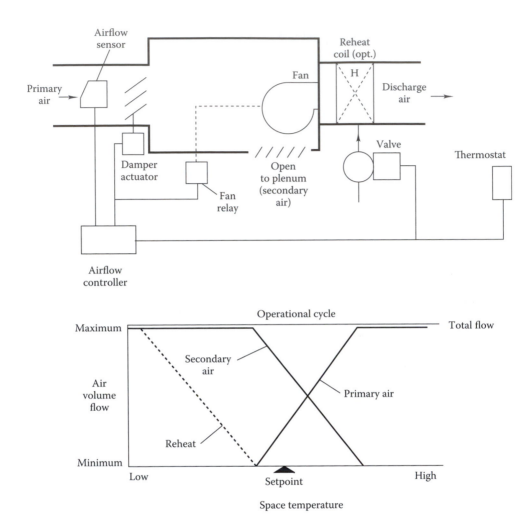

FIGURE 12.30
Series-type, fan-powered VAV box control subsystem and primary flow characteristic. The total box flow is constant at the level identified as maximum in the figure. The difference between primary and total airflow is the secondary air recirculated through the return air grille. Optional reheat coil requires airflow shown by the dashed line.

In addition to electrical specifications, the HVAC engineer often conveys electrical control logic by using a *ladder diagram*. An example is shown in Fig. 12.31 for the control of the supply and return fans in a central system. The electric control system is shown at the bottom and operates on low voltage (24 or 48 V ac) from the control transformer shown. The supply fan is started manually by closing the "start" switch. This activates the motor starter coil labeled 1M, thereby closing the three contacts labeled 1M in the supply fan circuit. The fourth 1M contact (in parallel with the start switch) holds the starter closed after the start button is released.

The manual-off-auto switch is typical and allows both automatic and manual operation of the return fan. When it is on the manual position, the fan starts. In the auto position, the fans will operate only when the adjacent 3M contacts are closed. Either of these actions activates the relay coil 2M, which in turn closes the three 2M contacts in the return fan motor starter. When either fan produces actual airflow, a flow switch is closed in the ducting,

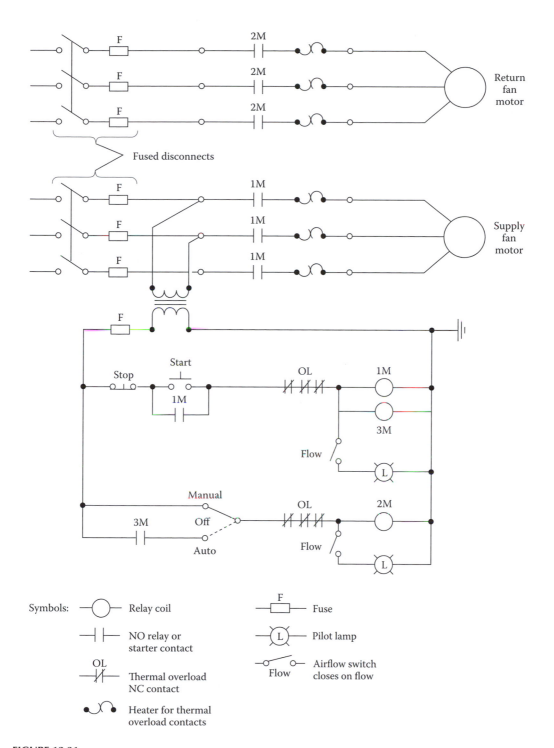

FIGURE 12.31
Ladder diagram for supply and return fan control. Manual-off-auto switch permits manual or automatic control of the return fan.

thereby completing the circuit to the pilot lamps (L). The fan motors are protected by fuses and thermal overload heaters. If motor current draw is excessive, the heaters (OL) shown in the figure produce sufficient heat to open the normally closed thermal overload contacts.

This example of a ladder diagram is primarily illustrative and is not typical of an actual design. In a fully automatic system, both fans would be controlled by 3M contacts actuated by the HVAC control system. In a fully manual system, the return fan would be activated by a fifth 1M contact, not by the 3M automatic control system.

12.5 Topics in Advanced Control System Design

The previous sections have described control system configurations and hardware design but have not described how one selects the type of control—P, PI, or PID—to be used or what the control constants K_p, K_i, and K_d should be. In this section, we will discuss some of the considerations that lead to selection of control algorithms. We will find Laplace transforms very useful for this purpose and will review this tool first. In this section, we will discuss primarily linear systems or systems that can be considered linear over a portion of their operating range.

12.5.1 Laplace Transforms

The Laplace transform is an integral transform that can be used to convert differential equations to algebraic equations. Solving the algebraic equation and inverting the result provide the solution to the differential equation. In HVAC control design, we are interested in solving the differential equations that model HVAC controls and the process being controlled. Even if the equations are not solved explicitly, the nature of the transform indicates how the control system will behave.

By definition, the Laplace transform $F(s)$ of a function of time $f(t)$ is found from

$$F(s) = \int_{0}^{\infty} e^{-st} f(t)\, dt \tag{12.20}$$

in which s is the complex variable

$$s = \sigma + j\omega \tag{12.21}$$

Wylie and Barrett (1982) show that the transforms of several common functions are expressed by the simple polynomials shown in Table 12.1. For HVAC system control analysis, we are most interested in the system response without reference to initial conditions. Therefore, we will drop the initial-condition terms from the transforms used in this section.

If a linear second-order system for an independent variable $y(t)$ has a forcing function $f(t)$, the transform of the equation with coefficients a, b, and c is

$$(as^2 + bs + c)Y(s) = F(s) \tag{12.22}$$

Solving for $Y(s)$ in the s domain is simply a matter of dividing both sides of the equation by the polynomial. The result is

$$Y(s) = \frac{F(s)}{as^2 + bs + c} \tag{12.23}$$

TABLE 12.1

Summary of Laplace Transforms

$F(s)$	$f(t)(t > 0)$	Notes
1	$\delta(t)$	Unit impulse
$\exp(-Ts)$	$\delta(t - T)$	Delayed impulse
$\dfrac{1}{s + a}$	$\exp(-at)$	Exponential decay
$\dfrac{1}{(s + a)^n}$	$\dfrac{t^{n-1} \exp(-at)}{(n-1)!}$	$n = 1, 2, 3, \ldots$
$\dfrac{1}{s}$	$u(t)$ or 1	Unit step
$\dfrac{1}{s^2}$	t	Unit ramp
$\dfrac{1}{s^n}$	$\dfrac{t^{n-1}}{(n-1)!}$	$0! = 1, n = 1, 2, 3, \ldots$
$\dfrac{\omega}{s^2 + \omega^2}$	$\sin \omega t$	Sine
$\dfrac{s}{s^2 + \omega^2}$	$\cos \omega t$	Cosine
$\dfrac{1}{(s + a)(s + b)}$	$\dfrac{e^{-at} - e^{-bt}}{b - a}$	$a \neq b$
$\dfrac{s}{(s + a)(s + b)}$	$\dfrac{ae^{at} - be^{-bt}}{a - b}$	$a \neq b$
$\dfrac{1}{s(s + a)}$	$\dfrac{1 - e^{-at}}{a}$	Exponential decay
$\dfrac{1}{s^2 + 2\zeta\omega_n s + \omega_n^2}$	$\dfrac{1}{\omega_d} \cdot e^{-\zeta\omega_n t} \sin \omega_d t$ $\left(\omega_d = \omega_n \sqrt{1 - \zeta^2}\right)$	$0 < \zeta < 1$
$\dfrac{s}{(s + a)^2}$	$(1 - at)e^{-at}$	
$\dfrac{a^2}{(s + a)^2 s}$	$1 - e^{-at}(1 + at)$	
$sF(s) - f(t = 0)$	$\dfrac{df(t)}{dt}$	Derivative
$\dfrac{F(s)}{s} + \dfrac{\int t(t)dt\vert_{t=0}}{s}$	$\int f(t)dt$	Integral

The *transfer function* TF is the ratio of output to forcing function. In this example,

$$\text{TF} = \frac{Y(s)}{F(s)} \tag{12.24}$$

To find the solution in the time domain, one uses results from the first and second columns of Table 12.1. If $F(s)$ in Eq. (12.23) were a constant, the solution would be exponential or sines and cosines or an exponential-sinusoidal combination.

Example 12.3: Laplace Transform

The Laplace transform of a differential equation is

$$Y(s) = \frac{s + 3}{s(s + 1)^2} \tag{12.25}$$

What is the inverse $y(t)$?

   Given: $Y(s)$

   Find: $y(t)$ if $y(0) = 0$

SOLUTION

First the given function must be rearranged so that the results in Table 12.1 can be used to invert the result. It is easy to show that Eq. 12.25 can be factored as follows:

$$Y(s) = \frac{3}{s} - \frac{3}{s+1} - \frac{2}{(s+1)^2} \tag{12.26}$$

Referring to Table 12.1, we see that each term type appears in the first column, and we can write the inverse function by inspection. The solution is

$$y(t) = 3 - 3e^{-t} - 2te^{-t} \tag{12.27}$$

COMMENTS

The solution is a constant with two exponential decay terms. For large times, note that $y(t) = 3$. By a theorem in Laplace transforms, it is known that this limit is the same as a related limit in the s domain:

$$\lim_{t \to \infty(t)} y = \lim_{s \to 0} sY(s) \tag{12.28}$$

From Eq. (12.28) we see that the long time limit can be found without inverting the Laplace transform.

   A complementary theorem also relates the initial value of a function to its transform:

$$\lim_{t \to 0(t)} y = \lim_{s \to \infty} sY(s) \tag{12.29}$$

12.5.2 Laplace Transforms for HVAC Equipment

A basic feedback controller can be represented by the schematic diagram in Fig. 12.32. The error e is processed by a controller which produces a signal V controlling the process. For a room during the heating season, the controlled variable is room temperature; and the controlling system is the thermostat sensor and its controller. The controlled process is the heat input into the room from the heating terminal device and the heat loss from the room.

   To write an equation for each block in the diagram, the appropriate Laplace transforms must be known. According to Letherman (1981), the characteristics of some common building components can be represented by the expressions given below. For the room itself, the transfer function TF is

$$\mathrm{TF}_{\mathrm{room}} = \frac{Ke^{-sL_{\mathrm{room}}}}{1 + \tau_{\mathrm{room}}s} \tag{12.30}$$

where
   $\mathrm{TF}_{\mathrm{room}} = $ room transfer function (the "process")
      $K = $ steady-state loss coefficient
   $L_{\mathrm{room}} = $ time delay (see Table 12.1, line 2)
   $\tau_{\mathrm{room}} = $ room time constant; the ratio of lumped room thermal capacitance to loss coefficient

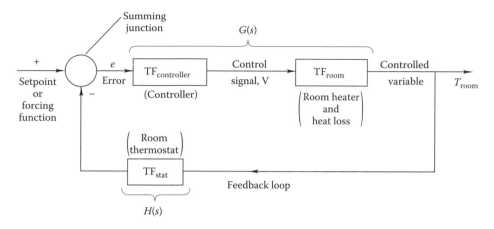

FIGURE 12.32
Simple feedback control loop for room-temperature control. Error *e* is processed by the controller, which produces a control signal to the controlled process (the room heating equipment and the room heat loss characteristic). The thermostat may have its own dynamics characterized by its transfer function TF_{stat}.

The transfer function of a PID controller is

$$\text{TF}_{\text{controller}} = \frac{V(s)}{E(s)} = K_p + \frac{K_i}{s} + K_d s \tag{12.31}$$

in which $E(s)$ is the transform of the error *e* and $V(s)$ is the transform of the controller output.

The closed-loop transfer function of the feedback controller shown in Figure 12.32 is

$$\text{TF}_{\text{loop}} = \frac{G(s)}{1 + G(s)H(s)} \tag{12.32}$$

in which

$$G(s) = \text{TF}_{\text{control}} \times \text{TF}_{\text{room}} \tag{12.33}$$

is the combined transfer function of the controller and process shown in Fig. 12.32, where $G(s)$ is sometimes called the *open-loop transfer function*. The corresponding *characteristic equation* is

$$G(s)H(s) + 1 = 0 \tag{12.34}$$

Shortly we shall see that the characteristic equation contains most of the information needed to assess the performance of linear control systems.

For a thermostat the following transfer function including both a lag and a first-order time constant is suggested:

$$\text{TF}_{\text{stat}} = \frac{e^{-sL_{\text{stat}}}}{1 + \tau_{\text{stat}} s} \tag{12.35}$$

The time lag L_{stat} is on the order of 30 to 40 s, and the time constant τ_{stat} is on the order of minutes for zone-mounted thermostats.

The first-order approximation to various lag terms, such as that resulting from the distance between a temperature measured at a coil and a controller at a remote chiller or those noted above, is given by the Padé approximation. This is convenient because it preserves a polynomial expression for a system's transfer function—a useful form when one is investigating stability, as we shall see below. To first order in sL, the transform of a lag is

$$e^{-sL} \approx \frac{2 - sL}{2 + sL} \tag{12.36}$$

Little is known of the transfer functions of other HVAC equipment and systems. It is therefore necessary to determine them experimentally, as we shall discuss shortly.

12.5.3 Control System Stability

One of the key results that Laplace transform analysis of building systems and controls can provide is a judgment as to whether the control system is stable when a step forcing function is imposed. As described in standard controls texts, a system is stable if all roots of the characteristic equation [Eq. (12.34)] have no positive real part (recall that s is complex and that roots of the characteristic equation are complex in general). For complicated systems it is laborious to find roots explicitly, but certain global characteristics of the equation can indicate stability. Two *necessary* conditions for stability are that

1. All powers of s must be present in the characteristic equation from zero to the highest order.
2. All coefficients in the characteristic equation must have the same sign.

These are necessary but not sufficient conditions. Routh's criterion is an algebraic method for further assessing stability. A simple algorithm is used to construct a matrix from the characteristic-equation coefficients. Then the first column of this matrix is examined for sign changes. The number of sign changes in the first column is equal to the number of positive roots. One positive (real-part) root or more indicates instability (Letherman, 1981). Examples 12.4 and 12.5 illustrate two methods for analyzing the stability of a simple proportional controller.

Example 12.4: Stability of a Second-Order Controller

Consider the stability of the heating coil controller shown in Fig. 12.2b with proportional-only control (that is, $K_i = 0.0$). Examine the behavior of the system as the proportional gain is varied. The transfer function of each component of the system is shown in parentheses in the figure. Note that the dynamics of the controlled device (i.e., valve and actuator) are not considered.

  Given: Component transfer functions

  Find: Stability conditions

  Figure: See Figs. 12.2b and 12.32.

  Assumptions: The coil system components and control systems are linear.

SOLUTION

The open-loop transfer function $G(s)$ is the product of the controller, device, and process transfer functions:

$$G(s) = K_p K_d \frac{\epsilon}{1 + \tau_{coil}s}$$

in which ϵ is the coil effectiveness and τ_{coil} is the coil time constant.[2] The sensor transfer function $H(s)$ is

$$H(s) = \frac{1}{1 + \tau_s s}$$

These two expressions can be combined to find the characteristic equation, Eq. (12.34), for this system:

$$G(s)H(s) + 1 = \frac{\tau_{coil}\tau_s s^2 + (\tau_{coil} + \tau_s)s + (K_p K_d \epsilon + 1)}{(1 + \tau_s s)(1 + \tau_{coil}s)} = 0$$

The denominator is nonzero; therefore, if any real root of the numerator has a positive part, the system is unstable. In this simple example, the quadratic formula can be used to find the roots r. We will then examine the result for stability. The roots of the characteristic equation are

$$r_1, r_2 = \frac{-(\tau_{coil} + \tau_s) \pm \sqrt{(\tau_{coil} + \tau_s)^2 - 4\tau_s \tau_{coil}(K_p K_d \epsilon + 1)}}{2\tau_{coil}\tau_s}$$

The real part of these roots cannot be positive; therefore, the system is stable. That is, for any physically realistic values of the five system parameters (of course, ϵ, the gains K, and the time constants τ must all be positive), the system is unconditionally stable if we include damped sinusoidal functions within the definition of stability. However, as the gain K_p increases, the frequency of system oscillation increases as the square root of the gain.

COMMENTS

The Routh criterion described in Example 12.5 could have been used to examine the stability of this system with the same result. It is an easier method of checking for instability than the approach used above. In addition, if the characteristic equation is of higher order, finding the roots is a much less convenient method than the Routh criterion.

Example 12.5: Proportional Controller Stability—Routh's Criterion

The transfer function of a heating coil (the process) TF_{proc} is determined experimentally to be[3]

$$TF_{proc} = \frac{1}{s(s + 4)(s + 6)}$$

and

$$TF_{control} = K_p$$

---

[2] In this example the coil is treated as a lumped capacitance.
[3] The terms in the denominator each relate to the dynamics of a piece of equipment in the process, e.g., the valve actuator, the valve, and the coil.

For simplicity, we ignore the dynamics of the duct temperature sensor. The coil is controlled by a proportional controller with gain K_p. For what values of gain is the system stable?

Given: TF_{proc}, $K_p > 0$

Find: Range of K_p for stability

Figure: See Fig. 12.25 (ignore ambient temperature reset feature) and Fig. 12.32.

Assumption: The system is linear.

SOLUTION

The characteristic equation, according to Eq. (12.34),[4] is

$$\frac{K_p}{s(s+4)(s+6)} + 1 = 0 \tag{12.37}$$

Clearing fractions gives

$$s^3 + 10s^2 + 24s + K_p = 0$$

It is seen that this equation satisfies the two conditions necessary for stability since all powers are present and since all coefficients have the same sign. Routh's criterion can be used to examine stability further. The Routh table for this equation is

$$
\begin{vmatrix}
1 & 24 \\
10 & K_p \\
\frac{240 - K_p}{10} & 0 \\
K_p & 0
\end{vmatrix}
$$

The first two rows of the table are constructed by entering the coefficients in "zigzag" order, starting at the upper left-hand corner, as is evident. The first term in the third row is the determinant of the four elements immediately above $(24 \times 10 - K_p \times 1)$ divided by the coefficient immediately above (10). The element in the fourth row is found in the same way from the determinant of rows 2 and 3:

$$\frac{K_p(240 - K_p)/10 - 10(0)}{(240 - K_p)/10} = K_p$$

All elements of the first column except the third are uniformly positive, as required by the Routh criterion for stability. For this term to be positive, the gain must be less than 240:

$$K_p < 240 \tag{12.38}$$

COMMENTS

Integral and derivative control can also be analyzed in this manner. In Prob. 12.29 we will see that integral control reduces stability since it increases the order of the characteristic equation by 1. On the other hand, derivative control can stabilize an unstable proportional controller since it reduces the order of the characteristic equation by 1. The control constants K_p, K_i, and K_d must therefore be selected with care. This is the final topic in this section.

---

[4] Since we are ignoring duct sensor dynamics, $H(s) = 1$.

12.5.4 Selection of Control Constants; Control System Simulation

There are two methods to empirically determine the proper control constants for an existing system that can be exercised and measured in the field. Either method can be used to finalize controller settings in the field from estimates by the HVAC designer. Figure 12.33a shows the temporary modifications to the controller needed to use the *reaction curve*

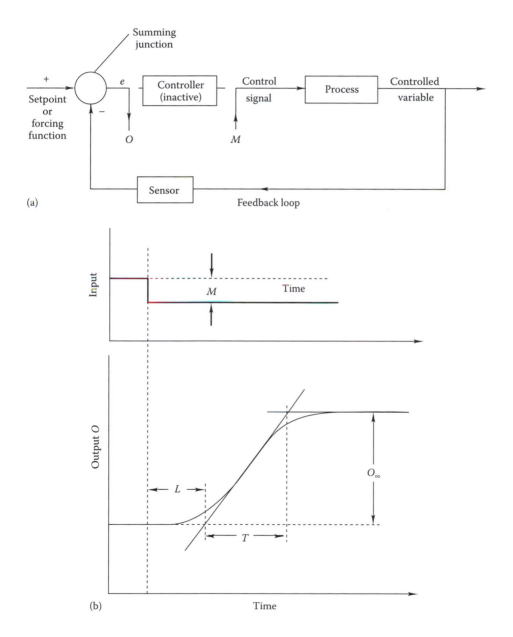

FIGURE 12.33

(a) Feedback control loop modified temporarily for reaction curve test. A control signal step change M is introduced into the process, and the system response (error) O is measured. (b) Typical reaction curve data showing steady output O_∞, lag L, and time constant T that result from an input step of M.

method. A step change of magnitude M is applied to the process. The response O at the output of the summing junction is measured and plotted, as shown in Fig. 12.33b. From the data three parameters are determined: a lag L, a time constant T, and the steady-state gain O_∞/M. These are used to construct an empirical first-order process transfer function

$$\text{TF}_{\text{proc}} = \frac{(O_\infty/M)e^{-sL}}{1 + Ts} \tag{12.39}$$

where O_∞/M is called the *system gain*. From this transfer function, acceptable values of the control constants can be found by using either a method due to Ziegler and Nichols (1942) or a more recent method due to Cohen and Coon (1953). Either method produces constants for P, PI, or PID controllers based on combinations of the three parameters shown in Fig. 12.33b. The constants so calculated give stable control with only a few oscillations whose amplitudes decrease by 75 percent per cycle. See the tables on the CD-ROM for details.

If this modest overshoot is undesirable, the *pole-zero cancellation* method of MacArthur et al. (1989) can be used to find an approach to control the setpoint that does not involve any overshoot. Briefly, this method applies the controller transfer function, Eq. (12.31) (without the derivative term), to the system transfer function, Eq. (12.39), to find the characteristic equation. The result involves a quotient of polynomials. By selecting the proportional and integral gains appropriately, terms in the numerator and denominator of the characteristic equation can be made to cancel, thereby lowering the order of the equation. Using this approach, one can find proportional and integral gains as a function of system time delay and system gain that result in the most rapid approach possible to the setpoint without any overshoot.

The *ultimate-frequency* method uses the undisturbed control shown in Fig. 12.32 but disables the integral and derivative terms by setting the integral and derivative gains to very small values. The proportional gain is then adjusted so that steady oscillations with neither increasing nor decreasing amplitude result. This specific value of K_p and the period of the ultimate steady-state frequency are used to find the recommended proportional, integral, and derivative gains by using rules in Letherman (1981). See the tables on the CD-ROM for details.

Both methods can be used to establish "good" control constants for the conditions under which the tests were conducted. However, most HVAC system controls have significant nonlinearities. As a result, settings at one condition will not necessarily be optimal at another. A recent solution to this problem uses the concept of *adaptive* control, whereby the control constants are adjusted in near real time as operating conditions change. The calculations needed for this method can be performed by a small computer in DDC systems. Underwood (1989) and Curtiss et al. (1993) report on HVAC applications of adaptive (self-tuning) controls.

Another method of establishing control system constants is to simulate both the HVAC system and its controls. Details of this method are beyond the scope of this book, but Example 12.6 indicates the approach.

Example 12.6: Heating Coil Control Simulation

Use the HCB software to study the response of a heating coil to a step change in the temperature setpoint with the following system characteristics. Is the control system stable?

Given: Coil inlet temperature: 30°F

Original coil air outlet setpoint: 55°F

New coil air outlet setpoint: 70°F

Coil throttling range: 50°F

Coil-to-sensor lag time: 0 s

Coil time constant: 120 s

Sensor time constant: 15 s

Proportional gain K_p: 1.0

Integral gain K_j: 0.038

Derivative gain K_d: 0.000

Find: Coil outlet air temperature, controller error, and valve position as a function of time by using the HCB software

Figure: See Fig. 12.2b.

Assumptions: The control system is linear, and the valve is quick-acting; i.e., there is no significant lag in valve response to the controller signal.

SOLUTION

Figure 12.34 shows the extended output of the HCB software for the conditions listed for a simulation period of 10 min (5 times the longest time constant in the system).

COMMENTS

The coil outlet temperature overshoots the 70°F setpoint by about 7°F. This occurs because the integral term gets "wound up" immediately after the setpoint change (Fig. 12.34b) and dominates the controller output signal, forcing the valve to stay open. Once the overshoot is

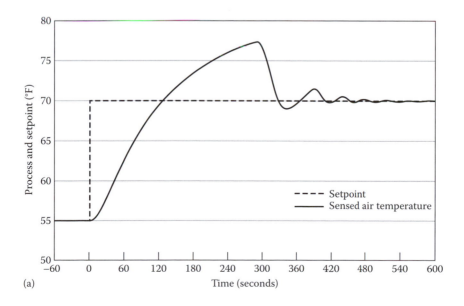

(a)

FIGURE 12.34

(a) Dynamic coil outlet temperature plot.

(continued)

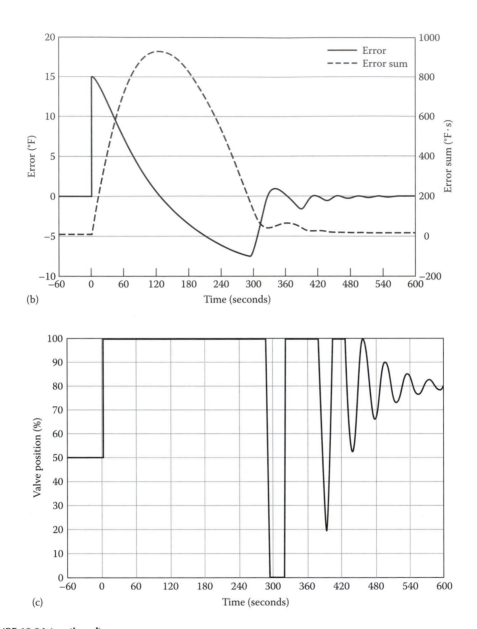

(b)

(c)

FIGURE 12.34 (continued)
(b) Corresponding controller error and error sum signals. (c) Valve position time history.

experienced for enough time the integral term and the proportional term have relatively equal effect on the process, creating the characteristic "ringing" response. After several cycles, the setpoint is achieved, and the error is essentially zero. The valve position shown in Fig. 12.34c is as expected: Immediately after the setpoint change, the valve opens fully where it remains for about 5 min. After 10 min has elapsed, the valve is near its final 80 percent position (the new setpoint is 40°F, or 80 percent of the valve throttling range of 50°F, above the coil air inlet temperature).

The reader is encouraged to try the following variations on Example 12.6:

1. Delete the integral control term. Note that the operation is stable but that an offset to the setpoint occurs after steady state is reached.

2. Add a time delay between the sensor and the controller. You will find that just a few seconds of delay here will cause control instability; therefore, one must place a sensor near the outlet of a coil.

3. Add a derivative term, and study a range of derivative values. Observe the nature of stability effects.

12.6 Summary

This chapter has introduced the important features of properly designed control systems for HVAC applications. Sensors, actuators, and control methods have been described. The method for determining control system characteristics either analytically or empirically has been discussed.

The following rules (ASHRAE, 1987) should be followed to ensure that the control system is as energy-efficient as possible:

1. Operate HVAC equipment only when the building is occupied or when heat is needed to prevent freezing. (In the morning, begin operation 30 min before occupancy to remove stale air.)

2. Consider the efficacy of night setback vis-à-vis building mass. Massive buildings may not benefit from night setback due to overcapacity needed for the morning pickup load.

3. Do not supply heating and cooling simultaneously. Do not supply humidification and dehumidification simultaneously.

4. Reset the heating and cooling air or water temperature to provide only the heating or cooling needed.

5. Use the most economical source of energy first; the most costly, last.

6. Minimize the use of outdoor air while complying with standards during the deep heating and cooling seasons.

7. Consider the use of "dead-band" or "zero-energy" thermostats.

8. Establish control settings for stable operation to avoid system wear and to achieve proper comfort.

Problems

The problems in this book are arranged by topic. The approximate degree of difficulty is indicated by a parenthetic italic number from 1 to 10 at the end of the problem. Problems are stated most often in USCS units; when similar problems are presented in SI units, it is

done with approximately equivalent values in parentheses. The USCS and SI versions of a problem are not exactly equivalent numerically. Solutions should be organized in the same order as the examples in the text: given, figure or sketch, assumptions, find, lookup values, solution. For some problems, the Heating and Cooling of Buildings (HCB) software included in this book may be helpful. In some cases it is advisable to set up the solution as a spreadsheet, so that design variations are easy to evaluate.

12.1 The thermostat inserted in the cooling system of an automotive engine controls the engine coolant temperature. Identify the sensor, controller, actuator, process, and controlled variable.   (3)

12.2 Residential natural gas–based heating system controls typically include a wall thermostat, a gas valve, and the furnace. Identify the sensor, controller, actuator, process, and controlled variable for the control system.   (3)

12.3 An electric duct heater has a capacity ranging from 0 to 10 kW. The proportional controller signal that controls heater power input ranges from 0 to 5 V dc. What are the throttling range and the proportional gain?   (3)

12.4 Write the equation relating heater output to sensed temperature for the controller described in Prob. 12.3 if the thermostat output voltage decreases linearly with temperature between 90°F (32°C) and 60°F (16°C) for a nominal thermostat setpoint of 75°F (24°C).   (4)

12.5 What is the steady-state error (in temperature) for the controller in Prob. 12.3 if the heat loss from the space heated by the electric heater is 7 kW? If it is 10 kW?   (4)

12.6 What is the resistance of a platinum RTD at -40 and 120°C? Its nominal resistance at 0°C is 100 Ω.   (3)

12.7 What is the resistance-temperature equation for a thermistor with the following measured data?   (5)

Temperature, K	283	288	293	298	303
Resistance, Ω	112,000	72,000	50,000	34,000	22,000

12.8 A thermometer inserted in a 55°C airstream has a time constant of 5.0 s. If the air temperature is rapidly increased to 75°C, when does the thermometer read 65°C? What is the thermometer reading after three time constants have passed?   (5)

12.9 What is the C_v of a linear water control valve that flows at 100 gal/min (6.3 L/s) with a pressure drop of 5 psi (35 kPa)? What is the gain of the valve, expressed in two ways: on the basis of its total stem travel [1.0 in (2.5 cm)] and on the basis of the control actuator voltage (0 to 10 V) that moves the stem over its full travel?   (4)

12.10 The flow through a coil and linear valve subsystem (Fig. 12.19a) is determined by the distribution of the available constant pressure difference (85 kPa) between the valve and the coil. Assume that the coil pressure drop is given by $\Delta P_{coil} = 2.5\dot{V}^2$ kPa, where $\dot{V}$ L/s is the volumetric flow rate. Even though the valve flow at constant imposed pressure across the valve would vary linearly with position, the loop flow does not because of the nonlinear nature of the coil. Plot the loop flow versus the percentage of valve travel for two values: $C_v = 0.5$ and $C_v = 1.3$. [The valve equation is $\dot{V}$ (L/s) = $C_v\sqrt{\Delta p \text{ (kPa)}}$]. Which value of C_v results in the most nearly linear valve travel compared to the loop flow characteristic?   (5)

12.11 What is the gain of the valve *as installed in the loop* for the configuration described in Prob. 12.10 for both C_v values? Note that the gain is not constant. If you had to pick one value of proportional gain for operation in the control midrange, what would it be? (6)

12.12 If the coil described in Prob. 12.10 is connected in a local loop with a linear three-way valve ($C_v = 0.5$), as shown in Fig. 12.19b, what pressure drop should be set across the balancing valve? What will the total loop flow be at valve midtravel? (7)

12.13 The flow characteristic of an equal-percentage valve is given by Eq. (12.12). A valve of this type with $K = 20$ and $k = 1.0$ is connected in the coil loop described in Prob. 12.10 with the valve having $C_v = 1.3$. Plot the valve flow versus position characteristic for the loop, and comment on its linearity. (5)

12.14 What is the C_v value of a linear water control valve that flows at 20 gal/min (1.2 L/s) with a pressure drop of 4 psi (28 kPa)? What is the gain of the valve expressed on the basis of its total stem travel [2.0 in (5 cm)] and on the basis of the control actuator voltage (0 to 5 V) that moves the stem over its full travel? (4)

12.15 Derive the steady-state transfer function of an air heat transfer coil that has two independent inputs—the inlet water and air temperatures. Draw a block diagram with the two inputs and the single output (the outlet air temperature) shown. (6)

12.16 Modify the model developed in Prob. 12.15 to include coil dynamics. Consider it to be a lumped-capacitance time constant τ_{hx} with the effectiveness ϵ. Modify the block diagram from Prob. 12.15 as appropriate. (7)

12.17 Select a steam control valve using saturated 80 psig steam to heat atmospheric pressure water for the following conditions. Water is to be heated to 20°F (11°C) at 82.5 gal/min (5.2 L/s). What is the value of C_v? Find both the critical flow pressure drop and the difference between supply and return steam pressure (atmospheric pressure) to find the proper entry in Eq. (12.19). (6)

12.18 The pressure drop through a water coil flowing at 15 gal/min (1 L/s) is 6 psi (42 kPa). What is the required value of C_v for good flow control as ensured by selecting a valve authority of 0.50? (3)

12.19 Determine the C_v value of a three-way valve used to control water flow through a cooling coil. The total coil and piping pressure drop under full-flow conditions in the coil loop is 3 ftWG. If the peak coil flow is 70 gal/min (4.2 L/s), what is the recommended C_v value and what type of valve should be selected? (4)

12.20 What is the inverse Laplace transform for this expression? (3)

$$\frac{s+3}{(s+4)(s-2)}$$

12.21 What is the Laplace transform of the following differential equation? Solve the equation if the initial conditions are $Y(t=0)=3$ and $Y'(t=0)$. (5)

$$\frac{d^2T}{dt^2} + 2\frac{dT}{dt} + 4 = 0$$

12.22 What is the transfer function for a temperature sensor (in a water stream) with mass m, convection heat transfer coefficient h, specific heat c, and surface area A? The transfer function expressed in the Laplace transform domain is the ratio of the transform of the sensed temperature to the transform of the fluid temperature.   (4)

12.23 A water heating tank is heated electrically with a 3-kW heater. The heat loss coefficient from the tank $UA = 40$ W/K, and the volume of water in the tank is 80 L. What is the open-loop transfer function relating the tank temperature to heat input (the tank-to-air temperature difference is a convenient variable here)?   (5)

12.24 A proportional controller with gain k is connected to the tank described in Prob. 12.23. If the open-loop transfer function is expressed as

$$\frac{K/(UA)}{1 + \tau s}$$

what is the time constant τ? At time zero, the setpoint is rapidly changed from a steady value of 15°C (also the environmental temperature) to 25°C. If the steady-state offset is found to be 1°C, what is the value of K? If the gain is double, what is the resulting offset?   (6)

12.25 A proportional controller is operated so that its setpoint is increased by 5 percent per minute. After an initial transient, the output increases by 1 percent per minute. What is the gain?   (3)

12.26 Write the transfer function for a baseboard room heating system including the dynamic effects of the room itself, the baseboard heating units, the thermostat, and the PI controller. The controlled variable is the room temperature.   (5)

12.27 What are the stability criteria for K, given these characteristic equations?   (4)

$$as^2 + bs + K = 0$$
$$as^3 + bs^2 + cs + K = 0$$

12.28 For what values of K is the following characteristic equation stable?   (4)

$$s^4 + 7s^3 + 11s^2 + 7s + (K + 1) = 0$$

12.29 A coil temperature PI controller with a fast-acting valve has a sensor time constant τ_s and a heating coil time constant τ_{coil}. Draw the system block diagram and write the characteristic equation if the coil is controlled by a PI controller. Determine the stability criteria on the proportional and integral gains for stability.   (7)

12.30 Is the system described in Prob. 12.29 with a proportional gain of 3 and integral gain of 0 stable if the sensor time constant is 3 s and the heating coil time constant is 60 s?   (7)

12.31 For the system described in Prob. 12.29, is a system with a proportional gain of 3 and air integral gain of 0.05 stable if the sensor time constant is 3 s and the heating coil time constant is 60 s?   (7)

12.32 The outlet temperature of a heating coil subjected to a step increase in inlet air temperature is shown in Fig. 12.2a. What is the process transfer function in the form of Eq. (12.39)?   (5)

12.33 Write the characteristic equation for a controller of the coil shown in Fig. 12.2a if the outlet temperature sensor has a time constant τ_s. Use the Padé approximation for the delay term, and ignore the dynamics of the control valve. If the sensor time constant is 10 s and a proportional-only control is used, what is the upper limit of the proportional gain for stability? (10)

12.34 Pneumatic controls can take advantage of small air pressures by varying the size of the diaphragm used in the actuator. What actuator force can a 15 psi control signal exert if the round diaphragm has a 3-in diameter? What if the diaphragm has a 6-in diameter? What if the diaphragm has a 12-in diameter? (2)

12.35 In the system schematic of Fig. 12.2a, what will be the effect on the controlled variable as the temperature sensor is moved farther away from the coil? (3)

12.36 The resistance of a platinum RTD is measured using a constant current of 1 mA. The nominal resistance of the RTD is 100 Ω at 0°C. The air around the RTD is at 20°C, but the current running through the RTD causes it to heat up.

(a) What is the theoretical resistance of the RTD at 20°C?

(b) What is the power dissipated by the RTD in the form of heat?

(c) If the heat loss from the RTD to the environment is 0.05 W/°C, what is the actual temperature of the RTD including the effect of self-heating? (5)

12.37 The temperature of an airstream is measured by connecting a platinum RTD to a building energy management system using 200 ft of 20-gauge wire. The wire has a resistance of 10 Ω per 1000 ft. How much error does the wire resistance introduce into the measurement if the nominal resistance of the RTD is 100 Ω at 32°F? How much error would the wire resistance add if the sensor were a thermistor with a nominal resistance of 25 kΩ at 32°F? (5)

12.38 An *analog-to-digital* converter is used to generate a computer-readable value from a measured voltage. Suppose a temperature sensor produces a 0 to 10-V dc signal over the range of −20 to 120°F. This signal is converted to an 8-bit value, which means that the total voltage range is divided into $2^8 = 256$ discrete levels. What is the minimum temperature change that can then be recorded using this temperature sensor? (3)

12.39 Air at 55°F enters a heating coil and comes out at 75°F. You measure the relative humidity with a handheld probe and find an inlet relative humidity of 59 percent and an outlet relative humidity of 31 percent. If there is no moisture added to the airstream, do these numbers make sense? (3)

12.40 An economizer mode attempts to mix outside air and building return air to minimize the amount of energy needed to condition the resulting mixed-air stream to match the desired supply air conditions. Suppose the conditions are outside air at 90°F and 40 percent relative humidity, the return air is at 80°F and 70 percent relative humidity, and the supply air setpoint it 55°F at 80 percent relative humidity. Should the economizer control use mostly outside air or mostly building return air? (4)

12.41 A heating coil with a throttling range of 20°F is controlled using proportional-only control. If the inlet temperature is 40°F, the setpoint is 55°F, and the controller gain is 0.1, what is the steady-state error? What if the gain is 1.0? What if the gain is 10.0? (4)

12.42 The hot water coil on an air handling unit has 120°F inlet water temperature, 110°F outlet water temperature, and a water flow rate of 30 gal/min. Also 10,000 ft^3/min of air enters this coil at 57°F and leaves at 73°F. If the temperature sensors are accurate to within ±1°F, does it appear that energy is conserved in this coil? Assume that the flow rates of air and water are measured without error.   (3)

12.43 A damper controller controls three sets of dampers as shown in Fig. P12.43. The three dampers are modulated to specify the amount of outdoor air entering the building. How can you calculate the fraction of outdoor air entering the air handling unit if you only know the return air temperature, the outdoor air temperature, and the mixed-air temperature? At what point does this relationship not make sense?   (8)

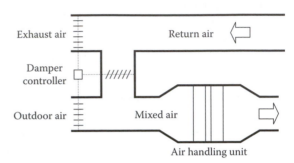

FIGURE 12.43

12.44 Airflow measurements are often made at several points across the face of an airstream so that the average airflow rate can be determined. A Pitot tube (see. Chap. 5) is often used for these kinds of measurements. If the dynamic pressure measurements at five points are 0.23, 0.20, 0.18, 0.25, and 0.20 (all in equivalent inches of water), what is the average velocity of the airstream?   (5)

12.45 Some standard automobiles will not start unless you depress the clutch. Create a simple ladder diagram that shows the start-up sequence for a car engine with this feature. Also include a rung on this ladder for the warning bell that sounds when you start the car without wearing a seat belt.   (6)

12.46 Suppose you run a reaction curve test on a heating coil. The valve of the coil is changed from 20 to 80 percent, resulting in a change of the coil outlet temperature of 48 to 63°F. The outlet temperature does not start increasing until about 10 s after the valve change and does not reach the final temperature until about 70 s after the valve change. What are the recommended PID control constants for this coil?   (6)

12.47 Most home thermostats include an *anticipator* that is used to shut off the heating system before the room actually reaches the setpoint value. The anticipator is a small electronic heating coil that fools the thermostat into thinking the room is warmer than it actually is. Describe why the thermostat would need such a feature. Most anticipators are adjustable. Why would you want to adjust the amount of anticipation?   (7)

12.48 Many Scandinavian residences are heated using radiant floor panels. A network of pipes is embedded in a concrete floor, and a pump circulates warm water through the pipes. The pipes warm the slab, and the slab then warms the room air by convection. Assuming a constant hot water temperature and a constant water flow rate when the pump is running, what is the transfer function that relates the hot water pump control and the temperature of the room?   (9)

12.49 In some air systems, airflow to a zone is measured using an array of Pitot tubes that measures the dynamic pressure of the moving air at many points. Most Pitot tube arrays use the *average* dynamic pressure to calculate the velocity of the air when in fact the proper way to make such measurement is to use the average of all the velocities calculated at each point. Imagine a Pitot tube array in which one-half of the Pitot tubes experience an actual velocity of 500 ft/min while the other half experience a velocity of 350 ft/min. What is the measurement error if the velocity is calculated using the average dynamic pressure?   (8)

References

ASHRAE (1987). *Handbook of HVAC Systems and Applications*. American Society of Heating, Refrigerating and Air-Conditioning Engineers, Atlanta.

Braun, J. E., J. W. Mitchell, S. A. Klein, and W. A. Beckman (1989). "Applications of Optimal Control to Chilled Water Systems without Storage." *ASHRAE Trans.*, vol. 95, pt. 1.

Coad, W. J. (1990). "Temperature Controls: The Industry's Dilemma." *Heating, Piping and Air Conditioning*, vol. 62 (October), pp. 103–104.

Cohen, G. H., and G. A. Coon (1953). "Theoretical Considerations for Retarded Control." *Trans. ASME*, vol. 75, pp. 827–834.

Curtiss, P. S., J. F. Kreider, and M. J. Brandemuehl (1993). "Adaptive Control of HVAC Processes Using Predictive Neural Networks." *ASHRAE Trans.*, vol. 99, pt. 1.

Grimm, N. R., and R. C. Rosaler (1990). *Handbook of HVAC Design*. McGraw-Hill, New York.

Haines, R. W. (1987). *Control Systems for Heating, Ventilating and Air Conditioning*, 4th ed. Van Nostrand Reinhold, New York.

Honeywell (1988). *Engineering Manual of Automatic Control*. Honeywell, Inc., Minneapolis.

Huang, P. H. (1991). "Humidity Measurements and Calibration Standards." *ASHRAE Trans.*, vol. 97, pt. 2.

Letherman, K. M. (1981). *Automatic Controls for Heating and Air Conditioning*. Pergamon, New York.

MacArthur, J. W., E. W. Grald, and A. F. Konar (1989). "An Effective Approach for Dynamically Compensated Adaptive Control." *ASHRAE Trans.*, vol. 95, pt. 2, pp. 411–423.

Madsen, T. L. (1989). "How Important Is the Location of the Room Thermostat?" *ASHRAE Trans.*, vol. 95, pt. 2.

Stoecker, W. F., and P. A. Stoecker (1989). *Microcomputer Control of Thermal and Mechanical Systems*. Van Nostrand Reinhold, New York.

Underwood, D. M. (1989). "Response of Self-Tuning Single Loop Digital Controllers to a Computer Simulated Heating Coil." *ASHRAE Trans.*, vol. 95, pt. 2.

Wylie, C. R., and L. C. Barrett (1982). *Advanced Engineering Mathematics*, 5th ed. McGraw-Hill, New York.

Ziegler, J. G., and N. B. Nichols (1942). "Optimum Settings for Automatic Controllers." *Trans. ASME*, vol. 64, pp. 759–763.

13

Lighting

Some 5 percent of the total U.S. energy consumption is used for lighting. In commercial buildings, lighting is the major consumer of electricity; in addition, it induces energy consumption for removing the heat of the lights. Clearly, lighting and the interaction between lighting and thermal loads deserve some attention by HVAC design engineers, even if in the past it may not have been considered as part of their domain. By the first law of thermodynamics, if a lamp consumes Q_{light} of electric power, it decreases the heating load and increases the cooling load by $(1-f)$ Q_{light}, where f is the fraction of Q_{light} that escapes to the outside. While the energy that escapes directly as light through the windows is usually negligible, a significant fraction of the heat can be carried off by the ventilation air and extracted to the outside. This is the case in commercial buildings where the lights are attached to the plenum and the air from a room is exhausted through the plenum. In such systems, f is approximately the product of the fraction of heat given off by convection and the fraction of the return airflow that is exhausted to the outside.

The HVAC design engineer is implicated even more if one wants to reduce the energy consumption by using daylight. Optimization of daylighting design is complicated because one has to trade off the solar heat gains due to increased daylight against the decreased electricity consumption for lights. Although the question of design optimization is deferred to Sec. 14.7, we present here a brief summary of the principles of lighting, both electric and natural.

13.1 Principles of Lighting

The quantity of light emitted, per unit time, by a source of radiation is called the *luminous flux*, and it is measured in units of *lumen* (abbreviated lm). The lumen is defined in terms of radiative power weighted by the spectral sensitivity of the human eye. For energy-conscious design, a quantity of special importance is the *luminous efficacy* of a light source, defined as the ratio of its light output (in lumens) to the necessary power input (in watts). The theoretical upper limit is 683 lm/W, corresponding to lossless conversion of the input to monochromatic light at the wavelength of 555 nm, where the eye is most sensitive. For other spectral distributions, the efficacy can only be lower. For instance, much of the solar spectrum is outside the visible range, and the luminous efficacy of daylight is only around 100 to 120 lm/W (relative to its radiative power). The luminous efficacy of incandescent lamps is around 10 to 20 lm/W, and that of fluorescent lamps is around 50 to 90 lm/W. The relatively high efficacy of daylight explains its interest for energy efficiency.

Of particular interest to the designer is the *illuminance* of the surface to be lighted. The illuminance is the luminous flux divided by the area on which it is incident; its unit is the *lux* (abbreviated lx):

$$1 \text{ lx} = 1 \text{ lm/m}^2 \qquad \qquad (13.1\text{SI})$$

In USCS units, the analogous quantity is the *footcandle* (abbreviated fc):

$$1 \text{ fc} = 1 \text{ lm/ft}^2 \qquad \qquad (13.1\text{US})$$

In addition to the flux, one may be interested in the angular distribution. The luminous flux emitted by a source into an angular region of unit solid angle [measured in steradians (sr)] is the *luminous intensity*; its unit is the *candela* (abbreviated cd):

$$1 \text{ cd} = 1 \text{ lm/sr} \qquad \qquad (13.2)$$

The apparent brightness of a surface depends on the flux received by the eye. This can be characterized as *luminance*, defined as the luminous flux emitted per unit area and per solid angle in a given direction.

Apart from energy consumption, the principal considerations for the design of lighting systems are

- Luminous flux
- Spectral composition (color rendition)
- Spatial and angular distribution of the radiation

As for the last consideration, a major concern is the possibility of extremely bright sources of light within the normal field of vision, something called *glare*. To understand this concept, consider, e.g., lightbulbs that are visible rather than hidden inside a luminaire. They cause less discomfort if they have a white coating rather than clear glass: Instead of coming from a small bright filament, the same amount of light is seen to emerge from the larger and less bright surface of the bulb, thus greatly reducing the contrast with the background. Quite generally, glare should be avoided.

But even if spectral and angular distributions can be quantified, lighting design choices are often based on criteria that defy easy analysis, such as prestige and aesthetics.

13.2 Electric lighting

There are a wide variety of electric light sources to choose from; some characteristics are summarized in Table 13.1. The data for efficacy and lifetime are approximate. Actual values can vary quite a bit from one model to another. The light output of a lamp decreases during its life, and in many applications one will replace a lamp long before it would burn out.

Incandescent lamps produce light by simple thermal radiation from a heated filament. Their efficacy is low because the temperature limitations of even the best filament material (tungsten) imply that only a small portion of the emitted power lies in the visible part of the electromagnetic spectrum; most of the power is given off as infrared radiation. In addition to the ordinary incandescent lamps that have been in use since Edison, a variant has

TABLE 13.1

Comparison of Light Sources*

Type	Efficacy, lm/W	Life, h	Comments
Incandescent			
Ordinary	10–20	1,000	Low first cost: best for seldom used spaces
Halogen	15–25	2,000	Excellent color rendition; output fairly constant over entire lifetime; good for task lights and special effects
Gaseous discharge			
Fluorescent	50–90	10,000	Offices and other commercial uses
Mercury vapor	50–60	10,000	Indoor commercial and outdoor uses
High-pressure sodium	100–150	15,000	Fairly good color rendition; indoor commercial and outdoor uses
Low-pressure sodium	200		Yellow; outdoor, especially roads

Source: Courtesy of Competitek, Inc., Lighting Report, Rocky Mountain Institute, Snowmass, CO, 1988.

* For a bit of historical perspective, note that artificial light used to be very expensive when the best source was a candle with 0.15 lm/W, lasting less than a day (Competitek, 1988).

become popular in recent years: the halogen lamp. It utilizes the tungsten-halogen cycle to counteract the principal cause of gradual degradation in ordinary incandescent lamps. This cycle makes the tungsten molecules that have evaporated return to the filament instead of condensing on the glass envelope. Furthermore, the filament can burn at higher temperature, thus improving both efficacy (by about 30 to 50 percent relative to ordinary incandescents) and color rendition.

Gaseous discharge lamps are based on quantum transitions between discrete energy levels, and they reach much higher efficacy. The wavelengths corresponding to these transitions do not, in general, lie in the desired region of the visible spectrum, but they can be shifted to longer wavelengths by means of special coatings. For instance, in fluorescent lamps, a gaseous discharge in a tube filled with Ar, Ar-Ne, or Kr produces ultraviolet radiation that is converted to light by phosphor coatings ("fluorescent") on the inside of the tube. Special coatings are available for improving the color rendition, although generally at increased cost and energy consumption. In recent years, compact fluorescent tubes have been developed that offer high efficacy in a size comparable to incandescent bulbs.

The output of a fluorescent lamp varies with the temperature of its environment. The temperature dependence for a typical tube is shown in Table 13.2; details can vary from one model to another. It is advisable to pay attention to the expected temperature of the lamp environment when one is selecting a fluorescent tube.

Gaseous discharge lamps must be connected to the mains via a special device, called the *ballast*. It prevents the current from exceeding the design value, it provides the starting voltage kick, and it corrects the power factor. The power consumption of the ballast can be a significant item, on the order of 10 percent of the total. More efficient, high-frequency ballasts have been developed in recent years.

Among commercially available lamps the record for luminous efficacy, around 200 lm/W, is held by the low-pressure sodium lamp. Unfortunately it produces an almost monochromatic yellow light, suitable only for outdoor applications such as roads. High-pressure sodium lamps yield white light with fairly good color rendition, but the efficacy is lower.

Only part of the light produced by a lamp reaches the work plane, with the rest being absorbed in the luminaire or by intervening objects. The *luminaire* is the housing of the lamp; it contains various reflective or refractive elements, designed to redistribute the light

TABLE 13.2

Lamp Wall Temperature (MLWT), Light Output (lm Index), and Luminous Efficacy (lm/W Index)
for Several Fluorescent System Configurations

Luminaire Configuration	MLWT, °C	m Index*	lm/W Index*
Nonairflow four-lamp lensed troffer			
No airflow	56.6	78.3	89.4
Airflow four-lamp lensed troffer			
No airflow	55.8	79.2	90.0
20 ft³/min	36.7	98.3	99.3
50 ft³/min	31.5	99.4	98.0
Nonairflow four-lamp louvered parabolic troffer			
No airflow	53.1	82.2	91.9
Airflow four-lamp parabolic troffer			
No airflow	51.8	83.8	93.1
20 ft³/min	40.9	95.6	98.8
50 ft³/min	35.7	99.0	99.8

Source: Courtesy of Competitek, Inc., Lighting Report, Rocky Mountain Institute, Snowmass, CO, 1988.
* As percent of output and efficacy under ANSI test conditions (open-air strip fixture at 25°C ambient).

in the desired fashion. The *luminaire efficiency*, defined as the ratio of the light leaving the luminaire to the light produced by the lamp, varies a great deal from one model to another. For instance, in a survey of luminaires for fluorescent tubes, efficiencies were found to range from 30 to 70 percent for devices that serve essentially the same purpose (direct illumination of a workspace below, with restricted exit angles to reduce glare) (Rabl, 1990). The low efficiencies were due to diffuse reflectors that trapped much of the radiation instead of sending it to the target area. Specular surfaces of high reflectivity perform far better, e.g., anodized or vacuum-deposited aluminum. Silver has the highest reflectivity, around 0.95, but it requires careful protection or it will tarnish. In recent years, silvered plastic films with durable coatings have been developed, and they are being used increasingly for luminaires (Competitek, 1988).

Most luminaires emit light not only straight toward the work plane but in other directions as well. Exact calculation of the amount of light that reaches the work plane would involve all the interreflections from walls, ceiling, etc., a daunting task that is usually simplified by various approximations. For hand calculations, the following method, known as the *lumen method*, is convenient and widely used. It determines the average illuminance E_w fc (lx) of the work plane by the formula

$$E_w = \frac{C_u F}{A_w} \tag{13.3}$$

where
    F = total light output of lamps (not counting absorption by luminaires), lm
  A_w = area of work plane
  C_u = coefficient of utilization

If all the light from the lamp reached the work plane, the average illuminance would be F/A_w. Thus the coefficient of utilization represents the efficiency with which the luminaire-room combination transfers the light to the work plane. It depends on the geometry and

TABLE 13.3

Typical Coefficient of Utilization Table for Particular Ceiling-Mounted Luminaire with Two Fluorescent Tubes, as a Function of Room Ratio [Eq. (13.4)] and of Reflectivities (Work Plane, Ceiling, and Wall)

Reflectivities of Work Plane	0.3					0.1								0.0
of Ceiling	0.7			0.5		0.7			0.5			0.3		0.0
of Wall	0.5	0.3	0.1	0.5	0.3	0.5	0.3	0.1	0.5	0.3	0.1	0.3	0.1	0.0
Room Ratio														
0.60	0.26	0.22	0.19	0.25	0.21	0.25	0.21	0.19	0.24	0.21	0.19	0.21	0.19	0.18
0.80	0.31	0.27	0.24	0.30	0.26	0.29	0.26	0.23	0.29	0.26	0.23	0.25	0.23	0.22
1.00	0.35	0.31	0.28	0.34	0.30	0.33	0.29	0.27	0.32	0.29	0.27	0.29	0.27	0.26
1.25	0.39	0.35	0.32	0.37	0.34	0.36	0.33	0.31	0.35	0.33	0.30	0.32	0.30	0.29
1.50	0.42	0.38	0.35	0.40	0.37	0.39	0.36	0.33	0.38	0.35	0.33	0.35	0.33	0.32
2.00	0.47	0.43	0.40	0.44	0.41	0.42	0.40	0.38	0.41	0.39	0.37	0.38	0.37	0.36
2.50	0.49	0.46	0.44	0.47	0.44	0.44	0.42	0.40	0.43	0.41	0.40	0.41	0.39	0.38
3.00	0.51	0.49	0.46	0.49	0.46	0.46	0.44	0.42	0.45	0.43	0.42	0.43	0.41	0.40
4.00	0.54	0.52	0.50	0.51	0.49	0.48	0.46	0.45	0.47	0.45	0.44	0.45	0.44	0.42
5.00	0.56	0.54	0.52	0.52	0.51	0.49	0.47	0.46	0.48	0.47	0.46	0.46	0.45	0.44

Source: Courtesy of data in Philips, *Philirama 89/90*, Catalog of products, Companie Philips Eclairage, Boulogne-Billancourt, France, 1989.

the reflectivities of the room, and on the luminaire; the geometry enters only via the room ratio (= 2 × ratio of floor area and wall area above work plane):

$$\text{Room ratio} = \frac{wl}{\Delta h(w + l)} \tag{13.4}$$

where

Δh = room height − work plane height
w = width
l = length of room

The coefficient of utilization of a luminaire is provided in tabular form by the manufacturer; an example is shown in Table 13.3. The method determines only the average illuminance over the work plane. The designer should verify that the room geometry and spacing of luminaires are such as to make the illuminance sufficiently uniform.

Example 13.1

A room is to be illuminated with fluorescent tubes, using the luminaire of Table 13.3. The room characteristics are

$$\text{Height } \Delta h = 6.3 \text{ ft } (1.9 \text{ m}) \quad \text{between ceiling and work plane}$$
$$\text{Width} = 40 \text{ ft } (12.2 \text{ m})$$
$$\text{Length} = 30 \text{ ft } (9.1 \text{ m})$$

and the reflectivities are 0.5 for the ceiling and walls and 0.2 for the work plane.

How many tubes and how many luminaires are needed to ensure a minimum illuminance of 50 fc (543 lx), averaged over the work plane, if each tube produces 3000 lm? What is the electric power if each tube, together with its ballast, requires 40 W?

Given: Luminaire of Table 13.3

$A_w = 1200$ ft^2 (111 m^2)

$E_w > 50$ fc (543 lx)

Find: Number of tubes, number of luminaires, and total electric power

Intermediate quantities: Room ratio $= 2.45$, from Eq. (13.4). For C_u one must interpolate between room ratios of 2.5 and 3.0 and between work plane reflectivities of 0.1 and 0.3. One finds $C_u = 0.46$, interpolated from Table 13.3 for wall and ceiling reflectivity of 0.5.

SOLUTION

$$\text{Required flux } F = \frac{E_w A_w}{C_u} \quad \text{from Eq. (13.3)}$$

$$= (50 \text{ lm/ft}^2)\left(\frac{1200 \text{ ft}^2}{0.46}\right)$$

$$= 130,435 \text{ lm}$$

$$\text{No. tubes} = 130,435 \text{ lm}/3000 \text{ lm/tube}$$

$$= 43.5 \text{ rounded up to } 44$$

$$\text{No. luminaries} = \frac{44 \text{ tubes}}{2 \text{ tubes/luminaire}} = 22 \text{ luminaires}$$

$$\text{Electric power} = 44 \text{ tubes} \times 40 \text{ W/tube} = 1760 \text{ W}$$

COMMENTS

The luminous efficacy of the tubes is 3000 lm/40 W $= 75$ lm/W. Only 45 percent of the light produced by the tubes actually reaches the work plane. The electric power per floor area is 1760 W/1200 ft^2 $= 1.47$ W/ft^2 (15.82 W/m^2), and if it all stays in the room, it imposes a cooling load of 1760 W$_t$ = 1760 W/(3.517 kW/ton) = 0.50 ton for this 1200 ft^2 (111 m^2) area. It will be interesting to compare this with the performance of the skylight in Example 13.2.

13.3 Principles of Daylighting

Daylighting (also known as *natural lighting*) is, of course, the oldest means of illumination, but it has gained new popularity since the oil crises. It has psychological appeal in addition to its promise of energy savings. The potential for savings is greatest in commercial buildings for a number of reasons: They are occupied during the day and use much electricity for lighting, the electricity during the day (peak period) is expensive, and the heat of the lights imposes a cooling load during much of the year. Having higher luminous efficacy, daylight is cooler than electric light. More than one-half of commercial space is located directly under a roof or adjacent to a window and thus is accessible to daylight. In residential buildings, by contrast, most of the light is needed when there is little daylight, and cooling loads are less important.

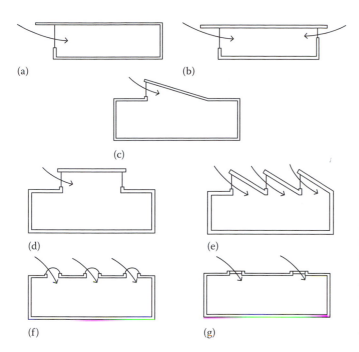

FIGURE 13.1
Design types for daylighting. (a) Unilateral lighting section; (b) bilateral lighting section; (c) clerestory lighting section; (d) roof monitor lighting section; (e) sawtooth lighting section; (f) domed skylight section; and (g) flat skylight section. (Courtesy of IES, *IES Lighting Handbook: Reference Volume*, Kaufman, J.E. (ed) and Christensen, J.F. (assoc ed), Illuminating Engineering Society of North America, New York, 1984a.)

There are a wide variety of architectural designs for daylighting. The most conventional types are shown in Fig. 13.1, including ordinary windows, clerestories, and skylights. Providing adequate daylight is a challenge because inside a building the transmitted light decreases rapidly with the distance from the opening. As a rule of thumb for ordinary windows, the maximum distance to which useful light can be brought into a building is 1.5 to 2 times the height of the upper window edge. The great advantage of skylights is their ability to spread the light quite evenly over a large work plane; their effectiveness for saving energy will be confirmed by the results in Sec. 14.7. However, skylights tend to be more expensive per unit area than vertical windows, because skylights require better waterproofing.

A major concern with daylighting is the avoidance of glare, i.e., excessive contrast, during periods of direct sunlight. A simple and effective countermeasure is to employ diffuse instead of clear glazing. Of course, that may not be an attractive solution if it interferes with a view to the outside. But diffuse glazing can be quite acceptable in skylights, clerestories, and other windows not used for vision. In any case, the simple measure of illuminance fails to capture all that is important about daylighting in practice. Aesthetics is an essential quality of daylighting, yet it eludes quantification. For example, if a horizontal skylight with diffuse glazing looks just like a fluorescent fixture, an architectural opportunity has been missed. As an alternative to diffuse glazing, one can avoid glare by interposing blinds or diffuse reflecting surfaces. Or one can tilt the aperture toward the north. The energy consequences will be examined in Sec. 14.7.

Figure 13.2 shows how skylights can be used to create an environment more pleasant and open than a conventional office space, and Fig. 13.3 shows a library lit by skylights. In recent years there has been some experimentation with innovative schemes based on light shelves or on tracking reflectors with light guides, with the goal of transporting daylight deeper into the building.

FIGURE 13.2
Roof structure with linear skylights. (Courtesy of Place, J.W. et al., *ASHRAE Trans.*, 93, 1, 1987.)

FIGURE 13.3
Boulder, Colorado, Public Library lighted by clerestories and skylights. (Courtesy of Lightforms, Inc.)

13.4 Analysis of Daylighting

For the analysis of daylighting, the following quantities need to be determined successively:

1. The daylight incident on the aperture, keeping track of three separate components: direct radiation from the solar disk, radiation from the sky, and radiation reflected by the ground

2. The fraction transmitted through the aperture, including losses due to dirt

3. The fraction transmitted to the work plane, taking into account multiple reflections at walls and other intervening surfaces (including the attenuation by a light well, if present)

4. The electricity that can be saved

Unlike the design conditions for thermal loads, the choice of design conditions for daylighting is not unique because daylighting is rarely designed as a standalone system for extreme conditions. In regions far from the equator, the seasonal change of day length makes it difficult or impossible to rely entirely on daylight for the duration of normal working hours. Complete backup with electric lamps is imperative in almost all situations. This reduces the role of design conditions from being an absolute minimum requirement to being merely a guide. One may want to require, somewhat arbitrarily, that no electric backup be necessary during certain periods, e.g., clear summer days, moderately overcast summer days, or winter days at noon.

The performance depends on the geometry of the daylighting system, on site and climate, and on the control of the backup. Calculations are difficult and tedious, if accuracy is desired. Fortunately, the chore can be performed by computer programs, such as DOE 2.1 (version b and higher) or WINDOWS or SUPERLIGHT. In this book we present only a simple method for design calculations, the *lumen method* of the IES (1989).

13.4.1 Daylight Data

The choice of data depends on what is obtainable and what accuracy is desired. At the crudest level, one uses just the design conditions, as presented below. Estimation of electricity savings is more difficult because it depends on the variability of the available radiation. At the level of hourly simulations, the simplest approach is to assume proportionality between daylight (in lumens) and solar radiation (in watts) and to treat the diffuse component as isotropic. But in reality the lumen distribution is not isotropic, even under a uniform clear or a uniform overcast sky, to say nothing of the effect of clouds. For instance, for an overcast sky, the IES model assigns a distribution proportional to

$$1 + 2\cos\theta_s$$

with zenith angle θ_s. Furthermore, the luminous efficacy of solar radiation varies as a function of meteorological conditions. These details are treated in more sophisticated simulation models. The key parameters needed for a detailed modeling of daylight availability are the astronomical coordinates of the sun (zenith, azimuth, solar time), the geographic location of the site (longitude, latitude, elevation above sea level), and the atmospheric conditions (turbidity, cloudiness).

The necessary geometric relationships were discussed already in Chap. 6. While the calculation of daylight availability is basically like the calculation of solar heat gain factors, with the replacement of watts by lumens, there is a difference in the detail to which angular distributions should be taken into account. For heat gains, it usually does not matter very much whether solar radiation is absorbed by a wall or by the floor, 1 or 3 m from a window; once a ray has entered, it can be counted as heat gain. But the daylighting effect is quite sensitive to the direction of the rays: Do they strike a wall or ceiling, or do they reach the target directly? Daylighting designers have developed their own solar radiation models, whose results may differ from models for thermal applications.

Even if a detailed modeling of the angular distribution of incident daylight is not feasible, it is recommended to carry out separate calculations for direct radiation (i.e., from the solar disk), for sky radiation, and for radiation reflected from the ground (the last two components are diffuse). Accordingly the design data of the IES Calculation Procedures Committee (IES, 1984b) are presented in a way that permits the separation of these three components. Direct illuminance is shown in Fig. 13.4, and sky illuminance in Fig. 13.5, all for the

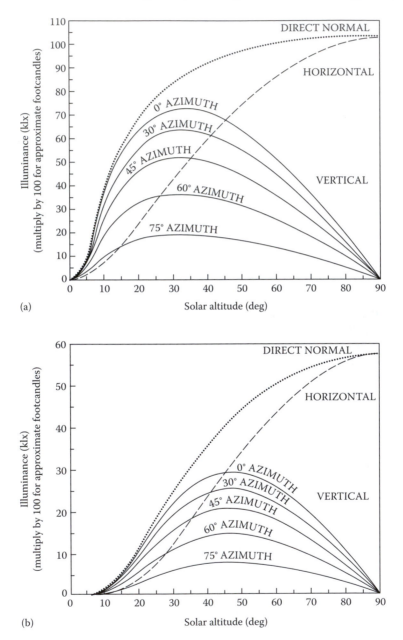

FIGURE 13.4

Direct component of design illuminance as a function of solar altitude ($90° - \theta_s$). Azimuth label indicates $|\phi_s - \phi_p|$ of azimuth angles of sun ϕ_s and of surface ϕ_p, respectively. HORIZONTAL = $E_{xh,dir}$, VERTICAL = $e_{xv,dir}$. (a) Clear sky; (b) partly cloudy sky. (Courtesy of IES, *IES Lighting Handbook: Reference Volume*, Kaufman, J.E. (ed) and Christensen, J.F. (assoc ed), Illuminating Engineering Society of North America, New York, 1984a.)

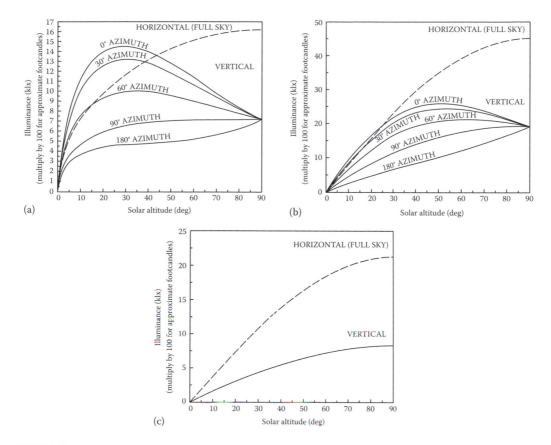

FIGURE 13.5

Diffuse sky component of design illuminance as a function of solar altitude ($90° - θ_s$). Azimuth label indicates $|φ_s - φ_p|$ of azimuth angles of sun $φ_s$ and of surface $φ_p$, respectively. HORIZONTAL $= E_{xh,s}$, VERTICAL $= E_{xv,s}/2$. (a) Clear sky, (b) partly cloudy sky, (c) overcast sky. (Courtesy of IES, *IES Lighting Handbook: Reference Volume*, Kaufman, J.E. (ed) and Christensen, J.F. (assoc ed), Illuminating Engineering Society of North America, New York, 1984a.)

horizontal plane and for several vertical planes. Finally, to help determine the illuminance received from the ground, Fig. 13.6 shows the diffuse illuminance on the horizontal, indicated by the label *horizontal (full sky)*. [In addition, Fig. 13.6 shows a set of curves labeled *horizontal (half sky)*; they will be needed as a supplementary variable to select a coefficient of utilization in Table 13.6].

Three sky conditions are considered: clear sky (part a), partly cloudy sky (part b), and overcast sky (part c). Since there is no direct radiation under overcast skies, there is no Fig. 13.4c. All graphs are plotted as function of the solar altitude angle (the complement $90° - θ_s$ of the zenith angle $θ_s$). The curves for vertical surfaces depend on the absolute value of the difference $φ_s - φ_p$ between the azimuth angles of the sun $φ_s$ and of the surface $φ_p$; in the graphs it is indicated by the label *azimuth*. For the overcast sky, the illuminance is independent of the azimuth. In using these data, one should not forget the enormous variability of the atmosphere; these design conditions are merely a representation of what might be typical.

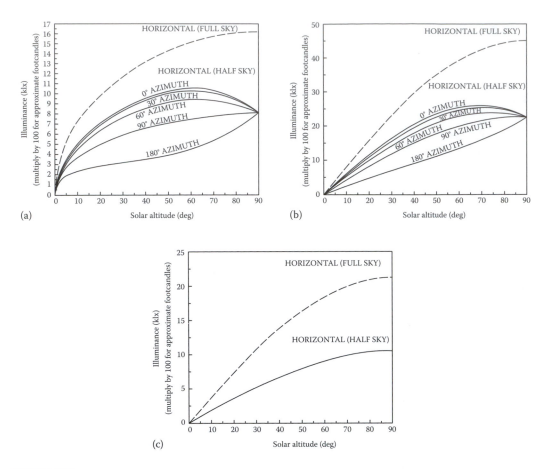

FIGURE 13.6
Diffuse horizontal half-sky component ($= E_{xh,s/2} =$ solid lines) of design illuminance on ground in front of a vertical surface as a function of solar altitude ($90° - \theta_s$). Azimuth label indicates $|\phi_s - \phi_p|$ of azimuth angles of sun ϕ_s and of surface ϕ_p, respectively. (a) Clear sky; (b) partly cloudy sky; (c) overcast sky. (Courtesy of IES, Recommended Practice for the Calculation of Daylight Availability. *J. Illuminating Eng. Soc.* (July), 381, 1984b.)

A quick look at the scales of these figures suggests that when direct illuminance can reach a surface, it is likely to dominate the other terms by far. For instance, when the altitude is 30° and the azimuth is 45° (for a vertical surface), Fig. 13.4a shows a direct illuminance of 50 klx under clear sky conditions. The corresponding diffuse sky illuminance is only 11.5 klx, from Fig. 13.5a. For an overcast sky, Fig. 13.5c shows that the surface receives only 4.5 klx.

13.4.2 Lumen Method for Toplighting

Skylights provide lighting from the top. So it comes as no surprise that the calculation is similar to the one described in Sec. 13.2 for electric lighting. For simplicity, we consider only horizontal apertures. Thus we need not worry about radiation reflected from the ground. Disregarding for the moment the directional dependence of the transmissivity of the glazing, we can state the basic formula for the average illuminance E_w on the indoor work plane in the form

$$E_w = E_{xh}\tau C_u \frac{A_h}{A_w} \tag{13.5}$$

where
  A_h = projected horizontal aperture area of skylights
  A_w = area of work plane (horizontal)
   τ = transmissivity of skylight (net, i.e., including losses due to dirt and due to frames, mullions and solar control devices, and, if appropriate, losses in light well)
  C_u = room coefficient of utilization (Table 13.4)
  E_{xh} = exterior illuminance on horizontal surface

The transmissivity τ is different for direct and for diffuse radiation. Adding appropriate subscripts, one can combine the contributions to the net transmissivity in the form

$$\tau_{dif} = T_{dif}T_{well}T_{control}R_A L_{dirt} \quad \text{for diffuse radiation} \tag{13.6a}$$

and

$$\tau_{dir} = T_{dir}T_{well}T_{control}R_A L_{dirt} \quad \text{for direct radiation} \tag{13.6b}$$

where
  T_{dif}, T_{dir} = transmissivity of glazing for diffuse and direct radiation (from manufacturer's brochure)
  T_{well} = transmissivity of light well, if present (= well efficiency of Fig. 13.7)
  $T_{control}$ = transmissivity of control devices such as louvers or diffusers, if present
  R_A = ratio of net to gross skylight area, to account for blocking by frame or mullions
  L_{dirt} = loss factor due to dirt (Table 13.5)

The dependence of the transmissivity of the glazing on the incidence angle necessitates a slight complication of Eq. (13.5). Under an overcast sky, there is only diffuse light, and one can write

$$E_w = E_{xh,s}\tau_{dif}C_u \frac{A_h}{A_w} \quad \text{for overcast sky} \tag{13.7}$$

where the exterior horizontal illuminance $E_{xh,s}$ from the sky can be found in Figure 13.5c and C_u comes from Table 13.4.

If a light well is present, the transmissivity is multiplied by an additional factor, the well efficiency, obtained from the ordinate of Fig. 13.7 as a function of well wall reflectivity and geometry. Of the light that has entered the room, the fraction that reaches the work plane is given by the room coefficient of utilization C_u, listed in Table 13.4 for several reflectivities and as a function of geometry, represented by the room cavity ratio

$$\text{Room cavity ratio} = \frac{5 \times \Delta h(w + l)}{wl} \tag{13.8}$$

where Δh = room height $-$ work plane height, w = width, and l = length of room [the room cavity ratio is 5 times the inverse of the room ratio of Eq. (13.4).

TABLE 13.4

Room Coefficient of Utilization C_u for Skylights, Based on Floor
Reflectivity $= 20$ Percent

Ceiling Reflectivity, Percent	RCR*	Wall Reflectivity, Percent		
		50	30	10
80	0	1.19	1.19	1.19
	1	1.05	1.00	0.97
	2	0.93	0.86	0.81
	3	0.83	0.76	0.70
	4	0.75	0.67	0.60
	5	0.67	0.59	0.53
	6	0.62	0.53	0.47
	7	0.57	0.49	0.43
	8	0.54	0.47	0.41
	9	0.53	0.46	0.41
	10	0.52	0.45	0.40
50	0	1.11	1.11	1.11
	1	0.98	0.95	0.92
	2	0.87	0.83	0.78
	3	0.79	0.73	0.68
	4	0.71	0.64	0.59
	5	0.64	0.57	0.52
	6	0.59	0.52	0.47
	7	0.55	0.48	0.43
	8	0.52	0.46	0.41
	9	0.51	0.45	0.40
	10	0.50	0.44	0.40
20	0	1.04	1.04	1.04
	1	0.92	0.90	0.88
	2	0.83	0.79	0.76
	3	0.75	0.70	0.66
	4	0.68	0.62	0.58
	5	0.61	0.56	0.51
	6	0.57	0.51	0.46
	7	0.53	0.47	0.43
	8	0.51	0.45	0.41
	9	0.50	0.44	0.40
	10	0.49	0.44	0.40

Source: Courtesy of IES, IES Recommended Practice for the Lumen Method of
Daylight Calculations, IES Report RP-23-1989, Illuminating
Engineering Society of North America, New York, 1989.
* RCR = room cavity ratio of Eq. (13.8).

The result is the average illuminance over the entire work plane. The illuminance will
approach uniformity to the extent that the distance between individual skylight openings
is small compared to Δh.

For clear and for partly cloudy days, Eq. (13.5) is generalized to contain two terms, one
for diffuse illuminance from the sky and one for direct illuminance from the sun:

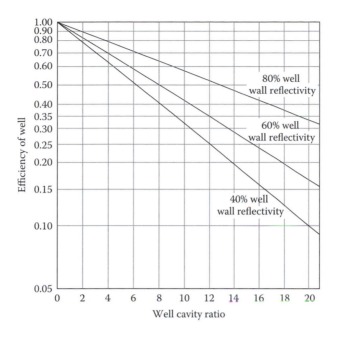

FIGURE 13.7
Well efficiency (transmissivity) of light well, for three values of wall reflectivity, as a function of height, length, and width of well, via the well cavity ratio [5 × height × (width + length)]/(length × width). (Courtesy of IES, IES Recommended Practice for the Lumen Method of Daylight Calculations, IES Report RP-23-1989, Illuminating Engineering Society of North America, New York, 1989.)

TABLE 13.5

Loss Factor Due to Dirt L_{dirt} as Function of Orientation of Glazing

	Orientation		
Location	**Vertical**	**Sloped**	**Horizontal**
Clean areas	0.9	0.8	0.7
Industrial areas	0.8	0.7	0.6
Very dirty areas	0.7	0.6	0.5

Source: Courtesy of IES, IES Recommended Practice for the Lumen Method of Daylight Calculations, IES Report RP-23-1989, Illuminating Engineering Society of North America, New York, 1989.

$$E_w = (E_{xh,s}\tau_{\text{dif}} + E_{xh,\text{dir}}\tau_{\text{dir}})C_u \frac{A_h}{A_w} \quad \text{for clear or partly cloudy skies} \qquad (13.9)$$

where
  $E_{xh,s}$ = diffuse illuminance from sky (Fig. 13.5a or b)
  $E_{xh,\text{dir}}$ = direct illuminance from sun (Fig. 13.4a or b)
  C_u = room coefficient of utilization (Table 13.4)

The skylight transmissivity τ_{dir} for direct radiation is different from τ_{dif} and varies with the angle of incidence. But the room coefficient of utilization is assumed independent of the angle of incidence, within the approximation of the lumen method.

Example 13.2

Find the illuminance for overcast and for clear days at noon, equinox, on the work plane of a room with horizontal skylights, without light well or control devices.

Given: Six evenly distributed horizontal skylights of total net area

$A_h = 6 \times 3$ ft $\times 3$ ft $(6 \times 0.9144$ m $\times 0.9144$ m)

Floor area (= work plane of area) $A_w = 30$ ft $\times 40$ ft $(9.144$ m $\times 12.192$ m)

Height difference between skylight and work plane $\Delta h = 6.3$ ft $(1.92$ m)

Latitude $\lambda = 40°$

$T_{dif} = 0.79$ $T_{dir} = 0.92$

Wall and ceiling reflectivity $= 0.50$

Work plane reflectivity $= 0.20$

Clean location

Find: E_w for overcast and for clear days

Assumption: Use lumen method for design conditions

Lookup values: Declination at equinox $\delta = 0°$ from Eq. (6.4)

Zenith angle $\theta_s = \lambda - \delta = 40°$ from Eq. (6.5)

Solar altitude $= 90° - \theta_s = 50°$

Azimuth $= 0$ at noon from Eq. (6.8)

$E_{xh,s} = 1600$ fc (≈ 16 klx) from Fig. 13.5c (overcast, horizontal)

$E_{xh,s} = 1400$ fc (≈ 14 klx) from Fig. 13.5a (clear sky, horizontal)

$E_{xh,dir} = 7500$ fc (≈ 75 klx) from Fig. 13.4a (clear sky, horizontal)

$L_{dirt} = 0.7$ from Table 13.5 for clean areas, horizontal

$\tau_{dif} = 0.79 \times 0.7 = 0.553$ (no area reduction because A_h is already net area)

$\tau_{dir} = 0.92 \times 0.7 = 0.644$

Room cavity ratio $= [5 \times 6.3 \times (40 + 30)]/(40 \times 30) = 1.84$ from Eq. (13.8)

$C_u = 0.89$ interpolated from Table 13.4 for wall and ceiling reflectivity of 0.5

Area ratio $A_h/A_w = 0.045$

Solution

For overcast days, the illuminance is, from Eq. (13.7),

$$E_w = 1600 \text{ fc} \times 0.553 \times 0.89 \times 0.045 = 36 \text{ fc} \ (\approx 360 \text{ lx})$$

For sunny days the illuminance is, from Eq. (13.9),

$$E_w = (1400 \text{ fc} \times 0.553 + 7500 \text{ fc} \times 0.644) \times 0.89 \times 0.045$$
$$= 31 + 192 \text{ fc} = 223 \text{ fc} \ (\approx 2230 \text{ lx})$$

Comments

This is the same room as in Example 13.1. For the sunny design day, the skylight, with a mere 4.5 percent of the floor area, provides more than 4 times the illuminance level of 50 fc (≈ 500 lx) that is typically required in offices. The total flux for 1200 ft^2 is 223 fc $\times$ 1200 ft$^2 = 268,000$ lm. Since sunlight has a luminous efficacy around 110 lm/W, the corresponding cooling load is $(268,000 \text{ lm})/(110 \text{ lm/W}) = 2440 \text{ W}_t = 2440 \text{ W}/(3.517 \text{ kW/ton}) = 0.69$ ton; this is not much larger than the cooling load of 1760 W_t for the electric system in Example 13.1, while providing more than 4 times as much light. The cooling load would be less than one-third of that imposed by the electric system, if the transmissivity of the skylight were controlled with an exterior shading device to give the same illuminance (an interior shading device would absorb much of the radiation—not good for reducing the cooling load). A selective coating could also help if it has high visible transmissivity and low shading coefficient.

This example illustrates why the optimization of daylighting is challenging. Under the right conditions or with the right control, daylight can indeed have a much higher efficacy than electricity; but for the average performance this advantage is less obvious. Without adjustable transmissivity, the illuminance will be either too high or too low most of the time, and finding the optimal design requires careful analysis of the annual performance. In Example 13.2, should the skylight be scaled up to yield 50 fc instead of 33 fc on the overcast day? Or should it be reduced to limit overheating during sunny days? The design conditions give no answer.

We will take up this question again in Sec. 14.7, where we will present some optimization results obtained by means of hourly computer simulations. They will show that the skylight area ratio $A_w/A_h = 0.045$ of Example 13.2 is not a bad guess, for a wide range of applications. And like most optimum values, there is fair tolerance to deviations from the optimum.

13.4.3 Lumen Method for Sidelighting

The lumen method for vertical apertures uses an equation similar to the one for toplighting. But there is a complication. Whereas with toplighting one can assume that all points in the work plane receive about the same amount of light, with sidelighting the illuminance varies strongly between different points in the room that are at different distances from the aperture. Therefore, the calculation should be done for several reference points located on a line perpendicular to the aperture. The procedure of IES (1989) recommends five reference points at 10, 30, 50, 70, and 90 percent of room depth, as indicated in Fig. 13.8.

The lumen method gives no credit for direct sunlight in the room; one assumes that it is kept out by overhangs, blinds, and shades in order to avoid problems with glare. Glare from direct sunlight can also be eliminated by using diffuse glazing, a technique that is

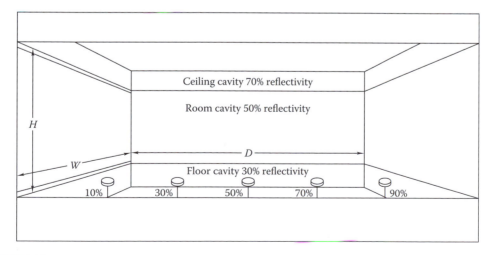

FIGURE 13.8
Standard conditions for the calculation of sidelighting. (Courtesy of IES, IES Recommended Practice for the Lumen Method of Daylight Calculations, IES Report RP-23-1989, Illuminating Engineering Society of North America, New York, 1989.)

convenient for transom windows above ordinary vision windows. Here we present only the simplest case: a window unshaded by plants, overhangs, or blinds.

Disregarding certain complications of angular distributions, the basic formula for the illuminance E_w on the work plane can be stated in the form

$$E_w = E_{xv}\tau C_u \tag{13.10}$$

where E_{xv} is the exterior vertical illuminance on the window and $\tau = \tau_{\text{dif}}$ is the diffuse transmissivity, calculated by Eq. (13.6a) as for skylights. No area ratio appears in the equation; it is implicit in the coefficient of utilization C_u. But for greater accuracy, the contributions from the sky and from the ground are calculated separately, with separate values for C_u, and then added:

$$E_w = \tau(E_{xv,s/2}C_{u,s} + E_{xv,g/2}C_{u,g}) \tag{13.11}$$

where $E_{xv,s/2}$ = vertical illuminance incident on window from the half sky in front of the window (Fig. 13.5), $E_{xv,g/2}$ = vertical illuminance reflected from the ground in front of the window, obtained from the horizontal illuminance E_{xh} by

$$E_{xv,g/2} = \frac{\rho_g}{2}E_{xh} \tag{13.12}$$

analogous to Eq. (6.24); ρ_g is the reflectivity of the ground. For overcast days E_{xh} is the full-sky value from Fig. 13.5c. For clear or for partly cloudy days, one takes E_{xh} as the sum of the direct horizontal component (Fig. 13.4a or b) and the horizontal diffuse sky component (Fig. 13.5a or b).

The coefficient of utilization C_u for sidelighting depends on the angular distribution of the light incident on the aperture. As a simple approximation for the values in Table 13.6, standard distributions have been assumed for sky radiation and for ground radiation, and separate tables are provided for different ratios of

$$E_{xv,s/2} = \text{vertical half-sky illuminance (Fig. 13.5)}$$

$$E_{xh,s/2} = \text{horizontal half-sky illuminance (Fig.13.6)}$$

Here we cite only a few excerpts to illustrate the method; for a full set of tables, including the effects of venetian blinds, we refer to IES (1989). Table 13.6 is arranged according to the ratio of room depth to window height (reading from top to bottom) and the ratio of window width to window height (reading from left to right). For each combination of these ratios, a set of five coefficients is listed, one for each of the five reference points.

Example 13.3

Suppose the room of Example 13.2 has a south-facing vertical window instead of the skylights. Find the illuminance in the center of the room for an overcast equinox day at noon.

Given: Floor area (= work plane area) $A_w = 30$ ft (9.144 m) (perpendicular to window) × 40 ft (12.192 m) (along window)

Room depth = 30 ft (9.144 m)   (*D* in Fig. 13.8)

Window width = 20 ft (6.096 m)   (*W* in Fig. 13.8)

TABLE 13.6

Coefficient of Utilization C_u for Window without Blinds, for Room of Figure 13.8

(a) Vertical/horizontal half-sky exterior illuminance ratio = $E_{xvs/2}/E_{xhs/2}$ = 0.75

Room Depth/Window Height	Percent D	Window Width/Window Height							
		0.5	1	2	3	4	6	8	Infinite
1	10	0.824	0.864	0.870	0.873	0.875	0.879	0.880	0.883
	30	0.547	0.711	0.777	0.789	0.793	0.798	0.799	0.801
	50	0.355	0.526	0.635	0.659	0.666	0.669	0.670	0.672
	70	0.243	0.386	0.505	0.538	0.548	0.544	0.545	0.547
	90	0.185	0.304	0.418	0.451	0.464	0.444	0.446	0.447
2	10	0.667	0.781	0.809	0.812	0.813	0.815	0.816	0.824
	30	0.269	0.416	0.519	0.544	0.551	0.556	0.557	0.563
	50	0.122	0.204	0.287	0.319	0.351	0.339	0.341	0.345
	70	0.068	0.116	0.173	0.201	0.214	0.223	0.226	0.229
	90	0.050	0.084	0.127	0.151	0.164	0.167	0.171	0.172
3	10	0.522	0.681	0.739	0.746	0.747	0.749	0.747	0.766
	30	0.139	0.232	0.320	0.350	0.360	0.366	0.364	0.373
	50	0.053	0.092	0.139	0.163	0.174	0.183	0.182	0.187
	70	0.031	0.053	0.081	0.097	0.106	0.116	0.116	0.119
	90	0.025	0.041	0.061	0.074	0.082	0.089	0.090	0.092
4	10	0.405	0.576	0.658	0.670	0.673	0.675	0.674	0.707
	30	0.075	0.134	0.197	0.224	0.235	0.243	0.243	0.255
	50	0.028	0.050	0.078	0.094	0.104	0.112	0.114	0.119
	70	0.018	0.031	0.048	0.059	0.065	0.073	0.074	0.078
	90	0.016	0.026	0.040	0.048	0.053	0.059	0.061	0.064
6	10	0.242	0.392	0.494	0.516	0.521	0.524	0.523	0.588
	30	0.027	0.054	0.086	0.102	0.111	0.119	0.120	0.135

(b) Vertical/horizontal half-sky exterior illuminance ratio = $E_{xvs/2}/E_{xhs/2}$ = 1.00

Room Depth/Window Height	Percent D	Window Width/Window Height							
		0.5	1	2	3	4	6	8	Infinite
1	10	0.671	0.704	0.711	0.715	0.717	0.726	0.726	0.728
	30	0.458	0.595	0.654	0.668	0.672	0.682	0.683	0.685
	50	0.313	0.462	0.563	0.589	0.598	0.607	0.608	0.610
	70	0.227	0.362	0.478	0.515	0.527	0.530	0.532	0.534
	90	0.186	0.306	0.424	0.465	0.481	0.468	0.471	0.472
2	10	0.545	0.636	0.658	0.660	0.661	0.665	0.666	0.672
	30	0.239	0.367	0.459	0.484	0.491	0.499	0.501	0.506
	50	0.121	0.203	0.286	0.320	0.335	0.348	0.351	0.355
	70	0.074	0.128	0.192	0.226	0.243	0.259	0.264	0.267
	90	0.058	0.101	0.156	0.188	0.207	0.215	0.221	0.223
3	10	0.431	0.561	0.607	0.613	0.614	0.616	0.615	0.631
	30	0.133	0.223	0.306	0.337	0.348	0.357	0.357	0.366
	50	0.058	0.103	0.155	0.183	0.197	0.211	0.213	0.218
	70	0.037	0.064	0.098	0.119	0.132	0.147	0.150	0.154
	90	0.030	0.051	0.079	0.098	0.110	0.122	0.126	0.129
4	10	0.339	0.482	0.549	0.560	0.563	0.566	0.565	0.593
	30	0.078	0.139	0.204	0.234	0.247	0.258	0.260	0.272
	50	0.033	0.060	0.094	0.114	0.126	0.139	0.143	0.150
	70	0.022	0.039	0.061	0.074	0.083	0.095	0.099	0.104
	90	0.019	0.032	0.050	0.061	0.070	0.080	0.084	0.089
6	10	0.211	0.343	0.433	0.453	0.458	0.461	0.461	0.518
	30	0.033	0.065	0.103	0.123	0.135	0.145	0.148	0.167

(continued)

TABLE 13.6 (continued)

Coefficient of Utilization C_u for Window without Blinds, for Room of Figure 13.8

(a) Vertical/horizontal half-sky exterior illuminance ratio = $E_{xv,s/2}/E_{xh,s/2}$ = 0.75

Room Depth/Window Height	Percent D	0.5	1	2	3	4	6	8	Infinite
	50	0.011	0.023	0.036	0.044	0.049	0.055	0.056	0.063
	70	0.009	0.018	0.027	0.032	0.035	0.040	0.041	0.046
	90	0.008	0.016	0.023	0.028	0.031	0.034	0.035	0.040
8	10	0.147	0.257	0.352	0.380	0.387	0.391	0.392	0.482
	30	0.012	0.026	0.043	0.054	0.060	0.067	0.070	0.086
	50	0.006	0.013	0.021	0.026	0.029	0.033	0.035	0.043
	70	0.005	0.011	0.017	0.021	0.023	0.026	0.027	0.034
	90	0.004	0.010	0.015	0.019	0.021	0.023	0.025	0.030
10	10	0.092	0.168	0.248	0.275	0.284	0.290	0.291	0.395
	30	0.006	0.014	0.026	0.032	0.036	0.041	0.044	0.059
	50	0.003	0.008	0.014	0.017	0.019	0.022	0.024	0.032
	70	0.003	0.007	0.012	0.014	0.016	0.018	0.019	0.026
	90	0.003	0.006	0.011	0.013	0.015	0.016	0.017	0.024

(b) Vertical/horizontal half-sky exterior illuminance ratio = $E_{xv,s/2}/E_{xh,s/2}$ = 1.00

Room Depth/Window Height	Percent D	0.5	1	2	3	4	6	8	Infinite
	50	0.015	0.029	0.047	0.057	0.064	0.073	0.077	0.086
	70	0.011	0.021	0.033	0.040	0.045	0.051	0.054	0.060
	90	0.010	0.019	0.028	0.034	0.038	0.044	0.046	0.052
8	10	0.135	0.238	0.326	0.353	0.362	0.366	0.367	0.452
	30	0.016	0.034	0.058	0.072	0.080	0.090	0.094	0.116
	50	0.008	0.017	0.027	0.034	0.039	0.045	0.048	0.059
	70	0.006	0.013	0.021	0.026	0.028	0.032	0.035	0.043
	90	0.005	0.012	0.019	0.023	0.025	0.029	0.031	0.038
10	10	0.090	0.165	0.244	0.272	0.283	0.290	0.291	0.395
	30	0.009	0.020	0.036	0.045	0.052	0.060	0.064	0.087
	50	0.005	0.010	0.019	0.023	0.026	0.030	0.033	0.044
	70	0.004	0.009	0.015	0.018	0.020	0.023	0.025	0.033
	90	0.003	0.008	0.014	0.016	0.018	0.020	0.022	0.030

(c) Vertical/horizontal half-sky exterior illuminance ratio = $E_{xv,s/2}/E_{xh,s/2}$ = 1.50

Room Depth/Window Height	Percent D	0.5	1	2	3	4	6	8	Infinite
1	10	0.503	0.528	0.536	0.541	0.544	0.557	0.558	0.559
	30	0.359	0.464	0.514	0.528	0.534	0.549	0.550	0.552
	50	0.261	0.384	0.471	0.499	0.508	0.524	0.526	0.527
	70	0.204	0.325	0.432	0.470	0.485	0.497	0.499	0.500
	90	0.179	0.295	0.412	0.456	0.475	0.474	0.477	0.478
2	10	0.412	0.477	0.490	0.492	0.493	0.498	0.499	0.505
	30	0.201	0.304	0.379	0.402	0.410	0.422	0.424	0.429
	50	0.115	0.192	0.269	0.304	0.320	0.339	0.343	0.347
	70	0.078	0.136	0.204	0.241	0.261	0.286	0.292	0.295

(d) Ground component

Room Depth/Window Height	Percent D	0.5	1	2	3	4	6	8	Infinite
1	10	0.105	0.137	0.177	0.197	0.207	0.208	0.210	0.211
	30	0.116	0.157	0.203	0.225	0.235	0.241	0.243	0.244
	50	0.110	0.165	0.217	0.241	0.252	0.267	0.269	0.270
	70	0.101	0.162	0.217	0.243	0.253	0.283	0.285	0.286
	90	0.091	0.146	0.199	0.230	0.239	0.290	0.292	0.293
2	10	0.095	0.124	0.160	0.178	0.186	0.186	0.189	0.191
	30	0.082	0.132	0.179	0.201	0.212	0.219	0.222	0.225
	50	0.062	0.113	0.165	0.189	0.202	0.214	0.218	0.220
	70	0.051	0.093	0.141	0.165	0.179	0.194	0.198	0.200

RCR	ρ																
	90	0.066	0.117	0.183	0.221	0.246	0.262	0.271	0.273	0.045	0.079	0.118	0.140	0.153	0.179	0.183	0.185
3	10	0.331	0.426	0.458	0.461	0.462	0.465	0.465	0.477	0.088	0.120	0.157	0.175	0.183	0.185	0.163	0.167
	30	0.121	0.202	0.275	0.304	0.316	0.327	0.329	0.337	0.059	0.107	0.154	0.176	0.187	0.198	0.193	0.198
	50	0.062	0.109	0.164	0.193	0.209	0.228	0.232	0.238	0.039	0.074	0.114	0.134	0.146	0.157	0.166	0.170
	70	0.041	0.073	0.114	0.138	0.154	0.176	0.183	0.188	0.031	0.055	0.085	0.101	0.111	0.122	0.127	0.130
	90	0.035	0.062	0.099	0.123	0.141	0.159	0.169	0.173	0.028	0.047	0.070	0.083	0.092	0.107	0.113	0.115
4	10	0.265	0.372	0.422	0.430	0.433	0.435	0.435	0.456	0.073	0.113	0.154	0.174	0.183	0.187	0.176	0.184
	30	0.077	0.137	0.199	0.229	0.243	0.256	0.259	0.272	0.040	0.082	0.127	0.148	0.159	0.170	0.177	0.185
	50	0.037	0.069	0.107	0.130	0.144	0.161	0.167	0.175	0.025	0.049	0.078	0.094	0.103	0.113	0.117	0.123
	70	0.026	0.046	0.073	0.089	0.101	0.119	0.126	0.132	0.020	0.036	0.054	0.065	0.071	0.079	0.083	0.087
	90	0.022	0.039	0.063	0.078	0.090	0.106	0.114	0.120	0.019	0.032	0.046	0.054	0.060	0.069	0.073	0.076
6	10	0.173	0.281	0.351	0.368	0.373	0.375	0.375	0.422	0.056	0.106	0.143	0.164	0.175	0.184	0.173	0.194
	30	0.037	0.073	0.115	0.137	0.151	0.164	0.168	0.189	0.021	0.050	0.081	0.098	0.107	0.117	0.123	0.138
	50	0.018	0.036	0.058	0.071	0.080	0.092	0.098	0.110	0.013	0.027	0.041	0.049	0.054	0.060	0.064	0.072
	70	0.013	0.026	0.040	0.049	0.056	0.064	0.069	0.078	0.011	0.021	0.029	0.033	0.035	0.039	0.041	0.046
	90	0.012	0.023	0.035	0.043	0.048	0.057	0.062	0.070	0.011	0.020	0.026	0.030	0.032	0.035	0.037	0.042
8	10	0.117	0.207	0.282	0.305	0.314	0.319	0.320	0.393	0.036	0.082	0.122	0.143	0.156	0.166	0.170	0.208
	30	0.020	0.042	0.071	0.087	0.098	0.111	0.116	0.143	0.011	0.029	0.050	0.062	0.070	0.078	0.082	0.101
	50	0.010	0.021	0.035	0.044	0.050	0.058	0.063	0.078	0.007	0.016	0.024	0.028	0.031	0.035	0.038	0.046
	70	0.007	0.016	0.026	0.032	0.036	0.041	0.045	0.055	0.006	0.013	0.018	0.020	0.021	0.023	0.025	0.030
	90	0.076	0.014	0.023	0.028	0.031	0.036	0.040	0.049	0.006	0.013	0.017	0.019	0.020	0.022	0.023	0.028
10	10	0.082	0.153	0.224	0.250	0.262	0.269	0.271	0.368	0.024	0.061	0.109	0.120	0.131	0.144	0.147	0.200
	30	0.012	0.026	0.047	0.059	0.068	0.078	0.084	0.114	0.006	0.017	0.034	0.040	0.046	0.053	0.056	0.076
	50	0.006	0.014	0.024	0.030	0.034	0.040	0.044	0.060	0.004	0.010	0.016	0.018	0.020	0.023	0.024	0.033
	70	0.005	0.011	0.019	0.022	0.025	0.029	0.032	0.043	0.004	0.009	0.013	0.014	0.015	0.016	0.016	0.022
	90	0.004	0.010	0.017	0.020	0.023	0.026	0.028	0.038	0.004	0.009	0.013	0.013	0.014	0.015	0.016	0.021

Source: Courtesy of IES, IES Recommended Practice for the Lumen Method of Daylight Calculations, IES Report RP-23-1989, Illuminating Engineering Society of North America, New York, 1989.

Window height = 5 ft (1.524 m) (H in Fig. 13.8)

Height difference between ceiling and work plane $\Delta h = 6.3$ ft (1.92 m)

Wall and ceiling reflectivity = 0.50

Work plane reflectivity = 0.20

Ground reflectivity = 0.20

Find: E_w at depth of 15 ft (= 50 percent point in Fig. 13.8)

Intermediate quantities: Solar altitude = $90° - \theta_s = 50°$ (from Example 13.2)

Azimuth of sun $\phi_s = 0°$ [from Eq. (6.8)]

Azimuth of window $\phi_p = 0°$ (south-facing)

Azimuth difference ("azimuth" of Fig. 13.4 to 13.6) $|\phi_s - \phi_p| = 0°$

$E_{xv,s/2} \approx 590$ fc exterior vertical illuminance on window, from Fig. 13.5c

$E_{xh} \approx 1500$ fc exterior horizontal illuminance, from Fig. 13.5c

$E_{xh,s/2} \approx 750$ fc horizontal half-sky exterior illuminance, from Fig. 13.6c

$L_{dirt} = 0.9$ from Table 13.5 for clean areas, vertical

$\tau_{dif} = 0.79 \times 0.9 = 0.711$ (no area reduction because A_h is already net area)

SOLUTION

The vertical illuminance reflected from the ground is, from Eq. (13.12),

$$E_{xv,g/2} = \frac{E_{xh}\rho_g}{2} = 1500 \text{ fc} \times \frac{0.2}{2} = 150 \text{ fc}$$

To determine the coefficients of utilization, we need several ratios:

$$\frac{\text{Window width}}{\text{Window height}} = \frac{20 \text{ ft}}{5 \text{ ft}} = 4.0$$

$$\frac{\text{Room depth}}{\text{Window height}} = \frac{30 \text{ ft}}{5 \text{ ft}} = 6.0$$

Horizontal half-sky exterior illuminance = $E_{xh,s/2} = 750$ fc
  Vertical/horizontal half-sky exterior illuminance ratio

$$\frac{E_{xv,s/2}}{E_{xh,s/2}} = \frac{590}{750} = 0.79$$

$$C_{u,s} = \begin{cases} 0.049 & \text{at } E_{xv,s/2}/E_{xh,s/2} = 0.75, \text{ from Table 13.6a} \\ 0.064 & \text{at } E_{xv,s/2}/E_{xh,s/2} = 1.00, \text{ from Table 13.6b} \end{cases}$$

By interpolation

$$C_{u,s} = 0.049 + (0.064 - 0.049)\left(\frac{0.79 - 0.75}{1.00 - 0.75}\right) = 0.051$$

For the ground component, $C_{u,g} = 0.054$, from Table 13.6d.
  Finally the illuminance at the work plane, from Eq. (13.11), is

$$E_w = 590 \times 0.711 \times 0.051 + 150 \times 0.711 \times 0.054 = 27 \text{ fc}$$

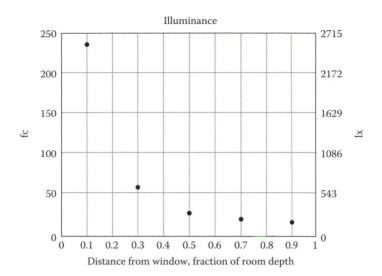

FIGURE 13.9
Illuminance on work plane of Example 13.3 as a function of distance from the window.

Repeating the calculation for other values of percentage of room depth, we obtain the results plotted in Fig. 13.9.

COMMENTS
The illuminance decreases rapidly over the first 10 ft ($\approx$3 m).

Problems

The problems in this book are arranged by topic. The approximate degree of difficulty is indicated by a parenthetic italic number from 1 to 10 at the end of the problem. Problems are stated most often in USCS units; when similar problems are presented in SI units, it is done with approximately equivalent values in parentheses. The USCS and SI versions of a problem are not exactly equivalent numerically. Solutions should be organized in the same order as the examples in the text: given, figure or sketch, assumptions, find, lookup values, solution. For some problems, the Heating and Cooling of Buildings (HCB) software included in this book may be helpful. In some cases it is advisable to set up the solution as a spreadsheet, so that design variations are easy to evaluate.

13.1 Define luminous efficacy. What are typical values for the most important sources of light? (*2*)

13.2 Consider a fluorescent tube that consumes 40 W to produce 3000 lm.

(a) What is the luminous efficacy of the tube?

(b) What is the system luminous efficacy if the tube is used in a luminaire with a coefficient of utilization of 0.45? (*3*)

13.3 Consider two options for the luminaires in a one-story office building with floor area 100 ft $\times$ 100 ft (30.48 m $\times$ 30.48 m). Each luminaire contains a fluorescent tube, consuming 40 W (including the ballast) and producing 3000 lm of light. For the illumination in the work plane, one luminaire has a coefficient of utilization of 0.60 while the other has 0.45; they are similar in other characteristics. The average illumination in the work plane is to be 50 fc (543 lx), after allowing for 10 percent degradation due to dirt and aging.

(a) How many luminaires are needed, and what is the electric power per floor area, for each of these options?

(b) If the utilization is 2500 h/yr and the electricity cost is $0.10/kWh, what is the difference in annual electricity cost?

(c) If the system life is 20 years and the highest permissible payback time is 10 years, how much more can one pay for the most efficient luminaire? Express the answer in dollars for the building and in dollars per luminaire. (6)

13.4 Design the lighting for the corridors of a commercial building, using fluorescent tubes in a luminaire with the coefficients of utilization of Table 13.2. Make the following assumptions:

- Average illumination on floor 50 fc (543 lx)
- Luminous efficacy of tubes 80 lm/W
- One 40-W tube per luminaire (including ballast)
- Corridor height of 8 ft (2.44 m)
- Corridor width of 8 ft (2.44 m)
- Corridor length large compared to height and width
- Ceiling reflectivity $= 0.7$
- Wall reflectivity $= 0.5$
- Floor reflectivity $= 0.2$

    (a) What is the spacing between luminaires?

    (b) What is the installed power in watts per square foot (meter)?

    (c) What is the annual electricity consumption if the lights are on 10 h/day for 6 days/week? (6)

13.5 As an alternative to the electric lighting of Prob. 13.4, consider horizontal skylights for the corridors that are on the top floor. Corridors are a good place for daylighting because wide variability of the illumination is usually acceptable. In this case, suppose that the system is considered adequate if it provides 50 fc (543 lx) under an overcast sky at noon of winter solstice. Assume the following values: latitude $= 40°N$, $T_{dif} = 0.79$, $L_{dirt} = 0.9$, $R_A = 0.85$.

(a) Estimate the area of the necessary skylights.

(b) How large are the electricity savings per square foot (meter) of corridor, relative to Prob. 13.4, if the electric backup is needed 10 percent of the time (full-power equivalent)?

(c) Estimate the concomitant savings in cooling energy if the cooling season lasts 1000 h and if the coefficient of performance (COP) of the cooling system is 3.0. Assume that 80 percent of the heat produced by the electric lights must be removed during the cooling season and that the luminous efficacy of daylight in this application is 90 lm/W, averaged over the cooling season. (8)

13.6 Suppose the room of Example 13.1 has a window of height 4 ft (1.22 m) running along its entire width. Assume the following values: latitude $= 40°N$, $T_{dif} = 0.79$, $L_{dirt} = 0.9, R_A = 0.85$.

(a) Evaluate the illumination due to daylight for a clear equinox sky, assuming no direct radiation on the window and reflectivity of ground $= 0.25$.

(b) If one does not take credit for any direct solar illumination in the room, to what extent does the result depend on the orientation of the room?

(c) Suppose that 60 percent of the electricity for the outer 15 ft (4.57 m) of the room can be saved, averaged over the year, if daylight sensors are used. How large are the savings? (10)

13.7 A compact fluorescent bulb can be used to replace a standard incandescent bulb. Suppose a particular task requires a total 200 lm. If a 50-W fluorescent bulb has a life of 7500 h and an efficacy of 70 lm/W while a 50-W incandescent bulb has a life of 1000 h and an efficacy of 15 lm/W, does it make more sense over the long run to use the compact fluorescent? Assume the compact fluorescent bulb costs $20, the incandescent bulb costs $1, and electricity costs 8¢/kWh. (4)

13.8 A 25 ft $\times$ 40 ft room at $35°$ latitude is to have an illumination level of 50 lm/ft^2 at the work plane. Two design options are to be evaluated: (i) 100% fluorescent lighting at the ceiling using the luminaire of Table 13.3 and lamps with 70 lm/W; (ii) horizontal skylights with ordinary single glazing and an aperture that provides 50 lm/ft^2 on June 21 at noon if the sky is clear. The ceiling is 6 ft above the work plane and the reflectivities of ceiling, wall, and work plane are 0.5, 0.3, and 0.2, respectively.

(a) What is the electric power of the fluorescent option?

(b) Using the CLFs of Chapter 7, estimate the corresponding cooling load at noon if the lights are turned on from 8 AM to 6 PM, and the 'a' and 'b' coefficients are 0.55 and D.

(c) How large are the skylights for the daylighting option?

(d) Using the CLFs of Chapter 7, estimate the corresponding cooling load at noon. Ignore any changes in conductive loads due to the replacement of ceiling by skylights. (7)

13.9 What is the instantaneous sensible cooling load for the conditions of Prob. 13.8 if the lighting level of 50 lm/ft^2 is to be achieved on an overcast day? Assume noon on June 21 and ignore any changes in conductive loads due to the replacement of ceiling by skylight. (5)

13.10 Which glazing, among the types listed in Table 6.6, minimizes the cooling load as calculated in Probs. 13.8 and 13.9? (5)

References

The standard references on lighting are the two parts of the *IES Lighting Handbook* (IES, 1984a and 1987).

Competitek, Inc. (1988). *Lighting Report*. Rocky Mountain Institute, 1739 Snowmass Creek Rd., Snowmass, CO 81654-9199.

IES (1984a). *IES Lighting Handbook: Reference Volume*. J. E. Kaufman, ed., and J. F. Christensen, assoc. ed. Published by IES (Illuminating Engineering Society of North America), 345 East 47th St., New York, NY 10017.

IES (1984b). "Recommended Practice for the Calculation of Daylight Availability." *J. Illuminating Eng. Soc.* (July), pp. 381–391.

IES (1987). *IES Lighting Handbook: Application Volume.* J. E. Kaufman, ed., and J. F. Christensen, assoc. ed. Published by IES (Illuminating Engineering Society of North America), 345 East 47th St., New York, NY 10017.

IES (1989). "IES Recommended Practice for the Lumen Method of Daylight Calculations." IES Report RP-23-1989. Illuminating Engineering Society of North America, 345 East 47th St., New York, NY 10017.

Philips (1989). *Philirama 89/90.* Catalog of products, Companie Philips Eclairage, Tour Vendôme Rond Point du Pont de Sèvres, 92516 Boulogne-Billancourt, France.

Place, J. W., J. P. Coutier, M. R. Fontoynont, R. C. Kammerud, B. Andersson, W. L. Carroll, M. A. Wahlig, F. S. Bauman, and T. L. Webster (1987). "The Impact of Glazing Orientation, Tilt, and Area on the Energy Performance of Room Apertures." *ASHRAE Trans., vol.* 93, p. 1.

Rabl, A. (1990). "Technologies d'éclairage: luminaires performants" (Lighting technologies: efficient luminaires). *Revue de l'énergie,* (October) no. 424, pp. 482–488.

14

Design for Efficiency

The search for the optimal design involves two kinds of analysis: calculation of performance and calculation of costs. The tools for calculating the performance have already been presented, and the calculation of costs is discussed in Chap. 15. Chapter 14 focuses on the process of the search and presents design recommendations and typical results. The recommendations attempt to distill the experience gained as result of the intensive research efforts on energy efficiency that have been stimulated by the oil crises of the 1970s.

In principle, we are faced with an optimization problem that could be solved by brute-force enumeration of all possible designs, evaluating the cost and performance of each. But in practice the number of design options is so enormous that the job would be hopeless without intelligent guidelines.

We begin by stating the design goal and describing search methods that can lead to efficient designs (Sec. 14.1). It is crucial that the designer have a good intuitive understanding of the elements of a building and their influence on energy consumption. For this purpose, rules and recommendations for energy-efficient design are presented in Sec. 14.2. Performance indices and theoretical limits are discussed in Sec. 14.3. Empirical performance data follow in Sec. 14.4. We proceed to specific design procedures and results for residential buildings in Sec. 14.5. Section 14.6 discusses HVAC systems for commercial buildings. The last section, Sec. 14.7, examines the energy savings that are achievable with daylighting.

To keep the discussion of energy efficiency on a firm and straightforward basis, we will take as a starting point whatever set of comfort conditions has been specified, assuming that all the designs under consideration will satisfy them. Thus we avoid questions such as "How much can be saved by lowering the thermostat set point 1°C?"—questions that are difficult and often controversial, involving tradeoffs between cost, comfort, and productivity.

With traditional design practice, it has not been easy to realize energy-efficient buildings. One obstacle is the still too frequent tendency to minimize initial cost rather than life cycle cost, especially in cases where the builder does not expect to be the occupant. It is hoped that rationality will prevail and life cycle costing will become common practice. (For help in that direction, see Chap. 15 on engineering economics. Another, often lamented, stumbling block has been the lack of communication between architects and engineers. Some engineers may even believe that it would be futile for them to worry about the design of the envelope of a building because those decisions have usually been dominated by the architect and the builder. That need not be the case: Many thermal characteristics of the envelope, e.g., insulation thickness, have little or no conflict with any architectural aspects.

Actually, an even greater obstacle on the road to efficiency has been the difficulty of carrying out sufficient analysis early enough in the design process. When accurate load calculations are not available until the design development phase (see the design phases in Table 1.1), it is too late for significant changes. However, in the future, advanced computer-aided design will enormously facilitate the process. If load calculations and cost estimation procedures are integrated into the drafting software used by the architect, the cost implications of each design element and of each design change can be observed immediately and interactively, beginning with the initial sketches. Thus the architect can directly benefit from the engineer's understanding of the energy aspects of the envelope. For these reasons, we do not hesitate to discuss envelope design options in a book for engineers.

14.1 The Road to Efficiency

14.1.1 The Design Goal

Quite generally, efficiency is easy to define for processes with a single input and a single output, both of which are quantifiable. Efficiency is a valuable concept, and we have already encountered it in the analysis of HVAC equipment. In Sec. 14.3 we describe some additional efficiency indices.

For a building as a whole, however, a precise definition of efficiency remains elusive. Already on the input side there is the complication of different forms of energy, even if one considers energy as the only input—quite apart from the question of whether the relevant input should be taken as money rather than energy. The output defies quantification since it involves all the services that a building is supposed to provide: protection from the elements, a comfortable and quiet indoor environment, good air quality, not to mention aspects such as prestige and aesthetics.

Therefore at the level of the building as a whole, we attempt only a qualitative formulation of efficiency: *An efficient building is one that provides the required conditions of comfort, convenience, etc., under the specified conditions of utilization for the lowest life cycle cost.*[1] It is in this sense that we describe the design goal as the search for efficiency.

We also note a further complication: the difficulty of foreseeing the future utilization of the building. For example, suppose that a building has been designed for internal gains of $15\,W/m^2$ but the occupants bring extra equipment that raises the gains to $30\,W/m^2$ (a common occurrence during the computer revolution of the last decade); surely the design would have been different, had one known in advance. Therefore, it is advisable to plan for flexibility and to avoid designs that are too vulnerable to changes in occupancy or utilization. A few additional sensitivity analyses during the design phase can help evaluate the flexibility.

14.1.2 The Search

How can designers know how close they have come to the optimal design, or whether and how they could reduce the energy consumption even further? In principle, the search for the optimal design is a straightforward optimization problem. But in practice the number of design options and variables is so large, especially for commercial buildings, that optimization by exhaustive search is simply not feasible, at least with the computers of the foreseeable future. Rather the search must be guided by intelligence. We recommend a combination of search (by parametric studies) and guidance (by performance indices). Experience and intuition are essential. To help develop an intuitive understanding of the influence of various building elements on energy consumption, we present general rules and recommendations in Sec. 14.2.

In *parametric studies*, one varies one parameter at a time and displays the effect on loads (peak and average) and on costs. Good graphic display of the results is important so the

---

[1] Many people, including the authors, have used the term *energy-efficient building* to express this idea. But in a literal sense, that term is not meaningful, for if one tried to define energy efficiency by substituting *energy* for *cost* in the statement of the design goal, the conclusion would be absurd: Without cost constraints one could add insulation, heat recovery, and the like, ad infinitum, reducing the energy consumption of a building to zero.

designer can see in which direction to pursue the search. It is instructive to try to identify, as far as possible, the contributions of individual elements of the design, e.g., the heat loss due to air exchange. This will be illustrated in Example 14.1.

For judging the performance of a design there are two types of criteria. One is to *compare the observed annual energy consumption per unit of floor area with that of other buildings* that serve the same function. Such data are presented in Sec. 14.4. Preferably the comparisons should be broken down by end uses (heating, cooling, lighting, ventilation, as well as peak loads), assuming, of course, that one has the necessary data.

But relying on observed data alone is a bit like the blind leading the blind: There is no guarantee that others are efficient. To go beyond that, one needs *theoretical guidelines*. There are two types of guidelines: systematic simulation studies and fundamental limits. A wide range of parametric studies of hypothetical buildings have been carried out, especially with the goal of formulating performance standards for buildings. The collective experience gained by such studies has been incorporated in standards such as the ASHRAE Standard 90.1-1989.

An example of a fundamental limit is the Carnot efficiency; it states the thermo-dynamic limit for heat pumps and air conditioners. Although we do not know how to define meaningful fundamental limits for the envelope of a building, we present in Sec. 14.3 certain efficiency indices for the HVAC system that are useful for the designer.

14.1.3 Parametric Studies

To illustrate the method of parametric studies, let us take a simple building like the one of Example 7.2 and ask how the energy consumption could be reduced. Starting from an initial design, called *reference*, we add various features to reduce the energy consumption. Since the ultimate goal is to find the design that is the most cost-effective, we also consider the incremental cost of each additional energy conservation feature. As a crude indicator of cost-effectiveness, we calculate the payback time; a more rigorous economic analysis would evaluate the life cycle savings or the rate of return, as explained in Chap. 15 on engineering economics.

Also, to highlight the essential features of a parametric study, we have simplified the example as much as possible. The building is a rectangular box, as shown in Fig. 7.7. We assume steady-state conditions, with all building parameters constant, in particular the indoor temperature, the air change rate, and the heat gains. We neglect heat exchange with the ground. We take the heating system efficiency constant at $\eta = 1.0$, as appropriate for electric resistance heating without distribution losses. And we consider only a small number of design variations of the envelope.

Example 14.1

Consider several envelope design variations for a house of boxlike shape and wood-frame construction. For each variation, calculate the annual heating energy as well as the payback time relative to the reference design.

   Given: House of Fig. 7.7 (as used for most of the examples of Chap. 7 but with different U values)

   Rectangular box 12 m × 12 m × 2.5 m (39.4 ft × 39.4 ft × 8.2 ft)

   Area of roof = 144 m^2

   Area of opaque walls = 96 m^2

   Area of glazing = 24 m^2, which is 20 percent of wall area

Heat gains $\dot{Q}_{gain} = 1$ kW (3412 Btu/h) constant year-round

Volume = 360 m^3

Outdoor air at 0.5 air change per hour, hence

$$\dot{V}_{\rho c_p} = 0.5 \times 360 \, m^3/h \times 1.2 \, kJ/(m^3 \cdot K) = 60 \, W/K \quad \text{[see Eq. (7.23)]}$$

$T_i = 20°C$ (68°F) constant year-round

Heating system efficiency $\eta = 1.0$

Energy price $p_{fuel} = \$19.44/GJ = 7¢/kWh$ ($20.50/MBtu), a typical value for electric heating on the east coast of the United States

Site: New York City

Reference design (design 1): U values of opaque surfaces equivalent to 0.10 m (4 in) of fiberglass for walls and 0.15 m (6 in) for roof; conductivity of fiberglass $k = 0.060$ W/(m $\cdot$ K) [0.10 Btu/(h $\cdot$ ft $\cdot$ °F)]; double-glazed windows with $U = 3.0$ W/(m^2 $\cdot$ K) [0.53 Btu/(h $\cdot$ ft^2 $\cdot$ °F)].

Variations to be considered: Add successively the following energy conservation features:

- Design 2: Increase roof insulation thickness from 0.15 to 0.25 m; incremental cost per volume of fiberglass is \$18/m^3.
- Design 3: Reduce U value of windows from 3.0 to 1.5 W/(m^2 $\cdot$ K) (low ϵ, triple-glazed); incremental cost per area \$140/m^2.
- Design 4: Heat recovery from exhaust ventilation with heat exchanger effectiveness $\epsilon = 0.75$; incremental cost \$1300.
- Design 5: Increase wall insulation thickness from 0.1 to 0.15 m; incremental cost per volume of fiberglass is \$18/m^3, plus incremental cost of changes in wall construction (0.15-m studs instead of 0.10-m) amounting to \$4.00/m^2 of wall area.

Assumptions: All parameters are constant; use the variable-base degree-day method. Calculate the U values of the wall and roof by considering only the fiberglass insulation, without film coefficients and support structure. There is no heat exchange with the ground.

Lookup values: Degree-day data from Fig. 8.3

SOLUTION

The problem has been solved with a spreadsheet, and the essential points are summarized in the tables below and in Fig. 14.1. We explain the details for the case of design 2.

The U values of the walls and roof are calculated as $U = k/t$, where t = thickness. For example, the U value of the roof is 0.060/0.15 W/(m^2 $\cdot$ K) = 0.40 W/(m^2 $\cdot$ K) for the reference design and is decreased to 0.060/0.25 W/(m$^2 \cdot$ K) = 0.24 W/(m$^2 \cdot$ K) for the improved design; hence the difference in U values is $\Delta U = 0.16$ W/(m$^2 \cdot$ K). The difference in the contribution to K_{tot} is $\Delta K_{tot} = 144$ m$^2 \times 0.16$ W/(m^2 $\cdot$ K) = 23.04 W/K, and the resulting K_{tot} for this design is 224.2 W/K.

The balance-point temperature is calculated according to Eq. (8.2) as

$$T_{bal} = T_i - \frac{\dot{Q}_{gain}}{K_{tot}}$$

$$= 20°C - \frac{1000 \, W}{224.2 \, W/K} = 15.95°C \tag{14.1}$$

The corresponding degree-days are $D(T_{bal}) = 2139$ K $\cdot$ days (read from the USCS scales of Fig. 8.3 and converted to SI).

From Eq. (8.6) the annual heating energy is

$$Q_h = \frac{K_{tot} D_h(T_{bal})}{\eta} \tag{14.2}$$

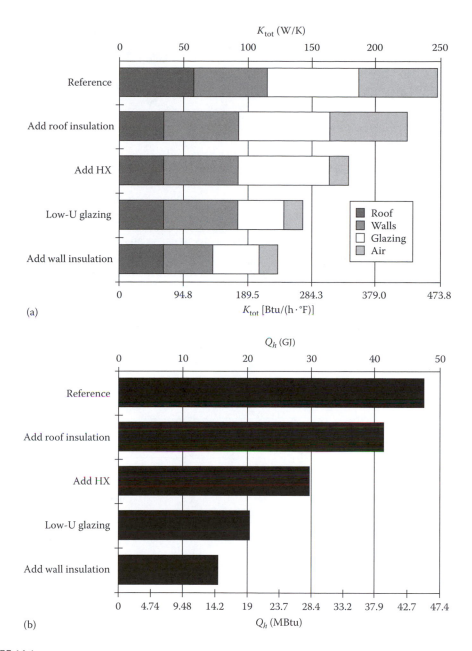

FIGURE 14.1
Parametric study of Example 14.1. (a) Contributions to total heat loss coefficient K_{tot}; (b) annual heating energy Q_h.

With $\eta = 1.0$, after converting to consistent units, we find

$$Q_h = 224.2 \, \text{W/K} \times 2139 \, \text{K} \cdot \text{days} = 41.42 \, \text{GJ}$$

Taking the difference from the consumption for the reference design, we find annual energy savings of $\Delta Q = 6.04$ GJ; they are worth $\Delta Q p_{fuel} = 6.04$ GJ $\times$ \$19.44/GJ $=$ \$117.40.

For the associated cost, one can assume that only the additional insulation material needs to be taken into account because neither additional structural support nor labor is necessary in new construction. Thus the incremental cost of the roof insulation is $\Delta C = 144 \text{ m}^2 \times 0.10 \text{ m} \times \$18/\text{m}^3 = \$259$.

Dividing the investment ΔC by the annual savings, we obtain the *payback time* as

$$\frac{\$259}{\$117.40/\text{yr}} = 2.2 \text{ yr}$$

This means that the extra cost of the roof insulation will have paid for itself after 2.2 years. The lifetime of the features is on the order of 20 years or longer. As a rule of thumb, if the lifetime is several times longer than the payback time and if there is negligible risk, the investment is very good for payback times shorter than 5 years and fairly good for payback times shorter than 10 years.

When several energy conservation features are under consideration, payback times could be evaluated for each feature separately or for any combination of features. For simplicity, we show only the cumulative cost (as the features are added on top of each other) and the payback time for the corresponding combinations.

It is instructive to plot, in Fig. 14.1, the components of K_{tot} and the heating energy Q_h; thus one sees which loss terms are the most important ones.

	Change			Contribution to K_{tot}				
Design	ΔU, W/(m² · K)	ΔK_{tot}, W/K	ΔC, \$	Roof	Walls	Glazing	Air	K_{tot}, W/K
1. Reference				57.6	57.6	72.0	60.0	247.2
2. Add roof insulation	0.16	23.04	259	34.6	57.6	72.0	60.0	224.2
3. Add HX		45.00	1300	34.6	57.6	72.0	15.0	179.2
4. Low-U glazing	1.50	36.00	3360	34.6	57.6	36.0	15.0	143.2
5. Add wall insulation	0.20	19.20	470	34.6	38.4	36.0	15.0	124.0

T_{bal}, °C	$D(T_{bal})$, K · days	Q, GJ	ΔQ, GJ	$\Delta Q p_{fuel}$, \$	Cumulative ΔC	Payback Time, yr
15.95	2222	47.46	0.00	0.00	0	
15.54	2139	41.42	6.04	117.40	259	2.2
14.42	1917	29.67	17.79	345.98	1559	4.5
13.01	1639	20.27	27.19	528.71	4919	9.3
11.93	1444	15.47	31.99	622.07	5390	8.7

COMMENTS

1. The reference design is representative of conventional construction during the 1980s. The features considered here do not require any unusual technologies, yet they reduce the consumption by a factor of 3.
2. In a real application, one should not limit oneself to envelope modifications. Other heating systems should be considered, in particular, a heat pump or a condensing furnace.
3. Considering only cumulative design modifications, one cannot identify and rank the cost-effectiveness of each individual feature. For instance, the reversal of payback times between designs 4 and 5 implies that extra wall insulation is more cost-effective than the low-U glazing, at least with the above cost assumptions. Also it may turn out that several features are cost-effective by themselves, but not in combination (the load remaining after adding one feature may be too small to justify the other).

4. There is a certain risk of being misunderstood or quoted out of context when one states specific numbers for payback times, since the results are highly variable with the assumptions. The cost of energy-saving features can vary quite a bit from one supplier to another. In some regions of the United States and in much of Canada, the number of degree-days is twice as large as in New York City; the corresponding savings would be twice as large and the payback time one-half as long. On the other hand, natural gas might be available at less than one-half the energy price assumed here, implying payback times twice as long.

In this example, it is easy to identify the heat loss contributions of individual design elements because the annual consumption is the product of degree-days and K_{tot}, and the contributions of the individual design elements to K_{tot} are explicit. When the conditions are not suitable for a steady-state analysis, the identification of individual contributions may be less obvious, especially if the computer simulation program does not list the contributions to K_{tot}. In such a case, it is advisable to begin by considering an upper and a lower bound of a parameter. For instance, setting the U value of the walls equal to zero, one obtains an upper limit on the savings achievable by wall insulation. The process of parametric search is all the more useful if it can be carried out interactively: One sees immediately in which sense to continue.

14.1.4 The Role of Building Type and Utilization

The design process depends on the type and utilization of the building in question. For instance, if skin loads dominate, one should concentrate one's efforts on improving the thermal performance of the envelope. High internal gains suggest that one should examine the efficiency of the lighting and pay special attention to the efficiency of the HVAC system. If fresh air requirements are large, a look at heat recovery is advisable. These matters depend on the type and utilization of the building. Therefore, we list in Table 14.1

TABLE 14.1

Some Typical Differences between Commercial and Residential Buildings

	Commercial Buildings	**Residential Buildings**
Occupancy	Mainly daytime, rarely nights/weekends	Almost always at night, often reduced or variable/irregular during day
Density (occupants per floor area)	Depends on building type; in office buildings relatively high density (around 0.05 to 0.1 person/m^2)	Usually low density (around 0.02 person/m^2 in wealthy countries)
Internal gains	Often relatively large	Relatively small
Ventilation	Usually central; outdoor air by central ventilation (in United States)	Outdoor air by infiltration (in United States)
HVAC system	Usually (in United States) large, central, and relatively complex systems	Simple; usually single-zone; air is not shared between residential multifamily units
Size	Often quite large	Small (in the case of houses)
Surface/volume ratio	Often fairly small	Usually quite large, even in apartment buildings
Important loads	Internal gains tend to be more important than envelope loads (lighting and cooling often dominant)	Envelope loads tend to be more important than internal gains (heating dominant in temperate and cold climates)
Design	Usually custom design	More or less mass-produced

some typical differences between the two major categories of buildings: commercial and residential. Of course, it should be understood that such a table is a gross simplification in view of the enormous variety of commercial buildings.

14.2 Design Elements and Recommendations

14.2.1 A List

The process of designing a building involves many choices, from the conception of the envelope to the specification of the HVAC equipment. As a general guide, we have listed in Table 14.2 the major elements, grouped in four categories:

- Environment
- Structure
- Equipment
- System and controls

As far as is practical, we have suggested simple rules to indicate how to choose each element for maximal energy efficiency *if cost were no object*. The rules are discussed below. Between some of these elements there is strong coupling. In particular, the interaction between the envelope and the HVAC system makes coordination between architect and engineer critical.

Obviously such a simple table cannot capture all the subtleties of building design (Crabb, 1988; Burt Hill Kosar Rittelmann Associates, 1985). There are an enormous variety of building types and utilization patterns, and in particular there are great differences between residential and commercial buildings (see Table 14.1). Nonetheless, we believe that a simple summary is a good starting point for honing one's intuition. The remainder of this section presents more specific comments on the elements of Table 14.2.

14.2.2 Environment

Here we have factors such as site and vegetation that are somewhat difficult to quantify. The site affects wind-induced heat transfer as well as solar gains due to shading by adjacent buildings or trees or due to reflected radiation. But in many situations, external constraints leave little freedom to influence these factors.

Shading should be carefully taken into account; it affects not only heating and cooling loads, but also daylighting. A particularly noticeable (and notorious) site effect occurs when a new structure is put up next to an existing building with large windows, casting either a shadow or a large amount of reflected radiation on it.

Wind patterns can affect infiltration and the surface heat transfer coefficient. Wind breaks can save heating energy, but they reduce the ventilation that is desirable for summer comfort in buildings without air conditioning. The tighter and better insulated a building, the lower its sensitivity to wind. Mechanical ventilation can also reduce the sensitivity to wind, as shown in Sec. 7.1.3. In current conventional construction with infiltration rates below 0.5 (air change per hour) and a reasonable amount of double glazing, the sensitivity

TABLE 14.2

Design Elements and Their Effect on Energy Consumption

Element	Variable	Recommendation (If Cost were No Object)
Environment		
Site and orientation		Pay attention to climate and to shading by surrounding objects (and to wind if natural ventilation is important).
Vegetation		Plants offer wind break, sound attenuation, shade, and evaporative cooling, plus beauty.
Structure		
Shape	S/V	If all else is fixed, heat loss is minimized by minimizing surface per volume.
Internal walls	Mc_p	Benefit of thermal mass depends on thermostat control (can store heat gains, but inhibits savings from thermostat setback).
Foundation		Do not overlook heat loss to ground.
		Ground coupling may be beneficial for cooling.
External walls, roof	U, K_{tot}	Should be as low as possible.
Doors		Use insulated doors with seals.
		If many door openings per day, install vestibule or revolving door.
	$\dot{V}_{inf}$	Should be no larger than needed for comfort. Use tight construction.
		Mechanical ventilation allows better control of air exchange than natural infiltration.
Windows		Use tight, insulating frames for windows.
		Movable night insulation for windows can be very effective.
Shading devices	$\dot{Q}_{sol}$	Use shading device as needed to control solar heat gains.
		Exterior shades are more effective than interior ones (especially if glass has high absorptivity).
		Solar gains are beneficial for heating, but undesirable for cooling. Beware of large east or west glazing without solar control.
Skylights	Daylight	Can save energy if lights controlled by daylight sensor (but be careful with heat gains, especially from large, unshaded south-facing skylights).
Sunspaces		Can save heating energy if temperature is allowed to float.
Equipment		
Service hot water		
Heating plant	η (COP)	Should be as high as possible.
		Heat pumps are more efficient than resistance heat, especially if source is not too cold.
Cooling plant		
Fans, pumps	Capacity	Variable speed/capacity for variable load.
Air outlets	Throw	Avoid stagnation of air.
Lights	Luminous efficacy	Aim for high luminous efficacy (lm/W).

(continued)

TABLE 14.2 (continued)

Design Elements and Their Effect on Energy Consumption

Element	Variable	Recommendation (If Cost were No Object)
System and control		
System		Minimize simultaneous heating and cooling.
Zoning		Match utilization, heat gains, and losses.
Ducts, pipes	ΔP	Should be as low as possible.
Airflow	$\dot{V}_{sup}$	Use variable rather than fixed.
Air supply temperature	T_{sup}	Use variable rather than fixed.
Thermostat	T_{min}, T_{max}	Use no more than needed for comfort.
Building management system		Use for diagnostics (compare actual and predicted performance).
		Detect degradation of equipment by comparison with historical records.
Sensors (daylight, occupancy)		Turn equipment off when not needed.
Economizer	$\dot{V}_{vent}$	Use instead of chiller whenever possible (enthalpy control is more efficient than temperature control).
Night ventilation		Consider if sensible cooling outweighs fan energy and latent gains; this is especially effective in massive buildings.
Cool storage		Look at possible savings from lower electricity cost at night (energy and demand charge) and from reduced chiller capacity.
Hot storage	$Q, \dot{Q}, \eta$	Look at possible savings from lower electricity cost at night (energy and demand charge) and from reduced heating equipment capacity.
		Also consider if free heat is available (e.g., from heat recovery) but mismatched in time from heating loads.
Heat recovery		Consider if free heat is available (e.g., from condenser of chiller or from exhaust ventilation).

Braces and arrows indicate connection between elements, variables, and recommendations. In practice, the recommendations must, of course, be tempered by cost and other constraints

of the total heat loss coefficient K_{tot} to wind does not appear very significant. For instance, K_{tot} of the house in Example 7.6 increases only 10 percent between the wind speeds of 7.5 mi/h (3.4 m/s) and 15 mi/h (6.7 m/s) of the ASHRAE summer and winter design conditions, and most of that variation is due to infiltration. Future construction is likely to be even less sensitive to wind. Wind patterns are most relevant where they are the most difficult to change—around old buildings.

Plants, especially large trees, can provide multiple benefits, breaking wind currents, attenuating noise, and offering shade and evaporative cooling, to say nothing about their beauty. A number of recent studies have emphasized the cooling benefit of plants (LBL, 1988). For example, by comparing air temperatures in neighboring sites with and without forest cover, differences around 4°F (2°C) have been observed by Akbari et al. (as reported in LBL, 1988). Meier (1990) has reviewed all the publications that report measured data for the effect of vegetation on cooling, and his findings are summarized in Table 14.3. Part a shows reductions in surface temperature; part b, reductions in energy consumption that can be attributed to the vegetation. Despite the enormous variety of

TABLE 14.3

Measured Data for Effect of Vegetation on Cooling

(a) Surface temperature reductions

Author and Year	Location and Climate	Type of Planting	Wall-Veg. Distance	Difference Measured	ΔT	Notes
J. Parker 1981	Miami, FL Hot/humid	Shrubs and trees	Shrubs < 1 m Trees < 10 m	Wall with and without plants	16°C	Westwall, 5 p.m., maximum value about 1 month apart
Hoyano 1988	Tokyo, Japan Hot/humid	Ivy covering	Touching	Wall with and without ivy	18°C	Westwall, 3 p.m., maximum value; 1 year apart
Hoyano 1988	Tokyo, Japan Hot/humid	Dense canopy ever-greens (Kaizuka hort)	0.2–0.6 m	Wall and inside plant surface	5–20°C	Westwall, 3 p.m. parallel measurements
McPherson 1989	Tucson, AZ Hot/dry, desert	18 shrubs and 5-cm decomposed granite	0.5 m	Wall with shrubs and no shrubs	17°C	Westwall, 3 p.m. different buildings
McPherson 1989	Tucson, AZ Hot/dry, desert	Turf, extending about 5 m from structure	Surrounding building	Wall with turf versus decomposed granite	6°C	Westwall, 3 p.m. different buildings
Makzoumi and Jaff, 1987	Baghdad, Iraq Hot-dry desert	Vine (*Luffa cylindrica*) on trellis	0.1–0.4 m	Wall with and without vines	17°C	Southwest wall, 3 p.m., maximum value, different building
Harazono 1989	Osaka, Japan Hot/humid	Rooftop hydroponic using lightweight planting substrates and mixed plants	0.1 m	One-half of roof with, one-half without	21°C	Average for 10 a.m. to 6 p.m. on clear August day
Halvorson 1984	Pullman, WA Temperate	Vertical vine canopy	n.a.	Wall with and without vine	20°C	

(continued)

TABLE 14.3 (continued)
Measured Data for Effect of Vegetation on Cooling

Author and Year	Location and Climate	Type of Planting	Energy Measurement	ΔE	ΔE, %	Notes
(b) Energy savings						
Air conditioning						
Dewalle et al. 1983	Central Pennsylvania Temperate	Forest site versus clear site	AC electricity for identical mobile homes	230 W	80	37-day test period
J. Parker 1983	Miami Hot/humid	Florida shrubs and trees	AC electricity with and without landscaping	5000 W	58	6-h (afternoon)
					24	test period 10-day periods
McPherson et al. 1989	Tucson AZ Hot/dry desert	Shrubs surrounding model house	AC electricity with and without shrubs	104 W	27	2-week period
McPherson et al. 1989	Tucson AZ Hot/dry desert	Turf surrounding model house	AC electricity with and without turf	100W	25	2-week period
D. Parker 1990	Palm Beach, FL Hot/humid	Miscellaneous trees and shrubs	Annual electricity for whole houses	1.8 W/m²	34	Inferred from regression of 25 houses, from landscape class 0→2,3
Heat gain						
Hoyano 1988	Tokyo, Japan Hot/humid	Vine-covered wall	Heat gain with and without vines	175 W/m²	75	Peak value at 4 p.m. on west wall
Hoyano 1988	Tokyo, Japan Hot/humid	Row of evergreens next to wall	Heat gain through wall	>60 W/m²	>50	Peak value at 4 p.m. on west wall for widely spaced trees
Harazono 1989	Osaka, Japan Hot/humid	Rooftop vegetation	One-half of roof with, one-half without	130 W/m²	90	Average from 10 a.m. to 4 p.m.

Source: Courtesy of Meier, A.K., *Measured Cooling Savings from Vegetative Landscaping*, vol. 4, American Council for an Energy Efficient Economy, 1990 Summer Study on Energy Efficiency in Buildings. Asilomar, CA, 1990, 133. With permission.

vegetative landscaping, the surface temperature reductions in Table 14.3a are quite similar, mostly in the range of 20 to 40°F (10 to 20 K) for peak conditions. The savings in cooling energy range from about 25 to 90 percent. The cooling benefits of plants can be very important!

On the other hand, compared to conventional equipment, the engineer may find the use of plants less predictable. Mechanical equipment, such as a chiller or a shading device, can usually be counted on to be ready in time for occupancy, and there is not much uncertainty about its performance. As for the landscaping, the engineer may have little influence, and then the plants may take many years to grow to full size. And, of course, plants impose certain maintenance costs, for watering, fertilizing, removal of weeds and dead leaves, etc. These obstacles may be more serious in practice than the fact that the heat transfer mechanisms are complex and difficult to analyze with precision. Again, factors beyond sheer energy may dominate the decision. In any case, it is important for the decision makers to be aware of the cooling benefits of plants.

14.2.3 The Structure and Envelope

For a given floor space, the environmental driving terms are minimized by minimizing the *surface-to-volume ratio*. Since spherical shapes are rarely suitable for buildings, the cube is the best according to this criterion. But in practice the shape is likely to be determined by other considerations. In particular, people like windows, and if one tries to maximize the number of rooms with windows, one ends up with a large surface-to-volume ratio.

The main effects of the structure on the thermal behavior of a building can be characterized by the following five variables, shown in Table 14.2:

- Total heat transmission coefficient K_{tot}
- Air exchange rate $\dot{V}$ due to infiltration and/or ventilation
- Thermal mass (= total effective heat capacity C_{eff})
- Solar heat gains $\dot{Q}_{sol}$
- Utilizable daylight

The last is relevant for energy efficiency only to the extent that daylight has a value, e.g., by virtue of appropriate control of the electric lighting.

The lower the *U values* of the building envelope, the lower the heating load and the lower the temperature-dependent component of the cooling load. The recommendation for low U values bears on comfort, too. The better the envelope, the lower the risk of discomfort caused by temperature gradients inside the building, by drafts from cold leaky windows, or by low radiant temperatures in perimeter zones. This is an instance where comfort and energy conservation go hand in hand.

A sizable part of the heating and cooling loads is proportional to the air exchange rate. Of course, a minimum of fresh outdoor air is required for comfort and health. But uncontrolled *infiltration* will rarely provide just what is needed. Infiltration depends on the details of the construction that are impossible to predict with precision, and it will vary with wind speed and temperature difference. The resulting air exchange is likely to be much higher than the optimum in winter, and in summer it may not be enough. In climates with large heating or cooling loads, airtight construction with mechanical ventilation is more conducive to energy efficiency than uncontrolled infiltration. If one opts for mechanical ventilation sufficient to supply the required flow of fresh air, then the envelope should be very tight. Swedish houses set a good example with rates around or below 0.1 air change per hour. Mechanical ventilation opens up a further opportunity for energy saving: heat recovery.

Windows demand the greatest attention. It is through the windows that the environmental driving terms affect the building most strongly. With conventional construction, conductive heat transfer and solar heat gains are an order of magnitude larger per unit area through glazing than through opaque walls. *Windows can make the difference between efficiency and waste.*

One very effective way of reducing heat losses through windows is to cover them with movable insulation at night. Although this is not common practice in the United States,[2] automated shutters could be sufficiently convenient to be accepted.

Architectural considerations tend to override U values when it comes to windows. Fortunately, the technology of windows is advancing. Windows with low heat loss coefficients are resulting from advances in low-emissivity coatings, aerogels, and evacuated panels. Combining low U values with control of solar transmissivity, via adjustable shading devices or electrochromic coatings, one can produce windows that are more energy-efficient than opaque walls. The recommendation, in Table 14.2, against large windows on east or west facades applies most strongly to conventional glazing. The better the control of solar heat gains, the more freedom one has in placing windows anywhere. But in any case, be careful with large, tilted, south-facing skylights without shading, for they would receive maximal solar heat gains.

Solar heat gains are beneficial only when they reduce the heating loads; in summer they can impose a severe penalty. It is crucial to pay attention to this term and to reduce it in summer, by using shading devices or glazing with a low shading coefficient. Solar heat gains tend to be less beneficial in commercial than in residential buildings because the gains coincide with the rather substantial heat gains from lights, office equipment, and occupants. Indeed, for many commercial buildings, cooling loads are more important than heating loads, even in cold climates.

As for *shading devices*, a basic choice is between interior and exterior shades. Interior devices can be lightweight and inexpensive, but they reduce heat gains only to the extent that they reflect radiation back out of the building. In practice, a sizable fraction of the solar radiation is absorbed by the interior shades or by the glazing (on the way in or on the way out). The ability to withstand wind loading makes exterior shades more costly, especially if they are adjustable or movable, but their thermal performance is better.

Since the 1970s, *sunspaces*, such as greenhouses and atria, have been in vogue. Even though they are frequently advertised as energy savers, the real savings depend on the details of design and control, in particular on occupant behavior. From a thermal point of view, a sunspace is very much like the airspace inside a double-glazed window, with extra surface for absorption of solar radiation and extra mass for its storage. *If* the temperature of the sunspace is allowed to float, then the air can act as a layer of insulation and solar heat gains can be stored by the thermal mass of the sunspace. However, if the sunspace is conditioned, the thermal boundary of the building is, in fact, moved to the outer glazing; the resulting load is larger than that in a building without this sunspace because the glazing provides less insulation than the opaque wall that would have been the thermal boundary.

---

[2] By contrast to Europe, where external or internal shutters are a common feature in residential buildings, and where people are well accustomed to the routine of closing them at night—often more for security than for comfort or energy conservation.

Therefore, *to save energy, a sunspace must remain unconditioned*. Like adding insulation, adding a sunspace can save energy; but unlike savings from insulation, savings from sunspaces are highly dependent on occupant behavior. Furthermore, in summer, a sunspace may become unbearable unless there is effective control of solar gains and ventilation. In winter, there is the risk of drafts and frost in the sunspace. In practice, the value of energy savings, if any, is often secondary to other considerations: the cost of the structure and its aesthetic appeal.

The light entering through windows, skylights, and other solar apertures can help save electricity for lighting—but only if lights are turned off in response to daylight. If properly designed and functioning correctly, a building with *daylighting* can save much energy, as discussed in Sec. 14.7.

Thermal mass[3] is a more complicated matter. The effects of interior mass and mass in the envelope differ somewhat, even though we have sometimes used a single variable for simplicity: the effective heat capacity C_{eff}. The effect of thermal mass depends on the control of the thermostat. If the thermostat is always kept at the same setting, the heat capacity in the interior of the building has almost no effect: Heat exchange with the mass is negligible. If one tries to save energy by setback during unoccupied periods, thermal mass is an obstacle because it slows the cooldown and necessitates greater capacity of the heating equipment; the setback recovery occurs during the early morning hours just at the time of the peak heating load. During the cooling season, this requirement for excess capacity is less of a problem because the peak cooling load is not coincident with setback recovery.

But thermal mass can also be desirable, especially during the cooling season or in buildings with passive solar heating. Peak cooling loads, as shown in Chap. 7, are significantly reduced by thermal mass. Even if the thermostat set point is constant, the temperature of the mass can differ somewhat from that of the air in the building, and the mass can soak up radiative energy from lights and solar radiation. The building shell can buffer the exterior heat pulse. In a climate where days are hot and nights are cool while the 24-h average stays within comfort limits, the interior of a massive building can remain very comfortable without any space conditioning whereas a lightweight structure would be intolerable. This principle can be applied in any building where there are substantial heat gains during part of the day, which can be stored by the thermal mass, to be released later.

Thus the recommendation for thermal mass involves a tradeoff between two opposing strategies to reduce heating energy: reduction of the indoor-outdoor temperature difference by thermostat setback, on one hand, and utilization of free heat gains, on the other hand. In addition, *thermal mass, especially thermal mass in the envelope, reduces peak cooling loads*; thus it can save equipment capacity. The bottom line depends on the climate, temporal distribution of heat gains, and occupancy pattern. Both heating and cooling should be taken into account. We present further analysis of this issue in Sec. 14.5.2, showing that thermal mass is beneficial for energy conservation in typical residential applications in the United States.

In practice, the potential savings may not be sufficient to justify a modification of customary construction methods for the sake of increasing or decreasing the thermal mass. However, in some cases, it may be fairly easy to modify the effective heat capacity by modifying the heat transfer between mass and air. Mass in external walls or the roof can be coupled (decoupled) permanently by placing the insulation on the exterior (interior)

---

[3] Also known as *thermal inertia or heat capacity*. See Sec. 7.3 for a discussion of the effective heat capacity.

surface. In commercial buildings with ceiling plenum and concrete floors between different stories, the effective mass can be increased by circulating ventilation air through the ceiling plenum, in direct contact with the concrete.

14.2.4 Equipment

In the third part of Table 14.2, the basic recommendation is very simple: *Choose equipment with high efficiency* [high coefficient of performance (COP) for the case of heat pumps and air conditioners]. Heat pumps are more efficient than electric resistance heating. Condensing furnaces are more efficient than noncondensing ones. Two-stage absorption chillers are more efficient than those with a single stage. Fluorescent lights are far more efficient than incandescent lights, etc. Always look for the most efficient equipment in the catalog, and do at least a quick evaluation; chances are, it will pay for itself. Even if it does not, one gets an idea of the relative importance of efficiency in a given situation. Information on efficient equipment can be found in several special publications, e.g., ACEEE (1989) for the residential sector and Usibelli et al. (1985) for the commercial sector. For electric equipment one can find excellent guidelines in the publications of the Electric Power Research Institute (EPRI).

The efficiency of *heat pumps* and *chillers* depends not only on the machine itself, but also on the method of extracting or rejecting the heat. Chiller COPs can range from about 2 for air-cooled window air conditioners to 5.5 for large centrifugal chillers with cooling towers. The smaller the temperature difference across which heat is to be pumped, the higher the COP. Furthermore, the COP increases somewhat with larger capacities. The COP varies with operating conditions; nominal or design values are usually quoted for full capacity and standard temperatures. When interpreting quoted performance values, one has to be careful of the operating conditions and how much of the ancillary equipment is included, such as fans and the cooling tower. Part-load characteristics must be considered for the calculation of annual energy consumption.

While air source heat pumps for space heating are very efficient in mild weather, they suffer a double handicap in cold weather: Both their COP and their capacity decrease, reaching the minimum when one needs the most heat. For that reason, air source heat pumps in the United States have been installed mostly to the south of the 40° latitude. Typical COPs have been around 1.5, averaged over a heating season. Much better COPs can be achieved during cold weather if groundwater or the earth can be used as a heat source, in places where this option is practical. The economics of heat pumps may be improved if they can do double duty as chillers in summer. For ground coupling that is also the best arrangement because it maintains, at least approximately, the long-term thermal balance of the ground. (See chap. 9 of ASHRAE, 1987).

Most HVAC loads are variable. For variable loads there is a further rule: Avoid penalties from poor part-load system efficiency. Thermal storage can be very effective for avoiding part-load operation of heating and cooling plants. For slowing down pumps and fans, *variable-speed* electric motor drives are a more efficient means than throttling or recirculating the flow.

The design and placement of *air diffusers* must not be neglected. Avoid short-circuiting or zones of stagnant air. The risk of stagnation is increased if cold air is blown into a room from below or hot air from above (be especially careful when heating tall rooms from above). The occupants may try to compensate for the lack of comfort by adjusting the thermostat and placing a greater load on the heating or cooling plant, but that does not solve the air quality problems. For best performance and comfort, perimeter zones should have two sets of diffusers, one for heating and one for cooling.

14.2.5 System and Controls

Here HVAC designers can test their mettle. The number of design options and variables is staggering, as one can see, e.g., from a look at the manual of the DOE2.1 computer code where a rather terse summary of system options sprawls over 300 pages. The choices begin with some basic questions: Does one want heating, does one want cooling, does one want mechanical ventilation, does one want humidity control? Sometimes the answer hinges more on custom than on climate; e.g., most older office buildings in Europe do not employ air conditioning or mechanical ventilation even where the cooling degree-days are comparable to those in the United States.

Another fundamental decision is the choice between *local* and *central* HVAC systems. Actually there is a spectrum of centralization. Air conditioning in a multi-family building may be supplied by window air conditioners, by air conditioners in each apartment, or by a central chiller for a building or group of buildings. *Centralization is cost-effective if economies and efficiencies of scale outweigh the cost and losses of the distribution system.* As an example of the cost crossover between the local and the central systems, Fig. 14.2 compares the life cycle cost of a central HVAC system with the local alternative, as a function of the number of apartments; in this particular case, the central system offers a lower life cycle cost (at 7 percent discount rate) if the number of apartments exceeds six. In office buildings, one usually opts for central systems. There are also considerations besides energy. Aesthetics, comfort, and noise can have a bearing, e.g., in the preference for central over window air conditioners. If several independent users share a system, the apportioning of costs may be troublesome because thermal energy is relatively difficult to meter.

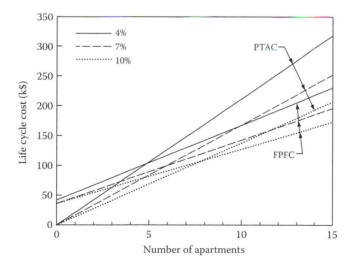

FIGURE 14.2
Life cycle cost versus number of apartments for local system and for central system (PTAC = packaged terminal air conditioners and FPFC = four-pipe fan coils) and three discount rates (4, 7, and 10 percent) in Chicago. (Courtesy of Byrne, S.J. and Fay, J.M., *ASHRAE Trans.*, 95, 2, 1989.)

The *distribution* of thermal energy is usually accomplished by *air* or *water*. In the past, steam was a common transport fluid for heating within buildings. But steam pipes undergo large temperature excursions under part-load conditions, and inside buildings the cacophony of contracting and expanding pipes can be quite a nuisance. Another problem arises from the frequent and notorious failures of steam traps and the resulting inefficiencies.

With central cooling, air is widely used as a distribution fluid, especially in the United States. Hydronic systems are also possible, distributing chilled water by pipes to local fan coils or induction coils; such systems are, in fact, quite common in Europe. In hydronic systems, one can avoid the need for an extra drainpipe for the condensate from the local cooling coils, if the outdoor air is supplied by a central ventilation system with central dehumidification and if the latent loads inside the building are not too large (this should be checked by calculation). One must pay attention to the pressure drop in ducts and pipes, because energy consumption increases with the pressure drop.

The system-and-control part of Table 14.2 states some apparently simple rules, but their implementation is not always obvious. *Minimization of simultaneous heating and cooling* is just plain common sense from the point of view of thermodynamics, although it is not always the solution with the lowest capital cost. The occurrence of simultaneous heating and cooling is especially likely in systems with *electric reheat*. If such a system is poorly adjusted, the electric reheat can easily compensate to maintain comfort, but the over-consumption of energy is costly and the problem may be difficult to detect.

For *zoning*, the general recommendation is to *match the distribution of heat gains and the utilization patterns of the building*. Along the perimeter, one should plan separate zones for facades that face in different directions, to accommodate different solar gains. In large buildings, at least one separate zone for the interior is advisable. Computer centers, used around the clock, and conference rooms, used occasionally, should be zoned apart from offices with normal working hours. And, of course, one invites trouble when one tries to control rooms with different loads by the same thermostat, e.g., a windowless laboratory and an office with windows.

The cooling energy can be reduced by resetting the supply air temperature in response to the loads. With regard to *thermostat set points*, the recommendation is to heat and cool no more than is necessary for comfort and safety. Certainly there should be a dead band between the onset of heating and the onset of cooling.

This brings us to a fundamental recommendation: *Turn equipment down or off when there is no need for it*. This applies to all equipment, from chillers to lights and office equipment. For instance, during the morning warm-up of an office building, do not open the outdoor air intake until occupancy begins. One of the advantages of a computerized building management system is its ability to ensure that the entire HVAC system operates just in time for each need and no more than necessary. Of course, there are items such as personal computers, printers, and copiers that one may not want to turn off each time one stops typing, simply because the start-up takes too long (with current designs and operating systems). Here one needs a controller that senses when equipment is not in use and turns it off automatically, followed by automatic and instantaneous restart the moment it is needed again (and, in the case of computers, without any loss of information).

For lights, the solution is automatic control by daylight sensors and occupancy sensors. The latter are particularly important for places such as conference rooms that are used intermittently by a large number of people. The lights are frequently left on in such places because nobody seems to feel an obligation to turn off the switch. Daylight sensors can save much electricity in offices with windows. Note that occupancy sensors can also bring important benefits as part of the security system of a building.

The *economizer*, rather than the chiller, should be used whenever the enthalpy of outdoor air is below the specified supply air enthalpy. Enthalpy control is somewhat more efficient than control based on the dry-bulb temperature of outdoor air (provided the enthalpy sensor works reliably).

Nighttime ventilation can reduce cooling loads in climates with cool nights. For air conditioned buildings in humid climates, one has to be careful not to increase the daytime latent load by moisture absorbed at the interior surfaces at night (Fairey and Kerestecioglu, 1985).

A few further design options are mentioned in Table 14.2. *Storage* and *heat recovery* deserve special attention. Cool storage can yield considerable savings if the total cost of electricity (energy charge plus demand charge, see Chap. 15 on economic analysis) at night is lower than that during the day, a common situation for large consumers. Just as important are capacity savings in buildings with daytime-only cooling loads; with storage, the same daily cooling energy can be produced by a smaller chiller running continuously. Compared to water tanks, cool storage with ice is more compact and can be bought as a prefabricated unit. For the analysis of storage, the most important variables are the energy capacity, charging and discharging rates, operating temperature range, and storage efficiency (fraction of stored energy that is recovered). Many publications on design and performance of thermal storage are available from the Electric Power Research Institute (for example, EPRI, 1984). Heat recovered from the exhaust air can be used for preheating, or it can be upgraded by heat pumps to supply service hot water or space heat.

14.2.6 Solar Energy

Solar energy for buildings has had its fashions, hopes, and disappointments. The oldest and so far most successful application has been service hot water in climates with little or no frost, while active systems for space heating and cooling have turned out to be too costly, in most cases (Duffie and Beckman, 1980; Kreider and Kreith, 1990). But passive solar heating, the approach by which the building itself is utilized as a collector, can be cost-effective. This topic is addressed in Sec. 14.5.3. Another successful application is solar preheating of ventilation air (see Chap. 9).

Another form of solar energy is daylight. In commercial buildings, daylighting holds much promise, especially in perimeter zones where people are sitting near the window anyway and only a light sensor is needed to reap the electricity savings. The savings achievable by daylighting are discussed in Sec. 14.7.

Designing for solar energy requires careful attention to the interrelation among solar radiation, outdoor temperature, and utilization of the building. For example, the heating load of a commercial building with high internal gains and a 40 h/week occupancy is likely to be so small that passive solar heating would not be appropriate. Solar energy utilization in buildings is more difficult than energy-efficient measures such as the use of insulation or the choice of equipment with high efficiency. If one adds too much insulation, the cost of the building will be higher than the optimum, but there is little risk of poor performance or discomfort. But if one makes the collector area of a building with passive solar heating too large, the consequences for comfort may be severe.

Much experience about passive solar design for commercial buildings has been gained from a series of 19 demonstration projects that were built around 1980 and followed by extensive performance evaluation. Detailed information about these buildings and their evaluation can be found in a book by Burt Hill Kosar Rittelmann Associates and Min Kantrowitz Associates (1987). These designs involved a variety of strategies, often in combination, especially direct gain or massive storage walls for passive heating, daylighting, shading devices, and night ventilation for cooling. While it is difficult to compress the conclusions to a few lines, one can say that these design approaches can be very successful. However, the design process is more difficult (and thus more costly), and the risk of problems is greater than with conventional design. There are risks when one uses unconventional technologies or when a design relies too much on specific behavior patterns of the occupants (e.g., manual adjustment of shading devices may not be followed as intended, or occupancy schedules may change). Nevertheless these 19 buildings were found to cost little, if any, more than conventional construction.

14.3 Performance Indices for Heating and Cooling

The concept of efficiency is of fundamental importance for the components of an HVAC system, and we have already made extensive use of it in previous chapters. In particular, we recall the importance of evaluating the annual efficiency for any equipment whose part-load efficiency is not constant.

In this section we take another look at efficiency by asking about theoretical limits for heating and cooling. At first sight, such a question may not seem meaningful. After all, in theory, one could just keep adding insulation and heat recovery until the heat load vanishes; such a limit is not very informative. However, the question is quite different if we take the heat loss coefficient and the heat gains of a building as a given starting point and then ask how much energy is needed if we have at our disposal a source of high thermodynamic quality such as electricity. The corresponding thermodynamic limit for the heating system efficiency is derived in the following sections, beginning with a simple idealized situation and then adding processes that are needed to accomplish the necessary heat transport into a real building. This leads to a definition of second-law efficiency. The equations are also relevant for cooling. Finally, in Sec. 14.3.5, we derive an efficiency index for multizone buildings where simultaneous heating and cooling loads are a frequent problem.

14.3.1 The Limit for Ideal Heat Pumps

Suppose that a building has a heating load $\dot{Q}_h$ and we have electricity available. We could use an electric resistance heater, but a heat pump would be more efficient. It could extract heat from the outdoor air at T_o and deliver heat at T_i, consuming work $\dot{W}_h$ (mechanical or electrical), as indicated in Fig. 14.3a. A heat pump is a heat engine running backward. For an ideal Carnot heat engine between a hot reservoir at T_i and a cold reservoir at T_o, the work output $\dot{W}_h$ and the heat input $\dot{Q}_h$ are related by

$$\dot{W}_h = \dot{Q}_h \frac{T_i - T_o}{T_i} \tag{14.3}$$

By running the process in reverse, $\dot{W}_h$ and $\dot{Q}_h$ change signs, but Eq. (14.3) remains unchanged. Therefore, in an ideal Carnot heat pump, $\dot{W}_h$ and $\dot{Q}_h$ are also related by

$$\dot{W}_h = \dot{Q}_h \frac{T_i - T_o}{T_i} \tag{14.4}$$

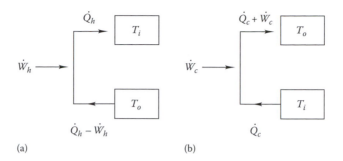

FIGURE 14.3
Thermodynamics of heat pump (a) for heating load $\dot{Q}_h$ and (b) for cooling load $\dot{Q}_c$.

Inserting typical temperatures, for example, $T_i = 20°C$ and $T_o = 0°C$, we find

$$\dot{W}_h = \dot{Q}_h \frac{20\,K}{(273 + 20)\,K} = 0.068\,\dot{Q}_h$$

This means that an ideal heat pump can deliver $1/0.068 = 14.7$ kWh$_t$ of heat for each 1 kWh$_e$ of electricity for these conditions (we have added subscripts t and e to distinguish thermal and electric energy forms).

Based on considerations such as these, some energy analysts have introduced the concept of the *second-law efficiency* of a process. It is defined as the ratio of the efficiency of the process to the efficiency of an ideal reversible process that satisfies the same demand (i.e., a process whose efficiency is as high as permitted by the second law of thermodynamics). For the above example of $T_i = 20°C$ and $T_o = 0°C$, it is very easy to find the second-law efficiency of electric resistance heating. For 1 kWh$_e$ of electric energy, the resistance heater delivers 1.0 kWh$_t$ of heat compared to 14.7 kWh$_t$ with an ideal heat pump. Hence the second-law efficiency of resistance heating is only $1/14.7 = 6.8$ percent, even though the ordinary (i.e., first-law efficiency) is 100 percent.

Next consider the case where $T_o > T_i$ and the building has a cooling load $\dot{Q}_c$. The corresponding energy flows are indicated in Fig. 14.3b. For a Carnot process, the load $\dot{Q}_c$ and the work input $\dot{W}_c$ are related by

$$\dot{W}_c = (\dot{Q}_c + \dot{W}_c)\frac{T_o - T_i}{T_o}$$

or

$$\dot{W}_c = \dot{Q}_c \frac{T_o - T_i}{T_i} \tag{14.5}$$

The ratio $\dot{Q}/\dot{W}$ of delivered thermal energy to work input is the efficiency relative to electricity; since it is a number greater than unity, it is usually called the *coefficient of performance* (COP):

$$COP = \frac{\dot{Q}}{\dot{W}} \tag{14.6}$$

Therefore, the COP of an ideal heat pump is

$$\text{COP}_{\text{Carn,h}} = \frac{T_i}{T_i - T_o} \tag{14.7}$$

For the above example of $T_i = 20°C$ and $T_o = 0°C$, we have a heating COP $= 14.7$.

The COP of an ideal air conditioner is

$$\text{COP}_{\text{Carn,c}} = \frac{T_i}{T_o - T_i} \tag{14.8}$$

For example, if $T_o = 35°C$ and $T_i = 25°C$, an ideal air conditioner has a COP of $(273 + 25)/10 = 29.8$; it delivers 29.8 kWh$_t$ of cooling for each 1 kWh$_e$ of electric input.

14.3.2 Finite Airflow Rates in Building

Since a Carnot heat pump does not use air as the working fluid, the heat $\dot{Q}_h$ must somehow be transferred to the air in the building; likewise, the heat drawn from the outdoor air must be transferred to the heat pump. That involves various heat exchanges which we shall now examine one by one. As a first step, let us consider an air heating system that supplies hot air at flow rate $\dot{V}$ and temperature T_{sup}. The heat supplied by this airstream is

$$\dot{Q}_h = \dot{V}\rho c_p(T_{\text{sup}} - T_i) \tag{14.9}$$

and it must be equal to the heating load $\dot{Q}_h = K_{\text{tot}}(T_i - T_o)$ if this heating system is to do its job. This implies the relation

$$T_{\text{sup}} - T_i = (T_i - T_o)\frac{K_{tot}}{\dot{V}\rho c_p} \tag{14.10}$$

The supply air temperature T_{sup} would be equal to T_i only in the limit of infinite flow rate. In practice, typical values of the supply air are 30 to 40°C (86 to 104°F) for heating and 12 to 16°C (53.6 to 60.8°F) for cooling. To find the resulting COP of the heat pump, we replace T_i by T_{sup} in Eq. (14.7). The corresponding results for cooling can be calculated in an analogous manner.

Example 14.2

Find the supply air temperature for a house with a typical value $K_{\text{tot}} = 250$ W/K [474 Btu/(h · °F)] if $T_i = 20°C$ (68°F), $T_o = 0°C$ (32°F), and the airflow rate of the heating system is $\dot{V} = 0.42$ m³/s.

Given: $T_i, T_o, K_{tot}, \dot{V}$

Find: T_{sup}

Lookup values: $\rho C_p = 1.2$ kJ/(m³ · K)

SOLUTION

The quantity $\dot{m}c_p = \dot{V}_p c_p = 0.42$ m³/s $\times$ 1.2 kJ/(m³ · K) $= 0.5$ kW/K. Inserting this into Eq. (14.10), we obtain

$$T_{\text{sup}} = 20°C + (20°C - 0°C)\frac{250\,\text{W/K}}{0.5\,\text{kW/K}} = 30°C = 303.15\,\text{K}\,(86°F)$$

The corresponding COP is

$$COP = \frac{T_{sup}}{T_{sup} - T_o} = \frac{303.15}{303.15 - 273.15} = 10.11$$

which is significantly less than the value of 14.7 with infinite flow rate.

14.3.3 The ΔT of Heat Exchangers

Further approaching realism, Fig. 14.4 joins heat exchangers to the heat pump, one between the outside air and the cold side (= evaporator) and one between the hot side (= condenser) and the indoor air. The working fluid is isothermal in the evaporator and in the condenser, since we are assuming either a Carnot cycle (for now) or a Rankine cycle (later). In terms of the heat exchanger effectiveness ε of Eq. (3.7) the heat flow is given by

$$\dot{Q}_h = \varepsilon_i \dot{C}_i (T_{cond} - T_{sup}) \tag{14.11}$$

where

ε_i = heat exchanger effectiveness on indoor air side
$\dot{C}_i = (\dot{m}c_p)_i$ = heat capacity–flow rate product on indoor air side
T_{sup} = supply air temperature of building
T_{cond} = temperature of condenser, which is higher than T_{sup} of Eq. (14.10) because ε_i is less than unity

For the outdoor side, one has an analogous equation

$$\dot{Q}_h = \dot{W}_h = \varepsilon_o \dot{C}_o (T_o - T_{evap}) \tag{14.12}$$

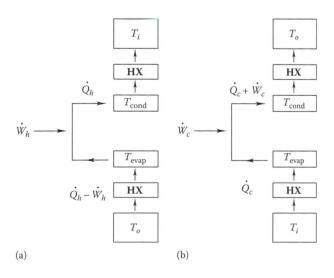

(a) (b)

FIGURE 14.4

Thermodynamics of heat pump with heat exchangers (HX) between heat pump and air. (a) For heating; (b) for cooling.

The heat exchanger effectiveness for these equations can be obtained from Fig. 2.15 by setting $\dot{C}_{min}/\dot{C}_{max} = 0$ (since $\dot{C}_{max}$ is, in effect, infinite in the condenser and in the evaporator, while $\dot{C}_{min}$ is the smaller $\dot{m}c_p$ of the respective airstreams). Still assuming the heat pump to be a Carnot heat pump, we can calculate the COP of this system by solving the last two equations for T_{cond} and T_{evap} and by using these temperatures instead of T_i and T_o in Eq. (14.7)

$$\text{COP} = \frac{T_{cond}}{T_{cond} - T_{evap}} \tag{14.13}$$

While T_{cond} can be obtained directly from Eq. (14.11), we first have to eliminate $\dot{W}_h$ from Eq. (14.12) before we can solve for T_{evap}. This can be done by setting

$$\dot{W}_h = \dot{Q}_h \frac{T_{cond} - T_{evap}}{T_{cond}}$$

according to Eq. (14.13). Inserting this into Eq. (14.12), one obtains the result for T_{evap}

$$T_{evap} = \frac{\varepsilon_o \dot{C}_o}{\varepsilon_o \dot{C}_o + \varepsilon_i \dot{C}_i (1 - T_{sup}/T_o)} T_o \tag{14.14}$$

Example 14.3

Find the COP of a Carnot heat pump for the conditions of Example 14.2 if $\dot{C}_i = \dot{C}_o = 0.5$ kW/K, and if the heat exchangers are of identical counterflow design, with areas $A_{HXi} = A_{HXo} = 4$ m^2, and heat transfer coefficients $U_{HXi} = U_{HXo} = 200$ W/(m$^2 \cdot$ K).

Given: $T_{sup} = 30°C$, $T_o = 0°C$,

$(K_{tot})_{HXi} = (K_{tot})_{HXo} = 0.80$ kW/K

$\dot{C}_i = \dot{C}_o = 0.50$ kW/K   on indoor and outdoor sides

Find: COP

Lookup values: Heat exchanger effectiveness from $C_{min}/C_{max} = 0$ curve of Fig. 2.15 for NTU = $(K_{tot})_{HX}/\dot{C} = 1.6$ is $\varepsilon = 0.80 = \varepsilon_i = \varepsilon_o$.

SOLUTION

The heating load of Example 14.2 is

$$\dot{Q}_h = (20°C - 0°C)(250 \text{ W/K}) = 5 \text{ kW}$$

Solving Eq. (14.11) for T_{cond}, we obtain

$$T_{cond} = T_{sup} + \frac{\dot{Q}_h}{\varepsilon_i \dot{C}_i} = 30°C + \frac{5 \text{ kW}}{0.80 \times 0.5 \text{ kW/K}} = 42.50°C$$

$$= 273.15 + 42.50 \text{ K} = 315.65 \text{ K}$$

From Eq. (14.14), noting that $\varepsilon_i \dot{C}_i = \varepsilon_o \dot{C}_o$ and that the temperature ratio must be evaluated in unit of Kelvins, we obtain

$$T_{evap} = \frac{1}{1 + 1 - T_{sup}/T_{cond}} T_o = \frac{1}{1 + 1 - 303.15/315.65} T_o = 0.962 \, T_o = 262.75 \, K$$
$$= -10.40°C$$

Finally the COP is

$$COP = \frac{T_{cond}}{T_{cond} - T_{evap}} = \frac{315.65 \, K}{315.65 \, K - 262.75 \, K} = 5.97$$

COMMENT

This is only about 40% of the value 14.7 with infinite flow rates and without heat exchanger penalties.

14.3.4 Real Heat Pumps

Practical heat pumps and air conditioners (of the compression type) are based on the Rankine cycle, as discussed in Chap. 3. The COP of an ideal Rankine heat pump with phase-change temperatures T_{cond} and T_{evap} can be written as a fraction f_{Rank} of the COP of the Carnot heat pump with upper and lower temperatures T_{cond} and T_{evap}:

$$COP_{Rank, ideal} = f_{Rank} \frac{T_{cond}}{T_{cond} - T_{evap}} \tag{14.15}$$

The factor f_{Rank} depends on the working fluid, but does not vary strongly over a typical range of operating temperatures. For heat pumps and air conditioners with Freon f_{Rank} is in the range of 0.7 to 0.8.

The COP of a real Rankine engine is further reduced by a factor f_{loss} because of various losses. The latter are dominated by friction in the compressor, so much so that one can approximate f_{loss} by the compressor efficiency. Typical values are also in the range of 0.7 to 0.8. Since this loss factor is relatively insensitive to temperature, one can approximate the efficiency of a real Rankine cycle as

$$COP_{Rank, real} = f_{Rank} f_{loss} \frac{T_{cond}}{T_{cond} - T_{evap}} \tag{14.16}$$

Comparison with results in Fig. 9.11 shows that this approximation is good to about 10 percent over the range of typical operating temperatures. Since typical values of $f_{Rank} f_{loss}$ are around 0.5, we can extract the rule of thumb that *the COP of real heat pumps and air conditioners is about one-half of the Carnot COP based on the actual temperatures of the evaporator and condenser.* Thus for the conditions of Example 14.3, the COP of a real heat pump can be expected to be around 3.0. If the areas or heat transfer coefficients of the heat exchangers are smaller than in Example 14.3, the COP is even lower.

14.3.5 Multizone Efficiency Index

A major cause of inefficiency in multizone[4] buildings is the simultaneous heating and cooling of different zones. To evaluate the magnitude of this problem, it is appropriate to compare the performance of the system in question with the performance of an "ideal" one-zone design where perfect mixing of the air ensures that all heat gains are utilized to offset heating loads before resorting to the economizer or chiller. All systems except the secondary HVAC system are kept the same in this comparison, in particular, the building envelope and the schedules for occupancy, thermostat, and minimum outdoor air requirement. Thus we define a *multizone efficiency index* $\eta_{\text{multiz},h}$ for heating as the ratio of the annual heating energy $Q_{h,1z}$ for the ideal one-zone system divided by the annual heating energy Q_h for the system in question, with analogous quantities for cooling:

$$\eta_{\text{multiz},h} = \frac{Q_{h,1z}}{Q_h} \quad \text{and} \quad \eta_{\text{multiz},c} = \frac{Q_{c,1z}}{Q_c} \tag{14.17}$$

The highest possible value of this index is unity. The smaller this index, the greater the overconsumption due to simultaneous heating and cooling. To illustrate the calculation and use of this index, consider the economizer operation of a multizone building.

Example 14.4

Consider a building that can be characterized by the following quantities and assumptions:

- Conductive heat loss coefficient $K_{\text{cond}} = 15$ kW/K.
- Total heat loss coefficient $K_{\text{tot}} = K_{\text{min}} = 15 + 10 = 25$ kW/K during cold weather (economizer off).
- $K_{\text{tot}} = K_{\text{max}} = 15 + 25 = 40$ kW/K during mild weather (economizer on).

(These numbers are representative of the Enerplex North, a 12,000-m^2 office building in New Jersey, which the authors have monitored in detail.)

To allow simple steady-state analysis, assume the building is occupied and conditioned around the clock, with $\dot{Q}_{\text{gain}} = 300$ kW and $T_i = 20°C$.

Assumptions: There is a single air handler, the economizer control is designed to maintain constant supply air temperature, and the return air temperature is equal to T_i.

Then the economizer equations of Sec. 3.2.2 (Example 3.6) are applicable, and the graph of $\dot{Q}$ versus T_o has a slope of K_{cond} when $T_{\text{min}} < T_o < T_{\text{max}}$ (economizer regime). When $T_o < T_{\text{min}}$, the economizer is turned off and the slope is K_{min}. This is shown in Fig. 14.5, where two consumption patterns are indicated, as labeled: economizer (between T_{min} and T_{max}, excess heat gains are vented by economizer) and "ideal" one-zone (assuming perfect mixing within building).

Find: $\eta_{\text{multiz},h}$ of Eq. (14.17)

Lookup values: Annual heating degree-hours for several bases (e.g., from bin data of HCB software):

$D_h(12.5°C) = 36,000$ K · h $D_h(8°C) = 20,440$ K · h

$D_h(1.25°C) = 5840$ K · h

---

[4] This discussion applies to any HVAC system for multiple zones (the "multizone system" described in Figure 11.22a is merely one particular, and rather inefficient, secondary HVAC system).

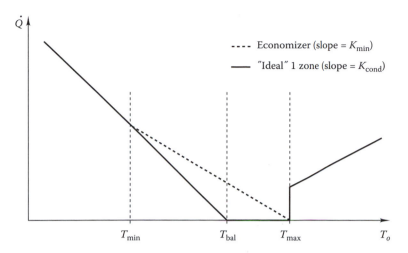

FIGURE 14.5
Energy use for heating and cooling of multizone building with economizer, $\dot{Q}$ versus T_o per Example 14.4.

SOLUTION

The economizer is turned off when T_o is below T_{min}:

$$T_{min} = \frac{K_{min} T_{bal} - K_{cond} T_{max}}{K_{min} - K_{cond}} \qquad (14.18)$$

The annual heating consumption $\dot{Q}_{h,1z}$ is easy to find because it corresponds to the straight line for $T_o < T_{bal}$. Hence it is given by

$$\dot{Q}_{h,1z} = K_{min} F_h(T_{bal}) \qquad (14.19)$$

where $D_h(T_{bal}) =$ annual heating degree-hours for base T_{bal} and

$$T_{bal} = T_i - \frac{\dot{Q}_{gain}}{K_{min}} \qquad (14.20)$$

The annual heating consumption $\dot{Q}_h$ with the economizer mode is a little more complicated because it corresponds to the solid line for $T_o < T_{min}$ and to the dashed line for $T_{min} < T_o < T_{max}$. It is given by

$$\dot{Q}_h = K_{cond} D_h(T_{max}) + (K_{min} - K_{cond}) D_h(T_{min}) \qquad (14.21)$$

The first term in this equation would be the answer if the consumption could follow the dashed line even below T_{min}; the second term adds the correction below T_{min}.

Inserting numerical values, we find

$$T_{bal} = 20 - \frac{300}{25} = 8°C \qquad D_h(8°C) = 20{,}440\ \text{K} \cdot \text{h}$$

and

$$T_{max} = 20 - \frac{300}{40} = 12.5°C \qquad D_h(12.5°C) = 36,000\,K \cdot h$$

The economizer is turned off when T_o is below T_{min}:

$$T_{min} = \frac{25 \times 8 - 15 \times 12.5}{25 - 15} = \frac{200 - 187.5}{10} = 1.25°C$$

The corresponding degree-hours are $D_h(1.25°C) = 5840\,K \cdot h$.
  The heating loads are

$$\dot{Q}_{h,1z} = 25\,kW/K \times 20,440\,k \cdot h = 511\,MWh$$

$$\dot{Q}_h = 15\,kW/K \times 36,000\,k \cdot h + (25 - 15)\,kW/K \times 5840\,k \cdot h = 598\,MWh$$

The multizone index for the economizer mode is

$$\eta_{multiz,h} = \frac{\dot{Q}_{h,1z}}{\dot{Q}_h} = \frac{511}{598} = 0.85$$

The overconsumption during the heating season is $598 - 511 = 87$ MWh, or $87/598 = 15$ percent.

COMMENTS

There is overconsumption due to multiple zones even though the chiller does not run at all. In effect, the economizer is a form of simultaneous heating and cooling because it increases the airflow for the entire building even though some zones require heating. This penalty could be avoided by means of heat recovery and thermal storage, and the overconsumption index quantifies the potential savings. A simple solution that avoids at least some of this overconsumption is to install two separate air handlers, one for the interior zone and one for the exterior zone.

14.4 Measured Performance

14.4.1 Data Sources

Measured performance data are crucial. They are the ultimate test of predictions, and they provide reference values to guide the design of new buildings. While we have already discussed in Sec. 8.5 the comparison between measured and predicted consumption, we bring up measured data again here to give the designer an appreciation of issues such as the dependence of energy consumption on building type and end use, the relation between construction cost and energy consumption, and the difficulties with the interpretation of data, e.g., due to ambiguities in the definition of floor area. There are a wide range of data types and data sources. At one end of the spectrum are the meter readings of the utility companies.

They are very reliable and accurate, but contain little information beyond the total consumption of a building during the billing interval (monthly in the United States, even longer time intervals in many other countries). For large consumers, the electricity consumption will be metered separately during peak (daytime weekdays), intermediate, and off-peak periods (nights and weekends), and the peak instantaneous power may be monitored as well.

Utility companies meter the consumption of each customer, but the data remain confidential between the company and the customer and are not available to others. The data become accessible only to the extent that they are communicated to organizations such as the Building Owners and Managers Association (BOMA) that publishes the data (BOMA, 1991). The most extensive compilations of consumption data have been published periodically by the Energy Information Administration (EIA, 1983). A summary, in Fig. 14.6, shows that the average consumption per unit floor area varies threefold between different building types, with residential and assembly being lowest and food sales and health care highest.

Such databases provide only aggregated information, such as the size of buildings, energy type, and annual energy, per building or unit of floor area. A designer would like far greater detail than that. In fact, the ideal database would be so detailed and precise that one could identify optimal strategies by correlating each design feature with measured performance and cost. In practice, the data are almost never sufficiently detailed and precise, and the correlations drown in noise. The greater detail one wants, the greater the effort and cost. Submetering is rare, even though it is most desirable to identify heating, cooling, lights, equipment, etc., and to correct for variations in weather or operating schedule. The separation of heating and cooling consumption is difficult in commercial

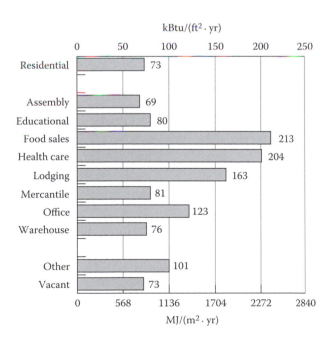

FIGURE 14.6
Annual energy consumption per unit of floor area, average by building type in the United States. (Courtesy of EIA, Nonresidential Buildings Energy Consumption Survey (NBECS): 1979 Consumption and Expenditures, Part 2, EIA Report DOE/EIA-0138(79)/2 (December), Energy Information Administration, U.S. Department of Energy, Washington, DC, 1983.)

buildings unless different fuels are used. In the future, the task of data collection and analysis will become increasingly easy thanks to advances in computerized building management systems. ASHRAE (1990) has established a standard for measuring and expressing building energy performance.

Note that the consumption observed during the during start-up year(s) is often higher than expected, but decreases as operators learn how to run a building more efficiently. The "bake-out" of moisture in building materials and the warm-up of the soil and basement can also cause abnormally high consumption during the first year.

To circumvent these limitations of purely empirical data, an interesting alternative has been developed by Crawley and Huang (1989). The building stock is characterized in terms of a small number of prototypical building descriptions, and these prototypes are then simulated by DOE2.1 to obtain a systematic database of disaggregated building loads. The multifamily buildings sector, e.g., has been characterized by 16 prototypes, and these data are available on diskette. With such data it is easy for researchers to evaluate the performance of modifications of a building or its equipment.

For measured data, the BECA (Buildings Energy Compilation and Analysis) database compiled by Lawrence Berkeley Laboratory (LBL) is a particularly valuable source. It covers both the residential sector and the commercial sector, and a brief summary is presented below. To be included in BECA, the data must meet certain requirements of detail and quality. For instance, both electric and fuel consumption must be reported so that different thermodynamic qualities can be taken into account.

It is interesting to compare buildings in a plot like Fig. 14.7 where electric consumption is on one axis and fuel on the other. Of course, such a comparison should be interpreted with some care in view of the wide variety of building types and climates. For example, even the most energy-efficient building will consume much electricity if it contains a high density of computing equipment. Also in Fig. 14.7 no explicit breakdown by end use is shown

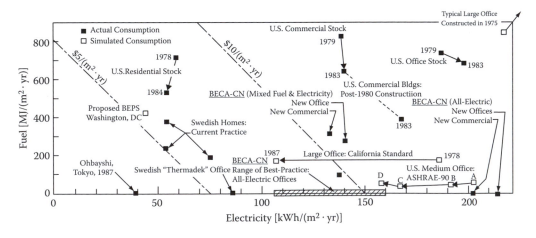

FIGURE 14.7
Annual fuel and electricity consumption (all end uses) per floor area, for several building types (1 kBtu/ft² = 11.36 MJ/m² = 3.156 kWh/m²). Vertical axis shows fuel; horizontal axis, electricity. Both scales are proportional to primary energy equivalents. The diagonal lines for total energy cost (at $5 and $10 per square meter per year) are based on $0.067 per kilowatthour for electricity and $6.25 per gigajoule for fuel. (Courtesy of Harris, Personal communication, 1990.)

(although fuel usually serves for space heating). To facilitate cost comparisons, lines of equal annual cost are drawn.

Figure 14.7 combines measured and calculated data, for several building types in different countries. One can see that the consumption by buildings in the United States has decreased markedly since the 1970s. Compared to U.S. homes, Swedish homes use only about one-half as much fuel despite the much colder climate. Commercial buildings consume much more electricity than homes, mostly because of higher lighting loads. That they also tend to require more fuel may appear perverse: With higher internal heat gains and a lower surface-to-volume ratio, their heating loads should be much lower. That is a manifestation of simultaneous heating and cooling loads in different zones of commercial buildings, a problem we already mentioned. Among commercial buildings the record for low consumption is held by the Ohbayashi Building, a demonstration project in Tokyo, Japan.

14.4.2 Residential Buildings

The BECA data for single-family houses are summarized in Fig. 14.8, which shows annual heating energy per unit floor area versus degree-days (from Ribot and Rosenfeld, 1982).

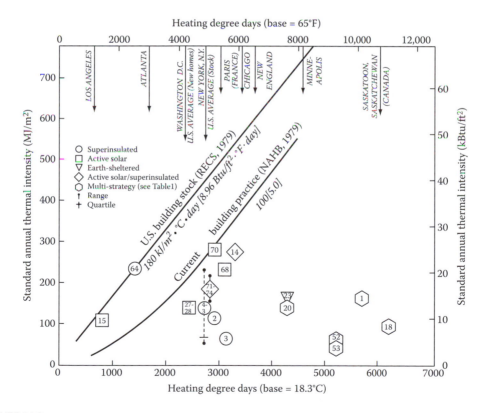

FIGURE 14.8

Annual heating energy consumption per unit of floor area for single-family houses (conventional construction and BECA). Data are normalized for standard interior temperature of 20°C and standard nonsolar heat gains of 1 kW. Further information on the 37 BECA houses, as identified by the numbers, can be found in Ribot and Rosenfeld (1982).

To make the comparison of different houses as meaningful as possible, the heating consumption has been normalized by assuming $T_i = 20°C$ and a standard nonsolar heat gain of 1 kW per house. The plot demonstrates vividly that a great deal of energy could be saved if the existing U.S. stock were replaced by the best available technology. While current building practice presents already a major advance, demonstration projects prove that much more is possible. For instance, the best Canadian homes in a climate with 6,000 K·days require no more than traditional homes in a climate with about one-tenth the number of degree-days.

14.4.3 Commercial Buildings

Piette and Riley (1986) and Piette et al. (1986) have published summaries of the BECA-CN (commercial new) data for new commercial buildings that were designed for energy efficiency. Their first report describes 133 buildings, followed by a second report with 152 buildings (103 office buildings, 21 schools, and a variety of other types). Let us look at their results.

Figure 14.9 shows performance for offices and schools, as site energy intensity in kBtu per square foot per year. In addition to all offices, a breakdown is shown by the following categories: large, small, with and without parking, with and without computer center. For each of these categories, several indicators of the statistical distribution are shown: median, average, ±1 standard deviation, maximum, and minimum. The average for all offices is 63 kBtu/(ft^2·yr), about one-half of the average of the existing stock

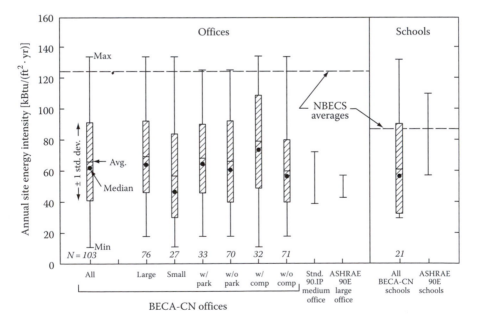

FIGURE 14.9

Site energy intensity for energy-efficient buildings (103 offices and 21 schools) of BECA database. The office category is further disaggregated according to large, small, with and without parking, and with and without computer center. The dashed line shows the average of existing U.S. stock in 1979. Simulation estimates for ASHRAE Standard 90 are also indicated. 1 kBtu/ft^2 = 11.36 MJ/m^2. (Courtesy of Piette, M.A. and Riley, R., Energy Use and Peak Power for New Commercial Buildings from the BECA-CN Data Compilation: Key Findings and Issues, Report LBL: 20896 (March), Lawrence Berkeley Laboratory, Berkeley, CA, 1986.)

(the NBECS survey of EIA, 1983), indicated by the dashed line at 124 kBtu/(ft² · yr). In fact, the performance of these buildings is close to that specified at that time by ASHRAE Standard 90. However, the range of energy intensities is large, and some buildings consume more than the NBECS average.

For schools, the performance of the existing stock is closer to the ASHRAE standard and to the average of the BECA-CN sample, but even here the BECA-CN sample suggests that there is much room for improvement.

For a slightly earlier and smaller sample, Table 14.4 from Piette and Riley (1986) presents another view of the BECA-CN data, broken down by different categories of commercial buildings. In addition to site energy, the resource energy is listed, calculated by multiplying electricity consumption by 3 to account for conversion losses.

The average annual energy cost for the BECA-CN buildings is $1.02 per square foot, much lower than the values of $1.50 and $1.93 per square foot reported, respectively, by the NBECS (EIA 1983) and by the BOMA (1991) surveys (the latter for large offices). Also of interest, because of demand charges, is the *electric load factor*, defined as the annual average load divided by the peak load. For the BECA-CN buildings it ranges from less than 0.3 to 0.7.

TABLE 14.4

Site Energy and Resource Energy Intensity for BECA-CN Buildings, and Comparison with NBECS Data for U.S. Stock in 1979 and with Simulations for ASHRAE Standard 90.1.
(1 kBtu/ft² = 11.36 MJ/m²)

Building Type	Site Intensity, kBtu/(ft² · yr)						
	BECA-CN				ASHRAE[†]	NBECS[‡]	
	N	Mean*	Median	Range	Std. 90.1	Mean	Range
Large office	69	61	59	15–129	43–57		
Small office	19	55	47	11–135	39–51	124	83–147
College	3	81	74	62–108		87	71–95
Secondary school	6	73	60	38–129	50–110	87	71–95
Elementary school	9	47	42	32–68		87	71–95
Retail store	8	77	69	39–134	57–75	87	82–92
Warehouse	4	42	40	26–62	43–89	108	53–158
Other	15	61	60	17–99		184	106–131
Total	133	60	56				

Source: Courtesy of Piette, M.A. and Riley, R., Energy Use and Peak Power for New Commercial Buildings from the BECA-CN Data Compilation: Key Findings and Issues, Report LBL: 20896 (March), Lawrence Berkeley Laboratory, Berkeley, CA, 1986.

* All BECA-CN means are unweighted; i.e., each building's energy intensity is weighted equally regardless of floor space.

[†] ASHRAE values are based on simulations for prototype buildings and are only rough approximations of how buildings would perform under the standard. The ranges include seven climate zones and two alternate HVAC systems (Battelle PNL, 1983). The range also incorporates Standard 90E both with and without daylighting. The values listed for a small office are based on a 49,500-ft² three-story building, which is near the BECA-CN size limit for small offices of 50,000 ft². The retail ranges are based on a mall department store, similar in type to most of the BECA-CN retail data. For educational buildings, the range is derived from a junior high prototype, although BECA-CN includes elementary schools, secondary schools, and colleges.

[‡] NBECS (EIA, 1981) does not distinguish between large and small offices; however, the average U.S. office from this sample has a floor area of 13,700 ft². NBECS does not distinguish between colleges, secondary schools, and elementary schools. All-electric buildings are averaged with those using mixed fuels. The ranges include subaverages across four U.S. census regions.

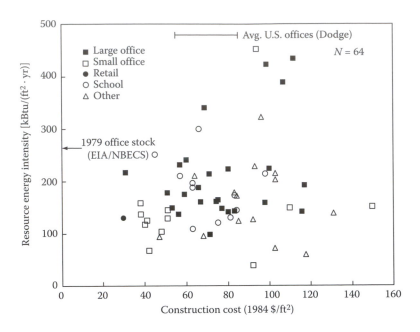

FIGURE 14.10

Resource energy intensity versus construction costs for new commercial buildings of BECA database. For comparison, national averages for construction costs ($55 to $85 per square foot) and for energy intensity are shown. Low-energy buildings do not necessarily have higher construction cost. 1 kBtu/ft² = 11.36 MJ/m² and $1/ft² = $10.76/m². (Courtesy of Piette, M.A. and Riley, R., Energy Use and Peak Power for New Commercial Buildings from the BECA-CN Data Compilation: Key Findings and Issues, Report LBL: 20896 (March), Lawrence Berkeley Laboratory, Berkeley, CA, 1986.)

Are energy-efficient buildings more expensive to construct? Certainly many efficiency features do increase the first cost. But as Figure 14.10 shows, this is not visible when one simply looks at total construction costs. The scatter is large, and no clear correlation between cost and energy performance is visible. The costs of these buildings range from $30 to $150 per square foot, compared to a national average between $55 and $85. The cost of a building is determined by many other factors beyond the design features that affect energy performance.

Piette and others have tried to correlate energy performance with many other variables. Most of them do not display any clear trends. Building size, e.g., might affect energy consumption. But as can be seen in Fig. 14.11, the main impression is a large dispersion of points. The only obvious pattern is an increased scatter for small buildings.

Collecting and interpreting such data are not without difficulties. For instance, a rather troubling ambiguity in all these performance data is the definition of the floor area. Is it gross area, net area, or conditioned area, and does it include or exclude the parking area (which may be partially conditioned)? Figure 14.12 shows that this point can easily change the reported energy-per-area values by more than 20 percent.

Another problem arises from year-to-year variations, due to weather or utilization. Rigorous weather correction requires submetered data for heating and cooling. While an energy signature model such as PRISM (Fels, 1986; Rabl and Rialhe, 1992) can try to identify these components from the total consumption, this is fraught with risk in commercial buildings with large base loads. In many commercial buildings, changes due to lighting or equipment have been found to overwhelm variations due to weather (Rabl et al., 1986; Piette et al., 1986).

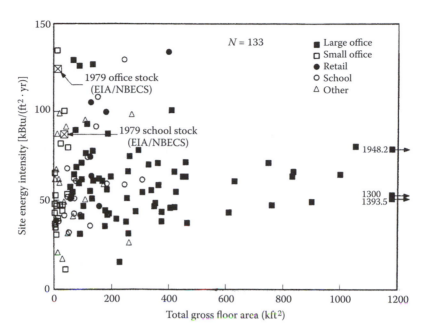

FIGURE 14.11
Site energy intensity versus building size for 133 new commercial buildings of BECA database. Small buildings are comparable on average to large buildings, but they show greater variation. 1 kBtu/ft^2 = 11.36 MJ/m^2 and 1 ft^2 = 0.0929 m^2. (Courtesy of Piette, M.A. and Riley, R., Energy Use and Peak Power for New Commercial Buildings from the BECA-CN Data Compilation: Key Findings and Issues, Report LBL: 20896 (March), Lawrence Berkeley Laboratory, Berkeley, CA, 1986.)

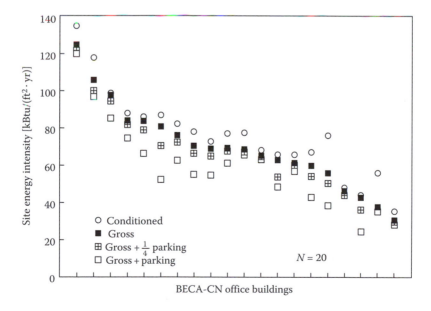

FIGURE 14.12
Sensitivity of site energy intensity to floor-area definitions. The different values resulting from four different definitions are shown for 20 large office buildings. (Courtesy of Piette, M.A. and Riley, R., Energy Use and Peak Power for New Commercial Buildings from the BECA-CN Data Compilation: Key Findings and Issues, Report LBL: 20896 (March), Lawrence Berkeley Laboratory, Berkeley, CA, 1986.)

14.5 Residential Buildings

Most residential HVAC systems are relatively simple because they are designed to serve a single zone. This tends to be true even for multifamily buildings, with each residence having its own single-zone HVAC system. The loads are dominated by the building envelope. Thus the basic recommendations for efficiency are insulation, airtightness, and efficient equipment, as we discussed at the start of this chapter. The prime example is Swedish housing. Their infiltration rate is so low that mechanical ventilation is used, with heat recovery from the exhaust airstream.

For residential buildings, the only major design complication arises from windows and the interaction with thermal mass. As the window area is increased, one gains free solar heat but at the price of increased conductive loss and cooling load. The tradeoffs are not obvious. In particular, they depend on the thermal mass of the building. The role of the relevant window variables (area, orientation, glazing type, and shading device) for lightweight houses is discussed next. Then we consider, in Sec. 14.5.2, the effect of varying the thermal mass of the walls. This will be followed by discussion of passive solar buildings where the heat capacity is greatly increased to improve the utilization of solar heat.

14.5.1 Lightweight Houses: Energy and Windows

Here we cite an interesting series of papers by Sullivan and Selkowitz (1985, 1987). Using the DOE2.1 simulation program with hourly weather data, these authors have analyzed the effects of varying the window size, glazing type, and building orientation for a single-family house of typical U.S. construction. The dimensions of this ranch-style house are shown in Fig. 14.13; the height is 8.0 ft (2.44 m), with an additional 5.35 ft (1.63 m) for the roof structure. The wall and roof are of wood-frame construction with insulation levels of $R = 11$ and 30 $(\text{h} \cdot \text{ft}^2 \cdot {}^\circ\text{F})/\text{Btu}$, respectively $[U = 0.52$ and 0.19 $\text{W}/(\text{m}^2 \cdot \text{K})]$. The major portion of the heat capacity lies in the slab-on-grade floor: a carpet-covered 4-in (0.1-m) thick slab of concrete on top of insulation and gravel.

The window sizes are fixed on three sides at a total of 8.65 percent of the floor area, while the size of a fourth window, called the *primary window*, is varied, such that the grand-total window area varies from 8.65 to 25.79 percent of the floor area. The glazing type is also varied, on all windows at the same time. To test the effect of orientation, the house is rotated in 45° increments, the primary window facing south, southwest, etc. The simulations were carried out for three locations: Madison, Wisconsin (cold continental climate); Lake Charles, Louisiana (hot and humid climate); and Phoenix, Arizona (hot and dry climate). Here we present an extract of their results.

Sullivan and Selkowitz employed an interesting technique to enhance the flexibility of presenting and using the results. Normally, each simulation yields results only for one set of input parameters, but here one would like the possibility of varying area A, conductance U, and shading coefficient SC continuously. The authors used linear least-squares regression analysis of simulation results at a few discrete values to obtain simple functional forms that summarize the essential trends. Thus they obtained polynomials of first and second order for the annual heating and for cooling energy as a function of AU and $A \times \text{SC}$ of the windows, as well as internal gains, air change, and conductance of opaque surfaces. The functional form is linear in all these variables, except for a quadratic term in $A \times \text{SC}$ for the primary window. The seasonal energy consumption includes the

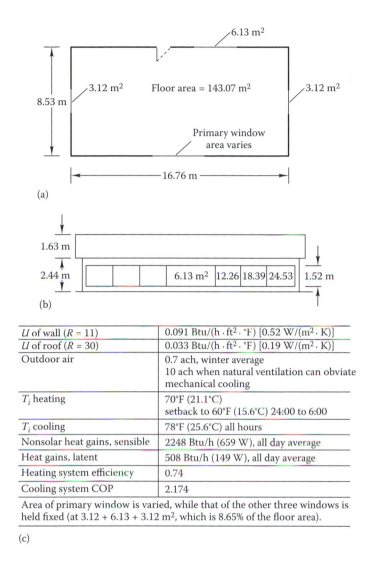

FIGURE 14.13
The house analyzed by Sullivan and Selkowitz (1985, 1987): (a) floor plan, (b) facade with primary window (placed symmetrically; numbers indicate sizes considered in this study), and (c) principal assumptions.

performance of the HVAC system: An efficiency of 0.74 for heating and a COP of 2.174 for cooling are assumed.

The resulting equation makes it easy to assess the role of the windows. In Figure 14.14, the annual energy for heating and cooling is plotted versus primary window area, with separate curves for different values of U and SC of the glazing. The curves shown here assume that the primary window faces south for a cold climate (Madison) and north for a hot climate (Lake Charles). The U values shown—1.006 Btu/(h·ft²·°F) [5.713 W/(m²·K)], 0.471 Btu/(h·ft²·°F) [2.675 W/(m²·K)], and 0.302 Btu/(h·ft²·°F) [1.715 W/(m²·K)]— correspond roughly to single, double, and triple glazing, while 0.094 Btu/(h·ft²·°F)

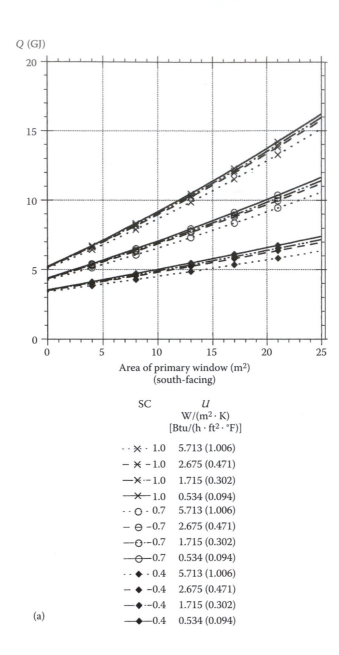

FIGURE 14.14

(a) Annual energy consumption for heating and cooling of the house analyzed by Sullivan and Selkowitz (1985) as a function of area of primary glazing, for several values of shading coefficient SC and of U value of glazing (same glazing on primary and nonprimary windows). Note different scales in different graphs. Cold climate (Madison, Wisconsin), cooling.

[0.534 W/(m$^2 \cdot$ K)] could be achieved with aerogel or evacuated windows. Interpolating between the curves, one can determine the consumption for any other window type.

The figures show that the heating load is very sensitive to the U value: The smaller U, the lower the heating load (even in Lake Charles, although here the magnitude is fairly small). The cooling load, by contrast, varies little with U; the slight decrease with increasing U is

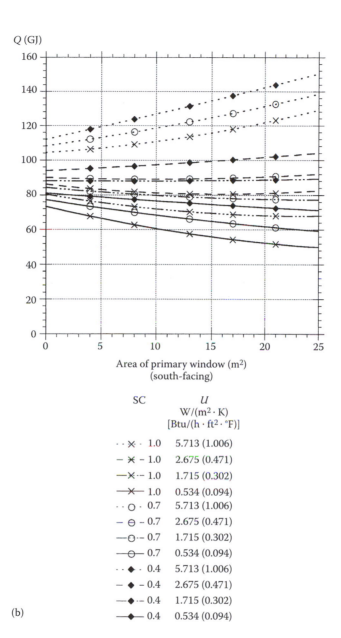

FIGURE 14.14 (continued)
(b) Cold climate (Madison, Wisconsin), heating.

(continued)

due to the fact that the average of T_o is slightly below that of T_i [$=78°F$ (25.6°C)] during the cooling season, even in Lake Charles. To understand these trends, note that heating loads are dominated by the indoor-outdoor temperature difference whereas cooling loads arise largely from heat gains, with the average indoor-outdoor temperature difference during the cooling season being rather small.

The variation with shading coefficient is as expected: A high SC reduces heating loads and increases cooling loads. Which load carries more weight in the total energy bill

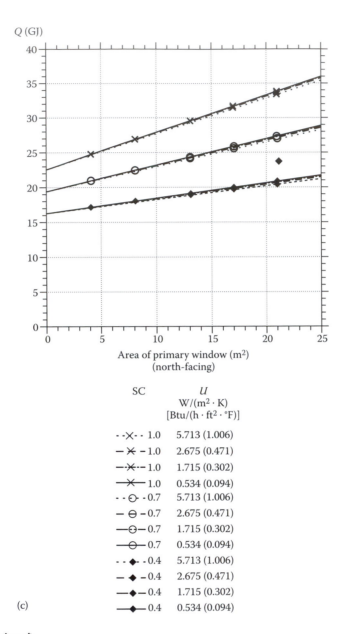

FIGURE 14.14 (continued)
(c) Hot climate (Lake Charles, Louisiana), cooling.

depends on the climate and on energy prices. Obviously, in hot climates a low SC is needed for low consumption.

Whether solar gains or conductive losses are more important can be seen from the variation of the heating load with the window area. With a single pane, the heating load increases with area; with a double pane, it is roughly constant; and when U is even lower, the windows bring a definite gain during the heating season (for the orientations and climates considered here). But note that the cooling load will invariably increase with the

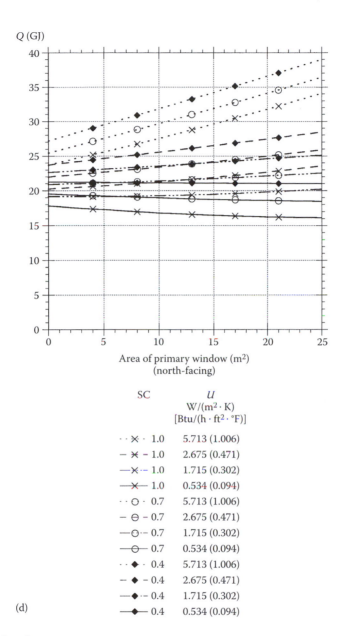

Q (GJ)

Area of primary window (m²)
(north-facing)

SC	U W/(m²·K) [Btu/(h·ft²·°F)]
··✕· 1.0	5.713 (1.006)
─✕─ 1.0	2.675 (0.471)
─✕·─ 1.0	1.715 (0.302)
─✕─ 1.0	0.534 (0.094)
··○· 0.7	5.713 (1.006)
─⊝─ 0.7	2.675 (0.471)
─○·─ 0.7	1.715 (0.302)
─⊖─ 0.7	0.534 (0.094)
··◆· 0.4	5.713 (1.006)
─◆─ 0.4	2.675 (0.471)
─◆·─ 0.4	1.715 (0.302)
─◆─ 0.4	0.534 (0.094)

(d)

FIGURE 14.14 (continued)
(d) Hot climate (Lake Charles, Louisiana), heating.

window area (unless it is compensated by a reduced lighting load, which is not considered here because it is unlikely in residential buildings). *Therefore, if windows are proposed as energy savers, evaluation of their consequences for cooling is imperative.*

To show the effect of orientation, Fig. 14.15 plots the incremental annual energy as a function of the orientation of the primary window, with the incremental annual energy being the change in annual heating and cooling energy as a result of increasing the primary window area from 0 to 198 ft² (18.39 m²). In this figure, specific glazing types are assumed, as listed in Table 14.5. For cooling in Madison, only two glazings are

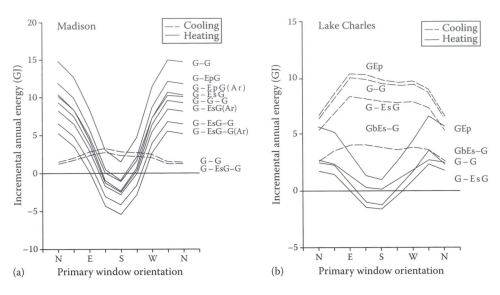

FIGURE 14.15

Change in annual heating and cooling energy as a result of increasing the primary window area from 0 to 18.39 m² (198 ft²) for the house analyzed by Sullivan and Selkowitz (1985) as a function of orientation of primary glazing, for glazing types of Table 14.5. (a) Cold climate (Madison, Wisconsin); (b) hot climate (Lake Charles, Louisiana).

TABLE 14.5

Glazing Properties Assumed for Fig. 14.15

Window Design	Gas Fill	Winter *U* Value	Summer *U* Value	Shading Coefficient	Solar Transmittance	Visible Transmittance
[†]G-G	Air	2.85 (0.50)	3.16 (0.56)	0.88	0.71	0.82
G-G-G	Air	1.86 (0.33)	2.20 (0.39)	0.79	0.61	0.74
G-EpG	Air	2.34 (0.41)	2.63 (0.46)	0.86	0.64	0.73
[†]G-EsG	Air	1.94 (0.34)	2.00 (0.35)	0.73	0.58	0.74
G-EpG	Argon	2.09 (0.37)	2.38 (0.42)	0.86	0.64	0.73
G-EsG	Argon	1.62 (0.28)	1.68 (0.30)	0.73	0.58	0.74
G-EsG-G	Air	1.32 (0.23)	1.53 (0.27)	0.71	0.52	0.71
G-EsG-G	Argon	1.11 (0.19)	1.30 (0.23)	0.72	0.52	0.71
[†]GEp	—	5.05 (0.89)	4.54 (0.80)	0.92	0.75	0.80
[†]GbEs-G	Air	1.94 (0.34)	2.05 (0.36)	0.37	0.26	0.46

Source: Courtesy of Sullivan, R. and Selkowitz, S. *ASHRAE Trans.*, 91, pt. 2A, 320, 1985.

Notes: 1. The units of the *U* value are W/(m² · °C) [Btu/(h · ft² · °F)].

2. G denotes glazing layer; Ep, a pyrolytic low-*e* coating (*e* = 0.35); and Es, a sputtered low-*e* coating (*e* = 0.15) on one side of the glazing; Gb denotes a bronze reflective coating. For the G-EsG-G and G-EpG units, the middle layer can be low-*e* coated glass or a low-*e* coated polyester film. Also, several other coatings on polyester film offer lower SCs with equivalent conductances to those shown.

3. Gap width between glazing layers is 12.7 mm (0.5 in).

[†] Windows examined in cooling-dominated locations.

shown because there is little difference between different types. For Lake Charles, only glazings of interest for hot climates are included. The low sensitivity of cooling loads to orientation may appear surprising, but it reflects the shade management assumed for these figures (solar gains reduced by 40 percent whenever direct solar gain on a window exceeded 63 W/m^2 during the cooling season). The heating loads do change significantly with orientation. For the glazings considered here, windows with northerly orientations increase the heating load, by contrast to Figure 14.15a and b where southern exposure was assumed.

Example 14.5

Find the annual savings for the house of Figs. 14.13 and 14.14 in Madison at a fuel price of $7/GJ if single-pane glazing with $U = 5.713\,\text{W}/(\text{m}^2 \cdot \text{K})$ is replaced by double-pane glazing with $U = 2.675\ \text{W}/(\text{m}^2 \cdot \text{K})$, the shading coefficient remaining the same at SC = 1.0. Assume a total window area of 12.37 m^2 ($= 0.0865 \times 143.06$ m^2); i.e., primary window area = 0. Express the result as savings per window area.

    Given: Figure 14.14b at $A = 0$, SC = 1.0

    Find: Difference between curves for $U = 5.713$ and 2.675 W/(m$^2 \cdot$ K)

SOLUTION

Reading the curves, we find $Q = 104$ GJ with single glazing and $Q = 86$ GJ with double glazing, a difference of 18 GJ, or roughly one-fifth of the load. This is worth

$$18 \text{ GJ} \times \$7/\text{GJ} = \$126 \text{ per year}$$

Dividing by 12.37 m^2, we obtain $10.19 per square meter of window per year.

COMMENTS

Suppose the incremental cost of double glazing is $50 per square meter. Then the payback time is $50/10.19 = 4.9$ years. *Savings due to good glazing are very important!*

When going to high-performance windows, one should not overlook the heat losses through window frames. Sullivan and Selkowitz have found that *the U value of the window frame is crucial in cold climates.* The above results assume good insulated frames with $U = 2.0\ \text{W}/(\text{m}^2 \cdot \text{K})$ and the frame area being 20 percent of the window area. The effect of the frame on the average U value, denoted U_{av}, can be seen from the equation [a simplified version of Eq. (6.45)]

$$U_{\text{av}} = \frac{U_g A_g + U_f A_f}{A_g + A_f} \tag{14.22}$$

where the subscripts g and f refer to glass and frame, respectively. If the frame area represents 20 percent of the window area, we have

$$U_{\text{av}} = 0.8 U_g + 0.2 U_f$$

For example, with $U_g = 2.0$ and $U_f = 5.2$ W/(m$^2 \cdot$ K) (conventional aluminum frame *with* thermal barrier), we would find

$$U_{av} = 0.8 \times 2.0 + 0.2 \times 5.2 = 2.64 \, W/(m^2 \cdot K)$$

a substantial increase over the value of the glazing itself. A conventional frame *without* thermal barrier has 12 W/(m$^2 \cdot$ K), and the penalty would be even larger.

14.5.2 Effects of Thermal Mass

The results that have just been presented are based on a fixed value of the heat capacity of the building, corresponding to the typical lightweight construction of U.S. houses. Now we consider how loads vary with the heat capacity. A large number of studies have been carried out to investigate this question (e.g., Christian, 1991; Byrne and Ritschard, 1985; and references cited in these articles). The general trends are clear; but due to the complex interplay of many different variables, it would be difficult to summarize the results in a form that is simple, accurate, and applicable in all situations. Therefore, we indicate only the general trends to guide the designer; for a precise analysis in a specific situation, a detailed dynamic calculation is recommended.

To begin, it is instructive to consider two extreme limits: zero heat capacity and infinite heat capacity. As pointed out by Mitchell and Beckman (1989), both cases can be analyzed with a simple static calculation, in terms of the balance-point temperature T_{bal} and heat loss coefficient K_{tot}. If the heat capacity is zero, there are no dynamic effects, and one can use the variable-base degree-day method of Sec. 8.1. In the limit of infinite heat capacity, the indoor temperature and the balance-point temperature approach constant values, and once again a steady-state method can be used. The difference in energy consumption between the two cases depends on the fluctuations of outdoor temperature T_o. There is no difference if T_o is always below T_{bal}. But if T_o fluctuates around T_{bal}, the zero-heat-capacity building needs heating whenever the *instantaneous* T_o is below T_{bal}, while the infinite-heat-capacity building needs heating only when the *average* T_o is below T_{bal}.

Mitchell and Beckman have evaluated these limits on a monthly time scale, i.e., interpreting *infinite capacity* to mean that heat gains are stored and reused during the course of each month. Some of their results are shown in Fig. 14.16, as daily heating requirement versus monthly average outdoor temperature, for three values of the heat capacity, keeping K_{tot} and T_{bal} fixed at values representative of conventional construction in Madison. The line labeled *large capacity* is a straight line, for $T_o < T_{bal}$, whose slope is the heat loss coefficient K_{tot} of the building; for $T_o > T_{bal}$, the heating requirement is zero. The zero-capacity line approaches the large-capacity line at low values of T_o, but lies significantly above it when T_o is comparable to T_{bal}. The behavior of a conventional house is indicated by the line labeled *typical capacity*. When the monthly average T_o is equal to T_{bal}, the mass effect is quite pronounced: the zero-capacity house needs 120 MJ/day, the typical-capacity house needs 60 MJ/day, and the large-capacity house needs nothing. At very low T_o, the three curves are almost the same.

This pattern remains true even when one accounts for more realistic conditions. For example, the DOE2.1 simulations of Byrne and Ritschard (1985) include nighttime thermostat setback during the heating season (from 21 to 16°C, midnight through 6 a.m.). Their results imply that as the heat capacity is increased, the energy savings from storage of free heat in the building mass more than compensate for the reduction of savings from (moderate) thermostat setback. Thermal mass brings significant energy savings in any case,

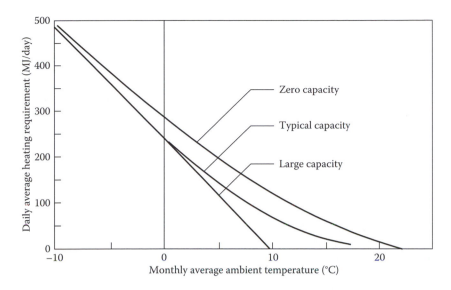

FIGURE 14.16

Daily average heating requirements as a function of monthly average outdoor temperature for a house in Madison, Wisconsin, for three values of the heat capacity. (Courtesy of Mitchell, J.W. and Beckman, W.A., *Solar Energy*, 42, 113, 1989.)

for heating and for cooling. The mass benefits for cooling are easy to understand because during much or most of the cooling season the daily average of T_o is below the thermostat set point for T_i. The mass averages in effect over the diurnal swings and causes the cooling load to follow the average rather than the instantaneous values of T_o.

For the practicing designer, it is helpful to show the benefit of thermal mass in terms of equivalent insulation. A table for this purpose has been prepared by the Council of American Building Officials (CABO, 1987), and we present here, in Table 14.6, an extract

TABLE 14.6

The U Value of a Massive Wall that is Equivalent in Terms of Annual Thermal Performance to a Lightweight Wall

°F Days (Base 65°F)	U of Light-Weight Wall Placement of Insulation	Equivalent U of Massive Wall								
		0.20			0.10			0.04		
		Exterior	Integral	Interior	Exterior	Integral	Interior	Exterior	Integral	Interior
<2000		0.28	0.28	0.25	0.16	0.15	0.12	0.08	0.07	0.04
2001–4000		0.27	0.27	0.24	0.15	0.14	0.12	0.08	0.06	0.04
4001–5501		0.25	0.26	0.23	0.14	0.13	0.11	0.07	0.06	0.04
5501–6501		0.23	0.24	0.22	0.12	0.12	0.11	0.06	0.05	0.04
6501–8000		0.22	0.22	0.21	0.11	0.11	0.10	0.05	0.05	0.04
>8000		0.20	0.20	0.20	0.10	0.10	0.10	0.04	0.04	0.04

Source: Courtesy of Christian, J.E., *ASHRAE Trans.*, 97, 2, 1991.

Notes: U is in Btu/(h · °F · ft^2) [1 Btu/(h · °F · ft^2) = 5.678 W/(K · m^2)]; heat capacity of massive wall is 6 Btu/(°F · ft^2) [123 kJ/(K · m^2)]; three placements of insulation relative to mass: exterior, integral, and interior; valid for residential applications.

of a more recent and refined version that has been recommended by Christian (1991). This table shows the U value of a massive wall that is equivalent to that of a lightweight wall, in the sense of yielding approximately the same annual energy use for typical residential applications. The massive wall has a heat capacity of 6 Btu/($°F \cdot ft^2$) [123 kJ/($K \cdot m^2$)] that corresponds to 2.5 in (6.5 cm) of heavyweight concrete, and three cases are considered for the placement of the insulation: external insulation, internal insulation, and insulation integral with the mass (e.g., insulation between two layers of brick or concrete). Since the savings depend on climate, separate entries are shown for different ranges of heating degree-days.

For example, in a climate with less than 2000°F·days, a lightweight wall with $U = 0.10$ Btu/($h \cdot °F \cdot ft^2$) is equivalent to a massive wall with $U = 0.16$ Btu/ ($h \cdot °F \cdot ft^2$) and insulation on the outside; in other words, it needs $\Delta R = 1/0.10 - 1/0.16 = 10 - 6.25 = 3.75$ ($h \cdot °F \cdot ft^2$)/Btu less of insulation. The difference between massive and lightweight walls is less important in colder climates, and above 8000°F·days the table shows no difference in U value. This is consistent with the explanation at the beginning of this section (see Fig. 14.16). For interior insulation the benefit is less pronounced than for exterior insulation, and integral insulation is intermediate between the interior and exterior cases.

To give a brief summary of thermal mass effects that have been discussed here or elsewhere in this book, we can say that thermal mass

- Is beneficial for cooling (it reduces both peak load and energy consumption)
- Reduces energy savings achievable by thermostat setback during the heating season
- Improves utilization of free heat gains when $T_{bal} - T_o$ changes sign during the day (especially important in sunny climates)

Also:

- The result is that on balance, in typical U.S. residential building designs, the addition of thermal mass can yield significant net energy savings, even during the heating season.
- In commercial buildings, the heating energy savings from a low-mass design with long and deep thermostat setback may outweigh the other benefits from thermal mass, and a detailed analysis is required.
- The extra capacity required for thermostat setback recovery in winter (the "pickup load") increases with thermal mass, but this can be avoided by reducing the depth or duration of the setback during the coldest weather.

14.5.3 Passive Solar Heating

Passive solar heating relies on the capacity of the building to collect a significant amount of solar heat during the day and to store some of it to offset heat losses during the night. Storage by phase-change materials would be interesting because they require less volume (which is at a premium in buildings). But given the difficulties of finding suitable phase-change materials, one usually is limited to sensible heat storage; brick or concrete is the most common choice, although water tanks have also been used. Thus a substantial temperature excursion, at least several degrees Celsius, is necessary if one wants to achieve a substantial solar contribution while keeping the amount of mass (and its cost) within reason.

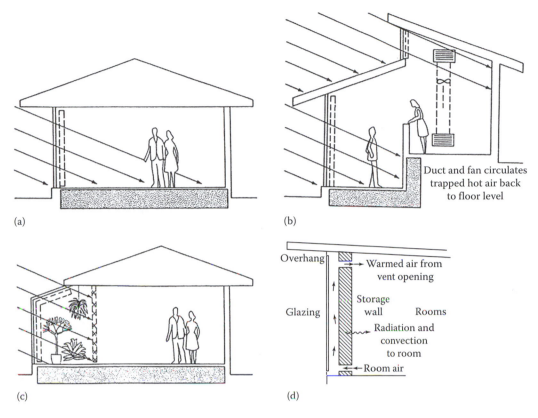

FIGURE 14.17
Schematic diagrams of passive solar heating methods. Additional option of movable insulation is shown as dotted line. (a) Direct gain; (b) direct gain with additional clerestory and fan; (c) sunspace (atrium, greenhouse); and (d) Trombe wall.

The principal approaches to passive heating are shown in Fig. 14.17. The *direct-gain* system is simply a sufficiently heavy floor (or other storage element) behind a large south-facing window. Flexibility can be gained by adding clerestory and/or circulating fan. The window should have a low U value, and movable night insulation greatly enhances the performance. *Sunspaces* such as attached greenhouses or atria are quite popular. As discussed in Sec. 14.2.3, the energy savings, if any, depend on whether the sunspace is conditioned.

The *storage wall* (also known as the *Trombe wall*, after its inventor), shown in Fig. 14.17d, is basically a vertical south-facing flat-plate collector with a heavy brick or concrete wall as absorber and storage. The wall thickness (on the order of 0.3 m) is sufficient to store the diurnal heat input and to release it to the room over the course of the night. To improve the control of the heat transfer, vents can be added at bottom and top, as indicated. Selective coatings for the absorber boost the performance.

Passive solar heating can be very effective at reducing energy bills, but it imposes certain constraints not compatible with everybody's preferences. The building must have a large window facing a southerly direction, with a shading device to avoid overheating in summer. A Trombe wall blocks at least part of the view to the south. With direct gain one can enjoy a full view, but furniture and carpets should not interfere with the absorption of radiation in the floor. The intense light may bleach textiles and artwork in the room. And then there are the above-mentioned temperature fluctuations.[5] Residents who want a steady 21°C might not be satisfied. Above all, the designer should pay close attention to summer performance. *Never recommend a passive solar design unless summer temperatures and cooling loads have been evaluated.* Otherwise the building may be uncomfortable in summer, or the problems will be covered up by using the air conditioner all the more.

To design a passive solar heating system, one has to choose appropriate values for the solar aperture (which depends on the size, orientation, and transmissivity of the glazing) and for the storage capacity (which depends on the amount, specific heat, and distribution of the mass). The choice is made on the basis of annual energy consumption—unlike conventional heating systems that are designed on the basis of peak loads. The reason lies in the fact that stand-alone solar energy systems would be far too expensive; a backup is almost always required for peak conditions. In most climates, the coldest weather may occur during overcast periods of such duration that it is advisable to choose the capacity of the backup with a conventional peak-load calculation.

Since temperature fluctuations in a passive solar building are important, steady-state methods such as the degree-day or bin methods of Sec. 8.1 are not appropriate for the prediction of annual performance. There are several dynamic simulation programs that are well suited to the analysis of passive solar systems, for example, BLAST (1986), SERIRES, or CALPAS3 (Berkeley Solar Group). Such programs have the advantage of providing, in addition to the calculation of annual energy, an evaluation of summer conditions. But they require a computer and hourly weather data. As an alternative, a variety of shorthand methods have been developed. The principal limitation of the existing shorthand methods is their inability to assess summer performance.

Perhaps best known among shorthand methods is the *solar load ratio method*, developed as correlation of a large number of hour-by-hour simulations (Jones, 1983; ASHRAE, 1984). The correlations are simple equations with a couple of coefficients that depend on the design and construction of the building. One calculation is needed for each month of the heating season; it can be programmed with a spreadsheet. Different designs, e.g., differing in thickness of storage floor, require different coefficients. Correlations have been developed for all the standard configurations and can be found in the cited references. Here we present only one particular example, designated sunspace D1; see Fig. 14.18. The key parameters are listed in Example 14.6, and the correlation is shown as a graph in Fig. 14.19.

To explain this method, one first has to address the interpretation of the performance of a passive solar element, such as a sunspace. By contrast to conventional heating systems, passive solar heating elements are an intrinsic part of the structure of a building. This is awkward if one wants to compare a building with to a building without passive solar heating. To resolve this difficulty, the solar load ratio method uses a heat loss coefficient that is based on the nonsolar parts of the envelope rather than on the total envelope of the building. This coefficient is called the *net load coefficient* (NLC) and has units of Btu per degree Fahrenheit per day:

$$\text{NLC} = \frac{24\,\text{h}}{1\,\text{day}} \left(\rho c_p \dot{V} + \sum_{\text{nonsolar}} UA \right) \qquad (14.23)$$

---

[5] For instance, the *solar load ratio* (SLR) design method, discussed below, allows a range from 18.3°C (65°F) to 23.9°C (75°F).

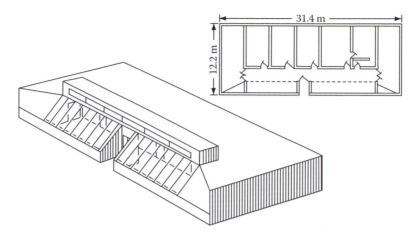

FIGURE 14.18
Sunspace design of Example 14.6. (Courtesy of ASHRAE, *Handbook of Fundamentals*, American Society of Heating, Refrigerating and Air-Conditioning Engineers, Atlanta, GA, 1989a. With permission.)

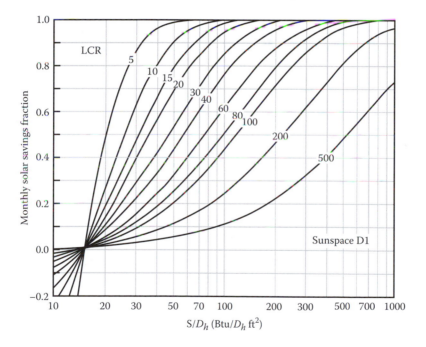

FIGURE 14.19
Correlation for solar savings fraction (SSF) for the sunspace design (D1) whose key parameters are listed in Example 14.6. (Courtesy of Jones, R.W. (ed), *Passive Solar Design Handbook*, vol. 3, American Solar Energy Society, Boulder, CO, 1983. With permission.)

It differs from K_{tot} of Eq. (7.24) by the exclusion of the solar elements of the envelope and by the factor 24 h/1 day. When comparing a solar building and a nonsolar building, one assumes that the solar parts of the envelope are replaced by an energy-neutral wall (i.e., an opaque wall with zero conductivity).

Another key parameter of this method is the *projected collector area* A_{proj}; it is the projection of the aperture area of the solar elements on the vertical south-facing plane.

The area is counted net, i.e., excluding mullions and shaded portions. For example, the building in Fig. 14.18 has a sunspace with south-facing glazing of net area 51 m^2, its normal being tilted at 50° from the vertical; its projected area is $A_{proj} = \sin 50° \times 51$ m$^2 = 39$ m^2.

As for weather data, the method requires the *monthly solar radiation S* on the collector area and the *monthly heating degree-days* $D_h(T_{bal})$ based on the balance-point temperature T_{bal}. The latter is calculated as in Eq. (8.2), but excluding the solar gains of the aperture. In terms of the NLC and daily solar gain $\dot{Q}_{gain,day}$ (in Btu per day), we have

$$T_{bal} = T_i - \frac{\dot{Q}_{gain,day}}{NLC} \tag{14.24}$$

For example, in Fig. 14.18 the solar gains from windows on the east, west, and north facades are included, but those from the sunspace are excluded from the calculation of T_{bal}. The solar radiation S on the aperture can be calculated from the common horizontal insolation data by using the method of Sec. 6.3.4.

Having determined these quantities for the design in question, one obtains the *solar savings fraction* (SSF) from a correlation such as the one in Fig. 14.19. It yields SSF as a function of two ratios, the ratio S/D_h and the load collector ratio (LCR):

$$LCR = \frac{NLC}{A_{proj}} \tag{14.25US}$$

Note that the correlation is dimensional and LCR must be stated with the dimensions of Btu per degree Fahrenheit per day per square foot. Finally, the auxiliary heating energy to be supplied by the backup is

$$Q_{aux} = NLC \times D_h \times (1 - SSF) \tag{14.26}$$

where SSF is the fraction of the load $NLC \times D_h$ that is supplied by the solar elements. This procedure is repeated for all months of the heating season to obtain the annual total.

Example 14.6

Find the auxiliary heating energy for the building of Fig. 14.18 in Denver, Colorado, in January.

Given: $T_i = 68°F$ (20°C)

$\dot{Q}_{gain} = 4200$ Btu/h (1231 W)

Sunspace design type D1 [12-in (0.3-m) masonry wall between sunspace and conditioned space]

Sunspace is unconditioned and semienclosed by conditioned space.

Aperture of sunspace is double-glazed with area 550 ft^2 (51.1 m^2), its normal being tilted at 50° from vertical.

South-facing, without night insulation

The envelope has the following characteristics:

	A or $\dot{V}$, ft^2 or ft^3/min	U, Btu/(h·°F·ft^2)	UA or $\rho c_p \dot{V}$, Btu/(h·°F)
Opaque walls	2000	0.040	80.0
Roof	3000	0.030	90.0
Floor (over crawl space)	3000	0.040	120.0
Windows (east, west, north)	100	0.550	55.0
Subtotal			345.0
Infiltration	200 ft^3/min (at 0.5 ach with $V = 24{,}000$ ft^3)	$\rho c_p = 0.015$ Btu/(ft^3·°F) in Denver	180.0
Subtotal			*525.0*
Sunspace	550	0.55	302.2
Total			827.2

The heat loss coefficient without the sunspace is indicated in italic as 525.0 Btu/(h·°F) (277 W/K). Hence the net load coefficient is

$$\text{NLC} = 24\,\text{h/day} \times 525.0\,\text{Btu/(h·°F)} = 12{,}600\,\text{Btu/(°F·day)}$$

Find: Q_{aux}

Lookup values: We need to look up solar radiation S and degree-days D_h for January in Denver. (For simplicity, we take these values as known; they could be calculated by using the methods of Secs. 6.3.4 and 8.1.2.)

$$S = 43{,}600\,\text{Btu/ft}^2 = \text{monthly solar radiation on aperture}$$

For the degree-days, we first determine the balance-point temperature from Eq. (14.24):

$$T_{\text{bal}} = 68°\text{F} - \frac{24\,\text{h} \times 4200\,\text{Btu/h}}{12{,}600\,\text{Btu/(°F·day)}} = 60°\text{F}$$

The corresponding degree-days are

$$D_h = 930°\text{F·days} = \text{monthly degree-days } D_h(T_{\text{bal}}) \text{ for base } T_{\text{bal}}$$

SOLUTION

The projected area is

$$A_{\text{proj}} = 550\,\text{ft}^2 \times \sin 50° = 421\,\text{ft}^2$$

By Eq. (14.25),

$$\text{LCR} = \frac{12{,}600\,\text{Btu/(°F·day·ft}^2)}{421\,\text{ft}^2} = 29.9\,\text{Btu/(°F·day)}$$

For the ratio S/D_h, we have

$$\frac{S}{D_h} = \frac{43.600\,\text{Btu/ft}^2}{930°\text{F} \cdot \text{day}} = 46.9\,\text{Btu/}(°\text{F} \cdot \text{day} \cdot \text{ft}^2)$$

Reading Fig. 14.19 at these values, we find

$$\text{SSF} = 0.51$$

This means that 51% of the heating requirement of the building is provided by the sunspace in January. Equation (14.26) yields the auxiliary heating requirement

$$\dot{Q}_{aux} = \text{NLC} \times D_h \times (1 - \text{SSF})$$
$$= 12{,}600\,\text{Btu/}(°\text{F} \cdot \text{day}) \times 930\,°\text{F} \cdot \text{days} \times (1 - 0.51) = 5.74\,\text{MBtu}$$

COMMENTS

In accordance with the assumptions of the method, this result presupposes that the thermostat dead band is 65 to 75°F for the conditioned part of the building and 45 to 95°F for the sunspace. The designer should check whether this is agreeable with the future occupants of the building.

As a general rule, one is likely to find that annual SSF values in the range of 0.4 to 0.7 are achievable with reasonable designs, except in the cloudiest regions of the United States; in other words, the passive solar approach can reduce the heating bill by about 40 to 70 percent.

Another shorthand method is based on the analytical solution of a simple thermal network. This approach has been developed by Gordon and Zarmi (1981) and further elaborated by Cowing and Kreider (1987). For greatest accuracy, the data for solar radiation and heating degree-days must be preprocessed in the form of a special frequency distribution. For the 26 SOLMET (1978) stations in the United States, these distributions have already been prepared by Cowing and Kreider.

As yet another method, we mention the un-utilizability method of Monsen et al. (1982). Here one calculates, for each month, the fraction of the solar radiation that is not utilizable because it is above the maximum level that would be useful for heating.

14.6 Commercial Buildings: HVAC Systems

To what extent the energy consumption of commercial buildings depends on the HVAC system can be seen from an interesting series of studies by Kao (1985). Using the program BLAST, Kao simulated four types of commercial buildings: a small office, a large office, a school, and a retail store. Building structure and utilization were kept fixed, while varying the HVAC system. A wide variety of HVAC systems were evaluated, about 20 of the most common types for each building. Here we present some selected results, as heating and cooling loads, excluding the efficiency of the primary system (or, to use the language of DOE2.1, only the performance of the "system" is considered, without "plant").

Six sites were considered, covering a broad range of climates; the heating degree-days ranged from 832 K · days (1498°F · days) in Lake Charles, Louisiana, to 4294 K · days

(7730°F · days) in Madison, Wisconsin, and the cooling degree-days from 47 K · days (84°F · days) in Santa Maria, California, to 1522 K · days (2739°F · days) in Lake Charles. The same envelope is assumed for all climates, even though different envelope designs might be optimal or required by building codes in different climates. The simulation results have been correlated with degree-days to highlight trends and to permit interpolation to other sites.

A summary description of buildings and operating conditions is presented in Table 14.7. The comfort conditions (outdoor airflow and thermostat set points for occupied periods) may appear a bit extreme today; they reflect standards imposed by the government during the energy crisis. Obeyed in public buildings but not in all private buildings, they have gradually been abandoned with the new energy affluence.

The results are shown as a function of degree-days in Figs. 14.20 to 14.23, with a brief description of the HVAC systems in each caption and with the numerical order corresponding roughly to the ranking in energy performance. *The highest and the lowest can easily differ by more than a factor of 2.* Generally the systems with the highest consumption are the ones that use constant volume, without economizer and with fixed supply air temperature. These conclusions for the economizer are in broad agreement with the simulations by Spitler et al. (1987), although the latter imply somewhat lower benefits for enthalpy control. The lowest consumption is achieved by VAV systems with enthalpy economizer, and with variable T_{sup}, reset according to zone load or outdoor temperature. Such reset lowers the amount of reheat.

VAV systems use far less than constant-volume systems for heating and almost always less for cooling (possibly with slight exceptions at low cooling loads). The economizer will always reduce the cooling load, although it may cause slightly increased heating.

TABLE 14.7

Summary of Building Characteristics for Figs. 14.20 to 14.23

	Units	Small Office	Large Office	School	Retail Store
Zones		2*(ESWN + Int)	ESWN + Int		2*(ESWN + Int)
Stories		3	12	1	2
A_{floor}	m^2 (ft^2)	2787 (30,000)	22,297 (240,000)	6136 (66,048)	14,270 (153,600)
A_{glass}/A_{floor}		0.15	0.14	0.04	0.01
Glass type		Double	Single, $\tau = 0.24$	Single	(Only glass doors)
A_{floor}/occupant (peak)	m^2 (ft^2)	11.6 (125)	9.3 (100)	2.3 (25)	3.7 (40)
Min. outdoor air/occupant	L/s (ft^3/min) per occupant	2.8 (6)	3.8 (8)	2.4 (5)	2.4 (5)
Min. outdoor air/ventil. air		0.1	0.1	0.2	0.2
K_{cond}/A	W/(K·m^2) [Btu/(h·ft^2)]	0.676 (0.119)	1.040 (0.183)	0.370 (0.066)	0.110 (0.019)
$T_{i,min}$	°C (°F)	18.9 (66.0)	20.0 (68.0)	20.0 (68.0)	22.2 (72.0)
$T_{i,max}$	°C (°F)	25.6 (78.0)	25.6 (78.0)	25.6 (78.0)	25.6 (78.0)
$T_{setback}$	°C (°F)	13.3 (56.0)	11.1 (52.0)	12.8 (55.0)	11.1 (52.0)
Heat gains (average during occupancy)	W/m^2 [Btu/(h·ft^2)]	34.7 (11.0)	49.2 (15.6)	63.4 (20.1)	55.8 (17.7)

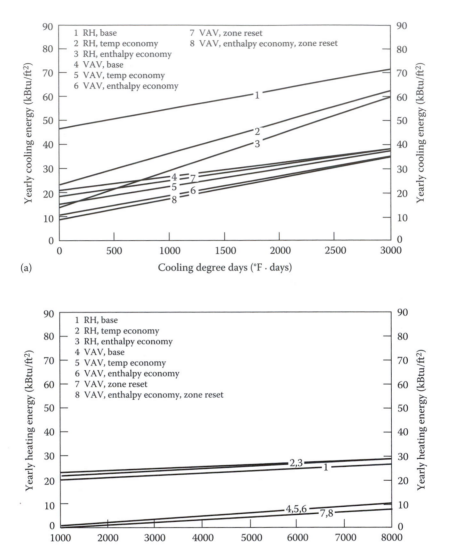

FIGURE 14.20
Heating and cooling loads for small office building with different HVAC systems as a function of degree-days. (1 = RH, base: One central air handler with preheat coil and cooling coil distributing conditioned air to each zone. Heating and cooling control in each zone accomplished by reheat coils in supply air ducts of each zone. No reheat coil in zone one (interior first two floors, with no surfaces to outside). T_{sup} set at 16.7°C (62°F) during working hours; 2 = RH, temperature economy: like 1 but with economizer controlled by temperature; 3 = RH, enthalpy economy: like 1 but with economizer controlled by enthalpy; 4 = VAV base: Central variable-volume fan (with variable-pitch inlet vanes) supplying air to all 10 zones. Each zone has VAV dampers, with minimal airflow below 24.4°C (76°F) and airflow increasing above this temperature to a maximum at 25.6°C (78°F). T_{sup} set at 12.8°C (55°F) during working hours; 5 = VAV, temperature economy: like 4 but with economizer controlled by temperature; 6 = VAV, enthalpy economy: like 4 but with economizer controlled by enthalpy; 7 = VAV, zone reset: like 4 but with T_{sup} variable between 12.8°C (55°F) and 16.7°C (62°F) during working hours according to zones with largest instantaneous cooling load; 8 = VAV, enthalpy economy, zone reset: like 7 but with economizer controlled by enthalpy.) (Courtesy of Kao, J.Y., *ASHRAE Trans.*, 91, pt. 2B, 810, 1985. With permission.)

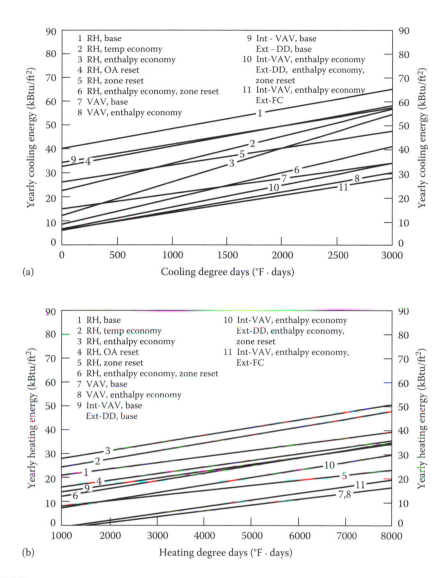

FIGURE 14.21

Heating and cooling loads for large office building with different HVAC systems as a function of degree-days. (1 = RH, base: Constant-volume system with terminal reheat. $T_{sup} = 15.0°C$ (59°F) including fan heat gain. Minimum outdoor air, during occupied hours only, at 10% of supply air; 2 = RH, temperature economy: like 1 but with economizer controlled by temperature; 3 = RH, enthalpy economy: like 1 but with economizer controlled by enthalpy; 4 = RH OA reset: like 1 but with T_{sup} of perimeter zones varied in linear fashion according to outdoor air T_o, between 15°C (at $T_o = 32.2°C$) and 17.2°C (at $T_o = 21.2°C$); 5 = RH zone reset: like 1 but with T_{sup} of perimeter and interior zones varied according to load of each zone. No limits on T_{sup} imposed; 6 = RH enthalpy economy, zone reset: like 1 but with economizer controlled by enthalpy and with T_{sup} of perimeter and interior zones varied according to load of each zone; 7 = VAV, base: VAV for interior and perimeter, flow variable from 20 to 100 percent of full flow. Constant flow of outdoor air during occupied hours. Constant $T_{sup} = 15°C$ (59°F). Reheat coils in perimeter zones only; 8 = VAV, enthalpy economy: like 7 but with economizer controlled by enthalpy; 9 = Int-VAV enthalpy economy, Ext-DD base: VAV base (like 7) for interior zones and DD (dual-duct) system for perimeter zones. Outdoor airflow constant during occupied hours. Cold duct $T_{sup} = 15°C$ (59°F) and $T_{hot} = 60°C$ (140°F) at all times; 10 = Int-VAV enthalpy economy, Ext-DD enthalpy economy, zone reset: like 9 but with economizer controlled by enthalpy for entire building, and cold and hot T_{sup} of dual-duct system reset according to zone loads; 11 = Int-VAV enthalpy economy, Ext-FC: like 8 for interior (VAV with enthalpy economizer) and four-pipe fan coil system for perimeter (with hot and cold water available year-round).) (Courtesy of Kao, J.Y., *ASHRAE Trans.*, 91, pt. 2B, 810, 1985. With permission.)

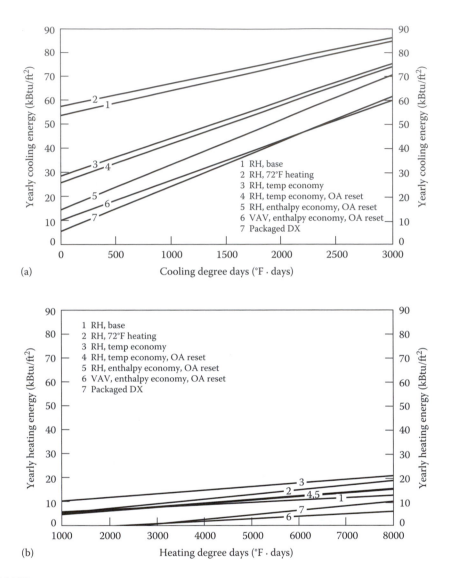

FIGURE 14.22
Heating and cooling loads for large retail store with different HVAC systems as a function of degree-days.
(1 = RH, base: constant volume for 10 zones, reheat coils for the 8 perimeter zones. $T_{sup} = 15.6°C$ (60°F) and
$T_i = 20°C$ (68°F) daytime; 2 = RH, 72°F heating: like 1 but with $T_i = 22.2°C$ (72°F) daytime instead of 20°C (68°F);
3 = RH, temperature economy: like 1 but with temperature economizer; 4 = RH, temperature economy, OA reset:
like 3 but with T_{sup} reset according to outdoor air temperature; 5 = RH, enthalpy economy: like 4 but with
enthalpy economizer; 6 = VAV, enthalpy economy, OA reset: VAV all zones, reheat in perimeter, enthalpy
economizer and T_{sup} reset according to outdoor air temperature; 7 = Packaged DX: packaged direct-expansion
units for 12 zones.) (Courtesy of Kao, J.Y., *ASHRAE Trans.*, 91, pt. 2B, 810, 1985. With permission.)

To minimize this heating penalty, *reset of T_{sup} is strongly advisable*. Enthalpy is more efficient
than temperature control for the economizer, attaining appreciably lower cooling load,
with little effect on heating. Both absolute and relative savings are larger for low cooling
degree-day climates (about 25 percent relative savings) than for high cooling degree-day
climates (about 5 percent relative savings).

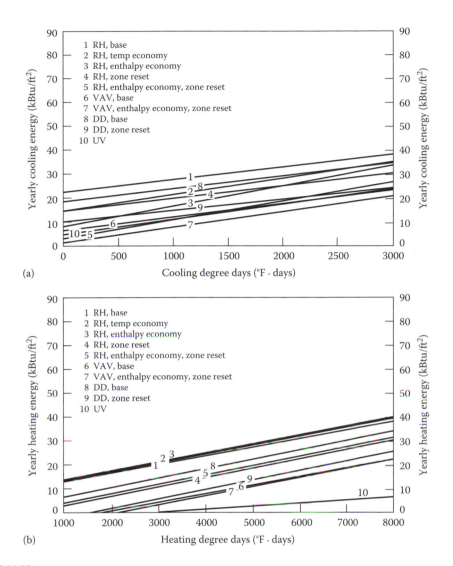

FIGURE 14.23

Heating and cooling loads for school with different HVAC systems as a function of degree-days. (1 = RH, base: constant volume, terminal reheat, T_{sup} = 15.0°C (59°F) fixed; 2 = RH, temperature economy: like 1 but with economizer controlled by temperature; 3 = RH, enthalpy economy: like 1 but with economizer controlled by enthalpy; 4 = RH zone reset: like 1 but with T_{sup} reset by zone load between 15.0°C (59°F) and 18.3°C (65°F); 5 = RH, enthalpy economy, zone reset: like 4 but with economizer controlled by enthalpy; 6 = VAV, base: variable inlet vanes, T_{sup} = 15.0°C (59°F) fixed, terminal reheat in perimeter; 7 = VAV, enthalpy economy, zone reset: like 6 but with enthalpy economizer and T_{sup} reset by zone load; 8 = DD, base: dual-duct system with fixed T_{sup} 15.0°C (59°F) cold and 48.9°C (120°F) hot.; 9 = DD, zone reset: like 8 but with hot and cold T_{sup} reset by zone load; 10 = UV: unit ventilators for heating and cooling, capable of admitting 100 percent outdoor air.) (Courtesy of Kao, J.Y., *ASHRAE Trans.*, 91, pt. 2B, 810, 1985. With permission.)

A number of other issues have not been addressed, in particular, the risk of discomfort and the flexibility in adjusting to changes in operating conditions or internal loads. Installation of reheat is the easiest protection against such eventualities, albeit at the price of increased energy consumption. Also, reheat systems render equipment failures more difficult to detect.

It would be interesting to evaluate the HVAC index of Eq. (14.4) for these systems, to see how much further improvement is possible. Unfortunately that would necessitate a rerun of the simulations, because in the presence of thermostat setback the loads cannot be calculated accurately with a static method. We suspect that a significant amount of reheat remains during periods with economizer operation, as we found in Example 14.4, and that it could be avoided by using *separate air handlers for zones whose loads differ significantly*. To see why, consider the economizer at a time when one zone calls for cooling and the other zone calls for heating. If both zones are supplied by the same air handler (with conventional VAV control), use of the economizer increases outdoor airflow for both; consequently the heating load becomes larger than it would have been if each zone had a separate air handler.

14.7 Design for Daylighting

In commercial buildings, the use of daylight is potentially the most important envelope design option for efficiency. There are numerous reasons. Energy consumption for lighting is large, and lights are used during daytime; thus daylight can displace expensive electricity during peak periods. Much of commercial floor space is sufficiently close to the outside, at the perimeter, or under the roof to enjoy easy access to daylight. The fraction directly under the roof is one-half by itself, according to Place et al. (1987). Furthermore, sunlight has high luminous efficacy, around 110 lm/W, about twice as high as conventional fluorescent lamps. Being cooler, sunlight offers the possibility of reducing the cooling loads of a building. And then there are, of course, important psychological benefits because daylighting can provide a view of the outside, a sense of openness, and a good spectral composition of the light.

By contrast, in residential buildings, the demand for lighting is usually quite limited during daytime; that is the reason why we discuss daylighting only in the commercial context. In residential buildings, it is more appropriate to consider the heat collection potential of windows, because residential heating loads tend to be relatively larger and better matched to the sun than those of commercial buildings.

Ordinary (vertical) windows are, of course, the simplest daylighting scheme; they do not call for any special ingredient beyond the installation of appropriate light controllers, but they are limited to the perimeter of a building. Skylights can provide daylight to any zone directly under the roof, but they imply higher construction cost than for a simple opaque roof. Finally, there are special schemes, based on light shelves or light guides. In this section, we discuss only the role of windows and of skylights, citing results from some important studies carried out at LBL. They are based on simulations of standard building designs. The results are indicative of general trends and provide valuable guidelines for the design of daylighted buildings, all the more so since many features of daylighting have been found to be fairly insensitive to climate. Thus, when carrying out simulations for a building under design, the designer knows in what range of parameters to look for the optimum.

The design variables to optimize are the aperture size, glazing type, and choice of shading device. The glazing is characterized by three variables: the shading coefficient SC, the transmissivity for visible radiation τ_v, and the conductance U. For the performance analysis of daylighting, it is convenient to group certain variables. One group is the *effective aperture* A_{eff}, normalized by the total area A_{tot} of the wall or roof,

$$A_{eff} = \frac{\tau_v A_{apert}}{A_{tot}} \tag{14.27}$$

The other is the ratio of visible to thermal transmittance

$$K_e = \frac{\tau_v}{SC} \tag{14.28}$$

It is proportional to the instantaneous luminous efficacy of the daylight transmitted into the building.

14.7.1 Windows

For the daylighting performance of ordinary (vertical) windows, we cite a study by Arasteh et al. (1986) and by Sweitzer et al. (1987). These authors have considered the perimeter zones of an intermediate floor of a standard office building, as shown in Fig. 14.24. They assume continuous dimmers for the control of the lights and, in the interest of thermal and visual comfort, the use of shades whenever the direct solar radiation transmitted through a window exceeds 63 W/m^2. Three levels of installed lighting power are evaluated (7.5, 18.3, and 29.1 W/m^2), all for the same design illumination of

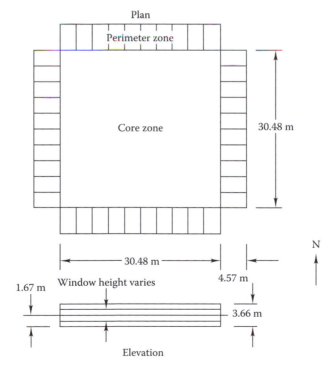

FIGURE 14.24
Diagram of building model for simulation of daylighting by Arasteh et al. (1986).

50 fc (540 lx). The HVAC system consists of a separate constant-volume system with fan coil and economizer for each zone, and the average COP of the cooling plant is 3.0.

The resulting energy consumption, as a function of the effective aperture, is shown in Fig. 14.25 for a hot and humid climate: Lake Charles, Louisiana. Total consumption is shown, as well as the components of lighting, cooling, office equipment, and fans. The dashed lines correspond to continuous dimming controls; the solid lines, to the absence of daylighting. With daylighting the total consumption drops appreciably as the aperture increases from 0 to about 0.1. Beyond that value, further gains in lighting are small while the cooling load continues to increase; thus the total consumption grows again.

Without dimming controls there is no initial drop; the consumption increases monotonically with aperture. This is most plausible: In a hot climate, the only energy benefit of windows could come from daylight—and without dimming controls there are no savings.

The magnitude of the savings depends on the desired illumination level, efficacy of the electric lighting system, lighting control, HVAC system, etc. The higher the installed lighting capacity, the higher the savings from daylight and the higher the optimal effective aperture, as shown in Fig. 14.26. The optimal aperture is typically in the range of 0.1 to 0.2.

Of course, the consumption is very sensitive to the choice of the glazing, especially the ratio $K_e = \tau_v/\text{SC}$. The higher, the better. The theoretical maximum is about 2.8. Most glazing for commercial buildings is in the range of 0.5 to 1.0, depending on its tint. Data for typical

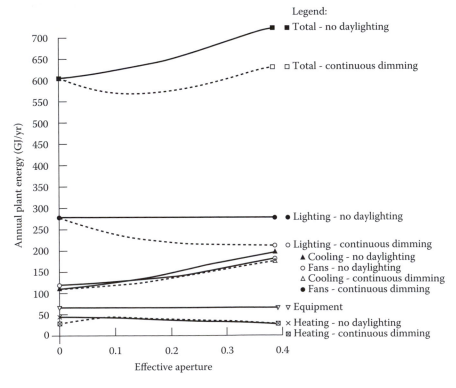

FIGURE 14.25

Annual plant energy requirements with daylighting (dashed lines) and without daylighting (solid lines) for the total building (five zones) of Fig. 14.24 as a function of the effective aperture, for Lake Charles, Louisiana. (Courtesy of Arasteh, D. et al., *Sunworld*, 10(4), 104, 1986.)

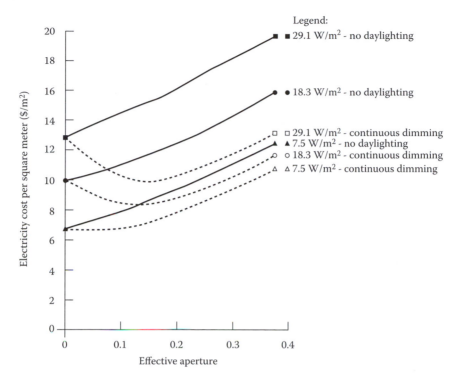

FIGURE 14.26

Annual total electricity cost versus A_{eff} for three levels of installed lighting capacity for the building of Fig. 14.24 as a function of the effective aperture, for Lake Charles, Louisiana. (Courtesy of Arasteh, D. et al., *Sunworld*, 10(4), 104, 1986.)

glazings are shown in Table 6.6. The total electricity consumption (excluding space heat) for a south zone in Madison, Wisconsin, and in Lake Charles, Louisiana, is plotted versus aperture in Fig. 14.27 for several values of K_e. The performance of specific glazings has been evaluated at a window/wall ratio of 0.75 and is indicated by the position of the numbers (1 to 9), referring to the types in Table 6.6.

The differences between different glazings are large, often on the order of 10 percent in total electricity. As for the benefit of increased K_e values, it is most pronounced when going from 0.5 to 1.0; beyond $K_e = 1.5$, further gains are negligible. As in the previous figures, the saturation of daylighting causes the consumption to increase at large apertures (beyond $A_e \approx 0.3$). No influence of U value is observable in this figure, because space heat is not included; the annual cooling energy is not very sensitive to the U value, the average temperature difference during the cooling season being small. It is interesting to note, both here and for the skylights in Fig. 14.28, that the optimal aperture seems to be quite insensitive to climate.

The savings due to daylight are worth on the order of $6 per square meter of floor area per year (Arasteh et al., 1986). This is to be compared with the cost of dimming controls, estimated around $10 to $16 per square meter in new construction, $43 to $54 per square meter in retrofits. In new construction, the economics are more favorable because this cost is partially offset by savings in cooling equipment. Thus the payback times for daylight controls with vertical windows appear to be on the order of a few years in new construction and around 8 years in retrofits.

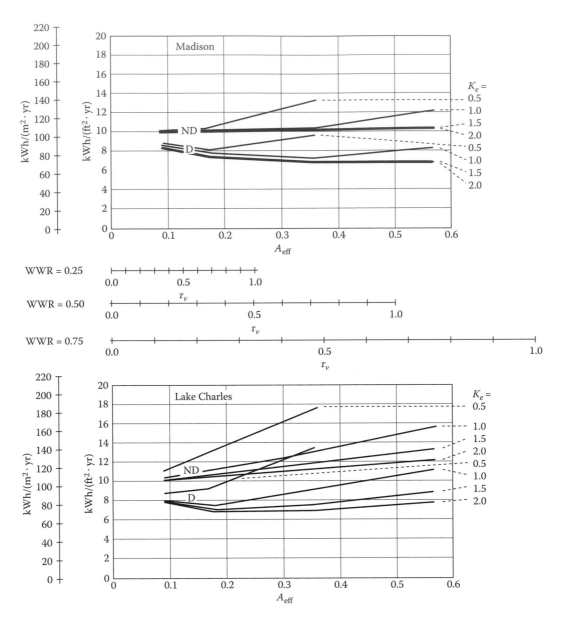

FIGURE 14.27
Total electricity use (lights, cooling, and fans, but not space heat) versus effective aperture A_{eff} with ($=$D) and without ($=$ND) dimming controls, for south zone of building of Fig. 14.24, for four values of $K_e = \tau v / SC$: 0.5, 1.0, 1.5, and 2.0 (some of the curves are almost on top of each other). *Top*: Madison, Wisconsin; *bottom*: Lake Charles, Louisiana. (Courtesy of Sweitzer, G. et al., *ASHRAE Trans.*, 93, 1, 1987. With permission.)

14.7.2 Roof Apertures (Skylights)

The potential of skylights can be appreciated from a study by Place et al. (1987). Using the simulation program BLAST, these authors evaluated a variety of roof apertures for the top floor of a standard office building of 100 ft × 100 ft (30 m × 30 m) base area.

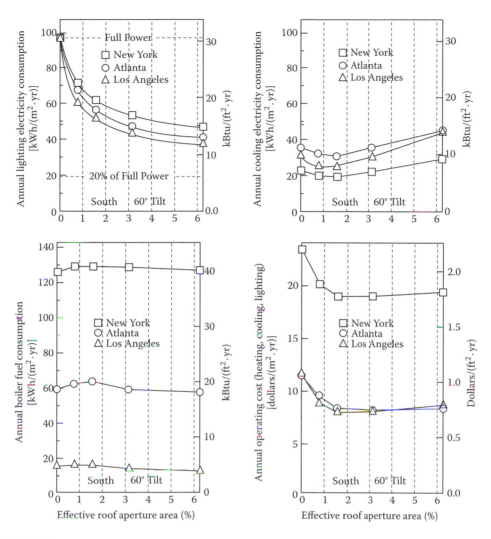

FIGURE 14.28

Annual consumption for office building with daylighting by skylights as a function of effective aperture area. (a) Electricity for lighting; (b) electricity for cooling; (c) boiler fuel; and (d) operating costs for cooling, heating, and lighting. (Courtesy of Place, J.W. et al., *ASHRAE Trans.*, 93, 1, 1987.)

They considered both linear apertures, as in Fig. 13.2, and localized apertures, with three tilts: horizontal, 60°, and vertical. Diffuse glazing or diffusers were assumed to prevent problems with glare. The design illumination level is 50 fc (540 lx) on the work plane. Power to the electric lights is reduced linearly in response to daylight while maintaining constant total illumination (except that the electricity cannot be reduced below a minimum of 20 percent of full power, typical of controllers available at the time of the study). The peak electric power for lighting is 2.5 W/ft^2, corresponding to a system luminous efficacy of 20 lm/W (i.e., relative to the illumination delivered on the work plane). The *system* luminous efficacy of daylight is taken as 72 lm/W (as opposed to the outdoor value of 110 lm/W).

The resulting annual consumption as a function of effective aperture is shown in Fig. 14.28, for lighting, cooling and heating, together with the total annual operating cost for these terms, at three sites: New York City, Los Angeles, and Atlanta. The basic pattern of Fig. 14.28 is similar to that in the preceding figures for vertical apertures. There is a rapid drop in total consumption as the aperture is increased from zero. Diminishing returns are reached with saturation of daylighting, and at large apertures the consumption goes up again. The main difference between vertical and horizontal apertures lies in the position of the optimum. While the optimal aperture was found to occur around 0.1 to 0.3 for vertical windows, it is an order of magnitude smaller, around 0.01 to 0.06, for skylights. Most of the savings are due to reduced consumption for lighting, but reductions in cooling loads are not to be overlooked. The curves are quite similar in shape for the three locations, even though the magnitudes are different. Likewise, the utility rates have a strong effect on the magnitude of the savings, but not on the choice of the design.

As for the performance of different skylight designs, horizontal apertures yield higher savings per aperture area, because they are most effective for beam daylighting, but vertical glazings (with orientations S, S+N, S+W+N+E, Sw+Se, or Nw+Ne+ Sw+Se) produce even lower total annual cost because they perform better at low sun angles; see Fig. 14.29. The optimal aperture for vertical skylights is larger, around 0.03 to 0.06, than that for horizontal skylights. A north aperture alone is not as good because it

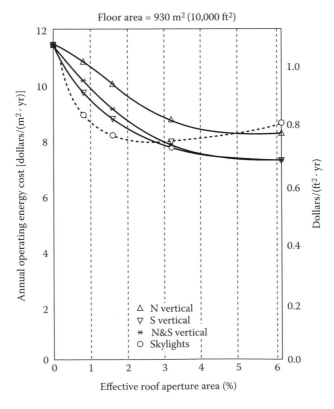

FIGURE 14.29
Annual operating energy cost for office building in Atlanta, Georgia, with daylighting by several types of skylights as a function of effective aperture area. (Courtesy of Place, J.W. et al., *ASHRAE Trans.*, 93, 1, 1987.)

cannot utilize beam radiation. Psychological considerations also play a role: The desire for sunlight is stronger in winter than in summer; hence skylights that are oriented toward the south may be preferable. All these designs presuppose that diffusing elements are used to prevent glare.[6]

The annual savings achievable with skylights can be as large as $4 per square meter of floor area for A_{eff} around 0.04 to 0.06. That is to be compared with the incremental first costs for dimmers and for skylights minus the savings from reduced cooling plant capacity. These costs are very much dependent on the details of construction and are difficult to estimate in general. To translate the $4 per square meter of floor area into permissible cost per skylight area, take the example of $A_{eff} = 0.04$ with a solar transmittance of $\tau_v = 0.50$, which implies that $A_{sky}/A_{floor} = 0.04/\tau_v = 0.08$; hence the annual savings are $4 per 0.08 m^2 = $50 per square meter of A_{sky}. While the savings per area of glazing are highest for horizontal skylights, this is not necessarily the appropriate criterion because of fixed costs associated with the installation of any skylight.

To conclude this chapter, note that *windows can be more efficient than opaque walls*. This is reassuring, since people will demand windows anyway (psychological benefits of daylight are an important bonus, even if an unquantifiable one). The challenge is laid down for the designer: Find energy-efficient solutions. Skylight designs have been tested in scale models and in real buildings; the results confirm that good illumination and high luminous system efficacy can indeed be achieved.

Problems

The problems in this book are arranged by topic. The approximate degree of difficulty is indicated by a parenthetic italic number from 1 to 10 at the end of the problem. Problems are stated most often in USCS units; when similar problems are presented in SI units, it is done with approximately equivalent values in parentheses. The USCS and SI versions of a problem are not exactly equivalent numerically. Solutions should be organized in the same order as the examples in the text: given, figure or sketch, assumptions, find, lookup values, solution. For some problems, the Heating and Cooling of Buildings (HCB) software included in this book may be helpful. In some cases it is advisable to set up the solution as a spreadsheet, so that design variations are easy to evaluate.

14.1 A central station converts fuel to electricity with a typical efficiency of 33 percent. How high must the COP of a heat pump be so that the overall fuel utilization is better with electric heating than with burning of fuel in local furnaces of 66 percent efficiency? Briefly state one or two reasons why this simple argument is not sufficient for a decision, quite apart from the cost considerations. (4)

14.2 For the house of Example 14.1 calculate the energy savings and the payback time, relative to the reference design, for each of the four design variations by itself. (4)

14.3 Calculate the annual heat lost per square meter (foot) of roof area as a function of the insulation thickness t. Even though, strictly speaking, the balance-point temperature changes with insulation, assume for simplicity that the number of degree-days

---

[6] And, of course, that attention will be paid to avoid practical problems such as leaks of rain and excessive snow accumulation.

remains constant at 2,000 K · days (3600°F · days). Using the cost data of Example 14.1, calculate the annual energy and the cost of the additional insulation. Plot the energy savings and the insulation cost versus insulation thickness from $t = 0.05$ to 0.5 m (0.164 to 1.64 ft). At what thickness is the payback time (relative to a thickness of 0.05 m) equal to 10 years?   (6)

14.4 Consider the comparison of HVAC systems for the small office building in Fig. 14.20, and try to estimate qualitatively how the comparison would change if the window/wall ratio (WWR) were changed? *Hint:* Consider the variability of zone loads for two extreme values of the window/wall ratio—0 and 1.   (5)

14.5 (This problem may take a fair amount of effort.) Consider a VAV system that supplies two zones: an interior zone and a perimeter zone. Both are maintained at $T_{int} = 23°C$ (73.4°F). Suppose the cooling load of the interior zone is constant at $\dot{Q}_1 = 1 kW_t$ (3412 Btu/h) while solar gains cause the load of the perimeter zone to vary from $\dot{Q}_1$ to $4\dot{Q}_1$ at a time when the outdoor temperature is at 33°C (91.4°F) and 50 percent relative humidity. These loads exclude the contribution of the outdoor air. The minimum outdoor air requirement is $\dot{V}_o = 34.7$ L/s (73.6 ft³/min), in each zone. Suppose that the supply air temperature is constant at $T_{sup} = 14°C$ (57.2°F) and 100 percent relative humidity. (a) Draw a schematic of the system and indicate how the airflow rates are controlled. (b) Find the maximum and minimum flow rates for the central supply airflow. (c) Find the maximum and minimum of the total cooling load, including outdoor air. (d) How much could the total cooling load be reduced if the zones were supplied by separate systems? (e) How would the results change if T_{sup} were variable? (f) Evaluate the multizone efficiency of Eq. (14.17).   (10)

14.6 Evaluate the contribution of outdoor air to the annual heating energy of a house for two design options: for natural infiltration and mechanical ventilation. Make the following assumptions. The house has volume $V = 360$ m³ (12,708 ft³) and conductive heat loss coefficient $K_{cond} = 150$ W/K [284.3 Btu/(h · °F)]. With mechanical ventilation there is a constant outdoor airflow of 10 L/s (20 ft²/min) per occupant, for an occupancy of 3. For natural infiltration use the LBL model of Sec. 7.1.4 with the coefficients for a two-story building at a site with shielding class 3; and choose the leakage area such that the infiltration rate is 3×10 L/s when the indoor-outdoor temperature difference $T_i - T_o = 5$ K (9°F) and the wind speed $= 5$ m/s (11.5 mi/h). Use the bin data for Chicago, and assume a constant wind speed of 5 m/s (11.5 mi/h), for simplicity. How does the heating energy reduction compare with the energy for the fan if the fan draws 100 W continuously?   (10)

14.7 (This problem may take a fair amount of effort.) Consider a VAV system that supplies two zones, an interior zone and a perimeter zone, each with 50-m² (538-ft²) floor area and height of 2.5 m (8.2 ft). Both are maintained at $T_i = 23°C$ (73.4°F). Suppose that the cooling load of the interior zone is constant at $\dot{Q}_1 = 1$ kW$_t$ (3412 Btu/h) while the perimeter zone has a heating load of $\dot{Q}_2 = (UA)_{per}(T_i - T_o)$. These loads exclude the contribution of the outdoor air. The minimum outdoor air requirement is $\dot{V}_o = 34.7$ L/s (73.6 ft³/min)(2.5 m × 50 m² × 1.0 ach) in each zone. The supply air temperature is 14°C (57.2°F). Neglect the effects of humidity. (a) Draw a schematic of the system, and indicate how the system is controlled. (b) Find the cutoff temperature of the economizer. (c) Find the supply airflow rate. (d) How much reheat is needed as a function of T_o? (e) Calculate and plot the total thermal power as a function of T_o. (f) Evaluate the multizone efficiency of Eq. (14.17).   (10)

14.8 Consider a building whose cooling load is proportional to the temperature difference between T_o and 65°F (18.3°C), reaching a peak of 200 tons (703 kW) at 100°F (37.8°C). Set up a spreadsheet with bin data for Albuquerque, New Mexico, to evaluate the annual energy consumption for these two options: a single 200-ton (703-kW$_t$) chiller with COP $= 4.5$, and two chillers with 100 tons (351.5 kW$_t$) each and COP $= 4.0$. Both have the same part-load efficiency curve, given by Eq. (10.29) with $A = 0.160$, $B = 0.316$, and $C = 0.519$ (the coefficients of Example 10.7). For the two-chiller system, you need to decide (you can use trial and error) at each temperature bin how to split the load among the two chillers to maximize the efficiency. (9)

14.9 As a simple approximation of a house with passive solar heating using a water Trombe wall (see Fig. 14.17d), assume that the only solar gains come from a south-facing window and that they are entirely absorbed in a perfectly mixed water tank. The house is characterized by the thermal network of Example 8.8 and Figure 8.11, and the storage tank is included in the heat capacity C_i; the other parameters are the same as in Example 8.8:

$$C_i = 1.0\,\text{kWh/K} + \rho V c_p \text{ of storage tank}$$
$$C_e = 5.0\,\text{kWh/K} \,(9.48\,\text{kBtu/°F})$$
$$R_i = 2.5\,\text{K/kW} \,[1.32\,(\text{h}\cdot°\text{F})/\text{Btu}] = R_o$$

Assume that the transmitted solar flux per area A of the window is given by

$$I(t) = \max\left\{0, -0.2 + \cos\left[(t - 12\,\text{h})\frac{\pi}{12\,\text{h}}\right]\right\} \times 800\,\text{W/m}^2$$

as function of time of day t. The outdoor temperature is

$$T_o(t) = \cos\left[(t - 16\,\text{h})\frac{\pi}{12\,\text{h}}\right] \times 5°\text{C}$$

(a) Write down the differential equations and specify all the input.

(b) Set up a spreadsheet or program to solve the equation for T_i numerically for the case where the only other energy input is a constant base level of 1.0 kW (you can use, e.g., the method illustrated in Example 8.9).

(c) Vary window area A and storage volume V until T_i will always be within the range of 20 to 24°C (68 to 75.2°F). (10)

14.10 Set up a spreadsheet for the transfer function method to solve Examples 7.17 to 7.19. For the outdoor temperature T_o, assume sinusoidal variation with average 75°F (23.9°C) and peak 89°F (31.7°C) (5 percent design conditions for Denver). Examine how much the peak cooling load and the daily cooling energy could be reduced if one ventilates the building with 1, 5, and 20 air changes per hour whenever T_o is below T_i. (10)

14.11 Suppose you have to design a daylighting system using skylights for a one-story office building with floor area 200 ft × 100 ft (60.96 m × 30.48 m) in Atlanta. How large would you choose the aperture area, and what type of glazing would you use? Approximately how large are the annual energy savings? (9)

14.12 Consider a house in Washington, D.C., with a heat loss coefficient $K_{tot} = 389$ Btu/ (h · °F), a constant indoor temperature of 68°F, and average internal gains of

3400 Btu/h. Compare the annual energy cost of heating this house with the following three options:

- A conventional natural gas furnace (efficiency of 75 percent)
- A condensing natural gas furnace (efficiency of 93 percent)
- A heat pump with these characteristics:

$$COP = \begin{cases} 2.6 & \text{for outdoor temperature greater than } 50°F \\ 2.2 & \text{for outdoor temperatures between 30 and } 50°F \\ 1.7 & \text{for outdoor temperatures between 15 and } 30°F \\ 1.0 & \text{for outdoor temperatures less than } 15°F \end{cases}$$

Bin temperature data or Washington can be found in the HCB software. Natural gas costs \$6/GJ and electricity costs 8¢/kWh.   (6)

14.13 The occupant of the house in Prob. 14.12 installs a thermostat that automatically reduces the set point temperature from 68 to 60°F during the hours of 8 p.m. through 8 a.m. What is the simple payback time if this thermostat costs \$100 and the house uses the conventional gas furnace? Neglect any transient behavior of the heating system or any thermal storage effects in the house.   (4)

14.14 Many university campuses in the United States are installing large cogeneration facilities to reduce their utility bills. Consider a hypothetical, unrealistically simple case where a university has a constant electric load of 10 MW year-round. The thermal load follows the *load duration curve* as shown in Fig. P14.14a. Assume the price of electricity is constant at \$0.035/kWh year-round and the price of gas is \$0.45/therm year-round. Suppose gas turbine cogeneration systems are available in any size and that of the energy contained in the gas, 30 percent is converted to electricity and 45 percent is available as usable heat. That is, the electrical and thermal efficiencies are 30 and 45 percent, respectively. What are the energy savings if (a) the system is sized to meet the electrical load (10 MW), (b) the system is sized to meet the peak thermal load (100 MMBtu/h), and (c) the system is sized to meet the base thermal load (20 MMBtu/h)?   (8)

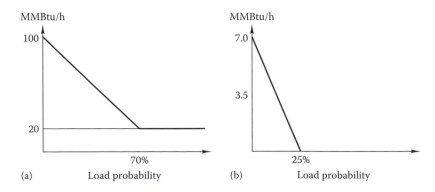

FIGURE P14.14
Load duration curve. The line represents the fraction of time (*X* axis) that the load is at or above the corresponding value on the *Y* axis. For example, for the load duration curve on the right, the dotted line shows that 12.5 percent of the time the load is above 3.5 MMBtu/h. Also note that these examples are highly simplified and that in practice the curves are less regular than shown here.

14.15 Consider an office building in New York that uses a 500-kW generator to produce power on site and reduce the overall peak demand. Recovered heat from the generator is used to satisfy the thermal load. The thermal load duration curve for this building is shown in Fig. 14.14b. Assume the electrical and thermal efficiencies of the generator are 30 and 50 percent, respectively, and that electricity costs 2¢/kWh with a demand charge of $20 per peak kilowatt and gas costs $6.50/MBtu. If the building electrical load is 200 kW during unoccupied hours and 1,500 kW during occupied hours, do most of the savings come from reduced electricity use or from reduced gas consumption used for heating? (9)

References

ACEEE (1989). *The Most Energy-Efficient Appliances*. Published annually by the American Council for an Energy-Efficient Economy (ACEEE), Suite 535, 1001 Connecticut Ave., NW, Washington, D.C. 20036.

Arastch, D., R. Johnson, S. Selkowitz, and D. Connell (1986). "Cooling Energy and Cost Savings with Daylighting in a Hot and Humid Climate." *Sunworld*, vol. 10, no. 4, p. 104.

ASHRAE (1984). *Passive Solar Heating Analysis: A Design Manual*. American Society of Heating, Refrigerating and Air-Conditioning Engineers, 1791 Tullie Circle NE, Atlanta, GA 30329.

ASHRAE (1985). *Design/Data Manual for Closed-Loop Ground Coupled Heat Pump Systems*. American Society of Heating, Refrigerating and Air-Conditioning Engineers, Atlanta.

ASHRAE (1987). *Handbook of HVAC Systems and Applications*. American Society of Heating, Refrigerating and Air-Conditioning Engineers, Atlanta.

ASHRAE (1989a). *Handbook of Fundamentals*. American Society of Heating, Refrigerating and Air-Conditioning Engineers, Atlanta.

ASHRAE (1989b). *Standard 90.1-1989: Energy Efficient Design of New Buildings, except Low-Rise Residential Buildings*. American Society of Heating, Refrigerating and Air-Conditioning Engineers, Atlanta.

ASHRAE (1990). *Standard 105-1990: Standard Methods of Measuring and Expressing Building Energy Performance*. American Society of Heating, Refrigerating and Air-Conditioning Engineers, Atlanta.

Bauman, F. S., J. W. Place, B. Andersson, J. Thornton, and T. C. Howard (1987). "The Experimentally Measured Performance of a Linear Roof Aperture Daylighting System." *ASHRAE Trans.*, vol. 93, p. 1.

Berkeley Solar Group (1981). *CALPAS3 User's Manual*. Berkeley, Calif.

Birdsall, B., W. F. Buhl, K. L. Ellington, A. E. Erdem, and F. C. Winkelmann (1990). "Overview of the DOE 2.1 Building Energy Analysis Program." Report LBL-19735, rev. 1. Lawrence Berkeley Laboratory, Berkeley, Calif.

BLAST (1986). "The Building Load Analysis and System Thermodynamics Program," Version 3. *User's Manual*. BLAST Support Office, University of Illinois, Champaign-Urbana.

BOMA (1991). Building Owners and Managers Association Experience Report (published annually).

Burt Hill Kosar Rittelmann Associates (1985). *Small Office Building Design Handbook: Design for Reducing First Costs and Utility Costs*. Van Nostrand Reinhold, New York.

Burt Hill Kosar Rittelmann Associates and Min Kantrowitz Associates (1987). *Commercial Building Design: Integrating Climate, Comfort and Cost*. Van Nostrand Reinhold, New York.

Byrne, S. J., and J. M. Fay (1989). "A Comparison of Central and Individual Systems for Space Conditioning and Domestic Hot Water in New Multifamily Buildings." *ASHRAE Trans.*, vol. 95, p. 2.

Byrne, S. J., and R. L. Ritschard (1985). "A Parametric Analysis of Thermal Mass in Residential Buildings." Lawrence Berkeley Laboratory Report LBL-20288, Berkeley, Calif.

CABO (1987). *Model Energy Code*. Council of American Building Officials, Falls Church, Va.

Christian, J. E. (1991). "Thermal Mass Credits Relating to Building Envelope Energy Standards." *ASHRAE Trans.*, vol. 97, p. 2.

Cowing, T., and J. F. Kreider (1987). "Solar Load Ratio Statistics for the United States and Their Use for Analytical Solar Performance Prediction." *ASME J. Solar Energy Eng.*, vol. 109, pp. 281–288.

Crabb, J. A. (1988). *An Approach to the Design of Energy Efficient Heated Buildings*. SWEG Report 41. Energy Studies Unit, Department of Physics, University of Exeter, Devon, U.K.

Crawley, D. B., and Y. J. Huang (1989). "Using the Office Building and Multifamily Data Bases in the Assessment of HVAC Equipment Performance." *ASHRAE Trans.*, vol. 95, p. 2.

Duffie, J. A., and W. A. Beckman (1980). *Solar Engineering of Thermal Processes*. Wiley, New York.

EIA (1981). "Nonresidential Buildings Energy Consumption Survey (NBECS): Fuel Characteristics and Conservation Practices." EIA Report (June). Energy Information Administration, U.S. Department of Energy, Washington.

EIA (1983). "Nonresidential Buildings Energy Consumption Survey (NBECS): 1979 Consumption and Expenditures, Part 2." EIA Report DOE/EIA-0138(79)/2 (December). Energy Information Administration, U.S. Department of Energy, Washington.

EPRI (1984). "Commercial Cool Storage Primer." Report EPRI EM-3371, Electric Power Research Institute, Palo Alto, CA 94303.

Fairey, P. W., and A. A. Kerestecioglu (1985). "Dynamic Modelling of Combined Thermal and Moisture Transport in Buildings: Effects on Cooling Loads and Space Conditions." *ASHRAE Trans.*, vol. 91, pt. 2A, p. 461.

Fels, M. F., ed. (1986). "Measuring Energy Savings: The Scorekeeping Approach." Special double issue of *Energy and Buildings*, vol. 9, nos. 1 and 2.

Gordon, J. M., and Y. Zarmi (1981a). "Analytic Model for Passively Heated Solar Houses: 1. Theory." *Solar Energy*, vol. 27, p. 331.

Gordon, J. M., and Y. Zarmi (1981b). "Analytic Model for Passively Heated Solar Houses: 2. User's Guide." *Solar Energy*, vol. 27, p. 343.

Grot, R. A., et al. (1985). "Evaluation of the Thermal Integrity of the Building Envelopes of Eight Federal Office Buildings." Center for Building Technology Report NBSIR 85-3147 (September). National Bureau of Standards, Gaithersburg, Md.

Jones, R. W., ed. (1983). *Passive Solar Design Handbook*, vol. 3. American Solar Energy Society, Boulder, Colo.

Kao, J. Y. (1982a). "Strategies for Energy Conservation in Small Office Buildings." Center for Building Technology Report NBSIR 84-2489 (June). National Bureau of Standards, Gaithersburg, Md.

Kao, J. Y. (1982b). "Strategies for Energy Conservation for a Large Retail Store." Center for Building Technology Report NBSIR 84-2580 (September). National Bureau of Standards, Gaithersburg, Md.

Kao, J. Y. (1983). "Strategies for Energy Conservation for a Large Office Building." Center for Building Technology Report NBSIR 84-2746. National Bureau of Standards, Gaitherburg, Md.

Kao, J. Y. (1984). "Strategies for Energy Conservation for a School Building." Center for Building Technology Report NBSIR 84-2831 (March). National Bureau of Standards, Gaitherburg, Md.

Kao, J. Y. (1985). "Control Strategies and Building Energy Consumption." *ASHRAE Trans.*, vol. 91, pt. 2B, pp. 810–817.

Kreider, J. F., and F. Kreith (1990). *Solar Design: Components, Systems and Economics*. Hemisphere, New York.

LBL (1988). "Energy and Environment Division: Annual Report." Report LBL-26585, Energy Efficient Buildings Program, Lawrence Berkeley Laboratory, Berkeley, Calif.

Meier, A. K. (1990). "Measured Cooling Savings from Vegetative Landscaping." *American Council for an Energy Efficient Economy, 1990 Summer Study on Energy Efficiency in Buildings*, vol. 4, p. 133, Asilomar, Calif.

Mitchell, J. W., and W. A. Beckman (1989). "Theoretical Limits for Storage of Energy in Buildings." *Solar Energy*, vol. 42, pp. 113–120.

Monsen, W. A., S. A. Klein and W. A. Beckman (1982). "The Un-Utilizability Design Method for Collector Storage Walls." *Solar Energy*, vol. 29(5), 421–429.

Piette, M. A., and R. Riley (1986). "Energy Use and Peak Power for New Commercial Buildings from the BECA-CN Data Compilation: Key Findings and Issues." Report LBL: 20896 (March), Lawrence Berkeley Laboratory, Berkeley, Calif.

Piette, M. A., L. W. Wall, and B. L. Gardiner (1986). "Measured Performance." *ASHRAE J.* (January), pp. 72–78.

Place, J. W., J. P. Coutier, M. R. Fontoynont, R. C. Kammerud, B. Andersson, W. L. Carroll, M. A. Wahlig, F. S. Bauman, and T. L. Webster (1987). "The Impact of Glazing Orientation, Tilt, and Area on the Energy Performance of Room Apertures." *ASHRAE Trans.*, vol. 93, p. 1.

Rabl, A., and A. Rialhe (1992). "Energy Signature Models for Commercial Buildings: Test with Measured Data and Interpretation." *Energy and Buildings*, vol. 19, pp. 143–154.

Rabl, A., L. K. Norford, and G. V. Spadaro (1986). "Steady State Models for the Analysis of Commercial Building Energy Data." *American Council for an Energy Efficient Economy, 1986 Summer Study*, vol. 9, p. 239, Santa Cruz, Calif.

Ribot, J., and A. Rosenfeld (1982). "Monitored Low-Energy Houses in North America and Europe: A Compilation and Economic Analysis." *American Council for an Energy Efficient Economy, 1982 Summer Study*, Santa Cruz, Calif.; and Report LBL-14788, Lawrence Berkeley Laboratory, Berkeley, Calif.

SERI (1984). "Passive Solar Performance: Summary of 1982–1983 Class B Results." Solar Energy Research Insitute, Report SERI/SP-271-2362, Golden, Colo.

SOLMET (1978). Volume 1: *User's Manual*. Volume 2: *Final Report, Hourly Solar Radiation Surface Meteorological Observations*. Report TD-9724. National Climatic Center, Asheville, N.C.

Spitler, J. D., D. C. Hittle, D. L. Johnson, and C. O. Pederson (1987). "A Comparative Study of the Performance of Temperature-Based and Enthalpy-Based Economy Cycles." *ASHRAE Trans.*, vol. 93, pt. 2, pp. 13–22.

Sullivan, R., and S. Selkowitz (1985). "Energy Performance Analysis of Fenestration in a Single-Family Residence." *ASHRAE Trans.*, vol. 91, pt. 2A, pp. 320–336.

Sullivan, R., and S. Selkowitz (1987). "Residential Heating and Cooling Energy Cost Implications Associated with Window Type." *ASHRAE Trans.*, vol. 93, p. 1.

Sweitzer, G., D. Arasteh, and S. Selkowitz (1987). "Effects of Low-Emissivity Glazings on Energy Use Patterns in Nonresidential Daylighted Buildings." *ASHRAE Trans.*, vol. 93, p. 1.

Usibelli, A., et al. (1985). "Commercial Sector Conservation Technologies." Lawrence Berkeley Laboratory Report LBL 18543 (February), Berkeley, Calif.

15

Costs

We do not live in paradise, and our resources are limited. Therefore it behooves us to try to reduce the cost of heating and cooling to a minimum—subject, of course, to the constraint of providing the desired indoor environment and services. But while capital costs and operating costs are readily stated in financial terms, other factors, such as comfort, convenience, and aesthetics, may be difficult or impossible to quantify. Furthermore, there is uncertainty: Future energy prices, future rental values, future equipment performance, future uses of a building, all are uncertain.

As a way around the difficulties, it is best to approach the design optimization in the following manner. First one evaluates the total cost for each proposed design or design variation, by properly combining all capital and operating costs. Then, knowing the cost of each design, one can select the "best," much like selecting the best product in a store where each product carries a price tag. Proceeding in this way, one separates the factors that can be quantified unambiguously (i.e., the calculation of the price tag) from those that are less tangible (e.g., aesthetics). The calculation of the price tag is the essence of engineering economics; it forms the main part of this chapter. Optimization and some effects of uncertainty are addressed at the end.

15.1 Comparing Present and Future Costs

15.1.1 The Effect of Time on the Value of Money

Before one can compare first costs (i.e., capital costs) and operating costs, one must apply a correction because a dollar (or any other currency unit) to be paid in the future does not have the same value as a dollar today. This time dependence of money is due to two, quite different causes. The first is inflation, the well-known and ever-present erosion of the value of our currency. The second reflects the fact that a dollar today can buy goods to be enjoyed immediately, or it can be invested to increase its value by profit or interest. Thus a dollar that becomes available in the future is less desirable than a dollar today; its value must be discounted. This is true even if there is no inflation. Both inflation and discounting are characterized in terms of annual rates.

Let us begin with inflation. To avoid confusion, it is advisable to add subscripts to the currency signs, indicating the year in which the currency is specified. For example, during the mid-1980s the inflation rate r_{inf} in Western industrial countries was around $r_{inf} = 4$ percent. Thus a dollar bill in 1986 is worth only $1/(1+0.04)$ as much as the same dollar bill 1 year before:

$$\$_{1986}1.00 = \$_{1985}\frac{1}{1+r_{inf}} = \$_{1985}\frac{1}{1+0.04} = \$_{1985}0.96$$

Actually the definition and measure of the inflation rate are not without ambiguities since different prices escalate at different rates and the inflation rate depends on the mix of goods assumed. Probably the most common measure is the *consumer price index* (CPI), an index that has arbitrarily been set at 100 in 1983. Its evolution is shown in Figure 15.1, along with another index of interest to the HVAC designer: the *Engineering News Record* construction

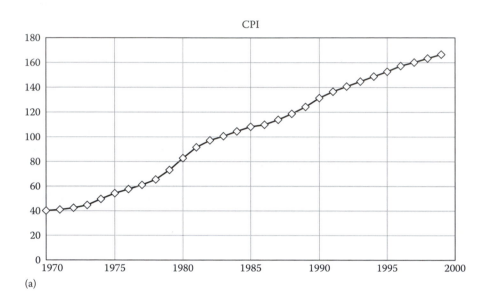

(a)

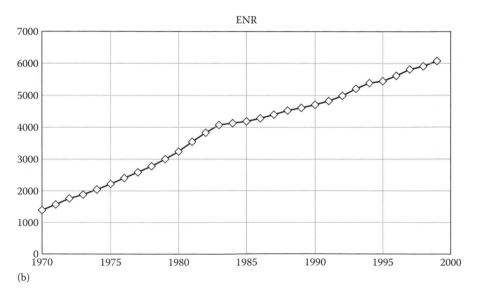

(b)

FIGURE 15.1

History of various cost indices: (a) CPI = consumer price index (from ftp://ftp.bls.gov/pub/special.requests/cpi/cpiai.txt). (b) ENR = *Engineering News Record* construction cost index.

cost index. In terms of the CPI, the average inflation rate from year ref to year ref $+ n$ is given by[1]

$$(1 + r_{\text{inf}})^n = \frac{\text{CPI}_{\text{ref}+n}}{\text{CPI}_{\text{ref}}} \tag{15.1}$$

Suppose $\$_{1985}1.00$ has been invested at an interest rate $r_{\text{int}} = 10$ percent, the *nominal* or *market* rate, as usually quoted by financial institutions. Then after 1 year this dollar has grown to $\$_{1986}1.10$, but it is worth only $\$_{1985}1.10/1.04 = \$_{1985}1.06$. To show the increase in the real value, it is convenient to define the real interest rate $r_{\text{int}0}$ by the relation

$$1 + r_{\text{int}0} = \frac{1 + r_{\text{int}}}{1 + r_{\text{inf}}} \tag{15.2}$$

or

$$r_{\text{int}0} = \frac{r_{\text{int}} - r_{\text{inf}}}{1 + r_{\text{inf}}}$$

The simplest way of dealing with inflation is to eliminate it from the analysis right at the start by using *constant currency* and expressing all growth rates (interest, energy price escalation, etc.) as real rates, relative to constant currency. After all, one is concerned about the real value of cash flows, not about their nominal values in a currency eroded by inflation. Constant currency is obtained by expressing the *current* or *inflating* currency of each year (i.e., the nominal value of the currency) in terms of equivalent currency of an arbitrarily chosen reference year ref. Thus the current dollar of year ref $+ n$ has a constant dollar value of

$$\$_{\text{ref}} = \frac{\$_{\text{ref}+n}}{(1 + r_{\text{inf}})^n} \tag{15.3}$$

A *real growth rate* r_0 is related to the *nominal growth rate* r in a way analogous to Eq. (15.2):

$$r_0 = \frac{r - r_{\text{inf}}}{1 + r_{\text{inf}}} \tag{15.4}$$

For low inflation rates one can use the approximation

$$r_0 \approx r - r_{\text{inf}} \quad \text{if } r_{\text{inf}} \text{ small} \tag{15.5}$$

As proved in Sec. 15.2.6, an analysis in terms of constant currency and real rates is exactly equivalent to one with inflating currency and nominal rates, if the investment is paid out of equity (i.e., without loan) and without tax deduction for depreciation or interest. Slight real differences between the two approaches can arise from the formulas for depreciation and for loan payments (in the United States, loan payments are usually arranged to have fixed amounts in current currency, and the real value of annual loan payments differs

---

[1] For simplicity we write the equations as if all growth rates were constant. Otherwise the factor $(1 + r)^n$ would have to be replaced by the product of factors for each year $(1 + r_1)(1 + r_2) \cdots (1 + r_n)$. Such a generalization is straightforward but tedious, and of dubious value in practice as it is chancy enough to predict average trends without trying to guess a detailed scenario.

between the two approaches). Therefore the inflating-dollar approach is commonly chosen in the U.S. business world.

However, when the constant-dollar approach is correct, it offers several advantages. Having one variable less, it is simpler and clearer. What is more important, the long-term trends of real growth rates are fairly well known even if the inflation rate turns out to be erratic. For example, from 1955 to 1980 the real interest rate on high-quality corporate bonds consistently hovered about 2.2 percent despite large fluctuations of inflation (Jones, 1982), while the high real interest rates of the 1980s have probably been a short-term anomaly. Riskier investments, such as the stock market, may promise higher returns, but they, too, tend to be more constant in constant currency.

Likewise, prices tend to be more constant when stated in terms of real currency. This is illustrated in Fig. 15.2 by comparing some energy prices in real and inflating dollars. For example, the market price ($=$ price in inflating currency) of crude oil reached a peak of \$36/barrel in 1981, which is 10 times higher than the market price during the 1960s, while in terms of constant currency the price increase over the same period was only a factor of 4. Crude oil during the oil crises is, of course, an example of extreme price fluctuations. For other goods the price in constant currency is far more stable (it would be exactly constant in the absence of relative price shifts among different goods). Therefore it is instructive to think in terms of real rates and real currency.

Example 15.1

Find the nominal and real escalation rates for residential electricity prices between 1970 and 1995.

    Given: Data in Fig. 15.2c. In these Fig. 15.2 the real prices are in 1992 dollars.

    Find: Real growth rate r_0 and nominal growth rate r

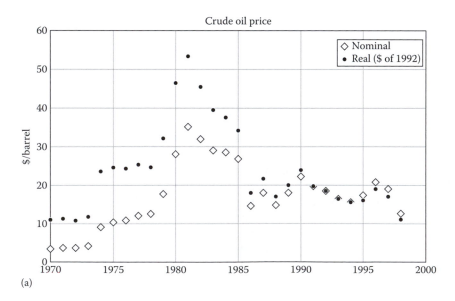

FIGURE 15.2
Energy prices, in constant dollars (solid symbols) and in inflating dollars (hollow symbols), from http://www.eia.doe.gov. (a) Crude oil.

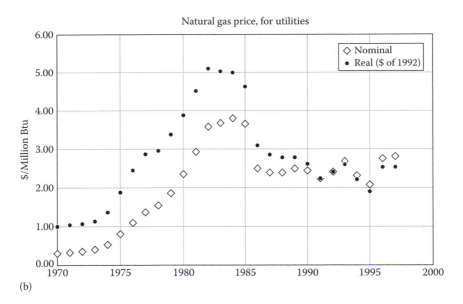

(b)

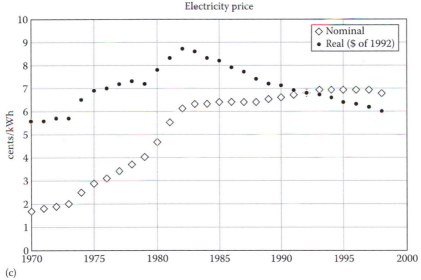

(c)

FIGURE 15.2 (continued)
(b) Natural gas (for utility companies). (c) Eectricity (average retail price).

SOLUTION

In 1970 the price was $p_{1970} = 1.7 ¢_{1970}/\text{kWh} = 5.6 ¢_{1992}/\text{kWh}$. In 1995 the price was $p_{1995} = 6.9 ¢_{1988}/\text{kWh} = 6.4 ¢_{1992}/\text{kWh}$. The number of years is $n = 1995 - 1970 = 25$. Hence the real growth rate is given by

$$r_0 = -1 + \sqrt[n]{\frac{p_{1985}}{p_{1970}}}$$

$$= -1 + \sqrt[25]{\frac{6.4}{5.6}} = -1 + 1.005 = 0.005$$

The nominal growth rate is

$$r = -1 + \sqrt[25]{\frac{6.9}{1.7}} = -1 + 1.058 = 0.058$$

COMMENTS

This highlights the importance of distinguishing between real and nominal growth rates. While the apparent price has grown by almost 6 percent per year, the real price increased only by 0.5 percent per year; the inflation rate averaged about 5.3 percent during this period.

15.1.2 Discounting of Future Cash Flows

As mentioned above, even if there were no inflation, a future cash amount F is not equal to its *present value* P; it must be discounted. The relation between P and its future value F_n in n years from now is given by the *discount rate* r_d, defined such that

$$P = \frac{F_n}{(1 + r_d)^n} \tag{15.6}$$

The higher the discount rate, the lower the present value of future transactions.

To determine the appropriate value of the discount rate, one has to ask, At what value of r_d one is indifferent between an amount P today and an amount $F_1 = P/(1 + r_d)$ a year from now? That depends on the circumstances and on individual preferences. Consider a consumer who would put his money in a savings account with 5 percent interest. His discount rate is 5 percent, because by putting the $1000 into this account he in fact accepts the alternative of $(1 + 5\%) \times \$1000$ a year from now. If instead he used it to pay off a car loan at 10 percent, then his discount rate would be 10 percent; paying off the loan is like putting the money into a savings account which pays at the loan interest rate. If the money allowed him to avoid an emergency loan at 20 percent, then his discount rate would be 20 percent. At the other extreme, if he hid the money in his mattress, his discount rate would be zero.

The situation becomes more complex when there are several different investment possibilities offering different returns at different risks, such as savings accounts, stocks, real estate, or a new business venture. By and large, if one wants the prospect of a higher rate of return, one has to accept a higher risk. Thus, as a more general rule, we can say that the appropriate discount rate for the analysis of an investment is the rate of return on alternative investments of comparable risk. In practice, that is sometimes quite difficult to determine, and it may be desirable to have an evaluation criterion that bypasses the need to choose a discount rate. Such a criterion is obtained by calculating the profitability of an investment in terms of an unspecified discount rate and then solving for the value of the rate at which the profitability goes to zero. That method, called the *internal rate of return method*, will be explained in Sec. 15.3.2.

Just as with other growth rates, one can specify the discount rate with or without inflation. If F_n is given in terms of constant currency, designated as F_{n0}, then it must be discounted with the real discount rate r_{d0}. The latter is, of course, related to the market discount rate r_d by

$$r_{d0} = \frac{r_d - r_{inf}}{1 + r_{inf}} \tag{15.7}$$

according to Eq. (15.4). Present values can be calculated with real rates and real currency, or with market rates and inflating currency; the result is readily seen to be the same because multiplying numerator and denominator of Eq. (15.6) by $(1 + r_{inf})^n$, one obtains

$$P = \frac{F_n}{(1 + r_d)^n} = \frac{F_n(1 + r_{inf})^n}{(1 + r_{inf})^n(1 + r_d)^n}$$

which is equal to

$$P = \frac{F_{n0}}{(1 + r_{d0})^n}$$

since

$$F_{n0} = \frac{F_n}{(1 + r_{inf})^n} \tag{15.8}$$

by Eq. (15.3).

The ratio P/F_n of present and future value is called the *present worth factor*. We shall designate it with the mnemonic notation

$$(P/F, r, n) = \frac{P}{F_n} = (1 + r)^{-n} \tag{15.9}$$

It is plotted in Fig. 15.3. Its inverse

$$(F/P, r, n) = \frac{1}{(P/F, r, n)} \tag{15.10}$$

is called the *compound amount factor*. These factors are the basic tool for comparing cash flows at different times. Note that we have chosen the so-called end-of-year convention by designating F_n as the value at the end of the nth year. Also, we have assumed annual intervals, which is generally an adequate time step for engineering economic analysis; accountants, by contrast, tend to work with monthly intervals, corresponding to the way most regular bills are paid. The basic formulas are the same, but the numerical results differ because of differences in the compounding of interest; this point will be explained more fully in Sec. 15.1.4 when we pass to the continuous limit by letting the time step approach zero.

Example 15.2

What might be an appropriate discount rate for analyzing the energy savings from a proposed new cogeneration plant for a university campus? Consider the fact that from 1970 to 1988 the endowment of the university has grown by a factor 8 (current dollars) due to profits from investments.

Given:

Growth factor in current dollars $= 8.0$

Increase in CPI $= \frac{118.3}{38.9} = 3.04$, from Fig. 15.1a over $N = 18$ years

Find: Real discount rate r_{d0}

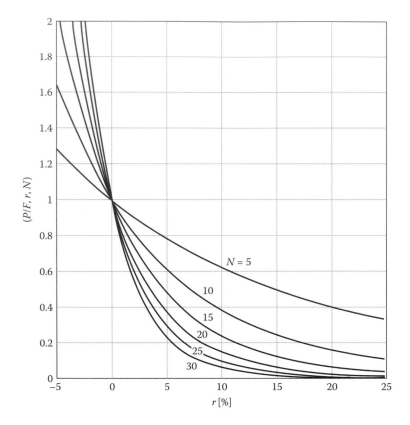

FIGURE 15.3
The present worth factor $(P/F, r, N)$ as function of rate r and number of years N.

SOLUTION

There are two equivalent ways of solving for r_{d0}. First, take the real growth factor, 8.0/3.04, and set it equal to $(1 + r_{d0})^N$. The result is $r_{d0} = 5.52$ percent.

Second, calculate market rate r_d by setting the market growth in current dollars equal to $(1 + r_d)^N$ and calculating the inflation by setting the CPI increase equal to $(1 + r_{inf})^N$. We find $r_d = 12.246$ percent and $r_{inf} = 6.371$ percent. Then we can solve Eq. (15.7) for r_{d0}, with the result

$$r_{d0} = \frac{0.12246 - 0.06371}{1 + 0.06371} = 5.52 \text{ percent}$$

the same as before.

COMMENTS

Choosing a discount rate is not without pitfalls. For this example the comparison with the real growth of other long-term investments seems appropriate; of course there is no guarantee that the endowment will continue growing at the same real rate in the future.

15.1.3 Equivalent Cash Flows and Levelizing

It will be convenient to express payments that are irregular or variable as equivalent equal payments in regular intervals; in other words, one replaces nonuniform series by equivalent uniform series. We shall refer to this technique as *levelizing*. It is useful because regularity facilitates understanding and planning. To develop the formulas, let us calculate the present value P of a series of N equal annual payments A. If the first payment occurs at the end of the first years, its present value is $A/(1+r_d)$. For the second year it is $A/(1+r_d)^2$, etc. Adding all the present values from years 1 to N, we find the total present value

$$P = \frac{A}{1+r_d} + \frac{A}{(1+r_d)^2} + \cdots + \frac{A}{(1+r_d)^N} \tag{15.11}$$

This is a simple geometric series, and the result is readily summed to

$$P = A\frac{1-(1+r_d)^{-N}}{r_d} \qquad \text{for } r_d \neq 0 \tag{15.12}$$

For zero discount rate this equation is indeterminate, but its limit $r_d \to 0$ is A/N, reflecting the fact that the N present values all become equal to A in that case. Analogous to the notation for the present worth factor, we designate the ratio of A and P by

$$(A/P, r_d, N) = \begin{cases} \frac{r_d}{1-(1+r_d)^{-N}} & \text{for } r_d \neq 0 \\ \frac{1}{N} & \text{for } r_d = 0 \end{cases} \tag{15.13}$$

It is called the *capital recovery factor*, and we have plotted it in Fig. 15.4. For the limit of long life $N \to \infty$, it is worth noting that $(A/P, r_d, N) \to r_d$ if $r_d > 0$.

The inverse is known as the *series present worth factor* since P is the present value of a series of equal payments A.

With the help of the present worth factor and capital recovery factor, any single expense C_n that occurs in year n, for instance, a major repair, can be expressed as an equivalent annual expense A that is constant during each of the N years of the life of the system. The present value of C_n is $P = (P/F, r_d, n)C_n$, and the corresponding annual cost is

$$\begin{aligned} A &= (A/P, r_d, N)(P/F, r_d, n)C_n \\ &= \frac{r_d}{1-(1+r_d)^{-N}}(1+r_d)^{-n}C_n \end{aligned} \tag{15.14}$$

Example 15.3

A system has salvage value of $1000 at the end of its useful life of $N=20$ years. What is the equivalent levelized annual value if the discount rate is 8 percent?

Given: $C_{20}=\$1000$, $N=n=20$, and $r_d=0.08$

Find: A

Lookup values:

$(A/P, r_d, N)=0.1019$ for Fig. 15.4 or Eq. (15.13)

$(P/F, r_d, n)=0.2145$ from Fig. 15.3 or Eq. (15.9)

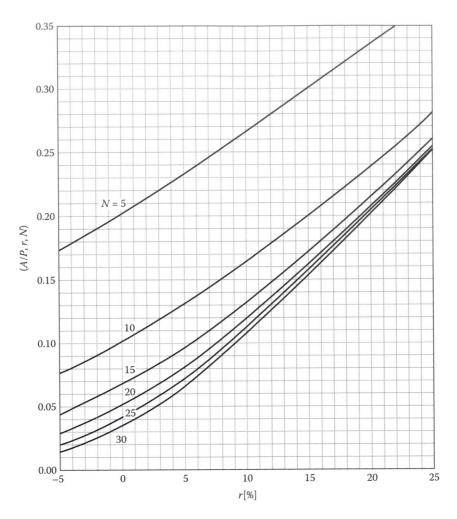

FIGURE 15.4
The capital recovery factor $(A/P, r, N)$ as function of rate r and number of years N.

SOLUTION

Insert into Eq. (15.14)

$$A = (A/P, r_d, N)(P/F, r_d, n)C_{20} = 0.1019 \times 0.2145 \times 1000\$/\text{yr} = 21.86\$/\text{yr}$$

A very important application of the capital recovery factor is the calculation of loan payments. In principle, a loan could be repaid according to any arbitrary schedule, but in practice the most common arrangement is based on constant payments in regular intervals. The portion of A due to interest varies, in a way to be calculated in Sec. 15.2.2; but to find the relation between A and the loan amount L, we need not worry about that. Let us first consider a loan of amount L_n that is to be repaid with a single payment F_n at the end of n years. With n years of interest, at loan interest rate r_l, the payment must be

$$F_n = L_n(1 + r_l)^n$$

Comparison with the present worth factor shows that the loan amount is the present value of the future payment F_n, discounted at the loan interest rate.

A loan that is to be repaid in N equal installments can be considered as the sum of N loans, the nth loan to be repaid in a single installment A at the end of the nth year. Discounting each of these payments at the loan interest rate and adding them, we find the total present value; it is equal to the total loan amount

$$L = P = \frac{A}{1 + r_l} + \frac{A}{(1 + r_l)^2} + \cdots + \frac{A}{(1 + r_l)^N} \qquad (15.15)$$

This is just the series of the capital recovery factor. Hence the relation between annual loan payment A and loan amount L is

$$A = (A/P, r_l, N)L \qquad (15.16)$$

Now the reason for the term *capital recovery factor* becomes clear: It is the rate at which a bank recovers its investment in a loan.

Example 15.4

A home buyer obtains a mortgage of $100,000 at an interest rate of 8 percent over 20 years. What are the annual payments?

    Given: $L = \$100,000$, $r_l = 8$ percent, $N = 20$ yr

    Find: A

SOLUTION

From Fig. 15.4 the capital recovery factor is 0.1019, and the annual payments are $10,190, approximately one-tenth of the loan amount.

Some payments increase or decrease at a constant annual rate. It is convenient to replace a growing or diminishing cost by an equivalent constant or *levelized* cost. Suppose the price of energy is p_e at the start of the first year, escalating at an annual rate r_e while the discount rate is r_d. If the annual energy consumption Q is constant, then the present value of all the energy bills during the N years of system life is

$$P_e = Q p_e \left[\left(\frac{1 + r_e}{1 + r_d} \right)^1 + \left(\frac{1 + r_e}{1 + r_d} \right)^2 + \cdots + \left(\frac{1 + r_e}{1 + r_d} \right)^N \right] \qquad (15.17)$$

(assuming the end-of-year convention described above). As in Eq. (15.3), we introduce a new variable $r_{d,e}$, defined by

$$1 + r_{d,e} = \frac{1 + r_d}{1 + r_e} \qquad (15.18)$$

or

$$r_{d,e} = \frac{r_d - r_e}{1 + r_e} \quad (\approx r_d - r_e \quad \text{if } r_e \ll 1) \qquad (15.19)$$

which allows us to write P_e as

$$P_e = (P/A, r_{d,e}, N)Qp_e \qquad (15.20)$$

Since $(A/P, r_d, N)$ is the inverse of $(A/P, r_d, N)$, we can write this as

$$P_e = (P/A, r_d, N)Q\left[\frac{(A/P, r_d, N)}{(A/P, r_{d,e}, N)}p_e\right]$$

If the quantity in brackets were the price, this would be just the formula without escalation. Let us call this quantity the *levelized* energy price $\bar{p}_e$, where

$$\bar{p}_e = \frac{(A/P, r_d, N)}{(A/P, r_{d,e}, N)}p_e \qquad (15.21)$$

It allows us to calculate the costs as if there were no escalation. Levelized quantities can fill a gap in our intuition which is ill prepared to gauge the effects of exponential growth over an extended period. The levelizing factor

$$\text{Levelizing factor} = \frac{(A/P, r_d, N)}{(A/P, r_{d,e}, N)} \qquad (15.22)$$

tells us in effect the average of a quantity that changes exponentially at a rate r_e while being discounted at a rate r_d, over a lifetime of N years. It is plotted in Fig. 15.5 for a wide range of the parameters.

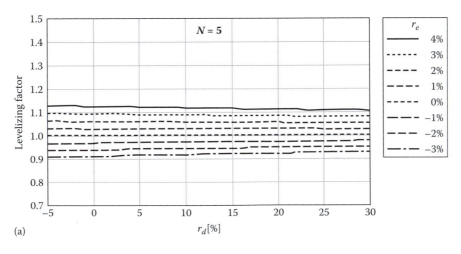

(a)

FIGURE 15.5
Levelizing factor $(A/P, r_d, N)/(A/P, r_{d,e}, N)$ as function of r_d and r_e for (a) $N = 5$ yr.

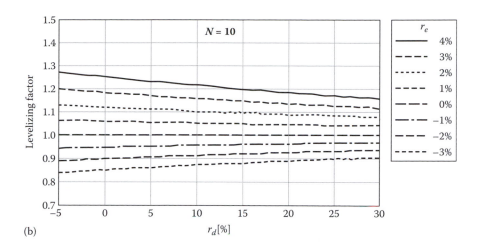

(b)

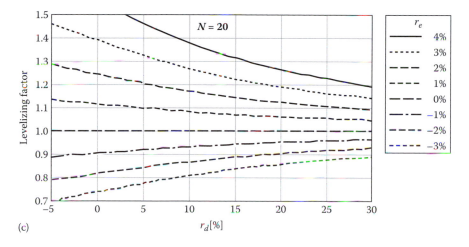

(c)

FIGURE 15.5 (continued)
(b) $N = 10$ yr, (c) $N = 20$ yr.

Example 15.5

The price of fuel is $p_e = \$5/GJ$ at the start of the first year, growing at a rate $r_e = 4$ percent while the discount rate is $r_d = 6$ percent. What is the equivalent levelized price over $N = 20$ years?

  Given: $p_e = \$5/GJ$, $r_e = 4$ percent, $r_d = 6$ percent, $N = 20$ yr
  Find: $\bar{P}_e$

SOLUTION

From Fig. 15.5 the levelizing factor is 1.44. Hence the levelized fuel price is $\bar{P}_e = 1.44 \times \$5/GJ = \$7.20$ GJ.

Several features may be noted in Fig. 15.5. First, the levelizing factor increases with cost escalation r_e, being unity if $r_e = 0$. Second, for a given escalation rate, the levelizing factor decreases as the discount rate increases, reflecting the fact that a high discount rate deemphasizes the influence of high costs in the future.

15.1.4 Discrete and Continuous Cash Flows

The above formulas suppose that all costs and revenues occur in discrete intervals. That is common engineering practice, in accord with the fact that bills are paid in discrete installments.Thus growth rates are quoted as annual changes even if growth is continuous. It is instructive to consider the continuous case.

Let us establish the connection between continuous and discrete growth by way of an apocryphal story about the discovery of e, the basis of natural logarithms. Before the days of compound interest, a mathematician who was an inveterate penny pincher thought about the possibilities of increasing the interest he earned on his money. He realized that if the bank gives interest at a rate of r per year, he could get even more by taking the money out after half a year and reinvesting it to earn interest on the interest as well. With m such compounding intervals per year, the money would grow by a factor

$$\left(1 + \frac{r}{m}\right)^m$$

and the larger m, the larger this factor. Of course, he looked at the limit $m \to \infty$ and found the result

$$\lim_{m \to \infty} \left(1 + \frac{r}{m}\right)^m = e^r \quad \text{with } e = 2.71828\ldots \tag{15.23}$$

At the end of 1 year the growth factor is $1 + r_{ann}$ with annual compounding at a rate r_{ann}, while with continuous compounding at a rate r_{cont} the growth factor is $\exp r_{cont}$. If the two growth factors are to be the same, the growth rates must be related by

$$1 + r_{ann} = \exp r_{cont} \tag{15.24}$$

With this replacement of rates, the continuous formulas in Table 15.1 yield the same results as the discrete ones. Similarly, with m compounding intervals at rate r_m continuous compounding is equivalent to annual compounding if one takes

$$1 + r_{ann} = \left(1 + \frac{r_m}{m}\right)^m \tag{15.25}$$

TABLE 15.1

Discrete and Continuous Formulas for Economic Analysis, with Growth Rate r and Time Horizon N. The Rates for the Discrete and Continuous Formulas are Related by Eq. (5.24)

Quantity Known	Quantity to be Found	Factor	Expression for Discrete Analysis	Expression for Continuous Analysis
P	F	$(F/P, r, N)$	$(1+r)^N$	$\exp(rN)$
F	P	$(P/F, r, N)$	$(1+r)^{-N}$	$\exp(-rN)$
P	A	$(A/P, r, N)$	$\dfrac{r}{1-(1+r)^{-N}}$	$\dfrac{\exp(r)-1}{1-\exp(-rN)}$
A	P	$(P/A, r, N)$	$\dfrac{1-(1+r)^{-N}}{r}$	$\dfrac{1-\exp(-rN)}{\exp(r)-1}$

Example 15.6

A bank quotes a nominal interest rate of 10 percent (i.e., annual growth without compounding). What is the equivalent annual growth rate with monthly, daily, and continuous compounding?

SOLUTION

For monthly compounding take $m = 12$ in Eq. (15.25), with the result

$$r_{12} = \left(1 + \frac{0.1}{12}\right)^{12} - 1 = 0.104713$$

With daily compounding we have

$$r_{365} = \left(1 + \frac{0.1}{365}\right)^{365} - 1 = 0.105155$$

and with continuous compounding

$$r_{\text{cont}} = \exp 0.1 - 1 = 0.105171$$

COMMENT

Beyond monthly compounding the differences are very small.

For small rates the the first three terms in the series expansion of the exponential give an approximation

$$r_{\text{ann}} \approx r_{\text{cont}}\left(1 + \frac{r_{\text{cont}}}{2}\right)$$

which is convenient if one does not have a calculator at hand.

15.1.5 The Rule of 70 for Doubling Times

Most of us do not have a good intuition for exponential growth. As a helpful tool we present therefore the rule of 70 for doubling times. The doubling time N_2 is related to the continuous growth rate r_{cont} by

$$2 = \exp N_2 r_{\text{cont}} \tag{15.26}$$

Solving the exponential relation for N_2, we obtain

$$N_2 = \frac{\ln 2}{r_{\text{cont}}} = \frac{0.693\ldots}{r_{\text{cont}}}$$

The product of doubling time and growth rate in units of percent is very close to 70 years:

$$N_2 r_{\text{cont}} \times 100 = 69.3\ldots \text{yr} \approx 70 \text{ yr} \tag{15.27}$$

In terms of annual rates the relation would be

$$N_2 = \frac{\ln 2}{\ln(1 + r_{\text{ann}})}$$

numerically close to Eq. (15.27) for small rates, but less convenient.

Example 15.7

Population growth rates average around 2 percent for the world as a whole and reach 4 percent in certain countries. What are the corresponding doubling times?

SOLUTION

$$\frac{70}{2} = 35 \text{ yr} \quad \text{for the world}$$

$$\frac{70}{4} = 17.5 \text{ yr} \quad \text{for countries with 4 percent growth}$$

Example 15.8

A consultant presents an economic analysis of an energy investment with $N = 20$ years, assuming a 10 percent escalation rate for energy prices without stating the inflation rate. Is that reasonable?

SOLUTION

A growth rate of 10 percent implies a doubling time of 7 years. There would be almost three doublings in 20 years, with a final energy price almost 8 times the original. In constant dollars that would clearly be an absurd hypothesis. A totally different conclusion emerges if the inflation were 6 percent: Then the real growth rate would be only $10 - 6 = 4$ percent, and the doubling time 17.5 years in constant currency; although extreme, that is not inconceivable given our experience since the 1973 oil shock.

COMMENT

There are two lessons:

1. Never state the growth rate or discount rate without indicating the corresponding inflation!
2. Be careful about assuming large growth rates over long time horizons! Use the rule of 70 to check whether the implications for the end of the time horizon make sense.

15.2 Life Cycle Cost

15.2.1 Cost Components

A rational decision is based on the true total cost. That is the sum of the present values of all cost components, and it is called *life cycle cost*. The cost components relevant for the HVAC engineer are

- Capital cost (= total initial investment)
- Energy costs
- Costs for maintenance, including major repairs
- Resale value
- Insurance
- Taxes

There is some arbitrariness in this assignment of categories. One could make a separate category for repairs, or one could include energy among O&M (operation and maintenance) cost, as it is done in some industries. There is, however, a good reason for keeping energy apart. In buildings, energy costs dominate the other O&M costs, and they can grow at a different rate. Electric rates usually contain charges for peak demand in addition to charges for energy (see Sec. 15.2.4). As a general rule, if an item is important, it merits separate treatment.

Rental income needs to be included if one wants to evaluate the profitability of the building, or if one wants to compare design options that would affect the rent. It can be left out of the picture if one is only concerned with comparing design options that do not differ in their effect on rental income. Likewise for cleaning, security, fire protection. *Quite generally when one is comparing two options, there is no need to include terms that would be the same for each.* For instance, when choosing between two chillers, one can focus on the costs associated directly with the chillers (capital cost, energy, maintenance), without worrying about the heating system if that is not affected. In some cases it becomes necessary to account for the effects of taxes, due to tax deductions for interest payments and depreciation; hence we discuss these items first, before presenting the equation for the complete system cost.

15.2.2 Principal and Interest

In the United States interest payments are deductible from the income tax, while payments for the reimbursement of the loan are not. A tax-paying investor therefore needs to know what fraction of a loan payment is due to interest. As explained in Sec. 15.1.3, we assume that a loan of duration N_l is repaid in N_l regular and equal payments A. (In this chapter we take N and N_l in years, but the formulas are valid for any choice of units, and for billing purposes the month is frequently chosen as payment period; slight numerical differences in the payments are due to compound interest.) Consider the nth payment, and let I_n = interest and P_n = principal (= loan reimbursement); their sum A is constant

$$I_n + P_n = A \tag{15.28}$$

Up to this point $n-1$ payments have been made, and so the debt remaining (on a loan of amount L) is

$$\text{Remaining debt} = L - P_1 - P_2 - \cdots - P_{n-1} \tag{15.29}$$

At a loan interest rate r_l the interest for the nth period is

$$I_n = r_l(L - P_1 - P_2 - \cdots - P_{n-1}) \tag{15.30}$$

Comparing I_{n+1} with I_n, one finds

$$I_{n+1} = I_n - r_l P_n \tag{15.31}$$

By means of Eq. (15.28) one can eliminate P_n with the result

$$I_{n+1} = (1 + r_l)I_n - r_l A \tag{15.32}$$

This recursion relation has the solution

$$I_n = (1 + r_l)^{n-1} r_l L + [1 - (1 + r_l)^{n-1}]A \tag{15.33}$$

as can readily be proved by mathematical induction. Since A and L are related by $A = (A/P, r_l, N_l)L$, where N_l is the duration of the loan and $(A/P, r_l, N_l)$ is the capital recovery factor of Eq. (15.13), this can be rewritten in the form

$$\frac{I_n}{A} = 1 - (1 + r_l)^{n-1-N_l} \tag{15.34}$$

It is worth noting the period n enters only in the combination $n - N_l$. This implies that the fractional allocation to principal and interest depends only on the number of periods $n - N_l$ left in the loan, not on the original life of the loan. A loan has no memory, so to speak.

In general, the loan interest rate r_l differs from the discount rate r used for the economic analysis, and the loan life N_l may be different from the system life N. Inserting $A = (A/P, r_l, N_l)$ into Eq. (15.34), we find that the interest payment I is related to loan amount L by

$$I_n = [1 - (1 + r_l)^{n-1-N_l}](A/P, r_l, N_l)L \tag{15.35}$$

The present value P_{int} of the total interest payments is found by discounting each I with the discount rate r and summing over n:

$$P_{\text{int}} = \sum_{n=1}^{N_l} \frac{1 - (1 + r_l)^{n-1-N_l}}{(1 + r_d)^n} (A/P, r_l, N_l)L \tag{15.36}$$

By using the formula for geometric series this can be transformed to

$$P_{\text{int}} = \left\{ \frac{(A/P, r_l, N_l)}{(A/P, r_d, N_l)} - \frac{(A/P, r_l, N_l) - r_l}{(1 + r_l)(A/P, r_{dl}, N_l)} \right\} L \tag{15.37}$$

with

$$r_{dl} = \frac{r_d - r_l}{1 + r_l}$$

If the incremental tax rate is τ, the total tax payments are reduced by τP_{int} (assuming a constant tax rate; otherwise the tax rate would have to be included in the summation).

Example 15.9

A solar water heating system costing \$2000 is financed with a 5-year loan at $r_l = 8$ percent. The tax rate is $\tau = 40$ percent. How much is the tax deduction for interest worth if the discount rate is $r_d = 8$ percent?

> *Given:* $L = \$2000$, $r_l = 8$ percent, $r_d = 8$ percent, $\tau = 40$ percent.
>
> *Find:* P_{int}
>
> *Lookup values:* $(A/P, r_{dl}, N_l) = 0.20$ and $(A/P, r_l, N_l) = 0.2505$ from Eq. (15.13)

SOLUTION

We have $r_{dl} = 0$, since $r_l = r_d$. Also $(A/P, r_{dl}, N_l) = 0.20$, and $(A/P, r_l, N_l) = 0.2505 = (A/P, r_d, N_l)$. Thus the present value of the interest payments is, from Eq. (15.37),

$$P_{int} = \left(1 - \frac{0.2505 - 0.08}{1.08 \times 0.20}\right)(\$2000) = \$421$$

At the stated tax rate that is worth $0.40 \times \$421 = \$168$.

15.2.3 Depreciation and Tax Credit

U.S. tax law allows business property to be depreciated. This means that for tax purposes the value of the property is assumed to decrease by a certain amount each year, and this decrease is treated as a tax-deductible loss. For the economic analysis one needs to express the depreciation as an equivalent present value. The details of the depreciation schedule have been changing with the tax reform of the 1980s. Instead of trying to present the full details, which can be found in the publications of the Internal Revenue Service, we merely note the general features. In any year n a certain fraction $f_{dep,n}$ of the capital cost (minus salvage value) can be depreciated. For example, in the simple case of straight-line depreciation over N_{dep} years

$$f_{dep,n} = \frac{1}{N_{dep}} \quad \text{for straight-line depreciation} \tag{15.38}$$

To obtain the total present value, one multiplies by the present worth factor and sums over all years from 1 to N:

$$f_{dep} = \sum_{n=1}^{N_{dep}} f_{dep,n}(P/F, r_d, n) \tag{15.39}$$

For straight-line depreciation the sum is

$$f_{dep} = \frac{(P/A, r_d, N_{dep})}{N_{dep}} \quad \text{for straight-line depreciation} \tag{15.40}$$

A further feature of some tax laws is the tax credit. For instance, in the United States for several years around 1980, tax credits were granted for certain renewable energy systems. If the tax credit rate is τ_{cred} for an investment C_{cap}, the tax liability is reduced by $\tau_{cred} C_{cap}$.

Example 15.10

A machine costs $10,000 and is depreciated with straight-line depreciation over 5 years, the salvage value after 5 years being $1000. Find the present value of the tax deduction for depreciation if the incremental tax rate is $\tau = 40$ percent, and the discount rate $r_d = 15$ percent.

Given: $C_{cap} = 10$ k$

$C_{salv} = 1$ k$

$N = 5$ yr

$\tau = 0.4$

$r_d = 0.15$

Find: $\tau \times f_{dep}(C_{cap} - C_{salv})$

Lookup values:

$$f_{dep} = \frac{(P/A, 0.15, 5 \text{ yr})}{5 \text{ yr}} = \frac{1/0.2983}{5} = \frac{3.3523}{5} = 0.6705, \quad \text{from Eq. (15.40)}$$

SOLUTION

For tax purposes the net amount to be depreciated is the difference

$$C_{cap} - C_{salv} = (10 - 1) \text{ k$} = 9 \text{ k$}$$

and with straight-line depreciation $1/N_{dep} = \frac{1}{5}$ of this can be deducted from the tax each year. Thus the annual tax is reduced by $\tau \times \frac{1}{5} \times 9$ k$ $= 0.40 \times 1.8$ k$ $= 0.72$ k$ for each of the 5 years. The present value of this tax reduction is

$$\tau \times f_{dep}(C_{cap} - C_{salv}) = 0.40 \times 0.6705 \times 9 \text{ k$} = 2.41 \text{ k$}$$

COMMENT

The present value of the reduction would be equal to 5×0.72 k$ $= 3.6$ k$ if r_d were zero. The discount rate of 15 percent reduces the present value by almost a one-third to $0.6705 = f_{dep}$.

15.2.4 Energy and Demand Charges

The cost of producing electricity has two major components: fuel and capital (for the power plant and distribution system). To the extent that it is practical, utility companies try to base the rate schedule on their production cost. Thus the rates for large customers contain four parts: the energy cost, a charge proportional to the peak demand, adjustments and surcharges, and taxes. Residential rates often omit the demand component because the cost of separate meters used to be considered too high.

For the energy charge a number of versions are commonly used. The simplest is a flat rate, with a single cost per kilowatthour consumed. In commercial and industrial rates, however, this is rarely the case. *Time-of-use rates* vary depending on the hour and type of day. Typically, the *on-peak* period occurs between 8 a.m. and 8 p.m. during weekdays, and the *off-peak* period is all other times. There may also be one or more *shoulder periods* between the on-peak and off-peak periods. The on-peak prices are usually 1.2 to 3 times higher than off-peak prices. *Block rates* vary the energy cost by the total amount of electricity used. For example, the first 1000 kWh may cost 6¢/kWh, the next 1000 kWh may cost 4¢/kWh, and any additional consumption would be 2¢/kWh. Note that there may be up to a half dozen blocks and that the rates do not necessarily decrease as the blocks "fill up." Furthermore, the amount paid for energy use in any given block may be adjusted by the total peak demand incurred while that block was in effect.

Time-of-use and block rates are also frequently applied to the demand component of commercial and industrial rates. For example, the first 100 kW of demand between 10 a.m. and 3 p.m. may be charged at one rate while the next 100 kW of demand during this time interval is charged at another rate, and so on. The demand charge may also be *ratcheted*; i.e., the highest demand incurred during a given month is used to calculate the demand charge for successive months. The ratchet period is typically 6 to 12 months and may use some percentage of the ratchet period peak demand or the current month's peak demand, whichever is higher.

The energy and demand rates can change seasonally. Most rates change between summer and winter, but a four-season split is not uncommon. Both electricity and gas bills are also adjusted, respectively, by the *energy cost adjustment* or the *purchased gas adjustment* and by any *surcharges*. They are essentially adders to the bill to account for monthly changes in the cost of fuel or energy transport and are applied either as a percentage of the total bill or as a function of the consumption component. Finally, state and local taxes are usually applied to the entire bill. There may also be a number of riders that dictate special fees and charges based on consumption and demand characteristics.

Several utilities now offer real-time pricing (RTP) electricity rates where the marginal costs of the utility are passed directly to the consumer. In these rates, the cost of electricity changes every hour of the day and can vary by a factor of 50 between the lowest and highest hourly costs. The highest costs occur only during 2 or 3 percent of the year.

As a simple example, consider a rate schedule with a monthly demand charge $p_{dem} = \$10/kW$ and the energy charge $p_e = \$0.07/kWh$. A customer with monthly energy consumption Q_m and peak demand P_{max} will receive a total bill of

$$\text{Monthly bill} = Q_m p_e + P_{max} p_{dem} \tag{15.41}$$

In most cases p_e and p_{dem} depend on time of day and time of year, being higher during the system peak than off peak. In regions with extensive air conditioning, the system peak occurs in the afternoon of the hottest days. In regions with much electric heating, the peak is correlated with outdoor temperature.

Example 15.11

A 100-ton electric chiller with COP = 3 is used for 8 months of the year (running at 100 percent capacity at least once per month during 4 months and at 50 percent capacity at least once per month during 4 months), and the total load is equivalent to 1000 h at peak capacity (a typical value around the belt from New York to Denver). What is the annual electricity bill if $p_e = \$0.10/kWh_e$ and $p_{dem} = \$10/kW_e$ per month?

Given: $P_{max,t} = 100$ tons $\times$ 3.516 kW$_t$/ton, with COP $= 3$ kW$_t$/kW$_e$

Annual energy $= P_{max} \times 1000$ h

Demand P_{max} for 4 months and $0.5 \times P_{max}$ for 4 months

$p_e = \$0.10/kWh_e$

$p_{dem} = \$10/kW_e$ per month

Find: Annual bill

SOLUTION

$$\text{Peak demand } P_{max} = \frac{100 \text{ tons} \times 3.516 \text{ kW}_t/\text{ton}}{3 \text{ kW}_t/\text{kW}_e} = 117.2 \text{ kW}_e$$

$$\text{Annual energy } Q = P_{max,e} \times 1000 \text{ h} = 117{,}200 \text{ kWh}_e$$

$$\text{Annual bill} = Qp_e + P_{max}p_{dem} \times (4 \times 1 + 4 \times 0.5)$$

$$= 117{,}200 \text{ kWh}_e \times \$0.10/\text{kWh}_e + 117.2 \text{ kW}_e \times \$10/\text{kW}_e \times 6$$

$$= \$11{,}720 + \$7032$$

$$= \$18{,}752$$

COMMENTS

In a real building the precise value of the peak demand may be difficult to predict because it depends on the coincidence of the demands of individual pieces of equipment.

The total cost per kilowatthour depends on the load profile. The more uneven the profile, the higher the cost. To take an extreme example, suppose the chiller were used only 1 h/yr, at full capacity. Then, with the rate structure of this example, the demand charge would be $1172 while the energy charge would be only $11.72, all that for consuming 1 kWh of energy. The total cost per kilowatthour would be $1172 + 11.72 = \$1183.72$ for 117.2 kWh, an effective electricity price of $10.10/kWh. This illustrates the interest of load leveling devices, such as cool storage for electric chillers.

15.2.5 The Complete Formula

The equations for a business investment can be stated in terms of before-tax cost or after-tax cost. Consider the purchase of fuel, with a market price of $5/GJ, by a business that is subject to an income tax rate $\tau = 40$ percent. Fuel, like all business expenditures, is tax-deductible. And, ultimately, it is paid by profits. To purchase 1 GJ one takes $5 of profits, before taxes; this reduces the tax liability by 5×40 percent $= \$2$, resulting in a net cost of only $3 after taxes.

We could do the accounting before or after taxes; the former counts the cash payments; the latter, their net (after-tax) values. The two modes differ by a factor $1 - \tau$, where τ is the income tax rate. For example, if the market price of fuel is $5/GJ and the tax rate $\tau = 40$ percent, then the before-tax cost of fuel is $5/GJ and the after-tax cost is $(1 - \tau)(\$5/\text{GJ}) = \$3/\text{GJ}$. Stated in terms of *after-tax cost*, the complete equation for the life cycle cost of an energy investment can be written in the form

C_{life}

$$
\begin{aligned}
= C_{\text{cap}}\{1 - f_l & \qquad\qquad\text{down payment} \\
+ f_l \frac{(A/P, r_l, N_l)}{(A/P, r_d, N_l)} & \qquad\qquad\text{cost of loan} \\
- \tau f_l \left[\frac{(A/P, r_l, N_l)}{(A/P, r_d, N_l)} - \frac{(A/P, r_l, N_l) - r_l}{(1 + r_l)(A/P, r_{d,l}, N_l)} \right] & \qquad\qquad\text{tax deduction for interest} \\
- \tau_{\text{cred}} & \qquad\qquad\text{tax credit} \\
- \tau f_{\text{dep}} \} & \qquad\qquad\text{depreciation} \\
- C_{\text{salv}} \left(\frac{1 + r_{\text{inf}}}{1 + r_d} \right)^N (1 - \tau) & \qquad\qquad\text{salvage} \\
+ Q p_e \frac{1 - \tau}{(A/P, r_{d,e}, N)} & \qquad\qquad\text{cost of energy} \\
+ P_{\text{max}} p_{\text{dem}} \frac{1 - \tau}{(A/P, r_{d,\text{dem}}, N)} & \qquad\qquad\text{cost of demand} \\
+ A_M \frac{1 - \tau}{(A/P, r_{d,M}, N)} \} & \qquad\qquad\text{cost of maintenance} \qquad (15.42)
\end{aligned}
$$

where

A_M = annual cost for maintenance, first-year \$
C_{cap} = capital cost, first-year \$
C_{salv} = salvage value, first-year \$
f_{dep} = present value of depreciation, as fraction of C_{cap}
f_l = fraction of investment paid by loan
N = system life, yr
N_l = loan period, yr
p_e = energy price, first-year \$/GJ
Q = annual energy consumption, GJ
r_d = market discount rate
r_e = market energy price escalation rate
$r_{d,e}$ = $(r_d - r_e)/(1 + r_e)$
r_{dem} = market demand charge escalation rate
$r_{d,\text{dem}}$ = $(r_d - r_{\text{dem}})/(1 + r_{\text{dem}})$
r_{inf} = general inflation rate
r_l = market loan interest rate
$r_{d,l}$ = $(r_d - r_l)/(1 + r_l)$
r_M = market escalation rate for maintenance costs
$r_{d,M}$ = $(r_d - r_M)/(1 + r_M)$
τ = incremental tax rate
τ_{cred} = tax credit

If there are several forms of energy, e.g., gas and electricity, the term $Q p_e$ is to be replaced by a sum over the individual energy terms. Many other variations and complications are possible; for instance, the salvage tax rate could be different from τ.

Example 15.12

Find the life cycle cost of the chiller of Example 15.11 under the following conditions:

Given:

System life $N = 20$ yr

Loan life $N_l = 10$ yr

Depreciation period $N_{dep} = 10$ yr, straight-line depreciation

Discount rate $r_d = 0.15$

Loan interest rate $r_l = 0.15$

Energy escalation rate $r_e = 0.01$

Demand charge escalation rate $r_{dem} = 0.01$

Maintenance cost escalation rate $r_M = 0.01$

Inflation $r_{inf} = 0.04$

Loan fraction $f_l = 0.7$

Tax rate $\tau = 0.5$

Tax credit rate $\tau_{cred} = 0$

Capital cost (at \$400/ton) $C_{cap} = 40$ k\$

Salvage value $C_{salv} = 0$

Annual cost of maintenance $A_M = 0.8$ k\$/yr ($= 2$ percent of C_{cap})

Capacity 100 tons $= 351.6$ kW$_t$

Peak electric demand 351.6 kW$_t$/COP $= 117.2$ kW$_e$

Annual energy consumption $Q = 100$ kton $\cdot$ h $= 351.6$ MWh$_t$

Electric energy price $p_e = 10$ ¢/kWh$_e = 100$ \$/MWh$_e$

Demand charge $p_{dem} = 10$ \$/kW$_e \cdot$ month, effective during 6 months of year

The rates are market rates.

Find: C_{life}

Lookup values:

$r_{d,l} = 0.0000$	$(A/P, r_l, N_l) = 0.1993$
$r_{d,e} = 0.1386$	$(A/P, r_d, N_l) = 0.1993$
$r_{d,dem} = 0.1386$	$(A/P, r_{d,l}, N_l) = 0.1000$
$r_{d,M} = 0.1386$	$(A/P, r_d, N) = 0.1598$
$(1 + r_{inf})/(1 + r_d) = 0.9043$	$(A/P, r_{d,e}, N) = 0.1498$
$f_{dep} = 0.502$ from Eq. (15.40)	$(A/P, r_{d,dem}, N) = 0.1498$
	$(A/P, r_{d,M}, N) = 0.1498$

SOLUTION

Components of C_{life} (all in k\$) per Eq. (15.42):

Down payment	12.0
Cost of loan	28.0
Tax deduction for interest	−8.0
Tax credit	0.0
Depreciation	−10.0
Salvage value	0.0

Cost of energy	39.1
Cost of demand charge	23.5
Cost of maintenance	2.7
Total $= C_{life}$	87.3

COMMENTS

A spreadsheet is recommended for this kind of calculation. The software provided with the book can also be used. The cost of energy and demand is higher than the capital cost.

15.2.6 Cost per Unit of Delivered Service

Sometimes it is interesting to know the cost per unit of delivered service (for instance, cost per ton-hour of cooling), analogous to the cost per driven mile for cars. This can be calculated as the ratio of levelized annual cost to annual delivered service. The levelized annual cost is obtained by multiplying the life cycle cost by the capital recovery factor for discount rate and system life. There appear two possibilities: the real discount rate r_{d0} and the market discount rate r_d. The quantity $(A/P, r_{d0}, N)C_{life}$ is the annual cost in constant dollars (of the initial year), whereas $(A/P, r_d, N)C_{life}$ is the annual cost in inflating dollars. The latter is difficult to interpret because it is an average over dollars of different real value. Therefore we levelize with the real discount rate because it expresses everything in first-year dollars, consistent with the currency of C_{life}. Thus we write the annual cost in initial dollars as

$$A_{life} = (A/P, r_{d0}, N)C_{life} \qquad (15.43)$$

The effective total cost per delivered service is therefore

$$\text{Effective cost per energy} = \frac{A_{life}}{Q} \qquad (15.44)$$

where $Q =$ annual delivered service (assumed constant, for simplicity).

The reader may wonder why we do not simply divide C_{life} by the service NQ delivered by the system over its lifetime. That would not be consistent because C_{life} is the present value, while NQ contains service flows (and thus monetary values) that are associated with future times. Really one must allocate service flows and costs within the same time frame. That is accomplished by dividing the levelized annual cost by the levelized annual service—and the latter is equal to Q because we have assumed that the consumption is constant from year to year.

Example 15.13

What is the cost per ton-hour for the chiller of Example 15.12?

Given: $C_{life} = 96.9$ k$, $Q = 100$ kton $\cdot$ h

Find: A_{life}/Q

Lookup values:

$r_{d0} = 0.1058$ from Eq. (15.14)

$(A/P, r_{d0}, N) = 0.1221$ from Eq. (15.13)

> **SOLUTION**
>
> Levelized annual cost in first-year dollars is
>
> $$A_{\text{life}} = (A/P, r_{d0}, N)C_{\text{life}} = 0.1221 \times 87.3 \text{ k} = \$10,659/\text{yr}$$
>
> $$\text{Cost per ton-hour} = \frac{A_{\text{life}}}{Q} = \$0.107/\text{ton} \cdot \text{h}$$

15.2.7 Constant Currency versus Inflating Currency

In the life cycle cost equation all cost components have been converted to equivalent present values (i.e., first-year costs). Let us see to what extent the result is the same whether one uses constant currency and real rates, or inflating currency and market rates. In the term for energy cost only the variable $r_{d,e} = (r_d - r_e)/(1 + r_e)$ depends on this choice. Inserting real rates according to

$$1 + r_{d0} = \frac{1 + r_d}{1 + r_{\text{inf}}} \tag{15.45}$$

and

$$1 + r_{e0} = \frac{1 + r_e}{1 + r_{\text{inf}}} \tag{15.46}$$

one finds that

$$r_{d,e} = \frac{(1 + r_{d0})(1 + r_{\text{inf}}) - (1 + r_{e0})(1 + r_{\text{inf}})}{(1 + r_{e0})(1 + r_{\text{inf}})} \tag{15.47}$$

and after canceling the factor $1 + r_{\text{inf}}$, one sees that this is equal to

$$r_{d,e} = \frac{1 + r_{d0} - (1 + r_{e0})}{1 + r_{e0}} = r_{d,e0} \tag{15.48}$$

The energy cost is the same, whether one uses real rates or market rates. The same holds for the maintenance cost term. The salvage term is also independent of this choice because

$$\frac{1 + r_{\text{inf}}}{1 + r_d} = \frac{1}{1 + r_{d0}}$$

By contrast, the ratio of capital recovery factors in the loan terms is not invariant, as one can see by inserting numerical values. For example, with $r_{d0} = 0.08$, $r_{l0} = 0.12$, and $r_{\text{inf}} = 0.05$ one finds, with $N = 20$ years,

$$\frac{(A/P, r_{l0}, N)}{(A/P, r_{d0}, N)} = 1.31$$

The corresponding market rates are $r_d = 0.134$ and $r_l = 0.176$, and the ratio becomes

$$\frac{(A/P, r_l, N)}{(A/P, r_d, N)} = 1.26$$

The difference arises from the fact that the cash flows are different. For a loan that is based on real rates, the annual payments are constant in constant currency, whereas for one based on market rates the payments are constant in inflating currency. Likewise the depreciation terms can depend on inflation.

It follows that the two approaches, constant currency and inflating currency, yield identical results for equity investments ($f_l = 0$) without depreciation. But if f_l or f_{dep} is not zero, there can be differences. Numerically the effect is not large, at most on the order of 10 percent, for inflation rates below 10 percent (Dickinson and Brown, 1979). The effect has opposite signs for the loan term and the depreciation term, leading to partial cancellation of the error.

15.3 Economic Evaluation Criteria

15.3.1 Life-Cycle Savings

Having determined the life cycle cost of each relevant design alternative, one can select the "best," i.e., the one that offers all desirable features at the lowest life cycle cost. Frequently one takes one design as reference and considers the difference between it and each alternative design. The difference is called *life cycle savings* relative to the reference case

$$S = -\Delta C_{life} \quad \text{with } \Delta C_{life} = C_{life} - C_{life, ref} \tag{15.49}$$

Often the comparison can be quite simple because only those terms that are different between the designs under consideration need to be considered. For simplicity we write the equations of this section only for an equity investment without tax. Then the loan fraction f_l in Eq. (15.39) is zero, and most of the complications of that equation drop out. Of course, the concepts of life cycle savings, internal rate of return, and payback time are perfectly general, and tax and loan can be readily included.

A particularly important case is the comparison of two designs that differ only in capital cost and operating cost: The one that saves operating costs has higher initial cost (otherwise the choice would be obvious, without any need for an economic analysis). Setting $f_l = 0$ and $\tau = 0$ in Eq. (15.38) and taking the difference between the two designs, one obtains the life cycle savings as

$$S = \frac{-\Delta Q p_e}{(A/P, r_{d,e}, N)} - \Delta C_{cap} \tag{15.50}$$

where

$\Delta Q = Q - Q_{ref} = $ difference in annual energy consumption

$\Delta C_{cap} = C_{cap} - C_{cap, ref} = $ difference in capital cost

$r_{d,e} = (r_d - r_e)/(1 + r_e)$

(If the reference design has higher consumption and lower capital cost, ΔQ is negative and ΔC_{cap} is positive with this choice of signs.)

Example 15.14

Compared to a one-stage model, a two-stage absorption chiller is more efficient, but its first cost is higher. Find the life cycle savings of a two-stage model for the followings situation.

Given:

Required chiller capacity 1000 kW$_t$ operating at 1000 h/yr full-load equivalent

A single-stage absorption chiller has COP $= 0.7$ and costs \$100/kW$_t$

($=$ reference system) while a two-stage absorption chiller has COP $= 1.1$ and costs \$130/kW$_t$.

Gas price $p_e = \$4/\text{GJ}$ at the start

Escalating at $r_e = 0$ percent (real)

Discount rate $r_d = 8$ percent (real)

Find: Life cycle savings for the two-stage chiller

Lookup value: $(A/P, r_d, N) = 0.1019$

SOLUTION

The annual energy consumption is

$$1000 \text{ kW}_t \times 1000 \text{ h}/\text{COP} = 1.0 \text{ MWh}_t/\text{COP}$$

$$= \begin{cases} 5.143 \times 10^3 \text{ GJ}_{\text{gas}} & \text{for COP} = 0.7 \\ 3.273 \times 10^3 \text{ GJ}_{\text{gas}} & \text{for COP} = 1.1 \end{cases}$$

Thus the difference in energy cost is

$$\Delta Q p_e = (3.273 - 5.143) \times 10^3 \text{ GJ} \times \$4/\text{GJ} = -\$7481 \text{ per year}$$

and the difference in capital cost is

$$\Delta C_{\text{cap}} = (130 - 100)(1000) = \$30{,}000$$

From Eq. (15.50) we find the life cycle savings

$$S = \frac{-\Delta Q p_e}{(A/P, r_{d,e}, N)} - \Delta C_{\text{cap}}$$

$$= \frac{\$7481}{0.1019} - 30{,}000 = \$73{,}415 - 30{,}000 = \$43{,}415$$

COMMENT

Even though the discount rate in this example is rather high (5 percent might be more appropriate), the life cycle savings are large. The investment certainly pays off.

15.3.2 Internal Rate of Return

The life cycle savings are the true savings if all the input is known correctly and without doubt. But future energy prices or system performance is uncertain, and the choice of the discount rate is not clear-cut. An investment in a building or its equipment is uncertain, and it must be compared with competing investments that have their own uncertainties. The limitation of the life cycle savings approach can be circumvented if one evaluates the profitability of an investment by itself, expressed as a dimensionless rate. Then one can rank different investments in terms of profitability and in terms of risk. General business experience can serve as a guide for expected profitability as function of risk level. Among investments of comparable risk, the choice can then be based on profitability.

More precisely, the profitability is measured as the *internal rate of return* r_r, defined as that value of the discount rate r_d at which the life cycle savings S are zero

$$S(r_d) = 0 \quad \text{at } r_d = r_r \tag{15.51}$$

For an illustration, take the case of Eq. (15.50) with energy escalation rate $r_e = 0$ (so that $r_{d,e} = r_d$), and suppose an extra investment ΔC_{cap} is made to provide annual energy savings $(-\Delta Q)$. The initial investment ΔC_{cap} provides an annual income from energy savings of

$$\text{Annual income} = (-\Delta Q)p_e \tag{15.52}$$

If ΔC_{cap} were placed in a savings account instead, bearing interest at a rate r_r, the annual income would be

$$\text{Annual income} = (A/P, r_r, N)\Delta C_{cap} \tag{15.53}$$

The investment behaves as a savings account whose interest rate r_r is determined by the equation

$$(A/P, r_r, N)\Delta C_{cap} = (-\Delta Q)p_e \tag{15.54}$$

Dividing by $(A/P, r_r, N)$, we see that right and left sides correspond to the two terms in Eq. (15.50) for the life cycle savings

$$S = \frac{-\Delta Q p_e}{(A/P, r_d, N)} - \Delta C_{cap} \tag{15.55}$$

and that r_r is indeed the discount rate r_d for which the life cycle savings are zero; it is the internal rate of return. Now the reason for the name is clear: It is the profitability of the project by itself, without reference to an externally imposed discount rate. When the explicit form of the capital recovery factor is inserted, one obtains an equation of Nth degree, generally not solvable in closed form. Instead one resorts to iterative or graphical solution. (There could be up to N different real solutions, and multiple solutions can indeed occur if there are sign changes in the stream of annual cash flows. But, not to worry, the solution is unique for the case of interest here: an initial investment that brings a stream of annual savings.)

Example 15.15

What is the rate of return for Example 15.14?

Given: $S = \frac{-\Delta Q p_e}{(A/P, r_{d,e}, N)} - \Delta C_{cap}$ with $r_{d,e} = r_d$ (because $r_e = 0$)

$(-\Delta Q)p_e = \$7481$

$\Delta C = \$30,000$

Find: r_r

SOLUTION

$S = 0$ for

$$(A/P, r_r, N) = \frac{-\Delta Q p_e}{\Delta C_{cap}}$$

$$= \frac{7481}{30,000} = 0.2494 \quad \text{with } N = 20$$

By iteration one finds $r_r = 0.246 = 24.6$ percent.

15.3.3 Payback Time

The *payback time* N_p is defined as the ratio of extra capital cost ΔC_{cap} to first-year savings

$$N_p = \frac{\Delta C_{cap}}{\text{first-year savings}} \tag{15.56}$$

(The inverse of N_p is sometimes called the *return on investment*.) If one neglects discounting, one can say that after N_p years the investment has paid for itself and any revenue thereafter is pure gain. The shorter N_p, the higher the profitability. As a selection criterion, the payback time is simple, intuitive, and obviously wrong because it neglects some of the relevant variables. Attempts have been made to correct for that by constructing variants such as a discounted payback time [by contrast to which Eq. (15.56) is sometimes called simple payback time], but the resulting expressions become so complicated that one might as well work directly with life cycle savings or internal rate of return.

The simplicity of the simple payback time is, however, irresistible. When investments are comparable to each other in terms of duration and function, the payback time can give an approximate ranking that is sometimes clear enough to be able to discard certain alternatives right from the start, thus avoiding the effort of detailed evaluation.

To justify the use of the payback time, let us recall Eq. (15.54) for the internal rate of return and note that it can be written in the form

$$(A/P, r_r, N) = \frac{1}{N_p}, \quad \text{or} \quad (P/A, r_r, N) = N_p \tag{15.57}$$

The rate of return is uniquely determined by the payback time N_p and the system life N. This equation implies a simple graphical solution for finding the rate of return if one plots $(P/A, r_r, N)$ on the x axis versus r_r on the y axis, as in Fig. 15.6. Given N and N_p, one simply looks for the intersection of the line $x = N_p$ (i.e., the vertical line through $x = N_p$) with the curve labeled by N; the ordinate (y axis) of the intersection is the rate of return r_r.

This graphical method can be generalized to the case where the annual savings change at a constant rate r_e. In that case Eq. (15.20) implies that the rate of return is replaced by $r_{r,e} = (r_r - r_e)/(1 + r_e)$, and Fig. 15.6 yields $r_{r,e}$ rather than r_r. In other words, Eq. (15.57) becomes

$$(P/A, r_{r,e}, N) = N_p \quad \text{with } r_{r,e} = \frac{r_r - r_e}{1 + r_e} \tag{15.58}$$

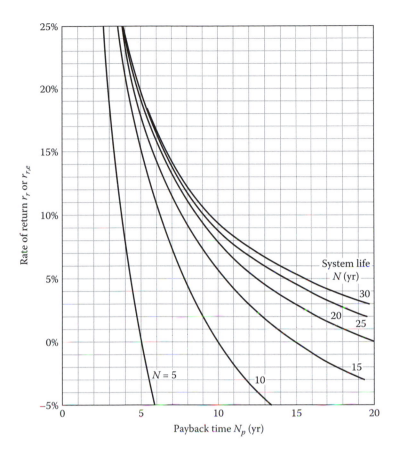

FIGURE 15.6
Relation between rate of return r_r, system life N, and payback time N_p. If r_e = escalation rate of annual savings $\neq 0$, the vertical axis is the variable $r_{r,e}$ from which r_r is obtained as $r_r = r_{r,e}(1 + r_e) + r_e$.

The graph yields $r_{r,e}$ which is readily solved for

$$r_r = r_{r,e}(1 + r_e) + r_e \qquad (15.59)$$

In particular, if r_e is equal to the general inflation rate r_{inf}, then $r_{r,e}$ is the real rate of return $r_{r,0}$.

Example 15.16

Find payback time for Example 15.15 and check the rate of return graphically.
   Given: First-year savings $(-\Delta Q)p_e = \$7481$
   Extra investment $\Delta C_{cap} = \$30,000$
   Find: N_p and r_r for $r_e = 0$ and 2 percent

SOLUTION

$$N_p = \frac{30,000}{7481} = 4.01 \ \text{yr}$$

it is independent of r_e. Then $r_r = 0.246$ for $r_e = 0$, from Fig. 15.6, and $r_r = 0.271$ for $r_e = 2$ percent, from Eq. (15.58).

Generally a real (i.e., corrected for inflation) rate of return above 10 percent can be considered excellent if there is low risk—a look at savings accounts, bonds, and stocks shows that it is difficult to find better. From the graph we see immediately that $r_{r,e}$ is above 10 percent if the payback time is shorter than 8.5 years (6 years), for a system life of 20 years (10 years). And $r_{r,e}$ is close to the real rate of return if the annual savings growth is close to the general inflation rate.

15.4 Complications of the Decision Process

In practice, the decision process is likely to bump into some obstacles. Suppose, e.g., that the annual operating cost of a proposed office building can be reduced by $1000 if one installs daylight sensors and dimmers for the lights, at an extra cost of $2000. The payback time is only 2 years. It looks like an irresistible investment opportunity, with a rate of return well above 25 percent, as shown by Fig. 15.6 (the exact value depends somewhat on lifetime and taxes, but that is beside the point). However, quite a few hurdles stand in the way.

First, to find out about this opportunity, the design engineer has to obtain the necessary information. Requesting catalogues, reading technical reviews of the equipment, carrying out the calculations of cost and performance all take time and effort. Under the pressures of the job, the engineer may not be willing to spend the extra time or to neglect other items that compete for her or his attention.

Suppose our engineer has done a good analysis and tries to convince the builder to spend the extra money. In the case of a speculative office building, the builder is likely to say, "Why should I pay a penny more, if only the future tenant will reap the benefit?" So the design engineer is forced to aim for lowest first cost.

Even if the builder is willing to spend a bit more for efficiency, with hopes that the prospect of reduced energy bills will make it easier to find tenants, the decision is not obvious. Can the builder trust the claims of the sales brochure or the calculations of the engineer? Daylight controls are relatively new, and perhaps the builder has heard that some of the first models did not live up to expectations. If malfunctions reduce the productivity of the workers, the hassle and the costs could more than nullify the expected savings. So the builder may refuse to take what he or she perceives as an excessive risk. The threat of a liability suit is a potent inhibiter; no wonder the building industry has a reputation for extreme conservatism.

This example illustrates the basic mechanisms that so frequently prevent the adoption of efficient technologies:

- Lack of information or excessive cost of obtaining the information
- Purchase decision made by someone who does not have to pay the operating costs
- Uncertainty (about future costs, reliability, etc.)

Any one of the hurdles can be sufficient cause to reject an investment. In the above example the decision to reject the lighting controller looks as if the discount rate were higher than 25 percent. Quite generally, these mechanisms have the effect of raising the apparent discount rate or foreshortening the time horizon. The resulting decisions appear irrational:

People do not spend as much for energy efficiency as would be optimal according to a life cycle cost analysis with the correct discount rate. In reality, this irrationality is but a reflection of other problems.

In the world of business, risk and uncertainty are all-pervasive, so much, so that most decision makers insist on very short payback times—almost always less than 5 years and frequently less than 2 years. But it depends on the nature of the business and on the circumstances. There are industries such as electric power plants where profits are sure (albeit moderate): Once a power plant has been built, it is expected to run smoothly for at least 30 years. Here the discount rates are low, and payback times are longer than 10 years. Governments, being charged with the long-term welfare of the citizens, also tend to have a long time horizon.

What does all this mean for the HVAC engineer? The more a design choice involves unproven technology or is dependent on occupant behavior for proper functioning, the more risky it is. For instance, a daylighting strategy that relies on manual control of shading devices by the occupants may not bring the intended savings because the occupants may not follow the intentions of the designer. Likewise, when one is considering a new design or a new piece of equipment without a track record, it is not irrational to demand short payback times.

By contrast, paying extra for an efficient boiler or chiller is a safe investment (assuming the equipment has a good reputation) because the occupant does not care how the heating or cooling is produced, as long as the environment is comfortable, pleasant, and healthy. Also, the building will certainly be heated and cooled over its entire life. Here a life cycle cost analysis with the correct discount rate is certainly in order, and it would be short-sighted to insist on payback times of less than 2 years.

Finally, what about the problem of the builder or landlord who refuses to pay for measures that would only reduce the energy bill of the tenant or of a future owner? This is a serious difficulty, indeed. In an ideal market the information about reduced energy cost would translate as higher rent or resale value, but in practice this process is slow and inefficient (there is a market failure, in the language of the economists). This justifies energy efficiency standards such as the ASHRAE Standard 90.1 and their enforcement by government regulations.

15.5 Cost Estimation

15.5.1 Capital Costs

For mass-produced consumer products, such as cars or cameras, the capital cost (i.e., the purchase price) is easy to determine, by looking at catalogues or newspaper ads or by calling the store. Even then there may be uncertainties: When you actually go the store, a discount may be offered on the spot to beat a competitor. Different prices can be found in different stores for identical products, not only because of differences in service or transportation but also because of the sheer difficulty of obtaining the price information.

And, of course, price is not the only criterion. Even more important, and more difficult to ascertain and compare, are the various characteristics of a good: the features it offers, the quality, the operating costs, and so on. Economists have even coined a special term, *cost of information*, so universal is the difficulty of finding the pertinent information.

For HVAC equipment the problems tend to be more complicated than for consumer goods. Transport and installation are an important item in addition to the cost of the equipment at the factory. The determination of the cost can become a major undertaking, especially for complex or custom-made systems. The capital cost of a system or component is known with certainty only when one has a firm contract from a vendor. But asking for bids on each design variation is simply not feasible—the cost of information would become prohibitive.

The more a design engineer wants to be sure of coming close to the optimal design, the more she or he needs to learn about the details of the cost calculation. Information on costs is available from a number of sources, e.g., Boehm (1987) and Konkel (1987). An important feature is the variation of the cost with size. Because of fixed costs and economies of scale, simple proportionality between cost and size is not the rule. But usually one can assume the following functional form over a limited range of sizes:

$$C = C_r \left(\frac{S}{S_r} \right)^m \qquad \text{for } S_{\min} < S < S_{\max} \tag{15.60}$$

where
C = cost at size S
C_r = cost at a reference size S_r
m = exponent

Typically m is in the range of 0.5 to 1.0; exponents less than unity are a reflection of economies of scale. On a logarithmic plot m is the slope of $\ln C$ versus $\ln S$. If m is not known, a value of 0.6 can be recommended as default. Tables on the CD-ROM summarize cost data for HVAC equipment in this form.

When interpreting such cost figures, one has to be careful about what is included and what is not. Is it the cost at the factory FOB (free on board, i.e., excluding transportation), the cost delivered to the site, or the cost installed? For items such as cool storage the space requirements may impose additional costs. And finally, what are the specific features and how is the quality?

Costs change not only with general inflation but also with the evolution of technology. The first models of a novel product tend to be expensive. Gradually mass production, technological advances, and competition combine to drive the prices down. General inflation or increases in the cost of some input, for instance, energy, will push in the opposite direction. The resulting evolution of the price of the product may be difficult to predict. Cost reductions due to technological advances are more likely with products of high technology (e.g., energy management systems) than with mature products that cannot be miniaturized (e.g., fans and motors). In some cases there is an improvement in a product rather than a reduction of its cost; variable-speed motors, for instance, are more expensive than constant-speed motors but allow better control or higher system efficiency.

Cost tabulations are based on sales or projects of the past, and they must be updated to the present by means of correction factors. For that purpose one could use general inflation (i.e., the CPI discussed in Sec. 15.1.1), but that is less reliable than specific cost indices for that class of equipment or that sector of the economy. The following two indices are particularly pertinent for buildings and HVAC equipment. One is the Marshall and Swift equipment cost index, values of which are published regularly in *Chemical Engineering*.

Another one is the construction cost index published by *Engineering News Record*. The latter is plotted in Fig. 15.1.

It is important during the design process to have a realistic understanding of all the relevant costs, yet the effort of obtaining these costs should not be prohibitive. Konkel (1987) describes a method that seems to be a good compromise between these conflicting requirements. The basic idea is to group certain portions of a project into what is called *unit operations*. The components of the unit operations, called *unit assemblies*, are itemized, priced, and plotted by size of unit operation. A boiler is an example of a unit operation; its unit assemblies include burner, air intake, flue, shutoff valves, piping, fuel supply, expansion tank; water makeup valves, and deaerator. The sizes and costs vary with the size of the boiler. Once the size-price relations have been found for each component, the size-price relation for the boiler as a whole is readily derived. Knowing the size-price relation for the unit operations, the designer can estimate the total cost of a project and its design variations without too much effort.

15.5.2 Maintenance and Energy

Maintenance cost and energy prices may evolve differently from general inflation and from each other. It is instructive to correct energy prices for general inflation, as in Fig. 15.2 where the prices of oil, gas, and electricity are shown in both current and constant dollars. One can see that some adjustments have occurred since the oil shocks of the 1970s. In fact, the real price of crude oil is almost the same in 1998 as during the 1960s.

What should we assume for the future? Projections of energy prices are published periodically by several organizations, for instance, the American Gas Association and the National Institute of Standards and Technology (Lippiat and Ruegg, 1990). Most analysts predict real escalation rates in the range of 0 to 3 percent, averaged over the next two decades. This is based on the gradual exhaustion of cheap oil and gas reserves and the fact that alternatives, i.e., coal, nuclear, and solar, are more expensive to utilize. Who knows? Further turmoil in the Middle East? What progress will be made in fusion, and how will public acceptance of nuclear power evolve? How much can be saved by improved efficiency, and at what cost? What constraints will be imposed by environmental concerns?

Data on maintenance costs can be obtained, e.g., from the "BOMA Experience Exchange Report," published annually by the Building Owners and Managers Association International (BOMA, 1987). Specifically for maintenance costs of HVAC equipment in office buildings, a succinct equation can be found in ASHRAE (1985). It states the annual cost A_M for maintenance, in dollars per floor area A_{floor}, is in the form

$$\frac{A_M}{A_{\text{floor}}} = C_{\text{base}} + an + h + c + d \tag{15.61}$$

where
  C_{base} = value for base system (fire-tube boilers for heating, centrifugal chillers for cooling, and VAV for distribution, during first year)
  n = age of equipment, years
  a = coefficient for age of equipment

and the coefficients h, c, and d allow the adjustment to other systems.

TABLE 15.2

HVAC Maintenance Costs of Eq. (15.61): Cost C_{base} of Base System
and Coefficients for Adjustment, 1983 U.S. Dollars

	$/ft^2	$/m^2
C_{base}	0.3335	3.590
Coefficient a for age per year	0.0018	0.019
Heating equipment, coefficient h		
Water-tube boiler	0.0077	0.083
Cast-iron boiler	0.0094	0.101
Electric boiler	−0.0267	−0.287
Heat pump	−0.0969	−1.043
Electric resistance	−0.1330	−1.432
Cooling equipment, coefficient c		
Reciprocating chiller	−0.0400	−0.431
Absorption chiller (single-stage)	0.1925	2.072
Water source heat pump	−0.0472	−0.508
Distribution system, coefficient d		
Single-zone	0.0829	0.892
Multizone	−0.0466	−0.502
Dual-duct	−0.0029	−0.031
Constant-volume	0.0881	0.948
Two-pipe fan coil	−0.0277	−0.298
Four-pipe fan coil	0.0580	0.624
Induction	0.0682	0.734

Source: Courtesy of ASHRAE, *Design/Data Manual for Closed-Loop Ground
　　　　　Coupled Heat Pump System*, American Society of Heating,
　　　　　Refrigerating and Air-Conditioning Engineers, Atlanta, GA, 1985.

Numerical values for C_{base}, a, h, c, and d are listed in Table 15.2. These values are in 1983 dollars. They still need to be adjusted to the year of interest by multiplication by the corresponding ratio of CPI values, as explained in Section 15.1. In using this equation one should keep in mind that it is based on a survey of office buildings originally published in 1986. Extrapolation to other building types or newer technologies may introduce large and unknown uncertainties.

Example 15.17

Estimate the annual HVAC maintenance cost for an office building that has a floor area of 1000 m^2 and is $n = 10$ years old in the year 2003. The system consists of an electric boiler, a reciprocating chiller, and a constant-volume distribution system. Suppose the CPI is 180 in 2003.

Given: $A_{floor} = 1000$ m^2

$n = 10$ yr

$\frac{CPC_{2003}}{CPI_{1983}} = \frac{180}{100} = 1.80$

Lookup values: From Table 15.2

$C_{base} = 3.59$, $a = 0.019$, $h = -0.287$, $c = -0.431$, $d = 0.948$ (in $\$_{1983}$/m^2)

Find: A_M

SOLUTION

Using Eq. (15.61), one finds

$$A_M = 1000\text{m}^2 \times [(3.59 + 0.019 \times 10 - 0.287 - 0.431 + 0.948)\$_{1983}/\text{m}^2]$$
$$= \$_{1983}4010/\text{m}^2$$

To convert to 2003 dollars, multiply by the CPI ratio

$$A_M = \$_{1983}4010/\text{m}^2 \times 1.80 = \$_{2003}7218/\text{m}^2$$

15.6 Optimization

In principle, the process of optimizing the design of a building is simple: Evaluate all possible design variations and select the one with the lowest life cycle cost. Who would not want to choose the optimum? In practice, it would be a daunting task to find the true optimum, among all conceivable designs. The difficulties, some of which have already been discussed, are

- The enormous number of possible design variations (building configuration and materials, HVAC systems, types and models of equipment, control modes)
- Uncertainties (costs, energy prices, reliability, occupant behavior, future uses of building)
- Imponderables (comfort, convenience, aesthetics)

Fortunately there is a certain tolerance for moderate errors, as we will show below. That facilitates the job greatly, because one can reduce the number of steps in the search for the optimum. Also, within narrow ranges some variables can be suboptimized without worrying about their effect on others.

Some quantities are easier to optimize than others. Optimizing the heating and cooling equipment, for a given building envelope, is less problematic than trying to optimize the envelope—the latter touches on the imponderables of aesthetics and image.

It is instructive to illustrate the optimization process with a very simple example: the thickness of insulation. The annual heat flow Q across the insulation is

$$Q = \frac{AkD}{t} \tag{15.62}$$

where
  $A =$ area, m^2
  $k =$ conductivity, $\text{W}/(\text{m} \cdot \text{K})$
  $D =$ annual degree-seconds, $\text{K} \cdot \text{s}$
  $t =$ thickness of insulation, m

The capital cost of the insulation is

$$C_{\text{cap}} = Atp_{\text{ins}} \tag{15.63}$$

with p_{ins} = price of insulation ($/m^3). The life cycle cost is

$$C_{life} = C_{cap} + Q\frac{p_e}{(A/P, r_{d,e}, N)} \qquad (15.64)$$

where p_e = first-year energy price and $r_{d,e}$ is related to discount rate and energy escalation rate as in Eq. (15.19). We want to vary the thickness t to minimize the life cycle cost, keeping all the other quantities constant. (This model is a simplification that neglects the fixed cost of insulation as well as possible feedback of t on D.) Eliminating t in favor of C_{cap}, one can rewrite Q as

$$Q = \frac{K}{C_{cap}} \qquad (15.65)$$

with a constant

$$K = A^2 k D p_{ins} \qquad (15.66)$$

Then the life cycle cost can be written in the form

$$C_{life} = C_{cap} + \frac{PK}{C_{cap}} \qquad (15.67)$$

where the variable

$$P = \frac{p_e}{(A/P, r_{d,e}, N)} \qquad (15.68)$$

contains all the information about energy price and discount rate. Here K is fixed, and the insulation investment C_{cap} is to be varied to find the optimum. Variable C_{life} and its components are plotted in Fig. 15.7. As t is increased, capital cost increases and energy cost decreases; C_{life} has a minimum at some intermediate value. Setting the derivative of C_{life} with respect to C_{cap} equal to zero yields the optimal value $C_{cap,o}$

$$C_{cap,o} = \sqrt{KP} \qquad (15.69)$$

Now here is an interesting question: What is the penalty for not optimizing correctly? In general, the following causes could prevent correct optimization:

- Insufficient accuracy of the algorithm or program for calculating the performance
- Incorrect information on economic data [e.g., the factor P in Eq. (15.69)]
- Incorrect information on technical data [e.g., the factor K in Eq. (15.69)]
- Unanticipated changes in the use of the building

Misoptimization would produce a design at a value C_{cap} different from the true optimum $C_{cap,o}$. For the example of insulation thickness, the effect on the life cycle cost can be seen

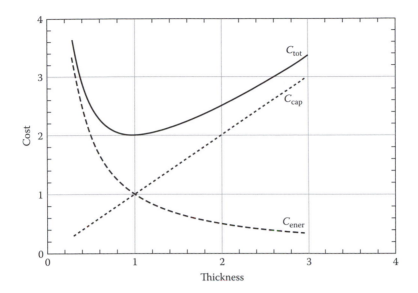

FIGURE 15.7

Optimization of insulation thickness. Insulation cost $= C_{cap}$, energy cost $= C_{ener}$, and life cycle cost $= C_{tot}$.

directly with the solid curve in Fig. 15.7. For instance, a $+10$ percent or -10 percent error in $C_{cap,o}$ would increase C_{life} by only $+1$ percent. Thus the penalty is not excessive for small errors.

This relatively large insensitivity to misoptimization is a feature much more general than the insulation model. As shown by Rabl (1985), the greatest sensitivity likely to be encountered in practice corresponds to the curve

$$\frac{C_{life,\,true}(C_{cap,\,o,\,guess})}{C_{life,\,true}(C_{cap,\,o,\,true})} = \frac{x}{1 + \log x} \quad (\text{"upper bound"}) \tag{15.70}$$

also shown in Fig. 15.8, with the label *upper bound*. Even here the minimum is broad; if the true energy price differs by $+10$ percent (-10 percent) from the guessed price, the life cycle cost increases only 0.4 percent (0.6 percent) over the minimum. Even when the difference in prices is 30 percent, the life cycle cost penalty is less than 8 percent.

Errors in the factor K (due to wrong information about price or conductivity of the insulation material) can be treated in the same way, because K and P play an entirely symmetric role in the above equations. Therefore curves in Fig. 15.8 also apply to uncertainties in other input variables.

The basic phenomenon is universal: Any smooth function is flat at an extremum. The only question is, How flat? For energy investments that question has been answered with the curves of Figs. 15.7 and 15.8. We can conclude that misoptimization penalties are definitely less than 1 percent (10 percent) when the uncertainties of the input variables are less than 10 percent (30 percent).

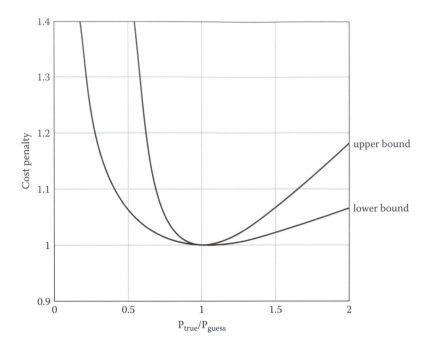

FIGURE 15.8
Life cycle cost penalty versus energy price ratio.

Problems

15.1 Estimate your personal discount rate, in both constant and inflating dollars.

15.2 Suppose you are offered the choice between $1000 today and $5000 in 10 years from now, and the average inflation rate is 4 percent. What are the real and the market discount rates at which you are indifferent between these two choices?

15.3 A bank advertises an interest rate of 10 percent (annual rate, no compounding). Use the rule of 70 (Sec. 15.1.5) to estimate how long will it take to double your money (a) in inflating currency and (b) in constant currency, if the inflation rate is 5 percent.

15.4 Suppose that market fuel prices escalate at a rate of $r_e = 8$ percent while the general inflation rate is $r_{inf} = 6$ percent.

(a) If fuel costs $5/MBtu ($5.275/GJ) today, what will be the price after 20 years, in both constant and inflating dollars?

(b) What is the corresponding levelized fuel price, if the discount rate is $r_{d,0} = 5$ percent (constant dollars)?

15.5 A building is to be heated with a furnace that costs $2500 and has a seasonal average efficiency of 90 percent. Suppose the annual heating load is 100 MBtu (105.5 GJ) and the fuel price is $6.00/MBtu ($6.33/GJ). Assume a lifetime of 20 years and a real discount rate of 5 percent.

(a) Calculate the life cycle cost of heating this building.

(b) Calculate the levelized total cost of delivered heat [in $/MBtu ($/GJ)]. What fraction is due to capital and what fraction due to energy?

15.6 Suppose a system of heat recovery from exhaust air costs $10,000 and reduces the annual heating bill by $2000. Assume a real discount rate of 5 percent.

(a) What is the payback time?

(b) Plot the life cycle savings versus the lifetime of the system for values from 5 to 20 years.

(c) Plot the rate of return versus the lifetime of the system for values from 5 to 20 years. Does the rate of return depend on the discount rate assumed for part *b*?

15.7 Compare two chillers, both rated for an output of 5 kW$_t$ (1.58-ton refrigeration):

- Model A has a COP of 2.0 and costs $1500.
- Model B has a COP of 2.5 and costs $2000.

They are used for 500 h/yr. Assume an electricity price of 10¢/kWh$_e$, a lifetime of 15 years, and a real discount rate of 5 percent.

(a) Calculate the life cycle costs.

(b) Find the payback time and rate of return for using model B instead of A.

(c) What is the highest price at which model B is competitive with A?

15.8 Consider two choices for the furnace to heat a building:

- An ordinary gas furnace with efficiency of 75 percent (seasonal average), costing $5000
- A condensing furnace with efficiency of 95 percent (seasonal average), costing $10,000

Suppose the annual heating load is 1000 MBtu (1055 GJ) and the fuel price is $5.00/MBtu ($5.275/GJ). Assume that both furnaces can be expected to last 15 years, and use a real discount rate of 5 percent.

(a) Calculate the life cycle cost for each.

(b) Calculate the levelized total cost of delivered heat for each. What fraction is due to capital, what fraction to energy?

(c) Find the payback time and rate of return for using model B instead of A.

15.9 A solar domestic water heater cost $C = \$2000$ and supplies $Q = 20$ GJ (18.96 MBtu) per year. The backup uses electric resistance heating with an efficiency of 100 percent. The price of electricity is $p_e = 6$¢/kWh, and the discount rate is 5 percent (real). Find the payback time, life cycle savings, and rate of return for the following cases:

(a) p_e and Q constant, life $N = 20$ years

(b) p_e and Q constant, life $N = 10$ years

(c) p_e and Q constant, life $N = 30$ years

(d) p_e escalates at 4 percent per year, Q constant, $N = 20$ years

(e) p_e constant, Q degrades at 1 percent per year, $N = 20$ years

15.10 Evaluate and compare the cost of generating electric power with the following three types of power plants, as a function of capacity factor ($=$ annual average power produced per peak power rating). Assume a system life of 30 years and a real discount rate of $r_{d,0} = 6$ percent; neglect O&M. Calculate the cost of electricity

($\not{c}$/kWh$_e$) for each of these two power plants and for each of these values of the capacity factor: 0.70 (= base load) and 0.10 (= peak load).

(a) Coal power plant: capital cost C_{cap} = \$1500/kW$_e$, efficiency (fuel to electric) η = 33.3 percent, fuel price p_f = \$2.50/GJ.

(b) Simple gas turbine: capital cost C_{cap} = \$250/kW$_e$, efficiency (fuel to electric) η = 20 percent, fuel price p_f = \$5.00/GJ.

(c) Solar power plant: capital cost C_{cap} = \$3000/kW$_e$, annual production corresponds to a capacity factor of 0.15 (this number depends on the amount of sunshine at the site).

(d) At what capital cost does the solar plant become competitive with the others?

(e) The total demand of an electric grid contains components that are relatively constant (typical of many industrial loads) and components that are more or less variable (typical of building loads, especially cooling). Which plant type would you use for demand peaks, which for constant demand?

References

ASHRAE TC1.8 (1985). "Analysis of Survey Data on HVAC Maintenance Costs." ASHRAE Technical Committee 1.8 Research Project 382.

Boehm, R. F. (1987). *Design, Analysis of Thermal Systems*. Wiley, New York.

BOMA (1987). "BOMA Experience Exchange Report: Income/Expense Analysis for Office Buildings." Published annually by Building Owners and Managers Association International. 1250 I Street, N. W., Suite 200, Washington, D.C.

Dickinson, W. C., and K. C. Brown (1979). "Economic Analysis for Solar Industrial Process Heat." Report UCRL-52814, Lawrence Livermore Laboratory, Livermore, Calif.

Jones, B. W. (1982). *Inflation in Engineering Economic Analysis*. Wiley Interscience, New York.

Konkel, J. H. (1987). *Rule of Thumb Cost Estimating for Building Mechanical Systems: Accurate Estimating and Budgeting Using Unit Assembly Costs*. McGraw-Hill, New York.

Lippiat, B. C., and R. T. Ruegg (1990). "Energy Prices and Discount Factors for Life-Cycle Cost Analysis 1990." Report NISTIR 85-3273-4, Annual Supplement to NBS Handbook 135 and NBS Special Publication 709. National Institute of Standards and Technology, Applied Economics Group, Gaithersburg, Md.

Rabl, A. (1985). "Optimizing Investment Levels for Energy Conservation: Individual versus Social Perspective and the Role of Uncertainty." *Energy Economics*, (October) p. 259.

Nomenclature

A	Annual payment
A	Area, ft^2 (m^2)
A_h	Projected horizontal aperture area of skylights, ft^2 (m^2)
A_{life}	Levelized annual cost
A_M	Annual cost for maintenance, first-year \$
$(A/P, r, N)$	Capital recovery factor
A_w	Area of work plane, m^2 (ft^2)
$a_{i,k}, a_{o,k}, a_{Q,k}$	Transfer function coefficients
a_s	Stack coefficient of Table 7.2, (ft^3/min)^2/(in^4 · °F) [(L/s)^2/(cm^4 · K)]
a_w	Wind coefficient of Table 7.3, (ft^3/min)^2/[in^4 · (mi/h)^2] {(L/s)^2/ [cm^4 · (m/s)^2]}}
AFUE	Annual fuel utilization efficiency, %
b_n	Conduction transfer function coefficients
C	Coefficient in Eq. (7.14) for number of people who pass door per hour, (ft^3/min)/(inWG)$^{1/2}$
C	Cost at size S
C	Pipe or duct-fitting pressure coefficient
C	Heat capacity with subscripts n for node n, e for envelope, and i for interior, Btu/°F (kJ/K)
C_{base}	Base level (energy other than heating or cooling, i.e., hot water, cooking, lights), Btu/yr (GJ/yr)
C_{cap}	Capital cost, first-year \$
C_d	Draft coefficient for resistance to airflow between floors
C_{eff}	Effective heat capacity of building, Btu/°F (J/K)
C_i	Concentration of pollutant in indoor air
C_i	Consumption in period i
C_{life}	Life cycle cost
C_o	Concentration of pollutant in outdoor air
C_p	Pressure coefficient
C_r	Cost at reference size S_r
C_{salv}	Salvage value, first-year \$
C_u	Coefficient of utilization
C_{yr}	Normalized annual consumption
CLF$_t$	Cooling load factor at time t
CLTD$_t$	Cooling load temperature difference at time t, °F (K)
COP	Coefficient of performance
CPI	Consumer price index
c	Flow coefficient in Eq. (7.1) for flow across opening, ft^3/(min · in WG^n) [m/(s · Pa^n)]
c_n	Conduction transfer function coefficients
c_p	Specific heat, Btu/(lb$_m$ · °F) [kJ/(kg · K)]
D	Diameter, ft (m)
$D_c(T_{\text{bal}})$	Cooling degree-days for base T_{bal}, °F · days (K · days)

$D_h(T_{bal})$	Heating degree-days for base T_{bal}, °F · days (K · days)
D_h	Hydraulic diameter, ft (m)
d_n	Conduction transfer function coefficients
E	Radiation emissive power, Btu/(h · ft²) (W/m²)
E_b	Blackbody emissive power, Btu/(h · ft²) (W/m²)
$E_{b\lambda}$	Spectral blackbody emissive power, Btu/(h · ft² · μ)[W/m² · μ)]
E_{xh}	Exterior horizontal total (direct + sky) illuminance, fc (lx)
$E_{xh,dir}$	Exterior horizontal direct illuminance (Fig. 13.4), fc (lx)
$E_{xh,s}$	Exterior horizontal sky illuminance (Fig. 13.5), fc (lx)
$E_{xh,s/2}$	Exterior half-sky illuminance on horizontal surface in front of window (Fig. 13.6), fc (lx)
$E_{xv,dir}$	Exterior vertical direct illuminance (Fig. 13.4), fc (lx)
$E_{xv,g/2}$	Vertical illuminance reflected from ground in front of window, fc (lx)
$E_{xv,s/2}$	Exterior vertical illuminance incident on window from half sky in front of window (Fig. 13.5), fc (lx)
E_t	Equation of time
E_w	Average illuminance of work plane, lx (fc)
ET*	Effective temperature (value of T_{ad} at 50 percent relative humidity), °F(°C)
F	Solar heat gain coefficient
F	Total flux produced by lamps (not counting absorption by luminaires), lm
f	Fluid friction factor
f_{dep}	Present value of total depreciation as fraction of C_{cap}
$f_{dep,n}$	Depreciation during year n as fraction of C_{cap}
f_l	Fraction of investment paid by loan
g	Acceleration due to gravity = 32 ft/s² (9.8 m/s²)
g_c	Conversion factor in USCS units, 32.17 lb$_m$ · ft/(lb$_f$ · s²)
H_0	Extraterrestrial daily irradiation, Btu/ft² (MJ/m²)
$H_{glo,hor}$	Daily global irradiation at earth's surface, Btu/ft² (MJ/m²)
$H_{glo,vert}$	Daily global irradiation on vertical surface, Btu/ft² (MJ/m²)
h	Enthalpy, Btu/lb (kJ/kg)
h	"Head" referring to pressure, ft (m, cm)
h_{c+r}	Combined heat transfer coefficient for convection and radiation (hcon + hrad), Btu/(h · ft² · °F) [W/(m² · K)]
h_{fg}	Enthalpy of liquid-gas phase change for water
h_{con}	Convection heat transfer coefficient, Btu/(h · ft² · °F) [W/(m² · K)]
h_i	Indoor surface heat transfer coefficient, Btu/(h · ft² · °F) [W/(m² · K)]
h_o	Outdoor surface heat transfer coefficient, Btu/(h · ft² · °F) [W/(m² · K)]
h_s	Heat transfer coefficient of space between panes of double glazing, Btu/(h · ft² · °F) [(W/(m² · K)]
I_0	Extraterrestrial irradiance, Btu/(h · ft²) (W/m²)
I_{dif}	Diffuse irradiance on horizontal surface, Btu/(h · ft²) (W/m²)
I_{dir}	Beam (direct) irradiance at normal incidence, Btu/(h · ft²) (W/m²)
$I_{glo,hor}$	Global horizontal irradiance, Btu/(h · ft²) (W/m²)
$I_{glo,p}$	Global irradiance on tilted plane, Btu/(h · ft²) (W/m²)
I_n	Interest payment during nth year
K	Coefficient in Eq. (7.15), ft³/(min · ft² · inWG$^{0.65}$); a constant
K_{cond}	Conductive heat transmission coefficient, Btu/(h · °F) (W/K)
K_d	Derivative controller constant
K_f	Pipe fitting pressure coefficient

K_i	Integral controller constant
K_p	Proportional controller constant
K_{max}	Total heat transmission coefficient of building with open windows or economizer, Btu/(h · °F) (W/K)
K_T	Daily solar clearness index [defined in Eq. (6.27) as ratio of terrestrial to extraterrestrial solar radiation]
$\bar{K}_T$	Monthly average solar clearness index
K_{tot}	Total heat transmission coefficient of building, Btu/(h · °F) (W/K)
k	Thermal conductivity, Btu/(h · ft · °F) [W/(m · K)]
k_T	Instantaneous or hourly clearness index
L	Loan amount
L	Longitude, deg
L_{dirt}	Loss factor due to dirt (Table 13.5)
l	Length of perimeter, ft (m)
l	Length of room, ft (m)
LMTD	Log mean temperature difference, °F (K)
M	Metabolic rate per surface area of body, met (W/m^2)
m	Exponent of relation between cost and size of equipment
$\dot{m}$	Mass flow rate, lb/h (kg/s)
N	Number of days in month
N	Number of nodes in thermal network
N	System life, yr
N_2	Doubling time
N_{bin}	Number of hours per bin of bin method
N_{dep}	Depreciation period, yr
n_h, n_j	Number of hours in a temperature bin
N_l	Loan period, yr
N_p	Payback time, yr
NTU	Number of heat transfer units in heat exchanger
n	Day of year ($n = 1$ for January 1)
n	Exponent of airflow equation
n	Polytropic process exponent
n	Year
n_i	Number of days in period i
p	Pressure, lb$_f$/in^2 (Pa)
$(P/F, r, n)$	Present worth factor
P_{int}	Present value of interest payments
P_{max}	Peak demand, kW
P_n	Principal during nth payment period
p_w	Partial vapor pressure of water vapor in air, Ib/in^2 (Pa)
p_{dem}	Demand charge, \$/(kW · month)
PF	Evaporative cooler performance factor
PLR	Part-load ratio
p_e	Energy price
$\bar{p}_e$	Levelized energy price
p_{ins}	Price of insulation, \$/m^3
Q	Energy consumption with subscripts c for cooling and h for heating, Btu (J)
$\dot{Q}$	Heat flow with subscripts c for cooling load, h for heating load, lat for latent load, floor for heat flow to floor, etc., Btu/h (W)

$\dot{q}$	Heat flux, Btu/(h·ft^2) (W/m^2)
R	Thermal resistance, (°F·h)/Btu (K/W) [$\equiv \Delta T/\dot{Q}$]
R_{th}	Thermal R value, (h·ft^2·°F)/Btu [(m^2·K)/W] [$\equiv \Delta T/\dot{q}$]
R^2	Correlation coefficient
R_A	Ratio of net to gross skylight area, to account for blocking by frame or mullions
$R_{n'n}$	Resistance between nodes n' and n
r	Radius, ft (m)
r_o	$(r - r_{inf})/(1 + r_{inf})$
$r_{d,e}$	$(r_d - r_e)/(1 + r_e)$
$r_{d,l}$	$(r_d - r_l)/(1 + r_l)$
$r_{d,M}$	$(r_d - r_M)/(1 + r_M)$
r_d	Market discount rate
$r_{dif}(\omega_{ss}, \omega)$	$= I_{dif}/H_{dif}$, ratio of hourly to daily solar radiation, h^{-1} (s^{-1})
r_e	Market energy price escalation rate
$r_{glo}(\omega_{ss}, \omega)$	$= I_{glo,hor}/H_{glo,hor}$, h^{-1} (s^{-1})
r_{inf}	General inflation rate
r_l	Market loan interest rate
r_M	Market escalation rate for maintenance costs
r_r	Internal rate of return
S	$= -C_{life} + C_{life,ref}$, life cycle savings
S	Size of equipment
SC	Shading coefficient
SEER	Seasonal energy efficiency ratio
SHGF	Solar heat gain factor, Btu/(h·ft^2) (W/m^2)
SHGF$_{max}$	Maximum solar heat gain factor, Btu/(h·ft^2) (W/m^2)
SPF	Seasonal performance factor
s	Annual savings
s	Entropy, Btu/(lb·°R) [kJ/(kg·K)]
T	Temperature, °R or °F (K or °C)
T_a	Dry-bulb air temperature, °F (°C)
T_{ad}	Adiabatic equivalent temperature, °F (°C)
T_{bal}	Balance-point temperature of building, °F (°C)
$T_{control}$	Transmissivity of control devices such as louvers or diffusers, if present
T_{dif}, T_{dir}	Transmissivities of glazing for diffuse and direct radiation
T_e	Temperature of envelope, °F (°C)
T_i	Indoor air temperature, °F (°C)
T_{max}	Temperature above which air conditioner is turned on, °F (°C)
T_n	Temperature of node n, °F (°C)
T_o	Outdoor air temperature °F (°C)
$T_{o,av}$	Average outdoor temperature on design day, °F (°C)
$T_{o,max}$	Design outdoor temperature, °F (°C)
T_{op}	Operative temperature, °F (°C)
T_{os}	Sol-air temperature, °F (°C)
$T_{os,t}$	Sol-air temperature of outside surface at time t, °F (°C)
$\bar{T}_{o,t}$	Average outdoor temperature for any hour t of month, °F (°C)
$\bar{T}_{o,yr}$	Annual average temperature, °F (°C)
T_r	Mean radiant temperature, °F (°C)
T_{sk}	Temperature of skin, °F (°C)

T_{vent}	Temperature above which air conditioner is turned on, °F (°C)
T_{well}	Transmissivity of light well, if present (well efficiency of Fig. 13.7)
t	Thickness of insulation, ft (m)
t_{DST}	Daylight saving time, h
t_{sol}	Solar time, h
t_{ss}	Sunset time, h
t_{std}	Standard time, h
U	Overall heat transfer coefficient, Btu/(h·ft^2·°F) [W/(m^2·K)]
TF	Transfer function
u	Internal energy, Btu/lb (kJ/kg)
V	Volume, ft^3/m^3
$\dot{V}$	Flow rate, ft^3/min (m^3/s or L/s)
$\dot{V}_o$	Outdoor airflow rate, L/s
$\dot{V}_{\text{pol}}$	Volumetric flow rate of pollution source, ft^3/min (m^3/s or L/s)
v	Wind speed, mi/h (ft/s, m/s); velocity
W	Work, ft·lb$_f$ (kJ)
W_i	Humidity ratio of indoor air
W_o	Humidity ratio of outdoor air
w	Thickness of wall, ft (m)
w	Width
x	Thermodynamic quality
x	Distance, ft (m)
y	Distance, ft (m)
z	Distance, ft (m)

Greek

α	Thermal expansion coefficient, °F^{-1}(°C^{-1})
α	Absorptivity for solar radiation
β	$= UA_{\text{tot}}/\eta$, ratio of total heat loss coefficient to efficiency of HVAC system, Btu/(h·°F) (W/K)
β_s	Altitude angle of sun ($= 90° - \theta_s$)
Δh	Room height − work plane height, ft (m)
Δh	Vertical distance from neutral pressure level, ft (m)
Δp	$= p_o - p_i$, pressure difference between outside and inside, lbft/ft^2, psi, or inWG (Pa); general pressure difference
ΔT	Indoor temperature difference $T_i - T_o$, °F (K)
Δt	Time step, h
Δx	Thickness of layer, ft (m)
δ	Declination
δq	$= \dot{Q}_{\text{ref}} - \dot{Q}$, difference in $\dot{Q}$ between two control modes
ϵ	Effectiveness of heat and mass transfer device
ϵ	Pipe roughness, ft (m)
η_h	Efficiency of heating system
θ_i	Incidence angle of sun on plane (angle between normal of plane and line to sun
θ_p	Zenith angle of plane (tilt from horizontal, up > 0)
θ_s	Zenith angle of sun
λ	Latitude

μ	Absolute viscosity, $lb_m/(ft \cdot s)$ $(Pa \cdot s)$
v	Kinematic viscosity, ft^2/s (m^2/s)
ρ	Density, lb_m/ft^3 (kg/m^3)
ρ	Reflectivity
ρ_g	Reflectivity of ground
σ	Stefan-Boltzmann constant, $Btu/(h \cdot ft^2 \cdot {}^{\circ}R^4)$ $[W/(m^2 \cdot K^4)]$
σ_{yr}	Annual standard deviation of monthly average outdoor temperatures, ${}^{\circ}F$ (K)
σ_m	Standard deviation of daily outdoor temperatures for each month, ${}^{\circ}F$ (K)
τ	Incremental tax rate
τ	Time constant
τ	Transmissivity
τ_{cred}	Tax credit
ϕ_p	Azimuth of plane (positive for orientations west of south)
ϕ_s	Azimuth of sun
ω	Solar hour angle
ω	Angular velocity, rad/s
ω_{ss}	Sunset hour angle

Subscripts

abs	Absolute pressure
ann	Annual growth rate
atm	Atmospheric pressure
c	Cooling loads
cont	Continuous growth rate
d	Dry
dew	Dew point
e	"Electric" for units of energy and power, e.g., kW_e
g	Gauge pressure
h	Heating loads
t	Thermal for units of energy and power, e.g., kW_t
wet	Wet bulb (temperature)
0	Real growth rate r_0

Appendix A

TABLE A.1

Properties of Saturated R-22

Temp., °F	Pressure, psia	Density, lb/ft³ Liquid	Volume, ft³/lb Vapor	Enthalpy, Btu/lb Liquid	Enthalpy, Btu/lb Vapor	Entropy, Btu/(lb · °F) Liquid	Entropy, Btu/(lb · °F) Vapor
−90	3.413	92.71	13.275	−12.921	94.572	−0.03271	0.25807
−80	4.778	91.75	9.7044	−10.355	95.741	−0.02587	0.25357
−70	6.555	90.79	7.2285	−7.783	96.901	−0.01919	0.24945
−60	8.830	89.81	5.4766	−5.201	98.049	−0.01266	0.24567
−50	11.696	88.83	4.2138	−2.608	99.182	−0.00627	0.24220
−45	13.383	88.33	3.7160	−1.306	99.742	−0.00312	0.24056
−40	15.255	87.82	3.2880	0.000	100.296	0.00000	0.23899
−35	17.329	87.32	2.9185	1.310	100.847	0.00309	0.23748
−30	19.617	86.81	2.5984	2.624	101.391	0.00616	0.23602
−25	22.136	86.29	2.3202	3.944	101.928	0.00920	0.23462
−20	24.899	85.77	2.0774	5.268	102.461	0.01222	0.23327
−15	27.924	85.25	1.8650	6.598	102.986	0.01521	0.23197
−10	31.226	84.72	1.6784	7.934	103.503	0.01818	0.23071
−5	34.821	84.18	1.5142	9.276	104.013	0.02113	0.22949
0	38.726	83.64	1.3691	10.624	104.515	0.02406	0.22832
5	42.960	83.09	1.2406	11.979	105.009	0.02697	0.22718
10	47.538	82.54	1.1265	13.342	105.493	0.02987	0.22607
15	52.480	81.98	1.0250	14.712	105.968	0.03275	0.22500
20	57.803	81.41	0.9343	16.090	106.434	0.03561	0.22395
25	63.526	80.84	0.8532	17.476	106.891	0.03846	0.22294
30	69.667	80.26	0.7804	18.871	107.336	0.04129	0.22195
35	76.245	79.67	0.7150	20.275	107.769	0.04411	0.22098
40	83.280	79.07	0.6561	21.688	108.191	0.04692	0.22004
45	90.791	78.46	0.6029	23.111	108.600	0.04972	0.21912
50	98.799	77.84	0.5548	24.544	108.997	0.05251	0.21821
55	107.32	77.22	0.5111	25.988	109.379	0.05529	0.21732
60	116.38	76.58	0.4715	27.443	109.748	0.05806	0.21644
65	126.00	75.93	0.4355	28.909	110.103	0.06082	0.21557
70	136.19	75.27	0.4026	30.387	110.441	0.06358	0.21472
75	146.98	74.60	0.3726	31.877	110.761	0.06633	0.21387
80	158.40	73.92	0.3451	33.381	111.066	0.06907	0.21302
85	170.45	73.22	0.3199	34.898	111.350	0.07182	0.21218
90	183.17	72.51	0.2968	36.430	111.616	0.07456	0.21134
95	196.57	71.79	0.2756	37.977	111.859	0.07730	0.21050
100	210.69	71.05	0.2560	39.538	112.081	0.08003	0.20965

(continued)

TABLE A.1 (continued)

Properties of Saturated R-22

Temp., °F	Pressure, psia	Density, lb/ft³ Liquid	Volume, ft³/lb Vapor	Enthalpy, Btu/lb Liquid	Enthalpy, Btu/lb Vapor	Entropy, Btu/(lb · °F) Liquid	Entropy, Btu/(lb · °F) Vapor
105	225.53	70.29	0.2379	41.119	112.278	0.08277	0.20879
110	241.14	69.51	0.2212	42.717	112.448	0.08552	0.20793
115	257.52	68.71	0.2058	44.334	112.591	0.08827	0.20705
120	274.71	67.89	0.1914	45.972	112.704	0.09103	0.20615
125	292.73	67.05	0.1781	47.633	112.783	0.09379	0.20522
130	311.61	66.17	0.1657	49.319	112.825	0.09657	0.20427
135	331.38	65.27	0.1542	51.032	112.826	0.09937	0.20329
140	352.07	64.33	0.1434	52.775	112.784	0.10220	0.20227
145	373.71	63.35	0.1332	54.553	112.692	0.10504	0.20119
150	396.32	62.33	0.1237	56.370	112.541	0.10793	0.20006
160	444.65	60.12	0.1063	60.145	112.035	0.11383	0.19757
170	497.35	57.59	0.0907	64.175	111.165	0.12001	0.19464
180	554.82	54.57	0.0763	68.597	109.753	0.12668	0.19102
190	617.53	50.62	0.0625	73.742	107.398	0.13432	0.18613
200	686.11	44.44	0.0478	80.558	102.809	0.14432	0.17805

Source: Courtesy of ASHRAE, *Handbook of Fundamentals*, American Society of Heating, Refrigerating and Air-Conditioning Engineers, Atlanta, GA, 1997. With permission.

TABLE A.2

Properties of Saturated R-134a

Temp., °F	Pressure, psia	Density, lb/ft³ Liquid	Volume, ft³/lb Vapor	Enthalpy, Btu/lb Liquid	Enthalpy, Btu/lb Vapor	Entropy, Btu/(lb · °F) Liquid	Entropy, Btu/(lb · °F) Vapor
−90	1.363	93.17	28.303	14.665	89.504	0.03717	0.24462
−80	1.997	92.21	19.783	11.755	91.005	0.02940	0.24125
−70	2.859	91.25	14.138	8.837	92.514	0.02182	0.23827
−60	4.006	90.28	10.310	5.907	94.026	0.01440	0.23563
−55	4.707	89.80	8.8656	4.437	94.783	0.01075	0.23443
−50	5.505	89.31	7.6569	2.963	95.539	0.00713	0.23331
−45	6.409	88.82	6.6405	1.484	96.295	0.00355	0.23225
−40	7.429	88.32	5.7819	0.000	97.050	0.00000	0.23125
−35	8.577	87.83	5.0533	1.489	97.804	0.00352	0.23032
−30	9.862	87.33	4.4325	2.984	98.556	0.00701	0.22945
−25	11.297	86.82	3.9014	4.484	99.306	0.01048	0.22863
−20	12.895	86.32	3.4452	5.991	100.054	0.01392	0.22786
−15	14.667	85.81	3.0519	7.505	100.799	0.01733	0.22714
−10	16.626	85.29	2.7116	9.026	101.542	0.02073	0.22647
−5	18.787	84.77	2.4161	10.554	102.280	0.02409	0.22584
0	21.162	84.25	2.1587	12.090	103.015	0.02744	0.22525
5	23.767	83.72	1.9337	13.634	103.745	0.03077	0.22470
10	26.617	83.18	1.7365	15.187	104.471	0.03408	0.22418

TABLE A.2 (continued)

Properties of Saturated R-134a

Temp., °F	Pressure, psia	Density, lb/ft³ Liquid	Volume, ft³/lb Vapor	Enthalpy, Btu/lb Liquid	Vapor	Entropy, Btu/(lb · °F) Liquid	Vapor
15	29.726	82.64	1.5630	16.748	105.192	0.03737	0.22370
20	33.110	82.10	1.4101	18.318	105.907	0.04065	0.22325
25	36.785	81.55	1.2749	19.897	106.617	0.04391	0.22283
30	40.768	80.99	1.1550	21.486	107.320	0.04715	0.22244
35	45.075	80.42	1.0484	23.085	108.016	0.05038	0.22207
40	49.724	79.85	0.9534	24.694	108.705	0.05359	0.22172
45	54.732	79.26	0.8685	26.314	109.386	0.05679	0.22140
50	60.116	78.67	0.7925	27.944	110.058	0.05998	0.22110
55	65.895	78.07	0.7243	29.586	110.722	0.06316	0.22081
60	72.087	77.46	0.6630	31.239	111.376	0.06633	0.22054
65	78.712	76.84	0.6077	32.905	112.019	0.06949	0.22028
70	85.787	76.21	0.5577	34.583	112.652	0.07264	0.22003
75	93.333	75.57	0.5125	36.274	113.272	0.07578	0.21979
80	101.370	74.91	0.4715	37.978	113.880	0.07892	0.21957
85	109.920	74.25	0.4343	39.697	114.475	0.08205	0.21934
90	119.000	73.57	0.4004	41.430	115.055	0.08518	0.21912
95	128.630	72.87	0.3694	43.179	115.619	0.08830	0.21890
100	138.830	72.16	0.3411	44.943	116.166	0.09142	0.21868
105	149.630	71.43	0.3153	46.725	116.694	0.09454	0.21845
110	161.050	70.68	0.2915	48.524	117.203	0.09766	0.21822
115	173.110	69.91	0.2697	50.343	117.690	0.10078	0.21797
120	185.840	69.12	0.2497	52.181	118.153	0.10391	0.21772
125	199.250	68.31	0.2312	54.040	118.591	0.10704	0.21744
130	213.380	67.47	0.2141	55.923	119.000	0.11018	0.21715
135	228.250	66.60	0.1983	57.830	119.377	0.11333	0.21683
140	243.880	65.70	0.1836	59.764	119.720	0.11650	0.21648
145	260.310	64.77	0.1700	61.727	120.024	0.11968	0.21609
150	277.570	63.80	0.1574	63.722	120.284	0.12288	0.21566
160	314.690	61.72	0.1345	67.823	120.650	0.12938	0.21463
170	355.510	59.42	0.1144	72.106	120.753	0.13603	0.21329
180	400.340	56.80	0.0965	76.636	120.493	0.14295	0.21151
190	449.550	53.70	0.0801	81.534	119.684	0.15029	0.20901
200	503.640	49.70	0.0646	87.088	117.906	0.15847	0.20519

Source: Courtesy of ASHRAE, *Handbook of Fundamentals*, American Society of Heating, Refrigerating and Air-Conditioning Engineers, Atlanta, GA, 1997. With permission.

Conversion Factors (* = exact)

Acceleration
$1 \text{ m/s}^2 = 3.281 \text{ ft/s}^2$

$1 \text{ ft/s}^2 = 0.3048 \text{ m/s}^2$

Area
$1 \text{ m}^2 = 10.76 \text{ ft}^2$
$1 \text{ km}^2 = 0.386 \text{ mi}^2$
$1 \text{ ha} = {}^*10^4 \text{ m}^2 = 2.47 \text{ acre}$

$1 \text{ ft}^2 = {}^*144 \text{ in}^2 = 0.0929 \text{ m}^2$
$1 \text{ mi}^2 = 2.590 \text{ km}^2$
$1 \text{ acre} = 43{,}560 \text{ ft}^2 = 4050 \text{ m}^2$

Costs
$1 \text{ ¢/kWh} = 2.778 \text{ \$/GJ} = 2.931 \text{ \$/MBtu}$
$1 \text{ \$/m}^2 = 0.0929 \text{ \$/ft}^2$
$1 \text{ \$/kg} = 0.45356 \text{ \$/lb}_m$
$1 \text{ \$/L} = 3.785 \text{ \$/gal}$

$1 \text{ \$/MBtu} = 0.948 \text{ \$/GJ} = 0.341 \text{ ¢/kWh}$
$1 \text{ \$/therm} = 9.48 \text{ \$/GJ}$
$1 \text{ \$/gal (No. 6 fuel oil @ } \eta = 1.0) = 6.32 \text{ \$/GJ}$
$1 \text{ \$/gal (No. 2 fuel oil @ } \eta = 1.0) = 6.77 \text{ \$/GJ}$

Density
$1 \text{ kg/m}^3 = 6.2430 \times 10^{-2} \text{ lb}_m/\text{ft}^3$

$1 \text{ lb}_m/\text{ft}^3 = 16.02 \text{ kg/m}^3$

Energy or work
$1 \text{ J} = {}^*1(\text{kg} \cdot \text{m}^2)/\text{s}^2 = {}^*10^7 \text{ erg}$
$\quad = 0.948 \times 10^{-3} \text{ Btu}$
$1 \text{ kWh} = {}^*3.6 \text{ MJ}$
$1 \text{ cal} = 4.187 \text{ J}$

$1 \text{ Btu} = 778.16 \text{ ft} \cdot \text{lb}_f = 1.055 \text{ kJ}$

$1 \text{ therm} = {}^*10^5 \text{ Btu} = 105.5 \text{ MJ}$
$1 \text{ quad} = {}^*10^{15} \text{ Btu} = 1.055 \times 10^{18} \text{ J}$
$1 \text{ ft} \cdot \text{lb}_f = 1.3558 \text{ J}$

Flow rate (mass)
$1 \text{ kg/s} = 2.2046 \text{ lb}_m/\text{s}$
$1 \text{ kg/s} = 132.3 \text{ lb}_m/\text{min}$
$1 \text{ kg/s} = 7937 \text{ lb}_m/\text{h}$

$1 \text{ lb}_m/\text{s} = 0.454 \text{ kg/s}$
$1 \text{ lb}_m/\text{min} = 7.56 \times 10^{-3} \text{ kg/s}$
$1 \text{ lb}_m/\text{h} = 0.1256 \times 10^{-3} \text{ kg/s}$

Flow rate (volume)
$1 \text{ m}^3/\text{s} = 2119 \text{ cfm}$
$1 \text{ m}^3/\text{s} = 1.585 \times 10^4 \text{ gpm}$

$1 \text{ cfm } (\text{ft}^3/\text{min}) = 0.4719 \text{ L/s}$
$1 \text{ gpm (gal/min)} = 2.228 \times 10^{-3} \text{ ft}^3/\text{s}$
$\qquad\qquad\qquad\quad = 0.0631 \text{ L/s}$

Flow rate (volume/area)

$1 \text{ cfm/ft}^2 = 5.01 \text{ L}/(\text{s} \cdot \text{m}^2)$

Force
$1 \text{ N} = {}^*1 \text{ (kg} \cdot \text{m)/s}^2 = {}^*10^5 \text{ dyn}$
$\quad = 0.2248 \text{ lb}_f$

$1 \text{ lb}_f = {}^*16 \text{ oz}_f = 4.4482 \text{ N}$

Heat flux
$1 \text{ W/m}^2 = 0.3170 \text{ Btu}/(\text{h} \cdot \text{ft}^2)$
$1 \text{ W/m}^2 = 0.0929 \text{ W/ft}^2$

$1 \text{ Btu}/(\text{h} \cdot \text{ft}^2) = 3.155 \text{ W/m}^2$
$1 \text{ W/ft}^2 = 10.76 \text{ W/m}^2$

Heat loss coefficient of building

$1 \text{ W/K} = 1.895 \text{ Btu/(h} \cdot {}^\circ\text{F)}$ $1 \text{ Btu/(h} \cdot {}^\circ\text{F)} = 0.528 \text{ W/K}$

Heat transfer coefficient

$1 \text{ W/(m}^2 \cdot \text{K)} = 0.1761 \text{ Btu/(h} \cdot \text{ft}^2 \cdot {}^\circ\text{F)}$ $1 \text{ Btu/(h} \cdot \text{ft}^2 \cdot {}^\circ\text{F)} = 5.678 \text{ W/(m}^2 \cdot \text{K)}$

Illuminance

$1 \text{ lx} = {}^*1 \text{ lm/m}^2 = 0.0929 \text{ fc}$ $1 \text{ fc (footcandle)} = {}^*1 \text{ lm/ft}^2 = 10.76 \text{ lx}$

Length

$1 \text{ m} = 3.281 \text{ ft} = 39.37 \text{ in} = 1.0936 \text{ yd}$ $1 \text{ ft} = {}^*12 \text{ in} = {}^*\frac{1}{3} \text{ yd} = {}^*0.3048 \text{ m}$

$1 \text{ km} = 0.622 \text{ mi}$ $1 \text{ mi} = 1.609 \text{ km}$

$1 \text{ cm} = {}^*0.01 \text{ m} = 0.3937 \text{ in}$ $1 \text{ in} = {}^*0.0254 \text{ m} = {}^*2.54 \text{ cm}$

Mass

$1 \text{ kg} = {}^*10^3 \text{ g} = 2.205 \text{ lb}_m$ $1 \text{ lb}_m = {}^*16 \text{ oz}_m = {}^*0.45356 \text{ kg}$

$1 \text{ ton (metric)} = {}^*10^3 \text{ kg}$ $1 \text{ grain} = {}^*1/7000 \text{ lb}_m = 0.0648 \text{ g}$

$1 \text{ g} = 0.0353 \text{ oz}_m$ $1 \text{ ton (U.S. long)} = {}^*2240 \text{ lb}_m = 1016 \text{ kg}$

$\qquad\qquad\qquad\qquad\qquad\quad$ $1 \text{ ton (U.S. short)} = {}^*2000 \text{ lb}_m = 907 \text{ kg}$

Power

$1 \text{ W} = 3.412 \text{ Btu/h}$ $1 \text{ Btu/h} = 0.2931 \text{ W}$

$1 \text{ kW} = 1.341 \text{ hp}$ $1 \text{ hp} = 550 \text{ (ft} \cdot \text{lb}_f)/\text{s} = 0.7457 \text{ kW}$

$1 \text{ kW} = 0.2844 \text{ ton refrigeration}$ $1 \text{ ton refrigeration} = 3.517 \text{ kW}$

$\qquad\qquad\qquad\qquad\qquad\quad$ $1 \text{ hp (boiler)} = 9.81 \text{ kW} = 33{,}475 \text{ Btu/h}$

Pressure

$1 \text{ Pa} = {}^*1 \text{ N/m}^2 = {}^*10^{-5} \text{ bar}$ $1 \text{ psi } (\text{lb}_f \cdot \text{in}^2) = 6.894 \text{ kPa} = 27.7 \text{ inWG}$

$\qquad = 1.450 \times 10^{-4} \text{ psi}$

$1 \text{ Pa} = 4.019 \times 10^{-3} \text{ inWG}$ $1 \text{ inWG} = 249.1 \text{ Pa} = 0.0361 \text{ psia}$

$1 \text{ atm (std. atmosphere)} = 101.325 \text{ kPa}$ $1 \text{ atm (std. atmosphere)} = 14.696 \text{ psi}$

$1 \text{ Pa} = 2.088 \times 10^{-2} \text{ lb}_f/\text{ft}^2$ $1 \text{ inHg} = 3.3772 \text{ kPa} = 0.49115 \text{ psia}$

$1 \text{ mmHg} = 0.01934 \text{ psia}$ $1 \text{ inWG}/100 \text{ ft} = 8.17 \text{ Pa/m}$

Specific enthalpy

$1 \text{ kJ/kg} = 0.4299 \text{ Btu/lb}_m$ $1 \text{ Btu/lb}_m = 2.3266 \text{ kJ/kg}$

Specific heat

$1 \text{ kg/(kg} \cdot \text{K)} = 0.2389 \text{ Btu/(lb}_m \cdot {}^\circ\text{F)}$ $1 \text{ Btu/(lb}_m \cdot {}^\circ\text{F)} = 4.1868 \text{ kJ/(kg} \cdot \text{K)}$

Temperature

${}^\circ\text{C} = {}^*({}^\circ\text{F} - 32) \times \frac{5}{9}$ ${}^\circ\text{F} = {}^*(9/5){}^\circ\text{C} + 32$

$\text{K} = {}^\circ\text{C} + 273.15$ ${}^\circ\text{R} = {}^\circ\text{F} + 459.67$

Thermal conductivity or resistance

$1 \text{ W/(m} \cdot \text{K)} = 0.5778 \text{ Btu/(h} \cdot \text{ft} \cdot {}^\circ\text{F)}$ $1 \text{ Btu/(h} \cdot \text{ft} \cdot {}^\circ\text{F)} = 1.731 \text{ W/(m} \cdot \text{K)}$

$1 \text{ W/(m} \cdot \text{K)} = 6.934 \text{ Btu} \cdot \text{in/(h} \cdot \text{ft}^2 \cdot {}^\circ\text{F)}$ $1 \text{ Btu} \cdot \text{in/(h} \cdot \text{ft}^2 \cdot {}^\circ\text{F)} = 0.144 \text{ W/(m} \cdot \text{K)}$

$1 \text{ K/W} = 0.5275 \text{ (}{}^\circ\text{F} \cdot \text{h)/Btu}$ $1 \text{ (}{}^\circ\text{F} \cdot \text{h)/Btu} = 1.896 \text{ K/W}$

Time
$1 \text{ yr} = 8760 \text{ h} = 3.1536 \times 10^7 \text{ s}$ $1 \text{ day} = 86{,}400 \text{ s}$

Velocity
$1 \text{ m/s} = 196.9 \text{ ft/min}$ $1 \text{ ft/min} = {}^*0.00508 \text{ m/s}$
$1 \text{ m/s} = 3.281 \text{ ft/s}$ $1 \text{ ft/s} = {}^*0.3048 \text{ m/s}$
$1 \text{ km/h} = 0.2778 \text{ m/s}$ $1 \text{ mi/h} = 0.4470 \text{ m/s}$
$1 \text{ m/s} = 2.2371 \text{ mi/h}$ $1 \text{ kn} = 1.152 \text{ mi/h} = 0.515 \text{ m/s}$
$1 \text{ rad/s} = 9.55 \text{ r/min}$ $1 \text{ r/min} = 0.1047 \text{ rad/s}$

Viscosity (absolute, dynamic)
$1 \text{ Pa} \cdot \text{s} = {}^*10 \text{ poise (P)} = 0.672 \text{ lb}_m/(\text{ft} \cdot \text{s})$ $1 \text{ lb}_m/(\text{ft} \cdot \text{s}) = 1.488 \text{ Pa} \cdot \text{s}$

Viscosity (kinematic)
$1 \text{ m}^2/\text{s} = {}^*10^4 \text{ stokes} = 10.76 \text{ ft}^2/\text{s}$ $1 \text{ ft}^2/\text{s} = 0.0903 \text{ m}^2/\text{s}$

Volume
$1 \text{ L} = {}^*10^{-3} \text{ m}^3 = 0.264 \text{ gal (U.S.)}$ $1 \text{ gal (U.S.)} = 3.785 \text{ L}$
$1 \text{ m}^3 = 35.3147 \text{ ft}^3$ $1 \text{ ft}^3 = 0.02832 \text{ m}^3 = 7.481 \text{ gal (U.S.)}$
          $1 \text{ bbl} = 42 \text{ gal} = 0.159 \text{ m}^3$

Decimal Multiples

Factor	SI Prefix	Symbol
10^{18}	exa	E
10^{15}	peta	P
10^{12}	tera	T
10^{9}	giga	G
10^{6}	mega	M
10^{3}	kilo	k
10^{2}	hecto	h
10^{1}	deca	da
10^{-1}	deci	d
10^{-2}	centi	c
10^{-3}	milli	m
10^{-6}	micro	μ
10^{-9}	nano	n
10^{-12}	pico	p

Some Useful Constants

Acceleration of gravity (standard value in United States): $g = 32.17$ ft/s^2 (9.80 m/s^2)

Air (standard design conditions, sea level): 14.696 psia (101,325 Pa), 70°F (21°C)
density $\rho = 0.075$ lb$_m$/ft^3 (1.20 kg/m^3)
specific heat $c_p = 0.24$ Btu/(lb$_m$ · °F) [1.00 kJ/(kg · K)]
sensible load $= \rho c_p \dot{V}(T_o - T_i)$, with $\rho c_p = 1.08$ [Btu/(h · °F)]/(ft^3/min) [120 (W/K)/(L/s)]
latent load $= \rho h_{fg} \dot{V} \, \Delta W$, with $\rho h_{fg} = 4840$ (Btu/h)/(ft^3/min) [3010 W/(L/s)]

Water
density $\rho = 62.32$ lb$_m$/ft^3 (998.2 kg/m^3) at 70°F (21°C)
specific heat $c_p = 1.000$ Btu/(lb$_m$ · °F)[4.186 kJ/(kg · K)]
latent heat of freezing $h_{if} = 143.8$ Btu/lb$_m$ (333.8 kJ/kg) at 32°F (0°C)
latent heat of vaporization $h_{fg} = 970.3$ Btu/lb$_m$ (2257 kJ/kg) at 212°F (100°C)

Gas constants:
universal $R_u = 1545$ ft · lb$_f$/(lb mol · °R) [8314 J/(kg mol · K)]
air $R_a = 53.35$ ft · lb$_f$/(lb$_m$ · °R) [287 J/(kg · K)]
water vapor $R_{H_2O} = 85.78$ ft · lb$_f$/(lb$_m$ · °R) [462 J/(kg · K)]

Stefan-Boltzmann Constant
$\sigma = 0.1714 \times 10^{-8}$ Btu/(h · ft^2 · °R^4) [5.67 × 10^{-8} W/(m^2 · K^4)]

Surface heat transfer coefficients (ASHRAE design values)
$h_i = 1.46$ Btu/(h · ft^2 · °F] [8.3 W/(m^2 · K)]
$h_o = 4.0$ Btu/(h · ft^2 · °F) [22.7 W/(m^2 · K)] at $v = 7.5$ mph (3.4 m/s) summer
$h_o = 6.0$ Btu/(h · ft^2 · °F) [34.0 W/(m^2 · K)] at $v = 15$ mph (6.7 m/s) winter

Other constants
$g_c = 32.17$ lb$_m$ · ft/(s^2 · lb$_f$)

Index